CHAPMAN & HALL/CRC APPLIED MATHEMATICS
AND NONLINEAR SCIENCE SERIES

Exact Solutions and Invariant Subspaces of Nonlinear Partial Differential Equations in Mechanics and Physics

CHAPMAN & HALL/CRC APPLIED MATHEMATICS AND NONLINEAR SCIENCE SERIES

Series Editors Goong Chen and Thomas J. Bridges

Published Titles

Computing with hp-ADAPTIVE FINITE ELEMENTS: Volume I One and Two Dimensional Elliptic and Maxwell Problems, Leszek Demkowicz

CRC Standard Curves and Surfaces with Mathematica®*: Second Edition,* David H. von Seggern

Exact Solutions and Invariant Subspaces of Nonlinear Partial Differential Equations in Mechanics and Physics, Victor A. Galaktionov and Sergey R. Svirshchevskii

Geometric Sturmian Theory of Nonlinear Parabolic Equations and Applications, Victor A. Galaktionov

Introduction to Fuzzy Systems, Guanrong Chen and Trung Tat Pham

Introduction to Partial Differential Equations with MATLAB®, Matthew P. Coleman

Mathematical Methods in Physics and Engineering with Mathematica, Ferdinand F. Cap

Optimal Estimation of Dynamic Systems, John L. Crassidis and John L. Junkins

Quantum Computing Devices: Principles, Designs, and Analysis, Goong Chen, David A. Church, Berthold-Georg Englert, Carsten Henkel, Bernd Rohwedder, Marlan O. Scully, and M. Suhail Zubairy

Forthcoming Titles

Computing with hp-ADAPTIVE FINITE ELEMENTS: Volume II Frontiers: Three Dimensional Elliptic and Maxwell Problems with Applications, Leszek Demkowicz, Jason Kurtz, David Pardo, Maciej Paszynski, Waldemar Rachowicz, and Adam Zdunek

Introduction to non-Kerr Law Optical Solitons, Anjan Biswas and Swapan Konar

Mathematical Theory of Quantum Computation, Goong Chen and Zijian Diao

Mixed Boundary Value Problems, Dean G. Duffy

Multi-Resolution Methods for Modeling and Control of Dynamical Systems, John L. Junkins and Puneet Singla

Stochastic Partial Differential Equations, Pao-Liu Chow

CHAPMAN & HALL/CRC APPLIED MATHEMATICS
AND NONLINEAR SCIENCE SERIES

Exact Solutions and Invariant Subspaces of Nonlinear Partial Differential Equations in Mechanics and Physics

Victor A. Galaktionov
University of Bath
U.K.

Sergey R. Svirshchevskii
Keldysh Institute of Applied Mathematics
Moscow, Russia

Chapman & Hall/CRC
Taylor & Francis Group
Boca Raton London New York

Chapman & Hall/CRC is an imprint of the
Taylor & Francis Group, an informa business

Chapman & Hall/CRC
Taylor & Francis Group
6000 Broken Sound Parkway NW, Suite 300
Boca Raton, FL 33487-2742

Printed in the United States of America on acid-free paper
10 9 8 7 6 5 4 3 2 1

International Standard Book Number-10: 1-58488-663-3 (Hardcover)
International Standard Book Number-13: 978-1-58488-663-1 (Hardcover)

Library of Congress Cataloging-in-Publication Data

Galaktionov, Victor A.
Exact solutions and invariant subspaces of nonlinear partial differential equations in mechanics and physics / Victor A. Galaktionov, Sergey R. Svirshchevskii.
p. cm. -- (Chapman & Hall/CRC applied mathematics and nonlinear science series ; 10)
Includes bibliographical references and index.
ISBN 1-58488-663-3 (alk. paper)
1. Differential equations, Partial--Numerical solutions. 2. Nonlinear theories--Methodology. 3. Invariant subspaces--Methodology. 4. Exact (Philosophy)--Mathematics. 5. Mathematical physics. 1. Svirshchevskii, Sergey R. II. Title.

QA377.G2217 2006
518'.64--dc22 2006049622

To our teacher, Sergey Pavlovich Kurdyumov

Contents

Introduction: Nonlinear Partial Differential Equations and Exact Solutions

Exact solutions: history, classical symmetry methods, extensions

One of the crucial problems in the theory of partial differential equations (PDEs) at its early stages in the eighteenth and nineteenth century was finding and studying classes of important equations that were integrable in closed form and, in particular, possessed explicit solutions. It seems that the first general type of explicit solutions were traveling waves in d'Alembert's formula for the linear wave equation. The method of separation of variables was developed by Fourier in the study of heat conduction problems, and was later generalized and extended by Sturm and Liouville in the 1830s. Many famous mathematicians, such as Euler, Lagrange, Liouville, Sturm, Laplace, Darboux, Bäcklund, Lie, Jacobi, Boussinesq, Goursat, and others developed various techniques for obtaining explicit solutions of a variety of linear and nonlinear models from physics and mechanics. Their methods included a number of particular transformations, symmetries, expansions, separation of variables, etc. Similarity solutions appeared in the works by Weierstrass around 1870, and by Bolzman around 1890. After the Blasius construction (1908) of the exact self-similar solution for the two-dimensional (2D) boundary layer equations proposed by Prandtl in 1904, similarity solutions of linear and nonlinear boundary-value problems became more common in the literature. General principles for finding solutions of systems of ODEs and PDEs by symmetry reductions date back to the famous Lie papers [389]–[393] published in the 1880s and 1890s.

In the first half of the twentieth century, the basic priorities in PDE theory were re-evaluated in light of the influence of mathematical physics. As a result of this, and possibly in view of the essential progress achieved in existence-uniqueness-regularity theory for classes of PDEs of different types, explicit solutions gradually began to lose their exceptional role. At that time, many results and techniques on explicit integration were forgotten. On the other hand, in the 1930s, and especially in the 1940s and 1950s, exact solutions and similarity reductions returned to the scene in the asymptotic and singularity analysis of difficult practical problems of gas and hydrodynamics which appeared in many fundamental technological, industrial, and military areas in different countries. In the 1930s, the first basic ideas and results in this area were due to von Mises, von Kármán, Bechert, Guderley, Sedov (in the 1940s), and others, who applied scaling and similarity techniques to the study of complicated nonlinear models and singularity phenomena. These gas and hydrodynamic models included systems of several nonlinear PDEs, for many of which a rigorous mathematical analysis remains elusive, even now. The exact similarity solutions were the only possible way to detect crucial features of nonstationary and singular evolution, such as focusing of spherical waves in gas dynamics and shock-wave phenomena. In light of this, it was no accident that the gas dynamic and hydrodynamic equations became the first applications of new general ideas and methods of the group analysis of the PDEs, which Ovsiannikov began to develop in the 1950s. On the basis of Lie groups, he proposed a general approach to invariant and partially invariant solutions of nonlinear PDEs. A notion of group-invariant solutions, including special cases of traveling waves and similarity patterns, was emphasized by Birkhoff on the basis of hydrodynamic problems in the 1940s.

In the second half of the twentieth century, the increase of interest in exact solutions and exactly solvable models was two-fold. Firstly, the applied areas related to modern physics, mechanics and technology induced more and more complicated models dealing with systems of nonlinear PDEs. In this context, it is worth mentioning the new theory of weak solutions of nonlinear degenerate porous medium equations initiated in the 1950s (uniqueness approaches dated back to classical Holmgren's method, 1901), and self-focusing in nonlinear optics described by blow-up solutions of the nonlinear Schrödinger equation in the beginning of the 1960s. Secondly, the effective development in the 1960s and 1970s of the method for the exact integration of nonlinear PDEs, such as the inverse scattering method and Lax pairs introduced an exceptional class of fully integrable evolution equations possessing countable sets of exact solutions, such as N-solitons.

It seems that the beginning of the twenty-first century may be characterized in a manner similar to the 1950s. At that time, the complexity of many nonlinear PDE models of principal interest rose so high that one could not expect a mathematically rigorous existence-regularity theory to be created soon. For instance, there are many fundamental open problems in the theory of higher-order multi-dimensional quasilinear thin film equations, higher-order KdV-type PDEs with nonlinear dispersion possessing compacton, peakon and cuspon-type solutions, quasilinear degenerate wave equations and systems including equations of general relativity. Modern PDE theory proposes a number of new canonical higher-order models, to which many classical techniques do not apply in principle. In these and other difficult areas of general PDE theory, exact solutions will continue to play a determining role and often serve as basic patterns, exhibiting the correct classes of existence, regularity, uniqueness and specific asymptotics.

The classical method for detecting similarity reductions and associated explicit solutions of various classes of PDEs is the Lie group method of infinitesimal transformations. These approaches and related extensions are explained in a series of monographs by L.V. Ovsiannikov, N.H. Ibragimov, G.W. Bluman and J.D. Cole, P.J. Olver, G.W. Bluman and S. Kumei amongst others. We refer to the "*CRC Handbook of Lie Group Analysis of Differential Equations*" [10] containing a large list of results and references on this subject.

Over the years, many generalizations of the concept of symmetry groups of nonlinear PDEs have been proposed. The first of these go back to Lie himself (contact transformations), to E. Cartan (dynamical symmetries, 1910), and to E. Noether (generalized symmetries, 1918). Other ideas that appeared in this period are discussed in Anderson–Kamran–Olver [11]. Many generalizations can be viewed as extensions of the classical *semi-inverse method* in Continuum Mechanics, which has a natural counterpart in symmetry methods (as was first noted by G. Birkhoff in the 1950s).

During the last fifty years, when more nonlinear models and applied PDEs began to attract the attention of mathematicians, many other fruitful attempts were made to extend the classical apparatus of Lie group symmetries for PDEs. A significant number of new classes of such generalized symmetries and corresponding exact solutions were found. Not pretending to completeness, precise statements, and the correct characterization of such ideas, we include in this list the following (specific power tools for integrable equations are not mentioned):

- the method of nonclassical symmetries (invariant surface conditions);
- the method of partially invariant solutions;
- the Bäcklund transformation method;
- the Baker–Hirota bilinear method;
- the direct and modified method;
- the conditional and generalized conditional symmetry method;
- the non-local symmetry method;
- the truncated Painlevé approach;
- the weak symmetries method;
- the side conditions method;
- the method of linear invariant subspaces for nonlinear operators;
- the method of linear determining equations;
- the method of B-determining equations;
- the nonlinear separation method;
- the functional separation method;
- the method of symmetry-preserving constraints;
- the symmetry-enhancing method;
- the differential constraint method.

We will present descriptions and references concerning most of the methods that are related to the techniques used in our analysis (some of the others can be traced out through use of the Index).

Most of the above methods can be reformulated by using the technicalities of the *method of differential constraints*. Such ideas initially appeared in the theory of first-order PDEs. In particular, Lagrange used differential constraints to determine total integrals of nonlinear equations with two independent variables

$$F(x, y, u, u_x, u_y) = 0.$$

Monge and Ampère proposed the technique of first integrals for solving the second-order PDEs

$$F(x, y, u, u_x, u_y, u_{xx}, u_{xy}, u_{yy}) = 0,$$

and in 1870, Darboux extended this approach by introducing an extra second-order PDE which is in involution with the original equation (this is what is now called a differential constraint). The history of this analysis and the detailed description of Darboux's method are given in Goursat [260] and Forsyth [196]. General theory of related overdetermined systems is due to many famous names, such as Riquier, Cartan, Ritt, and Spencer, as explained in Pommaret [468].

Systematic approaches to differential constraints, related symmetry and Lie group methods were proposed by Birkhoff in the 1940s (hydrodynamics and fluid dynamics) and by Yanenko in the 1960s (gas dynamics). A formal description of the method is not difficult: consider a PDE for solutions $u = u(x, t)$, with independent variables $(x, t) \in I\!R \times I\!R_+$, where t denotes the time-variable. Given a sufficiently smooth function $F(\cdot)$, consider the evolution PDE

$$F[u] \equiv F(u, Du, D^2u, ...) = 0, \tag{0.1}$$

where $Du = \{u_x, u_t\}$, $D^2u = \{u_{xx}, u_{xt}, u_{tt}\}$, etc. denote vectors of partial derivatives of arbitrary fixed finite orders. To find particular exact solutions consider, instead of the single PDE (0.1), a system of two (or possibly more) equations

$$\begin{cases} F_1[u] = 0, \\ \Phi[u] = 0, \end{cases} \tag{0.2}$$

where the second equation plays the role of an extra *differential constraint*. As usual, one can take $F_1 = F$ in the first equation, but, in general, these operators can be different under the hypothesis that the *consistency* of the system implies that such functions $u(x, t)$ also satisfy the original equation (0.1). For example, if $F[u] = F_1[u] - F_2[u]$, the following system may be considered:

$$\begin{cases} F_1[u] = \Phi[u], \\ \Phi[u] = F_2[u], \end{cases}$$

with an unknown operator Φ to be determined from the consistency condition.

The key ingredient of the differential constraint analysis is to find such suitable operator pairs $\{F_1, \Phi\}$ in (0.2). This is a difficult problem. Indeed, the consistency condition of the system leads to a PDE for Φ, which may be much more complicated than the original one (0.1) for u (to say nothing about PDEs in the multi-dimensional Euclidean space, where $x \in I\!R^N$). Nevertheless, there exists an essential advantage of this constraint analysis: one needs to find a *particular solution* of the compatibility equation.* In the methods listed above, the choice of suitable constraint operators was heavily affected by applying new additional ideas, including some results of classical group-invariant analysis and extensions, or those from neighboring areas of the theory and applications of the PDEs under consideration.

In sufficiently general settings (that do not deal with hard consistency of the PDEs for Φ) the scheme for the differential constraint method looks like using a practically random choice of consistent constraint operators Φ. In a natural sense, such a procedure does not essentially differ from *trial and error* dealing with *a priori* prescribed classes of functions $\{u(x, t, \alpha)\}$ (α is a parameter, possibly functional) to be substituted into the PDE (0.1) to check whether some of the functions by chance satisfy it.

* We do not mention the second important aspect of the method: how to find the solutions, corresponding to a consistent pair $\{F_1, \Phi\}$; this can also be extremely difficult.

The differential constraint may determine the possible class of solutions $\{u(x,t,\alpha)\}$, and, in many cases, this makes the procedure of seeking exact solutions algorithmic, rather than the trivial, random substitution of functions.

It is worth mentioning what is meant here by exact solutions. Indeed, the best opportunity is to detect the *explicit solutions* expressed in terms of elementary or, at least, known functions of mathematical physics (Euler's Gamma, Beta, elliptic, etc.), in terms of quadratures, and so on. But this is not always the case, even for simple semilinear PDEs. Therefore, *exact solutions* will mean those that can be obtained from some ODEs or, in general, from PDEs of *lower order* than the original PDE (0.1). For instance, such an extension of the notion of exact solutions was proposed by A.A. Dorodnitsyn in the middle of the 1960s.

In particular, our goal is to find a reduction of the PDEs to a finite number of ODEs representing a *dynamical system*.

Three-fold role of exact solutions: existence-uniqueness-asymptotics

Exact solutions of nonlinear models have always played a special role in the theory of nonlinear evolution equations. For difficult quasilinear PDEs or systems, exact solutions can often be the only possibility to formally describe the actual behavior of general, more arbitrary solutions. Furthermore, exact solutions are often crucial for developing general existence-uniqueness and asymptotic theory. There are many remarkable examples of important nonlinear models where an appropriate exact solution simultaneously reveals an optimal description of:

(i) local and global existence functional classes;

(ii) uniqueness classes; and,

(iii) classes of correct generic asymptotic behavior.

Actually, **(iii)** is well understood in rigorous or, more often, formal asymptotic analysis of nonlinear PDEs. The first two conclusions **(i)** and **(ii)** are harder to see and difficult to prove, even for reasonably simple evolution PDEs. Moreover, the particular space-time structure of such solutions may also detect useful features of the new methods and tools, which are necessary for studying general solutions. In the theory of parabolic reaction-diffusion equations, there exist seminal examples where the exact solutions determine the correct rescaled variables obtained via nonlinear transformations, in terms of which the Maximum Principle can be applied to extend regularity properties of these particular solutions to more general ones.

More and more often, modern theory of evolution PDEs deals with classes of extremely difficult, strongly nonlinear, higher-order equations with degenerate and singular coefficients. In particular, for at least twenty five years, a permanent source of such models is *thin film theory*, generating various fourth, sixth and higher-order thin film equations with non-monotone and non-divergent operators (essential parts of Chapters 3 and 6 are devoted to such equations). Bearing in mind the multi-dimensional setting in $I\!R^N$ for $N \geq 2$, it is unlikely that a rigorous, mathematically closed existence-uniqueness-regularity and singularity (blow-up) theory for these equations in different free-boundary settings will be developed soon. New exact solutions of thin film models will continue to supply us with a new regularity information that will be used to correct the existing methods in order to create a more general theory.

Linear invariant subspaces for nonlinear operators

As a key idea, we seek exact solutions of (0.1) on linear n-dimensional subspaces which in many cases are *invariant under the nonlinear operators* of the models. A formal general scheme for the approach is easy, though, as often happens, its abstract mathematical formulation leads to rather obscure explanations.

We define a subspace in terms of the linear span denoted by

$$W_n = \mathcal{L}\{f_1(x), ..., f_n(x)\},$$

with n unknown linearly independent basis functions $\{f_j(x)\}$. For instance, these functions are picked to be solutions of a given linear PDE

$$P[f] = 0, \tag{0.3}$$

where $P = P(D_x)$ is the *annihilator* of subspace W_n, in the sense that there holds $P : W_n \to \{0\}$. Then (0.1) is replaced by a system

$$\begin{cases} F[u] = 0, \\ \Phi[u] \in W_n, \end{cases} \tag{0.4}$$

where Φ is another unknown function (or, in general, a nonlinear operator). Using the annihilator (0.3), the second condition in (0.4) is written as a differential constraint

$$P[\Phi[u]] = 0.$$

Here the main difficulty appears: how to choose consistent pairs of operators Φ and P. We next can look for solutions in the form of finite expansions

$$\Phi[u(x,t)] = C_1(t)f_1(x) + ... + C_n(t)f_n(x) \in W_n \quad \text{for } t \in I\!R, \tag{0.5}$$

with unknown coefficients $\{C_j(t)\}$.

Finally, as the crucial step, assuming that the inverse Φ^{-1} exists, we demand the subspace W_n be *invariant* under the superposition of operators,

$$F \circ \Phi^{-1} : W_n \to W_n. \tag{0.6}$$

Then the operator $F \circ \Phi^{-1}$ is said to *preserve* or *admit* the subspace W_n. Substituting the expansion (0.5) into the PDE (0.1), most plausibly, leads to a low-dimensional reduction of the original PDE restricted to this invariant subspace.

In the case of first-order (in t) evolution PDEs with independent variables x and t,

$$u_t = F[u] \equiv F(u, u_x, u_{xx}, ...), \tag{0.7}$$

taking the identity $\Phi = I$ in (0.5), it follows that if W_n is invariant under F, then

$$F[u] = \Psi_1(C_1, ..., C_n)f_1 + ... + \Psi_n(C_1, ..., C_n)f_n \in W_n \quad \text{for } u \in W_n, \tag{0.8}$$

where $\{\Psi_j\}$ denote the expansion coefficients of $F[u]$ on W_n. Hence, (0.7) restricted to the invariant subspace W_n is the n-dimensional dynamical system (DS) for the expansion coefficients $\{C_j(t)\}$ in (0.5),

$$\begin{cases} C_1' = \Psi_1(C_1, ..., C_n), \\ \quad ... \quad ... \quad ... \\ C_n' = \Psi_n(C_1, ..., C_n). \end{cases} \tag{0.9}$$

For $n = 1$, 2, or 3, such DSs can often be studied on the phase-plane, or, at least, admit asymptotic analysis of some of their generic, stable orbits.

We will give several examples for which the above approach represents an easy way to predict such a linear structure of exact solutions. For instance, let us observe that, under the same invariance conditions, the second-order evolution equation

$$u_{tt} = F[u] \equiv F(u, u_x, u_{xx}, ...)$$

admits solutions (0.5), where $\Phi = I$, with a harder $2n$th-order DS,

$$\begin{cases} C_1'' = \Psi_1(C_1, ..., C_n), \\ \quad ... \quad ... \quad ... \\ C_n'' = \Psi_n(C_1, ..., C_n). \end{cases}$$

As a principal feature, this book can be viewed as a practical guide that introduces a number of techniques for constructing exact solutions of various nonlinear PDEs in $I\!R^N$ for arbitrary dimensions $N \geq 1$. Indeed, several such exact solutions can be obtained by other techniques including differential constraints which have been successfully developed algorithmically on the basis of computer symbolic manipulation techniques. Nevertheless, some other solutions, especially those of higher-order equations in $I\!R^N$, will be difficult to detect by such "purely computational" approaches. The ideas of linear invariant subspaces can play a decisive role in explaining such a geometric origin of invariant manifolds, the corresponding exact solutions, and extensions to other PDEs.

Examples: classic fundamental solutions belong to invariant subspaces

For linear homogeneous PDEs, the three-fold existence-uniqueness-asymptotics nature **(i)–(iii)** of exact solutions is straightforward in view of the classical concept of *fundamental* solutions of linear operators and convolution representations of general solutions. It is remarkable and surprising that, for a number of classical linear and quasilinear models, the *fundamental solutions are associated with linear subspaces invariant under nonlinear operators.*

The heat equation and linear subspace for its fundamental solution

Consider the canonical *heat equation* (HE)

$$u_t = \Delta u \quad \text{in} \;\; I\!R^N \times I\!R_+ \quad \left(\Delta = \sum_{i=1}^N \frac{\partial^2}{\partial x_i^2}\right). \tag{0.10}$$

Its *fundamental solution* denoted by $b(x, t)$ is given by the *Gaussian kernel*,

$$b(x, t) = (4\pi t)^{-\frac{N}{2}} \mathrm{e}^{-\frac{|x|^2}{4t}}, \tag{0.11}$$

and takes Dirac's delta $\delta(x)$ as initial data,

$$\lim_{t\to 0^+} b(x,t) = \delta(x), \tag{0.12}$$

where the convergence is understood in the sense of distributions.

As is well known in parabolic theory (see e.g., Friedman [205]), the structure of the Gaussian kernel in (0.11) illustrates *Tikhonov's uniqueness* (1935) [552] and local *existence* functional class of measurable functions,

$$\mathcal{U} = \big\{v(x) : \ \exists\ A > 0 \text{ and } a > 0, \ \text{ such that } |v(x)| \le A\mathrm{e}^{a|x|^2} \ \text{ in } I\!\!R^N\big\}.$$

Then the Cauchy problem for the HE with initial data $u_0(x) \in \mathcal{U}$ has a unique solution that is local in time and is given by the convolution

$$u(x,t) = b(\cdot,t) * u_0 \equiv (4\pi t)^{-\frac{N}{2}} \int_{I\!\!R^N} \mathrm{e}^{-\frac{|x-y|^2}{4t}}\, u_0(y)\,\mathrm{d}y. \tag{0.13}$$

By checking the convergence of the integral, it is easy to see that this formula guarantees the existence and uniqueness of the solution locally in time, at least for all $t < \frac{a}{4}$. In order to make the solution global in time, another growth condition should be imposed on the initial data, e.g., assuming that $|u_0(x)| \le A\mathrm{e}^{a|x|^{2-\varepsilon}}$, with an arbitrarily small constant $\varepsilon > 0$. In this case, the integral in (0.13) is finite for all $t > 0$.

The explicit formula (0.13) also determines the *asymptotic behavior* as $t \to \infty$ of global solutions. Namely, if initial data are integrable, $u_0 \in L^1(I\!\!R^N)$, and have unit mass, $\int u_0(x)\,\mathrm{d}x = 1$, as the fundamental solution does in (0.12), then

$$u(x,t) \approx b(x,t) \quad \text{for } t \gg 1. \tag{0.14}$$

It is convenient to express this asymptotic convergence in the rescaled sense by using the time-scaling factor $t^{N/2}$ as in (0.11). Then (0.14) reads

$$t^{\frac{N}{2}} \big|u(x,t) - b(x,t)\big| \to 0 \quad \text{as } t \to \infty \tag{0.15}$$

uniformly on expanding sets $\{|x| \le c\sqrt{t}\}$, where $c > 0$ is an arbitrary constant.

Invariant subspaces. The exponential structure of the fundamental solutions (0.11) suggests introducing the logarithmic variable

$$v(x,t) = \ln b(x,t) \equiv -\tfrac{N}{2}\ln(4\pi t) - \tfrac{1}{4t}\,|x|^2,$$

where the right-hand side belongs to the 2D subspace W_2 that is given by the span

$$W_2 = \mathcal{L}\{1, |x|^2\}. \tag{0.16}$$

The new function $v = \ln u$ satisfies the semilinear parabolic equation

$$v_t = \Delta v + |\nabla v|^2 \equiv F[v], \tag{0.17}$$

that contains the quadratic Hamilton–Jacobi operator $|\nabla v|^2$. Thus, the logarithmic change of variables leads to the nonlinear operator F in (0.17) that obviously preserves the subspace W_2. Substituting into (0.17) an arbitrary function

$$v(x,t) = C_0(t) + C_1(t)|x|^2 \in W_2 \quad \text{for } t \ge 0, \tag{0.18}$$

we find by calculating $\Delta|x|^2 = 2N$ and $|\nabla|x|^2|^2 = 4|x|^2$ that

$$C_0' + C_1'|x|^2 = F\big[C_0 + C_1|x|^2\big] \equiv 2NC_1 + 4C_1^2|x|^2.$$

This yields the dynamical system

$$\begin{cases} C_0' = 2NC_1, \\ C_1' = 4C_1^2, \end{cases} \tag{0.19}$$

which is easily integrated. The second equation implies that $C_1(t) = -\frac{1}{4t}$ up to translations in t. Therefore, $C_0' = -\frac{N}{2t}$, and this gives the fundamental solution (0.11) in terms of the original variable $u = \mathrm{e}^v$.

This analysis admits some easy and immediate extensions. Firstly, it is evident that the operator in (0.17) admits the $(N+1)$-dimensional invariant subspace

$$W_{N+1} = \mathcal{L}\{1, x_1^2, ..., x_N^2\} \implies v(x,t) = C_0(t) + C_1(t)x_1^2 + ... + C_N(t)x_N^2. \tag{0.20}$$

The DS then becomes $(N+1)$-dimensional,

$$\begin{cases} C_0' = 2\sum_{(i)} C_i, \\ C_j' = 4C_j^2, \quad j = 1, ..., N, \end{cases}$$

which can also be integrated. Secondly, one can consider the general invariant subspace of arbitrary quadratic polynomials

$$W_M = \mathcal{L}\{1, x_i, x_i x_j, \ i, j, = 1, ..., N\} \tag{0.21}$$

of dimension $M = \frac{N^2+3N+2}{2}$, where the expansion contains more coefficients generating an M-dimensional DS. Clearly, using the orthogonal transformations and translations, the exact solutions on the subspace (0.21) reduce to those on (0.20). But this is not the case for the corresponding second-order *hyperbolic* equation

$$v_{tt} = \Delta v + |\nabla v|^2$$

for which the family of solutions on the subspaces (0.21) and (0.20) differ essentially. In the corresponding DS, we have the second-order derivatives C_j'', and hence, both $\{C_j(0)\}$ and $\{C_j'(0)\}$ should be prescribed as initial data, so, for the subspace (0.21), it is a $2M$-dimensional DS.

The porous medium equation and linear subspace for its fundamental solution

For quasilinear parabolic equations for which convolution and eigenfunction expansion techniques are not applicable the determining features **(i)–(iii)** of exact solutions are not straightforward and demand different and difficult nonlinear mathematics. Consider the classic *porous medium equation* (PME)

$$u_t = \Delta u^m \quad \text{in } I\!R^N \times I\!R_+, \tag{0.22}$$

where $m > 1$ is a fixed exponent. By the Maximum Principle, the PME possesses nonnegative solutions $u(x,t)$, so that u^m makes sense for any non-integer value of m. The advanced theory of such degenerate parabolic equations that admit weak (generalized) solutions can be found in a number of monographs on parabolic PDEs;

see [148, 206, 245]. For $m = 1$, (0.22) reduces to the heat equation (0.10), so the PME can be viewed as its nonlinear extension.

Let us see if the quasilinear PME inherits some distinctive evolution properties available for the HE, and, especially, whether it admits a kind of *fundamental solution* to be understood, of course, in a different nonlinear way. The answer is yes, and the PME has the famous *Zel'dovich–Kompaneetz–Barenblatt* (ZKB, 1950) source-type self-similar solution that is again denoted by $b(x,t)$,

$$b(x,t) = t^{-kN} f(y), \quad y = \tfrac{x}{t^k}, \quad \text{where } k = \tfrac{1}{N(m-1)+2}\,. \tag{0.23}$$

The rescaled profile $f(y)$ is given explicitly,

$$f(y) = \big[A_0(a^2 - |y|^2)_+\big]^{\frac{1}{m-1}}, \quad \text{with the constant } A_0 = \tfrac{k(m-1)}{2m}\,, \tag{0.24}$$

where $(\cdot)_+$ denotes the positive part $\max\{(\cdot),0\}$. The constant $a > 0$ characterizes the preserved total mass of the solution. We want $b(x,t)$ to initially take Dirac's delta, as shown in (0.12). Direct computations yield the unique value of $a = a(m)$ (see e.g., [509, p. 21]),

$$\begin{gathered} 1 = \int_{I\!R^N} f(y)\,\mathrm{d}y \equiv N\,\omega_N \int_0^a z^{N-1}\big[A_0(a^2 - z^2)\big]^{\frac{1}{m-1}}\,\mathrm{d}z \\ \Longrightarrow \quad a^{\frac{2}{m-1}+N} = \pi^{-\frac{N}{2}} A_0^{-\frac{1}{m-1}} \frac{\Gamma(\frac{m}{m-1}+\frac{N}{2})}{\Gamma(\frac{m}{m-1})}\,, \end{gathered} \tag{0.25}$$

where Γ is Euler's Gamma function, and $\omega_N = \frac{2\pi^{N/2}}{N\Gamma(N/2)}$ denotes the volume of the unit ball in $I\!R^N$.

Returning to the rescaled fundamental profile (0.24), it follows that, unlike (0.11) for the heat equation, $b(x,t)$ is compactly supported in x for any $t > 0$. This is a striking property of the *finite propagation* for the quasilinear degenerate parabolic equation (0.22). At the free-boundary (interface), where $|y| = a$, the profile $f(y)$ has finite regularity, and $f^{m-1}(y)$ is just Lipschitz continuous.

Thus, it seems that the solutions of the HE and the PME correspond to entirely different functional settings. Nevertheless, a striking continuity with respect to the exponent m can be observed when passing to the limit as $m \to 1^+$ in (0.24). Then, using that, in (0.25), the ratio of Gamma functions is equal to $\big(\frac{m}{m-1}\big)^{N/2} + ...$, it is easy to conclude that, uniformly in y,

$$f(y) \to (4\pi)^{-\frac{N}{2}}\,\mathrm{e}^{-\frac{|y|^2}{4}} \quad \text{as } m \to 1^+,$$

where, on the right-hand side, there appears the rescaled Gaussian kernel of the fundamental solution (0.11). This means a continuous "branching" at $m = 1^+$ of the solution (0.24) from the fundamental solution (0.11) of the linear HE. Once more, this asserts using the term *fundamental solution* of the nonlinear PME.

It turns out that, in PME theory, the ZKB solution plays a similar three-fold role **(i)–(iii)**. Firstly, the inverse parabolic profile of the rescaled kernel $f(y)$ in (0.24) determines the class of uniqueness and local existence,

$$\mathcal{U} = \big\{v(x) \ge 0 : \ \exists\, A > 0 \text{ such that } v(x) \le A(1+|x|^2)^{\frac{1}{m-1}} \text{ in } I\!R^N\big\}.$$

It follows by comparison with the following separate variables blow-up solution:

$$u_*(x,t) = C_*|x|^{\frac{2}{m-1}}(T-t)^{-\frac{1}{m-1}}, \quad \text{with } C_* = \left[\tfrac{k(m-1)}{2m}\right]^{\frac{1}{m-1}},$$

that the weak solution exists, at least for all $t < T \sim A^{1-m}$. For global existence it suffices to restrict the growth rate at infinity, e.g., by assuming that $u_0(x) = O\big(|x|^{\frac{2}{m-1}-\varepsilon}\big)$ as $x \to \infty$ for some arbitrarily small $\varepsilon > 0$.

Secondly, in a similar manner, for nonnegative initial data $u_0 \in L^1(I\!R^N)$ with unit mass, $\int u_0(x)\,\mathrm{d}x = 1$, (0.14) holds. The asymptotic convergence (0.14) is again to be understood in the rescaled sense (0.15) with the time factor t^{kN}, instead of $t^{N/2}$. Note that $k = \frac{1}{2}$ for $m = 1$. The convergence is uniform on compact sets $\{|x| \le c\,t^k\}$, $c > 0$, corresponding to the new similarity variable y in (0.23).

Invariant subspaces. Though the ZKB-solution (0.23) is a classical example of self-similar solutions induced by a group of scaling transformations, let us now interpret it in terms of the same invariant subspace (0.16). The rescaled inverse parabolic profile (0.24) suggests using the new dependent variable

$$v = u^{m-1},$$

which is known as the *pressure* in the theory of filtration of liquids and gases in porous media. Most of the regularity results for the PME are formulated in terms of the pressure. Substituting $u = v^{\frac{1}{m-1}}$ into the PME yields the *pressure equation*

$$v_t = v\Delta v + \tfrac{1}{m-1}|\nabla v|^2 \equiv F[v]. \tag{0.26}$$

Similar to the transformed HE (0.17), the quadratic operator F in (0.26) preserves the subspace (0.16). Plugging (0.18) yields a slightly different dynamical system for the expansion coefficients,

$$\begin{cases} C_0' = 2NC_0C_1, \\ C_1' = \frac{2}{(m-1)k}\,C_1^2. \end{cases}$$

As in (0.19), the second equation is integrated independently, determining (0.23).

We easily reveal the dynamics on other extended subspaces of F in (0.26): the subspace (0.20) remains invariant, leading to the $(N{+}1)$-dimensional DS

$$\begin{cases} C_0' = 2C_0\sum_{(i)} C_i, \\ C_j' = 2C_j\sum_{(i)} C_i + \frac{4}{m-1}\,C_j^2, \quad j = 1, ..., N. \end{cases}$$

On the invariant subspace of arbitrary quadratic polynomials (0.21), the PME becomes an M-dimensional DS which is again reduced to that on the subspace (0.20) via rotations and translations. The quasilinear degenerate hyperbolic equation

$$v_{tt} = v\Delta v + \tfrac{1}{m-1}|\nabla v|^2$$

restricted to W_M becomes a $2M$th-order DS.

Elementary extensions to higher-order equations. We formally combine operators in (0.17) and (0.26), add extra operators, and create a fourth-order parabolic equation

$$v_t = F[v] \equiv -\alpha\Delta^2 v + \beta\Delta v + \gamma v\Delta v + \delta|\nabla v|^2 + \mu v + \nu, \tag{0.27}$$

with six arbitrary constants denoted by Greek letters. Such PDEs belong to the class of *Kuramoto–Sivashinsky equations* from flame propagation theory that will be studied in the subsequent chapters. Obviously, the fourth-order term $-\alpha \Delta^2 v$ vanishes on the invariant subspace (0.21), so (0.27) on W_M is an M-dimensional DS. The corresponding hyperbolic PDE is a *Boussinesq-type equation* from water-wave theory,

$$v_{tt} = -\alpha \Delta^2 v + \beta \Delta v + \gamma v \Delta v + \delta |\nabla v|^2 + \mu v + \nu,$$

which becomes a $2M$th-order DS on the same invariant subspace W_M.

Models, targets, prerequisites

On nonlinear models and PDEs to be considered

The underlying idea of invariant subspaces for nonlinear operators applies here to a large variety of nonlinear PDEs from many areas of mathematics, mechanics, and physics. Exact solutions on invariant subspaces arise in many quasilinear equations and various free-boundary problems from different applications. In this book, we will deal with various PDEs and models that exhibit some common nonlinear invariant features. Beyond this "invariant essence," many of the models have nothing in common and often belong to completely disjoint areas of mathematics.

We begin Chapter 1 with some history and present those classical and more recent examples of interesting solutions on invariant subspaces that were constructed in the twentieth century. In the rest of the book, we develop several techniques for constructing exact solutions that describe singularity behavior for various nonlinear PDEs, including (see Index for details and precise references)

- reaction-diffusion-absorption PDEs and combustion models;
- parabolic and hyperbolic PDEs with the p-Laplacian operators;
- gas dynamics models, including the *Kármán–Fal'kovich–Guderley equation*;
- fourth, sixth, and $2m$th-order *thin film equations*;
- fourth-order *Riabouchinsky–Proudman–Johnson equations*;
- free-boundary problems for the *Navier–Stokes equations* in $I\!R^2$;
- *Kuramoto–Sivashinsky equations* and extensions;
- *KdV-type equations* with blow-up, nonlinear dispersion PDEs with compactons;
- higher-order extensions of the *Rosenau–Hyman equation*;
- modifications of the *Fuchssteiner–Fokas–Camassa–Holm equations*;
- *Green–Naghdi equations*;
- *Harry Dym-type equations*;
- quasilinear pseudo-parabolic (*magma*) equations;
- quasilinear *wave* equations and dispersive *Boussinesq* models;
- *Zabolotskaya–Khokhlov-type equations*;
- *Zakharov–Kuznetsov equation with nonlinear dispersion*;
- quasilinear parabolic, hyperbolic, and *KdV-type* systems;
- *Maxwell equations* from nonlinear optics;
- *Monge-Ampère-type equations* of second and higher orders;
- *logarithmic Gauss curvature equations*;
- non-integrable PDEs admitting bilinear *Baker–Hirota* representations; etc.

In some cases, using exact solutions, we will describe interesting evolution properties that are related to singularity *blow-up* or *extinction* phenomena, *finite interface* propagation and regularity, with the special attention to *oscillatory, changing sign* behavior of weak solutions near interfaces. For several PDEs, this leads to many mathematical open problems, which we state when necessary. Most of the results are published for the first time.

Main problems and targets

There exist two main fundamental problems in invariant subspace theory:

• **Problem I, $\mathbf{F} \mapsto \{\mathbf{W_n}\}$**: *Given a nonlinear operator F, which invariant subspaces W_n does it preserve?*

• **Problem II, $\mathbf{W_n} \mapsto \{\mathbf{F}\}$**: *Given a subspace W_n, which nonlinear operators F admit it?*

In addition, there are a number of other practical questions, e.g.,

• *Which operators F admit higher-dimensional invariant subspaces as further extensions of the basic, simple invariant subspaces?*

• *Is there a well-defined procedure to detect invariant subspaces and their maximal dimensions* (*i.e., maximal dynamical systems that are restrictions of the PDE to the subspace*)*?*

Problem I is fundamental, and is key for the existence of lower-dimensional reductions of the PDEs. For arbitrary operators F, this does not admit a complete solution, but we will successfully study Problem I for many particular classes of nonlinear differential and discrete operators.

On the contrary, Problem II admits a complete algorithmic solution. It was solved for $N = 1$, i.e., for ordinary differential operators, by the second author of the book [544, 545] in terms of Lie–Bäcklund symmetries of linear ODEs. For general operators in $I\!R^N$, Problem II was solved in Kamran–Milson–Olver [312] by introducing a new approach to the annihilating differential operators. Nevertheless, as often happens in mathematics, a complete algorithmic solution does not assume easy practical applications of the results. It is said in [312, p. 316] that (for operators in $I\!R^N$) "The formulae for the affine annihilators and annihilators are often extremely complicated, even for relatively simple subspaces." Bearing in mind the practical aspects of calculations, the geometric concepts of invariant subspaces will continue to play an important role.

The general problem of finding invariant subspaces for wide classes of nonlinear operators in $I\!R^N$ is not completely solved here. We suspect that such a problem cannot be tackled with sufficient generality. Nevertheless, for quadratic and polynomial operators in $I\!R$, we present a complete classification of some types of invariant subspaces. We also introduce examples of invariant subspaces and exact solutions for classes of multi-dimensional quasilinear operators in $I\!R^N$.

Partially invariant subspaces: invariant sets

Another related direction of our analysis is the construction of *invariant sets* $M \subset W_n$ *on a linear subspace* W_n for operator F. This simply means that W_n is *partially invariant*, i.e., $F[W_n] \nsubseteq W_n$, but, for some part M of W_n,

$$F[M] \subseteq W_n. \tag{0.28}$$

The principal difference from the invariant subspaces for which $F : W_n \to W_n$ is that condition (0.28) leads to an *overdetermined* DS for the expansion coefficients $\{C_j(t)\}$ in (0.5).

Let us illustrate this for equation (0.7), assuming that W_n is not invariant under F in the sense of (0.6) with $\Phi = I$. Suppose, for instance, that F maps W_n onto an $(n+s)$-dimensional subspace, so that s new functions appear in the expansion

$$F : W_n \to W_{n+s} = \mathcal{L}\{f_1, ..., f_n, f_{n+1}, ..., f_{n+s}\},$$

and, instead of (0.8),

$$F[u] = \Psi_1(\cdot) f_1 + ... + \Psi_n(\cdot) f_n$$
$$+ \Psi_{n+1}(\cdot) f_{n+1} + ... + \Psi_{n+s}(\cdot) f_{n+s} \in W_{n+s}.$$

This leads to the same DS (0.9) accompanied by s extra algebraic conditions

$$\begin{cases} \Psi_{n+1}(C_1, ..., C_n) = 0, \\ \quad ... \qquad ... \qquad ... \\ \Psi_{n+s}(C_1, ..., C_n) = 0. \end{cases}$$

Such overdetermined DSs are not always consistent and are hard to study. The proof of the existence of the corresponding solutions on M becomes more involved, though we present a number of nonlinear evolution PDEs for which such overdetermined DSs are consistent.

Partial invariance as a manifestation of "partial integrability"

We discuss the principal link to integrable equations which admit countable sets of exact N-soliton and other solutions. We illustrate this by starting with the most classical integrable *Korteweg–de Vries* (KdV) *equation*

$$u_t + 6uu_x + u_{xxx} = 0, \tag{0.29}$$

which has been known since the 1870s and was first derived by J. Boussinesq. Following the standard scheme for integrable PDEs (see Newell [436, Ch. 4]), we apply the change $u = w_x$, yielding the *potential KdV equation*

$$w_t + 3(w_x)^2 + w_{xxx} = 0.$$

Next, setting

$$w = 2(\ln|v|)_x = \tfrac{2v_x}{v}, \quad \text{so that} \quad u = 2(\ln|v|)_{xx}, \tag{0.30}$$

reduces it to the homogeneous quadratic equation

$$F_*[v] \equiv vv_{xt} - v_x v_t + vv_{xxxx} - 4v_x v_{xxx} + 3(v_{xx})^2 = 0. \tag{0.31}$$

As a final step, the *Baker–Hirota bilinear method*† [284] is applied to derive a countable set of N-solitons $\{v_k(x,t)\}$, such that each solution $v_k(x,t)$ of (0.31) belongs to a *linear subspace* of exponential functions. We will use various linear subspaces to illustrate finite-dimensional dynamics, which exist for equation (0.31) and related models and correspond to well-known soliton-type solutions.

1-soliton on subspace $\mathbf{W_2^{exp}}$. This is the simplest *travelling wave* (TW) given by a single exponent,

$$v_1(x,t) = 1 + e^{\theta_1(x,t)}, \quad \text{where} \quad \theta_1(x,t) = p_1 x - p_1^3 t \tag{0.32}$$

and $p_1 \neq 0$ is a constant. Clearly, in this case, the 2D linear subspace (a module)

$$W_2^{\rm exp} = \mathcal{L}\{1, e^{p_1 x}\} \tag{0.33}$$

is invariant under the quadratic operator F_* in (0.31). Indeed, as usual, looking for solutions of (0.31) on $W_2^{\rm exp}$,

$$v(x,t) = C_1(t) + C_2(t) e^{p_1 x}, \tag{0.34}$$

and plugging it into (0.31) yields a single term, $(C_1 C_2' - C_2 C_1' + p_1^3 C_1 C_2) e^{p_1 x} = 0$, since the coefficient of the highest-degree exponential $e^{2p_1 x}$ vanishes, as the integrability demands. Therefore, the PDE (0.31) on $W_2^{\rm exp}$ reduces to the single ODE (an *underdetermined* DS)

$$C_1 C_2' - C_2 C_1' = -p_1^3 C_1 C_2 \quad \Longrightarrow \quad \left(\tfrac{C_2}{C_1}\right)' = -p_1^3 \tfrac{C_2}{C_1}, \tag{0.35}$$

so, on integration, $C_2(t) = A C_1(t) e^{-p_1^3 t}$, where A is a constant. Here, $C_1(t) \neq 0$ is an arbitrary smooth function that is eliminated by the differential change (0.30). Thus, up to an arbitrary multiplier $C_1(t)$, (0.34) represents the 1-soliton solution (0.32) belonging to the invariant subspace $W_2^{\rm exp}$. Notice that exact solutions (0.34) on $W_2^{\rm exp}$ can satisfy various PDEs involving operator F_*, e.g.,

$$\begin{aligned} \alpha v_{tt} + \beta v_t = {} & F_*[v] + \mu v + \nu + \sigma v_{xx} + \rho\big[v v_{xx} - (v_x)^2\big] \\ & + \varepsilon(v v_{xxx} - v_x v_{xx}) + \lambda\big[v v_{xxxx} - (v_{xx})^2\big] + \dots, \end{aligned} \tag{0.36}$$

with some linear and nonlinear operators preserving the subspace (0.33).

2-soliton on $\mathbf{W_4^{exp}}$. The 2-solitons are composed of three exponential patterns

$$v_2(x,t) = 1 + e^{\theta_1} + e^{\theta_2} + C_4 e^{\theta_1 + \theta_2}, \tag{0.37}$$

where, as in (0.32), $\theta_1 = p_1 x - p_1^3 t$, $\theta_2 = p_2 x - p_2^3 t$, and $p_1 \neq p_2$. In soliton theory [436, p. 123], applying the Baker–Hirota differential operator to (0.31) yields

$$C_4 = \left(\tfrac{p_1 - p_2}{p_1 + p_2}\right)^2.$$

As above, we can interpret (0.37) by using the linear subspace (module)

$$W_4^{\rm exp} = \mathcal{L}\{1, e^{p_1 x}, e^{p_2 x}, e^{(p_1 + p_2)x}\}.$$

† The bilinear differential operator in (0.31), transformations, such as (0.30), hierarchies of the KdV and KP equations, hyperelliptic representation of periodic solitons, etc., were introduced by H.F. Baker in 1903, [22]; see details in Athorne–Eilbeck–Enolskii [20, p. 275].

For instance, if $W_4^{\rm exp} = \mathcal{L}\{1, {\rm e}^x, {\rm e}^{2x}, {\rm e}^{3x}\}$ (this assumes more nonlinear interaction between terms than for the standard 2-soliton, see below), looking for

$$v(x,t) = C_1(t) + C_2(t){\rm e}^x + C_3(t){\rm e}^{2x} + C_4(t){\rm e}^{3x} \tag{0.38}$$

and substituting into (0.31) yields

$$\begin{aligned} &(C_1C_2' - C_2C_1' + C_1C_2){\rm e}^x + 2(C_1C_3' - C_3C_1' + 8C_1C_3){\rm e}^{2x} \\ &+ \left[C_2C_3' - C_3C_2' + 3(C_1C_4' - C_4C_1') + C_2C_3 + 81C_1C_4\right]{\rm e}^{3x} \\ &+ 2(C_2C_4' - C_4C_2' + 8C_2C_4){\rm e}^{4x} + (C_3C_4' - C_4C_3' + C_3C_4){\rm e}^{5x} = 0. \end{aligned} \tag{0.39}$$

Hence, for the given module $W_4^{\rm exp}$, there exists another module $\tilde{W}_5^{\rm exp}$ such that

$$F_* : W_4^{\rm exp} \to \tilde{W}_5^{\rm exp} = \mathcal{L}\{{\rm e}^x, {\rm e}^{2x}, {\rm e}^{3x}, {\rm e}^{4x}, {\rm e}^{5x}\}$$

(the coefficients of 1 and ${\rm e}^{6x}$ vanish). Equating the five coefficients in (0.39) to zero yields the *overdetermined* system of five equations for four functions

$$\begin{cases} C_1C_2' - C_2C_1' = -C_1C_2, \\ C_1C_3' - C_3C_1' = -8C_1C_3, \\ C_2C_3' - C_3C_2' + 3(C_1C_4' - C_4C_1') + C_2C_3 + 81C_1C_4 = 0, \\ C_2C_4' - C_4C_2' = -8C_2C_4, \\ C_3C_4' - C_4C_3' = -C_3C_4. \end{cases} \tag{0.40}$$

According to (0.28), the last two ODEs (projections onto ${\rm e}^{4x}$ and ${\rm e}^{5x}$) determine an *invariant set* M on $W_4^{\rm exp}$, in the sense that $F[u] \in W_4^{\rm exp}$ for all $u \in M$. Hence, the module $W_4^{\rm exp}$ is *partially invariant*. Writing all the ODEs (0.40), excluding the third one, in the form of (0.35) and integrating gives

$$C_2(t) = AC_1(t){\rm e}^{-t}, \quad C_3(t) = BC_1(t){\rm e}^{-8t}, \quad \text{and} \quad C_4(t) = DC_1(t){\rm e}^{-9t},$$

where, as above, $C_1(t)$ is arbitrary, and A, B, and D are constants. Plugging these expressions into the long third ODE in (0.40), rewritten in the form of

$$C_2^2\left(\tfrac{C_3}{C_2}\right)' + 3C_1^2\left(\tfrac{C_4}{C_1}\right)' + C_2C_3 + 81C_1C_4 = 0,$$

we obtain a single relation between constants, $AB = 9D$. This gives two exact solutions of 2-soliton type

$$v(x,t) = 1 \pm \left({\rm e}^{x-t} + 9b{\rm e}^{2(x-4t)}\right) + b{\rm e}^{3(x-3t)} \quad (b \in I\!R).$$

A similar interpretation of general N-soliton solutions means that, for the integrable equation (0.31),

$$\exists \text{ solutions on partially invariant modules } W_n^{\rm exp} \text{ for arbitrarily large } n.$$

In this sense, the fully integrable equations represent an exceptional limit case of evolution PDEs that possess exact solutions belonging to *an infinite number of invariant sets on linear exponential subspaces (modules) of arbitrarily large dimension.*

The invariance under the nonlinear operators can be treated as a kind of a *partial integrability property* (cf. "*...remnants of integrability*" [192, p. 573]), in the sense that we describe classes of nonlinear non-integrable PDEs for which only a *finite* *number* of invariant subspaces W_n, or sets with exact solutions, can be detected. In fact, for any arbitrarily large l, there exists a family of nonlinear non-integrable PDEs possessing at least l different types of solutions (looking like "N-solitons") on linear invariant subspaces W_n, or on sets, with n large enough (see Section 1.5.2). Such PDEs may be treated as *intermediate*, i.e., between general equations with no invariant properties at all, and the rather thin class of fully integrable PDEs.

Trigonometric subspace $\mathbf{W_3^{tr}}$: TWs. We next try solutions

$$v(x,t) = C_1 + C_2 \cos \gamma x + C_3 \sin \gamma x \in W_3^{\rm tr} = \mathcal{L}\{1, \cos \gamma x, \sin \gamma x\}, \qquad (0.41)$$

where $\gamma \in I\!R$ is a parameter. $W_3^{\rm tr}$ is *invariant* under F_*, so that restricting the PDE (0.31) to $W_3^{\rm tr}$ yields three ODEs

$$\begin{cases} C_2 C_3' - C_3 C_2' + 4\gamma^3 (C_2^2 + C_3^2) = 0, \\ C_1 C_3' - C_3 C_1' + \gamma^3 C_1 C_2 = 0, \\ C_2 C_1' - C_1 C_2' + \gamma^3 C_1 C_3 = 0. \end{cases}$$

The matrix of this first-order DS is singular and nontrivial solutions are possible for $C_1(t) \equiv 0$. This gives the TW

$$v(x,t) = \sin(\gamma x + 4\gamma^3 t).$$

In terms of the original function $u = 2(\ln |v|)_{xx}$, such solutions describe moving blow-up singularities with the following behavior near the poles:‡

$$u(x,t) \sim \frac{1}{(x - x_0(t))^2}, \quad \text{where} \quad x_0(t) = -4\gamma^2 t + \text{constant}. \qquad (0.42)$$

A slight modification of the KdV equation (0.31) produces another interesting evolution on $W_3^{\rm tr}$. For instance,

$$v(x,t) = 1 + \cos(x+t) \equiv 2\cos^2\left[\tfrac{1}{2}(x+t)\right] \quad \text{satisfies} \quad F_*[v] = 4.$$

This function, being extended by zero in $\{(x,t) : \frac{1}{2}|x+t| \ge \frac{\pi}{2}\}$, becomes a smooth *compacton*. Such compact structures entered nonlinear dispersion theory in the 1980s. We will discuss their mathematical well-posedness in Chapters 3–7.

Polynomial subspace $\mathbf{W_4^{p}}$: second rational solution. Similarly, equation (0.31) can be considered on the polynomial subspace such as

$$W_4^{\rm p} = \mathcal{L}\{1, x, x^2, x^3\}, \ \text{ i.e., } v(x,t) = C_1(t) + C_2(t)x + C_3(t)x^2 + C_4(t)x^3, \qquad (0.43)$$

which leads to a similar DS. Solving it yields

$$v(x,t) = 36t + b^2 x + 3bx^2 + 3x^3 \quad (b \in I\!R),$$

which, by (0.30), gives the second *rational* solution $u_2(x,t)$ of the KdV equation with the singular behavior (0.42) near poles (these are known since 1978, see survey [407]; the first rational solution is elementary, $u_1(x,t) = -\frac{2}{x^2}$).

‡ The study of the Schrödinger operator with the inverse square potential $U(x) \sim (x - x_0)^{-2}$ goes back to Hardy (1920), [280] (Hardy's inequality for embeddings of functional L^2 spaces with singular weights) and Friedrichs (1935), [208].

Polynomial-trigonometric subspace $W_4{}^{p,t}$: positons. Consider next the subspace $W_4^{p,t} = \mathcal{L}\{1, x, \cos x, \sin x\}$, composed of basis functions of subspaces in (0.43) and (0.41). Plugging the expansion on $W_4^{p,t}$ into equation (0.31) gives the solutions

$$v(x,t) = C_1(t) + C_2(t)x + C_3(t)\cos x + C_4(t)\sin x, \quad \text{where}$$
$$\begin{cases} C_1C_2' - C_2C_1' + C_3C_4' - C_4C_3' + 2\left(C_3^2 + C_4^2\right) = 0, \\ C_3C_1' - C_1C_3' + C_1C_4 - 3C_2C_3 = 0, \\ C_1C_4' - C_4C_1' + C_1C_3 + 3C_2C_4 = 0, \\ C_2C_4' - C_4C_2' + C_2C_3 = 0, \\ C_3C_2' - C_2C_3' + C_2C_4 = 0. \end{cases}$$

The first three ODEs are projections of the PDE onto 1, $\cos x$, and $\sin x$ respectively, while the last two represent the expansion coefficients of $x\cos x$ and $x\sin x$ that do not belong to $W_4^{p,t}$. Similar to (0.40), this DS yields two solutions

$$v(x,t) = \pm(3t + x) + \sin(x + t),$$

which are indeed the *positon solutions* of the KdV equation. Such positons, or harmonic breathers, have been recognized since the 1980s, [16, 418]. They exhibit the same type (0.42) of singularity (for continuous integrable models, all known positons have singularities), but a different behavior as $x \to \infty$. Similarly, the polynomial-exponential subspace $W_4^{p,e} = \mathcal{L}\{1, x, \cosh x, \sinh x\}$ leads to the *negatons*, that were first constructed in 1996, [485].

Exponential-trigonometric subspace $W_4{}^{e,t}$: complexitons. We now look for solutions of (0.31) on the trigonometric-exponential subspace,

$$v(x,t) = C_1(t)\cos x + C_2(t)\sin x + C_3(t)e^x + C_4(t)e^{-x}.$$

Substituting yields five ODEs being the projections of (0.31) onto 1, $e^x \sin x$, $e^x \cos x$, $e^{-x}\cos x$, and $e^{-x}\sin x$ respectively,

$$\begin{cases} -C_2C_1' + C_1C_2' + 2C_4C_3' - 2C_3C_4' + 4\left(C_1^2 + C_2^2\right) + 16C_3C_4 = 0, \\ -C_3C_1' - C_3C_2' + (C_1 + C_2)C_3' - 4C_2C_3 = 0, \\ -C_3C_1' + C_3C_2' + (C_1 - C_2)C_3' - 4C_1C_3 = 0, \\ C_4C_1' + C_4C_2' - (C_1 + C_2)C_4' - 4C_1C_4 = 0, \\ -C_4C_1' + C_4C_2' + (C_1 - C_2)C_4' - 4C_2C_4 = 0. \end{cases}$$

These are easily integrated by adding and subtracting two pairs of similar ODEs. Besides TWs, we obtain one more solution

$$v(x,t) = \cos(x - 2t) + \sinh(x + 2t), \tag{0.44}$$

which is determined up to an arbitrary smooth multiplier $C(t)$. This is precisely the *complexiton* solution that was constructed rather recently, [405]. Concerning a perturbed equation, note that $v(x,t) = \sin(x - 4t)$ solves $F_*[v] = 8$, and $v(x,t) = \cos(x - 2t) + \cosh(x + 2t)$ (cf. (0.44)) satisfies the equation $F_*[v] = 12$.

We have illustrated all types of known elementary soliton-type solutions of (0.31). All these solutions of the KdV equation can be constructed by the Wronskian method for integrable equations; a modern description is given in [407]. Similar DS reductions are also key for classes of non-integrable PDEs, though, of course, the consistency of DSs can be tricky and will be established in a few cases.

Sign-invariants for second-order parabolic equations (Chapter 8)

We also aim to emphasize a new interesting aspect of our exact solutions. It turns out that, for *second-order* parabolic equations, many solutions on invariant subspaces W_n may induce so-called *sign-invariants*, which are nonlinear differential operators $\mathcal{H}[u] = H(x, u, Du, D^2u, ...)$ preserving both their signs on evolution orbits. For the Cauchy problem in $I\!R^N \times I\!R_+$ with initial data $u_0(x)$, this means that

$$\mathcal{H}[u_0(x)] \le 0 \ (\ge 0) \text{ in } I\!R^N \implies \mathcal{H}[u(x,t)] \le 0 \ (\ge 0) \text{ in } I\!R^N \text{ for } t > 0. \quad (0.45)$$

Such *partial differential inequalities* are naturally associated with different *barrier techniques* in the theory of parabolic equations, where the Maximal Principle applies to control the operator signs on evolution orbits. Barrier approaches are the cornerstone of regularity and asymptotic theory of linear and nonlinear parabolic PDEs. For instance, classical Schauder and Bernstein estimates, as well as the Nash–Moser technique, are based on the Maximum Principle and use barrier analysis of parabolic differential inequalities. We refer to monographs [206, 245, 442, 472, 550].

In (0.45), the sign-invariant $\mathcal{H}$ preserves both signs, ≥ 0 and ≤ 0, on solutions of the parabolic PDE. The connection with invariant subspaces W_n is as follows:

$$\mathcal{H}[u] = 0 \quad \text{on } W_n \ \ (\text{or on a set } \ M \subset W_n). \qquad (0.46)$$

Vice versa, the equality (0.46) can be used to determine the corresponding sign-invariant $\mathcal{H}[u]$. We will show how to reconstruct such operators $\mathcal{H}$ by means of the structure of the invariant (or partially invariant) subspaces W_n. Of course, (0.46) is then a differential constraint generating solutions on W_n. It is important that, unlike just the constraint (0.46), the partial differential inequalities (0.45) characterize evolution properties of wider classes of solutions than simply those on W_n.

Discrete operators: applications to moving mesh methods and lattices (Chapter 9)

We will also deal with *discrete* nonlinear operators F for which we prove some results on the existence of linear invariant subspaces W_n and construct exact solutions of some discrete equations. As a further application, we describe invariant aspects of *moving mesh methods* (MMMs), which have become a powerful tool of numerical solutions of nonlinear PDEs possessing blow-up and other evolution singularities. We also introduce exact solutions on invariant subspaces for some *anharmonic lattices* associated with different evolution PDEs.

Prerequisites: a GUIDE *on models, nonlinear PDEs, and solutions*

The book is meant for advanced graduate level students and does not assume a knowledge of the fundamentals of the mathematical theory of PDEs and functional analysis, except the basics of the Maximum Principle for second-order parabolic equations in the theory of sign-invariants in Chapter 8 (though we have included necessary preliminary information). The knowledge of some standard aspects of ODE theory would be useful for performing some analytical manipulations and phase-plane diagrams. Sometimes our discussions around exact solutions on invariant subspaces include specific aspects of PDE theory. These parts can be omitted without causing any future confusion.

We hope that the present methods for parabolic, hyperbolic, KdV-type, and nonlinear dispersion PDEs, as well as discrete equations, will be useful for the readers with a mathematical background that is not necessarily applied or pure. We expect that several aspects of our analysis can be fruitful for researchers and students specializing in mechanics, physics, engineering, and those working with nonlinear PDEs.

Acknowledgements

This book is dedicated to our teacher, Sergey Pavlovich Kurdyumov, who taught us to detect, understand and respect exact solutions of nonlinear models, and how to extract the crucial information from each of the solutions that will be used in both theory and applications of PDEs. In the 1970s, S.P. Kurdyumov, with his former PhD students G.G. Elenin, A.P. Mikhailov, and N.V. Zmitrenko, created the methodology, and, eventually, a deep philosophy of blow-up singularities and heat localization in nonlinear media, starting with two simple, unusual exact solutions of the reaction-diffusion equations, exhibiting the paradoxical phenomena of regional blow-up (called then the *S-regimes*). This methodology determined several directions of blow-up theory for two decades to come, and is partially reflected in monographs [509], [245], and [226].

The authors would like to thank the following people for long-term collaboration: V.A. Dorodnitsyn, N.H. Ibragimov, J.R. King, S.I. Pohozaev, and S.A. Posashkov, who were the co-authors of some of the papers that we have used in the presentation of the results. During the last twenty-five years, the authors had the privilege of discussing various aspects of PDE theory, mathematical modeling, and group theory with many experts in mechanics, differential equations, and other areas of pure and applied mathematics. We would like to thank our colleagues and friends, G.I. Barenblatt, S. Kamin, R. Kersner, A.E. Shishkov, and J.L. Vazquez for fruitful discussions and useful comments or suggestions concerning nonlinear PDEs and their solutions that are used in this book. We especially thank J.D. Evans, E.V. Ferapontov, R.S. Fernandes, P.J. Harwin, O.V. Kaptsov, P.J. Olver, and P. Rosenau for discussions and valuable comments at the final stage.

This book was completed during regular visits of the second author to the Department of Mathematical Sciences, University of Bath, to which both authors are thankful for the permanent support and encouragement. The second author would like to thank the London Mathematical Society, The Royal Society, and the European INTAS for funding these visits and research.

Victor A. Galaktionov,
Sergey R. Svirshchevskii,
Bath–Moscow, November 2004

CHAPTER 1

Linear Invariant Subspaces in Quasilinear Equations: Basic Examples and Models

We begin this chapter with a few well-known and even classical examples of exact solutions of various nonlinear PDEs of mathematical physics with quadratic or cubic nonlinearities. We will treat these solutions from the point of view of the linear subspaces invariant under appropriate nonlinear operators. Indeed, ideas of low-dimensional reductions of evolution equations restricted to linear subspaces or manifolds have been known for a long time. Certainly there are other interesting solutions of a similar invariant nature, which we are not aware of. It would be interesting to detect more examples which date back to the first half of the twentieth and, hopefully, to the nineteenth century.

The rest of the chapter is devoted to further examples, in which we introduce empirical tools to study general properties of invariant subspaces, spaces of the corresponding nonlinear ordinary differential operators, and exact solutions. More systematic and advanced mathematics is developed in Chapter 2.

1.1 History: first examples of solutions on invariant subspaces

1.1.1 Five models from gas dynamics

Example 1.1 (**Ovsiannikov solutions**) In 1948, L.V. Ovsiannikov [455] showed that the study of spatial transonic flows of ideal polythropic gas leads to the following quasilinear elliptic-hyperbolic equation* in $I\!R^3$:

$$\boxed{\Delta u \equiv u_{xx} + u_{yy} = [(u-1)^2]_{zz} \equiv F[u],} \tag{1.1}$$

where $u = u(x, y, z)$ is the reduced projection of the flow velocity on the z-axis. Equation (1.1) is hyperbolic in the domain $\{(x, y, z) \in I\!R^3 : \ u(x, y, z) > 1\}$ and is elliptic in $\{u < 1\}$. Ovsiannikov detected its exact solutions in the following form:

$$u(x, y, z) = 1 + u_0(x, y) + u_1(x, y)z + \tfrac{1}{12}\, u_2(x, y)z^2. \tag{1.2}$$

Substituting this expression into (1.1) and equating the coefficients of 1, z, and z^2 (projections onto these functions) to zero yields that the functions $u_0(x, y)$, $u_1(x, y)$, and $u_2(x, y)$ satisfy the following system of elliptic PDEs in $I\!R^2$:

$$\begin{cases} \Delta u_0 = \frac{1}{3}u_2 u_0 + 2u_1^2, \\ \Delta u_1 = u_2 u_1, \\ \Delta u_2 = u_2^2. \end{cases} \tag{1.3}$$

Actually, existence of such solutions as (1.2), (1.3) reflects the straightforward fact

* We put boxes around the main PDEs possessing solutions on invariant subspaces.

that the linear subspace defined by the span $W_3 = \mathcal{L}\{1, z, z^2\}$ is *invariant* under the nonlinear operator F in (1.1), in the sense that

$$\text{for any } u \in W_3, \quad F[u] \in W_3 \quad (\text{or} \quad F[W_3] \subseteq W_3).$$

This invariance of W_3 under F is understood, as in standard linear algebra.

Of course, the system of three PDEs (1.3) is not easy to study in general, but it is a low-dimensional system for three functions defined in $I\!R^2$, unlike the original PDE (1.1) that is posed in $I\!R^3$. Moreover, the last equation for u_2 is independent of the others and can be studied separately.† Once this has been solved and a suitable function $u_2(x, y)$ has been determined, the rest of (1.3) yields a system of linear elliptic PDEs for u_0 and u_1 that can be studied by standard techniques.

This class of *Ovsiannikov's solutions*, as well as applied problems of analytical fluid mechanics [29] and other important applications in combustion theory [594], stimulated mathematical interests to such canonical semilinear elliptic PDEs

$$\Delta u = f(u), \tag{1.4}$$

with a given nonlinear function $f(u)$. The typical power nonlinearity is $f(u) = \pm u^p$, with the exponent $p > 1$, or $f(u) = |u|^{p-1}u$ for solutions u of changing sign. In elliptic theory, two classes of problems were most popular:

(i) the Dirichlet problem in a bounded domain $\Omega \subset I\!R^N$, with $u = 0$ on the boundary $\partial\Omega$, and

(ii) the problem in the whole space $I\!R^N$.

In the former case, for $f(u) = -u^p$, the famous *Sobolev critical exponent* occurs

$$p_S = \tfrac{N+2}{N-2} \quad \text{for } N \geq 3 \quad (p_S = \infty \text{ for } N = 1, \text{ or } 2).$$

The global and local properties of solutions are completely different in the subcritical, $p < p_S$, and the supercritical, $p > p_S$, ranges. In the critical case, $p = p_S$, there exists the explicit *Loewner–Nirenberg solution* [400]

$$u(x) = \Big[\frac{N(N-2)\lambda^{\frac{2}{N-2}}}{N(N-2)+\lambda^{\frac{4}{N-2}}|x|^2}\Big]^{(N-2)/2},$$

where $\lambda > 0$ is arbitrary. The questions of existence, nonexistence, and multiplicity of solutions for equation (1.4) have been actively studied in general elliptic theory during the last forty years. We refer to classical papers [331, 464] and to Mitidieri–Pohozaev [425] for history, references, and a systematic treatment of the nonexistence problem via the nonlinear capacity approach.

Example 1.2 (**Von Mises solutions**) Consider the potential equation of the 1D flow of a compressible gas

$$\boxed{\Phi_{tt} + 2\Phi_x\Phi_{xt} + a\Phi_{xx} + b\Phi_t\Phi_{xx} + c(\Phi_x)^2\Phi_{xx} = 0,} \tag{1.5}$$

where a, b, and c are constants. R. von Mises introduced the following class of exact

† L.V. Ovsiannikov was the supervisor, who proposed the equation $\Delta u = u^2$ in a bounded domain to S.I. Pohozaev in 1958 [467], that led to "Pohozaev's Identities" (1965) [464] in elliptic theory.

solutions of (1.5) (this is mentioned in Titov [553], we have not succeeded in tracing out the original von Mises work):

$$\Phi(x,t) = C_1(t) + C_2(t)x + C_3(t)x^2. \tag{1.6}$$

Plugging (1.6) into (1.5) yields an ODE system for the expansion coefficients $\{C_i\}$,

$$\begin{cases} C_1'' = -2C_2C_2' - 2aC_3 - 2bC_3C_1' - 2cC_2^2C_3, \\ C_2'' = -4(C_2C_3)' - 2bC_3C_2' - 8cC_2C_3^2, \\ C_3'' = -2(4+b)C_3C_3' - 8cC_3^3. \end{cases} \tag{1.7}$$

Similar to the example above, the finite expansion (1.6) indicates that the operator on the left-hand side of (1.5) composed of linear, quadratic, and cubic terms preserves the 3D subspace

$$W_3 = \mathcal{L}\{1, x, x^2\}.$$

The last equation for C_3 in (1.7) can be solved independently in terms of Jacobi elliptic functions.

In view of differential manipulations with expansion coefficients in square products on the right-hand side of (1.7), it is relevant to call W_3 an *invariant module*, which in Algebra [373, Ch. III] is used as a generalization of linear vector spaces with a field replaced by a ring; see Section 2.8. For simplicity, we sometimes keep using the term subspace if no confusion is likely.

Example 1.3 **(Guderley solutions)** Consider the potential equation for transonic flow written as

$$\boxed{\Phi_{yy} + \tfrac{N-1}{y}\,\Phi_y = (\gamma+1)\Phi_x\Phi_{xx} \ \text{ in } \{x>0,\ y>0\},} \tag{1.8}$$

where $N = 1$ or 2 and $\gamma = \frac{c_p}{c_v} > 1$ is the fixed constant, called the *adiabatic exponent*. From Guderley's book [267, p. 65]: "The solution of the exact potential equation of the flow in the throat of DE LAVAL nozzle has been obtained by MEYER [422] in the form of a series expansion. We shall show that the first term of this expansion represents the exact solution of the equation for transonic flow." K.G. Guderley presented two explicit solutions of (1.8),

$$\Phi(x,y) = \tfrac{c}{2}\,x^2 + \tfrac{c^2}{2}\,(\gamma+1)xy^2 + \tfrac{c^3}{24}(\gamma+1)^2y^4 \ \text{ for } \ N=1,$$
$$\Phi(x,y) = \tfrac{c}{2}\,x^2 + \tfrac{c^2}{4}(\gamma+1)xy^2 + \tfrac{c^3}{64}(\gamma+1)^2y^4 \ \text{ for } \ N=2;$$

see [267, p. 66, 69] (we keep the original notation).

These explicit *Guderley's solutions* belong to the subspace $W_3 = \mathcal{L}\{1, x, x^2\}$ which is invariant under the quadratic operator $F[\Phi] = \Phi_x\Phi_{xx}$ given on the right-hand side of (1.8). In addition, Guderley described properties of solutions

$$\Phi(x,y) = x^3 f(y)$$

belonging to the 1D invariant subspace $\mathcal{L}\{x^3\}$ of F. "The exponent of x could then be chosen such that the powers of x would cancel out from the equation," [267, p. 69]. Such solutions were also studied by H. Görtler [259].

These results altogether are expressed by saying that operator F admits the 4D invariant subspace

$$W_4 = \mathcal{L}\{1, x, x^2, x^3\}$$

with exact solutions

$$\Phi(x, y) = C_1(y) + C_2(y)x + C_3(y)x^2 + C_4(y)x^3$$

governed by the eighth-order DS

$$\begin{cases} C_1'' + \frac{N-1}{y}\, C_1' = 2(\gamma + 1)C_2C_3, \\ C_2'' + \frac{N-1}{y}\, C_2' = 2(\gamma + 1)\big(2C_3^2 + 3C_2C_4\big), \\ C_3'' + \frac{N-1}{y}\, C_3' = 18(\gamma + 1)C_3C_4, \\ C_4'' + \frac{N-1}{y}\, C_4' = 18(\gamma + 1)C_4^2. \end{cases}$$

Guderley's solutions correspond to $C_4(y) \equiv 0$. The last equation is the radial version of the quadratic elliptic PDE (1.4) with $f(u) = 18(\gamma + 1)u^2$.

Example 1.4 **(Titov's solutions)** It was shown by S.S. Titov [555] that the same quadratic operator $F[\Phi] = \Phi_x \Phi_{xx}$ admits another 3D subspace

$$W_3 = \mathcal{L}\big\{1, x^{\frac{3}{2}}, x^3\big\}.$$

This gives *Titov's solutions* of (1.8)

$$\Phi(x, y) = C_1(y) + C_2(y)x^{\frac{3}{2}} + C_3(y)x^3 \in W_3,$$

where the coefficients of the expansion satisfy the following ODE system:

$$\begin{cases} C_1'' + \frac{N-1}{y}\, C_1' = \frac{9}{8}(\gamma + 1)C_2^2, \\ C_2'' + \frac{N-1}{y}\, C_2' = \frac{45}{4}(\gamma + 1)C_2C_3, \\ C_3'' + \frac{N-1}{y}\, C_3' = 18(\gamma + 1)C_3^2. \end{cases}$$

Example 1.5 **(Ryzhov–Shefter solutions)** The *Lin–Reissner–Tsien* (LRT) *equation*

$$-\varphi_x\varphi_{xx} + \varphi_{yy} + \varphi_{zz} - 2\varphi_{xt} = 0 \quad \text{in } I\!R^3 \times I\!R \tag{1.9}$$

was discovered in 1948 [395] as a model for oscillation of a thin profile in transonic flow. O.S. Ryzhov and G.M. Shefter derived this equation later "...for the investigation of nonstationary processes in the vicinity of the surface of transition through the speed of sound in Laval nozzles when the dimensions and form of the critical cross section change with time sufficiently rapidly," [505, p. 939] In cylindrical coordinates

$$\begin{cases} z = r\cos\vartheta, \\ y = r\sin\vartheta, \end{cases}$$

(1.9) takes the form

$$\boxed{-\varphi_x\varphi_{xx} + \varphi_{rr} + \frac{1}{r}\,\varphi_r + \frac{1}{r^2}\,\varphi_{\vartheta\vartheta} - 2\varphi_{xt} = 0,} \tag{1.10}$$

and admits the following exact *Ryzhov–Shefter solutions*, 1959 (we keep the original notation from [505]):

$$\varphi = \lambda(t)x + \tfrac{1}{2}\,A(t)[x - \Delta(t)]^2 + h_1(\vartheta, t)[x - \Delta(t)]r^2 + h_2(\vartheta, t)r^4. \tag{1.11}$$

The expansion coefficients solve the following *PDE system*:

$$\begin{cases} \lambda_t + \frac{1}{2} A\lambda = A\Delta_t, \\ 2A_t + A^2 = h_{1\vartheta\vartheta} + 4h_1, \\ 2h_{1t} + h_1 A = h_{2\vartheta\vartheta} + 16h_2. \end{cases}$$

As shown in Example 1.3, solutions (1.11) are associated with the invariant subspace $W_3 = \mathcal{L}\{1, x, x^2\}$ of the operator $\varphi_x \varphi_{xx}$. There exists its 4D invariant extension $W_4 = \mathcal{L}\{1, x, x^2, x^3\}$. There are other more detailed invariant interpretations. For instance, taking the subspace $W_6 = \mathcal{L}\{1, x, r^2, x^2, xr^2, r^4\}$ and hence solutions

$$\varphi(x, r, \vartheta, t) = C_1 + C_2 x + C_3 x^2 + C_4 r^2 + C_5 x r^2 + C_6 r^4$$

yields the following system of PDEs for the coefficients $\{C_i(\vartheta, t)\}$:

$$\begin{cases} 2C_2C_3 = 4C_4 + C_{4\vartheta\vartheta} - 2C_{2t}, \\ 4C_3^2 = 4C_5 + C_{5\vartheta\vartheta} - 4C_{3t}, \\ 2C_3C_5 = 16C_6 + C_{6\vartheta\vartheta} - 2C_{5t}, \\ C_{1\vartheta\vartheta} = 0, \;\; C_{2\vartheta\vartheta} = 0, \;\; C_{3\vartheta\vartheta} = 0. \end{cases} \tag{1.12}$$

Solutions (1.11) then correspond to

$$C_1 = \tfrac{1}{2} A\Delta^2, \;\; C_2 = \lambda - A\Delta, \;\; C_3 = \tfrac{1}{2} A, \;\; C_4 = -h_1\Delta, \;\; C_5 = h_1, \;\; C_6 = h_2.$$

The general solution of (1.12) is as follows:

$$C_1 = a_1(t)\vartheta + b_1(t), \;\; C_2 = a_2(t)\vartheta + b_2(t), \;\; C_3 = a_3(t)\vartheta + b_3(t),$$

$$\begin{aligned} C_4 &= K_1 \cos 2\vartheta + K_2 \sin 2\vartheta + \alpha\vartheta^2 + \beta\vartheta + \gamma, \\ C_5 &= \tilde{K}_1 \cos 2\vartheta + \tilde{K}_2 \sin 2\vartheta + \tilde{\alpha}\vartheta^2 + \tilde{\beta}\vartheta + \tilde{\gamma}, \\ C_6 &= (\mu_1 + \nu_1\vartheta)\cos 2\vartheta + (\mu_2 + \nu_2\vartheta)\sin 2\vartheta, \end{aligned}$$

where $K_{1,2}(t)$, $\tilde{K}_{1,2}(t)$ are arbitrary functions, and other coefficients $\alpha(t)$, $\beta(t)$, ... are expressed by functions $\{a_i(t), b_i(t)\}$ by substituting into the PDE (1.10). Other PDE systems occur by studying (1.10) on the 3D invariant subspace $\mathcal{L}\{1, r^2, r^4\}$.

1.1.2 Nonlinear wave equation

Example 1.6 (Quadratic wave equation) Ovsiannikov [456, p. 286] performed a classification of group-invariant solutions of the following system:

$$\begin{cases} u_y = v_x, \\ u u_x = v_y, \end{cases}$$

which also describes transonic gas flows. This is equivalent to the *quadratic wave equation* $u_{yy} = (uu_x)_x$, or, replacing $y \mapsto t$,

$$\boxed{u_{tt} = F[u] \equiv \tfrac{1}{2}(u^2)_{xx} \quad \text{in } \mathbb{R} \times \mathbb{R}} \tag{1.13}$$

Olver and Rosenau introduced the following explicit solution of (1.13):

$$u(x, t) = \alpha t^2 + at + b \pm \sqrt{2\alpha}\, x, \quad \text{where } \alpha > 0, \tag{1.14}$$

that is "...not obtainable by partial invariance by appending the second order side condition

$$u_{tt} = 2\alpha, \tag{1.15}$$

where α is a constant," [448, p. 112].

These solutions belong to the 3D invariant subspace $W_3 = \mathcal{L}\{1, x, x^2\}$ preserved by operator F in (1.13). Plugging

$$u(x,t) = C_1(t) + C_2(t)x + C_3(t)x^2 \in W_3 \tag{1.16}$$

into the PDE yields the following DS:

$$\begin{cases} C_1'' = C_2^2 + 2C_1C_3, \\ C_2'' = 6C_2C_3, \\ C_3'' = 6C_3^2. \end{cases} \tag{1.17}$$

The solutions (1.14) then correspond to the particular case $C_3(t) \equiv 0$, where the second ODE is $C_2'' = 0$. Choosing $C_2(t) = \pm\sqrt{2\alpha}$ yields $C_1'' = 2\alpha$, whence come solutions (1.14). On the other hand, taking $C_2(t) = \alpha t$ ($\alpha \neq 0$) leads to the new polynomial solution

$$u(x,t) = \tfrac{\alpha^2}{12}\,t^4 + at + b + \alpha t x.$$

Fixing now a nontrivial solution $C_3(t) = \frac{1}{t^2}$ of (1.17) yields Euler's ODE for C_2,

$$t^2 C_2'' = 6C_2 \quad \Longrightarrow \quad C_2(t) = At^3 + \tfrac{B}{t^2}, \tag{1.18}$$

where A and B are arbitrary constants of integration. Finally, solving the first ODE yields a more general family of solutions on W_3 ($D,\ E \in I\!R$),

$$u(x,t) = \tfrac{A^2t^8}{54} + \tfrac{ABt^3}{2} + \tfrac{B^2}{4t^2} + Dt^2 + \tfrac{E}{t} + \left(At^3 + \tfrac{B}{t^2}\right)x + \tfrac{1}{t^2}\,x^2.$$

For $\alpha = 0$ in the side condition (1.15), the explicit solution [448, p. 112] is

$$u(x,t) = \pm(t+a)\sqrt{x+b},$$

which, after translation, belongs to the 1D invariant subspace $W_1 = \mathcal{L}\{\sqrt{x}\}$. The dynamics on W_1 with solutions $u(x,t) = C(t)\sqrt{x}$ is described by the ODE

$$C'' = 0.$$

1.1.3 *Quadratic Boussinesq-type equations*

Example 1.7 (**Olver–Rosenau solution**) In 1986, Olver and Rosenau [448] considered the following *Boussinesq-type equation*:

$$\boxed{u_{tt} = F[u] \equiv u_{xx} + \beta(u^2)_{xx} + \gamma\, u_{xxtt} \quad \text{in } I\!R \times I\!R,} \tag{1.19}$$

which was introduced by Boussinesq in 1871 [74] for studying long waves in shallow water. This equation also describes longitudinal waves in solid rods with effects of lateral inertia included. In [448], the following *Olver–Rosenau solution* of (1.19) was constructed:

$$u(x,t) = -\tfrac{1}{2\beta} + \tfrac{3\gamma}{2\beta t^2} + \tfrac{1}{2\beta t^2}\,x^2, \tag{1.20}$$

where parameters of translations in x and t are not included. Therefore, this is a two-parameter family of solutions.

Such a simple solution initiated a discussion on general invariant group origins of exact solutions. Written in the form of

$$u(x,t) = -\tfrac{1}{2\beta} + \varphi(t)\psi(x), \quad \text{with } \varphi(t) = \tfrac{1}{2\beta t^2} \text{ and } \psi(x) = x^2 + 3\gamma,$$

the solution looks like a standard affine version of a separable solution (i.e., becoming separable after shifting in u by $-\frac{1}{2\beta}$), and hence is expected to be obtained by local group approaches dealing with groups of scaling or other non-classical methods. Nevertheless, it was proved that "...the entire two-parameter family could not have come from a single local group," [448, p. 111].

Concerning the invariant subspace treatment of (1.20), it is easy to observe the subspace $W_2 = \mathcal{L}\{1, x^2\}$ that is invariant under the quadratic operator F in (1.19). As done in Example 1.6, we take solutions (1.16) on the extended subspace W_3, and, on substitution into the PDE, obtain the system

$$\begin{cases} C_1'' = 2C_3 + 2\beta C_2^2 + 4\beta C_1 C_3 + 2\gamma\, C_3'', \\ C_2'' = 12\beta C_2 C_3, \\ C_3'' = 12\beta C_3^2. \end{cases}$$

As far as explicit solutions are concerned, the last equation gives

$$C_3(t) = \tfrac{1}{2\beta t^2}.$$

Substituting into the second ODE yields Euler's equation (1.18). Finally, the following solutions of the Boussinesq-type equation (1.19) are obtained:

$$u(x,t) = -\tfrac{1}{2\beta} + \tfrac{\beta A^2 t^8}{27} + \beta A B t^3 + \left(\tfrac{3\gamma}{2\beta} + \tfrac{\beta B^2}{2}\right)\tfrac{1}{t^2}$$
$$+ Dt^2 + \tfrac{E}{t} + \left(At^3 + \tfrac{B}{t^2}\right)x + \tfrac{1}{2\beta t^2}\,x^2.$$

Bearing in mind translations in x and t, this is a six-parameter family of solutions which, for $A = B = D = E = 0$, gives the Olver–Rosenau solution (1.20).

1.1.4 Examples from reaction-diffusion-absorption theory

We next turn the attention to nonlinear reaction-diffusion-absorption PDEs which have given a record number of various exact solutions, including those on invariant subspaces. The basic nonlinear diffusion operator in such parabolic equations was already derived by J. Boussinesq [77] , who, in 1904, studied non-stationary flows of soil water under the presence of *free surface*, and derived the PDE

$$u_t = \gamma\,(uu_x)_x. \tag{1.21}$$

Here, $\gamma = \frac{k}{m}$ is a positive constant, where k is the *filtration coefficient* and m is the *porosity* of soil. The function $u = u(x,t)$ is the pressure of the ground water. Here, (1.21) is the quadratic *porous medium equation* (PME). Boussinesq also derived the exact solution of the PME (1.21) in separate variables

$$u(x,t) = X(x)T(t).$$

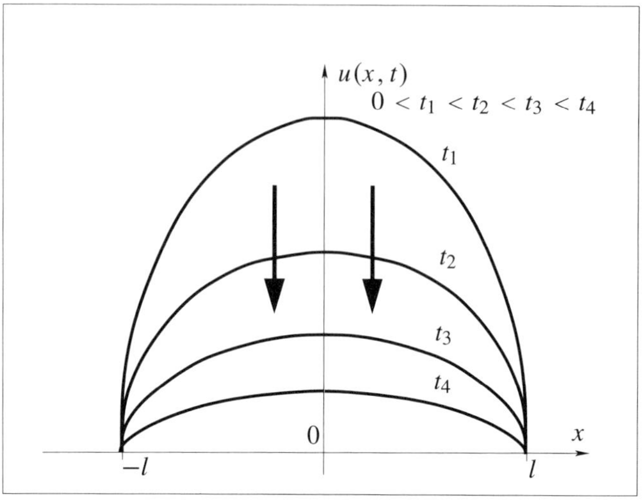

Figure 1.1 Evolution described by the Boussinesq solution (1.22).

Plugging into (1.21) yields two independent ODEs for functions $T(t)$ and $X(x)$,

$$\frac{T'}{T^2} = \gamma \, \frac{(XX')'}{X} = -\lambda,$$

where $\lambda > 0$ is the parameter of separation. Solving the first equation leads to the so-called *Boussinesq solution*

$$u(x,t) = \frac{X(x)}{\lambda t}, \tag{1.22}$$

where $X \geq 0$ is a solution of the ODE

$$\gamma \, (XX')' = -\lambda X.$$

Solving this ODE on a bounded interval $x \in (-l, l)$ with the zero Dirichlet boundary conditions

$$X(-l) = X(l) = 0$$

yields the *Boussinesq ordered regime* that describes the time decay of solutions of the initial-boundary value problem for the PME on a bounded interval. See Figure 1.1. The fact that the Boussinesq solution (1.22) is asymptotically stable and that the corresponding decay rate $O\left(\frac{1}{t}\right)$ for $t \gg 1$ is correct for general solutions of the PME for arbitrary bounded initial data $u(x,0) = u_0(x) \geq 0$ was proved much later in the 1970s; see details and references in [245, Ch. 2].

For the PME in the whole space, i.e., for $x \in I\!R$ (the Cauchy problem), the famous *Zel'dovich–Kompaneetz–Barenblatt* (ZKB) *solution* is key for stability analysis as $t \to \infty$. We have discussed the ZKB solution in the Introduction (see (0.23)) and refer to a great amount of literature in [245] concerning the foundation of PME theory.

More complicated spatio-temporal patterns can occur for the PME with extra low-order operators, such as reaction or absorption ones. There are many models of this type. For instance, the PME with a nonlinear *convection* term

$$u_t = \gamma\,(uu_x)_x + \beta uu_x,$$

also known as the *diffusion-convection Boussinesq equation*, occurs in the various fields of petroleum technology and ground water hydrology. Let us begin with another example, where the interesting exact solutions on invariant subspaces arise.

Example 1.8 (**PME with absorption: Kersner's solution**) Consider the exact solution constructed by R. Kersner in the middle of the 1970s; see references in [333, 334]. At that time, Kersner was a PhD student supervised by A.S. Kalashnikov, who performed in the 1960s-70s the pioneering research of localization-extinction phenomena for nonlinear degenerate parabolic PDEs, including equations from diffusion-absorption theory. Key results are reflected in his fundamental survey [309]. Among Kalashnikov's other PDE models, there is a famous diffusion-absorption equation with the *critical* absorption exponent

$$v_t = \left(v^{\sigma} v_x\right)_x - v^{1-\sigma}, \tag{1.23}$$

where $\sigma > 0$ is a parameter. In filtration theory, according to G.I. Barenblatt, absorption power-like terms $-v^p$ describe the phenomenon of seepage on a permeable bed. The Cauchy problem for equation (1.23) admits weak nonnegative compactly supported solutions. The first explicit localized solutions of such diffusion-absorption equations were constructed by L.K. Martinson and K.B. Pavlov in 1972; see details and references in [509, p. 21].

Let us derive explicit solution of (1.23) using the invariant subspaces. Introducing the *pressure* variable from filtration theory, $u = v^{\sigma}$, yields a PDE with the quadratic differential operator and a constant sink,

$$\boxed{u_t = F[u] \equiv uu_{xx} + \tfrac{1}{\sigma}(u_x)^2 - \sigma.} \tag{1.24}$$

Clearly, operator $F[u]$ preserves the 2D subspace $W_2 = \mathcal{L}\{1, x^2\}$, since

$$F[C_1 + C_2x^2] = 2C_1C_2 - \sigma + 2\big(1 + \tfrac{2}{\sigma}\big)C_2^2x^2 \in W_2.$$

Therefore, (1.24) admits solutions

$$u(x,t) = C_1(t) + C_2(t)x^2, \tag{1.25}$$

with the expansion coefficients $C_1(t)$ and $C_2(t)$ satisfying the dynamical system

$$\begin{cases} C_1' = 2C_1C_2 - \sigma, \\ C_2' = 2\big(1 + \tfrac{2}{\sigma}\big)C_2^2. \end{cases} \tag{1.26}$$

Integrating the uncoupled second ODE and substituting

$$C_2(t) = -\tfrac{\sigma}{2(\sigma+2)t}$$

into the first equation yields *Kersner's solution* (1976)

$$u(x,t) = \left[A_0 t^{-\frac{\sigma}{\sigma+2}} - \tfrac{\sigma(\sigma+2)}{2(\sigma+1)}\,t - \tfrac{\sigma}{2(\sigma+2)t}\,x^2\right]_+,$$

where A_0 is an arbitrary constant. Despite its elementary structure, the solution is

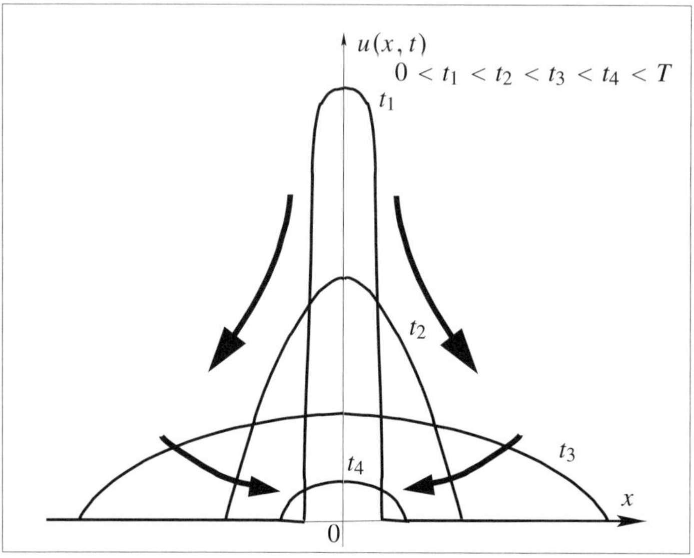

Figure 1.2 Finite-time extinction for the PME with absorption (1.23) described by Kersner's solutions (1.25); T is the extinction time, so $u(x, T) \equiv 0$.

not group-invariant if $A_0 \neq 0$. The positive part $[\cdot]_+$ determines weak solutions of (1.24) with finite interfaces, so they describe interesting and principal phenomena of non-Darcy interface propagation with turning points, extinction patterns, quenching, etc. Figure 1.2 shows this unusual extinction behavior. Similar explicit solutions also exist for the multi-dimensional PME with absorption in $I\!R^N \times I\!R_+$ (Martinson, 1979, [414])

$$u_t = \nabla \cdot (u^\sigma \nabla u) - u^{1-\sigma},$$

and for other extended PME-type models, see [509, p. 103].

Example 1.9 (**Oron–Rosenau solution**) In 1986, A. Oron and P. Rosenau considered the following fast diffusion equation with absorption [453]:

$$v_t = (\sqrt{v}\,)_{xx} - \sqrt{v}, \tag{1.27}$$

which, in plasma physics, describes energy diffusion in a strong magnetic field in the presence of energy sinks due to plasma radiation. It was shown that (1.27) admits the *Oron–Rosenau solution*

$$v(x,t) = B^2(x)\big(C_0 \textstyle\int \frac{\mathrm{d}x}{B^2(x)} - t\big)^2, \tag{1.28}$$

where C_0 is a constant and $B(x)$ satisfies the ODE

$$B'' + 2B^2 - B = 0.$$

Bearing in mind the idea of invariant subspaces, we derive the quadratic version of (1.27) by setting $v = u^2$, to obtain the PDE

$$F[u] \equiv 2uu_t = u_{xx} - u. \tag{1.29}$$

In the space of smooth functions of the time-variable t, operator F in (1.29) admits the 2D subspace

$$W_2 = \mathcal{L}\{1, t\}.$$

Since

$$F[C_1 + C_2 t] = 2C_1C_2 + 2C_2^2 t \in W_2,$$

there exist the corresponding solutions

$$u(x,t) = C_1(x) + C_2(x)t \in W_2. \tag{1.30}$$

On substitution into (1.29), we obtain the following fourth-order DS:

$$\begin{cases} C_1'' - C_1 = 2C_1C_2, \\ C_2'' - C_2 = 2C_2^2. \end{cases}$$

Since $C_2C_1'' = C_1C_2''$, on integration, we have

$$C_2C_1' = C_1C_2' + C_0,$$

with a constant C_0. Integrating again yields

$$C_1(x) = C_0C_2(x) \int \frac{\mathrm{d}x}{C_2^2(x)},$$

which yields the solution (1.28) with $B = -C_2$.

Example 1.10 (**Dyson–Newman solution**) In 1980, W.I. Newman [437] considered the following quasilinear parabolic equation:

$$\boxed{u_t = F[u] \equiv \tfrac{1}{2}(uu_x)_x + u(1-u).} \tag{1.31}$$

It is a quasilinear extension of the *Kolmogorov–Petrovskii–Piskunov–Fisher* (KPPF) *equation* of population genetics,

$$u_t = \tfrac{1}{2}\, u_{xx} + u(1-u),$$

which, since the 1930s, induced several fundamental directions in mathematical theory of nonlinear parabolic PDEs. The original KPP-paper (1937) [353] contains a number of famous mathematical ideas and results.

As stated in [437], using the idea from a personal communication with F. Dyson (1978), Newman looked for solutions composed of the hyperbolic cosine. To be precise, in terms of invariant subspaces, solutions take the form

$$u(x,t) = C_1(t) + C_2(t)\cosh x, \tag{1.32}$$

belonging to the subspace $W_2 = \mathcal{L}\{1, \cosh x\}$ which is invariant under the quadratic operator F in (1.31). Then the expansion coefficients satisfy the DS

$$\begin{cases} C_1' = -C_1^2 - \frac{1}{2}\, C_2^2 + C_1, \\ C_2' = -\frac{3}{2}\, C_1C_2 + C_2. \end{cases} \tag{1.33}$$

Unlike a simpler quadratic DS (1.26), system (1.33) cannot be solved explicitly, but is integrated in quadratures, giving interesting properties of finite-front propagation and evolution to traveling waves in such nonlinear media. In particular, this

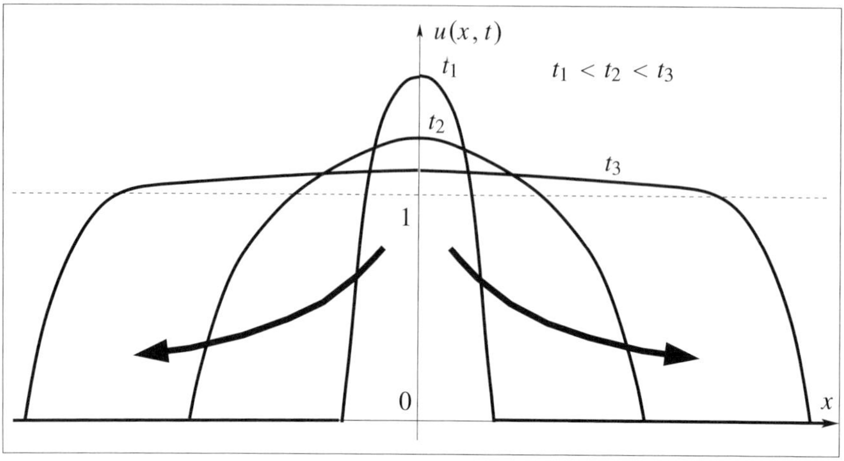

Figure 1.3 Formation of a traveling wave in the quasilinear model (1.31) described by Dyson–Newman's solution (1.32).

Dyson–Newman's solution propagates for $t \gg 1$ with the asymptotic speed $\frac{1}{2}$. See Figure 1.3. There are other applications of such solutions in the theory of reaction-absorption PDEs; see [509, p. 106] and references therein.

Example 1.11 **(Blow-up: Galaktionov's solution)** The semilinear heat equation

$$\boxed{u_t = F[u] \equiv u_{xx} + (u_x)^2 + u^2 \quad (u > 0),} \tag{1.34}$$

which was introduced to PDE theory in 1979 (see [245, Ch. 9] for history), plays a decisive role in blow-up combustion problems. This is the only *semilinear* reaction-diffusion equation of the second order that generates the *regional blow-up* (*S-regime*) for which bell-shaped solutions blow up on spatial intervals of the length 2π, [509, p. 294]. The change $u = \ln v$ transforms (1.34) into a semilinear heat equation,

$$v_t = v_{xx} + v \ln^2 v, \tag{1.35}$$

where the reaction term, $q(v) = v \ln^2 v$, is "almost" linear as $v \to +\infty$, but, nevertheless, satisfies the *Osgood criterion* of blow-up,

$$\int^{\infty} \tfrac{ds}{q(s)} < \infty.$$

Therefore, solutions of (1.35) with sufficiently large initial data blow-up in finite time creating unusual localized blow-up patterns. Mathematical analysis of such blow-up localization phenomena uses specific stability techniques from singular perturbation theory and exact solutions; see details in books [509, Ch. 4] and [245, Ch. 9].

Operator $F[u]$ in (1.34) preserves the 2D subspace $W_2 = \mathcal{L}\{1, \cos x\}$ [232, 217]. Thus, for arbitrary C_1 and C_2,

$$F[C_1 + C_2 \cos x] = C_1^2 + C_2^2 + (2C_1 - 1)C_2 \cos x \in W_2.$$

This gives the exact solutions of (1.34) of the form

$$u(x,t) = C_1(t) + C_2(t) \cos x, \tag{1.36}$$

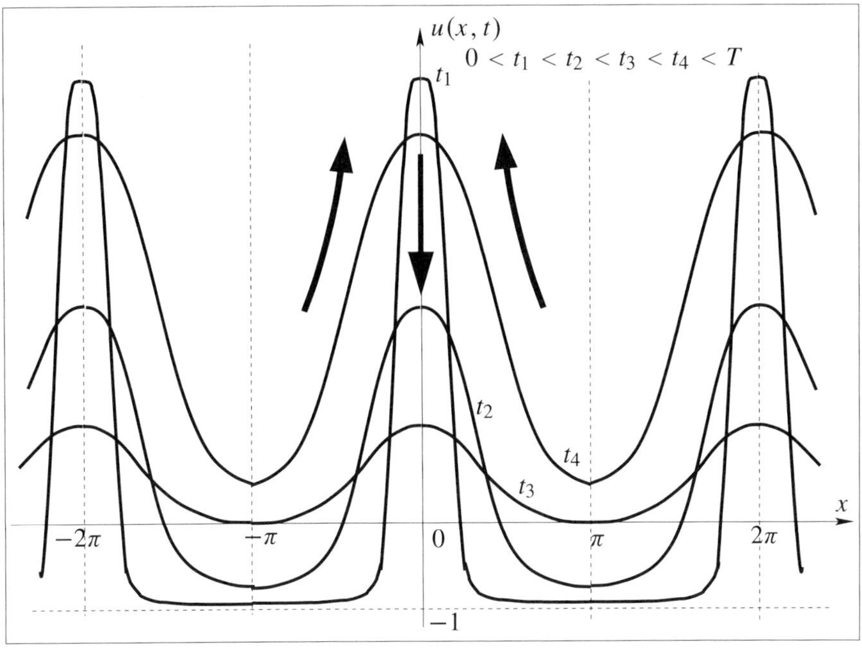

Figure 1.4 Non-monotone blow-up evolution of the invariant solutions (1.36), (1.37).

where the coefficients $C_1(t)$ and $C_2(t)$ satisfy the DS

$$\begin{cases} C_1' = C_1^2 + C_2^2, \\ C_2' = (2C_1 - 1)C_2. \end{cases} \tag{1.37}$$

This is not integrated explicitly and is studied on the phase-plane. In Figure 1.4 the non-monotone with time behavior of such explicit solutions is shown. These describe two singularities: the initial collapse of Dirac's delta-type initial data posed at points $\pm 2\pi k$, and finite-time blow-up afterwards. It is curious that this exact 2π-periodic (in x) *Galaktionov's solution* (1.36), (1.37) [217, 232] is not localized and blow-up globally as $t \to T^-$ at any point $x \in I\!R$. The blow-up rate is strikingly *non-uniform* [245, p. 242]: as $t \to T^-$, at maxima $x = 0$ and minima points $x = \pm\pi$, respectively,

$$u(0, t) = \tfrac{1}{T-t}\,(1 + o(1)) \to +\infty \quad \text{and}$$

$$u(\pm\pi, t) = \tfrac{1}{2}\,|\ln(T - t)|(1 + o(1)) \to +\infty.$$

Nevertheless, the intersection comparison with such exact solutions guarantees that any bell-shaped blow-up solution of (1.34) is spatially *effectively* localized as $t \to T^-$ on intervals of length 2π, [245, p. 258].

Example 1.12 (**Parabolic system: King's first solution**) The following system of two second-order PDEs:

$$\boxed{\begin{cases} v_t = (w v_x - v w_x)_x, \\ w_t = v_{xx}\,, \end{cases}} \tag{1.38}$$

is a simple model for the solid-state diffusion of a substitutional impurity by a vacancy mechanism; see King [340] and references therein. In this paper, among other results on explicit and similarity solutions, it was shown that (1.38) admits exact polynomial *King's first solution*

$$v(x,t) = C_1(t) + C_2(t)x + C_3(t)x^2 + C_4(t)x^3,$$
$$w(x,t) = D_1(t) + D_2(t)x + D_3(t)x^2 + D_4(t)x^3,$$

where the expansion coefficients solve the DS

$$\begin{cases} C_1' = 2(D_1C_3 - C_1D_3), \\ C_2' = 2(D_2C_3 - C_2D_3) + 6(D_1C_4 - C_1D_4), \\ C_3' = 6(D_2C_4 - C_2D_4), \\ C_4' = 4(D_3C_4 - C_3D_4), \\ D_1' = 2C_3, \;\; D_2' = 6C_4, D_3' = 0, \;\; D_4' = 0. \end{cases}$$

Here, operators in the right-hand sides of (1.38) preserve the 4D subspace $W_4 = \mathcal{L}\{1, x, x^2, x^3\}$. For the operator $F_1[v, w] = (wv_x - vw_x)_x$ from the first equation, this means that $F_1 : W_4 \times W_4 \to W_4$.

The second polynomial expansion detected in [340] is as follows:

$$v(x,t) = C_1(t) + C_2(t)x + C_3(t)x^2 + C_4(t)x^3 + C_5(t)x^4,$$
$$w(x,t) = D_1(t) + D_2(t)x + D_3(t)x^2,$$

with the resulting DS

$$\begin{cases} C_1' = 2(D_1C_3 - C_1D_3), \\ C_2' = 2(D_2C_3 - C_2D_3) + 6D_1C_4, \\ C_3' = 6D_2C_4 + 12D_1C_5, \\ C_4' = 4D_3C_4 + 12D_2C_5, \\ C_5' = 10D_3C_5, \;\; D_1' = 2C_3, \\ D_2' = 6C_4, \;\; D_3' = 12C_5. \end{cases}$$

Note that components v and w belong to different subspaces,

$$v \in W_5 = \mathcal{L}\{1, x, x^2, x^3, x^4\}, \;\; w \in W_3 = \mathcal{L}\{1, x, x^2\}, \;\; \text{so} \;\; F_1 : W_5 \times W_3 \to W_5.$$

Example 1.13 (Fast diffusion equation: King's second solution) The following construction is also due to King [342]. Using in the fast diffusion equation

$$v_t = \left(v^{-\frac{3}{2}} v_x\right)_x$$

the pressure transformation $u = v^{-3/2}$ reduces it to the equation with quadratic nonlinearities

$$\boxed{u_t = F[u] \equiv u u_{xx} - \tfrac{2}{3}(u_x)^2.} \tag{1.39}$$

This possesses exact *King's second solution*

$$u(x,t) = C_1(t) + C_2(t)x + C_3(t)x^2 + C_4(t)x^3.$$

Plugging this into the PDE yields the DS

$$\begin{cases} C_1' = 2C_3C_1 - \frac{2}{3}\,C_2^2, \\ C_2' = 6C_1C_4 - \frac{2}{3}\,C_2C_3, \\ C_3' = 2C_2C_4 - \frac{2}{3}\,C_3^2, \\ C_4' = 0. \end{cases}$$

This means that the quadratic operator F in (1.39) admits the 4D subspace

$$W_4 = \mathcal{L}\{1, x, x^2, x^3\} \quad \big(\text{and } F : W_4 \to W_3 = \mathcal{L}\{1, x, x^2\}\big).$$

The final two examples represent some remarkable invariant subspaces of the *maximal* dimension (a crucial theoretical aspect to be studied in the next chapter).

Example 1.14 (**Reaction-diffusion equation: 5D polynomial subspace**) Consider now the quasilinear equation with the negative exponent $\sigma = -\frac{4}{3}$, corresponding to the case of fast diffusion and a specific superlinear reaction term:

$$v_t = \left(v^{-\frac{4}{3}} v_x\right)_x + v^{\frac{7}{3}}. \tag{1.40}$$

Using the pressure transformation $u = v^{-4/3}$ yields the quadratic PDE

$$\boxed{u_t = F[u] \equiv u u_{xx} - \tfrac{3}{4}(u_x)^2 - \tfrac{4}{3}.} \tag{1.41}$$

It was shown in Galaktionov [220] that operator F preserves the 5D subspace

$$W_5 = \mathcal{L}\{1, x, x^2, x^3, x^4\},$$

so (1.41) admits the solution

$$u(x,t) = C_1(t) + C_2(t)x + C_3(t)x^2 + C_4(t)x^3 + C_5(t)x^4,$$

with the coefficients $\{C_i(t)\}$ satisfying the DS

$$\begin{cases} C_1' = 2C_1C_3 - \frac{3}{4}\,C_2^2 - \frac{4}{3}, \\ C_2' = 6C_1C_4 - C_2C_3, \\ C_3' = 12C_1C_5 + \frac{3}{2}\,C_2C_4 - C_3^2, \\ C_4' = 6C_2C_5 - C_3C_5, \\ C_5' = 2C_3C_5 - \frac{3}{4}\,C_4^2. \end{cases}$$

This fifth-order DS is not easy to study, but some particular features of such exact solutions can be obtained and used for comparison with general solutions of (1.40). The equation (1.40) admits single point blow-up, and the exact solutions describe interesting generic blow-up patterns.

Example 1.15 (**Reaction-absorption equation: 5D trigonometric subspace**) Consider an equation with the same fast diffusion and a different absorption term,

$$v_t = \left(v^{-\frac{4}{3}} v_x\right)_x - v^{-\frac{1}{3}}. \tag{1.42}$$

The pressure transformation $u = v^{-4/3}$ now yields the quadratic PDE

$$\boxed{u_t = F[u] \equiv u u_{xx} - \tfrac{3}{4}(u_x)^2 + \tfrac{4}{3}\,u^2.} \tag{1.43}$$

Here, F admits the 5D subspace spanned by trigonometric functions,

$$W_5 = \mathcal{L}\{1, \cos(\lambda x), \sin(\lambda x), \cos(\tfrac{\lambda x}{2}), \sin(\tfrac{\lambda x}{2})\}, \quad \text{where } \lambda = \tfrac{4}{\sqrt{3}}.$$

Therefore, the PDE (1.43) has exact solutions on W_5 [220]

$$u(x,t) = C_1 + C_2 \cos(\lambda x) + C_3 \sin(\lambda x) + C_4 \cos(\tfrac{\lambda x}{2}) + C_5 \sin(\tfrac{\lambda x}{2}),$$

where the coefficients $\{C_i(t)\}$ solve the DS

$$\begin{cases} C_1' = \frac{4}{3}\, C_1^2 - 4(C_2^2 + C_3^2) - \frac{1}{2}\, C_3^2, \\ C_2' = -\frac{8}{3}\, C_1 C_2 + \frac{1}{2}(C_4^2 - C_5^2), \\ C_3' = -\frac{8}{3}\, C_1 C_3 + C_4 C_5, \\ C_4' = \frac{4}{3}\, C_1 C_4 - 4(C_3 C_5 + C_2 C_4), \\ C_5' = \frac{4}{3}\, C_1 C_5 - 4(C_2 C_5 - C_3 C_4). \end{cases}$$

This DS is more difficult, though some key asymptotic properties of orbits can be detected that describe interface and extinction phenomena for (1.42).

1.2 Basic ideas: invariant subspaces and generalized separation of variables

1.2.1 Invariant subspaces

Following the above examples, consider a general first-order evolution PDE

$$u_t = F[u], \tag{1.44}$$

where F is a kth-order ordinary differential operator,

$$F[u] \equiv F\big(x, u, u_x, \ldots, D_x^k u\big).$$

Here, $F(\cdot)$ is a given sufficiently smooth function and D_x denotes $\frac{\partial}{\partial x}$.

Let $\{f_i(x),\ i = 1, \ldots, n\}$ be a finite set of $n \geq 1$ linearly independent functions, and let W_n denote their linear span,

$$W_n = \mathcal{L}\{f_1(x),\ \ldots,\ f_n(x)\}.$$

W_n is an n-dimensional linear subspace consisting of their linear combinations with real coefficients,

$$u = \sum_{i=1}^{n} C_i f_i, \quad \text{for any vector } \mathbf{C} = \{C_i\} \in I\!R^n.$$

The subspace W_n is said to be *invariant* under the given operator F if

$$F[W_n] \subseteq W_n,$$

and then F is said to *preserve* or *admit* W_n. As in linear algebra, this means

$$F\Big[\sum_{i=1}^{n} C_i f_i(x)\Big] = \sum_{i=1}^{n} \Psi_i(C_1,\ \ldots,\ C_n) f_i(x) \quad \text{for any } \mathbf{C} \in I\!R^n,$$

where $\{\Psi_i\}$ are the expansion coefficients of $F[u] \in W_n$ in the basis $\{f_i\}$.

It follows that *if the linear subspace W_n is invariant under F, then equation* (1.44) *has solutions of the form*

$$u(x,t) = \sum_{i=1}^{n} C_i(t) f_i(x), \tag{1.45}$$

where the coefficients $\{C_i(t)\}$ satisfy the dynamical system

$$C_i'(t) = \Psi_i(C_1(t), \dots, C_n(t)), \quad i = 1, \dots, n.$$

The PDE (1.44), which is an infinite-dimensional DS, being restricted to the invariant subspace W_n becomes an n-dimensional dynamical system.

1.2.2 First extension: second-order hyperbolic equations

A first extension is obvious: for the second-order evolution PDE

$$u_{tt} = F[u], \tag{1.46}$$

there exist solutions (1.45) governed by the $2n$th-order DS

$$C_i''(t) = \Psi_i(C_1(t), \dots, C_n(t)), \quad i = 1, \dots, n. \tag{1.47}$$

There are other easy generalizations to higher-order PDEs. For instance, if operator P is a linear *annihilator* of the subspace W_n, i.e., $P : W_n \to \{0\}$, then, for arbitrary operators F_1, the PDE

$$u_{tt} = F[u] + (P[u])\, F_1[u]$$

on W_n reduces to the same DS (1.47).

1.2.3 Second extension: invariant subspaces for delay-PDEs

If, for a given operator F, an invariant subspace W_n has been detected, one can find other types of equations of differential, integral, or functional types, which can be restricted to W_n. Another simple extension is to consider the functional *delay-PDE* corresponding to (1.44),

$$u_t(t) = F[u(t-1)], \tag{1.48}$$

where the right-hand side is defined for the solution $u(\cdot, t-1)$ with the 1-retarded time-argument. The discrete evolution mechanism of such equations is well-suited for various applications. Differential delay models appear in population genetics, bioscience problems, control theory, electrical networks with lossless transmission lines, etc.; see Remarks. Theory of functional delay-ODEs, to say nothing of the delay-PDEs, is not as advanced as that of standard differential equations. In particular, the questions of symmetries, constraints, reductions, and exact solutions are less developed, and, in many cases, it is not clear how to translate related notions to non-local-in-time functional operators.

In the present case, we arrive at the same invariant conclusion: if $F : W_n \to W_n$, then (1.48) admits exact solutions (1.45) for which the expansion coefficients solve the following system of delay-ODEs:

$$C_i'(t) = \Psi_i(C_1(t-1), \dots, C_n(t-1)), \quad i = 1, \dots, n.$$

Delay-ODEs are infinite-dimensional DSs, which are difficult to study, but are simpler than delay-PDEs (1.48).

1.2.4 *Generalized separation of variables: first simple example*

Let us present next an example explaining some features of the main problem of determining invariant subspaces for a given nonlinear operator. Consider the standard quadratic ordinary differential operator from reaction-diffusion theory

$$F[u] = \alpha(u_{xx})^2 + \beta u u_{xx} + \gamma (u_x)^2 + \delta u^2, \tag{1.49}$$

with arbitrary real parameters α, β, γ, and δ. Such operators occur in several applications that will be discussed later on. Consider a 2D subspace

$$W_2 = \mathcal{L}\{1, f(x)\}, \tag{1.50}$$

where the first basic function is constant 1, and the set $\{1, f(x)\}$ is assumed to be linearly independent. Obviously, the simplest 1D subspace $W_1 = \mathcal{L}\{1\}$ is invariant under F, since

$$F[1] = \delta \in W_1.$$

Therefore, we need to determine a single second function $f(x)$ from the invariance condition

$$F[W_2] \subseteq W_2. \tag{1.51}$$

Substituting into (1.49)

$$u = C_1 + C_2 f \in W_2,$$

where C_1 and C_2 are arbitrary constants, yields

$$F[u](x) = \delta C_1^2 + 2\delta C_1 C_2 f(x) + \beta C_1 C_2 f''(x) + C_2^2 F[f](x).$$

The first two terms belong to W_2. Consider the last two terms. Since $C_1 C_2$ and C_2^2 are independent, (1.51) is valid iff there exist parameters $\mu_{1,2}$ and $\nu_{1,2}$ such that f satisfies the following *overdetermined* system of ODEs:

$$\begin{cases} f'' = \mu_1 + \nu_1 f, \\ F[f] = \mu_2 + \nu_2 f. \end{cases} \tag{1.52}$$

The second equation implies that $\hat{W}_1 = \mathcal{L}\{f\}$ is also invariant if such an f exists for $\mu_2 = 0$ (and $\nu_2 \neq 0$). If $\mu_2 \nu_2 \neq 0$, then $F : \hat{W}_1 \to W_2$ and, in a natural sense, the element f *generates* the 2D invariant subspace (1.50).

This procedure of determining admissible basis functions $f(x)$ from an overdetermined system of ODEs with several free parameters is called the *generalized separation of variables* (GSV). In the present case, the GSV can be performed easily, since the first equation is linear and, clearly, for various values of parameters, there are six types of functions,

$$f(x) \in \{x, x^2, \cos \lambda x, \sin \lambda x, \cosh \lambda x, \sinh \lambda x\}, \quad \text{with } \lambda = \text{constant} \neq 0. \tag{1.53}$$

Substituting each of the functions f into the second equation in (1.52), we obtain the set (a linear subspace) of quadratic operators preserving subspace (1.50). We do not do this here; however, we do present the results of more general computations in the next section.

For 3D and multi-dimensional subspaces, the GSV leads to complicated overdetermined systems of ODEs that do not admit a simple treatment. Even for a general 2D subspace $\mathcal{L}\{f_1, f_2\}$ with two unknown basis functions, the GSV becomes essentially more involved. In our further study of invariant subspaces in Chapter 2, we will use another approach associated with Lie–Bäcklund symmetries of linear ODEs, and will return to the general theory of GSV in Section 7.3.

The above GSV reveals typical basis functions (1.53) of the invariant subspaces (1.50) for quadratic operators. These are:

(i) polynomial,
(ii) trigonometric, and
(iii) exponential subspaces,

which will be studied later on.

On related aspects of finite commutative rings. Consider the operator F in (1.49) in the linear space K of real analytic functions of the single variable x. The quadratic polynomial structure of (1.49) suggests introducing the *commutative product*

$$u * v = \alpha u_{xx} v_{xx} + \tfrac{\beta}{2}\big(u v_{xx} + v u_{xx}\big) + \gamma\, u_x v_x + \delta u v \tag{1.54}$$

for any $u, v \in K$. In this case, K becomes a *commutative ring* with the product (1.54), which is not associative in general.

It is interesting to interpret *nilpotents* and *idempotents* of this ring. To this end, for instance, consider the corresponding hyperbolic PDE (1.46). Then a nilpotent $\varepsilon(x)$ satisfying

$$\varepsilon * \varepsilon = 0, \quad \text{i.e.,} \quad F[\varepsilon] = 0,$$

is indeed a stationary solution of (1.46). On the other hand, any idempotent $e(x)$ satisfying

$$e * e = e, \quad \text{i.e.,} \quad F[e] = e,$$

is associated with the separate variables solution

$$u(x,t) = \varphi(t) e(x), \quad \text{where} \quad \varphi''(t) = \varphi^2(t).$$

For instance, the blow-up function $\varphi(t) = 6(T-t)^{-2}$ can be chosen.

We are now looking for 2D subrings A of K, and will describe where a link to overdetermined systems of ODEs is coming from. Assume that, in a subring A, there exists a generating element p such that p and $p * p$ are linearly independent. Actually, it can be shown that this is the case for any subring; see references in Remarks. This implies that p satisfies the system of two ODEs

$$\begin{cases} p * (p * p) = \mu_1 + \nu_1 (p * p), \\ (p * p) * (p * p) = \mu_2 + \nu_2 (p * p), \end{cases}$$

with four free parameters, as above. It is a system of two fourth-order nonlinear ODEs for p, which is difficult to study for general quadratic operators F.

1.3 More examples: polynomial subspaces

In the last three sections we present further examples of PDEs with quadratic, cubic, and other polynomial operators preserving linear subspaces of various dimensions. These results are introductory to more advanced theory developed in the subsequent chapters.

1.3.1 Classification and first examples of polynomial subspaces

We study second-order ($k = 2$) quadratic and cubic operators admitting subspaces that are composed of polynomials of the fixed order n,

$$W_n = \mathcal{L}\{1, x, ..., x^{n-1}\}, \quad \text{with } n \geq 2. \tag{1.55}$$

Operators preserving such a given subspace form a linear space. In the next propositions, the bases of such linear spaces of nonlinear operators are described.

Proposition 1.16 *Subspace* (1.55) *is invariant under the general quadratic operator of the second order*

$$F[u] = b_6(u_{xx})^2 + b_5 u_x u_{xx} + b_4 u u_{xx} + b_3 (u_x)^2 + b_2 u u_x + b_1 u^2 \tag{1.56}$$

only in the following four cases:

(i) $\mathbf{n} = \mathbf{2}$ *with a 5D space spanned by operators*

$$F_1[u] = (u_{xx})^2, \;\; F_2[u] = u_x u_{xx}\,,$$
$$F_3[u] = u u_{xx}\,, \;\; F_4[u] = (u_x)^2, \;\; F_5[u] = u u_x,$$

i.e., $b_1 = 0$ *in* (1.56);

(ii) $\mathbf{n} = \mathbf{3}$ *with a 4D space spanned by*

$$F_1[u] = (u_{xx})^2, \;\; F_2[u] = u_x u_{xx}\,, \;\; F_3[u] = u u_{xx}\,, \;\; F_4[u] = (u_x)^2,$$

i.e., $b_1 = b_2 = 0$;

(iii) $\mathbf{n} = \mathbf{4}$ *with a 3D space spanned by*

$$F_1[u] = (u_{xx})^2, \;\; F_2[u] = u_x u_{xx}\,, \;\; F_3[u] = u u_{xx} - \tfrac{2}{3}(u_x)^2,$$

i.e., $b_1 = b_2 = 0$ *and* $b_3 = -\frac{2}{3} b_4$;

(iv) $\mathbf{n} = \mathbf{5}$ *with a 2D space spanned by*

$$F_1[u] = (u_{xx})^2 \quad \text{and} \quad F_2[u] = u u_{xx} - \tfrac{3}{4}(u_x)^2,$$

i.e., $b_1 = b_2 = b_5 = 0$ *and* $b_3 = -\frac{3}{4} b_4$.

For $\mathbf{n} \geq \mathbf{6}$, *no nontrivial operators* (1.56) *preserving subspace* (1.55) *exist.*

Proof. For $n \leq 5$, the proof is straightforward by plugging the finite sum expansion

$$u = C_1 + C_2 x + ... + C_n x^{n-1}$$

into operator (1.56) and equating the coefficients of the expansion of $F[u]$, corresponding to higher-degree terms x^l with $l \geq n$, to zero. Any computer codes on algebraic manipulations in Maple, Matematica, MatLab, or Reduce, etc., are suitable for this analysis. The last negative statement for $n \geq 6$ will follow from a more general result to be proved in Section 2.2 (Theorem 2.8). □

A similar approach applies to other propositions presented below for various operators and subspaces.

Proposition 1.17 *Subspace* (1.55) *is invariant under the general cubic operator of the second order*

$$\begin{aligned} F[u] &= b_{10}(u_{xx})^3 + b_9(u_{xx})^2 u_x + b_8(u_{xx})^2 u + b_7 u_{xx}(u_x)^2 \\ &+ b_6 u_{xx} u_x u + b_5 u_{xx} u^2 + b_4(u_x)^3 + b_3(u_x)^2 u + b_2 u_x u^2 + b_1 u^3 \end{aligned} \tag{1.57}$$

only for the following three cases:

(i) $\mathbf{n} = \mathbf{2}$ *with an 8D space spanned by*

$$\begin{gathered} F_1[u] = (u_{xx})^3, \quad F_2[u] = u_x(u_{xx})^2, \quad F_3[u] = u(u_{xx})^2, \\ F_4[u] = (u_x)^2 u_{xx}, \quad F_5[u] = u u_x u_{xx}, \quad F_6[u] = u^2 u_{xx}, \\ F_7[u] = (u_x)^3, \quad F_8[u] = u(u_x)^2; \end{gathered}$$

(ii) $\mathbf{n} = \mathbf{3}$ *with a 6D space spanned by*

$$\begin{gathered} F_1[u] = (u_{xx})^3, \quad F_2[u] = u_x(u_{xx})^2, \\ F_3[u] = u(u_{xx})^2, \quad F_4[u] = (u_x)^2 u_{xx}, \\ F_5[u] = u_x[2uu_{xx} - (u_x)^2], \quad F_6[u] = u[2uu_{xx} - (u_x)^2]; \end{gathered}$$

(iii) $\mathbf{n} = \mathbf{4}$ *with a 2D space spanned by*

$$F_1[u] = (u_{xx})^3 \quad \text{and} \quad F_2[u] = u_{xx}\big[uu_{xx} - \tfrac{2}{3}(u_x)^2\big].$$

For $\mathbf{n} \geq \mathbf{5}$, *no nontrivial cubic operators* (1.57) *preserving subspace* (1.55) *exist.*

Example 1.18 (**Quadratic PDEs**) As an illustration of case (iii) in Proposition 1.16, we consider a fully nonlinear PDE

$$u_t = \alpha(u_{xx})^2 + \beta u_x u_{xx} + \gamma\big[uu_{xx} - \tfrac{2}{3}(u_x)^2\big]. \tag{1.58}$$

Nonlinearities $(u_{xx})^2$ and $(u_{xx})^3$ are typical for the *dual porous medium equations* in filtration theory; see references in [49]. In this case, (1.58) has solutions

$$u(x,t) = C_1(t) + C_2(t)x + C_3(t)x^2 + C_4(t)x^3,$$

$$\begin{cases} C_1' = 2\gamma C_1 C_3 - \frac{2}{3}\gamma C_2^2 + 2\beta C_2 C_3 + 4\alpha C_3^2, \\ C_2' = 6\gamma C_1 C_4 + 6\beta C_2 C_4 - \frac{2}{3}\gamma C_2 C_3 + 4\beta C_3^2 + 24\alpha C_3 C_4, \\ C_3' = 2\gamma C_2 C_4 - \frac{2}{3}\gamma C_3^2 + 18\beta C_3 C_4 + 36\alpha C_4^2, \\ C_4' = 18\beta C_4^2. \end{cases}$$

The last equation with $\beta > 0$ and $C_4(0) > 0$ implies finite-time blow-up,

$$C_4(t) = \tfrac{C_4(0)}{1 - 18\beta C_4(0)t} \to +\infty \quad \text{as } t \to T^-, \tag{1.59}$$

where $T = [18\beta C_4(0)]^{-1}$. If $C_4(0) < 0$, then

$$C_4(t) = -\tfrac{|C_4(0)|}{1 + 18\beta |C_4(0)| t} \to 0 \quad \text{as } t \to +\infty$$

is well-defined for all $t > 0$. For the delay-PDE (1.48), the corresponding delay-ODE

$$C_4'(t) = 18\beta C_4^2(t-1) \tag{1.60}$$

always has global solutions, and, for $\beta > 0$ and $C_4(0) > 0$, instead of blow-up (1.59), a *super-exponential* growth occurs

$$C_4(t) \sim \tfrac{2\ln 2}{9\beta}\, 2^t \mathrm{e}^{2^t} \quad \text{for } t \gg 1. \tag{1.61}$$

For the "hyperbolic" PDE (it is actually hyperbolic in $\{\alpha u_{xx} + \beta u_x + \gamma u > 0\}$)

$$u_{tt} = \alpha (u_{xx})^2 + \beta u_x u_{xx} + \gamma \left[u u_{xx} - \tfrac{2}{3}(u_x)^2 \right],$$

the eighth-order DS with the same right-hand sides appears. Adding any linear differential operator with constant coefficients to the right-hand side of the PDE does not affect the invariant property. For example, similar solutions on W_4 exist for the following fourth-order semilinear hyperbolic PDE:

$$u_{tt} = -u_{xxxx} + u u_{xxx} + \alpha (u_{xx})^2 + \beta u_x u_{xx} + \gamma \left[u u_{xx} - \tfrac{2}{3}(u_x)^2 \right].$$

Example 1.19 **(Cubic hyperbolic equation)** It follows from Proposition 1.17(ii) that the quasilinear cubic hyperbolic PDE

$$\boxed{u_{tt} = u^2 u_{xx} - \tfrac{1}{2}\, u (u_x)^2 \quad \text{on } W_3}$$

has the solutions

$$u(x,t) = C_1(t) + C_2(t)x + C_3(t)x^2, \tag{1.62}$$

where the equivalent DS takes the form

$$C_i'' = \left(2C_1C_3 - \tfrac{1}{2}\, C_2^2\right) C_i, \quad i = 1, 2, 3.$$

In the next example, the quadratic operator includes derivatives in t.

Example 1.20 **(Gibbons–Tsarev equation)** Consider the PDE in $I\!R \times I\!R$

$$\boxed{u_{tt} = F_\beta[u] + \mu u + \nu, \quad \text{where } F_\beta[u] = u_x u_{tx} - \beta u_t u_{xx}.} \tag{1.63}$$

For $\beta = 1$, $\mu = 0$, and $\nu = 1$, this is the *Gibbons–Tsarev* (GT) *equation* [252]

$$\boxed{u_{tt} = F_1[u] + 1 \equiv u_x u_{tx} - u_t u_{xx} + 1} \tag{1.64}$$

from the theory of Benney moment equations (a system of hydrodynamic type). Equation (1.64) is Galilean invariant; see extra details in Example 1.23.

For F_β, the *basic* subspace (an invariant module), admitted for any β, is indeed $W_3 = \mathcal{L}\{1, x, x^2\}$, and looking for the solution (1.62) of (1.63) yields

$$\begin{cases} C_1'' = C_2 C_2' - 2\beta C_1' C_3 + \mu C_1 + \nu, \\ C_2'' = 2(1-\beta) C_2' C_3 + 2 C_2 C_3' + \mu C_2, \\ C_3'' = 2(2-\beta) C_3 C_3' + \mu C_3. \end{cases}$$

Concerning extensions of the basic subspace, operator F_β admits

$$W_4 = \mathcal{L}\{1, x, x^2, x^3\} \text{ for } \beta = \tfrac{3}{2} \text{ and } \hat{W}_3 = \mathcal{L}\{1, x^2, x^4\} \text{ for } \beta = \tfrac{4}{3}.$$

Consider the first case of $\beta = \frac{3}{2}$, where

$$u = C_1 + C_2 x + C_3 x^2 + C_4 x^3 \in W_4 \quad \Longrightarrow \tag{1.65}$$
$$F_{\frac{3}{2}}[u] = 3\big(C_3 C_4' - C_4 C_3'\big)x^3 + \big(3C_2 C_4' - 6C_4 C_2' + C_3 C_3'\big)x^2$$
$$+ \big(2C_2 C_3' - 9C_4 C_1' - C_3 C_2'\big)x - 3C_3 C_1' + C_2 C_2'.$$

For the second case $\beta = \frac{4}{3}$, choosing the PDE

$$\boxed{u_{tt} = u_x u_{tx} - \tfrac{4}{3}\, u_t u_{xx} + 1,}$$

we find solutions from the second invariant subspace,

$$u(x,t) = C_1(t) + C_2(t)x^2 + C_3(t)x^4 \in \hat{W}_3, \tag{1.66}$$
$$\begin{cases} C_1'' = -\frac{8}{3}\, C_2 C_1' + 1, \\ C_2'' = \frac{4}{3}\, C_2 C_2' - 16 C_3 C_1', \\ C_3'' = -8C_3 C_2' + \frac{16}{3}\, C_2 C_3'. \end{cases}$$

It is curious that, unlike in most of the previous cases, for $\beta = \frac{4}{3}$, the full subspace of the fourth-order polynomials

$$\hat{W}_5 = \mathcal{L}\{1, x, x^2, x^3, x^4\}$$

is not invariant under $F_{4/3}[u]$. The computations are easy:

$$u = C_1 + C_2 x + C_3 x^2 + C_4 x^3 + C_5 x^4 \in \hat{W}_5 \quad \Longrightarrow$$
$$F_{\frac{4}{3}}[u] = 4\big(C_4 C_5' - C_5 C_4'\big)x^5 + \tfrac{1}{3}\big(16C_3 C_5' - 24 C_5 C_3' + 3C_4 C_4'\big)x^4$$
$$+ \tfrac{2}{3}\big(6C_2 C_5' - 18 C_5 C_2' + 5C_3 C_4' - 3C_4 C_3'\big)x^3$$
$$+ \tfrac{1}{3}\big(-48 C_5 C_1' + 9C_2 C_4' - 15 C_4 C_2' + 4C_3 C_3'\big)x^2$$
$$+ \tfrac{2}{3}\big(-12 C_4 C_1' + 3C_2 C_3' - C_3 C_2'\big)x + \tfrac{1}{3}\big(-8C_3 C_1' + 3C_2 C_2'\big).$$

The projection onto x^5 is $4(C_4 C_5' - C_5 C_4') \neq 0$ in general, so no invariance of $\hat{W}_5$ exists. A unified approach to $\frac{\partial}{\partial t}$-dependent operators is developed Section 2.7.

Example 1.21 (**Chaplygin gas equation**) The following system in $I\!R \times I\!R$:

$$\begin{cases} \rho_t + \rho_x u_x + \rho u_{xx} = 0, \\ u_t + \frac{1}{2}(u_x)^2 = \frac{\lambda}{\rho^2}, \end{cases} \tag{1.67}$$

consists of the equation of continuity for the density $\rho(x,t)$, and Euler's force equation for an ideal fluid of zero vorticity with the velocity potential $u(x,t)$, in which the pressure P is related to the density by

$$P = -\frac{2\lambda}{\rho}, \quad \text{where } \lambda = \text{constant} \neq 0.$$

Expressing ρ from the first equation in (1.67) and substituting into the second yields the *Chaplygin gas equation* (1904) [103]

$$\left(\frac{1}{\sqrt{u_t + \frac{1}{2}(u_x)^2}}\right)_t + \left(\frac{u_x}{\sqrt{u_t + \frac{1}{2}(u_x)^2}}\right)_x = 0. \tag{1.68}$$

Its solutions are expressed via those of the linear wave equations $X_{tt} = X_{ss}$; see [80]. On differentiation, we obtain from (1.68) a PDE with quadratic operators,

$$\boxed{F[u] \equiv u_{tt} + 2u_x u_{xt} - 2u_{xx}u_t = 0.} \tag{1.69}$$

The basic *polynomial subspace* for F (a module) is still $W_3 = \mathcal{L}\{1, x, x^2\}$, so that, for the solution (1.62) of (1.69), there occurs the DS

$$\begin{cases} C_1'' = -2C_2C_2' + 4C_3C_1', \\ C_2'' = -4C_2C_3', \\ C_3'' = -4C_3C_3'. \end{cases}$$

For $C_3(t) = \frac{1}{4t}$, explicit solutions can be found (cf. Example 1.6).

Similarly to the previous example, for the more general equation

$$\boxed{F_{\alpha\beta}[u] \equiv u_{tt} + \alpha u_x u_{xt} + \beta u_{xx}u_t = 0,} \tag{1.70}$$

operator $F_{\alpha\beta}$ admits

$$W_4 = \mathcal{L}\{1, x, x^2, x^3\}, \quad \text{if } 3\alpha + 2\beta = 0;$$
$$\hat{W}_3 = \mathcal{L}\{1, x^2, x^4\}, \quad \text{if } 4\alpha + 3\beta = 0.$$

In the first case, $3\alpha + 2\beta = 0$, for solutions (1.65) of (1.70), the DS is

$$\begin{cases} C_1'' = \alpha(3C_3C_1' - C_2C_2'), \\ C_2'' = \alpha(9C_4C_1' + C_3C_2' - 2C_2C_3'), \\ C_3'' = \alpha(6C_4C_2' - C_3C_3' - 3C_2C_4'), \\ C_4'' = 3\alpha(C_4C_3' - C_3C_4'). \end{cases}$$

In the second case, $4\alpha + 3\beta = 0$, the DS for solutions (1.66) takes the form

$$\begin{cases} C_1'' = \frac{8}{3}\alpha C_2C_1', \\ C_2'' = \frac{4}{3}\alpha(12C_3C_1' - C_2C_2'), \\ C_3'' = \frac{8}{3}\alpha(3C_3C_2' - 2C_2C_3'). \end{cases}$$

Concerning other subspaces, we take the *trigonometric* subspace

$$u(x,t) = C_1(t) + C_2(t)\cos x + C_3(t)\sin x,$$

and find that equation (1.70) with $\beta = -\alpha$ (this is true for (1.69)) admits such solutions with the DS

$$\begin{cases} C_1'' = -\alpha\big(C_2C_2' + C_3C_3'\big), \\ C_2'' = -\alpha C_2C_1', \\ C_3'' = -\alpha C_3C_1'. \end{cases}$$

Example 1.22 (**Born–Infeld equations**) Consider now the *Born–Infeld* (BI) *equations* in $I\!R \times I\!R$

$$\begin{cases} \rho_t + \left(\sqrt{\frac{\rho^2c^2+a^2}{c^2+(u_x)^2}}\right)_x = 0, \\ u_t + \rho c^2\sqrt{\frac{c^2+(u_x)^2}{\rho^2c^2+a^2}} = 0, \end{cases} \tag{1.71}$$

which were introduced in 1934 as a nonlinear correction to the linear Maxwell equations for electromagnetism, [73]. At the limit $c \to \infty$ (c is the speed of light), this relativistic Born–Infeld model reduces to the above non-relativistic Chaplygin gas equation with $\lambda = \frac{1}{2}a^2$. The BI equations belong to the family of 6×6 systems of hyperbolic conservation laws, together with two solenoidal constraints on the magnetic field and electric displacements; see [80] and [266] for details on physics and mathematics. Excluding ρ from the system (1.71) yields a single PDE

$$\boxed{\left(\frac{u_t}{\sqrt{c^4+c^2(u_x)^2-(u_t)^2}}\right)_t + c^2\left(\frac{u_x}{\sqrt{c^4+c^2(u_x)^2-(u_t)^2}}\right)_x = 0,}$$

which gives a cubic polynomial equation for $u(x,t)$,

$$\boxed{c^2\big[c^2+(u_x)^2\big]u_{tt} + c^2\big[c^4+(c^2-1)(u_x)^2-(u_t)^2\big]u_{xx} - (c^2-1)u_t u_x u_{xt} = 0.}$$

In particular, exact solutions exist on $\mathcal{L}\{1, x\}$, i.e., $u(x,t) = C_1(t) + C_2(t)x$.

Example 1.23 (**Galilean invariant PDEs**) Another similar quadratic operator occurs in general *Galilean invariant* PDEs

$$F[u] \equiv u_{xx}u_t - u_x u_{tx} = G[u] \equiv G(u, u_x, u_{xx}, \ldots), \tag{1.72}$$

where $G[u]$ is an arbitrary (possibly elliptic) operator. This equation is invariant under the *Galilean transformation* of the reference frame,

$$x \mapsto x + \upsilon t \quad \text{for any } \upsilon = \text{constant}. \tag{1.73}$$

So, if $u(x,t)$ is a solution of (1.72), $u(x+\upsilon t, t)$ is also a solution. Galilean transformations and invariance under generalized Galilean algebras play an important role in the analysis of various PDEs, and especially in many physical and mechanical models; see Remarks for references. Furthermore, besides (1.73), equation (1.72) admits a stronger symmetry generating an exceptionally wide class of solutions: if $u(x,t)$ is a solution of (1.72), then

$$u(x + f(t), t) \text{ is a solution for any } C^1\text{-function } f(t). \tag{1.74}$$

Without extra hypotheses (say, symmetry), the Cauchy problem for (1.72) makes no sense, since by (1.74) for given initial data, it admits an infinite-dimensional set of solutions. Equations such as (1.72) are fully nonlinear PDEs with unknown concepts of proper solutions and local regularity properties. Exact solutions may help to clarify some evolution characteristics and possible singularities of such PDEs. Operator F is related to the *remarkable* operator $F_{\rm rem}[u] = uu_{xx} - (u_x)^2$ to be studied later on; see Examples 1.36 and 2.45. Equation (1.72) may admit exact solutions on various 2D invariant modules of F, such as $\mathcal{L}\{1, \cos\gamma x\}$ and $\mathcal{L}\{1, \cosh\gamma x\}$ for any $\gamma \neq 0$.

Consider the Galilean invariant equation with the porous medium operator on the right-hand side (for convenience, the original equation was divided by u_{xx})

$$\boxed{u_t - \frac{1}{u_{xx}} u_x u_{tx} = (uu_x)_x \quad \text{in } I\!R \times I\!R_+.} \tag{1.75}$$

The invariant subspace for both left and right-hand sides is $W_3 = \mathcal{L}\{1, x, x^2\}$. Bearing in mind the translational nonuniqueness, we take the symmetric expression

$$u(x,t) = C_1(t) + C_2(t)x^2,$$

which, similar to the ZKB solution, may be expected to exhibit the properties of the source-type pattern for (1.75), and obtain the DS

$$\begin{cases} C_1' = 2C_1C_2, \\ C_2' = -6C_2^2. \end{cases}$$

Integrating yields the following compactly supported solution:

$$u(x,t) = \left[A(T-t)^{\frac{1}{3}} - \tfrac{1}{6(T-t)}\, x^2\right]_+ \equiv (T-t)^{\frac{1}{3}}\left(A - \tfrac{1}{6}\, y^2\right)_+, \qquad (1.76)$$

where $y = x/(T-t)^{2/3}$ is the spatial rescaled variable and $A > 0$ is a constant. The positive part in (1.76) formally mimics a typical *finite propagation* property for the PME and needs a special setting of a *free boundary problem* (FBP). We will deal with several problems like that for more mathematically reliable PDEs. Hence, (1.76) is a self-similar solution of (1.75), exhibiting another nonlinear phenomenon, such as the finite-time *extinction*: $u(x,t)$ vanishes identically as $t \to T^-$.

Stability analysis of these unusual singular patterns uses the corresponding rescaled solution given by

$$u(x,t) = (T-t)^{\frac{1}{3}} v(y,\tau) \quad \text{with} \quad \tau = -\ln(T-t),$$

satisfies the following rescaled PDE:

$$v_\tau - \tfrac{v_y}{v_{yy}}\left(v_{\tau y} + \tfrac{2}{3}\, v_{yy} y + \tfrac{1}{3}\, v_y\right) = (v v_y)_y - \tfrac{2}{3}\, v_y y + \tfrac{1}{3}\, v. \qquad (1.77)$$

The profile $g(y) = (A - \frac{1}{6}\, y^2)_+$ that is obtained in (1.76) from the exact solution may be treated as a "weak" stationary solution of (1.77), possibly, in an FBP framework, which is not known. For general rescaled solutions, the passage to the limit $\tau \to \infty$ in (1.77) is an OPEN PROBLEM (recall that the general well-posedness of such problems is also obscure). Local regularity and any potential or gradient properties of such flows are unknown. Inserting other operators on the right-hand side of (1.75) may yield a different type of solutions on invariant subspaces, including trigonometric or exponential ones, leading to non-similarity solutions (see further examples below for better justifies models).

1.3.2 General polynomial operators

Let $\nu = \{\nu_1, ..., \nu_K\} \in I\!R^K$ for $2 \le K \le n-1$ be a row (a multiindex) of nonnegative integers of fixed length $M = |\nu| = \nu_1 + ... + \nu_K \ge 2$. The corresponding *primitive monomial* operator is given by

$$F_\nu[u] = \textstyle\prod_{j=1}^K \left(D_x^j u\right)^{\nu_j}. \qquad (1.78)$$

We next define the polynomial operator with arbitrary constants $a_\nu \in I\!R$,

$$F[u] = \textstyle\sum_{(|\nu|=M)} a_\nu F_\nu[u]. \qquad (1.79)$$

Proposition 1.24 (i) *Each operator* (1.78) *and, hence* (1.79), *admits the subspace* W_n *in* (1.55) *if*

$$\textstyle\sum_{j=1}^K (n-1-j)\nu_j \le n-1 \quad \textit{for all} \quad \nu \textit{ such that } \{|\nu| = M\}. \qquad (1.80)$$

(ii) *If* (1.80) *is not valid for* ν *from some subset* $\mathcal{N} \subset I\!R^K$ *and*

$$\sum_{(\nu\in\mathcal{N})}(n-1-j)\nu_j = n,$$

then W_n *is invariant under the extra condition*

$$\sum_{(\nu\in\mathcal{N})} a_\nu \prod_{(j)}\big[(n-1)(n-2)...(n-j)\big]^{\nu_j} = 0. \tag{1.81}$$

Proof. Fix an arbitrary

$$u = C_1 + C_2 x + ... + C_n x^{n-1} \in W_n.$$

(i) Taking into account the higher-degree terms only, due to $(1.80)^{\ddagger}$,

$$F_\nu[x^{n-1}] \sim x^\rho \in W_n, \quad \text{with } \rho = \sum(n-1-j)\nu_j.$$

(ii) For any $\nu \in \mathcal{N}$, we need to control the coefficient of x^n in each monomial,

$$F_\nu[... + C_n x^{n-1}] = ... + C_n^M \prod_{(j)}\big[(n-1)(n-2)...(n-j)\big]^{\nu_j} x^n.$$

Summing up over all $\nu \in \mathcal{N}$ yields that the condition (1.81) cancels the term with x^n in $F[u]$. □

Remark 1.25 **(Extensions)** It is not difficult to extend the result in (ii) to the case

$$\sum_{(\nu\in\mathcal{N}_1)}(n-1-j)\nu_j = n+1 \quad \text{for some } \mathcal{N}_1 \subset I\!R^K.$$

Then $F[u]$ will contain terms with x^{n+1} having the common multipliers C_n^M, and the next lower-order term on $\mathcal{L}\{x^n\}$ with the common multiplies $C_{n-1}C_n^{M-1}$. Therefore, we will need two hypotheses that guarantee that W_n is still invariant. The first hypothesis is similar to (1.81), and the second one is slightly more complicated but still easy to derive in general. This analysis can be continued to include three or more hypotheses on the operator.

We finish with the following application to more general equations.

Remark 1.26 **(Polynomial PDEs)** Now let $\nu = \|\nu_{i,j}\|$ be a $K\times K$ matrix of non-negative integers, with $2 \le K \le n-1$, and let the corresponding monomial

$$F_\nu[u] = \prod_{(i,j)}(D_t^i D_x^j u)^{\nu_{i,j}}$$

be defined in the space of sufficiently smooth functions $u = u(x,t)$ of two independent variables x and t. The polynomial operator is then

$$F[u] = \sum_{(\nu)} a_\nu(t) F_\nu[u].$$

Assume that, similar to (1.80),

$$\sum_{(i,j)}(n-1-j)\nu_{i,j} \le n-1 \quad \text{for all } \nu.$$

Then the PDE

$$F[u] = L[u] + f(x,t) \quad \text{on } W_n,$$

‡ Here, "$\sim$" means equality up to a non-zero constant multiplier.

where $f(\cdot, t) \in W_n$ for all t and L is a linear operator

$$L = \sum b_{i,j}(x,t)\, D_t^i D_x^j,$$

with $b_{i,j}(\cdot, t) \in W_{j+1}$ for $1 \le i, j \le K$ and $t \in I\!R$, reduces to an ODE system.

1.3.3 One-dimensional polynomial subspaces

The 1D invariant subspaces are easy to deal with. For instance, consider the second-order operator of nonlinear diffusion

$$F_2[u] = (k(u)u_x)_x, \quad \text{with} \quad k(u) = \alpha u^p + \beta u^{2(p+1)}, \tag{1.82}$$

where $p \neq -1$ is a constant. In this case, (1.82) admits the 1D subspace

$$W_1 = \mathcal{L}\{x^\mu\}, \quad \text{with} \quad \mu = \tfrac{1}{p+1} \quad (x > 0) \tag{1.83}$$

(and $\mathcal{L}\{e^x\}$ if $p = -1$), since the first term $(\alpha u^p u_x)_x$, corresponding to αu^p, vanishes. We will plug this operator into some PDEs.

Example 1.27 (**Parabolic, KdV, and Cahn–Hilliard PDEs**) First, we add to F_2 a general first-order operator preserving subspace (1.83),

$$u_t = F_2[u] + \gamma u + \textstyle\sum_{(j \ge 0)} a_j u^j (u_x)^{-\frac{\mu(j-1)}{\mu-1}} \quad (\mu \neq 1).$$

Then there exists the solution

$$u(x,t) = C(t)x^\mu, \tag{1.84}$$

where $C(t)$ satisfies the ODE

$$C' = \beta\mu(\mu+1)C^{2p+3} + \gamma C + \sum a_j \mu^{-\frac{\mu(j-1)}{\mu-1}} C^{\frac{\mu-j}{\mu-1}}.$$

For $p = -\frac{1}{2}$, i.e., $\mu = 2$, we introduce a third-order *KdV-type equation* (odd-order nonlinear PDEs are the subject of Chapter 4)

$$\boxed{u_t = u_{xxx} + \big[\big(\tfrac{\alpha}{\sqrt{u}} + \beta u\big)u_x\big]_x + \gamma u + \delta u u_{xx},}$$

where the ODE takes the form $C' = 2(3\beta + \delta)C^2 + \gamma C$. For the fourth-order *Cahn–Hilliard equation* (see Section 3.1 for details), we take $\mu = 3$ for $p = -\frac{2}{3}$, so

$$\boxed{\begin{aligned} u_t = -u_{xxxx} + \big[\big(\alpha u^{-\frac{2}{3}} + \beta u^{\frac{2}{3}}\big)u_x\big]_x \\ + \gamma u + \delta u_x u_{xx} + \varepsilon u u_{xxx} + \ldots \end{aligned}} \tag{1.85}$$

admits solutions (1.84) with $\mu = 3$.

Example 1.28 (**Thin film models**) In a similar fashion, the fourth-order thin film operator (see Section 3.1)

$$F_4[u] = -[(\alpha u^p + \beta u^q)u_{xxx}]_x, \quad \text{where} \quad q = \tfrac{4(p+1)}{3},$$

preserves subspace (1.83) with $\mu = \frac{3}{p+1}$. Hence, the *thin film equation* (TFE)

$$u_t = F_4[u] + \gamma u + \delta u^5 (u_x)^{-\frac{4\mu}{\mu-1}} \quad (p \neq 2)$$

has the solution (1.84) with the ODE

$$C' = -\mu(\mu^2 - 1)(\mu - 2)\beta C^{q+1} + \gamma C + \delta \mu^{-\frac{4\mu}{\mu-1}} C^{\frac{\mu-5}{\mu-1}}.$$

1.3.4 Operators of the same total order: criterion of invariance

Quadratic case. Fix a natural $k \geq 1$, and consider a general quadratic operator of the following form:

$$F[u] = \sum_{(i+j=k)} a_{i,j}\, D_x^i u\, D_x^j u, \tag{1.86}$$

where the coefficients satisfy $a_{i,j} = a_{j,i}$ for all $\{i, j \geq 0 : i + j = k\}$. The total differential order of each monomial is equal to k. Note that, for polynomial invariant subspaces, it is actually suffices to consider operators of the form (1.86); see Section 2.6 for more details and references. The particular case $k = 4$ is associated with thin film operators

$$F[u] = -u u_{xxxx} + \beta u_x u_{xxx} + \gamma (u_{xx})^2 \quad (\alpha,\ \beta \in I\!R),$$

which will be systematically studied in Chapter 3. The symmetric form of this operator is as follows:

$$F[u] = -\tfrac{1}{2}\, u u_{xxxx} + \tfrac{1}{2}\, \beta u_x u_{xxx} + \gamma (u_{xx})^2 + \tfrac{1}{2}\, \beta u_{xxx} u_x - \tfrac{1}{2}\, u_{xxxx} u.$$

Taking first u from the standard polynomial subspace,

$$u = \sum_{\mu=0}^{n-1} C_\mu x^\mu \in W_n = \mathcal{L}\{1, x, ..., x^{n-1}\}, \tag{1.87}$$

substituting into (1.86), and introducing the notation $\sigma_\mu^i = \mu(\mu - 1)...(\mu - i + 1)$, we obtain

$$\begin{aligned} F[u] &= \sum_{(i,j)} a_{i,j} \big(\sum_{(\mu)} C_\mu x^\mu\big)^{(i)} \big(\sum_{(\nu)} C_\nu x^\nu\big)^{(j)} \\ &= \sum_{(i,j)} a_{i,j} \big(\sum_{(\mu)} C_\mu \sigma_\mu^i x^{\mu-i}\big)\big(\sum_{(\nu)} C_\nu \sigma_\nu^j x^{\nu-j}\big) \\ &= \sum_{(\mu,\nu)} C_\mu C_\nu \big(\sum_{(i,j)} a_{i,j}\, \sigma_\mu^i \sigma_\nu^j\big)\, x^{\mu+\nu-k}. \end{aligned}$$

Since constants $\{C_\mu\}$ are independent, we conclude that $F : W_n \to W_n$ iff

$$\sum_{(i+j=k)} a_{i,j}\, \sigma_\mu^i \sigma_\nu^j = 0 \quad \text{for all } \mu \leq \nu \text{ such that } \mu + \nu - k > n - 1. \tag{1.88}$$

Similarly, choosing any polynomial subspace with “gaps”,

$$u = \sum_{(\mu \in \mathcal{N})} C_\mu x^\mu \in W_n, \tag{1.89}$$

where the summation is performed over some subset of nonnegative integers $\mathcal{N}$, the invariance condition in (1.88) must be valid for all $\mu + \nu - k \notin \mathcal{N}$. Dealing with invariant subspaces for quadratic operators (1.86) pose a number of interesting mathematical problems. Later on, we perform a detailed study of operators like (1.86) on polynomial (Section 2.6.4), exponential (see Example 1.49, and Section 7.1 for more general operators), and some other subspaces.

Cubic case. Consider the cubic operators from thin film theory (Chapter 3)

$$F[u] = \sum_{(i+j+l=k)} a_{i,j,l}\, D_x^i u\, D_x^j u\, D_x^l u, \tag{1.90}$$

where $\{a_{i,j,l}\}$ are elements of an absolutely symmetric tensor of rank 3 (symmetric relative to any pair of indices). Substituting (1.87) yields

$$F[u] = \sum_{(\mu,\nu,\omega)} C_\mu C_\nu C_\omega \left(\sum_{(i,j,l)} a_{i,j,l}\, \sigma^i_\mu \sigma^j_\nu \sigma^l_\omega\right) x^{\mu+\nu+\omega-k}.$$

Therefore, the invariance criterion of W_n is

$$\sum_{(i,j,l)} a_{i,j,l}\, \sigma^i_\mu \sigma^j_\nu \sigma^l_\omega = 0 \quad \text{for all } \mu \le \nu \le \omega, \;\; \mu+\nu+\omega-k > n-1. \tag{1.91}$$

Similar conditions occur for quartic, quintic and higher-degree operators.

1.4 Examples: trigonometric subspaces

1.4.1 Some classification results and examples

We next move to quadratic and cubic operators preserving

$$W_3 = \mathcal{L}\{1, \cos x, \sin x\}. \tag{1.92}$$

This 3D subspace and its 5D extensions induce finite Fourier expansions of solutions of several nonlinear PDEs. For such subspaces, typical computations are more involved than those for the polynomial expansions. We restrict our attention to the two most important cases $n = 3$ and $n = 5$. The general results in Section 2.2 will be devoted to subspaces of arbitrary dimensions.

Proposition 1.29 *Subspace* (1.92) *is invariant under the quadratic operator* (1.56) *iff the coefficients* $\{b_j\}$ *satisfy the linear system*

$$\begin{cases} b_4 = b_6 + b_1 - b_3, \\ b_5 = b_2. \end{cases}$$

The set of such operators is a 4D linear space spanned by

$$F_1[u] = u_{xx}(u_{xx} + u), \;\; F_2[u] = u_x(u_{xx} + u),$$
$$F_3[u] = u(u_{xx} + u), \;\; F_4[u] = (u_x)^2 + u^2.$$

Proposition 1.30 *Subspace* (1.92) *is invariant under the cubic operator* (1.57) *iff*

$$\begin{cases} b_1 = \frac{1}{2}(b_5 + b_7 - b_{10}), \;\; b_2 = \frac{1}{2}\, b_6, \\ b_3 = \frac{1}{2}(-b_5 + 3b_7 + 2b_8 - 3b_{10}), \;\; b_4 = b_9 - \frac{1}{2}\, b_6. \end{cases}$$

The set of such operators is a 6D linear space spanned by

$$F_1[u] = u_{xx}[(u_{xx})^2 + (u_x)^2], \;\; F_2[u] = u_x[(u_{xx})^2 + (u_x)^2],$$
$$F_3[u] = u[(u_{xx})^2 + (u_x)^2], \;\; F_4[u] = u_{xx}[2uu_{xx} - (u_x)^2 + u^2],$$
$$F_5[u] = u_x[2uu_{xx} - (u_x)^2 + u^2], \;\; F_6[u] = u[2uu_{xx} - (u_x)^2 + u^2].$$

Next, consider the 5D trigonometric subspace

$$W_5 = \mathcal{L}\{1, \cos x, \sin x, \cos 2x, \sin 2x\}. \tag{1.93}$$

Proposition 1.31 *Subspace* (1.93) *is invariant under the quadratic operator* (1.56) *iff the coefficients* $\{b_j\}$ *satisfy*

$$\begin{cases} b_2 - 2b_5 = 0, \;\; b_2 - 4b_5 = 0, \\ b_1 - 4b_3 - 4b_4 + 16b_6 = 0, \\ 2b_1 - 4b_3 - 5b_4 + 8b_6 = 0. \end{cases}$$

The set of such operators is a 2D linear space spanned by

$$F_1[u] = uu_{xx} - \tfrac{3}{4}(u_x)^2 + u^2 \quad \text{and} \quad F_2[u] = (u_{xx} + 4u)^2.$$

Proposition 1.32 *No nontrivial cubic operators* (1.57) *admit subspace* (1.93).

Proof. Subspace (1.93) is invariant under (1.57) iff

$$\begin{gathered} b_2 = b_4 = b_6 = b_9 = 0,\ b_3 = 4b_7,\ \ b_5 = b_8 - 9b_5 + 8b_7 = 0, \\ b_{10} - b_1 + 3b_7 = 0,\ \ -3b_1 + b_5 + 8b_7 = 0, \\ 15b_1 - 5b_5 - 44b_7 = 0,\ 21b_1 - 4b_5 - 64b_7 = 0. \end{gathered}$$

This system admits the trivial solution only. □

Some examples of PDEs with standard basic trigonometric subspaces have been studied in Section 1.1. Let us consider another example motivated by the structure of the quadratic operator of the GT equation from Example 1.20.

Example 1.33 **(Periodic solutions of the Gibbons–Tsarev equation)** Similar to Example 1.20, the quadratic operator of the GT equation (1.64) is convenient to consider on the module (1.92). Hence, there exist exact solutions

$$u(x,t) = C_1(t) + C_2(t)\cos x + C_3(t)\sin x,$$

$$\begin{cases} C_1'' = C_2C_2' + C_3C_3' + 1, \\ C_2'' = C_1'C_2, \\ C_3'' = C_1'C_3. \end{cases}$$

Further extensions of trigonometric subspaces are possible for a three-parameter family of such quadratic operators,

$$F[u] = u_xu_{tx} - \beta u_tu_{xx} + \gamma uu_{txx} + \delta uu_t \quad (\beta, \gamma, \delta \in I\!R). \tag{1.94}$$

Proposition 1.34 *The only operator* (1.94) *associated with the 5D module* (1.93) *is as follows:*

$$F[u] = u_xu_{tx} - \tfrac{2}{3}\,u_tu_{xx} - \tfrac{2}{3}\,uu_{txx} - \tfrac{4}{3}\,uu_t. \tag{1.95}$$

Proof. We begin with the necessity. Using the subspace of cos-functions,

$$u = C_1 + C_2\cos x + C_3\cos 2x \in W_3,$$

we give the results of final calculations, keeping only the suspicious terms that may not belong to W_3,

$$\begin{aligned} F[u] = {} & \big[4\sin^2 2x + (4\beta - 4\gamma + \delta)\cos^2 2x\big]C_3C_3' \\ & + \big[(4\beta - \gamma + \delta)\cos x\cos 2x + 2\sin x\sin 2x\big]C_3C_2' \\ & + \big[(\beta - 4\gamma + \delta)\cos x\cos 2x + 2\sin x\sin 2x\big]C_2C_3' + \dots . \end{aligned}$$

These three terms belong to W_5 for arbitrary values $C_{2,3}$ and $C_{2,3}'$ iff

$$\begin{cases} 4\beta - 4\gamma + \delta - 4, \\ 4\beta - \gamma + \delta = 2, \\ \beta - 4\gamma + \delta = 2, \end{cases}$$

from which come $\beta = -\gamma = \frac{2}{3}$ and $\delta = -\frac{4}{3}$ yielding operator (1.95).

To see the sufficiency, we write down the operator (1.95) in the form

$$F[u] = -\tfrac{2}{3}\big[uu_{xx} - \tfrac{3}{4}(u_x)^2 + u^2\big]_t.$$

The operator in square brackets, up to scaling $x \mapsto \frac{\sqrt{3}}{2}x$, coincides with the operator in Example 1.15 and admits the subspace (1.93), so the same subspace is available for F, where W_5 becomes a module. □

Example 1.35 (**Fast diffusion equation**) Consider the following quasilinear fast diffusion equation with source and absorption:

$$v_t = \left(v^{-\frac{4}{3}} v_x\right)_x - \tfrac{3}{4}\, v^{-\frac{1}{3}} - \tfrac{3}{2}\, av - \tfrac{3}{2}\, bv^{\frac{5}{3}}, \tag{1.96}$$

where a and b are constants. Setting $u = v^{-2/3}$ (notice that u *is not* the standard pressure) yields a PDE with a cubic operator,

$$\boxed{u_t = u^2 u_{xx} - \tfrac{1}{2}\, u(u_x)^2 + \tfrac{1}{2}\, u^3 + au + b.}$$

It follows from Proposition 1.30 (see operator F_6 therein) that (1.96) has solutions

$$v(x,t) = \big[C_1(t) + C_2(t)\cos x + C_3(t)\sin x\big]^{-\frac{3}{2}}, \tag{1.97}$$

$$\begin{cases} C_1' = \tfrac{1}{2}\, C_1(C_1^2 - C_2^2 - C_3^2) + aC_1 + b, \\ C_2' = \tfrac{1}{2}\, C_2(C_1^2 - C_2^2 - C_3^2) + aC_2, \\ C_3' = \tfrac{1}{2}\, C_3(C_1^2 - C_2^2 - C_3^2) + aC_3. \end{cases}$$

These solutions can be derived by using a quadratic representation of (1.96). Setting $u = v^{-4/3}$ (the correct pressure transformation) yields a family of solutions on the 5D subspace (1.93) that includes those given by (1.97); cf. Example 1.15.

Example 1.36 (**Riabouchinsky–Proudman–Johnson (RPJ) equation**) Consider the *RPJ equation* [488, 473]

$$\boxed{u_{xxt} = \varepsilon u_{xxxx} - uu_{xxx} + u_x u_{xx}\,,} \tag{1.98}$$

with the third-order quadratic operator $F[u] = -uu_{xxx} + u_x u_{xx}$. This PDE arises in the so-called *Berman problem* and is derived from the *Navier–Stokes equations* in $I\!R^2$ describing the unsteady plane flow of a viscous incompressible fluid confined between parallel walls. In this case, $\varepsilon = \frac{1}{\mathrm{Re}}$, where Re is the *Reynolds number*. The stationary solutions of (1.98) were studied by A.S. Berman in 1953, [43].

Equation (1.98) can be integrated once to reveal the special quadratic operator

$$u_{xt} = \varepsilon u_{xxx} - \big[uu_{xx} - (u_x)^2\big] + g(t),$$

where $g(t)$ is an arbitrary smooth function of integration related to the fluid pressure. We will refer to

$$F_{\mathrm{rem}}[u] = uu_{xx} - (u_x)^2 \tag{1.99}$$

as the *remarkable operator*, in view of its outstanding invariant properties that will be used a few times later on. Namely, F_{rem} admits the 3D subspace

$$W_3 = \mathcal{L}\{1, \cos\gamma x, \sin\gamma x\} \quad \text{for any } \gamma \neq 0. \tag{1.100}$$

Using this subspace (1.100) gives

$$u(x,t) = C_1(t) + C_2(t)\cos\gamma x + C_3(t)\sin\gamma x,$$

with the following system of an algebraic and two differential equations:

$$\begin{cases} g = -\gamma^2(C_2^2 + C_3^2), \\ C_2' = -\varepsilon\gamma^2 C_2 - \gamma C_1 C_3, \\ C_3' = -\varepsilon\gamma^2 C_3 + \gamma C_1 C_2. \end{cases}$$

Concerning polynomial subspaces, for the third-order quadratic operator in (1.98), the basic subspace is 4D, $W_4 = \mathcal{L}\{1, x, x^2, x^3\}$. The 5D, $W_5 = \mathcal{L}\{1, x, x^2, x^3, x^4\}$, is available for the modified operator

$$F[u] = -uu_{xxx} + \tfrac{1}{2}\,u_x u_{xx} : W_5 \to W_5.$$

Example 1.37 (**Partial invariance in time-dependent cases**) Let us return to the original RPJ equation (1.98). Clearly, the operator there preserves the subspace of linear functions $W_2^{\rm lin} = \mathcal{L}\{1, x\}$, as well as the trigonometric subspace (1.100) for *arbitrary* $\gamma \in I\!R$. These two properties make it possible to construct solutions of (1.98) of the following form:

$$u(x,t) = C_1(t) + C_2(t)x + C_3(t)\cos(\gamma(t)x) + C_4(t)\sin(\gamma(t)x), \tag{1.101}$$

where now $\gamma(t)$ is a function to be determined. Such solutions admit simple invariant motivations; see also Remarks on relations to *positons* of integrable PDEs. Here, the basis functions depend on the independent variable t, so a proper setting includes partially invariant modules to be treated in the next chapter, Section 2.8.

Now, using a natural scaling idea, we consider the whole differential operator

$$\hat{F}[u] = u_{xxt} - F[u] \equiv u_{xxt} + uu_{xxx} - u_x u_{xx}, \tag{1.102}$$

which is defined in the space of smooth functions $u = u(x,t)$ of both independent variables. First, we get rid of the $\gamma(t)$-dependence in the spatial variable in (1.101) by introducing the new independent variable

$$\hat{x} = \gamma(t)x.$$

Then (1.102) reads

$$\hat{F}[u] = \gamma\gamma'\hat{x}u_{\hat{x}\hat{x}\hat{x}} + 2\gamma\gamma' u_{\hat{x}\hat{x}} + \gamma^2 u_{\hat{x}\hat{x}t} - \gamma^3(u_{\hat{x}}u_{\hat{x}\hat{x}} - uu_{\hat{x}\hat{x}\hat{x}}),$$

while the exact solutions (1.101) take the form

$$u(\hat{x},t) = C_1 + \tfrac{C_2}{\gamma}\,\hat{x} + C_3\cos\hat{x} + C_4\sin\hat{x}. \tag{1.103}$$

Second, for convenience, let us introduce the new function v by

$$u = v + \tfrac{C_2}{\gamma}\,\hat{x},$$

so that

$$\hat{F}[u] \mapsto \hat{\mathcal{F}}[v] = \gamma\,(\gamma' + \gamma C_2)\hat{x}v_{\hat{x}\hat{x}\hat{x}} + \gamma\,(2\gamma' - \gamma C_2)v_{\hat{x}\hat{x}} \\ + \gamma^2\big[v_{\hat{x}\hat{x}t} - \gamma\,(v_{\hat{x}}v_{\hat{x}\hat{x}} - vv_{\hat{x}\hat{x}\hat{x}})\big].$$

$\hat{\mathcal{F}}[v]$ preserves the standard trigonometric subspace

$$\hat{W}_3 = \mathcal{L}\{1, \cos \hat{x}, \sin \hat{x}\}, \quad \text{if} \;\; \gamma' + \gamma C_2 = 0, \tag{1.104}$$

and the following action of the operator is obtained:

$$\hat{\mathcal{F}}[v] = 3\gamma\gamma' v_{\hat{x}\hat{x}} + \gamma^2 v_{\hat{x}\hat{x}t} - \gamma^3(v_{\hat{x}} v_{\hat{x}\hat{x}} - v v_{\hat{x}\hat{x}\hat{x}}). \tag{1.105}$$

Actually, $\hat{W}_3$ is invariant under all three operators on the right-hand side of (1.105). This is seen from the identity in the original independent variables,

$$\begin{aligned} \hat{F}[u] = {} & \big[-(C_3\gamma^2)' - C_1C_4\gamma^3 + C_2C_3\gamma^2\big] \cos \gamma x \\ & + \big[-(C_4\gamma^2)' + C_1C_3\gamma^3 + C_2C_4\gamma^2\big] \sin \gamma x \\ & + \gamma^2(\gamma' + \gamma C_2)\big(C_3 x \sin \gamma x - C_4 x \cos \gamma x\big), \end{aligned} \tag{1.106}$$

where the last bad term not belonging to the subspace vanishes by (1.104).

Using the expansion (1.106), we obtain for solutions (1.101) of (1.98) three ODEs for five coefficients

$$\begin{cases} C_3' = -\varepsilon\gamma^2 C_3 + 3C_2C_3 - \gamma C_1 C_4, \\ C_4' = -\varepsilon\gamma^2 C_4 + 3C_2C_4 + \gamma C_1 C_3, \\ \gamma' = -C_2\gamma. \end{cases} \tag{1.107}$$

This undetermined DS allows us to control the pressure and pose suitable boundary conditions. For instance, the homogeneous Dirichlet conditions,

$$u(\pm 1, t) = 0 \quad \text{for} \;\; t > 0,$$

imply by (1.101) that

$$C_1 = -C_3 \cos \gamma \quad \text{and} \quad C_2 = -C_4 \sin \gamma.$$

Plunging these into (1.107) yields a well-posed DS for three coefficients,

$$\begin{cases} C_3' = -\varepsilon\gamma^2 C_3 + (\gamma \cos \gamma - 3 \sin \gamma) C_3 C_4, \\ C_4' = -\varepsilon\gamma^2 C_4 - 3C_4^2 \sin \gamma - \gamma C_3^2 \cos \gamma, \\ \gamma' = \gamma \sin \gamma \, C_4. \end{cases}$$

Similar computations can be performed for hyperbolic functions $\cosh(\gamma(t)x)$ and $\sinh(\gamma(t)x)$ in (1.101); see further examples below.

1.4.2 Non-local models from Navier–Stokes equations in $I\!R^2$

Example 1.38 (**Nonstationary von Kármán solutions**) Consider the *Navier–Stokes equations* in $I\!R^2$

$$\begin{cases} u_t + uu_x + vu_y = -\frac{1}{\rho} p_x + \nu(u_{xx} + u_{yy}), \\ v_t + uv_x + vv_y = -\frac{1}{\rho} p_y + \nu(v_{xx} + v_{yy}), \\ u_x + v_y = 0, \end{cases} \tag{1.108}$$

where $(u, v) = \mathbf{u}$ is the velocity field, p is the pressure, $\rho > 0$ is the constant density, and $\nu > 0$ is the constant kinematic viscosity. We are looking for solutions

$$u = \int_0^x f(z,t)\,\mathrm{d}z, \quad v = -y\, f(x,t), \quad \text{and} \quad p = h(x,t). \tag{1.109}$$

The structure of such solutions corresponds to a nonstationary plane version of the *von Kármán solutions* [327], which have been applied to various problems of fluid dynamics; see [42] and references in [13, Ch. 7].

Plugging (1.109) into (1.108) yields a semilinear non-local parabolic equation for the function $f(x,t)$, replaced now by $u = u(x,t)$:

$$\boxed{u_t = u_{xx} - u^2 + u_x \int_0^x u(z,t)\,\mathrm{d}z.} \tag{1.110}$$

Consider for (1.110) a free-boundary problem on $(0, s(t)) \times I\!R_+$ with the symmetry condition at the origin, $u_x(0,t) = 0$, and the following conditions at the free boundary $x = s(t)$:

$$u_x(s(t),t) = 0, \quad s' = -\int_0^s u(z,t)\,\mathrm{d}z.$$

For derivation, references, and further mathematical results, see [245, Ch. 9]. Using the same approach, after time-dependent scaling, as shown in (1.104), we obtain solutions of the type (1.101), with $\gamma(t) = \frac{\pi}{s(t)}$,

$$u(x,t) = C_1(t) + C_2(t)\cos\left(\tfrac{\pi x}{s(t)}\right),$$

$$\begin{cases} C_1' = -C_1^2 - C_2^2, \\ C_2' = -2C_1C_2 - (\frac{\pi}{s})^2 C_2, \\ s' = -sC_1. \end{cases}$$

Example 1.39 **(General solutions on $\mathbf{W_2}$)** In connection with the previous example, let us perform more general invariant analysis of the Navier–Stokes equations (1.108). The spatial structure of von Kármán solutions (1.109) expresses the fact that (1.108) can posses solutions on the 2D subspace

$$u, v, p \in W_2 = \mathcal{L}\{1, y\},$$

which is not invariant under all the nonlinear operators. So, we set

$$\begin{cases} u = C_1(x,t) + C_2(x,t)y, \\ v = D_1(x,t) + D_2(x,t)y, \\ p = E_1(x,t) + E_2(x,t)y. \end{cases}$$

Substituting into (1.108) yields the following system of eight equations (projections of the three PDEs onto the vectors 1, y, and y^2 respectively) for six coefficients:

$$\begin{cases} C_{1t} + C_1C_{1x} + D_1C_2 = -\frac{1}{\rho}E_{1x} + \nu C_{1xx}, \\ C_{2t} + (C_1C_2)_x + C_2D_2 = -\frac{1}{\rho}E_{2x} + \nu C_{2xx}, \\ C_2C_{2x} = 0, \\ D_{1t} + C_1D_{1x} + D_1D_2 = -\frac{1}{\rho}E_2 + \nu D_{1xx}, \\ D_{2t} + C_2D_{1x} + C_1D_{2x} + D_2^2 = \nu D_{2xx}, \\ C_2D_{2x} = 0, \\ C_{1x} + D_2 = 0, \\ C_{2x} = 0. \end{cases}$$

The shortest equations imply that

$$\text{either } C_2 = 0, \quad \text{or} \quad C_{2x} = 0 \text{ and } D_{2x} = 0.$$

Consider for instance the first case. The second equation then yields that $E_2 = E_2(t)$ is arbitrary. Next, plugging

$$C_1 = -\int D_2\,\mathrm{d}x$$

from the seventh equation into the previous one (with $C_2 = 0$), we obtain the independent governing equation of the type (1.110) for the coefficient D_2. Solving a suitable FBP for this parabolic equation yields $D_2(x,t)$, and then the fourth equation becomes a linear parabolic PDE for D_1 with an arbitrary function $E_2(t)$. Finally, the first equation can be treated as an ODE for $E_1(x,t)$ with a parameter t.

Example 1.40 **(Polynomial subspace)** The following non-local semilinear heat equation on $(0,1) \times I\!R_+$ is also associated with the Navier–Stokes equations in $I\!R^2$:

$$\boxed{u_t = u_{xx} + \beta(u_x)^2 + u^2 - \mu u_x \int_0^x u(z,t)\,\mathrm{d}z - (1+\mu)\int_0^1 u^2(z,t)\,\mathrm{d}z.}$$

For $\mu = \frac{3}{2}$, it admits solutions on the subspace of quadratic polynomials,

$$u(x,t) = C_1(t) + C_2(t)x + C_3(t)x^2,$$

$$\begin{cases} C_1' = -\frac{3}{2}C_1^2 - \frac{5}{2}(C_1C_2 + \frac{2}{3}C_1C_3) + (\beta - \frac{5}{6})C_2^2 - \frac{5}{4}C_2C_3 - \frac{1}{2}C_3^2 + 2C_3, \\ C_2' = -\frac{1}{2}C_1C_2 + 4\beta C_2C_3, \\ C_3' = -C_1C_3 + \frac{1}{4}C_2^2 + 4\beta C_3^2. \end{cases}$$

1.4.3 Criterion of invariance for polynomial operators

Consider a general quadratic operator of the form

$$F[u] = \textstyle\sum_{(i,j)} a_{i,j}\, D_x^i u D_x^j u, \tag{1.111}$$

where $a = \|a_{i,j}\|$ $(i,j = 0, ..., k)$ is a given constant real symmetric $(k+1)\times(k+1)$ matrix. By P, we denote the corresponding polynomial,

$$P(X,Y) = \textstyle\sum a_{i,j} X^i Y^j, \quad \text{so} \quad F[u] = P(D_x, D_y)u(x,t)u(y,t)\big|_{y=x}. \tag{1.112}$$

Consider the function

$$u = \textstyle\sum_{(\mu\in\mathcal{N})}\big[C_\mu \cos(\mu x) + D_\mu \sin(\mu x)\big], \tag{1.113}$$

where $\mathcal{N}$ is a subset of nonnegative integers. If $\mathcal{N} = \{0, 1, ..., n\}$, then u belongs to the standard trigonometric subspace $W_{2n+1} = \mathcal{L}\{1, \cos x, \sin x, ..., \cos(nx), \sin(nx)\}$. Differentiating in each term and performing simple manipulations, we find the action

$$\begin{aligned} F[u] = {} & \textstyle\sum_{(\mu,\nu)}\Big\{\sum_{(i,j)} a_{i,j}\,\mu^i\nu^j\big[C_\mu\cos(\mu x + \frac{\pi}{2}i) + D_\mu\sin(\mu x + \frac{\pi}{2}i)\big] \\ & \times\big[C_\nu\cos(\nu x + \frac{\pi}{2}j) + D_\nu\sin(\nu x + \frac{\pi}{2}j)\big]\Big\} = \tfrac{1}{2}\textstyle\sum_{(\mu,\nu)}\Big\{\sum_{(i,j)} a_{i,j} \\ & \times\Big(C_\mu C_\nu \mu^i\nu^j\big[\cos((\mu+\nu)x + \tfrac{\pi}{2}(i+j)) + \cos((\mu-\nu)x + \tfrac{\pi}{2}(i-j))\big] \\ & + C_\mu D_\nu(\mu^i\nu^j + \nu^i\mu^j)\big[\sin((\mu+\nu)x + \tfrac{\pi}{2}(i+j)) - \sin((\mu-\nu)x + \tfrac{\pi}{2}(i-j))\big] \\ & + D_\mu D_\nu \mu^i\nu^j\big[\cos((\mu-\nu)x + \tfrac{\pi}{2}(i-j)) - \cos((\mu+\nu)x + \tfrac{\pi}{2}(i+j))\big]\Big)\Big\}. \end{aligned}$$

In the second sum with $C_\mu D_\nu$, we obtained two similar terms with $C_\mu D_\nu$ and $C_\nu D_\mu$, and in the latter, we changed the summation $(\mu, \nu) \mapsto (\nu, \mu)$ to get a single term. The first term with $C_\mu C_\nu$ has the coefficient

$$\tfrac{1}{2} \sum_{(i+j \text{ is even})} a_{i,j}\, \mu^i \nu^j \big[(-1)^{\frac{i+j}{2}} \cos((\mu+\nu)x) + (-1)^{\frac{i-j}{2}} \cos((\mu-\nu)x)\big]$$
$$+ \tfrac{1}{2} \sum_{(i+j \text{ is odd})} a_{i,j}\, \mu^i \nu^j \big[(-1)^{\frac{i+j+1}{2}} \sin((\mu+\nu)x) + (-1)^{\frac{i-j+1}{2}} \cos((\mu-\nu)x)\big].$$

Since all the products $C_\mu C_\nu$, $C_\mu D_\nu$, and $D_\mu D_\nu$ (latter two generate similar sums) are linearly independent, taking into account cos and sin functions of $(\mu+\nu)x$, we obtain first two groups of invariance conditions:

$$\sum_{(i+j \text{ is even})} a_{i,j}\, \mu^i \nu^j (-1)^{\frac{i+j}{2}} = 0, \qquad \sum_{(i+j \text{ is odd})} a_{i,j}\, \mu^i \nu^j (-1)^{\frac{i+j+1}{2}} = 0 \tag{1.114}$$

for any $\mu, \nu \in \mathcal{N}$ with $\mu \le \nu$, such that $\mu + \nu \notin \mathcal{N}$.

Similarly, the terms with cos and sin of the argument $(\mu - \nu)x$ yield two more groups of invariance conditions (1.114) in which, everywhere in both sums, $i + j$ should be replaced by $i - j$. These apply for $\nu - \mu \notin \mathcal{N}$, $\mu \le \nu$. For the standard subspace W_{2n+1} without "gaps", these two groups of conditions do not appear.

For operators (1.86) with fixed $i + j = k$ and the subspace W_{2n+1}, the invariance conditions (1.114) are simplified and involve the polynomial (1.112),

$$P(\mu, \nu) = 0 \quad \text{for all } \mu \le \nu \text{ such that } \mu + \nu > n. \tag{1.115}$$

We will study such linear systems on the coefficients $\{a_{i,j}\}$ in Section 1.5.2 devoted to exponential subspaces.

Similar invariance analysis can be performed for the cubic operators (1.90) with analogous trigonometric manipulations and conclusions.

1.5 Examples: exponential subspaces

1.5.1 Classification and examples

We begin with the 3D subspace

$$W_3 = \mathcal{L}\{1, \mathrm{e}^x, \mathrm{e}^{-x}\}. \tag{1.116}$$

Proposition 1.41 *Subspace* (1.116) *is invariant under operator* (1.56) *iff*

$$\begin{cases} b_1 + b_2 + b_3 + b_4 + b_5 + b_6 = 0, \\ b_1 - b_2 + b_3 + b_4 - b_5 + b_6 = 0. \end{cases}$$

The set of such operators is a 4D linear space spanned by

$$F_1[u] = u(u_{xx} - u), \;\; F_2[u] = u_x(u_{xx} - u),$$
$$F_3[u] = u(u_{xx} - u), \;\; F_4[u] = (u_x)^2 - u^2.$$

Example 1.42 (**Blow-up for fast diffusion equation with source**) The quasilinear fast diffusion equation with a quadratic reaction term

$$v_t = \left(\tfrac{1}{v} v_x\right)_x + v^2 \quad \text{in } I\!R \times I\!R_+ \tag{1.117}$$

admits blow-up solutions for sufficiently large initial functions $v_0(x) > 0$. By the pressure transformation $u = \frac{1}{v}$, (1.117) reduces to the quadratic PDE

$$\boxed{u_t = F_{\rm rem}[u] - 1 \equiv uu_{xx} - (u_x)^2 - 1.} \tag{1.118}$$

By Proposition 1.41, the remarkable operator $F_{\rm rem}$ in (1.118) (note that $F_{\rm rem}$ equals to $F_3 - F_4$) preserves subspace (1.116) or, which is the same, the subspace of hyperbolic functions

$$W_3 = \mathcal{L}\{1, \cosh x, \sinh x\}. \tag{1.119}$$

Studying the evolution of the symmetric in x functions yields the solutions of (1.117)

$$u(x,t) = C_1(t) + C_2(t)\cosh x \in W_2 = \mathcal{L}\{1, \cosh x\},$$

$$\begin{cases} C_1' = C_2^2 - 1, \\ C_2' = C_1 C_2. \end{cases} \tag{1.120}$$

The DS admits the first integral

$$C_2^2 - C_1^2 - 2\ln|C_2| = \text{constant},$$

and can be integrated in quadratures. The exact blow-up asymptotics of the DS (1.120) describe some interesting features of localized, single point blow-up patterns for the original PDE (1.117).

It is curious that "trigonometric" Proposition 1.29 also yields solutions

$$u(x,t) = C_1(t) + C_2(t)\cos x \in W_2 = \mathcal{L}\{1, \cos x\},$$

where the DS is slightly different,

$$\begin{cases} C_1' = -C_2^2 - 1, \\ C_2' = -C_1 C_2. \end{cases}$$

Such solutions describe distinct features of blow-up of 2π-periodic solutions.

We now introduce the exponential analog of Example 1.37.

Example 1.43 (Partially invariant module). Consider the $\frac{\partial}{\partial t}$-dependent operator

$$\hat{F}[u] = u_t - F_{\rm rem}[u] \equiv u_t - \left[uu_{xx} - (u_x)^2\right] \quad \text{for } u = u(x,t) \text{ given by}$$

$$u(x,t) = C_1(t) + C_2(t)x + C_3(t)e^{\gamma(t)x}. \tag{1.121}$$

This case of "time-dependent" subspaces admits a natural invariant treatment after scaling $\hat{x} = \gamma(t)x$, as shown in Example 1.37. We omit further related computations. In the original variables, the expansion is

$$\hat{F}[u] = C_1' + C_2^2 + C_2'x + \left(C_3' - C_1C_3\gamma^2 + 2C_2C_3\gamma\right)e^{\gamma x} + C_3\left(\gamma' - C_2\gamma^2\right)xe^{\gamma x}.$$

For instance, for the semilinear Kuramoto–Sivashinsky-type equation (see Section 3.8 for details)

$$\boxed{u_t = -u_{xxxx} + uu_{xx} - (u_x)^2 + \alpha u + \beta,}$$

looking for solutions in the form (1.121) leads to the following DS:

$$\begin{cases} C_1' = -C_2^2 + \alpha C_1 + \beta, \\ C_2' = \alpha C_2, \\ C_3' = \gamma C_3(\gamma C_1 - 2C_2) + (\alpha - \gamma^4)C_3, \\ \gamma' = C_2 \gamma^2. \end{cases}$$

The second ODE gives $C_2(t) = \mathrm{e}^{\alpha t}$ if $\alpha \neq 0$, and $C_2(t) = 1$ if $\alpha = 0$. Substituting $C_2(t)$ into the last equation, we obtain two functions $\gamma(t)$ for which such solutions exist,

$$\gamma(t) = \begin{cases} \alpha(\mathrm{e}^{\alpha T} - \mathrm{e}^{\alpha t})^{-1} & \text{for } \alpha \neq 0, \\ \frac{1}{T-t} & \text{for } \alpha = 0, \end{cases}$$

where T is a fixed blow-up time of the solution.

Omitting straightforward analysis of cubic operators (1.57) admitting (1.116), consider the 5D exponential subspace

$$W_5 = \mathcal{L}\{1, \mathrm{e}^x, \mathrm{e}^{-x}, \mathrm{e}^{2x}, \mathrm{e}^{-2x}\}. \tag{1.122}$$

Proposition 1.44 *Subspace* (1.122) *is invariant under* (1.56) *iff*

$$\begin{cases} 16b_6 + 8b_5 + 4b_4 + 4b_3 + 2b_2 + b_1 = 0, \\ 16b_6 - 8b_5 + 4b_4 + 4b_3 - 2b_2 + b_1 = 0, \\ 8b_6 + 6b_5 + 5b_4 + 4b_3 + 3b_2 + 2b_1 = 0, \\ 8b_6 - 6b_5 + 5b_4 - 4b_3 - 3b_2 + 2b_1 = 0. \end{cases}$$

The set of such operators is a 2D linear space spanned by

$$F_1[u] = uu_{xx} - \tfrac{3}{4}(u_x)^2 - u^2 \quad \text{and} \quad F_2[u] = (u_{xx} - 4u)^2.$$

Proposition 1.45 *No nontrivial cubic operators* (1.57) *admit the subspace* (1.122).

The proof is the same as that of Proposition 1.32. A full classification of 5D invariant subspaces for quadratic second-order operators will be the main subject of Theorem 2.22.

Example 1.46 In a similar manner, the operator of the GT equation (1.64) is associated with the hyperbolic subspace (1.119). The extension to subspace (1.122) for the operator family (1.94) is performed as in Proposition 1.34.

1.5.2 General polynomial operators

We now deal with the general quadratic kth-order operators (1.111) with the polynomial (1.112).

Set $p_1 = 0$ and fix other $n-1$ real exponents, introducing the set

$$\Gamma = \{p_i,\ i = 1, ..., n\}, \quad \text{where } p_i \neq p_j \text{ for } i \neq j.$$

Consider the n-dimensional linear subspace

$$\begin{gathered} W_n = \mathcal{L}\{\mathrm{e}^{p_i x},\ i = 1, ..., n\}, \quad \text{and denote} \\ \Gamma' = \{p_m + p_l :\ p_m, p_l \in \Gamma,\ p_m + p_l \notin \Gamma\}. \end{gathered} \tag{1.123}$$

Proposition 1.47 *Subspace* (1.123) *is invariant under the operator* (1.111) *iff*

$$P(p_m, p_l) = 0 \quad \text{for all pairs } (p_m, p_l) \text{ such that } p_m + p_l \in \Gamma'. \tag{1.124}$$

Proof. Taking arbitrary

$$u = \sum_{(i)} C_i \mathrm{e}^{p_i x} \in W_n \tag{1.125}$$

and plugging into (1.111) yields

$$\begin{aligned} F[u] &= \sum_{(i,j)} \sum_{(m,l)} a_{i,j}\, p_m^i p_l^j C_m C_l\, \mathrm{e}^{(p_m+p_l)x} \\ &\equiv \sum_{(m,l)} P(p_m, p_l) C_m C_l \mathrm{e}^{(p_m+p_l)x}. \end{aligned} \tag{1.126}$$

Since $\{C_i\}$ are independent, it follows from (1.126) that the coefficients of $\mathrm{e}^{(p_m+p_l)x}$ must vanish for any $p_m + p_l \in \Gamma'$. □

Dealing with invariant subspaces, in this proof, we do not take into account a possible correlation of similar terms in (1.126) that can give rise to cancellation of some of the terms. Actually, for *arbitrary* values of expansion coefficients $\{C_i\}$ involved, this is not possible. Assuming existence of some algebraic relations between expansion coefficients $\{C_i\}$, meaning *partial* invariance of W_n, this topic will be under scrutiny in Section 7.1. Evidently, the linear system (1.124) on coefficients $\{a_{i,j}\}$ means existence of an infinite number of quadratic (non-integrable) PDEs with solutions on exponential subspaces W_n of arbitrary finite dimension n (e.g., soliton-type solutions).

Corollary 1.48 *Given an arbitrarily large integer $l > 1$, there exists a polynomial operator* (1.111) *of the order k large enough, admitting at least l exponential invariant subspaces W_n,* (1.123)*, of any dimension $n = 1, 2, ..., l$.*

The corresponding first-order evolution PDEs

$$u_t = F[u] \equiv \sum_{(i,j)} a_{i,j}\, D_x^i u D_x^j u \quad \text{on } W_n \tag{1.127}$$

are n-dimensional DSs. The second-order PDEs, $u_{tt} = F[u]$ on W_n, are $2n$th-order DSs. Such classes of PDEs can be treated as *intermediate* relative to known integrable equations admitting infinitely many (partially) invariant subspaces.

Let us return to systems (1.124). The general operator (1.111) contains $\frac{(k+1)(k+2)}{2}$ free coefficients $\{a_{i,j}\}$. So the system (1.124) cannot have more than $\frac{(k+1)(k+2)}{2} - 1$ linearly independent equations to generate such nontrivial operators. The total number of elements in Γ' (its *cardinal number*) satisfies $\sharp\Gamma' \le \frac{(k+1)(k+2)}{2} - 1$. In an example below, we illustrate solvability of such systems in a particular setting.

Example 1.49 (**Operators with monomials of the same total order**) Consider operators (1.86) that contain $\left[\frac{k}{2}\right] + 1$ unknown coefficients $\{a_{i,j}\}$. Without loss of generality, we assume that k is even and $\sharp\,\Gamma' = s = \frac{k}{2}$. By $(p_{l_1}, p_{m_1}), ..., (p_{l_s}, p_{m_s})$, we denote the pairs of exponents such that $p_{l_j} + p_{m_j} \in \Gamma'$ for all $j = 1, ..., s$. The system (1.124) is then composed of precisely s equations,

$$\begin{aligned} & a_{0,k}\, \tfrac{1}{2}\big(p_{l_j}^k + p_{m_j}^k\big) + a_{1,k-1}\, \tfrac{1}{2}\big(p_{l_j}^{k-1} p_{m_j} + p_{l_j} p_{m_j}^{k-1}\big) + ... \\ & \quad + a_{\frac{k}{2},\frac{k}{2}}\, p_{l_j}^{\frac{k}{2}} p_{m_j}^{\frac{k}{2}} = 0 \ \text{ for } j = 1, ..., s. \end{aligned} \tag{1.128}$$

(If k is odd, the last term is replaced by $a_{\frac{k-1}{2},\frac{k+1}{2}}\, p_{l_j}^{\frac{k-1}{2}}\, p_{m_j}^{\frac{k+1}{2}}$, and the analysis is similar.) Dividing each equation in (1.128) by $p_{l_j}^k$ and denoting $r_j = \frac{p_{m_j}}{p_{l_j}}$ yields the equivalent system

$$\begin{aligned} &a_{0,k}\left(1+r_j^k\right)+a_{1,k-1}\left(r_j+r_j^{k-1}\right)+\dots \\ &+a_{\frac{k}{2}-1,\frac{k}{2}+1}\left(r_j^{\frac{k}{2}-1}+r_j^{\frac{k}{2}+1}\right)+2a_{\frac{k}{2},\frac{k}{2}}\, r_j^{\frac{k}{2}}=0. \end{aligned} \tag{1.129}$$

For mutually distinct ratios $\{r_j\}$, we then obtain a *unique* operator (1.86), up to a scalar factor. It is not very difficult to solve the system (1.129),

$$\begin{aligned} &a_{l,k-l}=(-1)^l\sigma_l\left(r_1,\dots,r_s,\tfrac{1}{r_1},\dots,\tfrac{1}{r_s}\right) \;\text{ for }\; l=0,\dots,\tfrac{k}{2}-1; \\ &a_{\frac{k}{2},\frac{k}{2}}=\tfrac{1}{2}(-1)^{\frac{k}{2}}\sigma_{\frac{k}{2}}(\cdot) \;\text{ for }\; l=0. \end{aligned} \tag{1.130}$$

Here, $\sigma_l(\cdot)$ are the *elementary symmetric function* of $2s=k$ arguments,

$$\sigma_l(y_1,\dots,y_{2s})=\textstyle\sum_{(1\le i_1<\dots<i_l\le 2s)} y_{i_1}\cdots y_{i_l} \quad (\sigma_0(\cdot)=1).$$

For instance, for $k=4$, the unique thin film operator (1.86) has the form

$$u_{xxxx}u-\left(r_1+r_2+\tfrac{1}{r_1}+\tfrac{1}{r_2}\right)u_{xxx}u_x+\tfrac{1}{2}\left(r_1r_2+2+\tfrac{r_1}{r_2}+\tfrac{r_2}{r_1}+\tfrac{1}{r_1r_2}\right)(u_{xx})^2.$$

For the space $W_3=\mathcal{L}\{1,\mathrm{e}^x,\mathrm{e}^{2x}\}$, i.e., $p_1=0$, $p_2=1$, and $p_3=2$, we have two pares of exponents, $(1,2)$ and $(2,2)$, generating $\Gamma'=\{3,4\}$. Hence, we obtain $r_1=2$ and $r_2=1$, and the operator

$$F[u]=u_{xxxx}u-\tfrac{9}{2}\,u_{xxx}u_x+\tfrac{7}{2}(u_{xx})^2.$$

The corresponding parabolic and hyperbolic PDEs,

$$u_t=F[u] \quad\text{and}\quad u_{tt}=F[u],$$

have finite-dimensional dynamics on W_3. An ODE reduction exists for the fifth-order *nonlinear dispersion* PDE $u_t=(F[u])_x$. We will study such models in Section 4.2.

Example 1.50 (**TFEs**) Consider a fourth-order equation from thin film theory (see more on such degenerate parabolic models in Section 3.1),

$$\boxed{u_t=F[u]\equiv -uu_{xxxx}+\alpha u_xu_{xxx}+\beta(u_{xx})^2+\gamma uu_{xx}.} \tag{1.131}$$

Let $\Gamma=\{0,1,2,3\}$, so the PDE is restricted to the subspace

$$W_4=\mathcal{L}\{1,\mathrm{e}^x,\mathrm{e}^{2x},\mathrm{e}^{3x}\}, \tag{1.132}$$

and let us look for a solution in the form of expansion

$$u(x,t)=C_1(t)+C_2(t)\mathrm{e}^x+C_3(t)\mathrm{e}^{2x}+C_4(t)\mathrm{e}^{3x}. \tag{1.133}$$

Since $1+2=3$, the structure of this solution corresponds to the 2-solitons. The polynomial of operator F in (1.131) is

$$P=-\tfrac{1}{2}\left(X^4+Y^4\right)+\tfrac{\alpha}{2}\left(XY^3+X^3Y\right)+\beta X^2Y^2+\tfrac{\gamma}{2}\left(X^2+Y^2\right),$$

$\Gamma' = \{4, 5, 6\}$, so, by Proposition 1.47, subspace (1.132) is invariant under F iff

$$P(1,3) = P(2,3) = P(3,3) = 0.$$

This yields a linear system for parameters,

$$\begin{cases} 15\alpha + 9\beta + 5\gamma = 41, \\ 78\alpha + 72\beta + 13\gamma = 97, \\ 9\alpha + 9\beta + \gamma = 9. \end{cases}$$

This has a unique solution, so that there exists a single thin film operator

$$F[u] = -uu_{xxxx} + 10\, u_x u_{xxx} - \tfrac{74}{9}(u_{xx})^2 - 7uu_{xx}$$

preserving (1.132). The PDE (1.131) on W_4 then reduces to the fourth-order DS

$$\begin{cases} C_1' = 0, \\ C_2' = 2P(0,1)C_1C_2, \\ C_3' = 2P(1,1)C_2^2 + 2P(0,2)C_1C_3, \\ C_4' = 2P(0,3)C_1C_4 + 2P(1,2)C_2C_3, \end{cases}$$

that can be solved explicitly. We may take $C_1 = 1$ from the first ODE, and, from the second, $C_2(t) = A_2 e^{2P(0,1)t}$, with the constant $A_2 = C_2(0)$. The rest of the ODEs can be explicitly solved step by step. The resulting exact solution (1.133) is not always composed of elementary solitonic exponential terms, such as $e^{px+\omega t}$. For instance, linear (and quadratic) polynomials in t appear in the resonance case $P(0,2) = 2P(0,1)$. For solutions not having the 2-soliton structure, e.g.,

$$u(x,t) = C_1(t) + C_2(t)e^x + C_3(t)e^{2x} + C_4(t)e^{4x},$$

where $\Gamma' = \{3, 5, 6, 8\}$, we need four conditions of invariance of the corresponding subspace

$$P(1,2) = P(1,4) = P(2,4) = P(4,4) = 0.$$

The maximal number of such invariance conditions of general exponential spaces W_4 is six for the case where no correlation between products of exponential terms in (1.126) is available.

Example 1.51 (**5D subspaces**) Consider a general second-order PDE

$$\boxed{u_{tt} = F[u] = \textstyle\sum_{(i,j)} a_{i,j}\, D_x^i u D_x^j u}$$

on W_5 with the expansion (cf. the canonical 5D subspace (1.122))

$$u(x,t) = C_1(t) + C_2(t)e^x + C_3(t)e^{-x} + C_4(t)e^{\frac{1}{3}x} + C_5(t)e^{2x}.$$

Then $\Gamma = \{-1, 0, \frac{1}{3}, 1, 2\}$ and $\Gamma' = \{-2, -\frac{2}{3}, \frac{2}{3}, \frac{4}{3}, \frac{7}{3}, 3, 4\}$, with seven conditions $P(-1,-1) = P(-1,\frac{1}{3}) = P(\frac{1}{3},\frac{1}{3}) = P(1,\frac{1}{3}) = P(\frac{1}{3},2) = P(1,2) = P(2,2) = 0$. The DS is

$$\begin{cases} C_1'' = P(0,0)C_1^2 + 2P(1,-1)C_2C_3, \\ C_2'' = 2P(0,1)C_1C_2 + 2P(-1,2)C_3C_5, \\ C_3'' = 2P(0,-1)C_1C_3, \\ C_4'' = 2P(0,\frac{1}{3})C_1C_4, \\ C_5'' = 2P(0,2)C_1C_5 + P(1,1)C_2^2. \end{cases}$$

For solutions on the general subspace (1.123), the maximal number of invariant conditions in (1.124) is $\frac{n(n-1)}{2}$ (recall that $p_1 = 0$ is fixed *a priori*). For proper operators F with higher-degree polynomials, there exist subspaces of higher dimension, and hence, solutions via DSs can be constructed for corresponding evolution PDEs.

On exponential-trigonometric subspaces. If complex exponents $p_j = a_j + \mathrm{i} b_j$, $b_j \neq 0$, are involved in (1.123), this gives the exact solutions for a system of PDEs stated for the real and imaginary parts of $u = U + \mathrm{i}V$,

$$\boxed{\begin{cases} U_t = F[U] - F[V], \\ V_t = 2P[U, V]. \end{cases}} \tag{1.134}$$

The real case $V = 0$ yields the single PDE $U_t = F[U]$. Since

$$\mathrm{e}^{p_j x} = \mathrm{e}^{a_j x}(\cos b_j x + \mathrm{i} \sin b_j x),$$

the expansion on W_n consists of exponential-trigonometric functions, and algebraic manipulations become more technical.

On quadratic $\frac{\partial}{\partial t}$-dependent operators. Consider more general operators that include differentiating in t,

$$F[u] = \textstyle\sum_{(i,j,r,s)} a_{i,j,r,s} \big(D_x^i D_t^r u\big)\big(D_x^j D_t^s u\big), \tag{1.135}$$

where, for each pair of fixed $\{r, s\}$, the real matrix $\|a_{i,j,r,s}\|$ is symmetric relative to indices $\{i, j\}$. Such operators occur in the theory of fully integrable and other equations; cf. the quadratic representation (0.31) of the KdV equation in the Introduction and the GT equation (1.64). The action of F to functions (1.125) on W_n is similar,

$$F[u] = \textstyle\sum_{(m,l)} \big(\sum_{(r,s)} \sum_{(i,j)} a_{i,j,r,s}\, p_m^i p_l^j C_m^{(r)} C_l^{(s)}\big) \mathrm{e}^{(p_m + p_l)x},$$

where the internal sum can be viewed as the action of the polynomial operator

$$F_{r,s}[u] \equiv P_{r,s}(p_m, p_l) C_m^{(r)} C_l^{(s)}.$$

Assuming that all the derivatives $\{C_m^{(r)}\}$ are arbitrary, the subspace (module) W_n is invariant if all terms like that vanish for any $p_m + p_l \in \Gamma'$, so that Proposition 1.47 remains valid. We postpone analysis of such solutions until Section 7.1, where using partially invariant linear subspaces will provide us with more flexibility.

On cubic and other operators. Consider cubic operators with polynomials

$$F[u] = \textstyle\sum_{(i,j,k)} a_{i,j,k}\, D_x^i u D_x^j u D_x^k u, \quad P(X, Y, Z) = \sum a_{i,j,k}\, X^i Y^j Z^k, \tag{1.136}$$

where $\{a_{i,j,k}\}$ is an absolutely symmetric (i.e., relative to any pair of indices) tensor of rank 3. In this case, the analogy of formula (1.126) includes all triple products forming the terms $\mathrm{e}^{(p_m + p_l + p_s)x}$, with invariance conditions

$$P(p_m, p_l, p_s) = 0, \quad \text{if} \quad p_m + p_l + p_s \in \Gamma'.$$

The approach is extended to polynomial operators of higher algebraic degree.

1.5.3 On a relation between exponential and polynomial subspaces

Consider a class of PDEs possessing solutions of the form

$$\bar{u}(x,t) = C(t)\mathrm{e}^{\varphi(t)x}, \tag{1.137}$$

with two expansion coefficients $C(t) > 0$ and $\varphi(t)$. The basic nonlinearity is now

$$g(u) = u \ln u \quad (u > 0), \quad \Longrightarrow \quad g(\bar{u}) = (C \ln C)\, \mathrm{e}^{\varphi x} + (C\varphi)\, x\mathrm{e}^{\varphi x}.$$

The time-derivative also satisfies $u_t : \mathcal{L}\{\mathrm{e}^{\varphi x}\} \to \mathcal{L}\{\mathrm{e}^{\varphi x}, x\mathrm{e}^{\varphi x}\}$. Moreover, differential operators $D_x^k(\ln u)$ with any $k \geq 2$ are annihilators of the subspace $\mathcal{L}\{\mathrm{e}^{\varphi x}\}$ for any φ. Therefore, solutions (1.137) are available for the PDEs

$$u_t = \textstyle\sum_{(j\geq 1)} D_x^j\big[a_j u \ln u + b_j (\ln u)_x\big] + a_0 u \ln u, \quad \text{with the DS} \tag{1.138}$$

$$\begin{cases} \varphi' = \sum_{(j\geq 0)} a_j \varphi^{j+1}, \\ C' = \sum_{(j\geq 0)} a_j \varphi^j (j + \ln C)C. \end{cases}$$

Setting $u = \mathrm{e}^v$ in (1.138) (so, in view of (1.137), $\bar{v} = \ln C + \varphi x$) yields a PDE with the leading polynomial operator,

$$\boxed{v_t = \mathrm{e}^{-v} \textstyle\sum_{(j\geq 1)} a_j D_x^j(\mathrm{e}^v v) + \mathrm{e}^{-v} \sum_{(j\geq 1)} b_j D_x^{j+1} v + a_0 v.}$$

The right-hand side admits the subspace of linear functions $\mathcal{L}\{1, x\}$, from whence come the equivalent solutions

$$\bar{v}(x,t) = C_1(t) + C_2(t)x$$

(the second term vanishes).

Example 1.52 (**Cahn–Hilliard-type equation**) The following parabolic PDE:

$$u_t = -u_{xxxx} + \big[\big(\alpha \ln u + \tfrac{\beta}{u} + \gamma\big)u_x\big]_x + \delta u \ln u \tag{1.139}$$

admits solutions (1.137), where

$$\begin{cases} \varphi' = \delta\varphi + \alpha\varphi^3, \\ C' = [-\varphi^4 + (\alpha \ln C + \alpha C + \gamma C)\varphi^2 + \delta \ln C]C. \end{cases}$$

Hence,

$$\varphi(t) = \mathrm{e}^{\delta t}[\tfrac{\alpha}{\delta}(A - \mathrm{e}^{2\delta t})]^{-\frac{1}{2}} \quad \text{for } \delta \neq 0,$$

and

$$\varphi(t) = [2\alpha(-t)]^{-\frac{1}{2}} \quad \text{for } \delta = 0.$$

1.5.4 On subspaces of irrational functions: dipole-type solutions

The quadratic operator from PME theory

$$F_2[u] = uu_{xx} + \gamma\,(u_x)^2, \quad \text{where} \quad \gamma \neq -1, \tag{1.140}$$

admits a simple 2D subspace (here $x > 0$)

$$W_2 = \mathcal{L}\{x^\alpha, x^2\}, \quad \text{with} \quad \alpha = \tfrac{1}{1+\gamma}. \tag{1.141}$$

The origin of such a remarkable subspace is the well-known *Barenblatt–Zel'dovich* (BZ) *dipole solution* (1957) [28] of the PME

$$v_t = (v^\sigma v_x)_x \quad \text{in } I\!R_+ \times I\!R_+ \quad \text{and} \quad v(0,t) \equiv 0 \quad (\sigma > 0).$$

The pressure variable $u = v^\sigma$ then solves $u_t = F_2[u]$ containing operator (1.140) with $\gamma = \frac{1}{\sigma}$. The BZ solution is self-similar (see (3.146) for $\sigma = 1$) and is induced by a group of scalings. The explicit integrability of the resulting ODE follows from the conservation law of the *first moment* of solutions of the PME, meaning that $\int_0^\infty x v(x,t)\,\mathrm{d}x$ is preserved with time if it is finite at $t = 0$. See more details on such solutions in Section 3.6 devoted to higher-order diffusion equations. Consider another related example.

Example 1.53 (**Quadratic wave equation**) The quasilinear quadratic "forced" (an extra term μu on the right-hand side) and "damped" (the friction term βu_t on the left-hand side with $\beta > 0$) wave equation

$$\boxed{u_{tt} + \beta u_t = (u u_x)_x + \mu u \equiv F_2[u] + \mu u \quad (\gamma = 1)} \tag{1.142}$$

admits exact dipole-type solutions on the subspace (1.141),

$$u(x,t) = C_1(t)\sqrt{x} + C_2(t)x^2,$$
$$\begin{cases} C_1'' + \beta C_1' = \frac{15}{4} C_1 C_2 + \mu C_1, \\ C_2'' + \beta C_2' = 6C_2^2 + \mu C_2. \end{cases}$$

These solutions are not connected with a scaling group, leaving (1.142) invariant.

1.5.5 *Another special subspace*

Various non-standard invariant subspaces will be studied in the next chapter, but we present a simple example here.

Example 1.54 (**Quasilinear wave equation with source**) The hyperbolic PDE

$$\boxed{u_{tt} = u_x u_{xx} + u^2 \quad (u_x \geq 0)} \tag{1.143}$$

admits solutions

$$u(x,t) = C_1(t) + C_2(t) f(x) \in \mathcal{L}\{1, f(x)\},$$

where f satisfies the ODE for elliptic functions

$$f' f'' + f^2 = \mu f + \nu,$$

with arbitrary fixed parameters μ and ν. The DS is (see applications in Section 5.2)

$$\begin{cases} C_1'' = C_1^2 + \nu C_2^2, \\ C_2'' = \mu C_2^2 + 2C_1 C_2. \end{cases}$$

1.5.6 *On reduction to quadratic equations*

It is no accident that the main classification results on invariant subspaces are associated with evolution equations having quadratic or cubic nonlinearities. First of all, the majority of the nonlinear PDEs of mathematical physics fall into this class. Secondly, it will be shown in Section 2.2 that, in several important cases, linear invariant subspaces of the *maximal* dimension appear precisely for quadratic operators only. Therefore, it is important to specify classes of nonlinear evolution PDEs (say, of parabolic type) which can be reduced to those having quadratic operators.

Here, we consider a class of 1D *quasilinear heat* or *reaction-diffusion* equations

$$u_t = k(u)u_{xx} + q(u), \tag{1.144}$$

which are widely used in many areas of mechanics, combustion theory, and biology. Let us perform a smooth change

$$u = g(v)$$

and describe those PDEs (1.144) which can be reduced to equations with quadratic operators. It is not difficult to see that, up to a linear change, there exist three types of such equations:

$$\begin{aligned}
&\text{(I)} \quad u_t = \tfrac{u^n}{\mu u^n+\nu}\, u_{xx} + \tfrac{\delta u^n+\kappa+\lambda u^{-n}}{\mu u^n+\nu}\, u, \\
&\text{(II)} \quad u_t = \tfrac{e^u}{\mu e^u+\nu}\, u_{xx} + \tfrac{\delta e^u+\kappa+\lambda e^{-u}}{\mu e^u+\nu}, \\
&\text{(III)} \quad u_t = \tfrac{1}{\mu \ln u+\nu}\, u_{xx} + \tfrac{\delta \ln^2 u+\kappa \ln u+\lambda}{\mu \ln u+\nu}\, u,
\end{aligned}$$

where μ and ν are constants satisfying $\mu^2 + \nu^2 \neq 0$ and $n \neq 0$. In the first PDE (I), the change $u = v^{1/n}$ reduces it to

$$\text{(I)} \quad (\mu v + \nu)v_t = v v_{xx} + \left(\tfrac{1}{n} - 1\right)(v_x)^2 + n(\delta v^2 + \kappa v + \lambda),$$

where the quadratic operator also arises on the left-hand side. This class of PDEs will be more systematically used in Chapter 7 dealing with partially invariant subspaces.

In the second equation (II), the change is logarithmic, $u = \ln v$, where v solves

$$\text{(II)} \quad (\mu v + \nu)v_t = v v_{xx} - (v_x)^2 + \delta v^2 + \kappa v + \lambda.$$

In the third PDE (III), we use the exponential transformation $u = e^v$, yielding

$$\text{(III)} \quad (\mu v + \nu)v_t = v_{xx} + (v_x)^2 + \delta v^2 + \kappa v + \lambda.$$

For $\mu = 0$, the left-hand sides become linear and the exact solutions of these three types of quadratic PDEs can be found by using invariant subspaces.

Example 1.55 (**Semilinear cubic heat equation**) Consider equation (I) with parameters $\mu = 1$, $\nu = \kappa = 0$, and $n = -1$,

$$u_t = u_{xx} + \delta u + \lambda u^3.$$

Setting $u = v^{1/n} = \frac{1}{v}$ yields the following quadratic PDE (see Section 7.4.1):

$$\boxed{v v_t = v v_{xx} - 2(v_x)^2 - \delta v^2 - \lambda.}$$

Remarks and comments on the literature

§ 1.1. In all chapters we consider various quasilinear second-order parabolic PDEs. Such models occur in diffusion and combustion theory; see descriptions and earlier references in the classic book [594]. Modern filtration theory goes back to the beginning of the twentieth century to works by N.Ye. Zhukovskii, who is better known for his fundamental research in aerodynamics, hydrodynamics, and ODE theory (on his non-oscillation test in 1892, see [226, p. 19]). His contribution to "theory of ground waters" is explained in Kochina's paper [350]. Parabolic PDE models in filtration theory of liquids and gases in porous media were already derived by Leibenzon in the 1920s and 1930s [378], by Richard's (1931) [489], and Muskat (1937) [429]. More information on parabolic PDEs, systems, and various aspects of existence-uniqueness and blow-up singularity theory can be found in [9, 148, 149, 164, 205, 206, 226, 245, 403, 509, 530, 533, 578].

§ 1.2. First systematic studies of invariant subspaces for nonlinear operators were performed in [217, 220, 232], where some earlier references can be found. Theory and applications of functional differential equations (Section 1.2.3) are available in many books; see e.g., [276, 351]. Concerning Lie-group symmetry analysis for delay-differential equations, see [549].

General aspects of the GSV for evolution PDEs (Section 1.2.4) were developed in [238] (a similar term "nonlinear separation of variables" was suggested in [445]). Finite rings were used in [555], where a full description of 2D commutative subrings for the product

$$u * v = (uv)_{xx},$$

associated with the PME operator in

$$u_t = (u^2)_{xx},$$

was given. This reiterated the well-known solutions, including the dipole Barenblatt–Zel'dovich solution [28], see (3.146), on the invariant subspace

$$\mathcal{L}\{\sqrt{x}, x^2\}$$

(i.e., (1.141) for $\gamma = 1$) of the operator $(u^2)_{xx}$.

For linear differential operators, the problem of characterizing those operators that leave a given finite-dimensional subspace of polynomials invariant has been known as the *generalized Bochner problem* [67]. Some partial solutions have connections with *Burnside's theorem* on polynomial generator representations of endomorphisms of an irreducible module for Lie algebra. Differential operators (i.e., Schrödinger operators) preserving a finite-dimensional subspace W_n of a given space of smooth functions are said to be *partially* (or *quasi-exactly*) *solvable*. See basic ideas in Turbiner [563], [188], references in [312], and discussion in the more recent paper [257]. In mathematical physics, there are several self-adjoint 1D operators admitted countable sets of polynomial (monomial) eigenfunctions; e.g. the classic second-order self-adjoint one

$$F_2 = D_x^2 - \tfrac{1}{2}\, x D_x \quad \text{in } I\!R,$$

with eigenfunctions being Hermite polynomials, or its higher-order non-self-adjoint analogies

$$F_{2m} = (-1)^{m+1} D_x^{2m} - \tfrac{1}{2m}\, x D_x \quad \text{in } I\!R \text{ for any } m = 2, 3, \ldots$$

for which polynomial eigenfunctions are known to be complete and closed in a weighted L^2-space [163] (in fact, any order, $2m \mapsto k \geq 3$, fits, with more delicate topology for odd k). In general, the problem on finite-dimensional invariant subspaces for such linear operators, especially for operators in $I\!R^N$, is known as intractable.

§ 1.3–1.5. Several results are given in [239]. Galilean invariant PDEs in Example 1.23 were introduced in [213], symmetries of some nonlinear heat conduction models were studied in [112]; see also references therein. (The mathematics of such fully nonlinear models, including existence and nonuniqueness is not clear, so some conclusions there are not justified.) Exact solutions of (1.64) on 3D polynomial and exponential subspaces were obtained in [323] by detecting the so-called *defining equations* and linear differential constraints. Exact solutions on invariant subspaces for the quasilinear heat equation with source and convection,

$$u_t = (D(u)u_x)_x + P(u)u_x + Q(u),$$

were studied in [476]; see also [481, p. 145] for extensions. Proposition 1.24 was proved in [553]. The exact solutions of some particular equations, such as (1.85) and (1.139), were derived in [246] via nonclassical symmetries. It is well known that, for $b = 0$, equation (1.96) admits a five-parameter Lie group of transformations and several explicit invariant solutions; see [150] and [228]. Solutions in Example 1.36 were constructed in [220]. See [160, 159] for some other applications of invariant subspaces to parabolic PDEs. Solutions in Example 1.37 were studied in detail in [346], where further information on the Berman problem can be found. Solutions on subspaces of the mixed polynomial-trigonometric or exponential type (cf. (1.101))

$$\mathcal{L}\{1, x, \cos x, \sin x\}, \quad \text{or} \quad \mathcal{L}\{1, x, \cosh x, \sinh x\}$$

are typical for the *positons* or the *negatons* solutions of the integrable PDEs, such as the KdV or KP equations; see Remarks to Section 4.1.

Derivation of 1D non-local parabolic models in Example 1.38 (and partially in Example 1.40) from the Navier–Stokes equations in $I\!R^2$ can be found in [13, Ch. 7]. Solutions in Example 1.38 were constructed in [475] (blow-up singularities were rigorously described in [244]). Example 1.40 is given in [220].

Solutions (1.121) in Example 1.43 were detected in [110] by using a linear ordinary differential constraint with coefficients, depending on t (no other new exact solutions of quadratic parabolic PDEs were obtained there by this approach). For evolution PDEs (1.127) with general quadratic operators (1.111), some solutions on exponential partially invariant subspaces were obtained in [397]; see further extensions and references in Section 7.1.

Notice another classical quadratic *Prandtl's equation* occurring in *Prandtl's boundary layer theory* (proposed in 1904, [470]; the first exact similarity solution in $I\!R^2$ is due to Blasius (1908) [63]) for incompressible fluids,

$$\psi_{yyy} + \psi_x\psi_{yy} - \psi_y\psi_{xy} - \psi_{yt} + uu_x + U_t = 0, \tag{1.145}$$

where $\psi = \psi(x, y, t)$ is the stream function and $U(x, t)$ is the given external far-field (at $y = \infty$) velocity distribution; see further mathematical details in Oleinik–Samokhin [444]. This PDE has the same quadratic operator as that in the Gibbons–Tsarev equation (1.64). Exact solutions of (1.145) are studied in [402], where further references can be found. Another quadratic operator arises in the *symmetric regularized long wave equation*

$$u_{tt} + u_{xx} + uu_{xt} + u_x u_t + u_{xxtt} = 0;$$

see exact solutions in [180] constructed by an interesting algebraic method using some linear polynomial structures.

Some other solutions on invariant subspaces can be explained by using various differential constraints and determining equations; see [322, 447, 592] and [13, Ch. 3]. More general quasilinear wave models (cf. (1.143)),

$$u_{tt} = (u_x)^m u_{xx} + f(u),$$

appear in applications to gas dynamics and shallow water wave theory; see [18], where group properties are studied, and Chapter 5 for more results.

CHAPTER 2

Invariant Subspaces and Modules: Mathematics in One Dimension

This chapter is devoted to mathematical theory of invariant subspaces for ordinary differential operators. We establish a general representation of operators preserving a given linear subspace (the Main Theorem), and prove other properties that are necessary for further application to nonlinear PDEs in the subsequent chapters. In particular, we obtain the sharp estimate of the dimension of linear subspaces invariant under a nonlinear operator of a given differential order (the Theorem on maximal dimension).

2.1 Main Theorem on invariant subspaces

There exist two main problems in the context of invariant subspaces:

Problem I, $\mathbf{F} \mapsto \{\mathbf{W_n}\}$: *given operator F, find all invariant subspaces W_n*, and

Problem II, $\mathbf{W_n} \mapsto \{\mathbf{F}\}$: *given invariant subspace W_n, find all operators F.*

In this section, Problem II is studied. We assume that the subspace

$$W_n = \mathcal{L}\{f_1(x), ..., f_n(x)\} \tag{2.1}$$

is defined as the space of solutions of a linear nth-order ODE,

$$L[y] \equiv y^{(n)} + a_1(x)y^{(n-1)} + \ ... \ + a_{n-1}(x)y' + a_n(x)y = 0, \tag{2.2}$$

for which the functions $\{f_i(x)\}$ form a *fundamental set of solutions*, an FSS; see e.g., [132, 434]. Let $\mathcal{F}_{n-1}(W_n)$ (or simply $\mathcal{F}_{n-1}$) denote the whole set of differential operators

$$F[y] = F(x, y, y', \ ..., y^{(k)}) \tag{2.3}$$

of the order k not greater than $n-1$, leaving the subspace (2.1), i.e., equation (2.2), invariant. The function $F(\cdot)$ in (2.3) is assumed to be smooth enough. The *invariance condition* of the subspace W_n with respect to F takes the form

$$L[F[y]]\big|_{L[y]=0} \equiv 0. \tag{2.4}$$

We will use *first integrals* of the equation (2.2) (i.e., expressions that are constant on solutions). There exists the following set of first integrals:

$$I_i[y] = \alpha_{i,1}(x)y^{(n-1)} + \ ... \ + \alpha_{i,n-1}(x)y' + \alpha_{i,n}(x)y \quad \text{for } i = 1, ..., n, \tag{2.5}$$

where the coefficients $\alpha_{i,1}(x), \ ..., \alpha_{i,n}(x)$ are defined recursively by

$$\alpha_{i,1} = g_i, \ \alpha_{i,2} = a_1 g_i - (\alpha_{i,1})', \ ..., \ \alpha_{i,n} = a_{n-1} g_i - (\alpha_{i,n-1})' \tag{2.6}$$

via an arbitrary FSS $\{g_i(x)\}$ of the equation that is conjugated (adjoint) to (2.2).

We now present a general description of $\mathcal{F}_{n-1}(W_n)$. This result is referred to as the *Main Theorem* on invariant subspaces.

Theorem 2.1 **("Main Theorem")** *The set $\mathcal{F}_{n-1}(W_n)$ consists of all operators of the form*

$$F[y] = \sum_{i=1}^{n} A^i(I_1, \ldots, I_n) f_i(x), \tag{2.7}$$

where $A^i(I_1, \ldots, I_n)$ for $i = 1, \ldots, n$ are arbitrary smooth functions of the complete set of first integrals of equation (2.2).

Proof. Consider the invariance condition (2.4),

$$D^n F + a_1(x) D^{n-1} F + \ldots + a_{n-1}(x) DF + a_n(x) F = 0, \tag{2.8}$$

where D is the operator of the full derivative by virtue of equation (2.2), i.e.,

$$\begin{aligned} D = \frac{\mathrm{d}}{\mathrm{d}x}\Big|_{(2.2)} = \frac{\partial}{\partial x} + y' \frac{\partial}{\partial y} + \ldots + y^{(n-1)} \frac{\partial}{\partial y^{(n-2)}} \\ -\big[a_1 y^{(n-1)} + \ldots + a_{n-1} y' + a_n y\big] \frac{\partial}{\partial y^{(n-1)}} . \end{aligned} \tag{2.9}$$

In the extended space of variables $\{x, y, y', \ldots, y^{(n-1)}\}$, we introduce the new variables $\{\tilde{x}, I_1, \ldots, I_n\}$ by formulae

$$\tilde{x} = x \quad \text{and} \quad I_i = I_i(x, y, y', \ldots, y^{(n-1)}) \;\; \text{for} \;\; i = 1, \ldots, n.$$

Then operator D in (2.9) takes the form $\tilde{D} = \partial/\partial \tilde{x}$, and the invariance condition (2.8) is transformed into

$$\tilde{D}^n F + a_1(\tilde{x}) \tilde{D}^{n-1} F + \ldots + a_{n-1}(x) \tilde{D} F + a_n(\tilde{x}) F = 0. \tag{2.10}$$

The general solution of (2.10) is given by the FSS $\{f_i(x)\}$ of (2.2),

$$F = A^1(I_1, \ldots, I_n) f_1(\tilde{x}) + \ldots + A^n(I_1, \ldots, I_n) f_n(\tilde{x}),$$

where $A^i(I_1, \ldots, I_n)$ for $i = 1, \ldots, n$ are arbitrary sufficiently smooth functions. In the original variables, this is the equality (2.7). □

Remark 2.2 Condition (2.4) is actually the invariance criterion for equation (2.2) with respect to the *Lie-Bäcklund operator* $X = F[y]\frac{\partial}{\partial y}$, and, therefore, all the results can be interpreted in terms of symmetries of linear ODEs; see more details in [12, 297, 446] and [10, Ch. 5].

Remark 2.3 If the FSS $\{f_i(x)\}$ is known, the FSS $\{g_i(x)\}$ of the conjugate equation can be obtained without integrating by the standard formulae

$$g_i = (-1)^{n+i} \frac{W[f_1, \ldots, f_{i-1}, f_{i+1}, \ldots, f_n]}{W[f_1, \ldots, f_n]} \quad \text{for } i = 1, \ldots, n, \tag{2.11}$$

where

$$W[h_1, \ldots, h_l] = \begin{vmatrix} h_1 & h_2 & \cdots & h_l \\ h_1' & h_2' & \cdots & h_l' \\ \cdots & \cdots & \cdots & \cdots \\ h_1^{(l-1)} & h_2^{(l-1)} & \cdots & h_l^{(l-1)} \end{vmatrix}$$

denotes the Wronskian of the given functions $\{h_1, \ldots, h_l\}$.

Remark 2.4 (**Application to PDEs**) For the first integrals (2.5) constructed by the FSS (2.11), the following *duality* condition holds: for all $i,\ j = 1, ..., n$,

$$I_i[g_j] = \delta_{ij} = \begin{cases} 1 \ \text{for}\ i = j, \\ 0 \ \text{for}\ i \neq j, \end{cases} \quad (\delta_{ij} \ \text{is Kronecker's delta}). \tag{2.12}$$

Therefore, given the evolution PDE

$$\boxed{u_t = F[u]}$$

with the operator F defined by (2.7) via first integrals $\{I_i\}$ satisfying (2.12), setting

$$u(x,t) = \textstyle\sum_{i=1}^{n} C_i(t) f_i(x)$$

leads to the DS

$$C_i' = A^i(C_1, ..., C_n), \quad i = 1, ..., n.$$

The right-hand sides of these ODEs are precisely the coefficients in the operator representation (2.7).

Theorem 2.1 is illustrated by a few examples, where we take $g_i = f_i$ if the conjugate operator L^* coincides with L or $-L$.

Example 2.5 (**2D subspace**) Choose the 2D subspace $W_2 = \mathcal{L}\{e^x, e^{-x}\}$ with the ODE

$$y'' - y = 0.$$

This equation coincides with its conjugate, so we set $g_1 = f_1 = e^x$, $g_2 = f_2 = e^{-x}$. By (2.7), all the operators F admitting W_2 are

$$F[y] = A^1(I_1, I_2)e^x + A^2(I_1, I_2)e^{-x}, \tag{2.13}$$

where first integrals are $I_1 = (y' - y)e^x$ and $I_2 = (y' + y)e^{-x}$.

We can use this expression for describing operators satisfying some additional conditions. For instance, we now find all the operators that do not depend explicitly on x (these are especially important in applications). To this end, we use the notation

$$A_i^k = \tfrac{\partial A^k}{\partial I_i}, \quad A_{ij}^k = \tfrac{\partial^2 A^k}{\partial I_i \partial I_j}, \quad \dots .$$

The translation-invariant operators satisfy

$$\tfrac{\partial F}{\partial x} \equiv 0, \quad \text{or} \tag{2.14}$$

$$\big(I_1 A_1^1 - I_2 A_2^1 + A^1\big)e^x + \big(I_1 A_1^2 - I_2 A_2^2 - A^2\big)e^{-x} \equiv 0.$$

Equating the coefficients of e^x and e^{-x} to zero, and solving the resulting system yields $A^1 = I_2 B^1(I_1 I_2)$ and $A^2 = I_1 B^2(I_1 I_2)$, so that by (2.13),

$$F[y] = (y' + y)B^1(J) + (y' - y)B^2(J),$$

where $J = I_1 I_2 = (y')^2 - y^2$ and $B^1(J)$ and $B^2(J)$ are arbitrary functions. Some particular cases are as follows:

$$F_1[y] = \tfrac{1}{y'-y}, \ F_2[y] = \tfrac{1}{y'+y}, \ F_3[y] = (y' - y)(y' + y)^2,$$
$$F_4[y] = y'\big[(y')^2 - y^2\big], \quad \text{and} \quad F_5[y] = y\big[(y')^2 - y^2\big].$$

These operators yield a number of evolution PDEs with solutions on W_2,

$$u(x,t) = C_1(t)e^x + C_2(t)e^{-x}; \tag{2.15}$$

for example, the following reaction-diffusion equations:

$$\boxed{u_t = u_{xx} + \tfrac{1}{u_x+u}, \quad u_t = u_{xx} + u[(u_x)^2 - u^2], \quad u_t = [(u_x)^2 - u^2]u_{xx}\,,}$$

etc. (Similarly, hyperbolic PDEs with $u_{tt} = ...$ can be treated.) Clearly, W_2 is invariant under the linear operator u_{xx}. For the non-polynomial operator $F[u] = \frac{1}{u_x+u}$ in the first equation, W_2 is invariant, since

$$F\big[C_1e^x + C_2e^{-x}\big] = \tfrac{1}{2C_1}\,e^{-x} \in W_2 \quad (C_1 \neq 0).$$

Therefore, solutions (2.15) generate the DS $C_1' = C_1$, $C_2' = C_2 + \frac{1}{2C_1}$.

Example 2.6 **(3D subspace)** Consider $W_3 = \mathcal{L}\{1, e^x, e^{-x}\}$ defined by the ODE

$$y''' - y' = 0.$$

Up to the sign, this equation coincides with its conjugate, so that we take the following functions $g_1 = f_1 = 1$, $g_2 = f_2 = e^x$, and $g_3 = f_3 = e^{-x}$. By Theorem 2.1, the general representation of operators F preserving W_3 is

$$F[y] = A^1(I_1, I_2, I_3) + A^2(I_1, I_2, I_3)e^x + A^3(I_1, I_2, I_3)e^{-x}, \tag{2.16}$$

where A^1, A^2, and A^3 are arbitrary smooth functions of the first integrals

$$I_1 = y'' - y, \quad I_2 = (y'' - y')e^x, \quad \text{and} \quad I_3 = (y'' + y')e^{-x}.$$

As above, let us detect the translational invariant operators F satisfying (2.14). This leads to a system for functions A^1, A^2, and A^3,

$$I_2A^1_2 - I_3A^1_3 = 0, \quad I_2A^2_2 - I_3A^2_3 + A^2 = 0, \quad I_2A^3_2 - I_3A^3_3 - A^3 = 0.$$

Solving it and using (2.16) yields

$$F[y] = B^1(J_1, J_2) + (y'' + y')B^2(J_1, J_2) + (y'' - y')B^3(J_1, J_2),$$

where $J_1 = I_1 = y'' - y$, $J_2 = I_2I_3 = (y'')^2 - (y')^2$, and $B^1(J_1, J_2)$, $B^2(J_1, J_2)$, and $B^3(J_1, J_2)$ are arbitrary functions.

If, in addition to (2.14), we assume that F is of the first order, i.e.,

$$\frac{\partial F}{\partial y''} \equiv 0,$$

we obtain the 4D linear space of operators spanned by

$$F_1[y] = (y')^2 - y^2, \quad F_2[y] = y', \quad F_3[y] = y, \quad \text{and} \quad F_4[y] = 1.$$

These operators make it possible to construct evolution PDEs with solutions on W_3,

$$u(x,t) = C_1(t) + C_2(t)e^x + C_3(t)e^{-x}.$$

In particular, this is the case for the semilinear heat equation with absorption

$$\boxed{u_t = u_{xx} + (u_x)^2 - u^2.}$$

Example 2.7 (**Radial geometry**) Consider the following ODE for the subspace:

$$y'' = \tfrac{\beta-1}{x}\, y', \quad \text{where} \quad \beta \neq 0, 1. \tag{2.17}$$

This representation is related to the radial Laplace operator

$$\Delta_N = \tfrac{\mathrm{d}^2}{\mathrm{d}x^2} + \tfrac{N-1}{x}\,\tfrac{\mathrm{d}}{\mathrm{d}x}$$

for the (non-integer and, possibly, negative) dimension $N = 2 - \beta$. Here $x > 0$ denotes the radial space variable. The FSS of (2.17) is

$$f_1(x) = 1 \quad \text{and} \quad f_2(x) = x^{\beta}.$$

By (2.11), we readily find that, up to a constant factor, the FSS of the conjugate equation is

$$g_1(x) = x \quad \text{and} \quad g_2(x) = x^{1-\beta}.$$

Then Theorem 2.1 implies that the linear space of nonlinear operators admitting $W_2 = \mathcal{L}\{1, x^{\beta}\}$ has the form

$$F[y] = A^1(I_1, I_2) + A^2(I_1, I_2)x^{\beta},$$

with arbitrary functions A^1 and A^2 of the first integrals

$$I_1 = xy' - \beta y \quad \text{and} \quad I_2 = x^{1-\beta} y'.$$

The translation-invariant operators satisfying (2.14) are spanned by

$$F_1[y] = (y')^{\frac{\beta}{\beta-1}}, \quad F_2[y] = y, \quad \text{and} \quad F_3[y] = 1.$$

The operator F_1 generates the following semilinear higher-order PDEs ($m \geq 2$):

$$u_t = \tfrac{\partial^m u}{\partial x^m} + \lambda (u_x)^{\frac{m}{m-1}} \quad (u_x > 0), \tag{2.18}$$

where λ is a constant. The operator on the right-hand side has the invariant subspace $W_2 = \mathcal{L}\{1, x^m\}$, and hence, (2.18) admits solutions

$$u = C_1(t) + C_2(t)x^m.$$

Substituting into the PDE yields $C_1' + C_2' x^m = m!\, C_2 + \lambda (mC_2)^{\frac{m}{m-1}} x^m$, which leads to the following dynamical system:

$$\begin{cases} C_1' = m!\, C_2, \\ C_2' = \lambda\, (mC_2)^{\frac{m}{m-1}}. \end{cases}$$

It is solved explicitly and gives the solutions of (2.18) of the form

$$u(x,t) = \tfrac{(m-1)^{m-1}}{m^m} \Big[\tfrac{m!}{\lambda (m-2)} \tfrac{1}{(a_1-\lambda t)^{m-2}} + \tfrac{x^m}{(a_1-\lambda t)^{m-1}} + a_2 \Big] \text{ for } m \neq 2,$$
$$u(x,t) = \tfrac{1}{4}\big[-\tfrac{2}{\lambda} \ln|a_1 - \lambda t| + \tfrac{x^2}{a_1-\lambda t} + a_2 \big] \text{ for } m = 2,$$

where a_1 and a_2 are arbitrary constants. Note that the change $x = \tilde{x}^{1/\beta}$ transforms (2.17) into a simpler equation, $y_{\tilde{x}\tilde{x}} = 0$.

2.2 The optimal estimate on dimension of invariant subspaces

We consider the question of the maximal dimension of invariant subspaces for nonlinear operators of a fixed differential order. This is an important, crucial aspect of Problem I: $F \mapsto \{W_n\}$. It is especially key for applications, since this establishes the maximally possible order of dynamical systems as restrictions of the PDEs.

In all the previous examples, the dimensions of invariant subspaces were less than or equal to three for the first-order operators, and less than or equal to five for the second-order ones. The following theorem establishes the maximal dimension of invariant subspaces for arbitrary kth-order nonlinear ordinary differential operators.

Theorem 2.8 **("Theorem on maximal dimension")** *If a linear subspace W_n is invariant under a nonlinear ordinary differential operator* (2.3) *of the order k, then*

$$\boxed{n \le 2k+1.} \tag{2.19}$$

Proof. Arguing by contradiction, assume that there exists an invariant subspace W_n of (2.3), and (2.19) is not valid, i.e., $n \ge 2k+2$. Let us show that F is then a linear operator. Let W_n be defined by the linear ODE

$$y^{(n)} = a_1 y^{(n-1)} + \,...\, + a_{n-1} y' + a_n y. \tag{2.20}$$

By the condition (2.4), the following identity must be true on solutions of (2.20):

$$D^n F \equiv a_1 D^{n-1} F + \,...\, + a_{n-1} DF + a_n F, \tag{2.21}$$

where D denotes the operator of the total derivative in x. For convenience, performing algebraic manipulations with the function $F(\cdot)$, the following notation is used:

$$y_j = y^{(j)} \quad \text{for all } j \ge 0 \ \ (y_0 = y), \quad \text{and} \ \ F_{y_j} = \tfrac{\partial F}{\partial y_j}.$$

Then (2.20) reads $y_n = a_1 y_{n-1} + ... + a_n y_0$. Differentiating $F(x, y_0, y_1, ..., y_k)$, and keeping leading linear and quadratic terms yields

$$DF = y_{k+1} F_{y_k} + ..., \quad D^2 F = y_{k+2} F_{y_k} + (y_{k+1})^2 F_{y_k y_k} + ...,$$
$$D^3 F = y_{k+3} F_{y_k} + 3 y_{k+1} y_{k+2} F_{y_k y_k} + ...,$$
$$D^4 F = y_{k+4} F_{y_k} + \big[4 y_{k+1} y_{k+3} + 3 (y_{k+2})^2\big] F_{y_k y_k} + ...,$$

etc. By induction, after any $p \ge 4$ steps, we find

$$\begin{aligned} D^p F = {} & y_{k+p} F_{y_k} + \big[\textstyle\sum_{i=1}^{[\frac{p}{2}]-1} C_p^i \, y_{k+p-i} y_{k+i} \\ & + \nu C_p^{[\frac{p}{2}]} y_{k+p-[\frac{p}{2}]} y_{k+[\frac{p}{2}]}\big] F_{y_k y_k} + ..., \end{aligned} \tag{2.22}$$

where $\big[\frac{p}{2}\big]$ denotes the integer part, C_p^i for $i = 1, ..., \big[\frac{p}{2}\big]$ are binomial coefficients, and $\nu = \frac{1}{2}$ for p even, $\nu = 1$ for odd. Here, we separate the first linear term that contains the highest-order derivative y_{k+p}. The sum in square brackets in (2.22) is composed of the quadratic (in higher-order derivatives) summands that exhibit the maximal total order of both the derivatives: $(k+p-i)+(k+i) = 2k+p$. For $p = n$, keeping in (2.22) only the quadratic terms, containing at least one derivative of the order not less than $n-1$, gives

$$D^n F = \big[\textstyle\sum_{i=1}^{k+1} \alpha_i \, y_{k+n-i} y_{k+i}\big] F_{y_k y_k} + ..., \tag{2.23}$$

where $\alpha_i = C_n^i$ for $i = 1, ..., k$, $\alpha_{k+1} = C_n^{k+1}$ if $n > 2k + 2$, and $\alpha_{k+1} = \frac{1}{2} C_n^{k+1}$ if $n = 2k+2$. Here, we do not display the linear term. The total order of the summands in the square brackets is equal to $2k + n$. By (2.20), all the derivatives y_{k+n-i} for $i = 1, 2, ..., k$ can be linearly expressed in terms of $y_{n-1}, ..., y_0$. Keeping in the square brackets in (2.23) the terms containing y_{n-1}, we find

$$D^n F = \left[\sum_{i=1}^k \tilde{\alpha}_i y_{k+i} + \alpha_{k+1} y_{2k+1}\right] y_{n-1} F_{y_k y_k} + \dots , \tag{2.24}$$

where $\tilde{\alpha}_i = \alpha_i \chi_i$ and χ_i are expressed via $a_1, ..., a_n$ and their derivatives. There exists a single quadratic term in (2.24) of the maximal total order $2k + n$, namely,

$$\alpha_{k+1} y_{2k+1} y_{n-1} F_{y_k y_k}.$$

It is easy to see that such terms do not appear in the derivatives $D^p F$ with $p < n$. In order for (2.21) to be valid, we should then set

$$F_{y_k y_k} = 0. \tag{2.25}$$

Taking into account (2.25), we derive similar to (2.23) that

$$D^n F = \left[\sum_{i=1}^{k+1} \beta_i y_{k+n-i} y_{k-1+i}\right] F_{y_k y_{k-1}} + \dots ,$$

where β_i are some positive coefficients. In the summands, containing at least one derivative of the order not less than $n - 1$, the total order of the derivatives is now $2k + n - 1$. Excluding, as above, the derivatives y_{k+n-i} for $i = 1, 2, ..., k$ by using (2.20) yields the unique quadratic term of the maximal total order

$$\beta_{k+1} y_{n-1} y_{2k} F_{y_k y_{k-1}}.$$

Such summands do not appear in any other derivative $D^p F$ for $p < n$. Equating this coefficient to zero implies

$$F_{y_k y_{k-1}} = 0.$$

Assume that the equalities

$$F_{y_k y_k} = F_{y_k y_{k-1}} = \dots = F_{y_k y_{k-(j-1)}} = 0$$

have been proved, i.e., F_{y_k} depends on the arguments $x,\ y_0,\ ...,\ y_{k-j}$ only. Similarly to the above calculus,

$$D^n F = \left[\sum_{i=1}^{k+1} \gamma_i y_{k+n-i} y_{k-j+i}\right] F_{y_k y_{k-j}} + \dots , \tag{2.26}$$

where $\gamma_i > 0$ and the total order of the quadratic summands relative to higher derivatives is $2k + n - j$. Expressing y_{k+n-j} again in terms of $y_{n-1}, ..., y_0$ yields that, in (2.26), there exists the unique quadratic summand of the total order $2k + n - j$,

$$\gamma_{k+1} y_{n-1} y_{2k+1-j} F_{y_k y_{k-j}}.$$

No other derivatives $D^p F$ with $p < k$ produce such a term. Hence,

$$F_{y_k y_{k-j}} = 0. \tag{2.27}$$

By induction, it follows that (2.27) holds for all $j = 0, 1, ..., k$, i.e.,

$$F[y] = f_k(x) y_k + \widetilde{F}(x, y_0, \dots , y_{k-1}).$$

Since the summand $f_k(x)y_k$ generates the terms in (2.21) that are linear in y_0, ..., y_{n-1}, the above conclusions based on the analysis of the quadratic terms can literally be repeated for operator $\widetilde{F}(x, y_0, \ldots, y_{k-1})$, replacing k by $k-1$ in all instances. This yields

$$\widetilde{F}[y] = f_{k-1}(x)y_{k-1} + \widetilde{\widetilde{F}}(x, y_0, \ldots, y_{k-2}).$$

Acting in a similar fashion yields that the function $F(x, y_0, y_1, \ldots, y_k)$ is linear relative to all the arguments y_i for $i = k, ..., 0$. Hence,

$$F[y] = f_k(x)y_k + \ldots + f_0(x)y_0 + f(x),$$

completing the proof. □

For arbitrary k, there exist nonlinear *quadratic* operators admitting invariant subspaces of the maximal dimension $2k+1$ as the following example shows.

Example 2.9 (**Subspace of maximal dimension**) The kth-order quadratic operator

$$F[y] = \big(y^{(k)}\big)^2 \tag{2.28}$$

admits the $(2k{+}1)$th-dimensional subspace

$$W_{2k+1} = \mathcal{L}\{1, x, \ldots, x^{2k}\}.$$

In the next three remarks, we comment on some aspects of our analysis.

Remark 2.10 Let operator F be admitted by the linear ODE (2.2) that defines the invariant subspace (2.1). Let

$$F[y] = \textstyle\sum_{(i)} F_i[y],$$

where the sum is finite or infinite, and each $F_i[y] = F_i(x, y, y', ..., y^{(k_i)})$ is a homogeneous polynomial in y, y', ..., and $y^{(k_i)}$ of degree i. In this case, every operator F_i is also admitted by this ODE.

This directly follows from the invariance criterion (2.4) that is written down as

$$\textstyle\sum_{(i)} \big(L[F_i[y]]\big|_{L[y]=0} \big) \equiv 0.$$

Hence, in view of the homogenuity of all the summands, this splits into the system

$$L[F_i[y]]\big|_{L[y]=0} \equiv 0 \quad \text{for all} \quad i.$$

Remark 2.11 Any change of variables

$$y = \alpha(x)\widetilde{y}, \quad \widetilde{x} = \beta(x), \tag{2.29}$$

where $\alpha(x)$, $\beta(x)$ are given functions, transforms any homogeneous polynomial in $y, y', ..., y^{(k)}$ into a homogeneous polynomial of the same degree in $\widetilde{y}$, $\widetilde{y}'$, ..., $\widetilde{y}^{(k)}$.

Two operators $F[y] = F(x, y, y', ..., y^{(k)})$ and $\widetilde{F}[\widetilde{y}] = \widetilde{F}(\widetilde{x}, \widetilde{y}, \widetilde{y}', ..., \widetilde{y}^{(k)})$ are called *equivalent*, if there exists the change of variables (2.29) such that

$$\widetilde{F}[\widetilde{y}] = \tfrac{1}{\alpha(x)}\, F[y]. \tag{2.30}$$

If operator F preserves the given subspace

$$W_n = \mathcal{L}\{f_1(x), \ldots, f_n(x)\},$$

then the equivalent operator $\widetilde{F}$ admits the invariant subspace

$$\widetilde{W}_n = \mathcal{L}\{\widetilde{f}_1(\widetilde{x}), \ldots, \widetilde{f}_n(\widetilde{x})\},$$

where $\widetilde{f}_i(\widetilde{x}) = f_i(x(\widetilde{x}))/\alpha(x(\widetilde{x}))$ for all $i = 1, \ldots, n$.

Remark 2.12 Subspace W_n, invariant under the operator $F[y]$, is also invariant under $\widetilde{F}[y] = F[y + f]$, with any function $f(x) \in W_n$.

Example 2.13 **(Exponential $\tilde{\mathbf{W}}_{\mathbf{2k+1}}$)** The change of variables

$$x = \mathrm{e}^{\tilde{x}}, \quad y = \mathrm{e}^{2k\tilde{x}}\,\tilde{y},$$

transforms (2.28) into the following quadratic operator

$$\tilde{F}[\tilde{y}] = \mathrm{e}^{-2k\tilde{x}}\big[\big(\mathrm{e}^{-\tilde{x}}\tfrac{\mathrm{d}}{\mathrm{d}\tilde{x}}\big)^k\big(\mathrm{e}^{2k\tilde{x}}\,\tilde{y}\big)\big]^2.$$

The invariant subspace is now of the exponential type rather than polynomial,

$$\tilde{W}_{2k+1} = \mathcal{L}\big\{\mathrm{e}^{-2k\tilde{x}}, \ldots, \mathrm{e}^{-\tilde{x}}, 1\big\}.$$

Note that $\tilde{F}[\tilde{y}]$ is a translation-invariant operator.

2.3 First-order operators with subspaces of maximal dimension

In view of Theorem 2.8, for nonlinear first-order operators, the dimension of invariant subspaces cannot exceed three. We now describe all first-order operators

$$F[y] = F(x, y, p), \quad \text{where } p = y', \tag{2.31}$$

admitting 3D subspaces given by the linear ODEs

$$y''' = a_1(x)y'' + a_2(x)y' + a_3(x)y, \tag{2.32}$$

with sufficiently arbitrary coefficients $a_k(x)$ for $k = 1, 2, 3$. The invariance criterion (2.4) then takes the form

$$D^3F - \big(a_1D^2F + a_2DF + a_3F\big) \equiv 0, \tag{2.33}$$

where D is the operator of the full derivative via (2.32),

$$D = \tfrac{\partial}{\partial x} + y'\tfrac{\partial}{\partial y} + y''\tfrac{\partial}{\partial y'} + \big[a_1(x)y'' + a_2(x)y' + a_3(x)y\big]\tfrac{\partial}{\partial y''}.$$

Lemma 2.14 *Any operator (2.31) preserving 3D subspaces can be represented in the following form:*

$$F[y] = F_2[y] + F_1[y] + F_0[y], \quad \textit{where} \tag{2.34}$$

$$F_2[y] = Lp^2 + Myp + Ny^2, \quad F_1[y] = Rp + Sy, \quad F_0[y] = f, \tag{2.35}$$

and L, M, N, R, S, and f are arbitrary functions of x.

Proof. It is easy to check that the left-hand side of identity (2.33) is a third-degree polynomial in the variable y'', while the coefficient of $(y'')^3$ is $\frac{\partial^3 F}{\partial p^3}$. Equating this coefficient to zero and solving the resulting equation implies

$$F(x, y, p) = A_2(x, y)p^2 + A_1(x, y)p + A_0(x, y), \tag{2.36}$$

where $A_2(x, y)$, $A_1(x, y)$, and $A_0(x, y)$ are some functions. Rewriting the coefficients of $(y'')^s$ for $s = 2, 1$, and 0 by using (2.36), we obtain polynomials in p with coefficients, depending only on x and y. Equating the coefficients of the leading powers of p in these polynomials to zero yields

$$\frac{\partial A_2}{\partial y} = 0, \quad \frac{\partial^2 A_1}{\partial y^2} = 0, \quad \frac{\partial^3 A_0}{\partial y^3} = 0.$$

In view of (2.36), this completes the proof. □

From the representation (2.34) by using Remark 2.10, we see that this operator is admitted by the equation (2.32) iff each operator in (2.35) is also admitted by this ODE. Note that the admissibility of F_0 implies that $f(x) \in W_3$. In what follows, we consider all operators up to such trivial summands, and up to multiplication by an arbitrary non-zero number.

Let us consider the quadratic operator $F_2[y]$ in (2.35).

Lemma 2.15 *Any operator $F_2[y]$ admitting a 3D subspace is equivalent to*

$$F[y] = (y')^2 + Ny^2, \quad \textit{where } N = -1, \ 0, \ \textit{or } 1. \tag{2.37}$$

The 3D subspace invariant with respect to operator (2.37) *is given by the ODE*

$$y''' + Ny' = 0 \tag{2.38}$$

and has the form

$$W_3 = \mathcal{L}\{1, \mathrm{e}^{-x}, \mathrm{e}^{x}\} = \mathcal{L}\{1, \cosh x, \sinh x\} \quad \textit{for } N = -1; \tag{2.39}$$

$$W_3 = \mathcal{L}\{1, x, x^2\} \quad \textit{for } N = 0; \ \textit{and} \tag{2.40}$$

$$W_3 = \mathcal{L}\{1, \cos x, \sin x\} \quad \textit{for } N = 1. \tag{2.41}$$

Proof. Let us use the first transformation in (2.29). By (2.30), operator F_2 reads

$$\begin{aligned} \widetilde{F}[\widetilde{y}] = \tfrac{1}{\alpha(x)} F_2\big(x, \alpha\widetilde{y}, (\alpha\widetilde{y})'\big) = L\alpha(\widetilde{y}')^2 \\ + (2L\alpha' + M\alpha)\widetilde{y}\,\widetilde{y}' + \tfrac{1}{\alpha}\big[L(\alpha')^2 + M\alpha\alpha' + N\alpha^2\big]\widetilde{y}^2. \end{aligned} \tag{2.42}$$

Consider the following two cases: (i) $L(x) \neq 0$, and (ii) $L(x) \equiv 0$, $M(x) \neq 0$. In the former case of (i), choosing $\alpha(x)$ from the condition

$$\alpha' = -\frac{M(x)}{2L(x)}\,\alpha, \tag{2.43}$$

the operator (2.42) is reduced to

$$\widetilde{F}[\widetilde{y}] = \alpha L(\widetilde{y}'_x)^2 + \alpha\big(N - \tfrac{M^2}{4L}\big)\widetilde{y}^2, \tag{2.44}$$

where the sign of α is picked so that $\alpha(x)L(x) > 0$.

Similarly, in the latter case of (ii), $\alpha(x)$ is chosen from the condition $\alpha' = -\frac{N(x)}{M(x)}\,\alpha$. Then operator (2.42) takes the form

$$\widetilde{F}[\widetilde{y}] = \alpha M\,\widetilde{y}\,\widetilde{y}'_x. \tag{2.45}$$

Let us apply the second transformation in (2.29). Since $\frac{\mathrm{d}}{\mathrm{d}x} = \beta'(x)\,\frac{\mathrm{d}}{\mathrm{d}\widetilde{x}}$, by choosing β from the condition

$$\beta'(x) = \big[L(x)\alpha(x)\big]^{-\frac{1}{2}} \tag{2.46}$$

in the case of (2.44), and from $\beta' = [M(x)\alpha(x)]^{-1}$ in the case of (2.45), we derive, respectively, that

$$\widetilde{\widetilde{F}}[\widetilde{y}(\widetilde{x})] = (\widetilde{y}'_{\widetilde{x}})^2 + \widetilde{N}(\widetilde{x})\widetilde{y}^2, \tag{2.47}$$

where $\widetilde{N}(\widetilde{x}) = \alpha(x)\big[N(x) - \frac{M(x)^2}{4L(x)}\big]$, and

$$\widetilde{\widetilde{F}}[\widetilde{y}(\widetilde{x})] = \widetilde{y}\,\widetilde{y}'_{\widetilde{x}}. \tag{2.48}$$

Assume now that, for the operator F_2, there exists a 3D subspace. Then the equivalent operators (2.47) and (2.48) must have the same property. A direct checking via the invariance condition (2.33) shows that, in the case of (2.47), this is possible only for $\widetilde{N} = \text{const}$. The subspace is then defined by an equation of the form (2.38), while, in the case of (2.48), 3D subspaces do not exist. It remains to note that, if $\widetilde{N} = \text{const} \neq 0$, (2.47) is reduced by extensions along $\widetilde{x}$ to the form (2.37) with $|N| = 1$. $\square$

The proof implies that the first operator in (2.35) can preserve 3D subspaces iff

$$L(x) \neq 0 \tag{2.49}$$

and $\widetilde{N}(\widetilde{x}) = \alpha(x)\big[N(x) - \frac{M^2(x)}{4L(x)}\big] = \text{const}$. In view of (2.43), after differentiating in x, one can rewrite down the last condition in terms of the coefficients of the operator only. Namely,

$$\frac{\mathrm{d}}{\mathrm{d}x}\left[N(x) - \frac{M^2(x)}{4L(x)}\right] \equiv \frac{M(x)}{2L(x)}\left[N(x) - \frac{M^2(x)}{4L(x)}\right]. \tag{2.50}$$

Conditions (2.49) and (2.50) make it possible to determine at once, by the coefficients of operator F_2, whether it admits 3D subspaces or not. The functions $\alpha(x)$ and $\beta(x)$ defining the variable change (2.29) are found from equations (2.43) and (2.46), respectively. This gives the following straightforward consequence for operators with constant coefficients.

Lemma 2.16 *Up to a constant multiplier, there exist exactly two translation-invariant quadratic operators F_2 in (2.35) that preserve 3D subspaces,*

$$F[y] = (y')^2 + Ny^2 \quad \textit{and} \tag{2.51}$$

$$F[y] = (y' + Cy)^2, \tag{2.52}$$

where $C \neq 0$ and N are arbitrary constants. By the change of variables

$$y = \mathrm{e}^{-Cx}\widetilde{y} \quad \textit{and} \quad \widetilde{x} = \mathrm{e}^{\frac{C}{2}x}, \tag{2.53}$$

operator (2.52) is transformed into (2.51) with $N = 0$.

Note that the 3D subspace of the operator $\tilde{F}[\widetilde{y}] = (\widetilde{y}'_{\widetilde{x}})^2$ has the form $\tilde{W}_3 = \mathcal{L}\{1, \widetilde{x}, \widetilde{x}^2\}$ (see (2.40)). By Remark 2.11, the 3D subspace of (2.52) is

$$W_3 = \mathrm{e}^{-Cx}\mathcal{L}\{1, \mathrm{e}^{\frac{C}{2}x}, \mathrm{e}^{Cx}\} = \mathcal{L}\{1, \mathrm{e}^{-\frac{C}{2}x}, \mathrm{e}^{-Cx}\}.$$

Let us now consider the linear summand $F_1[y]$ in (2.34). By Lemma 2.15, it suffices to describe all such operators for which equation (2.38) is invariant. The invariance criterion implies the following result.

Lemma 2.17 *The linear operator F_1 in (2.35) is admitted by equation (2.38) iff*

$$R(x) \in W_3 \quad \text{and} \quad S(x) = -R'(x) + C,$$

where C is an arbitrary constant.

Combining Lemmas 2.14, 2.15, and 2.17 gives the complete description of operators (2.31) preserving 3D subspaces.

Theorem 2.18 *Any nonlinear first-order operator (2.31) admitting 3D subspaces is reduced by transformations (2.29), (2.30) to the form*

$$F[y] = (y')^2 + Ny^2 + R(x)y' + [C - R'(x)]y, \tag{2.54}$$

where $N = -1$, 0 or 1, $R(x)$ satisfies (2.38), and C is a constant. The corresponding 3D subspaces are defined by equation (2.38) and are given by (2.39)–(2.41).

Following Remark 2.12, let us replace y by $y + f$ with a function $f(x) \in W_3$ to obtain, up to trivial summands $\widetilde{f}(x) \in W_3$, the operator

$$\widetilde{F}[y] = (y')^2 + Ny^2 + [R(x) + 2f']y' + [S(x) + 2Nf]y, \tag{2.55}$$

where $S(x) = -R(x)' + C$. Note that $R(x) \in W_3$ implies that $S(x) \in W_3$. If $N \neq 0$, setting $f(x) = -\frac{S(x)}{2N}$, we see that the coefficient of y vanishes, and the coefficient of y' becomes constant. Consequently, in this case, the operator (2.54) is reduced to

$$F[y] = (y')^2 + Ny^2 + Ry',$$

where $N = \pm 1$ and $R = \text{const}$. If $N = 0$, the coefficients of operator (2.54) are

$$R(x) = -C_1x^2 + C_2x + C_3 \ \text{ and } \ S(x) = -R(x)' + C = -2C_1x + C_4,$$

where $C_1, \dots, C_4$ are constants. Putting $f(x) = -\frac{1}{2}\left(\frac{1}{2}C_2x^2 + C_3x\right)$ into (2.55) implies that operator (2.54) is reduced to

$$F[y] = (y')^2 + R_1x^2y' + (-2R_1x + R_2)y$$

with constants R_1 and R_2.

Since subspaces W_3, which are invariant under operators (2.51) and (2.52), are also invariant with respect to any linear operators with constant coefficients, Lemma 2.16 implies the following result.

Theorem 2.19 *The set of nonlinear translation-invariant first-order operators that admit 3D subspaces, up to a constant multiplier, is exhausted by*

$$F[y] = (y')^2 + Ny^2 + Ry' + Sy \quad \text{and} \tag{2.56}$$

$$F[y] = (y' + y)^2 + Ry' + Sy, \tag{2.57}$$

where N, R, S, and C are arbitrary constants.

As follows from Lemma 2.16, the quadratic parts of operators (2.57) and (2.56) for $N = 0$ are related by the change of variables (2.53). It is easy to check, however, that this is not valid for full operators containing linear terms. Indeed, by the change (2.53), the operator (2.57) for $R \neq 0$ is reduced to

$$\widetilde{F}[\widetilde{y}] = \tfrac{C^2}{4}\left(\widetilde{y}'_{\widetilde{x}}\right)^2 + \tfrac{RC}{2}\,\widetilde{x}\,\widetilde{y}'_{\widetilde{x}} + (S - RC)\widetilde{y},$$

depending explicitly on $\widetilde{x}$.

2.4 Second-order operators with subspaces of maximal dimension

We now study subspaces of maximal dimension for nonlinear operators

$$F[y] = F(x, y, p, q), \quad \text{where} \quad p = y' \text{ and } q = y''. \tag{2.58}$$

By Theorem 2.8, the dimension cannot exceed five, so we will deal with 5D linear subspaces given by the ODEs

$$y^{(5)} = a_1(x)y^{(4)} + a_2(x)y''' + a_3(x)y'' + a_4(x)y' + a_5(x)y. \tag{2.59}$$

In this case, the invariance criterion leads to a result similar to Lemma 2.14.

Lemma 2.20 *Any operator* (2.58) *preserving a 5D subspace can be represented in the form of*

$$\begin{gathered} F[y] = F_2[y] + F_1[y] + F_0[y], \quad \textit{where} \\ F_2[y] = b_{2,2}q^2 + b_{1,2}pq + b_{0,2}yq + b_{1,1}p^2 + b_{0,1}yp + b_{0,0}y^2, \\ F_1[y] = b_2 q + b_1 p + b_0 y, \quad \textit{and} \quad F_0[y] = b, \end{gathered}$$

where $b_{i,j}$ and b_i for $i, j = 0$, 1, and 2, and b are functions of x.

Proof. This result is based on analyzing the condition of invariance of equation (2.59) under the operator (2.58). In this case, the condition takes the form

$$D^5 F - \big(a_1 D^4 F + a_2 D^3 F + a_3 D^2 F + a_4 DF + a_5 F\big) \equiv 0, \quad \text{where} \tag{2.60}$$

$$\begin{gathered} D = \tfrac{\partial}{\partial x} + y_1 \tfrac{\partial}{\partial y} + y_2 \tfrac{\partial}{\partial y_1} + y_3 \tfrac{\partial}{\partial y_2} + y_4 \tfrac{\partial}{\partial y_3} \\ + \big[a_1(x)y_4 + \ ... \ + a_4(x)y_1 + a_5(x)y\big] \tfrac{\partial}{\partial y_4}. \end{gathered}$$

As above, we have set

$$y_k = \tfrac{\mathrm{d}^k y}{\mathrm{d}x^k} \quad \text{for } k = 1, 2, \ ... \ .$$

The left-hand side of the identity (2.60) is a polynomial in the variables y_3 and y_4, with coefficients that should be equal to zero. Consider the resulting conclusions that have been obtained by using the package Reduce [282] for analytic manipulations.

Coefficients of $y_3 y_4^2$ and $y_3^2 y_4$ lead to conditions

$$F_{y_2 y_2 y_2} = F_{y_1 y_2 y_2} = 0, \tag{2.61}$$

where the lower indices of F denote the corresponding partial derivatives.

In view of (2.61), the *coefficient of y_4^2* implies the equation, which by differentiating in y_1 takes the form

$$2F_{y_1 y_1 y_2} + 3F_{y y_2 y_2} = 0.$$

Simultaneously, *the coefficient of y_3^3* yields

$$3F_{y_1 y_1 y_2} + 2F_{y y_2 y_2} = 0.$$

These equalities imply

$$F_{y y_2 y_2} = 0 \quad \text{and} \tag{2.62}$$

$$F_{y_1y_1y_2} = 0. \tag{2.63}$$

It follows from (2.61) and (2.62) that $F_{y_2y_2} = C(x)$, and hence, F has the form

$$F = \tfrac{1}{2} C y_2^2 + \phi(x, y, y_1) y_2 + \psi(x, y, y_1). \tag{2.64}$$

Later on, we suppose that C is a constant, since the general case $C = C(x)$ is reduced to the constant case by the change (2.29).

Using (2.64), the condition of vanishing the *coefficient of* y_4^2 is

$$10(\phi_{y_1} + a_1 C) = 3a_1 C, \quad \text{from whence come}$$

$$\phi_{y_1} \equiv F_{y_1y_2} = -\tfrac{7}{10} a_1 C \tag{2.65}$$

$$\text{and} \quad F_{yy_1y_2} = 0. \tag{2.66}$$

Then the *coefficient of* y_3y_4 leads to

$$-60a_1' C + 20F_{y_1y_1} + 30F_{yy_2} - 5a_1^2 C + 24a_2 C = 0,$$

which, on differentiation in y_1 and y, implies, respectively, that

$$F_{y_1y_1y_1} = 0 \quad \text{and} \quad 2F_{yy_1y_1} + 3F_{yyy_2} = 0. \tag{2.67}$$

The *coefficient of* y_3^2, on differentiation in y_1, yields $5F_{yy_1y_1} + 4F_{yyy_2} = 0$ that, by the previous equality, means

$$F_{yy_1y_1} = F_{yyy_2} = 0. \tag{2.68}$$

By differentiating in y, the same coefficient implies that

$$F_{yyy_1} = 0. \tag{2.69}$$

Finally, consider the *coefficient of* y_4. Equating it to zero and differentiating in y twice the resulting relation, we find that

$$F_{yyyy} y_1 + F_{xyyy} = 0.$$

Differentiating next in y_1 and using (2.69) yields

$$F_{yyyy} = F_{xyyy} = 0. \tag{2.70}$$

We now apply the above results for defining the coefficients in (2.64). It follows from (2.63), (2.66), and (2.68) that the function ϕ is linear in y and y_1,

$$\phi = \phi_1(x) y_1 + \phi_2(x) y + \phi_3(x).$$

Similarly, (2.67)–(2.70) imply that function ψ takes the form

$$\psi = C_1 y^3 + \widetilde{\psi}(x, y, y_1),$$

where $C_1 =$ constant, and $\widetilde{\psi}$ is a second-degree polynomial in y and y_1. Hence, the right-hand side of (2.64) is a sum of homogeneous polynomials of the zero, first, and second degree in variables y, y_1, y_2, and of the third-degree polynomial $C_1 y^3$. Each of those polynomials determines a differential operator, and, clearly, equation (2.59) is invariant with respect to any of them, and, in particular, with respect to $\widetilde{F} = C_1 y^3$ (see Remark 2.10). However, this is possible, provided that $C_1 = 0$, since, by Theorem 2.8, a nonlinear operator of the zero order cannot preserve a subspace of dimension exceeding one. This completes the proof. □

Likewise, for the case of a first-order operator, we shall carry out further analysis by studying the *quadratic* operators

$$F[y] = b_{2,2}q^2 + b_{1,2}pq + b_{0,2}yq + b_{1,1}p^2 + b_{0,1}yp + b_{0,0}y^2, \tag{2.71}$$

restricting ourselves to the case of constant coefficients,

$$b_{i,j} = \text{constant for } i, j = 0, 1, 2. \tag{2.72}$$

In this case, the following result holds:

Lemma 2.21 *If equation* (2.59) *is invariant under the operator* (2.71), (2.72), *then the coefficients of the equation do not depend on* x.

Proof. In (2.71), let $b_{22} = \frac{C}{2}$ and $C^2 + b_{12}^2 + b_{02}^2 \neq 0$. We deduce from (2.65) that

$$b_{12} = -\tfrac{7}{10}\, a_1 C, \tag{2.73}$$

and, hence, it suffices to consider the following two cases:

Case (i): $C = 2,\ a_1 = -\frac{5}{7}\, b_{12} = \text{constant}$, and

Case (ii): $C = 0,\ b_{12} = 0$.

Let us again concentrate on the invariance condition (2.60). In case of (i), equating the coefficients of y_3y_4, y_2y_4, y_1y_4, and yy_4 to zero, we have

$$\begin{gathered} -5a_1^2 + 24a_2 + 15b_{02} + 20b_{11} = 0, \\ 20a_2' - 7a_1^3 - 14a_1a_2 + 20a_1b_{02} + 10a_1b_{11} + 100a_3 + 75b_{01} = 0, \\ 2a_2'b_{12} + a_1^2b_{02} + a_1b_{01} + 2a_2b_{02} + 20a_4 + 10b_{00} = 0, \\ a_2'b_{02} + 10a_5 = 0. \end{gathered}$$

From these equalities, we derive successively that the coefficients $a_2, \ldots, a_5$ are constant, and, moreover, it turns out that $a_5 = 0$.

In the second case of (ii), equating the coefficients of y_3y_4, y_2y_4, y_1y_4, and y_3^2 to zero yields

$$\begin{gathered} 3b_{02} + 4b_{11} = 0, \quad 4a_1b_{02} + 2a_1b_{11} + 15b_{01} = 0, \\ a_1'(5b_{02} + 2b_{11}) + a_1^2b_{02} + a_1b_{01} + 2a_2b_{02} + 10b_{00} = 0, \\ -2a_1b_{02} - 3a_1b_{11} + 5b_{01} = 0. \end{gathered} \tag{2.74}$$

The first equation gives $b_{11} = -\frac{3}{4}\, b_{02}$, and the second and fourth equation then take, respectively, the form

$$a_1b_{02} + 6b_{01} = 0 \quad \text{and} \quad a_1b_{02} + 20b_{01} = 0,$$

whence comes $a_1b_{02} = b_{01} = 0$. Since, in this case, b_{02} does not vanish (we consider second-order operators), it follows that $a_1 = 0$, and then the third equation of the system (2.74) yields $a_2 = \text{const}$. By these conditions, studying the coefficients of y_1y_3, y_1y_2, and yy_2 implies that $a_3 = 0$, $a_4 = \text{constant}$, and $a_5 = 0$, which completes the proof. □

The next theorem gives a full description of operators (2.71), (2.72) preserving 5D subspaces. All operators are shown up to scaling in x and multiplication by an arbitrary non-zero number. In cases of **(1)–(3)** below, where the same subspace is invariant under arbitrary linear combinations

$$F[y] = C_1F_1[y] + C_2F_2[y], \quad \text{where } C_1, C_2 = \text{constant}, \tag{2.75}$$

of linearly independent operators F_1 and F_2, only these basis operators are given. In all the cases, the corresponding invariant subspaces and (in parentheses) their defining equations (2.59) are indicated.

Theorem 2.22 *The set of operators* (2.71) *with constant coefficients preserving 5D subspaces is exhausted by:*

(1) $F_1[y] = y''y - \frac{3}{4}(y')^2, \quad F_2[y] = (y'')^2,$ *with the subspace*

$$W_5 = \mathcal{L}\{1, x, x^2, x^3, x^4\} \quad \big(y^{(5)} = 0\big);$$

(2) $F_1[y] = y''y - \frac{3}{4}(y')^2 + y^2, \quad F_2[y] = \big(y'' + 4y\big)^2,$ *with*

$$W_5 = \mathcal{L}\{1, \cos x, \sin x, \cos 2x, \sin 2x\} \quad \big(y^{(5)} = -5y''' - 4y'\big);$$

(3) $F_1[y] = y''y - \frac{3}{4}(y')^2 - y^2, \quad F_2[y] = \big(y'' - 4y\big)^2,$ *with*

$$W_5 = \mathcal{L}\{1, \cosh x, \sinh x, \cosh 2x, \cosh 2x\} \quad \big(y^{(5)} = 5y''' - 4y'\big);$$

(4) $F[y] = \big(y'' - \frac{7}{2}y' + 3y\big)^2,$ *with*

$$W_5 = \mathcal{L}\big\{1, e^{\frac{1}{2}x}, e^x, e^{\frac{3}{2}x}, e^{2x}\big\} \quad \big(4y^{(5)} = 20y^{(4)} - 35y''' + 25y'' - 6y'\big);$$

(5) $F[y] = \big(5y'' - \frac{27}{2}y' + 7y\big)\big(y'' - \frac{3}{2}y' - y\big),$ *with*

$$W_5 = \mathcal{L}\big\{1, e^{-\frac{1}{2}x}, e^{\frac{1}{2}x}, e^x, e^{2x}\big\} \quad \big(4y^{(5)} = 12y^{(4)} - 7y''' - 3y'' + 2y'\big);$$

(6) $F[y] = (y'' - 5y' + 6y)(y'' - 2y' - 3y),$ *with*

$$W_5 = \mathcal{L}\big\{1, e^{-x}, e^x, e^{2x}, e^{3x}\big\} \quad \big(y^{(5)} = 5y^{(4)} - 5y''' - 5y'' + 6y'\big).$$

Note that operators F_2 in **(1)** and F in **(4)** are equivalent; see Example 2.13, where $k = 2$ and $\tilde{x} \mapsto -\frac{1}{2}\tilde{x}$. The equivalence transformation for operators F_1 in **(1)** and F_1 in **(3)** is $y = x^2\tilde{y}$, $\tilde{x} = \ln x$, i.e., $\alpha = x^2$ and $\beta = \ln x$ in Remark 2.11.

Proof. We consider in greater detail cases (i) and (ii) that appeared in the proof of Lemma 2.21.

Case (i): $C = 2, \ a_1 = -\frac{5}{7}b_{12}$. Then, as has been shown above, $a_5 = 0$. Consider two possibilities: $b_{12} \neq 0$ and $b_{12} = 0$.

(i.1) In the subcase where $b_{12} \neq 0$, using the scaling in x implies $b_{12} = -7$. The first coefficient on the right-hand side of (2.59) is then $a_1 = 5$.

Equating the coefficients of y_3y_4, y_3^2, y_2y_4, y_1y_4, y_2y_3, y_1y_3, y_2^2, y_1y_2, and y_1^2 to zero in the identity (2.60) (the rest of the coefficients vanish in view of the assumed hypotheses), we obtain the system

$$\begin{gathered}
15b_{02}+20b_{11}+24a_2-125=0,\\
-10b_{02}-15b_{11}+5b_{01}-37a_2+4a_3=0,\\
20b_{02}+10b_{11}+15b_{01}-14a_2+20a_3-175=0,\\
(2a_2+25)b_{02}+5b_{01}+10b_{00}+20a_4=0,\\
7a_2b_{02}+4a_2b_{11}-50b_{01}+20b_{00}-35a_2-116a_3+10a_4=0,\\
(5a_2+3a_3)b_{02}+2a_2b_{01}-40b_{00}-95a_4=0,\\
10a_3b_{02}+8a_3b_{11}-3a_2b_{01}-30b_{00}-35a_3-35a_4=0,\\
5(a_3+3a_4)b_{02}+10a_4b_{11}+3a_3b_{01}-6a_2b_{00}-35a_4=0,\\
5a_4b_{02}+5a_4b_{01}-2a_3b_{00}=0.
\end{gathered}$$

Using the first four equations yields the expressions for the coefficients of operator (2.71), with $b_{22}=\frac{C}{2}$, in terms of the coefficients of (2.59),

$$\begin{gathered}
b_{02}=\tfrac{1}{35}(-124a_2-32a_3-675),\\
b_{11}=\tfrac{1}{35}(51a_2+24a_3+725),\quad b_{01}=\tfrac{1}{35}(164a_2-20a_3+825),\\
b_{00}=\tfrac{1}{175}(124a_2^2+32a_2a_3+1815a_2+450a_3-350a_4+6375).
\end{gathered}\tag{2.76}$$

The fifth equation then implies the following:

$$a_4=-\tfrac{1}{25}\big(4a_2^2+95a_2+30a_3+375\big).\tag{2.77}$$

Using these conditions, we reduce the remaining four equations to

$$\begin{gathered}
428a_2^2+276a_2a_3+5415a_2+32a_3^2+1875a_3+175a_4+17000=0,\\
1236a_2^2+964a_2a_3+13365a_2+128a_3^2+4875a_3-875a_4+38250=0,\\
372a_2^3+96a_2^2a_3+5445a_2^2+1670a_2a_3+2325a_2a_4\\
+19125a_2+550a_3^2+600a_3a_4+2250a_3+10250a_4=0,\\
124a_2^2a_3+32a_2a_3^2+1815a_2a_3-500a_2a_4\\
+450a_3^2+300a_3a_4+6375a_3-1875a_4=0.
\end{gathered}$$

Resolving the first three equations with respect to the quadratic terms yields:

$$\begin{gathered}
a_2^2=\tfrac{5}{36}(-93a_2+2a_3-275),\\
a_2a_3=\tfrac{5}{36}(30a_2-41a_3+125),\\
a_3^2=\tfrac{5}{36}(-165a_2-46a_3-875).
\end{gathered}\tag{2.78}$$

It is not difficult to check that, in view of these relations, the last equation becomes the identity. From the first equation of system (2.78), it follows that

$$a_3=\tfrac{1}{2}\big(\tfrac{36}{5}\,a_2^2+93a_2+275\big),\tag{2.79}$$

and the second and third equations then take the form

$$144a_2^3+2680a_2^2+15925a_2+30625=0\quad\text{and}\tag{2.80}$$

$$1296a_2^4+33480a_2^3+317525a_2^2+1310750a_2+1990625=0.\tag{2.81}$$

The roots of the cubic equation (2.80) are

$$-\tfrac{35}{4},\quad -\tfrac{175}{36},\quad\text{and}\quad -5,$$

simultaneously satisfying equation (2.81).

For each of the obtained values of a_2, using the formulae (2.79), (2.77), and (2.76), we determine the rest of the coefficients,

$$\begin{gathered} a_2=-\tfrac{35}{4}, \quad a_3=\tfrac{25}{4}, \quad a_4=-\tfrac{3}{2}, \\ b_{02}=6, \quad b_{11}=\tfrac{49}{4}, \quad b_{01}=-21, \quad b_{00}=9; \end{gathered} \tag{2.82}$$

$$\begin{gathered} a_2=-\tfrac{175}{36}, \quad a_3=\tfrac{125}{36}, \quad a_4=-\tfrac{725}{162}, \\ b_{02}=-\tfrac{110}{21}, \quad b_{11}=\tfrac{1345}{84}, \quad b_{01}=-\tfrac{25}{21}, \quad b_{00}=\tfrac{9950}{567}; \end{gathered} \tag{2.83}$$

$$\begin{gathered} a_2=-5, \quad a_3=-5, \quad a_4=6, \\ b_{02}=3, \quad b_{11}=10, \quad b_{01}=3, \quad b_{00}=-18. \end{gathered} \tag{2.84}$$

Collections of the coefficients (2.82) and (2.84) correspond, respectively, to items **(4)** and **(6)** in the statement of the theorem, while (2.83) corresponds to item **(5)** up to scaling in x.

(i.2) In the subcase where $b_{12}=0$, from (2.73), it follows that $a_1=0$. Equating the coefficients of y_3y_4, y_2y_4, y_1y_4, y_3^2, y_2y_3, y_1y_3, y_2^2, y_1^2, and y_1y_2 in (2.60) to zero (the rest of the coefficients are identically zero), we find

$$\begin{gathered} 24a_2+15b_{02}+20b_{11}=0, \quad 4a_3+3b_{01}=0, \\ a_2b_{02}+10a_4+5b_{00}=0, \quad 4a_3+5b_{01}=0, \\ 7a_2b_{02}+4a_2b_{11}+10a_4+20b_{00}=0, \quad 2a_2b_{01}+3a_3b_{02}=0, \\ -3a_2b_{01}+10a_3b_{02}+8a_3b_{11}=0, \quad -2a_3b_{00}+5a_4b_{01}=0, \\ -6a_2b_{00}+3a_3b_{01}+15a_4b_{02}+10a_4b_{11}=0. \end{gathered}$$

The second and the fourth equations yield $a_3=b_{01}=0$, so the system is simplified and has the form

$$\begin{gathered} 24a_2+15b_{02}+20b_{11}=0, \\ a_2b_{02}+10a_4+5b_{00}=0, \\ 7a_2b_{02}+4a_2b_{11}+10a_4+20b_{00}=0, \\ -6a_2b_{00}+15a_4b_{02}+10a_4b_{11}=0. \end{gathered}$$

Taking b_{11} and b_{00} from the first two equations and plugging into the rest of them, a single additional condition is obtained $a_4=-\frac{4}{25}a_2^2$. The final solution of the original system is then written as

$$b_{11}=-\tfrac{3}{5}a_2-\tfrac{3}{4}b_{02}, \quad b_{00}=-\tfrac{1}{5}a_2b_{02}+\tfrac{4}{25}a_2^2, \quad a_4=-\tfrac{4}{25}a_2^2,$$

with arbitrary parameters a_2 and b_{02}. Further analysis of these equations leads to those operators (2.75) from **(1)–(3)** of the theorem that are singled out by the condition $C_2 \neq 0$.

Case (ii): If $C=0$, $b_{12}=0$, $b_{02}\neq 0$, then $a_1=a_3=a_5=0$. See the proof of Lemma 2.21. Studying the coefficient of y_2y_4 in (2.60) yields $b_{01}=0$, while vanishing the coefficients of y_3y_4, y_1y_4, and y_2y_3, y_1y_2 (other coefficients are cancelled by the above conditions) leads to the system

$$3b_{02} + 4b_{11} = 0, \quad a_2 b_{02} + 5b_{00} = 0,$$
$$7a_2 b_{02} + 4a_2 b_{11} + 20b_{00} = 0, \quad -6a_2 b_{00} + 15a_4 b_{02} + 10a_4 b_{11} = 0.$$

The solutions of this system are

$$b_{11} = -\tfrac{3}{4}\, b_{02}, \quad b_{00} = -\tfrac{1}{5}\, a_2 b_{02}, \quad a_4 = -\tfrac{4}{25}\, a_2^2$$

that lead to the last case of operators (2.75) with $C_2 = 0$. The theorem is proved. □

Notice that the description of *all* translation-invariant second-order operators that preserve 5D subspaces is made by adding arbitrary linear operators of the second order with constant coefficients.

By Theorem 2.22, it is easy to reconstruct a large number of quasilinear and fully nonlinear heat and wave equations composed of linear combinations of operators,

$$u_t = F[u] \quad \text{and} \quad u_{tt} = F[u], \tag{2.85}$$

exhibiting interesting finite-dimensional evolution on the corresponding invariant subspaces. Various singularity formation phenomena, such as quenching, extinction, blow-up, and propagation of finite interfaces, can be traced out by using such exact solutions. We postpone more detailed singularity analysis until the next chapter, where we begin the study of invariant subspaces and solutions of fourth-order thin film equations with similar quadratic DSs.

2.5 First and second-order quadratic operators with subspaces of lower dimensions

In this section, we consider quadratic operators

$$F[y] = b_{11}(y')^2 + 2b_{10}y'y + b_{00}y^2 \quad \text{and} \tag{2.86}$$

$$F[y] = b_{22}(y'')^2 + 2b_{21}y''y' + 2b_{20}y''y + b_{11}(y')^2 + 2b_{10}y'y + b_{00}y^2 \tag{2.87}$$

that admit invariant subspaces of the dimension which is less than maximal. The coefficients of the operators, as well as the equations of invariant subspaces

$$y^{(n)} = r_{n-1}y^{(n-1)} + ... + r_1 y' + r_0 y$$

(notice the change in notation), are assumed to be constant. Such operators are plugged into a number of nonlinear evolution PDEs of different types, including reaction-diffusion, combustion, flame propagation, Boussinesq equations of water wave interaction, and others, which typically take the evolution form (2.85). Notice that many partial differential operators in $I\!R^N$ reduce to ordinary differential operators in radial geometry, so that they also fall into the scope of the present analysis.

2.5.1 First-order operators

We begin with the operators (2.86). Omitting most of technical calculus details obtained by Reduce, let us present the final results. For completeness, the 3D subspaces are also included.

2D subspaces. Subspaces W_2 are defined by the ODE

$$y'' = r_1 y' + r_0 y.$$

The invariance condition for (2.86) gives the following two cases:

Case I: $r_0 = 0$, i.e.,

$$y'' - r_1 y' = 0, \quad \text{so that}$$

$$W_2 = \mathcal{L}\{1, x\} \text{ if } r_1 = 0, \quad \text{or} \quad W_2 = \mathcal{L}\{1, \mathrm{e}^{r_1 x}\} \text{ if } r_1 \neq 0. \tag{2.88}$$

This equation (and the corresponding subspace) is invariant under a 2D linear space of operators spanned by

$$F_1[y] = y(y' - r_1 y) \quad \text{and} \quad F_2[y] = y'(y' - r_1 y). \tag{2.89}$$

For $r_1 \neq 0$, one can choose another basis, such as

$$F_2[y] + r_1 F_1[y] = (y')^2 - r_1^2 y^2 \quad \text{and} \quad F_2[y] - r_1 F_1[y] = (y' - r_1 y)^2.$$

Case II: $r_0 = -\frac{2}{9} r_1^2 \neq 0$. Then, up to scaling in x, the ODE and the subspace are

$$y'' - 3y' + 2y = 0 \quad \text{and} \quad W_2 = \mathcal{L}\{\mathrm{e}^x, \mathrm{e}^{2x}\}. \tag{2.90}$$

The 1D space of operators is spanned by

$$F[y] = (y' - 2y)^2. \tag{2.91}$$

3D subspaces. These are given by

$$y''' = r_2 y'' + r_1 y' + r_0 y.$$

There are two cases:

Case I: $r_0 = r_2 = 0$, i.e.,

$$y''' - r_1 y' = 0.$$

Up to scaling, we can consider $r_1 = 0, \pm 1$ and obtain the subspaces

$$W_3 = \mathcal{L}\{1, x, x^2\} \quad (r_1 = 0),$$
$$W_3 = \mathcal{L}\{1, \cos x, \sin x\} \quad (r_1 = -1),$$
$$W_3 = \mathcal{L}\{1, \cosh x, \sinh x\} \quad (r_1 = 1).$$

The basis of operators is

$$F[y] = (y')^2 - r_1 y^2. \tag{2.92}$$

Case II: $r_0 = 0$ and $r_1 = -\frac{2}{9} r_2^2 \neq 0$. As above, we set $r_2 = 3$, $r_1 = -2$ to obtain

$$y''' - 3y'' + 2y' = 0 \quad \text{and} \quad W_3 = \mathcal{L}\{1, \mathrm{e}^x, \mathrm{e}^{2x}\}.$$

The basis is

$$F[y] = (y' - 2y)^2. \tag{2.93}$$

It is easy to reduce operator (2.93) to (2.92) with $r_1 = 0$ (see Lemma 2.15).

2.5.2 Second-order operators

Consider now the operators (2.87).

2D subspaces. The ODE for W_2 is

$$y'' = r_1 y' + r_0 y.$$

For arbitrary r_1 and r_0, there exists the basis

$$F_1[y] = y(y'' - r_1 y' - r_0 y), \quad F_2[y] = y'(y'' - r_1 y' - r_0 y), \\ \text{and} \quad F_3[y] = y''(y'' - r_1 y' - r_0 y), \tag{2.94}$$

composed of trivial operators annihilating W_2. The subspace of operators is extended in two cases:

Case I: $r_0 = 0$. The ODE is

$$y'' - r_1 y' = 0,$$

and the subspaces are as shown in (2.88). In addition to (2.94), two operators are included:

$$F_4[y] = y'(y' - r_1 y) \quad \text{and} \quad F_5[y] = y(y' - r_1 y).$$

As above, for $r_1 \neq 0$, we can choose the basis

$$F_4[y] + r_1 F_5[y] = (y')^2 - r_1^2 y^2 \quad \text{and} \quad F_4[y] - r_1 F_5[y] = (y' - r_1 y)^2.$$

Case II: $r_0 = -\frac{2}{9} r_1^2 \neq 0$ ($r_1 = 3$, $r_0 = -2$), i.e., (2.90) holds. The additional operator is (2.91). Denoting $f[y] = y' - 2y$, we can write down the above ODE and the operator as follows:

$$f' - f = 0 \quad \text{and} \quad F_4[y] = (f[y])^2.$$

3D subspaces. Such subspaces W_3 are given by

$$y''' = r_2 y'' + r_1 y' + r_0 y.$$

There are two cases:

Case I: $r_0 = 0$, i.e.,

$$y''' - r_2 y'' - r_1 y' = 0.$$

For arbitrary r_2 and r_1, there exist three linear independent operators

$$F_1[y] = y(y'' - r_2 y' - r_1 y), \quad F_2[y] = y'(y'' - r_2 y' - r_1 y), \\ \text{and} \quad F_3[y] = y''(y'' - r_2 y' - r_1 y).$$

The extensions are:

(i) If $r_2 = 0$, then $F_4[y] = (y')^2 - r_1 y^2$;
(ii) If $r_2 = 3$ and $r_1 = -2$, the extension is given by $F_4[y] = (y' - 2y)^2$.
These results are similar to those given above. The next case is new

Case II: $r_0 \neq 0$. There exists a family of invariant subspaces given by the ODE with the parameter $\alpha \in I\!R$:

$$f' - f \equiv y''' - \alpha y'' + (3\alpha - 7)y' - 2(\alpha - 3)y = 0, \quad \text{where}$$

$$f[y] = y'' + (1-\alpha)y' + 2(\alpha-3)y.$$

The corresponding subspaces are:

$$\begin{aligned} \alpha \neq 4,\, 5: &\quad W_3 = \mathcal{L}\{e^x, e^{2x}, e^{(\alpha-3)x}\}, \\ \alpha = 4: &\quad W_3 = \mathcal{L}\{e^x, xe^x, e^{2x}\}, \\ \alpha = 5: &\quad W_3 = \mathcal{L}\{e^x, e^{2x}, xe^{2x}\}. \end{aligned}$$

For any $\alpha \in I\!R$, the following operator is admitted:

$$F_1[y] = (f[y])^2 = [y'' + (1-\alpha)y' + 2(\alpha-3)y]^2.$$

The space of operators is extended for:

(i) If $\alpha = 2$, then the additional operator is

$$F_2[y] = (y'' - y' - 2y)(y' - 5y) + 2[(y')^2 - yy' - 2y^2];$$

(ii) If $\alpha = 6$, then $F_2[y] = (y'' - 5y' + 6y)(y' - 3y)$; and
(iii) If $\alpha = 7$, then $F_2[y] = (y'' - 5y' + 4y)^2$.
The case $\alpha = 3$ coincides with case **I**(ii).

4D subspaces. Invariant subspaces W_4 are given by the ODE

$$y^{(4)} - r_3 y''' - r_2 y'' - r_1 y' - r_0 y = 0.$$

There are several cases:

Case I: $r_0 = 0$.

I.1. $r_1 = r_3 = 0$. For any $r_2 \in I\!R$, there exists the operator $F_1[y] = (y'')^2 - r_2(y')^2$. For $r_2 = 0$, there exist additional operators

$$F_2[y] = y'y'' \quad \text{and} \quad F_3[y] = yy'' - \tfrac{2}{3}(y')^2.$$

I.2. A one-parameter family of invariant subspaces is given by

$$f'' - f' \equiv y^{(4)} - \alpha y''' + (3\alpha - 7)y'' - 2(\alpha-3)y' = 0, \tag{2.95}$$

where $f[y] = y'' + (1-\alpha)y' + 2(\alpha-3)y$. For any $\alpha \in I\!R$, there exists the operator

$$F_1[y] = (f[y])^2 = [y'' + (1-\alpha)y' + 2(\alpha-3)y]^2.$$

The invariant subspaces are:

$$\begin{aligned} \alpha \neq 3,\, 4,\, 5: &\quad W_4 = \mathcal{L}\{1, e^x, e^{2x}, e^{(\alpha-3)x}\}, \\ \alpha = 3: &\quad W_4 = \mathcal{L}\{1, x, e^x, e^{2x}\}, \\ \alpha = 4: &\quad W_4 = \mathcal{L}\{1, e^x, xe^x, e^{2x}\}, \\ \alpha = 5: &\quad W_4 = \mathcal{L}\{1, e^x, e^{2x}, xe^{2x}\}. \end{aligned}$$

Extra extensions are available in the following cases:

(i) If $\alpha = 2$, there exist

$$F_2[y] = (y'' - y' - 2y)(y'' - 3y' + 2y), \quad F_3[y] = (y'' - y' - 2y)(y' - 2y);$$

(ii) If $\alpha = 6$, then $F_2[y] = (y'' - 5y' + 6y)(y' - 3y)$; and

(iii) If $\alpha = 7$, then $F_2[y] = (y'' - 5y' + 4y)^2$.

The above cases partially coincide with similar ones above. The next case, as well as others below, are new.

(iv) If $\alpha = 1$, then $F_2[y] = (y'' - y')^2 - 4(y' - y)^2$.

I.3. Another one-parameter family of subspaces given by

$$y^{(4)} - y''' - \beta y'' + \beta y' = 0 \quad \left(\beta \ne \tfrac{1}{4},\, 4\right)$$

is invariant under the operator $F[y] = (y'' - y')^2 - \beta(y' - y)^2$. The corresponding subspaces are given by

$$\begin{aligned} \beta \ne 0,\, 1: &\quad W_4 = \mathcal{L}\{1, e^x, e^{\sqrt{\beta}x}, e^{-\sqrt{\beta}x}\},\\ \beta = 0: &\quad W_4 = \mathcal{L}\{1, x, x^2, e^x\},\\ \beta = 1: &\quad W_4 = \mathcal{L}\{1, e^x, xe^x, e^{-x}\}. \end{aligned}$$

For $\beta = \frac{1}{4}$ and $\beta = 4$, this yields equation (2.95) with $\alpha = 2$ (up to scaling $x \mapsto 2x$) and $\alpha = 1$, respectively. We also obtain one of the operators admitted by (2.95).

I.4. The one-parameter family of subspaces given by

$$y^{(4)} - 2y''' + (\gamma + 1)y'' - \gamma y' = 0 \quad \left(\gamma \ne -2,\, \tfrac{2}{9}\right)$$

admits the operator $F[y] = (y'' - 2y' + y)(y'' - y') + \gamma (y' - y)^2$. For $\gamma = \frac{2}{9}$ and $\gamma = -2$, these yield equation (2.95) with $\alpha = 6$ (up to scaling $x \mapsto 3x$) and $\alpha = 2$ respectively, and one of the admitted operators. Subspaces are

$$\begin{aligned} \gamma \ne 0,\, \tfrac{1}{4}: &\quad W_4 = \mathcal{L}\{1, e^x, e^{\frac{1}{2}(1-\sqrt{1-4\gamma})x}, e^{\frac{1}{2}(1+\sqrt{1-4\gamma})x}\},\\ \gamma = 0: &\quad W_4 = \mathcal{L}\{1, x, e^x, xe^x\},\\ \gamma = \tfrac{1}{4}: &\quad W_4 = \mathcal{L}\{1, e^x, e^{\frac{x}{2}}, xe^{\frac{x}{2}}\}. \end{aligned}$$

Case II: $r_0 \ne 0$. The only subspace W_4 and the operator are given by

$$\begin{aligned} f'' - 3f' + 2f \equiv y^{(4)} - 10y''' + 35y'' - 50y' + 24y = 0,\\ F[y] = (y'' - 7y' + 12y)^2 \equiv (f[y])^2. \end{aligned}$$

The ODE is easily integrated and yields $W_4 = \mathcal{L}\{e^x, e^{2x}, e^{3x}, e^{4x}\}$.

5D subspaces. This case corresponds to the maximal dimension of invariant subspaces admitted by nonlinear quadratic second-order differential operators studied in the previous section. For completeness, we now briefly comment on these conclusions using a slightly different representation of the results. As a new feature, some lower-dimensional invariant subspaces from W_5 are also presented. This study of linear invariant manifolds on invariant subspaces is important for future applications to nonlinear evolution PDEs. In general, such questions are not straightforward at all.

Subspaces W_5 are given by the ODE

$$y^{(5)} - r_4 y^{(4)} - r_3 y''' - r_2 y'' - r_1 y' - r_0 y = 0.$$

Invariant subspaces are known to occur only if $r_0 = 0$. There are six cases that have already been indicated in Theorem 2.22. We present these up to scaling in x.

Cases (1)–(3). The ODE is $y^{(5)} - 5ry''' + 4r^2y' = 0$, where, up to scaling, we set $r = 0, \pm 1$ and obtain the subspaces

$$W_5 = \mathcal{L}\{1, x, x^2, x^3, x^4\} \ \ (r = 0),$$
$$W_5 = \mathcal{L}\{1, \cos x, \sin x, \cos 2x, \sin 2x\} \ \ (r = -1),$$
$$W_5 = \mathcal{L}\{1, \cosh x, \sinh x, \cosh 2x, \sinh 2x\} \ \ (r = 1).$$

The basis of operators is $F_1[y] = yy'' - \frac{3}{4}(y')^2 - ry^2$ and $F_2[y] = (y'' - 4ry)^2$.

Case (4). The ODE is

$$f''' - 3f'' + 2f' \equiv y^{(5)} - 10y^{(4)} + 35y''' - 50y'' + 24y' = 0,$$

where $f[y] = y'' - 7y' + 12y$ and $W_5 = \mathcal{L}\{1, \mathrm{e}^x, \mathrm{e}^{2x}, \mathrm{e}^{3x}, \mathrm{e}^{4x}\}$. The operator is $F[y] = (y'' - 7y' + 12y)^2 \equiv (f[y])^2$. Note that the subspace $\mathcal{L}\{\mathrm{e}^{3x}, \mathrm{e}^{4x}\}$ is the kernel (the null-set) of F.

There exist other lower-dimensional subspaces from W_5, such as

$$W_4 = \mathcal{L}\{\mathrm{e}^x, \mathrm{e}^{2x}, \mathrm{e}^{3x}, \mathrm{e}^{4x}\} \ \text{ and } \ W_3 = \mathcal{L}\{\mathrm{e}^{2x}, \mathrm{e}^{3x}, \mathrm{e}^{4x}\},$$

as well as the 2D ones $W_2 = \mathcal{L}\{\mathrm{e}^{3x}, \mathrm{e}^{4x}\}$ and $W_2' = \mathcal{L}\{\mathrm{e}^{2x}, \mathrm{e}^{4x}\}$. Such subspaces are easily found by the change

$$\tilde{x} = \mathrm{e}^{-x}, \ \ \tilde{y} = \mathrm{e}^{-4x}y, \ \ \text{implying that} \ \ \tilde{F}[\tilde{y}] = \mathrm{e}^{-4x}F[y] = (\tilde{y}_{\tilde{x}\tilde{x}})^2.$$

Therefore, the subspaces W_5, W_4, W_3, W_2, and W_2' are transformed into $\tilde{W}_5 = \mathcal{L}\{1, \tilde{x}, \tilde{x}^2, \tilde{x}^3, \tilde{x}^4\}$, $\tilde{W}_4 = \mathcal{L}\{1, \tilde{x}, \tilde{x}^2, \tilde{x}^3\}$, $\tilde{W}_3 = \mathcal{L}\{1, \tilde{x}, \tilde{x}^2\}$, $\tilde{W}_2 = \mathcal{L}\{1, \tilde{x}\}$, and $\tilde{W}_2' = \mathcal{L}\{1, \tilde{x}^2\}$, respectively.

Case (5). The ODE and the subspace are

$$y^{(5)} - 6y^{(4)} + 7y''' + 6y'' - 8y' = 0 \ \text{ and } \ W_5 = \mathcal{L}\{1, \mathrm{e}^{-x}, \mathrm{e}^x, \mathrm{e}^{2x}, \mathrm{e}^{4x}\},$$

with the operator $F[y] = (5y'' - 27y' + 28y)(y'' - 3y' - 4y)$. Here, the kernel is $\ker F = \mathcal{L}\{1, \mathrm{e}^{-x}, \mathrm{e}^{4x}\}$. It is easy to see that $W_4 = \mathcal{L}\{1, \mathrm{e}^{-x}, \mathrm{e}^x, \mathrm{e}^{4x}\} \subset W_5$ is also invariant.

Case (6). The ODE and the subspace are

$$y^{(5)} - 5y^{(4)} + 5y''' + 5y'' - 6y' = 0 \ \text{ and } \ W_5 = \mathcal{L}\{1, \mathrm{e}^{-x}, \mathrm{e}^x, \mathrm{e}^{2x}, \mathrm{e}^{3x}\}.$$

The operator is $F[y] = (y''-5y'+6y)(y''-2y'-3y)$. Then $\ker F = \mathcal{L}\{1, \mathrm{e}^{-x}, \mathrm{e}^{3x}\}$ and $W_4 = \mathcal{L}\{1, \mathrm{e}^x, \mathrm{e}^{2x}, \mathrm{e}^{3x}\} \subset W_5$ is invariant.

This completes the classification of invariant subspaces for the nonlinear second-order operators, since the subspaces of the dimension higher than five can be preserved by linear differential operators only (Section 2.2).

2.6 Operators preserving polynomial subspaces

In this section, we study ordinary differential operators preserving polynomial subspaces. Our goal is a complete description of the whole set of such operators, as well as of subsets of operators of a given order and translation-invariant operators.

2.6.1 Main Theorem for polynomial subspaces

From now on, we concentrate on subspaces of the polynomial type,

$$W_n = \mathcal{L}\{1, x, \, ..., x^{n-1}\}, \tag{2.96}$$

defined by the simplest linear ODE

$$y^{(n)} = 0. \tag{2.97}$$

As explained in Section 2.1, we introduce the first integrals via solutions g_i of the conjugate equation $(-1)^n z^{(n)} = 0$. Choosing $g_i(x) = f_i(x) = x^{i-1}$ for $i = 1, ..., n$ yields the complete set of first integrals in the form of

$$I_i = \sum_{j=1}^{i} (-1)^{j-1} \tfrac{(i-1)!}{(i-j)!} \, x^{i-j} \, y^{(n-j)} \quad \text{for } i = 1, ..., n. \tag{2.98}$$

Therefore, Theorem 2.1 reads as follows:

Theorem 2.23 *The set $\mathcal{F}_{n-1}$ of nonlinear operators F preserving the polynomial subspace* (2.96) *is generated by the operators*

$$F[y] = \textstyle\sum_{i=1}^{n} A^i(I_1, \, ..., I_n) x^{i-1}, \tag{2.99}$$

where $A^i(I_1, \, ..., I_n)$, $i = 1, ..., n$, are any smooth functions of first integrals (2.98).

As was mentioned above, every operator (2.3) preserving subspace (2.1) defines a Lie-Bäcklund symmetry operator $X = F[y] \frac{\partial}{\partial y}$ of the corresponding linear ODE (2.2). The whole set $\mathcal{F}_{n-1}$ of such operators forms an infinite-dimensional Lie algebra completely described by Theorem 2.1. In its turn, Theorem 2.23 provides us with all Lie-Bäcklund symmetries for equation (2.97) that were found in [12].

Along with the algebra $\mathcal{F}_{n-1}$, we consider its linear subspaces $\mathcal{F}_m$ with $m \le n-2$, consisting of operators (2.3) of the order not greater than m. Operators of the order not greater than $n-2$ give a linear space $\mathcal{F}_{n-2}$. We call these the *operators* or *symmetries of submaximal order*. One purpose of this section is to obtain a general representation for the sets $\mathcal{F}_m$, with $m \le n-2$, in the case of the subspace (2.96) (equation (2.97)). In particular, we prove that the dimension of the linear space $\mathcal{F}_{n-2}$ of the operators of submaximal order is expressed as

$$\dim \mathcal{F}_{n-2} = C_{2n-1}^{n} \tag{2.100}$$

via the binomial coefficient $C_n^k = \frac{n!}{k!(n-k)!}$.

For $n = 3$, this has been actually shown by Lie [393], who found that the maximal dimension of the algebra of contact symmetries on the plane cannot exceed $C_5^3 = 10$, and that this maximal value is attained on the equation $y''' = 0$. A generalization of this result for arbitrary ODEs of the order $n = 4$ was obtained in [298]. This, together with Theorem 2.24, proved below, makes plausible the following conjecture (this is an OPEN PROBLEM).

Conjecture 2.1 *For any nth-order ODEs (linear or not), the maximal dimension of the linear space $\mathcal{F}_{n-2}$ of symmetries of submaximal order is equal to C_{2n-1}^{n}.*

Bearing in mind applications to PDEs, we also consider the subspace $\widehat{\mathcal{F}}_{n-1} \subset \mathcal{F}_{n-1}$ of translation invariant quadratic operators and give its complete description. Finally, we perform a description of all the translation-invariant operators preserving (2.96).

In this section, we refer to operators (2.3) admitted by equations (2.2) or (2.97) as *symmetry operators*, or simply *symmetries* of these ODEs.

2.6.2 Operators of submaximal and lower orders

A symmetry operator (2.3) for (2.2) is defined by the invariance condition (2.4),

$$L[F[y]] \,\big|_{L[y]=0} \equiv 0. \tag{2.101}$$

We will assume that the function F in (2.3) is analytic and is represented by a convergent power series in variables $y, y', ..., y^{(n-1)}$, with coefficients depending on x. Considering this series as an infinite sum of homogeneous polynomials in $y, y', ..., y^{(n-1)}$, and using the fact that the ODE (2.2) is linear, we obtain from (2.101) that every such homogeneous polynomial defines a symmetry as well (see Remark 2.10). Hence, we can suppose beforehand that operators (2.3) have the homogeneous form

$$F[y] = \sum_{i_0+...+i_{n-1}=k} \beta_{i_0,...,i_{n-1}}(x)\, y^{i_0}(y')^{i_1}\cdot ... \cdot(y^{(n-1)})^{i_{n-1}},$$

where $k = 0, 1, 2, ...$ and $0 \le i_0, ..., i_{n-1} \le k$. Using (2.5), one can express y and its derivatives as linear homogeneous functions of $I_1, ..., I_n$, thus obtaining the representation

$$F[y] = \sum_{i_1+...+i_n=k} \gamma_{i_1,...,i_n}(x)\, (I_1)^{i_1}\cdot ... \cdot(I_n)^{i_n}, \tag{2.102}$$

where, in accordance with Theorem 2.1, the coefficients $\gamma_{i_1,...,i_n}(x)$ are linear combinations of the fundamental set of solutions $\{f_i(x)\}$.

Let us return to the equation (2.97). Symmetries of the order $n-2$ are singled out by the condition

$$\tfrac{\partial F}{\partial y^{(n-1)}} \equiv \textstyle\sum_{i=1}^{n} x^{i-1}\tfrac{\partial F}{\partial I_i} = 0, \tag{2.103}$$

where F is considered as a function of $x, I_1, ..., I_n$. The first integrals of the corresponding characteristic system are

$$x \quad \text{and} \quad J_i = xI_i - I_{i+1} \ \text{ for } i = 1, ..., n-1,$$

and therefore, the general expression for F is given by $F = \widetilde{F}(x, J_1, ..., J_{n-1})$ with an arbitrary function $\widetilde{F}$. Passing in (2.102) from I_i, $i = 1, ..., n$, to the new variables

$$J_0 = I_1, \quad J_i = xI_i - I_{i+1} \quad (i = 1, ..., n-1),$$

so that $I_1 = J_0$ and $I_{i+1} = x^i J_0 - \sum_{j=1}^{i} x^{i-j} J_j$, we obtain

$$F[y] = \sum_{i_1+...+i_{n-1}=k} \delta_{i_1,...,i_{n-1}}(x)\, (J_1)^{i_1}\cdot ... \cdot(J_{n-1})^{i_{n-1}}, \tag{2.104}$$

where we have taken into account that F does not depend on J_0.

Notice that $\delta_{i_1,\dots,i_{n-1}}(x)$ and $(J_1)^{i_1}\cdot\ldots\cdot(J_{n-1})^{i_{n-1}}$ are some polynomials in x of degrees $s_{i_1,\dots,i_{n-1}}$ and k respectively, with coefficients, depending on I_i, with $i = 1, \dots, n$. The leading term of x in (2.104) can be written as

$$x^{s+k} \sum_{i_1+\ldots+i_{n-1}=k} C_{i_1,\dots,i_{n-1}}(I_1)^{i_1}\cdot\ldots\cdot(I_{n-1})^{i_{n-1}},$$

where $C_{i_1,\dots,i_{n-1}}$ are some constants, and $s = \max s_{i_1,\dots,i_{n-1}}$. It follows from (2.99) that $s + k \le n-1$, providing us with estimates $s_{i_1,\dots,i_{n-1}} \le n-1-k$ for $k \le n-1$. By Theorem 2.23, every monomial

$$x^{i_0}(J_1)^{i_1}\cdot\ldots\cdot(J_{n-1})^{i_{n-1}}, \tag{2.105}$$

with $i_0, \dots, i_{n-1} \ge 0$ and $i_0 + \ldots + i_{n-1} \le n-1$, defines a symmetry of the ODE (2.97). Therefore, $\mathcal{F}_{n-2}$ is a linear span of these monomials. Calculating the number of operators (2.105) yields

Theorem 2.24 *The linear space $\mathcal{F}_{n-2}$ of symmetry operators of submaximal order for equation* (2.97) *is spanned by the operators* (2.105), *and* (2.100) *holds.*

Remark 2.25 The theorem remains valid without the analyticity assumption on F. Indeed, applying the condition (2.103) to the general representation (2.99) yields

$$\textstyle\sum_{i=1}^{n}\left(\sum_{j=1}^{n} A^i_j x^{j-1}\right)x^{i-1} \equiv 0, \quad \text{where } A^i_j = \frac{\partial A_i}{\partial I_j}.$$

Equating the coefficients of the different powers of x to zero gives the system

$$\textstyle\sum_{(i+j=k)} A^i_j = 0 \quad \text{for } k = 2, \dots, 2n, \tag{2.106}$$

that leads to the necessary result.

In the next examples, we apply Theorems 2.23 and 2.24 to the cases $n = 2$ and 3.

Example 2.26 **(n = 2)** For $n = 2$, equation (2.97) is

$$y'' = 0, \tag{2.107}$$

and, for the spaces $\mathcal{F}_1$ and $\mathcal{F}_0$, we obtain the following results:

(i) The whole Lie-Bäcklund algebra $\mathcal{F}_1$ of the equation (2.107) is generated by operators

$$F[y] = A^1(I_1, I_2) + A^2(I_1, I_2)x, \tag{2.108}$$

where A^1 and A^2 are arbitrary functions of the first integrals $I_1 = y'$ and $I_2 = xy' - y$.

(ii) The linear space $\mathcal{F}_0$ is 3D spanned by operators $x^{i_0}(J_1)^{i_1}$ $(i_0, i_1 \ge 0, i_0+i_1 \le 1)$, or, explicitly,

$$1, \ x, \ J_1, \tag{2.109}$$

where $J_1 = xI_1 - I_2 = y$. The operators (2.109) define a 3D subalgebra of the algebra of Lie point symmetries of equation (2.107).

Illustrating Remark 2.25 above, note that, in this case, system (2.106) reads

$$A^1_1 = 0, \ A^1_2 + A^2_1 = 0, \ A^2_1 = 0.$$

Its solutions are $A^1 = C_1 - C_3 I_2$ and $A^2 = C_1 + C_3 I_2$, that after substitution into (2.108) yield $F[y] = C_1 + C_2 x + C_3(xI_1 - I_2)$. This coincides with (2.109).

Example 2.27 (n = 3) For $n = 3$, equation (2.97) is

$$y''' = 0, \tag{2.110}$$

and, for the spaces $\mathcal{F}_2$ and $\mathcal{F}_1$, Theorems 2.23 and 2.24 yield the following.

(i) The whole Lie-Bäcklund algebra $\mathcal{F}_2$ of the equation (2.110) is generated by

$$F[y] = A^1(I_1, I_2, I_3) + A^2(I_1, I_2, I_3)x + A^3(I_1, I_2, I_3)x^2,$$

where A^1, A^2, and A^3 are arbitrary functions of the first integrals

$$I_1 = y'', \quad I_2 = xy'' - y', \quad \text{and} \quad I_3 = x^2 y'' - 2xy' + 2y. \tag{2.111}$$

(ii) The subspace $\mathcal{F}_1$ (coinciding with the subalgebra of contact symmetries) is 10D spanned by the operators $x^{i_0}(J_1)^{i_1}(J_2)^{i_2}$ $(i_0, i_1, i_2 \geq 0,\ i_0 + i_1 + i_2 \leq 2)$, or,

$$1,\ x,\ x^2,\ J_1,\ xJ_1,\ J_2,\ xJ_2,\ (J_1)^2,\ J_1J_2,\ (J_2)^2, \quad \text{where} \tag{2.112}$$

$$J_1 = xI_1 - I_2 = y' \quad \text{and} \quad J_2 = xI_2 - I_3 = xy' - 2y. \tag{2.113}$$

Plugging the last expressions into (2.112) yields a basis of the well-known 10D algebra of contact symmetries of (2.110) described by Lie in [393].

Using a general representation of the operators $F \in \mathcal{F}_{n-2}$ given by Theorem 2.24, and successively applying conditions $\partial F/\partial y^{(p)} = 0$ with $p = n-2,\ n-3,\ \dots,\ 1$, one can obtain a complete description of the linear spaces $\mathcal{F}_{n-1-k}$ for arbitrary $k = 1, \dots, n-1$. The result is expressed in terms of differences defined recursively by

$$J_i^k = xJ_i^{k-1} - J_{i+1}^{k-1} \quad \text{for } i = 1, \dots, n-k, \tag{2.114}$$

with $J_i^0 = I_i$ for $i = 1, \dots, n$, and $J_i^1 = J_i$ for $i = 1, \dots, n-1$. Taking into account that J_i^k for $i = 1, \dots, n-k$ are polynomials in x of the degree k with coefficients, linearly depending on $I_1, \dots, I_n$, we arrive at the following theorem which includes Theorem 2.24 as a special case.

Theorem 2.28 *The linear space $\mathcal{F}_{n-1-k}$, $1 \leq k \leq n-1$, of symmetry operators of the order not greater than $n-1-k$ for equation (2.97) is spanned by the operators*

$$x^{i_0}\,(J_1^k)^{i_1} \cdot \dots \cdot (J_{n-k}^k)^{i_{n-k}},$$

where $i_0, i_1, \dots, i_{n-k}$ are nonnegative integers satisfying the condition $i_0 + k(i_1 + \dots + i_{n-k}) \leq n-1$.

Example 2.29 (n = 4) Let $n = 4$, so equation (2.97) is

$$y^{(4)} = 0. \tag{2.115}$$

Theorems 2.24 and 2.28 yield the following conclusions:

(i) ($k = 1$) The subspace $\mathcal{F}_2$ of the operators of submaximal order is 35D spanned by the operators

$$x^{i_0}(J_1)^{i_1}(J_2)^{i_2}(J_3)^{i_3} \quad \text{for } i_0, i_1, i_2, i_3 \geq 0, \quad i_0 + i_1 + i_2 + i_3 \leq 3,$$

where $J_i = xI_i - I_{i+1}$ for $i = 1, 2$, and 3, and the first integrals are written as

$$I_1 = y''', \quad I_2 = xy''' - y'', \quad I_3 = x^2y''' - 2xy'' + 2y', \\ \text{and} \quad I_4 = x^3y''' - 3x^2y'' + 6xy' - 6y. \tag{2.116}$$

Substituting (2.116) into the expressions for J_i, one obtains

$$J_1 = y'', \quad J_2 = xy'' - 2y', \quad \text{and} \quad J_3 = x^2y'' - 4xy' + 6y. \tag{2.117}$$

(ii) $(k = 2)$ Space $\mathcal{F}_1$, corresponding to the algebra of point symmetries, is spanned by $x^{i_0}(K_1)^{i_1}(K_2)^{i_2}$, $i_0, i_1, i_2 \geq 0$, $i_0 + 2(i_1 + i_2) \leq 3$, or, explicitly,

$$1, \ x, \ x^2, \ x^3, \ K_1, \ xK_1, \ K_2, \ xK_2,$$

where $K_1 = xJ_1 - J_2 = 2y'$ and $K_2 = xJ_2 - J_3 = 2xy' - 6y$. (Here we set $J_1^2 = K_1$ and $J_2^2 = K_2$.)

(iii) $(k = 3)$ Space $\mathcal{F}_0$ is spanned by $x^{i_0}(L_1)^{i_1}$, $i_0, i_1 \geq 0$, $i_0 + 3i_1 \leq 3$, or

$$1, \ x, \ x^2, \ x^3, \ L_1,$$

where $L_1 = xK_1 - K_2 = 6y$. (We denote $J_1^3 = L_1$.)

Example 2.30 $(\mathbf{n} \geq 4)$ The subspace $\mathcal{F}_1 = \mathcal{F}_{n-1-(n-2)}$ $(k = n-2)$ for equation (2.97) with $n \geq 4$, coinciding with the algebra of point symmetries, is spanned by operators

$$x^{i_0}(J_1^{n-2})^{i_1}(J_2^{n-2})^{i_2}, \quad i_0, i_1, i_2 \geq 0, \quad i_0 + (n-2)(i_1 + i_2) \leq n-1, \quad \text{or}$$

$$1, \ x, \ \ldots, \ x^{n-1}, \ J_1^{n-2}, \ xJ_1^{n-2}, \ J_2^{n-2}, \ xJ_2^{n-2}. \tag{2.118}$$

Then, for the expressions J_i^k, $k = 0, \ldots, n-1$, the following representation holds:

$$J_i^k = k!\, x^{i+k}\left(x^{-1-k}y^{(n-i-k)}\right)^{(i-1)} \quad \text{for } i = 1, \ldots, n-k, \tag{2.119}$$

and, therefore, one obtains

$$J_1^{n-2} = (n-2)!\, x^{n-1}(x^{1-n}y') = (n-2)!\, y' \quad \text{and} \\ J_2^{n-2} = (n-2)!\, x^n(x^{1-n}y)' = (n-2)!\,[xy' + (1-n)y].$$

Plugging into (2.118) yields the well-known $(n+4)$-dimensional algebra of Lie point symmetries of (2.97) with $n \geq 4$, [374].

2.6.3 Translation-invariant operators

In the context of applications to evolution PDEs, it makes sense to obtain a description of operators admitted by the equation (2.97) which do not depend explicitly on x. We denote by $\widehat{\mathcal{F}}_{n-1}$ the linear subspace of such translation-invariant operators of orders not greater than $n-1$. In this case, $\widehat{\mathcal{F}}_{n-1}$ is a subalgebra of the symmetry algebra $\mathcal{F}_{n-1}$. Using the representation (2.99) for the operators $F \in \mathcal{F}_{n-1}$, the translation-invariant operators are singled out by the condition (2.14), i.e.,

$$\frac{\partial F}{\partial x} \equiv 0.$$

Example 2.31 (**n = 2**) Let us start with the case $n = 2$, i.e., with the ODE $y'' = 0$. Applying condition (2.14) to operator (2.108) yields the identity

$$(I_1 A_2^1 + A^2) + x\, I_1 A_2^2 = 0,$$

where $A_i^k \equiv \partial A^k/\partial I_i$, so that $I_1 A_2^1 = -A^2$ and $A_2^2 = 0$. This system has the general solution $A^2 = I_1 A(I_1)$, $A^1 = -I_2 A(I_1) + B(I_1)$ with arbitrary functions $A(I_1)$ and $B(I_1)$. Plugging these expressions into (2.108) yields

$$F[y] = (xI_1 - I_2)A(I_1) + B(I_1) = yA(y') + B(y').$$

Example 2.32 (**n = 3**) Acting in a similar way in the case $n = 3$ with the ODE $y''' = 0$, one obtains the following general description of operators F belonging to $\widehat{\mathcal{F}}_2$:

$$F[y] = (y')^2 A(\hat{I}_1, \hat{I}_2) + y' B(\hat{I}_1, \hat{I}_2) + C(\hat{I}_1, \hat{I}_2),$$

where A, B, and C are arbitrary functions of the expressions $\hat{I}_1 = I_1$ and $\hat{I}_2 = I_1 I_3 - (I_2)^2$ via the first integrals (2.111) which imply $\hat{I}_1 = y''$ and $\hat{I}_2 \sim 2yy'' - (y')^2$.

Generalizing these examples yields the following result.

Theorem 2.33 *The set $\widehat{\mathcal{F}}_{n-1}$ of translation-invariant operators admitted by the equation (2.97) is generated by*

$$F[y] = \textstyle\sum_{i=1}^n \alpha_i(\hat{I}_1, \ldots, \hat{I}_{n-1})(y^{(n-2)})^{i-1},$$

where α_i are arbitrary functions of the homogeneous polynomials

$$\begin{gathered} \hat{I}_1 = I_1, \quad \hat{I}_2 = I_1 I_3 - (I_2)^2, \\ \hat{I}_k = (I_1)^{k-1} I_{k+1} - (I_2)^k - \textstyle\sum_{i=1}^{k-2} C_k^i \hat{I}_{k-i}(I_2)^i, \;\; 3 \le k \le n-1, \end{gathered} \tag{2.120}$$

of the first integrals (2.98).

Proof. We apply condition (2.14) to operator (2.99). Using that $\partial I_1/\partial x = 0$ and $\partial I_k/\partial x = (k-1)I_{k-1}$ for $k = 2, \ldots, n$ yields the identity

$$\sum_{i=1}^{n-1}\Big[\sum_{k=2}^{n}(k-1)I_{k-1}A_k^i + iA^{i+1}\Big]x^{i-1} + \Big[\sum_{k=2}^{n}(k-1)I_{k-1}A_k^n\Big]x^{n-1} \equiv 0,$$

which is reduced to the following system of linear equations:

$$XA^i = -iA^{i+1} \quad (i = 1, \ldots, n-1), \quad XA^n = 0 \tag{2.121}$$

with $X \equiv \sum_{k=2}^n (k-1)I_{k-1}\partial/\partial I_k$. One can verify that, for functions (2.120), the relations $X\hat{I}_i = 0$ $(i = 1, \ldots, n-1)$ hold, while $X(I_2/I_1) = 1$, and, hence, in variables

$$t_i = \hat{I}_i \quad (i = 1, \ldots, n-1) \quad \text{and} \quad t_n = \tfrac{I_2}{I_1},$$

the operator X is transformed into $\partial/\partial t_n$. The system (2.121) is then rewritten as $\partial A^i/\partial t_n = -iA^{i+1}$ $(i = 1, \ldots, n-1)$, $\partial A^n/\partial t_n = 0$, or, in the equivalent form,

$$A^k = \tfrac{(-1)^{k-1}}{(k-1)!}\,\tfrac{\partial^{k-1}A^1}{\partial {t_n}^{k-1}} \quad (k = 2, \ldots, n), \qquad \tfrac{\partial^n A^1}{\partial {t_n}^n} = 0.$$

This provides us with general expressions for A^k,

$$A^k = (-1)^{k-1} \sum_{i=k}^{n} C_{i-1}^{k-1} (t_n)^{i-k} \widetilde{\alpha}_i(t_1, \dots, t_{n-1}) \quad (k = 1, \dots, n),$$

with arbitrary functions $\widetilde{\alpha}_i$ $(i = 1, \dots, n)$. Substituting these into (2.99) yields

$$\begin{aligned} F[y] &= \sum_{k=1}^{n} \sum_{i=k}^{n} C_{i-1}^{k-1} (-x)^{k-1} (t_n)^{i-k} \widetilde{\alpha}_i \\ &= \sum_{i=1}^{n} \big[\sum_{k=1}^{i} C_{i-1}^{k-1} (-x)^{k-1} (t_n)^{i-k}\big] \widetilde{\alpha}_i \\ &= \sum_{i=1}^{n} (t_n - x)^{i-1} \widetilde{\alpha}_i = \sum_{i=1}^{n} (x I_1 - I_2)^{i-1} \alpha_i, \end{aligned}$$

where $\alpha_i = \widetilde{\alpha}_i/(-I_1)^{i-1}$. Since $x I_1 - I_2 = y^{(n-2)}$, this completes the proof. □

Applying Theorem 2.33 to the case $n = 4$ leads to

Example 2.34 **(n = 4)** For equation (2.115), $y^{(4)} = 0$, the set $\widehat{\mathcal{F}}_3$ consists of operators

$$F[y] = \sum_{i=1}^{4} (y'')^{i-1} \alpha_i(\hat{I}_1, \hat{I}_2, \hat{I}_3),$$

where $\alpha_i(\cdot)$ are arbitrary functions of the homogeneous polynomials

$$\hat{I}_1 = I_1, \quad \hat{I}_2 = I_1 I_3 - (I_2)^2 \text{ and } \hat{I}_3 = (I_1)^2 I_4 - (I_2)^3 - 3\hat{I}_2 I_2.$$

Substituting the expressions (2.116) for I_1, I_2, and I_3 yields

$$\hat{I}_1 = y''', \quad \hat{I}_2 = 2y'y''' - (y'')^2, \text{ and } \hat{I}_3 = -6y(y''')^2 + 6y'y''y''' - 2(y'')^3.$$

2.6.4 *Translation-invariant quadratic operators*

We return to the study of quadratic operators and present a more detailed description of operators admitting polynomial subspaces. Namely, we study the special class $\hat{\mathcal{F}}_{n-1}^{\mathrm{q}} \subset \hat{\mathcal{F}}_{n-1}$ of translation-invariant quadratic operators preserving the subspace (2.96) (the ODE (2.97)). For exponential subspaces, a related problem was solved earlier in Section 1.5.2 (Example 1.49). We will again use the notation $y_i = y^{(i)}$.

Let Q be the linear span of all the monomials $y_i y_j$ for $i, j = 0, \dots, n-1$. In what follows, we represent Q as a direct sum of the form

$$Q = \oplus_{k=0}^{n-1} Q_{n-1-k}, \tag{2.122}$$

where Q_{n-1-k} denotes a linear space of quadratic operators of the order $n-1-k$, with the basis that is constructed below. The linear space $\hat{\mathcal{F}}_{n-1}^{\mathrm{q}} \subset Q$ will then be expressed in terms of Q_{n-1-k}'s.

Theorem 2.28 and representation (2.102) imply that the set $\mathcal{F}_{n-1-k}^{\mathrm{q}}$ of *all* quadratic operators of the order not greater than $n-1-k$ admitted by the equation (2.97) is generated by

$$F = \sum_{i_0=0}^{n-1-2k} \big(x^{i_0} \sum_{(i,j=1, i \le j)}^{n-k} \alpha_{i_0,i,j} J_i^k J_j^k\big), \tag{2.123}$$

where $\alpha_{i_0,i,j}$ are arbitrary constants and k satisfies the inequality

$$0 \le k \le \tfrac{n-1}{2}. \tag{2.124}$$

The subset $\hat{\mathcal{F}}_{n-1-k}^{\mathrm{q}} \subset \mathcal{F}_{n-1-k}^{\mathrm{q}}$ of *translation-invariant* operators is singled out from (2.123) by the condition (2.14).

The construction of the basis of Q_{n-1-k} is fulfilled in a few steps.

Step 1. We start with a particular case of operators of the form

$$F = \sum_{(i,j=1,i\le j)}^{n-k} \alpha_{i,j}\, J_i^k J_j^k. \tag{2.125}$$

Here, instead of (2.124), we assume that $0 \le k \le n-1$. In this case, the invariance of the subspace (2.96) can be violated, since the definition of J_i^k (see (2.114)) implies that the right-hand side of (2.125) is a polynomial of x of the degree $2k$, whose coefficients depend on first integrals of the equation (2.97). Therefore, the operator (2.125) maps W_n given by (2.96) onto $W_{2k+1} = \mathcal{L}\{1, ..., x^{2k}\}$, so that there is no invariance for $k > \frac{n-1}{2}$. Nevertheless, at the first step, we construct all the operators of the form (2.125) that are translation-invariant. The linear space of such operators is denoted by Q_{n-i-k}^{2k}.

From (2.119), the following identities are obtained:

$$\frac{\partial J_1^k}{\partial x} = 0 \;\; \text{and} \;\; \frac{\partial J_i^k}{\partial x} = (i-1)J_{i-1}^k \;\; \text{for} \;\; i = 2, ..., n-k. \tag{2.126}$$

Then (2.114) and (2.126) imply that it suffices to consider operators (2.125) with $i + j = s$, $2 \le s \le 2(n-k)$ that can be written as

$$F = \begin{cases} \sum_{i=1}^{[\frac{s}{2}]} \alpha_i J_i^k J_{s-i}^k & \text{for } s = 2, ..., n-k+1, \\ \sum_{i=s-n+k}^{[\frac{s}{2}]} \alpha_i J_i^k J_{s-i}^k & \text{for } s = n-k+2, ..., 2(n-k). \end{cases}$$

Condition (2.114) leads to the system of linear homogeneous equations for the coefficients $\{\alpha_i\}$. It can be shown that a nontrivial solution exists only for even s that belong to the first interval, i.e., in the case of the operators

$$F_{\frac{s}{2}}^{2k} = \sum_{i=1}^{\frac{s}{2}} \alpha_i J_i^k J_{s-i}^k \quad \text{for even } s \in \{2, ..., n-k+1\}. \tag{2.127}$$

Applying (2.119) and (2.126) to (2.127) yields (for $s \ge 4$) the identity

$$\begin{aligned} 0 = \tfrac{\partial}{\partial x} F_{\frac{s}{2}}^{2k} &= \sum_{i=1}^{\frac{s}{2}} \alpha_i \big[(i-1)J_{i-1}^k J_{s-i}^k + (s-i-1)J_i^k J_{s-i-1}^k\big] \\ &= \sum_{i=1}^{\frac{s}{2}} \big[i\alpha_i + (s-i-1)\alpha_i\big] J_i^k J_{s-i-1}^k + \big(\tfrac{s}{2}-1\big)\alpha_{\frac{s}{2}} J_{\frac{s}{2}}^k J_{\frac{s}{2}-1}^k. \end{aligned}$$

This gives the following system for the coefficients:

$$\begin{cases} \alpha_{i+1} = -\frac{s-i-1}{i}\,\alpha_i \;\; \text{for } i = 1, ..., \frac{s}{2}-2 \;\; (s \ge 6), \\ \alpha_{\frac{s}{2}} = -\frac{s}{2(s-2)}\,\alpha_{\frac{s}{2}-1}. \end{cases}$$

Setting $\alpha_1 = 1$ yields

$$\alpha_i = (-1)^{i-1} C_{s-2}^{i-1} \;\; \text{for } i = 1, ..., \tfrac{s}{2}-1,$$

$$\alpha_{\frac{s}{2}} = (-1)^{\frac{s}{2}-1} C_{s-3}^{\frac{s}{2}-2}.$$

Substituting these expressions into (2.127) and taking into account the obvious case $s = 2$, we obtain a basis of Q_{n-1-k}^{2k},

$$\begin{gathered} F_1^{2k} = (J_1^k)^2, \\ F_{\frac{s}{2}}^{2k} = \sum_{i=1}^{\frac{s}{2}-1} (-1)^{i-1} C_{s-2}^{i-1} J_i^k J_{s-i}^k + (-1)^{\frac{s}{2}-1} C_{s-3}^{\frac{s}{2}-2} \big(J_{\frac{s}{2}}^k\big)^2, \end{gathered} \tag{2.128}$$

where $s \in \{4, ..., n-k+1\}$ is even.

Step 2. Operators (2.128) provide us with a part of the basis of the linear space Q_{n-1-k}. The rest of the basis can be constructed as follows. Note that, for $k = n-1$, the only operator that remains in (2.128) is $F_1^{2(n-1)} = (J_1^{n-1})^2$. Let $k \le n-2$. Replacing k by $k+1$ in (2.128) leads to the $(n-2-k)$th-order operators $F_{\frac{s}{2}}^{2k+2}$ for even $s \in \{2, ..., n-k\}$. Using operator D of the total derivative in x, we introduce the following $(n-1-k)$th-order operators:

$$F_{\frac{s}{2}}^{2k+1} = DF_{\frac{s}{2}}^{2k+2} \quad \text{for even } s \in \{2, ..., n-k\}. \tag{2.129}$$

From (2.119), one can derive that

$$DJ_i^{k+1} = (k+1)J_i^k \quad \text{for } i = 1, ..., n-k-1.$$

Applying this allows us to rewrite (2.129) in terms of $\{J_i^k\}$ and $\{J_i^{k+1}\}$ as

$$\begin{aligned} F_1^{2k+1} &= DF_1^{2k+2} = 2(k+1)J_1^{k+1}J_1^k, \\ F_{\frac{s}{2}}^{2k+1} &= DF_{\frac{s}{2}}^{2k+2} = (k+1)\big[\textstyle\sum_{i=1}^{\frac{s}{2}-1}(-1)^{i-1}C_{s-2}^{i-1}\big(J_i^k J_{s-i}^{k+1} \\ &\quad + J_i^{k+1}J_{s-i}^k\big) + 2(-1)^{\frac{s}{2}-1}C_{s-3}^{\frac{s}{2}-2}J_{\frac{s}{2}}^{k+1}J_{\frac{s}{2}}^k\big] \end{aligned} \tag{2.130}$$

for even $s \in \{2, ..., n-k\}$. Operators (2.130) map W_n onto $W_{2k+2} = \mathcal{L}\{1, ..., x^{2k+1}\}$. Let Q_{n-1-k}^{2k+1} denote the linear envelope of these operators.

Notice that the set of operators (2.128) and (2.130) is linearly independent and that every operator $F \in \hat{\mathcal{F}}_{n-1-k}^{\text{q}}$ is represented as a linear combination of these operators and operators of less orders. (This is readily seen from the explicit representation of the operators given below.) The subspace Q_{n-1-k} is then defined as the linear span of the operators (2.128) and (2.130). Obviously, the representation $Q_{n-1-k} = Q_{n-1-k}^{2k} \oplus Q_{n-1-k}^{2k+1}$ holds.

Step 3. In order to obtain the explicit representation of the operators (2.128), we use the formulae (2.119). It follows that

$$J_i^k = (-1)^{i-1}(i+k-1)!\, y_{n-i-k} + ... \quad \text{for } i = 1, ..., n-k,$$

whereby "..." we denote the x-dependent summands that are mutually cancelled by plugging into (2.128). Substituting these expressions into (2.128) yields

$$\begin{aligned} &F_1^{2k} \sim (y_{n-1-k})^2, \quad F_{\frac{s}{2}}^{2k} \sim \textstyle\sum_{i=1}^{\frac{s}{2}-1}(-1)^{s-i-1}C_{i+k-1}^k \\ &\times C_{s-i+k-1}^k y_{n-i-k}y_{n-s+i-k} + \tfrac{1}{2}(-1)^{\frac{s}{2}-1}\big(C_{\frac{s}{2}+k-1}^k\big)^2\big(y_{n-\frac{s}{2}-k}\big)^2 \end{aligned} \tag{2.131}$$

for even $s \in \{4, ..., n-k+1\}$. Using (2.129) (or (2.130)) yields

$$\begin{aligned} &F_1^{2k+1} \sim y_{n-1-k}y_{n-2-k}, \\ &F_{\frac{s}{2}}^{2k+1} \sim C_{s-1+k}^{k+1}y_{n-1-k}y_{n-s-k} + \textstyle\sum_{i=1}^{\frac{s}{2}-1}(-1)^{s-i-1} \\ &\times \big(C_{i+k}^{k+1}C_{s-i+k}^{k+1} - C_{i+1+k}^{k+1}C_{s-i-1+k}^{k+1}\big)y_{n-i-k-1}y_{n-s+i-k} \end{aligned} \tag{2.132}$$

for even $s \in \{2, ..., n-k\}$. Leaving in (2.131) and (2.132) only the terms containing the leading order derivative y_{n-1-k} yields

$$F^{2k}_{\frac{s}{2}} \sim y_{n-1-k}y_{n-s+1-k} + \dots \quad \text{for even } s \in \{2, ..., n-k+1\}, \text{ and}$$

$$F^{2k+1}_{\frac{s}{2}} \sim y_{n-1-k}y_{n-s-k} + \dots \quad \text{for even } s \in \{2, ..., n-k\}.$$

Here, all the monomials of the form $y_{n-1-k}y_{n-i-k}$ for $i = 1, ..., n-k$ are available and moreover each of these appear in a single operator only. Hence, operators (2.131), and (2.132) are linearly independent, and their envelope Q_{n-1-k} contains all the operators of the order $n-1-k$ up to adding lower order operators. The dimensions of the subspaces are given by the following integer parts: $\dim Q^{2k}_{n-1-k} = \left[\frac{n-1-k}{2}\right]$ and $\dim Q^{2k+1}_{n-1-k} = \left[\frac{n-k}{2}\right]$. Hence,

$$\dim Q_{n-1-k} = \dim Q^{2k}_{n-1-k} + \dim Q^{2k+1}_{n-1-k} = n-k.$$

For $Q' = \oplus_{k=0}^{n-1} Q_{n-1-k}$, we have

$$\dim Q' = \textstyle\sum_{k=0}^{n-1}(n-k) = \frac{n(n+1)}{2} = \dim Q.$$

Therefore, $Q' = Q$, completing the construction of the representation (2.122).

Noting that subspace (2.96) is invariant with respect to $F \in Q^r_{n-1-k}$ ($r = 2k$ or $r = 2k+1$) iff $r \le n-1$ and that all such operators form the set $\hat{\mathcal{F}}^{\rm q}_{n-1}$, we arrive at the following result.

Theorem 2.35 *The linear space $\hat{\mathcal{F}}^{\rm q}_{n-1}$ of translation-invariant quadratic operators of the order not greater than $n-1$, which preserve subspace (2.96), is represented as:*

$$\text{(i) } \hat{\mathcal{F}}^{\rm q}_{n-1} = Q^{n-1}_{\frac{n-1}{2}} \oplus \left(\oplus_{k=0}^{\frac{n-3}{2}} Q_{n-1-k}\right), \quad \dim \hat{\mathcal{F}}^{\rm q}_{n-1} = \frac{3n^2+4n+9}{8} \quad \text{for odd } n,$$

and

$$\text{(ii) } \hat{\mathcal{F}}^{\rm q}_{n-1} = \oplus_{k=0}^{\frac{n-2}{2}} Q_{n-1-k}, \quad \dim \hat{\mathcal{F}}^{\rm q}_{n-1} = \frac{n(3n+2)}{8} \quad \text{for even } n.$$

It is seen that the set of translation-invariant quadratic operators of the order k, which preserve the polynomial subspace (2.96) of the maximal dimension $n = 2k+1$, coincides with the subspace Q^{2k}_k. Setting $n = 2k+1$ in (2.131) yields

Theorem 2.36 *The basis of the linear subspace Q^{2k}_k of translation-invariant kth-order operators preserving the polynomial subspace of the maximal dimension $n = 2k+1$ is given by*

$$F^{2k}_1 \sim (y_k)^2, \; F^{2k}_{\frac{s}{2}} \sim \textstyle\sum_{i=1}^{\frac{s}{2}-1} (-1)^{s-i-1} C^k_{i+k-1} C^k_{s-i+k-1} y_{k+1-i} y_{k+1-s+i} + \frac{1}{2}(-1)^{\frac{s}{2}-1}\left(C^k_{\frac{s}{2}+k-1}\right)^2 \left(y_{k+1-\frac{s}{2}}\right)^2 \tag{2.133}$$

for even $s \in \{4, ..., k+2\}$. The dimension is $\dim Q^{2k}_k = \left[\frac{k+2}{2}\right]$.

Example 2.37 (**n = 5**) For $n = 5$, the subspace and its equation are

$$W_5 = \mathcal{L}\{1, ..., x^4\} \quad \text{and} \quad y_5 = 0. \tag{2.134}$$

Applying (2.128), (2.131), and (2.130), we obtain Table 2.1, where the boldface subspaces $\mathbf{Q^j_i}$ are composed of operators admitted by the ODE in (2.134). Second-order

Table 2.1 *Quadratic operators for* $W_5 = \mathcal{L}\{1, ..., x^4\}$ $(y_5 = 0)$

		Basis of operators	$F[W_5]$
Q_0	Q_0^8	$F_1^8 = (J_1^4)^2 \sim (y_0)^2$	$W_9 = \mathcal{L}\{1, ..., x^8\}$
Q_1	Q_1^7	$F_1^7 = DF_1^8 \sim y_1 y_0$	$W_8 = \mathcal{L}\{1, ..., x^7\}$
	Q_1^6	$F_1^6 = (J_1^3)^2 \sim (y_1)^2$	$W_7 = \mathcal{L}\{1, ..., x^6\}$
Q_2	Q_2^5	$F_1^5 = DF_1^6 \sim y_2 y_1$	$W_6 = \mathcal{L}\{1, ..., x^5\}$
	$\mathbf{Q_2^4}$	$F_1^4 = (J_1^2)^2 \sim (y_2)^2$, $F_2^4 = J_1^2 J_3^2 - (J_2^2)^2 \sim 4y_2 y_0 - 3(y_1)^2$	$W_5 = \mathcal{L}\{1, ..., x^4\}$
Q_3	$\mathbf{Q_3^3}$	$F_1^3 = DF_1^4 \sim y_3 y_2$, $F_2^3 = DF_2^4 \sim 2y_3 y_0 - y_2 y_1$	$W_4 = \mathcal{L}\{1, ..., x^3\}$
	$\mathbf{Q_3^2}$	$F_1^2 = (J_1^1)^2 \sim (y_3)^2$, $F_2^2 = J_1^1 J_3^1 - (J_2^1)^2 \sim 3y_3 y_1 - 2(y_2)^2$	$W_3 = \mathcal{L}\{1, ..., x^2\}$
Q_4	$\mathbf{Q_4^1}$	$F_1^1 = DF_1^2 \sim y_4 y_3$, $F_2^1 = DF_2^2 \sim 3y_4 y_1 - y_3 y_2$	$W_2 = \mathcal{L}\{1, x\}$
	$\mathbf{Q_4^0}$	$F_1^0 = (J_1^0)^2 \sim (y_4)^2$, $F_2^0 = J_1^0 J_3^0 - (J_2^0)^2 \sim 2y_4 y_2 - (y_3)^2$, $F_3^0 = J_1^0 J_5^0 - 4J_2^0 J_4^0 + 3(J_3^0)^2$ $\sim 2y_4 y_0 - 2y_3 y_1 + (y_2)^2$	$W_1 = \mathcal{L}\{1\}$

operators preserving the subspace (2.134) (it is of the maximal dimension for such operators) form the subspace $\mathbf{Q_2^4}$.

Remark 2.38 If $F[W_n] = W_k$ for $k < n$, the evolution equation

$$u_t = F[u]$$

admits exact solutions on W_n,

$$u(x, t) = \textstyle\sum_{i=0}^{n-1} C_i(t) x^i,$$

where the corresponding DS contains $n - k$ trivial equations

$$C_i' = 0 \quad \text{for} \quad i = k + 1, ..., n.$$

For an illustration, see Example 1.13.

Table 2.2 *Quadratic operators for* $W_4 = \mathcal{L}\{1, ..., x^3\}$ $(y_4 = 0)$

		Basis of operators	$F[W_4]$
Q_0	Q_0^6	$F_1^6 = (J_1^3)^2 \sim (y_0)^2$	$W_7 = \mathcal{L}\{1, ..., x^6\}$
Q_1	Q_1^5	$F_1^5 = DF_1^6 \sim y_1 y_0$	$W_6 = \mathcal{L}\{1, ..., x^5\}$
	Q_1^4	$F_1^4 = (J_1^2)^2 \sim (y_1)^2$	$W_5 = \mathcal{L}\{1, ..., x^4\}$
Q_2	$\mathbf{Q_2^3}$	$F_1^3 = DF_1^4 \sim y_2 y_1$	$W_4 = \mathcal{L}\{1, ..., x^3\}$
	$\mathbf{Q_2^2}$	$F_1^2 = (J_1^1)^2 \sim (y_2)^2$, $F_2^2 = J_1^1 J_3^1 - (J_2^1)^2 \sim 3y_2 y_0 - 2(y_1)^2$	$W_3 = \mathcal{L}\{1, x, x^2\}$
Q_3	$\mathbf{Q_3^1}$	$F_1^1 = DF_1^2 \sim y_3 y_2$, $F_2^1 = DF_2^2 \sim 3y_3 y_0 - y_2 y_1$	$W_2 = \mathcal{L}\{1, x\}$
	$\mathbf{Q_3^0}$	$F_1^0 = (J_1^0)^2 \sim (y_3)^2$, $F_2^0 = J_1^0 J_3^0 - (J_2^0)^2 \sim 2y_3 y_1 - (y_2)^2$	$W_1 = \mathcal{L}\{1\}$

Using Table 2.1, it is easy to obtain a similar description of the set of quadratic operators for the subspace (2.96) (equation (2.97)) with any $n < 5$. Denote expressions (2.119), related to the equation (2.97), by $J_i^{n,k}[y]$ for $i = 1, ..., n-k$. Let $J_i^{n+1,k}[y]$ for $i = 1, ..., n+1-k$ be analogous expressions for the equation $y_{n+1} = 0$. It directly follows from (2.119) that the first $n-k$ of those are obtained from $\{J_i^{n,k}[y]\}$ by the change $y \mapsto y_1$, i.e., $J_i^{n+1,k}[y] = J_i^{n,k}[y_1]$ for $i = 1, ..., n-k$, and that only the last one, $J_{n+1-k}^{n+1,k}[y] = k!x^{n+1}(x^{-1-n}y)^{(n-k)}$, contains y_0.

Example 2.39 (n = 4) This observation makes it possible, starting from the basis of operators for equation $y_{n+1} = 0$, to obtain a similar basis for equation (2.97). To this end, it is necessary to delete operators that contain y_0 and to perform the substitution $y_i \mapsto y_{i-1}$ in the operators that are left. Doing so, from Table 2.1, we obtain Table 2.2 that represents the set of operators, corresponding to the equation

$$y_4 = 0. \tag{2.135}$$

Remark 2.40 We make some comments concerning the set $\mathcal{F}$ of operators admitted by a given linear ODE $L[y] = 0$.

1). If $F_1, F_2 \in \mathcal{F}$, then $F_1 F_2 \in \mathcal{F}$, with the standard definition of the superposition, $(F_1 F_2)[y] = F_1[F_2[y]]$.

2). Let L have constant coefficients. Then:
(i) The operator D of the total derivative in x belongs to $\mathcal{F}$;
(ii) If $F \in \mathcal{F}$, both DF and FD belong to $\mathcal{F}$ and have the order that is larger by one than the order of F; and
(iii) If $F[y]$ is a homogeneous polynomial of y and its derivatives, then $DF[y]$ and $FD[y]$ are homogeneous polynomials of the same degree.

Using these remarks, it is possible to obtain a basis of the linear space $\hat{\mathcal{F}}^{\rm q}_{n-1}$ of translation-invariant operators admitted by the ODE (2.97), starting with the operators of the lowest orders that form the basis of $Q^{n-1}_{\frac{n-1}{2}}$ for odd n and $Q_{\frac{n}{2}}$ for even n. For example, in the case of the equation (2.134), the basis of $\mathbf{Q^4_2}$,

$$F^4_1[y] \sim (y_2)^2 \quad \text{and} \quad F^4_2[y] \sim y_2 y_0 + ...$$

(only the leading-order terms are indicated), and operators

$$F^4_1 D[y] \sim (y_3)^2, \;\; DF^4_1[y] \sim y_3 y_2, \;\; F^4_2 D[y] \sim y_3 y_1 + ..., \;\; DF^4_2[y] \sim y_3 y_0 + ... \, ,$$
$$F^4_1 D^2[y] \sim (y_4)^2, \;\; DF^4_1 D[y] \sim y_4 y_3 + ... \, , \;\; D^2 F^4_1[y] \sim y_4 y_2 + ... \, ,$$
$$DF^4_2 D[y] \sim y_4 y_1 + ... \, , \;\; D^2 F^4_2[y] \sim y_4 y_0 + ...$$

form a basis of $\hat{\mathcal{F}}^{\rm q}_4$.
In the case of the equation (2.135), the basis of Q_2,

$$F^2_1[y] \sim (y_2)^2, \;\; F^3_1[y] \sim y_2 y_1, \;\; \text{and} \;\; F^2_2[y] \sim y_2 y_0 + ... \, ,$$

and the additional operators

$$F^2_1 D[y] \sim (y_3)^2, \;\; DF^2_1[y] \sim y_3 y_2, \;\; F^2_2 D[y] \sim y_3 y_1 + ..., \;\; DF^2_2[y] \sim y_3 y_0 + ... \, ,$$

form a basis in $\hat{\mathcal{F}}^{\rm q}_3$. In general, such bases differ from those in Tables 2.1 and 2.2.

2.7 Extensions to $\frac{\partial}{\partial t}$-dependent operators

Here, we describe a generalization of the main results to operators, including the time derivative u_t,

$$F[u] = F(x, u, u_1, ..., u_k, u_t, u_{1,t}, ..., u_{k,t}), \;\; \text{where} \; u_i = \frac{\partial^i u}{\partial x^i} \tag{2.136}$$

and $u_0 = u$. Such applications are associated with the GT equation in Example 1.20 having the quadratic operator of the form

$$F_1[u] = u_x u_{tx} - u_t u_{xx} \, , \tag{2.137}$$

and with several other models.

For a given set of functions $f_1(x)$, ..., $f_n(x)$ that form a fundamental set of solutions of the ODE (2.2), we consider the set of their linear combinations,

$$W_n = \left\{ \textstyle\sum_{i=1}^n C_i(t) f_i(x) \right\}, \tag{2.138}$$

which is actually a module [373, Ch. III]. Without fear of confusion, we also keep the notation (2.1) for modules. The coefficients $\{C_i(t)\}$, as well as functions $\{f_i(x)\}$, are assumed to be sufficiently smooth. The module (2.138) is supposed to be a collection of solutions of the linear equation

$$L[u] \equiv u_n + \sum_{i=1}^{n} a_i(x)u_{n-i} = 0 \tag{2.139}$$

for $u = u(x,t)$. The condition for W_n to be invariant with respect to operator F, $F[W_n] \subseteq W_n$, has the form of the identity

$$L[F[u]] \equiv 0 \quad \text{for every solution of (2.139).} \tag{2.140}$$

We use the notation $\mathcal{F}_{n-1}(W_n)$ for the set of operators (2.136) with $k \leq n-1$ that leave the module (2.138) (or equation (2.139)) invariant. The formulation of the Main Theorem that describes this module and its proof are similar to those of Theorem 2.1.

2.7.1 Main Theorem

Theorem 2.41 **("Main Theorem")** *The set $\mathcal{F}_{n-1}(W_n)$ of operators* (2.136) *that leave module* (2.138) *invariant is given by*

$$F[u] = \sum_{i=1}^{n} A^i(I[u], I[u_t]) f_i(x), \tag{2.141}$$

where $I[u] = (I_1[u], \ldots, I_n[u])$ is the complete set of first integrals of the ODE (2.139) *and A^i for $i = 1, \ldots, n$ are arbitrary smooth functions.*

Proof. The invariance condition (2.140) is written down as

$$\begin{aligned} &D^n F + \ldots + a_{n-1}(x) DF + a_n(x) F = 0, \quad \text{with} \\ &D = \tfrac{\partial}{\partial x} + \sum_{i=0}^{n-2} u_{i+1} \tfrac{\partial}{\partial u_i} - \left(\sum_{i=1}^{n} a_i u_{n-i}\right) \tfrac{\partial}{\partial u_{n-1}} \\ &\quad + \sum_{i=0}^{n-2} u_{i+1,t} \tfrac{\partial}{\partial u_{i,t}} - \left(\sum_{i=1}^{n} a_i u_{n-i,t}\right) \tfrac{\partial}{\partial u_{n-1,t}}. \end{aligned} \tag{2.142}$$

Note that $I[u_t]$ is a complete set of first integrals of the equation $L[u_t] = 0$. Therefore, $DI_i[u] = 0$ and $DI_i[u_t] = 0$ for all $i = 1, \ldots, n$. Passing from $(x, u, u_1, \ldots, u_{n-1}, u_t, u_{1,t}, \ldots, u_{n-1,t})$ to the new variables

$$\tilde{x} = x, \quad I_i = I_i[u], \quad \text{and} \quad J_i = I_i[u_t] \ \text{for} \ i = 1, \ldots, n,$$

we transform D into $\tilde{D} = \frac{\partial}{\partial \tilde{x}}$ and equation (2.139) into

$$\tilde{D}^n \tilde{F} + \sum_{i=1}^{n} a_i(x) \tilde{D}^{n-i} \tilde{F} = 0.$$

Then the general solution is

$$\tilde{F}[u] = \sum_{i=1}^{n} A^i(I_1, \ldots, I_n, J_1, \ldots, J_n) f_i(\tilde{x}),$$

where A^i are arbitrary functions. Returning to the original variables yields (2.141) that completes the proof. □

Remark 2.42 If operators (2.136) also contain higher-order derivatives $\frac{\partial^i u}{\partial t^i}$ for $i = 2, \ldots, m$, instead of (2.141), we find

$$F[u] = \sum_{i=1}^{n} A^i\left(I[u], I[u_t], \ldots, I[\tfrac{\partial^m u}{\partial t^m}]\right) f_i(x). \tag{2.143}$$

Remark 2.43 If $t = (t_1, ..., t_p) \in I\!R^p$, we set in (2.143)

$$u_t = \{u_{t_i}\}, \quad I[u_t] = \{I[u_{t_i}]\};$$
$$u_{tt} = \{u_{t_i t_j}\}, \quad I[u_{tt}] = \{I[u_{t_i t_j}]\} \ (i, j = 1, ..., p); \ \text{ etc.}$$

2.7.2 *Theorem on the maximal dimension*

Theorem 2.8 is also extended to operators (2.136).

Theorem 2.44 ("Theorem on maximal dimension") *If the module* (2.138) *is invariant under the operator* (2.136) *or its generalizations in Remarks* 2.42, 2.43, *and the operator is nonlinear, then* (2.19) *holds.*

Proof. As in the proof of Theorem 2.8, it is convenient to use slightly different equations of the W_n,

$$u_n = a_1 u_{n-1} + ... + a_{n-1} u_1 + a_n u, \tag{2.144}$$

and the corresponding invariance condition,

$$D^n F \equiv a_1 D^{n-1} F + ... + a_{n-1} DF + a_n F \quad \text{on} \ \ (2.144), \tag{2.145}$$

where D is the operator of the total derivative in x. We replace

$$u_t \mapsto v, \quad \text{ so that } \ u_{j,t} \mapsto v_j.$$

Arguing by contradiction, let $n \ge 2k + 2$. Following the lines of the proof of Theorem 2.8, for functions $F(x, u, u_1, ..., u_k, v, v_1, ..., v_k)$, we find the formula that is similar to (2.22),

$$\begin{aligned} D^p F = {} & u_{k+p} F_{u_k} + v_{k+p} F_{v_k} + \big[\textstyle\sum_{i=1}^{[\frac{p}{2}]-1} C_p^i \, u_{k+p-i} u_{k+i} \\ & + \nu C_p^{[\frac{p}{2}]} u_{k+p-[\frac{p}{2}]} u_{k+[\frac{p}{2}]}\big] F_{u_k u_k} + \big[\textstyle\sum_{i=1}^{[\frac{p}{2}]-1} C_p^i \, v_{k+p-i} v_{k+i} \\ & + \nu C_p^{[\frac{p}{2}]} v_{k+p-[\frac{p}{2}]} v_{k+[\frac{p}{2}]}\big] F_{v_k v_k} + \big[\textstyle\sum_{i=1}^{p-1} C_p^i \, u_{k+p-i} v_{k+i}\big] F_{u_k v_k} + ..., \end{aligned} \tag{2.146}$$

where we keep the linear terms containing higher-order derivatives of the order $k + p$, and also the quadratic terms of the total order $2k + p$.

Applying the arguments used in the proof of Theorem 2.8 first to F as a function of the variables $u, u_1, ..., u_k$, and second of $v, v_1, ..., v_k$, immediately gives

$$F_{u_i u_j} = 0 \ \text{ and } \ F_{v_i v_j} = 0 \quad \text{for all } \ i, j = 0, 1, ..., k.$$

Therefore, omitting the linear terms, (2.146) takes the form

$$D^p F = \big[\textstyle\sum_{i=1}^{p-1} C_p^i \, u_{k+p-i} v_{k+i}\big] F_{u_k v_k} + ... \tag{2.147}$$

For $n \ge 2k + 2$, we select in the square brackets in (2.147) the derivatives in u of the order not less than $n - 1$, so

$$D^n F = \big(\textstyle\sum_{i=1}^{k+1} a_i u_{k+n-i} v_{k+i}\big) F_{u_k v_k} + ... \tag{2.148}$$

Using the linear equation (2.144) of W_n, we express all the derivatives u_{k+n-i} for $i = 1, 2, ..., k$ in terms of $u_{n-1}, ..., u$. Then (2.148) will contain a single quadratic term with $u_{n-1} v_{2k+1}$, i.e., $C_n^{k+1} u_{n-1} v_{2k+1} F_{u_k v_k}$, that has the maximal total order

$2k+p$. Such terms cannot appear in the derivatives D^pF for $p < n$. Hence, in order to satisfy the invariance condition, (2.145) we need $F_{u_k v_k} = 0$. Taking this into account, similar to (2.148), we find

$$D^n F = \left[\sum_{i=1}^{k+1} \beta_i u_{k+n-i} v_{k-1+i}\right] F_{u_k v_{k-1}} + \ldots,$$

where $\beta_i > 0$. All the indicated summands have the total order of the derivatives $2k+n-1$. Excluding the derivatives u_{k+n-i} for $i = 1, 2, \ldots, k$ gives the unique quadratic term with $u_{n-1} v_{2k}$, namely

$$\beta_{k+1} u_{n-1} v_{2k} F_{u_k v_{k-1}},$$

which verifies the maximal total order $2k+n-1$ (such summands cannot occur in D^pF for $p < n$). Equating this coefficient to zero yields $F_{u_k v_{k-1}} = 0$. Assuming next that

$$F_{u_k v_k} = F_{u_k v_{k-1}} = \ldots = F_{u_k v_{k-(j-1)}} = 0,$$

we obtain

$$D^n F = \left[\sum_{i=1}^{k+1} \gamma_i u_{k+n-i} v_{k-j+i}\right] F_{u_k v_{k-j}} + \ldots, \quad \text{where } \gamma_i > 0.$$

Hence, $F_{u_k v_{k-j}} = 0$ for any $j = 0, 1, \ldots, k$. Since the variables u and v are equivalent, by a similarity argument, this implies that $F_{u_{k-j} v_k} = 0$ for $j = 0, 1, \ldots, k$.

It follows that

$$F[y] = f_k(x) u_k + g_k(x) v_k + \widetilde{F}(x, u, \ldots, u_{k-1}, v, \ldots, v_{k-1}).$$

Repeating the same speculations for the function $\widetilde{F}$ yields

$$\widetilde{F} = f_{k-1}(x) u_{k-1} + g_k(x) v_{k-1} + \widetilde{\widetilde{F}}(x, u, \ldots, u_{k-2}, v, \ldots, v_{k-2}),$$

etc. Finally,

$$F = \sum_{i=1}^{k} \left[f_i(x) u_i + g_i(x) v_i\right] + h(x),$$

i.e., $F[u]$ is a linear operator. □

2.7.3 Examples

Example 2.45 (W_3 for the operator (2.137)) According to Theorem 2.44, the operator (2.137) from the GT equation can admit W_n of dimension n not exceeding five. The analysis of the invariance conditions implies that the maximal dimension is in fact three. Up to scalings in x, all 3D modules are described by the equation

$$u''' + a u' = 0, \quad \text{with } a = -1,\ 0, \text{ and } 1,$$

and have the form

$$W_3 = \mathcal{L}\{1, e^x, e^{-x}\}, \quad W_3 = \mathcal{L}\{1, x, x^2\}, \quad \text{and } W_3 = \mathcal{L}\{1, \cos x, \sin x\}.$$

Symmetric restrictions of operators. Given a $\frac{\partial}{\partial t}$-dependent operator (2.136), we introduce its *symmetric restriction*

$$\hat{F}[u] = F(x, u, u_1, \ldots, u_k, u, u_1, \ldots, u_k), \tag{2.149}$$

which is obtained from (2.136) by replacing $u_t \mapsto u$. For example, the GT-operator (2.137) has the symmetric restriction

$$\hat{F}_1[u] = (u_x)^2 - uu_{xx} \equiv -F_{\rm rem}[u],$$

where the remarkable operator (1.99) appears. The relation between operators and their symmetric restrictions is as follows:

Proposition 2.46 *If the module* (2.138) *is invariant under the operator* (2.136), *then it is invariant under its symmetric restriction* (2.149).

In general, nonlinear operators can have *linear* symmetric restrictions, or even the *null-restriction* $\hat{F} = 0$.

Example 2.47 (Quadratic operators with polynomial W_3, W_4, and W_5) Let us describe all the quadratic operators with constant coefficients of the second order in x and of the first order in t (as in (2.137)), preserving polynomial modules of dimensions 3, 4, and 5. Analyzing the invariance conditions yields the following bases of the linear space of such operators:

(i) $W_3 = \mathcal{L}\{1, x, x^2\}$:

$$\underline{F_1[u] = (u_{xx})^2}, \;\; F_2[u] = (u_{txx})^2, \;\; F_3[u] = u_{xx}u_{txx};$$
$$\underline{F_4[u] = u_x u_{xx}}, \;\; F_5[u] = u_{tx}u_{txx}, \;\; F_6[u] = u_{tx}u_{xx}, \;\; F_7[u] = u_x u_{txx};$$
$$\underline{F_8[u] = uu_{xx}}, \;\; F_9[u] = u_t u_{txx}, \;\; F_{10}[u] = u_t u_{xx}, \;\; F_{11}[u] = uu_{txx};$$
$$\underline{F_{12}[u] = (u_x)^2}, \;\; F_{13}[u] = (u_{tx})^2, \;\; F_{14}[u] = u_x u_{tx};$$
$$F_{15}[u] = uu_{tx} - u_t u_x.$$

The underlined operators are the symmetric restrictions of the next ones. The last operator, F_{15}, has the null-projection, i.e., $\hat{F}_{15} = 0$.

(ii) In a similar fashion, for $W_4 = \mathcal{L}\{1, x, x^2, x^3\}$:

$$\underline{F_1[u] = (u_{xx})^2}, \;\; F_2[u] = (u_{txx})^2, \;\; F_3[u] = u_{xx}u_{txx}, \;\; F_4[u] = u_{tx}u_{txx};$$
$$\underline{F_5[u] = u_x u_{xx}}, \;\; F_6[u] = u_x u_{txx}, \;\; F_7[u] = u_{tx}u_{xx};$$
$$\underline{F_8[u] = uu_{xx} - \tfrac{2}{3}(u_x)^2}, \;\; F_9[u] = u_t u_{txx} - \tfrac{2}{3}(u_{tx})^2,$$
$$F_{10}[u] = u_t u_{xx} - \tfrac{2}{3}\, u_x u_{tx}, \;\; F_{11}[u] = uu_{txx} - \tfrac{2}{3}\, u_x u_{tx}.$$

(iii) For the 5D subspace $W_5 = \mathcal{L}\{1, x, x^2, x^3, x^4\}$, we find the following operators:

$$\underline{F_1[u] = (u_{xx})^2}, \;\; F_2[u] = (u_{txx})^2, \;\; F_3[u] = u_{xx}u_{txx};$$
$$\underline{F_4[u] = uu_{xx} - \tfrac{3}{4}(u_x)^2}, \;\; F_5[u] = u_t u_{txx} - \tfrac{3}{4}(u_{tx})^2,$$
$$F_6[u] = uu_{txx} + u_t u_{xx} - \tfrac{3}{2}\, u_x u_{tx}; \;\; F_7[u] = u_x u_{txx} - u_{tx}u_{xx}.$$

Note that $\hat{F}_7 = 0$. As an illustration, let us present the computations concerning the last three operators for functions

$$u = C_1 + C_2 x + C_3 x^2 + C_4 x^3 + C_5 x^4 \in W_5.$$

The following holds:

$$F_5[u] = 12(C_4C_5' - C_4'C_5)x^4 + 16(C_3C_5' - C_3'C_5)x^3 + 6(2C_2C_5'$$
$$- 2C_2'C_5 + C_3C_4' - C_3'C_4)x^2 + 6(C_2C_4' - C_2'C_4)x + 2(C_2C_3' - C_2'C_3);$$

$$F_6[u] = \tfrac{1}{4}\big[8C_3'C_5' - 3(C_4')^2\big]x^4 + \big(6C_2'C_5' - C_3'C_4'\big)x^3 + \tfrac{1}{2}\big[24C_1'C_5'$$
$$+ 3C_2'C_4' - 2(C_3')^2\big]x^2 + \big(6C_1'C_4' - C_2'C_3'\big)x + 4C_1'C_3' - \tfrac{3}{4}(C_2')^2;$$

$$F_7[u] = \tfrac{1}{2}\big(4C_3C_5' + 4C_3'C_5 - 3C_4C_4'\big)x^4 + \big(6C_2C_5' + 6C_2'C_5 - C_3C_4'$$
$$- C_3'C_4\big)x^3 + \tfrac{1}{2}\big(24C_1C_5' + 24C_1'C_5 + 3C_2C_4' + 3C_2'C_4 - 4C_3C_3'\big)x^2$$
$$+ 2\big(6C_1C_4' + 6C_1'C_4 - C_2C_3' - C_2'C_3\big)x + 4C_1C_3' + 4C_1'C_3 - 3C_2C_2'.$$

By these expressions, the PDEs $F_k[u] = g \in W_5$ reduce to ODE systems.

2.7.4 Operators $F : W_n \to \tilde{W}_m$

It is convenient to use the class of $\frac{\partial}{\partial t}$-dependent operators (2.136) to describe the next generalization, where we move away from the notion of invariant modules, but keep the main mathematical tools and results.

Consider two different modules, W_n and $\tilde{W}_m$, given by two different ODEs, written in the form of

$$\begin{aligned} W_n &= \big\{\textstyle\sum_{i=1}^n \alpha_i(t) f_i(x)\big\}, \quad u_n = \textstyle\sum_{j=1}^n a_j(x) u_{n-j}; \\ \tilde{W}_m &= \big\{u = \textstyle\sum_{i=1}^m \beta_i(t) \tilde{f}_i(x)\big\}, \quad u_m = \textstyle\sum_{j=1}^m b_j(x) u_{m-j}, \end{aligned} \tag{2.150}$$

where $u_j = \frac{\partial^j u}{\partial x^j}$. We are looking for operators $F : W_n \to \tilde{W}_m$ and denote by $\mathcal{F}_{n-1}(W_n, \tilde{W}_m)$ the whole set of such operators. In the case where $W_n = \tilde{W}_m$, we return to the concept of invariant modules. If $F \in \mathcal{F}_{n-1}(W_n, \tilde{W}_m)$, then

$$\text{for any } \; u = \textstyle\sum_{i=1}^n C_i(t) f_i(x), \;\; F[u] = \textstyle\sum_{i=1}^m \Psi_i(C_1, ..., C_n, C_1', ..., C_n')\tilde{f}_i(x).$$

In particular, the PDE

$$F[u] = 0 \quad \text{on } \; W_n$$

is equivalent to the system of ODEs

$$\begin{cases} \Psi_1(C_1, ..., C_n, C_1', ..., C_n') = 0, \\ \quad \cdots \quad \cdots \quad \cdots \\ \Psi_m(C_1, ..., C_n, C_1', ..., C_n') = 0. \end{cases}$$

The necessary and sufficient condition for operator F to map any solution of the first ODE, $L[u] = 0$, in (2.150) into a solution of the second one, $\tilde{L}[u] = 0$, is

$$\tilde{L}[F[u]]\big|_{L[u]=0} \equiv 0,$$

which looks similar to (2.4). By the same analysis, we arrive at the following result, where the same notation, as in Theorem 2.41, is kept.

Theorem 2.48 (**"Main Theorem"**) *The set $\mathcal{F}_{n-1}(W_n, \tilde{W}_m)$ consists of operators*

$$F[u] = \textstyle\sum_{i=1}^m A^i(I[u], I[u_t])\tilde{f}_i(x),$$

where $I[u] = (I_1[u], \ldots, I_n[u])$ *is the complete set of first integrals of the first ODE in* (2.150), A^i *for* $i = 1, \ldots, m$ *are any smooth functions, and* $\{\tilde{f}_1(x), \ldots, \tilde{f}_m(x)\}$ *is a fundamental set of solutions of the second ODE in* (2.150).

Remarks 2.42 and 2.43 are also applied to mappings $W_n \to \tilde{W}_m$.

Theorem 2.49 ("Theorem on maximal dimension") *Let* $F : W_n \to \tilde{W}_m$, *where* $m \le n$, *be a nonlinear operator* (2.136) *of the differential order* k. *Then* $n \le 2k + 1$.

Proof. Let $m = n$ and W_n and $\tilde{W}_n$ be prescribed by the linear equations

$$u_n = a_1 u_{n-1} + \ldots + a_{n-1} u_1 + a_n u, \tag{2.151}$$

$$u_n = b_1 u_{n-1} + \ldots + b_{n-1} u_1 + b_n u,$$

respectively. Then, for F, the following holds:

$$D^n F \equiv b_1 D^{n-1} F + \ldots + b_{n-1} DF + b_n F \quad \text{on} \ \ (2.151),$$

where D is the operator of the full derivative in x. The result of the theorem follows from the arguments that have been used in the proof of Theorem 2.44.

In the case of $m < n$, $\tilde{W}_m$ can be extended to $\tilde{W}_n$ by adding $n - m$ arbitrary functions to the basis that form, together with $\{\tilde{f}_i(x),\ i = 1, \ldots, m\}$, a linearly independent set. Then $F : W_n \to \tilde{W}_n$ and the previous argument works. □

Corollary 2.50 *A nonlinear* k*th-order differential operator* $F : W_n \to \tilde{W}_m$, *where* $n \ge 2k + 2$, *can exist only if* $m > n$.

These results are valid in the cases indicated in Remarks 2.42 and 2.43, as well as for the operators that do not contain derivatives in t.

In the following example, we consider the operator (1.94),

$$F[u] = u_x u_{tx} - \beta u_t u_{xx} + \gamma u u_{txx} + \delta u u_t.$$

For some particular choices of the coefficients, we find 3D modules W_3 and $\tilde{W}_3$ such that $F : W_3 \mapsto \tilde{W}_3$.

Example 2.51 ($\tilde{\mathbf{W}}_3 \ne \mathbf{W}_3$) Let $\gamma = \beta \ne 0$ and $\delta = 0$, i.e.,

$$F[u] = u_x u_{tx} + \beta(u u_{txx} - u_t u_{xx}).$$

There exist two cases:

1. The modules are given by

$$W_3 = \mathcal{L}\{1, x, \ln x\} \quad \left(u''' = -\tfrac{2}{x}\, u''\right),$$

$$\tilde{W}_3 = F[W_3] = \mathcal{L}\left\{1, \tfrac{1}{x}, \tfrac{1}{x^2}\right\} \quad \left(u''' = -\tfrac{6}{x}\, u'' - \tfrac{6}{x^2}\, u'\right),$$

and the action of the operator is

$$F[C_1 + C_2 x + C_3 \ln x] = C_2 C_2'$$

$$+ \left[(C_2 C_3)' + \beta\left(C_3 C_2' - C_2 C_3'\right)\right] \tfrac{1}{x} + \left[C_3 C_3' + \beta\left(C_3 C_1' - C_1 C_3'\right)\right] \tfrac{1}{x^2}.$$

As an application, taking the PDE

$$F[u] \equiv u_x u_{tx} + \beta(u u_{txx} - u_t u_{xx}) = \tfrac{1}{2},$$

we find the exact solution

$$u(x,t) = C_1(t) + C_2(t)x + C_3(t)\ln x,$$
$$\begin{cases} C_2C_2' = \frac{1}{2}, \\ (C_2C_3)' + \beta\big(C_3C_2' - C_2C_3'\big) = 0, \\ C_3C_3' + \beta\big(C_3C_1' - C_1C_3'\big) = 0. \end{cases}$$

The first equation of this DS yields $C_2(t) = \sqrt{t}$, and the rest of equations are also explicitly integrated, so, for $\beta \neq 0,\ 1$, the solutions are

$$u(x,t) = \tfrac{1}{\beta}(k_2 - \mu \ln t)k_1 t^{\mu} + \sqrt{t}\, x + k_1 t^{\mu} \ln x,$$

where $\mu = \frac{\beta+1}{2(\beta-1)}$, and $k_{1,2}$ are arbitrary constants.

2. The module and its image are given by

$$W_3 = \mathcal{L}\{1, \ln x, \ln^2 x\} \quad \big(u''' = -\tfrac{3}{x}\, u'' - \tfrac{1}{x^2}\, u'\big),$$
$$\tilde{W}_3 = F[W_3] = \mathcal{L}\big\{\tfrac{1}{x^2}, \tfrac{\ln x}{x^2}, \tfrac{\ln^2 x}{x^2}\big\} \quad \big(u''' = -\tfrac{9}{x}\, u'' - \tfrac{19}{x^2}\, u' - \tfrac{8}{x^3}\, u\big).$$

The action is

$$F[C_1 + C_2 \ln x + C_3 \ln^2 x] = \big[C_2C_2' + \beta\big(2C_1C_3' - 2C_3C_1' - C_1C_2' + C_2C_1'\big)\big]\,\tfrac{1}{x^2}$$
$$+\,2\big[(C_2C_3)' + \beta\big(C_2C_3' - C_3C_2' - C_1C_3' + C_3C_1'\big)\big]\,\tfrac{\ln x}{x^2}$$
$$+\big[4C_3C_3' + \beta\big(C_3C_2' - C_2C_3'\big)\big]\,\tfrac{\ln^2 x}{x^2}.$$

Example 2.52 ($\tilde{\mathbf{W}}_3 = \mathbf{W}_3$) Let $\beta = -\gamma = \frac{2}{3}$ and $\delta = -\frac{4}{3}$, i.e.,

$$F[u] = u_x u_{tx} - \tfrac{2}{3}\, u_t u_{xx} - \tfrac{2}{3}\, u u_{txx} - \tfrac{4}{3}\, u u_t. \tag{2.152}$$

There are two cases:

1. The module and action of the operator are

$$W_3 = \mathcal{L}\{1, \cos 2x, \sin 2x\} \quad (u''' = -4u'),$$
$$F[C_1 + C_2 \cos 2x + C_3 \sin 2x] = 4(C_2C_2' + C_3C_3')$$
$$-\,\tfrac{4}{3}\, C_1C_1' + \tfrac{4}{3}(C_1C_2)' \cos 2x + \tfrac{4}{3}(C_1C_2)' \sin 2x.$$

2. The module and the action are

$$W_3 = \mathcal{L}\{1, \cos x, \cos 2x\} \quad \big(u''' = 3\cot x\, u'' - (1 + 3\cot^2 x)u'\big),$$
$$F[C_1 + C_2 \cos x + C_3 \cos 2x] = \big(2C_3^2 + \tfrac{1}{4}C_2^2 - \tfrac{2}{3}C_1^2\big)'$$
$$+\,2\big[C_2\big(C_3 - \tfrac{1}{3}C_1\big)\big]' \cos x + \big(\tfrac{4}{3}C_1C_3 - \tfrac{1}{4}C_2^2\big)' \cos 2x.$$

Both 3D modules are sub-modules of the 5D one; see Proposition 1.34.

2.8 SUMMARY: Basic types of equations and solutions

For the reader's convenience, we present here a summary of equations, solutions, and main schemes of our analysis to be used later on for constructing exact solutions in the subsequent chapters. Almost all models and PDEs to be studied therein fall into the scope of the following classification:

Here, we use the notation $\mathbf{C} = (C_1, ..., C_n)^T \in I\!R^n$ for constant vectors, and $\mathbf{C}(t) = (C_1(t), ..., C_n(t))^T$ for sufficiently smooth vector-functions. For all the equations, we are looking for solutions in the standard form

$$u(x,t) = \sum_{i=1}^n C_i(t) f_i(x), \tag{2.153}$$

where the functional set $\{f_i(x)\}$ is linearly independent.

2.8.1 Invariant and partially invariant subspaces

• **Invariant subspaces.** Consider the equation

$$l[u] = F[u]. \tag{2.154}$$

Here, l is a linear operator that depends on t and includes only derivatives in t, while F is a nonlinear operator that depends on x and includes only derivatives in x. A linear subspace is

$$W_n = \left\{ \sum_{i=1}^n C_i f_i(x) : \ \mathbf{C} = (C_1, ..., C_n)^T \in I\!R^n \right\}. \tag{2.155}$$

If W_n is invariant with respect to F, i.e., $F[W_n] \subseteq W_n$, or, which is the same,

$$F\left[\sum_{i=1}^n C_i f_i(x)\right] = \sum_{i=1}^n \Psi_i[\mathbf{C}] f_i(x) \quad \text{for any } \mathbf{C},$$

where $\Psi_i[\mathbf{C}] = \Psi_i(C_1, ..., C_n)$ are some algebraic functions, then (2.154) possesses solutions (2.153) with the coefficients satisfying the dynamical system

$$l[C_i(t)] = \Psi_i[\mathbf{C}(t)], \quad i = 1, ..., n. \tag{2.156}$$

This is the most frequently occurring instance throughout this text.

• **Partially invariant subspaces.** For the subspace (2.155), there exists an (invariant) set $M \subset W_n$ that is defined by a system of algebraic equations

$$\chi_k[\mathbf{C}] = 0 \ \text{ for } \ k = 1, ..., r,$$

such that $F[M] \subset W_n$. Solutions (2.153) are determined by

$$\begin{cases} l[C_i(t)] = \Psi_i[\mathbf{C}(t)], \ i = 1, ..., n, \\ \chi_k[\mathbf{C}(t)] = 0, \ k = 1, ..., r. \end{cases}$$

Example 2.53 The operator

$$F[u] = u_{xx} + (u_x)^2 + u^2 - 1$$

admits the invariant subspace $W_3 = \mathcal{L}\{1, \cos x, \sin x\}$. It is checked that, for any constants $\{C_i\}$,

$$F[C_1 + C_2 \cos x + C_3 \sin x] = C_1^2 + C_2^2 + C_3^2 - 1$$
$$+ (2C_1 - 1) C_2 \cos x + (2C_1 - 1) C_3 \sin x \in W_3.$$

The subspace $W_2 = \mathcal{L}\{\cos x, \sin x\}$ is partially invariant with respect to F:

$$F[C_1 \cos x + C_2 \sin x] = C_1^2 + C_2^2 - 1 - C_1 \cos x - C_2 \sin x \in W_2$$

iff $C_1^2 + C_2^2 = 1$.

Example 2.54 For the operator

$$F[u] = u_{xx} + \tfrac{1}{2}\left(1 - \tfrac{1}{u}\right)(u_x)^2 + 2\left(1 + \tfrac{1}{u}\right),$$

there exists the partially invariant subspace $W_2 = \mathcal{L}\{1, x^2\}$:

$$F[C_1 + C_2x^2] = 2C_1 + 2C_2^2 + \tfrac{2(C_1C_2+1)}{C_1+C_2x^2} \in W_2 \quad \text{iff} \quad C_1C_2 = -1.$$

2.8.2 *Invariant and partially invariant modules*

• **Invariant modules.** The equation takes the form

$$F_*[u] = 0, \tag{2.157}$$

where F_* is a nonlinear operator that depends on x and t and includes derivatives in these variables. The set of linear combinations

$$W_n = \left\{\textstyle\sum_{i=1}^n C_i(t) f_i(x)\right\} \tag{2.158}$$

is a module, [373, Ch. III]. If W_n is invariant with respect to F_*, $F_*[W_n] \subseteq W_n$, or

$$F_*\left[\textstyle\sum_{i=1}^n C_i(t) f_i(x)\right] = \textstyle\sum_{i=1}^n \Psi_i[\mathbf{C}(t)] f_i(x) \quad \text{for any } \mathbf{C}(t),$$

where $\Psi_i[\mathbf{C}(t)] = \Psi_i(t, \mathbf{C}(t), \mathbf{C}'(t), ...)$ are some differential operators, then, for solutions (2.153) of equation (2.157), there occurs the system

$$\Psi_i[\mathbf{C}(t)] = 0, \quad i = 1, ..., n. \tag{2.159}$$

If the module W_n is invariant with respect to operator F_*, the PDE

$$\tfrac{\partial^q u}{\partial t^q} = F_*[u],$$

where q is greater than the maximal order of the time derivatives that are available in F_*, possesses solutions (2.153) with coefficients that solve the DS

$$C_i^{(q)}(t) = \Psi_i[\mathbf{C}(t)], \quad i = 1, ..., n.$$

• **Partially invariant modules.** If there exists a set $M \subset W_n$ that is defined by

$$\chi_k[\mathbf{C}(t)] = 0 \quad \text{for} \quad k = 1, ..., r,$$

such that $F[M] \subset W_n$, then, instead of (2.159), solutions (2.153) are obtained from the system

$$\begin{cases} \Psi_i[\mathbf{C}(t)] = 0, \ i = 1, ..., n, \\ \chi_k[\mathbf{C}(t)] = 0, \ k = 1, ..., r, \end{cases}$$

with $\Psi_i[\mathbf{C}] = \Psi_i(t, \mathbf{C}(t), \mathbf{C}'(t), ...)$ and $\chi_k[\mathbf{C}] = \chi_k(t, \mathbf{C}(t), \mathbf{C}'(t), ...)$ being differential operators.

Example 2.55 For the operator

$$F[u] = u_{xt}u_x - \tfrac{3}{2}\, u_{xx}u_t,$$

there exists the invariant module $W_4 = \mathcal{L}\{1, x, x^2, x^3\}$:

$$\begin{aligned} F[C_1 + C_2x + C_3x^2 + C_4x^3] &= C_2C_2' - 3C_3C_1' + \left(2C_2C_3' - C_3C_2' - 9C_4C_1'\right)x \\ &+ \left(3C_2C_4' - 5C_3C_3' - 6C_4C_2'\right)x^2 + \left(3C_3C_4' - 3C_4C_3'\right)x^3 \in W_4. \end{aligned}$$

Example 2.56 For the operator

$$F[u] = u_{xt}u_x - u_{xx}u_t, \tag{2.160}$$

there exists the partially invariant module $W_3 = \mathcal{L}\{1, e^x, xe^x\}$:

$$F[C_1 + C_2 e^x + C_3 x e^x] = -(C_2 + 2C_3)C_1' e^x \\ - C_3 C_1' x e^x + (C_2 + C_3)C_3' e^{2x} \in W_3$$

iff $(C_2 + C_3)C_3' = 0$.

2.8.3 Further generalizations

• As above, consider the equation (2.157) and the module (2.158). Let there exist another module

$$\tilde{W}_m = \left\{\sum_{i=1}^m \tilde{C}_i(t)\tilde{f}_i(x)\right\},$$

with basis $\{\tilde{f}_i(x)\}$ and expansion coefficients $\{\tilde{C}_i(t)\}$. If $F[W_n] \subseteq \tilde{W}_m$, i.e., for any $\mathbf{C}(t)$, there exist differential operators $\Psi_i[\mathbf{C}(t)] \equiv \Psi_i(t, \mathbf{C}(t), \mathbf{C}'(t), ...)$, such that

$$F_*\left[\sum_{i=1}^n C_i(t) f_i(x)\right] = \sum_{i=1}^m \Psi_i[\mathbf{C}(t)]\tilde{f}_i(x),$$

then, for solutions (2.153), there occurs the system

$$\Psi_i[\mathbf{C}(t)] = 0, \quad i = 1, ..., m.$$

• If there exists a set $M \subset W_n$ that is defined by

$$\chi_k[\mathbf{C}(t)] = 0 \quad \text{for } k = 1, ..., r,$$

such that $F[M] \subset \tilde{W}_n$, then, for the coefficients of solutions (2.153), there occurs the system

$$\begin{cases} \Psi_i[\mathbf{C}(t)] = 0, & i = 1, ..., m, \\ \chi_k[\mathbf{C}(t)] = 0, & k = 1, ..., r. \end{cases}$$

• Another straightforward generalization is achieved when the bases of modules W_n and $\tilde{W}_m$ are functions of both independent variables x and t, i.e.,

$$W_n = \left\{\sum_{i=1}^n C_i(t) f_i(x, t)\right\} \quad \text{and} \quad \tilde{W}_m = \left\{\sum_{i=1}^m \tilde{C}_i(t)\tilde{f}_i(x, t)\right\}.$$

The variables x and t can also be vectors, $x = (x_1, ..., x_\mu)$ and $t = (t_1, ..., t_\nu)$.

Example 2.57 For the operator (2.160), there exist the modules

$$W_3 = \mathcal{L}\{e^x, e^{-x}, e^{3x}\} \quad \text{and} \quad \tilde{W}_3 = \mathcal{L}\{1, e^{2x}, e^{4x}\}, \qquad \text{such that}$$

$$F[C_1 e^x + C_2 e^{-x} + C_3 e^{3x}] \\ = -2(C_1C_2)' e^x - \left[3(C_2C_3)' + C_2C_3' + 9C_3C_2'\right]e^{2x} \\ + \left[3(C_1C_3)' - C_1C_3' - 9C_3C_1'\right]e^{4x} \in \tilde{W}_3.$$

Example 2.58 For the operator

$$F[u] = u_{xxt} + uu_{xxx} - u_x u_{xx}$$

(see Example 1.37), we have

$$W_3 = \mathcal{L}\{x, \cos(\gamma(t)x), \sin(\gamma(t)x)\} \quad \text{and}$$

$$\tilde{W}_4 = \mathcal{L}\{\cos(\gamma(t)x), \sin(\gamma(t)x), x\cos(\gamma(t)x), x\sin(\gamma(t)x)\},$$

where $\gamma(t)$ is a smooth function. In this case,

$$F[C_1 x + C_2 \cos(\gamma x) + C_3 \sin(\gamma x)]$$
$$= \left[\gamma^2 C_1 C_2 - (C_2 \gamma^2)'\right] \cos(\gamma x) + \left[\gamma^2 C_1 C_3 - (C_3 \gamma^2)'\right] \sin(\gamma x)$$
$$+ \gamma^2(\gamma' + \gamma C_1)\, x [C_2 \sin(\gamma x) - C_3 \cos(\gamma x)] \in \tilde{W}_4.$$

Notice that, at the same time, W_3 can be considered as a partially invariant module:

$$F[W_3] \subseteq W_3 \quad \text{iff} \quad C_1 = -\tfrac{\gamma'}{\gamma}.$$

Remarks and comments on the literature

§ 2.1. We follow [541, 544]. A survey on some results connected with applications of invariant subspaces to evolution PDEs is given in [546]. Extended surveys on this and other methods of reduction of differential equations can be found in [127, 312, 445].

§ 2.2. A general scheme for main proofs is explained in [543]; see [544, 545] for the statements of the results and discussion.

§ 2.3–2.4. Concerning these results, see [545].

§ 2.6. We follow [547]. A detailed analysis of the structure of the set of polynomial operators, preserving polynomial subspaces, based on the notion of deficiency, is fulfilled in [257]. The description of translation-invariant quadratic operators presented in Section 2.6.4 is similar to that given in [257, Sect. 6].

Open problems

- Given an arbitrary subspace (2.1), describe the set $\widehat{\mathcal{F}}_{n-1}$ of translation-invariant operators preserving W_n. [In Section 2.6.3, this is done for polynomial subspaces. The whole set of operators (2.3) preserving W_n is given by Theorem 2.1.]
- Perform a general description of the set of translation-invariant homogeneous polynomial operators of a given degree $p \geq 3$ that preserve the polynomial subspace (2.96). [For quadratic operators, see Theorem 2.35 and [257, Sect. 6].]
- The same problem for quadratic and higher-degree operators preserving subspaces of trigonometric and exponential types. [For quadratic operators with exponential subspaces, see Example 1.49; other particular cases are considered in Propositions 1.29–1.32, 1.41, 1.44, and 1.45.]
- Prove that a nonlinear differential operator (2.3) preserving a subspace of the maximal dimension is necessarily quadratic. [In Sections 2.3 and 2.4 this was proved for operators of the first and second order.]
- Perform a full description of second-order operators (2.3) preserving invariant subspaces of the maximal dimension. [For translation-invariant operators, see Section 2.4.]
- Perform the previous description for operators of orders greater than two.
- See Conjecture 2.1 in Section 2.6.1.

CHAPTER 3

Parabolic Equations in One Dimension: Thin Film, Kuramoto-Sivashinsky, and Magma Models

In the following three chapters, we use invariant subspaces for various higher-order operators and PDEs in one space dimension. More attention is now paid to models of physical importance and evolution properties of the exact solutions. In the current chapter, we will deal mainly with quasilinear parabolic PDEs, including the well-known thin film and Kuramoto–Sivashinsky models. We also consider the lesser known magma equation, generating an interesting family of formally pseudo-parabolic models. These classes of PDEs are associated with different areas of mechanics and physics. Using the exact solutions makes it possible to detect new and unusual properties of nonlinear models. Mathematical existence, uniqueness, and regularity theory of many higher-order PDEs to be studied is still a ways from being complete. Their solutions on finite-dimensional invariant subspaces yield an exceptional opportunity for describing key evolution aspects of such problems, including singularity formation phenomena of blow-up and extinction. Whenever possible and reasonable, we discuss some mathematical aspects of these nonlinear problems slightly more rigorously, and state, when it seems necessary and interesting, OPEN PROBLEMS.

For most 1D higher-order ordinary differential operators, the results on invariant subspaces can be extracted from the Main Theorem in Section 2.1 and from other results of Chapter 2. Nevertheless, dealing with simple polynomial or trigonometric subspaces for typical fourth or higher-order operators, sometimes we include direct simple calculations which help to reveal some additional features of the associated PDEs of parabolic type. Such an easy geometric analysis often gives straightforward extensions of the models, admitting invariant subspaces, and thus avoiding the use of technical manipulations via the Main Theorem.

3.1 Thin film models and solutions on polynomial subspaces

We first deal with a class of quasilinear fourth and higher-order operators, which, since the 1980s, began to play a determining role in the theory of nonlinear degenerate parabolic PDEs. It is not an exaggeration to say that nowadays higher-order nonlinear degenerate *thin film equations* are of the same fundamental importance as the second-order *porous medium equations* used in the 1950s-80s.

3.1.1 Typical quasilinear models from thin film theory

Fourth-order thin film equations. Thin film flows correspond to the case where the size of the flow domain in one direction is essentially smaller in comparison with that in the other directions. This enables to derive simplified models from the Navier–

Stokes equations. Namely, coupling to boundary conditions on the fluid-substrate interface and fluid surface yields a single scalar equation for the fluid height. The quasilinear parabolic fourth-order *thin film equation* (TFE)

$$u_t = -\big(u^n u_{xxx}\big)_x, \quad \text{with a fixed exponent } n \neq 0, \tag{3.1}$$

is key in thin film theory. For $n = 0$, this is the well-known linear *bi-harmonic equation*

$$u_t = -u_{xxxx}. \tag{3.2}$$

The connection and certain similarities between (3.1) and (3.2) will be important in what follows.

The quasilinear TFE (3.1) with $n = 3$, as well as

$$(h^3)_y = -3\big(h^3 h_{xxx}\big)_x, \tag{3.3}$$

arise from *Reynolds' equation* for *Stokes' flow*

$$u_t = (u^n p_x)_x, \quad \text{or} \quad u_t = \nabla_x \cdot (u^n \nabla_x p) \quad (n = 3). \tag{3.4}$$

This describes slow sliding or spreading of thin Newtonian liquid drops (film) over a horizontal surface in the presence of high surface tension. Here, $u = u(x,t) \geq 0$ denotes the height of the droplet, and $p = p(x,t)$ stands for the pressure of the fluid. For $n = 3$, this is precisely Reynolds' equation from lubrication theory. It also applies to the case $n = 2$, where the fluid is able to "slip" over the solid.

The case $n = 1$ corresponds to flows in a porous medium or in a Hele–Shaw cell which is formed by two immiscible fluids that are separated by a thin interface with thickness $2u$. The dependence of p upon u (and its spatial derivatives) is of importance in establishing the force driving the spreading of the droplet. The fourth-order TFE (3.1) corresponds to capillary driven flow of a thin droplet of thickness $u(x,t)$, so that, under suitable rescaling, the pressure is

$$p = -u_{xx}. \tag{3.5}$$

Equation (3.3) corresponds to steady capillary-gravity driven flows down a substrate at an angle α to the horizontal in a special (x, y)-geometry.*

Illustrating a typical derivation of thin film models, consider a Hele–Shaw flow between two parallel plates separated by a fixed distance h_0. This is governed by two PDEs,

$$\begin{cases} \text{conservation of mass:} \quad u_t + (uv)_x = 0, \quad \text{and} \\ \text{Darcy's law:} \quad v = -\frac{h_0^2}{12\mu}\, p_x, \end{cases} \tag{3.6}$$

where v is the average velocity of the fluid in the film, μ is the fluid viscosity, and p is the pressure. If the flow is governed by the surface tension, then

$$p_x = -\gamma\kappa \equiv -\gamma\, u_{xx}$$

(cf. (3.5)), where γ is the surface tension and κ is the curvature of the surface. Substituting p_x into the second equation in (3.6) and the resulting v into the first equation

* As usual, we put extra references on this subject in the Remarks.

yields the PDE

$$u_t + \tfrac{\gamma h_0^2}{12\mu}(u u_{xxx})_x = 0.$$

This is (3.1) with $n = 1$, where the constant is scaled out. The case $n = 2$ represents Navier-slip-dominated Stokes flows of thin film. Then the analog of (3.3) has the form

$$(h^2)_y = -2\big(h^2 h_{xxx}\big)_x.$$

Taking into account other monomial differential terms occurring after differentiating in the fourth-order operators in the above models, let us introduce generalized TFEs comprising a family of homogeneous operators

$$u_t = F[u] \equiv -\big[u^n u_{xxx} + \beta u^{n-1} u_x u_{xx} + \gamma\, u^{n-2}(u_x)^3\big]_x, \tag{3.7}$$

where β and γ are constants. These PDEs share the scaling and homogenuity properties of the standard TFE, which is important for our invariant subspace analysis.

More models containing other terms and operators appear in *lubrication theory*, where the long-wave unstable *lubrication equations* take the form

$$u_t = -\big(u^n u_{xxx}\big)_x - \big(u^m u_x\big)_x, \quad \text{with } n > 0,\ m > 0. \tag{3.8}$$

In general, the lubrication approximation that describes the motion of long-wave unstable thin films deals with equations of the same type (3.4) for which the homogeneous nonlinearity u^n (with typically $n = 3$) is replaced by a perturbed function $u^n + \beta u^k$ with $\beta > 0$ and $k \in (0, 3)$. This corresponds to classes of *slip* boundary conditions, while the above case, $\beta = 0$, reflects the *no-slip* conditions. Furthermore, instead of (3.5), the pressure p in the thin film is given by

$$p = -u_{xx} + G(u),$$

where the term $G(u)$ describes forces exerted on the film, e.g., hydrostatic body forces or intermolecular forces due to *van der Waals interaction*. Such models lead to a more general class of the fourth-order parabolic PDEs, such as

$$u_t = -(f(u)u_{xxx})_x - (g(u)u_x)_x \tag{3.9}$$

with given functions f and g, or to extended models containing other terms. As was mentioned, a typical nonlinearity in the main higher-order term is

$$f(u) = \alpha u^n + \beta u^k$$

with an extra exponent $k > 0$ and positive constants α and β. Such PDEs also occur as stable and unstable *Cahn–Hilliard* models in phase transition, fluid interfaces, and rupture of thin films. They can describe thin jets in Hele–Shaw cells and fluid droplets hanging from a ceiling. In the semilinear case, $n = 0$, these equations are of the modified Kuramoto–Sivashinsky type that describe flame propagation and several other phenomena, such as solidification of a hyper-cooled melt

More lower-order terms enter the *Benney equation* that describes the nonlinear dynamics of the interface of 2D liquid films flowing on a fixed inclined plane,

$$h_t + \tfrac{2R}{3}(h^3)_x + \varepsilon\big[\big(\tfrac{8R^2}{15}\, h^6 - \tfrac{2R}{3}\cot\theta\, h^3\big)h_x + \Sigma\, h^3 h_{xxx}\big]_x = 0, \tag{3.10}$$

where R is the unit-order *Reynolds number* of the flow driven by gravity, σ is the rescaled *Weber number* (related to surface tension σ), θ is the angle of plane inclination to the horizontal, and $\varepsilon = \frac{d}{\lambda} \ll 1$, with d being the average thickness of the film and λ the wavelength of the characteristic interfacial disturbances.

Another class of TFEs in 1D includes a two-parameter family of models, such as

$$u_t = -\big(u^n (u^s)_{xxx}\big)_x + \text{(lower-order terms)}, \tag{3.11}$$

where n and s are some fixed exponents. We will deal with some of these PDEs which can be reduced to quadratic or cubic representations.

Concerning multi-dimensional TFEs with non-power nonlinearities, as a typical example, consider the PDE

$$h_t + \nabla \cdot \big[\big(-G\,h^3 + \tfrac{B\,M\,h^2}{2P(1+B\,h)^2}\big)\nabla h\big] + S\,\nabla \cdot (h^3 \nabla \Delta h) = 0 \tag{3.12}$$

that describes, in the dimensionless form, the dynamics of a film in $I\!R^3$ subject to the actions of thermocapillary, capillary, and gravity forces. Here, G, M, P, B, and S are the gravity, Marangoni, Prandtl, Biot, and inverse capillary numbers respectively.

Another important related fourth-order model with similar operators in $I\!R^2$ or $I\!R^3$ is the *Cahn–Hilliard equation* applied in the study of phase separations in cooling binary solutions, such as alloys, glasses, and polymer mixtures, where the pattern formation phenomena occur,

$$u_t = -\nabla \cdot (f(u) \nabla \Delta u + g(u) \nabla u) \quad \text{in } I\!R^N \times I\!R_+ .$$

Here, $f(u)$ and $g(u)$ are given coefficients which are typically quadratic or cubic functions. Here, u measures the difference in mass fractions of the two components of the alloy.

The 1D TFE (3.1) and other similar models exhibit some exceptional mathematical properties. In particular, some free-boundary problems (FBPs) for (3.1) admit nonnegative solutions, which is a natural and desirable property for thin film applications. Recall that, typically, $u(x,t)$ measures the height of the film and is not intended to change sign (at least, significantly). Mathematical theory of such fourth and higher-order TFEs has been originated in the pioneering work by Bernis and Friedman [44], who developed new ideas, techniques, and initiated further systematic mathematical study of weak solutions of the higher-order quasilinear degenerate parabolic PDEs. Further references are put into the Remarks.

Sixth-order thin film equations. Thin film theory generates other higher-order nonlinear PDEs. For instance, another pressure dependence in (3.4) given by

$$p = u_{xxxx}$$

corresponds to the case where an elastic plate covers the droplet surface. This leads to the sixth-order parabolic equation

$$u_t = \big(u^n u_{xxxxx}\big)_x + \text{(lower-order terms)}. \tag{3.13}$$

According to Landau–Lifshitz [372], $m = -u_{xx}$ is the *bending moment* on the overlaying plate, and $\Sigma = u_{xxx}$ is the *shearing force*. As in the fourth-order case, $n = 3$

corresponds to Newtonian viscous fluids that do not slip along the substrate, while $n = 2$ is associated with the Navier-slip dominated case. In addition, $n = 3$ arises in a model of the oxidation of silicon in semiconductor devices. The parameter $n = 1$ corresponds to thin film flows in a Hele–Shaw cell. The linear counterpart of this TFE for $n = 0$ is the *tri-harmonic equation*

$$u_t = u_{xxxxxx}.$$

Setting $n = 1$ gives the leading quadratic operator to be considered later on,

$$\boxed{u_t = (uu_{xxxxx})_x + (\text{lower-order terms}).}$$

On free-boundary conditions. For one-dimensional TFEs, the free-boundary conditions at the interface $x = s(t)$, where $u(s(t),t) = 0$, are of principal importance. For the fourth-order PDE (3.1), the most physically relevant are the *zero contact angle* and *zero-flux* (mass conservation) conditions, i.e.,

$$u = u_x = u^n u_{xxx} = 0 \quad \text{at} \quad x = s(t). \tag{3.14}$$

These make sense also for $n = 0$, i.e., for the bi-harmonic equation (3.2). In general, for the fourth-order TFEs, these three conditions on the *a priori* unknown free boundary (together with standard Dirichlet or Neumann ones at fixed boundary points and given initial data) are expected to be sufficient to specify a solution. A rigorous proof of uniqueness (to be discussed later) is always a difficult mathematical problem, remaining open for many FBPs.

Other types of free-boundary conditions can also be considered, e.g., those of the *one-phase Florin* type

$$u_x = S[u] \quad \text{at} \quad x = s(t),$$

where the operator S may depend on other derivatives of u at the interface. For the heat equation, $u_t = u_{xx}$, the case $S[u] =$ constant gives *Florin's* boundary condition which has been known since the 1950s, [190]. See [454] for details.

In both cases, it is important to derive the *dynamic interface equation* representing the dependence of the interface speed $s'(t)$ on the corresponding interface "slopes" that are determined in terms of the derivatives u_x, u_{xx},... at $x = s(t)$. For instance, for the zero-flux Stefan FBP with conditions

$$u = u^n u_{xxx} = 0 \quad \text{and} \quad u_x = S[u] \quad \text{at} \quad x = s(t),$$

for sufficiently smooth solutions, we have, by differentiating, that

$$\tfrac{\mathrm{d}}{\mathrm{d}t}\, u(s(t),t) = u_x s' + u_t = 0,$$

yielding the following formal Stefan-type free-boundary condition, which is the interface equation:

$$s' = -\tfrac{1}{u_x}\, u_t \equiv -\tfrac{1}{S[u]}\, (u^n u_{xxx})_x \big|_{u=0} \, .$$

Here, the right-hand side may contain an indeterminacy (if, say, $S[u] = 0$ on the interface). Existence of such a limit as $x \to s(t)$ is very difficult to justify for general weak solutions. Using particular explicit solutions, we reveal the structure of dynamic interface equations and discuss related mathematical aspects. In some cases, we will need more general free-boundary conditions for some exact solutions.

For the sixth-order TFEs, such as (3.13), a standard FBP setting includes four free-boundary conditions

$$u = u_x = u_{xx} = u^n u_{xxxxx} = 0 \quad \text{at} \quad x = s(t), \tag{3.15}$$

meaning zero-height, zero contact angle, *zero-curvature*, and zero-flux. Some other conditions of Stefan–Florin's type can also be introduced to treat solutions on invariant subspaces.

2mth-order thin film equations in $I\!R$ and $I\!R^N$. The $2m$th-order 1D TFEs for $n = 1$ take the form

$$\boxed{u_t = (-1)^{m+1} D_x \big(u D_x^{2m-1} u\big) + \text{(lower-order terms)},} \tag{3.16}$$

where $D_x = \frac{\partial}{\partial x}$, and, as usual, by lower-order terms, we mean operators of lower differential order. The sign multiplier $(-1)^{m+1}$ in front of the higher-order operator guarantees that, for any $m = 1, 2, ...,$ the PDE is of parabolic type in the positivity domain $\{u > 0\}$ of the solution. This ensures local existence and regularity properties of strictly positive classical solutions that sometimes can be extended to compactly supported solutions by using special types of regularization; see Remarks. A full mathematical theory of such PDEs concerning existence, uniqueness, and differential properties of weak, strong, or maximal solutions of the Cauchy problem, or FBPs is still not fully justified, especially for higher orders where $m \geq 3$.

All of the above models admit a natural formulation in the N-dimensional geometry. For instance, (3.16) in $I\!R^N \times I\!R_+$ reads

$$\boxed{u_t = (-1)^{m+1} \nabla \cdot (u \nabla \Delta^{m-1} u) + \text{(lower-order terms)}.}$$

Free-boundary conditions should also be adapted to this $2m$th-order, N-dimensional case and should include m+1 conditions on the free-boundary surface.

3.1.2 Quadratic models: polynomial subspaces

As usual, for construction of solutions on linear invariant subspaces, we need to rewrite the PDE in a form having principle algebraically homogeneous polynomial operators, or, even better, including a quadratic or a cubic one. The thin film operator F in (3.1) is already quadratic for $n = 1$ and is cubic for $n = 2$. Some other thin film-type models can be reduced to quadratic PDEs. For instance, consider the quasilinear fourth-order TFE of the type (3.11), with $n = 1$,

$$v_t = -(v(v^s)_{xxx})_x.$$

Setting $u = v^s$ yields a homogeneous quadratic operator in the PDE

$$\boxed{u_t = -s\big(u u_{xxxx} + \tfrac{1}{s}\, u_x u_{xxx}\big).}$$

Let us next introduce a more general class of quadratic operators in the TFE

$$\boxed{u_t = F[u] \equiv -u u_{xxxx} + \beta u_x u_{xxx} + \gamma\, (u_{xx})^2.} \tag{3.17}$$

In (3.17), we include three quadratic *primitive monomial* operators $F_{1,2,3}$ with similar actions on elementary polynomials, i.e., for any monomial x^l $(l \geq 4)$,

$$F_j[x^l] = c_{jl}\, x^{M(l)}, \quad \text{where } M(l) = 2l - 4$$

for $j = 1, 2$, or 3 with constants $c_{jl} \neq 0$. We then write $F_j[x^l] \sim x^{M(l)}$. Obviously (this is a general principle), for any $M(l) \leq l$, the polynomial subspace $W_{l-1} = \mathcal{L}\{1, x, ..., x^{l-1}\}$ is invariant under F_j. Adding other quadratic terms with $M(l)$ less than $2l-4$, such as $u_x u_{xxxx}$, $u_{xx}u_{xxx}$ ($M = 2l-5$ for both), $u_{xx}u_{xxxx}$, $(u_{xxx})^2$ ($M = 2l-6$), etc., will not affect the dimension of polynomial subspaces, although this can change other aspects of mapping. The first result is straightforward and establishes the dimension of the *basic* invariant subspace that exists for arbitrary β and γ.

Proposition 3.1 *Operator* F *in* (3.17) *preserves the 5D subspace*

$$W_5 = \mathcal{L}\{1, x, x^2, x^3, x^4\}. \tag{3.18}$$

Proof. Indeed, it suffices to check mappings of the higher-degree term, $F[x^l] \sim x^{2l-4}$. For invariance, one needs $2l - 4 \leq l$, which yields $l \leq 4$. □

For solutions on W_5,

$$u(x,t) = C_1(t) + C_2(t)x + C_3(t)x^2 + C_4(t)x^3 + C_5(t)x^4,$$

the TFE (3.17) is a fifth-order DS

$$\begin{cases} C_1' = -24C_1C_5 + 6\beta C_2C_4 + 4\gamma\, C_3^2, \\ C_2' = 24(\beta - 1)C_2C_5 + 12(\beta + 2\gamma)C_3C_4, \\ C_3' = 24(2\beta + 2\gamma - 1)C_3C_5 + 18(\beta + 2\gamma)C_4^2, \\ C_4' = 24(4\beta + 6\gamma - 1)C_4C_5, \\ C_5' = 24(4\beta + 6\gamma - 1)C_5^2. \end{cases}$$

The last ODE is solved independently, and then the rest of the equations become linear and give the explicit solutions.

The next question is about possible extensions of the basic subspace (3.18). These and further results can be extracted from Section 2.6.4. We present direct proofs in view of their simplicity.

Proposition 3.2 *Operator* F *in* (3.17) *preserves the following subspaces:*

(i) $W_6 = \mathcal{L}\{1, x, x^2, x^3, x^4, x^5\}$, *if* $15\beta + 20\gamma = 6$;

(ii) $W_7 = \mathcal{L}\{1, x, x^2, x^3, x^4, x^5, x^6\}$, *if* $4\beta + 5\gamma = 2$;

(iii) $W_8 = \mathcal{L}\{1, x, x^2, x^3, x^4, x^5, x^6, x^7\}$, *if* $\beta = \frac{16}{7}$, $\gamma = -\frac{10}{7}$;

(iv) $W_9 = \mathcal{L}\{1, x, x^2, x^3, x^4, x^5, x^6, x^7, x^8\}$, *if* $\beta = \frac{5}{2}$, $\gamma = -\frac{45}{28}$.

(v) F *does not admit the 10D subspace*

$$W_{10} = \mathcal{L}\{1, x, x^2, x^3, x^4, x^5, x^6, x^7, x^8, x^9\},$$

or polynomial subspaces of higher dimension.

Proof. (i) For arbitrary parameters β and γ, $F: W_6 \to W_7$, and the given invariance condition ensures vanishing the coefficient of x^6 in the expansion of $F[u]$ for any $u \in W_6$.

(ii) Similarly, $F: W_7 \to W_9$, and the prescribed condition guarantees that the coefficients of x^7 and x^8 simultaneously vanish.

(iii) In this case, one needs to check the invariance conditions implying vanishing the coefficients of x^8, x^9, and x^{10}. For functions $u = C_1 + ... + C_6 x^5 + C_7 x^6 + C_8 x^7 \in W_8$, the following holds:

$$F[u] = ... + \big[180(4\beta + 5\gamma - 2)C_7^2 + 30(49\beta + 56\gamma - 32)C_6 C_8\big]x^8 \\ + 60(35\beta + 42\gamma - 20)C_7 C_8 x^9 + 42(35\beta + 42\gamma - 20)C_8^2 x^{10}.$$

In view of the independence of the terms C_7^2 and C_6C_8 in the first square bracket, this leads to the following linear system for coefficients β and γ:

$$\begin{cases} 4\beta + 5\gamma = 2, \\ 49\beta + 56\gamma = 32, \\ 35\beta + 42\gamma = 20. \end{cases}$$

The system has a unique solution indicated in (iii).

(iv) The result is straightforward for even polynomials $u = C_1 + C_3 x^2 + C_5 x^4 + C_7 x^6 + C_9 x^8 \in W_9$ for which

$$F[u] = ... + 24(124\beta + 140\gamma - 85)C_7 C_9 x^{10} + 112(24\beta + 28\gamma - 15)C_9^2 x^{12},$$

so that, for invariance, we need

$$\begin{cases} 124\beta + 140\gamma = 85, \\ 24\beta + 28\gamma = 15. \end{cases}$$

This yields the operator in (iv). For arbitrary $u \in W_9$, we use Reduce.

(v) The negative conclusion follows from Theorem 2.8 on the maximal dimension, since the optimal estimate (2.19) yields

$$n \le (2k+1)\big|_{k=4} = 9, \tag{3.19}$$

where n is the dimension of the invariant subspace W_n of the kth-order operator. □

As an illustration of the proof of (v), it is not difficult to check that plugging $u = C_1 + C_2 x + ... + C_8 x^7 + C_9 x^8 + C_{10} x^9 \in W_{10}$ leads to an inconsistent linear system for the parameters $\{\beta, \gamma\}$.

Operators preserving subspaces of maximal dimension. According to Theorem 2.8, the maximal dimension of subspaces satisfies (3.19), i.e., it is nine for arbitrary nonlinear fourth-order operators. Proposition 3.2(iv) shows a single such operator from the family (3.17). Clearly, the maximal dimension is achieved for the quadratic fourth-order fully nonlinear operator

$$F_{\rm m1}[u] = (u_{xxxx})^2, \tag{3.20}$$

which has the *minimal* degree $M(l) = 2l - 8$, and admits subspace W_9 in Proposition 3.2. Concerning general quadratic fourth-order operators (here $u' = u_x$)

$$\begin{aligned} F[u] = {} & \alpha_1 (u^{(4)})^2 + \alpha_2 u''' u^{(4)} + \alpha_3 u'' u^{(4)} + \alpha_4 u' u^{(4)} + \alpha_5 u u^{(4)} \\ & + \alpha_6 (u''')^2 + \alpha_7 u'' u''' + \alpha_8 u' u''' + \alpha_9 u u''' \\ & + \alpha_{10} (u'')^2 + \alpha_{11} u' u'' + \alpha_{12} u u'' + \alpha_{13} (u')^2 + \alpha_{14} u u' + \alpha_{15} u^2, \end{aligned} \tag{3.21}$$

where $\alpha_1^2 + ... + \alpha_5^2 \neq 0$, the following complete classification result holds:

Proposition 3.3 *There exist precisely three linearly independent operators from the family* (3.21) *preserving subspace* W_9 *in Proposition* 3.2. *These are* (3.20) *and*

$$F_{\rm m2}[u] = -uu^{(4)} + \tfrac{5}{2}\,u'u''' - \tfrac{45}{28}(u'')^2,$$
$$F_{\rm m3}[u] = u''u^{(4)} - \tfrac{5}{6}(u''')^2.$$

The same operators follow from the general formulae in Theorem 2.36. $F_{\rm m2}$ is given in Proposition 3.2(iv), and $F_{\rm m3}$ is the only new one. As in classification of second-order operators in Section 2.4, the ODE

$$u^{(9)} = 0 \quad \text{that defines} \quad W_9.$$

is invariant under the 3D operator space spanned by $\{F_{\rm m1}, F_{\rm m2}, F_{\rm m3}\}$.

Example 3.4 (**Fourth-order p-Laplacian operator**) The dimension of polynomial subspaces may also increase for the higher-order p-Laplacian operators. For instance, parabolic PDEs

$$u_t = -\nabla \cdot \left(u^n |\nabla \Delta u|^l \nabla \Delta u\right) \tag{3.22}$$

appear in thin film flows in the capillary-driven case (the pressure is $p = \Delta u$), in which the viscosity of the spreading liquid is temperature-dependent; see Remarks. Setting $n = 0$ and $l = 1$, consider the 1D operator

$$F[u] = -\left[(u_{xxx})^2\right]_x.$$

Since here $M(l) = 2l - 7$, this admits W_8 from Proposition 3.2(iii).

3.1.3 Cubic operators

In the next example, consider the cubic operators from (3.7) with $n = 2$,

$$F[u] = -\left[u^2 u_{xxx} + \beta u u_x u_{xx} + \gamma\,(u_x)^3\right]_x, \tag{3.23}$$

where each monomial operator has $M(l) = 3l - 4$. Then the basic subspace for any β and γ is $W_3 = \mathcal{L}\{1, x, x^2\}$ ($l = 3l - 4$, i.e., $l = 2$), on which the principal fourth-order operator in (3.23) annuls identically. This becomes nontrivial on some extended subspaces.

Proposition 3.5 *Operator* (3.23) *preserves the following subspaces:*

$$\begin{aligned} &\text{(i)} \quad W_4 = \mathcal{L}\{1, x, x^2, x^3\}, \quad \textit{if} \quad 6\beta + 9\gamma = -2; \\ &\text{(ii)} \quad W_3^{\rm even} = \mathcal{L}\{1, x^2, x^4\}, \quad \textit{if} \quad \beta = -\tfrac{3}{2}, \ \gamma = \tfrac{3}{4}. \end{aligned} \tag{3.24}$$

(iii) $W_5 = \mathcal{L}\{1, x, x^2, x^3, x^4\}$ *is not invariant for any* β *and* $\gamma \in I\!R$.

Proof is straightforward. In (iii), we derive an inconsistent linear system for β, γ.

The monomial $F[u] = (u_{xxxx})^3$ satisfies, in our notation, $F[x^l] \sim x^{3l-12}$ and has $W_7 = \mathcal{L}\{1, x, ..., x^6\}$. Most plausibly, this seven is the maximal dimension that is achieved by cubic operators of this type.

3.2 Applications to extinction, blow-up, free-boundary problems, and interface equations

We begin with a simple quadratic TFE using the basic polynomial subspace W_4. It reveals some interesting evolution properties of such models, which are not easy to justify rigorously for more general classes of solutions.

3.2.1 Invariant subspace and exact solutions

We introduce a quadratic TFE with the constant negative absorption term,

$$\boxed{u_t = -(uu_{xxx})_x - 1.} \tag{3.25}$$

It is common to have thin film models in the divergent form, so the mass conservation holds. On the other hand, under some circumstances, TFEs may contain non-divergent operators. For instance, source-like terms may be relevant for a monolayer film in coexistence with vapor due to the well-studied phenomena of adsorption (interaction of gases and liquids with solid surfaces) and condensation; see [83] and [559] for more recent references. Homogeneous nucleation phenomena are natural in phase transition theory (see Lifshitz–Pitaevskii [394]), and condensation of liquid droplets from a supersaturated vapor is an example. Absorption terms can occur in view of the *evaporation* phenomenon, or due to the *permeability* of the surface. Such general non-divergent TFEs are derived in [451]; see also Remarks.

Evaporation and condensation phenomena have been a subject of research and debate for more than a century. Classical nucleation theory dates back to Laplace's work (1806) on surface energy and tension, to Thompson's (Lord Kelvin) theory establishing the *Thompson formula* (1870) for the dependence of the critical radius r of a droplet on the vapor pressure p, the famous condensation theory by Hertz (1882) and Knudsen (1915), Becker–Döring's equations of nucleation (1935), Zel'dovich–Frenkel's equation (1942), Lifschitz–Slyozov's theory of coarsening (1961), and other ideas and results. Papers [518, 394] contain detailed overviews of the history.

Equation (3.25) represents a formal mathematical model explaining key features of extinction phenomena in Stefan–Florin FBPs for thin film equations. For the second-order PMEs, similar absorption models are

$$u_t = (u^\sigma u_x)_x - u^p, \quad \text{with } \sigma > 0 \text{ and } p > -(\sigma + 1), \tag{3.26}$$

where $p = 0$ gives the constant absorption -1. These models exhibit interfaces, finite-time extinction, and quenching in the strong absorption range $p < 1$, and are key for general existence, uniqueness, interface propagation, and asymptotic theory; see references and results in [245, Ch. 4, 5] and [226, Sect. 7.11].

For (3.25), a compactly supported nonnegative continuous initial function $u_0(x)$ is taken. Then, formally, in the Cauchy problem, we want the constant absorption term to act only in the positivity domain $\{u > 0\}$, since otherwise, the solution will immediately take negative values. This implies that, instead of -1 everywhere, one needs to have the absorption term $-1\chi_{\{u>0\}}(x)$, where $\chi_A(x)$ is the characteristic function of the set $A \in I\!R$, so $\chi_A(x) = 1$ if $x \in A$, and zero otherwise. Then the

Heaviside function should be put into the right-hand side of (3.25) instead of -1,

$$H(u) = \begin{cases} 1 & \text{if } u > 0, \\ 0 & \text{if } u \le 0, \end{cases}$$

representing a strong discontinuous nonlinearity. We keep the use of the term -1 in the equation, since this cannot affect the construction of the explicit solutions. For FBPs, the Heaviside function is not necessary. (By regular approximations, this can be avoided in the CP as well; see below, though solutions are of changing sign.)

According to Proposition 3.1, restricting our attention to symmetric in x functions, consider the solutions

$$u(x,t) = C_1(t) + C_2(t)x^2 + C_3(t)x^4 \in W_3 = \mathcal{L}\{1, x^2, x^4\}. \tag{3.27}$$

Substituting into (3.25) yields a simple DS

$$\begin{cases} C_1' = -24C_1C_3 - 1, \\ C_2' = -72C_2C_3, \\ C_3' = -120C_3^2. \end{cases} \tag{3.28}$$

The last ODE is integrated independently, $C_3(t) = \frac{1}{120t}$, and the first two equations then become linear and give the explicit solutions

$$u(x,t) = \left(A_0 t^{-\frac{1}{5}} - \tfrac{5}{6}t - E_0 t^{-\frac{3}{5}} x^2 + \tfrac{1}{120t} x^4\right)_+, \tag{3.29}$$

where $A_0 > 0$ and $E_0 \ge 0$ are arbitrary constants, and $(\cdot)_+$ denotes the positive part. For these sufficiently smooth Lipschitz continuous in x solutions (3.29), the zero-flux condition is valid,

$$u = uu_{xxx} = 0 \quad \text{at } x = s(t), \tag{3.30}$$

but $u_x \ne 0$ at the interface in general. The zero contact angle condition fails, and hence, there occurs another FBP to be specified next. The zero contact angle version of the FBP is considered later on in Example 3.10.

3.2.2 Free-boundary setting

It is easy to identify the dynamic free boundary equation governing propagation of the interface. Since exact solutions $u(x,t)$ are smooth functions (excluding $t = 0$ and the extinction time $t = T$ to be studied separately), differentiating gives

$$u(s(t),t) = 0 \quad \Longrightarrow \quad u_x s' + u_t = 0 \quad \Longrightarrow \quad s' = -\tfrac{1}{u_x} u_t\,. \tag{3.31}$$

From (3.25), using that, for sufficiently smooth C^4-solutions, $uu_{xxxx} = 0$ on the interface, we infer that, at $x = s(t)$, $u_t = -u_x u_{xxx} - 1$, and this determines the *regularity* dynamic interface equation

$$s' = S[u] \equiv u_{xxx} + \tfrac{1}{u_x} \quad \text{at } x = s(t) \text{ for } t \in (0,T). \tag{3.32}$$

The interface operator $S[u]$ on the right-hand side is of the third order and consists of two *interface operators*. The second operator is non-positive, since $u_x \le 0$ at the right-hand interface, while the first one is positive, since

$$u_{xxx}(s(t), t) = \tfrac{1}{5t}\, s(t) > 0, \quad \text{provided that } s(t) > 0.$$

This leads to a competition between terms of different signs in the interface equation (3.32), and makes it possible to observe a complicated non-monotone behavior of interfaces, which can have *turning* points to be studied below.

Apparently, equation (3.32) holds for any smooth solution $u(x, t)$ satisfying two free-boundary conditions (3.30), so it is just a manifestation of sufficient regularity of solutions. We are still short of a third condition to create a well-posed FBP, which is assumed to admit a unique solution (given, of course, by (3.29)).

Let us now specify the actual *governing* dynamic interface equation generating exact solutions (3.27). First of all, $C_3(t) = \frac{1}{120t} = \frac{1}{4!}\, u_{xxxx}$, which yields the following time-parameterization on the invariant subspace:

$$t = \frac{1}{5u_{xxxx}}.$$

Next, calculating the interface position $s(t)$ from (3.27) in terms of the expansion coefficients yields

$$s(t) = R(C_1(t), C_2(t), C_3(t)) \quad \text{for } t \in (0, T), \tag{3.33}$$

where R is an irrational function which is obtained from the bi-quadratic equation $C_1 + C_2 R^2 + C_3 R^4 = 0$. Differentiating (3.33) and using the DS (3.28), we finally arrive at the governing interface equation of the form

$$s' = \tilde{R}(C_1(t), C_2(t), C_3(t)), \quad \text{where} \quad t = \frac{1}{5u_{xxxx}}. \tag{3.34}$$

The interface operator $\tilde{R}$ is now of the fourth order, so it differs from the already known regularity equation (3.32). On the other hand, two equalities, (3.32) and (3.34), imply a stationary Neumann-type condition on the interface,

$$u_{xxx} + \frac{1}{u_x} = \tilde{R}(C_1(t), C_2(t), C_3(t))\big|_{t=1/5u_{xxxx}} \quad \text{at } x = s(t), \tag{3.35}$$

which can replace the dynamic condition (3.34).

Thus we have posed the FBP, which can be locally analyzed by using the *von Mises transformation* $(x, t, u) \mapsto (u, t, X)$, where, close to the interface,

$$X(u(x, t), t) \equiv x.$$

Then X satisfies a quasilinear fourth-order PDE degenerated at the interface, which, in the new variables, is fixed at the origin,

$$x = s(t) \quad \Longleftrightarrow \quad u = 0.$$

Therefore, two boundary conditions are given at $u = 0$:

(i) (3.32) that fixes a class of sufficiently smooth solutions, and

(ii) (3.34) or (3.35) (both should be rewritten in terms of the new variables).

Such FBPs are difficult from a mathematical point of view. It is worth recalling the fundamental results in linear theory, where the well-posedness of boundary-value problems are governed by the *Lopatinskii–Shapiro conditions* [401]. It is important to characterize necessary general conditions for the arising nonlinear FBPs, to ensure existence of a unique local-in-time solution for small $u > 0$ constructed by semi-group methods of parabolic PDE theory. In general, these are OPEN PROBLEMS. We postpone an example of such an analysis for a simpler, but more physically relevant FBP with the zero contact angle condition, as in (3.14).

More on second-order PME with absorption: interface equation. The mathematics of such FBPs for the TFE are difficult, but, relative to some key features, are similar to those for the second-order parabolic PDEs which are much better understood. This counterpart of (3.25) is the *PME with absorption* (3.26) having $p = 0$ and $\sigma = 1$,

$$\boxed{u_t = (uu_x)_x - 1.} \tag{3.36}$$

For the Cauchy problem, or, which is the same, for the *maximal solutions* constructed by regular approximations (see [226, Ch. 7] for details), the derivation of the free-boundary condition as the analogy of (3.32) is similar and leads to the dynamic equation with the first-order operator S,

$$s' = S[u] \equiv -u_x + \tfrac{1}{u_x} \quad \text{at} \quad x = s(t) \quad \text{for} \quad t \in (0, T). \tag{3.37}$$

It is known (see examples in [226, Sect. 7.11]) that (3.37) holds almost everywhere (a.e.) for any weak solution of the Cauchy problem for (3.36) with bounded nonnegative compactly supported initial data. The proof relies on the Maximum Principle and intersection comparison arguments based on Sturm's Theorem on zero sets, [226, Ch. 7], so we do not need the von Mises transformation creating a non-standard parabolic problem with dynamic boundary conditions. There are other well-posed FBPs for (3.36) with various higher-order dynamic free-boundary conditions that differ from (3.37). Typical existence and regularity properties of such *non-maximal* solutions are also driven by Sturmian intersection comparison arguments. Such general FBP theory is developed in [226, Ch. 8]. All of these MP and the intersection comparison techniques fail for higher-order TFEs.

3.2.3 Initial Dirac's mass

We now begin to describe evolution properties of exact solutions on W_3 that, actually, do not critically depend on the particular FBP setting that was revealed above.

It follows that explicit solutions (3.29) create a singular short-time behavior as $t \to 0^+$ that reflects the *initial singularity* phenomenon, i.e., *blow-up* at $t = 0^+$. We choose special values $A_0 = 1$ and $E_0 = \frac{1}{\sqrt{30}}$ for which (3.29) takes the form

$$u(x,t) = \left[-\tfrac{5}{6}\,t + t^{-\frac{1}{5}} f_0(y)\right]_+, \quad \text{where} \quad f_0(y) = \left(1 - \tfrac{1}{\sqrt{120}}\, y^2\right)^2 \tag{3.38}$$

and $y = x/t^{1/5}$. It is important that the profile $f_0(y)$ itself satisfies the zero contact angle condition at its interface point, i.e.,

$$f_0(y_0) = f_0'(y_0) = 0 \quad \text{at} \quad y_0 = (120)^{\frac{1}{4}}.$$

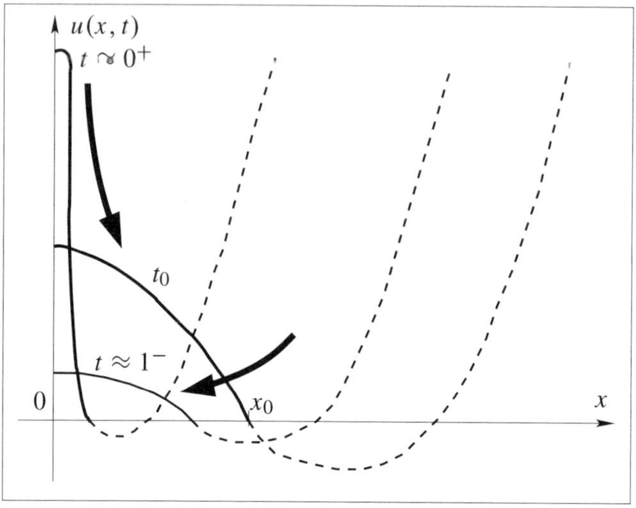

Figure 3.1 Evolution properties of solutions (3.38): (i) Dirac's delta for $t \approx 0^+$, (ii) interface turning point at $t = t_0$, and (iii) extinction as $t \to 1^-$.

This is the rescaled similarity profile of the pure quadratic TFE

$$u_t = -(uu_{xxx})_x, \quad \text{with the similarity solution } u(x,t) = t^{-\frac{1}{5}} f_0\big(\tfrac{x}{t^{1/5}}\big)$$

that has been recognized since the 1980s, [531]. Taking into account the spatial hump concentrated near the origin inside the interval $\{|y| \le y_0\}$, solution (3.38) satisfies a remarkable initial condition given by a measure: as $t \to 0^+$,

$$u(x,t) \to M_0 \delta(x), \quad \text{with} \quad M_0 = 2 \int_0^{y_0} f_0(y)\,\mathrm{d}y = \tfrac{16}{15}(120)^{\frac{1}{4}},$$

where $\delta(x)$ is Dirac's delta concentrated at $x = 0$. Figure 3.1 for $t \approx 0^+$ illustrates such a singular *initial blow-up* behavior.

3.2.4 Extinction patterns

Thus, explicit solutions (3.29) are generated by measures as initial data. As $t > 0$ increases, the phenomenon of *finite-time extinction* occurs. Setting $A_0 = \frac{5}{6}$ now for convenience, (3.29) gives at the origin $x = 0$ the following behavior:

$$\sup_x u(x,t) \equiv u(0,t) = \tfrac{5}{6}\big(t^{-\frac{1}{5}} - t\big).$$

Hence, $u(x,t)$ vanishes at the extinction time $T = 1$, so that $u(x,t) \equiv 0$ for all $t \ge 1$. The *extinction* and *blow-up* behavior are the main singularity formation phenomena in nonlinear PDEs, especially in reaction-diffusion-absorption theory. There exists a large amount of mathematical literature on these subjects, representing a practically complete understanding of second-order parabolic PDEs; see key references in [245, 509]. For higher-order diffusion equations, the results are rare, and many types of singularity patterns are not well-described and await rigorous treatment.

Using (3.29) with $A_0 = \frac{5}{6}$, we will describe the finite-time extinction as $t \to 1^-$ for the given FBP by using the corresponding rescaled independent variables

$$\xi = \tfrac{x}{\sqrt{1-t}} \quad \text{and} \quad \tau = -\ln(1-t) \to +\infty \quad \text{as } t \to 1^-. \tag{3.39}$$

Then, rescaling solution (3.29) yields, up to exponentially smaller terms,

$$u(x,t) = \mathrm{e}^{-\tau}\big(1 - E_0\xi^2 + \tfrac{1}{120}\,\mathrm{e}^{-\tau}\xi^4\big)_+ + \dots . \tag{3.40}$$

For any constant $E_0 > 0$, we obtain a pattern which is expected to be asymptotically stable in the rescaled sense, meaning simply that

$$\mathrm{e}^{\tau} u(x,t) \to \big(1 - E_0\xi^2\big)_+ \equiv g(\xi) \quad \text{as} \quad t \to 1. \tag{3.41}$$

Indeed, this illustrates stability of self-similar solutions of (3.25) of the form

$$u_{\mathrm{s}}(x,t) = (1-t)g(\xi), \ \xi = \tfrac{x}{\sqrt{1-t}}, \quad \text{where} \tag{3.42}$$

$$-(gg''')' - \tfrac{1}{2}\, g'\xi + g - 1 = 0, \tag{3.43}$$

and g also satisfies the corresponding free-boundary conditions. Therefore, using explicit solutions (3.29) implies that the self-similar behavior is stable, at least on the subspace W_3, where almost all solutions (excluding those with $E_0 = 0$ to be analyzed below) asymptotically take the form of the self-similar ones.

The extension of stability analysis beyond the invariant subspace is a difficult OPEN PROBLEM, which has the following asymptotic setting: Given a general solution $u(x,t)$ of the FBP with extinction at $t = 1$ and $x = 0$, the rescaled function $v(\xi,\tau) = (1-t)u(x,t)$, $x = \xi\sqrt{1-t}$, satisfies the non-stationary rescaled PDE

$$v_\tau = -(v v_{\xi\xi\xi})_\xi - \tfrac{1}{2}\, v_\xi \xi + v - 1, \tag{3.44}$$

with the operator from (3.43). Thus, we arrive at the asymptotic stabilization problem as $\tau \to +\infty$. For the second-order PDEs, such as (3.36), the asymptotic extinction behavior is well understood and proved rigorously (see [245, Ch. 5]), where the analysis uses the MP, comparison, and intersection comparison approaches that do not apply to higher-order models.

3.2.5 On interface turning points

The interfaces of the explicit solution (3.38) are not monotone with time. The position of the right-hand interface is

$$s(t) = (120)^{\frac{1}{4}} t^{\frac{1}{5}} \left[1 - \left(\tfrac{5}{6}\right)^{\frac{1}{2}} t^{\frac{3}{5}}\right]^{\frac{1}{2}},$$

so that $s(0) = s(T) = 0$, where $T = \left(\frac{6}{5}\right)^{\frac{5}{6}}$, and $s(t)$ attains its maximum at

$$t_0 = \left(\tfrac{24}{125}\right)^{\frac{5}{6}}.$$

At the moment $t = t_0 \in (0,T)$, the solution has the *turning point* of the interface at $x = x_0 = s(t_0)$. Therefore, close to the maximum point of $s(t)$, the interface has a specific quadratic behavior

$$s(t) - x_0 = -a_1(t-t_0)^2 + \dots, \quad \text{with some } a_1 > 0. \tag{3.45}$$

As we have seen, the turning point occurs due to the presence of two terms of different signs in the regularity interface equation (3.32). This is a delicate phenomenon in FBP theory. Even for the second-order PMEs with absorption, such as (3.36), there are some mathematical open questions concerning the whole countable family of different turning patterns. For instance, it is known that, for (3.36), the turning kth pattern takes spatial shape related to the kth-order *Hermite polynomial* being the eigenfunction of the corresponding linearized operator [241]. We are going to develop a similar "spectral theory" for the TFE and detect corresponding kth-order polynomials.

Solution (3.38) gives an extremely rare opportunity to explicitly study the turning asymptotic pattern, which it is convenient to describe in coordinates that are rescaled about the point (x_0, t_0) by setting

$$y = x_0 - x \geq 0, \quad \zeta = \tfrac{y}{t_0 - t} \geq 0, \quad \text{and} \quad \tau = -\ln(t_0 - t). \tag{3.46}$$

In view of the turning point characterization (3.45), using Taylor's expansion, it is not difficult to obtain from (3.38) that, for small $y > 0$ and $\tau \gg 1$,

$$u(x,t) = \left[c_0 y + \mathrm{e}^{-2\tau}\left(-c_1 + c_2\zeta + c_3 \mathrm{e}^{-\tau}\zeta^2 + c_4 \mathrm{e}^{-2\tau}\zeta^3 + \ldots\right)\right]_+, \tag{3.47}$$

where c_k denote some fixed constants with $c_{0,1} > 0$. Solving approximately the equation $u(x,t) = 0$ yields the interface position $y_0(\tau) = \frac{c_1}{c_0}\,\mathrm{e}^{-2\tau} + \ldots$, coinciding with (3.45) if $a_1 = \frac{c_1}{c_0}$. The first linear term in (3.47) is the asymptotic trace of the *stationary profile* $G(y)$, about which the turning effect occurs via a certain *focusing* phenomenon. Hence, G solves the stationary TFE

$$-(GG''')' - 1 = 0, \quad G(y) = a_0 y + a_1 y^2 + a_2 y^3 + \ldots \text{ as } y \to 0^+, \tag{3.48}$$

where the expansion corresponds to the necessary free-boundary conditions at $y = 0$. Thus, $a_2 = -\frac{1}{6a_0}$ $(a_0 = c_0)$, meaning that $s'(t_0) = 0$ by (3.32). The explicit formula (3.47) is the asymptotic expansion about the stationary profile $G(y)$, so we set $u(y,t) = G(y) + Y(y,t)$ to get the linearized PDE

$$Y_t = -(G(y)Y_{yyy})_y - (YG'''(y))_y - (YY_{yyy})_y, \tag{3.49}$$

where the inhomogeneous term $-(GG_{yyy})_y - 1$ has vanished due to (3.48) and the coefficients are $G(y) = a_0 y + \ldots$ and $G'''(y) = 6a_2 + \ldots$ as $y \to 0$. Let us write (3.49) in terms of the new spatial rescaled variable

$$Y(y,t) = \mathrm{e}^{-2\iota}\, w(\eta, \tau), \quad \text{where } \eta = \tfrac{y}{\sqrt{t_0 - t}}, \tag{3.50}$$

that is associated with the already known behavior (3.47), to obtain the following exponentially perturbed equation:

$$w_\tau = \mathbf{B}w - \mathrm{e}^{-\frac{\tau}{2}} 6a_2 w_\xi - \mathrm{e}^{-3\tau}(w w_{\eta\eta\eta})_\eta + \ldots, \tag{3.51}$$

where $\mathbf{B}$ is the linear degenerate fourth-order operator

$$\mathbf{B}w = -a_0(\eta w_{\eta\eta\eta})_\eta - \tfrac{1}{2}\, w_\eta \eta + 2w. \tag{3.52}$$

This operator is not symmetric and most plausibly does not admit a self-adjoint extension in any weighted L^2-spaces. In general, the spectral properties of such non-self-adjoint operators are difficult. Let us determine the point spectrum of the special extension. Notice first that solutions of the eigenvalue equation

$$\mathbf{B}\psi = \lambda\psi$$

exhibit the following generic exponential growth near the infinite singular point:

$$\psi(\eta) \sim e^{\mu\eta} \quad \text{as } \eta \to +\infty, \text{ with } \mu = (2a_0)^{-\frac{1}{3}}\left(\tfrac{1}{2} \pm \mathrm{i}\,\tfrac{\sqrt{3}}{2}\right).$$

Therefore, a proper functional setting for $\mathbf{B}$ involves the weighted space L^2_ρ with the exponential weight $\rho(\eta) = e^{-b\eta}$, where $b \in \left(0, \frac{1}{2}(2a_0)^{-1/3}\right)$ is a constant. It turns out that $\mathbf{B}$ possesses the real point spectrum in spaces of functions that are sufficiently smooth at $\eta = 0$ (e.g., such as H^4_ρ),

$$\sigma_p(\mathbf{B}) = \left\{\lambda_k = -\tfrac{k}{2} + 2,\ k = 0, 1, 2, ...\right\},$$

and each eigenfunction $\psi_k(\eta)$ is a kth-degree polynomial

$$\psi_k(\eta) = b_k\left[\eta^k + \tfrac{2k(k-1)(k-2)^2}{3}\, a_0\eta^{k-1} + ...\right],$$

where b_k are some normalization constants. In a proper setting, the resolvent of $\mathbf{B}$ is compact in L^2_ρ (see Remarks for extra details), so the whole spectrum $\sigma(\mathbf{B})$ can be made discrete. These polynomials are complete and closed in $L^2_\rho(I\!R)$; see Kolmogorov–Fomin [352, p. 431]. So, studying (3.51), eigenfunction expansions can be used. To this end, let us introduce the adjoint operator

$$\mathbf{B}^* v = -a_0(\eta v')''' + \tfrac{1}{2}\, v'\eta + \tfrac{5}{2}\, v$$

with the same spectrum and find the bi-orthogonal (in $L^2(I\!R)$) set $\{\psi_k^*\}$ of eigenfunctions satisfying, after normalization, $\langle \psi_i, \psi_j^* \rangle = \delta_{ij}$. Then, in L^2_ρ, we can look for solutions of (3.51) in the form of

$$w(\eta, \tau) = \textstyle\sum_{(k \geq 0)} c_k(\tau)\psi_k(\eta), \quad \text{where } c_k(\tau) = \langle w(\cdot, \tau), \psi_k^* \rangle,$$

and study the DS for the expansion coefficients $\{c_k(\tau)\}$ that describes the asymptotic behavior. It follows that the turning patterns satisfying the rescaled equation (3.51) correspond to the evolution on the stable and center manifolds that are tangent to the corresponding subspaces of $\mathbf{B}$, i.e., as $\tau \to \infty$, $w(\eta, \tau) \sim e^{\lambda_k \tau}\psi_k(\eta)$, with $\lambda_k \in \sigma_p(\mathbf{B})$ for some $k \geq 4$, so $\lambda_k \leq 0$. In particular, according to the scaling (3.50), taking $k = 4$ with $\lambda_4 = 0$ gives the asymptotic behavior for the *first turning pattern* (most probably, generic)

$$Y(y, t) \sim e^{-2\tau}\,\psi_4(\eta) + \dots, \quad \text{where} \quad \psi_4(\eta) = b_4\left(\eta^4 + 32a_0\eta\right)$$

is the corresponding eigenfunction. This determines the behavior on sets defined as $y = O(\sqrt{t_0 - t})$ to be matched with the behavior (3.47) on smaller sets in the variable ζ in (3.46), i.e., for $y = O(t_0 - t)$, to get a global structure of this first turning pattern. A rigorous justification of such asymptotic and matching analysis is OPEN for higher-order TFEs. The exact solutions remain the only tool to test such a curious singularity formation phenomenon.

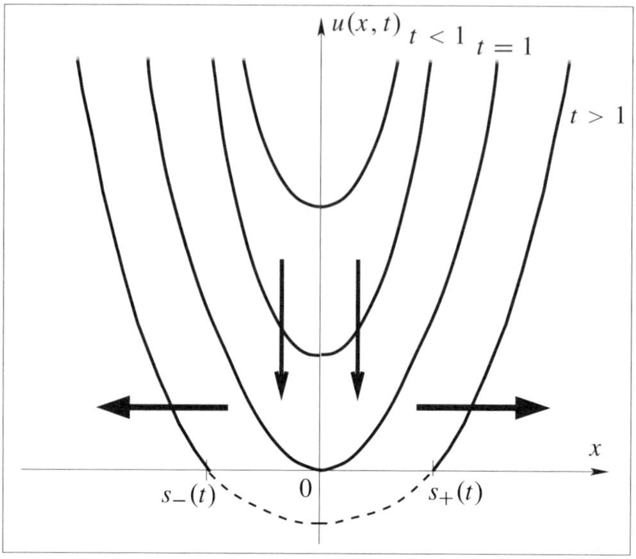

Figure 3.2 The exact solution (3.29) with $A_0 = \frac{5}{6}$, $E_0 < 0$: (i) quenching at $t = 1$, and (ii) two interfaces appear for $t > 1$.

3.2.6 *Quenching patterns*

Take $A_0 = \frac{5}{6}$ again, and now fix an arbitrary $E_0 \geq 0$ in (3.29), giving the single point *quenching*, where the strictly positive solution first touches the singularity zero level $\{u = 0\}$ at $t = 1$, as shown in Figure 3.2. If $E_0 < 0$ for the asymptotic description of quenching, we still can use the same rescaled variables (3.39) to observe convergence (3.41), where the self-similar profile is now strictly positive,

$$g(\xi) = 1 + |E_0|\xi^2.$$

The asymptotic quenching phenomenon is described by the same rescaled equation (3.44) with, plausibly, a generic stabilization to the similarity profile; proof is an OPEN PROBLEM. After quenching, for $t > 1$, two interfaces appear with the non-Lipschitz behavior

$$s_{\pm}(t) = +\frac{1}{\sqrt{|E_0|}}\sqrt{t-1} + \dots \quad \text{as } t \to 1^+, \tag{3.53}$$

so that a smooth flow, corresponding to the uniformly parabolic TFE for $t < 1$, is replaced by the FBP for the degenerate equation for $t > 1$, with a quenching transition at $t = 1^-$. In general, the questions of solution extensions beyond singular quenching remain OPEN. For similar second-order equations, such as (3.36), this determines extensions of order-preserving semigroups for various types of singularities, [226, Sect. 6.2].

A different quenching pattern occurs for $E_0 = 0$ in (3.29), and a distinct spatial rescaled variable η is necessary for describing the limit $t \to 1^-$,

$$e^{\tau}u(x,t) \to 1 + \tfrac{1}{120}\eta^2, \quad \text{where } \eta = \tfrac{x}{(1-t)^{1/4}}. \tag{3.54}$$

Then we introduce another rescaled function, $v(\eta, \tau) = e^{\tau} u(x, t)$, that is defined according to (3.54) and satisfies the following equation:

$$v_\tau = -\tfrac{1}{4} v_\eta \eta + v - 1 - e^{-\tau}(v v_{\eta\eta\eta})_\eta. \qquad (3.55)$$

Unlike the rescaled parabolic PDE (3.44), the thin film operator now reduces to a fourth-order perturbation that is exponentially small as $\tau \to \infty$. This gives a *singular perturbation problem* for the linear autonomous Hamilton–Jacobi equation

$$v_\tau = -\tfrac{1}{4} v_\eta \eta + v - 1. \qquad (3.56)$$

The convergence as $\tau \to \infty$ of some classes of solutions of (3.55) and (3.56) represents an interesting OPEN PROBLEM; though such behavior is not expected to be generic, unlike the above similarity with $E_0 < 0$. For PMEs with absorption (3.36) for which, in (3.55), the fourth-order operator is replaced by the diffusion one $(v v_\eta)_\eta$, such behavior is proved to be stable; see results in [245, Ch. 5].

After quenching at $t = 1$, the interfaces also exhibit a different behavior (cf. (3.53))

$$s_\pm(t) = \pm\big[120(t-1)\big]^{\frac{1}{4}} + \ldots \quad \text{as } t \to 1^+.$$

We expect that it is not generic and corresponds to very flat initial data $u(x, 1^-)$ created by such a peculiar unstable quenching.

3.2.7 On extinction patterns with zero contact angle, the Cauchy problem

In the case of the standard FBP having conditions as in (3.14), it seems that the present TFE with absorption does not possess exact solutions on invariant subspaces (for other parameters, such solutions do exist; see Example 3.10 below). On the other hand, self-similar solutions of the form (3.42) may be introduced, where $g(\xi)$ is an even positive solution of the ODE (3.43) on some interval $\xi \in (0, \xi_0)$ with conditions

$$g(\xi_0) = g'(\xi_0) = (gg''')(\xi_0) = 0, \quad \text{and } g'(0) = g'''(0) = 0, \qquad (3.57)$$

corresponding to the *symmetry* at the origin $\xi = 0$. The behavior close to the interface at $\xi = \xi_0$ is

$$g(\xi) = (\xi_0 - \xi)^2 \sqrt{2|\ln(\xi_0 - \xi)| + C} + \ldots \quad (C \in I\!R).$$

Existence (or nonexistence) of such $g(\xi)$ leads to two-parameter, $\{\xi_0, C\}$, shooting to satisfy two symmetry conditions at $\xi = 0$ and remains an OPEN PROBLEM.

In the Cauchy problem for such degenerate PDEs to be discussed systematically later on, solutions typically are of changing sign, so the PDE and the similarity ODE are modified as follows:

$$u_t = -(|u|u_{xxx})_x - \operatorname{sign} u \implies -(|g|g''')' - \tfrac{1}{2} g'\xi + g - \operatorname{sign} g = 0. \qquad (3.58)$$

Then, (3.57) are also valid at interfaces, but the solutions $g(\xi)$ are smoother and are oscillatory with a different behavior (corresponding to $(|g|g''')' + \operatorname{sign} g = 0$)

$$g(\xi) = (\xi_0 - \xi)^2 \varphi(s), \quad s = \ln(\xi_0 - \xi), \quad \text{as } \xi \to \xi_0^-,$$

where the *oscillatory component* $\varphi(s)$ is a periodic solution of a nonlinear ODE.

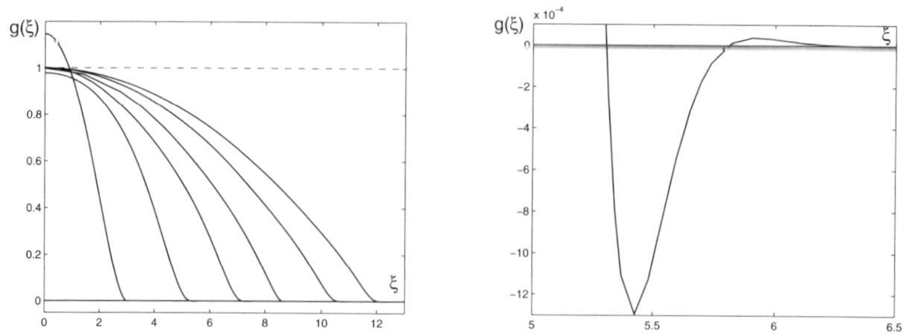

(a) six profiles (b) oscillations, enlarged

Figure 3.3 Similarity extinction profiles satisfying the ODE (3.58) with the conditions (3.57); six profiles (a) and oscillations near the interface for the second profile (b).

Details will be presented in Section 3.7. Figure 3.3(a) shows six similarity profiles for $\xi > 0$. In (b), these numerical results reveal the oscillatory character of solutions near interfaces, an intriguing part of our future analysis.

Example 3.6 (TFE with cubic operator) Take the unique operator from Proposition 3.5(ii) and consider the TFE with constant absorption

$$\boxed{u_t = -\big[u^2 u_{xxx} - \tfrac{3}{2}\, u u_x u_{xx} + \tfrac{3}{4}(u_x)^3\big]_x - 1.}$$

Taking the subspace of even fourth degree polynomials yields solutions

$$u(x,t) = C_1(t) + C_2(t)x^2 + C_3(t)x^4,$$

$$\begin{cases} C_1' = 3C_1C_2^2 - 24C_1^2C_3 - 1, \\ C_2' = -9(C_2^2 + 2C_1C_3)C_2, \\ C_3' = 30(4C_1C_3 - C_2^2)C_3. \end{cases}$$

This DS is more difficult and explicit solutions do not exist. A simpler system occurs by setting $C_2(t) \equiv 0$, which corresponds to the subspace $W_2 = \mathcal{L}\{1, x^4\}$, but it is not clear whether the resulting extinction behavior is generic. The asymptotics of extinction as $t \to T^-$ is easy to detect, since, on any bounded orbit of the above DS,

$$C_1(t) = T - t + O\big((T-t)^2\big),$$

while $C_{2,3}(t)$ remain almost constant for $t \approx T$. This gives the necessary asymptotics of the extinction (or quenching for positive solutions) behavior. The corresponding rescaled equations can be formulated in a similar fashion. Other features of the FBPs can also be described, though some of the computations are not explicit or easy.

3.2.8 *Quartic operators: applications to extinction and blow-up*

As a new application, consider the fourth-order TFE with source or absorption

$$v_t = -\big(v^n v_{xxx}\big)_x + g(v). \tag{3.59}$$

Introducing the new independent variable

$$v = u^{\mu}, \quad \text{where} \quad \mu = \tfrac{3}{n}, \tag{3.60}$$

splits the thin film operator into five primitive monomials of the algebraic homogenuity four,

$$\begin{aligned} u_t = -\big[&u^3 u_{xxxx} + (4\mu - 1)u^2 u_x u_{xxx} \\ &+ 3(\mu - 1)u^2 (u_{xx})^2 + 3(\mu - 1)(2\mu - 1)u(u_x)^2 u_{xx} \\ &+ \mu(\mu - 1)(\mu - 2)(u_x)^4\big] + \tfrac{n}{3}\, u^{1-\mu} g(u^{\mu}). \end{aligned} \tag{3.61}$$

Assume that the last zero-order term is a linear function, i.e.,

$$\tfrac{n}{3}\, u^{1-\mu} g(u^{\mu}) = a + bu. \tag{3.62}$$

Then polynomial subspaces will depend on the differential operator in (3.61). Firstly, this admits the trivial subspace of linear functions $W_2 = \mathcal{L}\{1, x\}$, which does not provide us with interesting solutions. In particular, these are traveling waves that are unbounded as $x \to \infty$. We are interested in solutions of typical bell-shaped forms localized on a bounded interval in x. Secondly, extensions of W_2 are possible in the following cases:

Proposition 3.7 *The operator given in* (3.61) *and* (3.62) *preserves* $W_3 = \mathcal{L}\{1, x, x^2\}$ *iff* $n = 3$, $n = 6$, *or* $n = -2$.

Proof. Plugging $u = C_1 + C_2 x + C_3 x^2 \in W_3$ into the quartic operator yields that the terms on $\mathcal{L}\{x^3, x^4\}$ vanish iff

$$4(\mu - 1)[3 + 6(2\mu - 1) + 4\mu(\mu - 2)] = 0,$$

which yields either $\mu = 1$ or $4\mu^2 + 4\mu - 3 = 0$, i.e., $\mu = \frac{1}{2}$ or $\mu = -\frac{3}{2}$. □

Example 3.8 (**Extinction behavior**) Let us choose $a = -1$ and $b = 0$ in (3.62), giving a constant absorption term in (3.61).

Case $n = 3$ ($\mu = 1$). This is quite simple, since the resulting PDE

$$\boxed{u_t = F_4[u] - 1 \equiv -u^3 u_{xxxx} - 3u^2 u_x u_{xxx} - 1} \tag{3.63}$$

does not contain lower differential terms, so that, being restricted to W_3, it is equivalent to the linear ODE

$$u_t = -1. \tag{3.64}$$

Curiously, this trivial evolution on W_3 describes the actual general phenomenon of asymptotic degeneracy of the PDEs near extinction. To show this, consider nonnegative and even in x (symmetric) solutions on $W_2 = \mathcal{L}\{1, x^2\}$. Noticing that the general solution of (3.64) is

$$u(x, t) = h(x) - t,$$

where $h(x)$ is arbitrary (initial data), we choose the parabolic profile $h(x) = T - d\,x^2$ with positive constants T and d. Next, bearing in mind the FBP, take the positive part to obtain the following pattern near the extinction time T:

$$u(x, t) = \big(T - t - d\,x^2\big)_+ \equiv (T - t)g(\xi) = (T - t)\big(1 - d\,\xi^2\big)_+, \tag{3.65}$$

where ξ denotes the spatial rescaled variable $x/\sqrt{T-t}$. For general even bell-shaped solutions of (3.63), this asymptotic behavior on W_2 suggests introducing the rescaled solution

$$u(x,t) = (T-t)w(\xi,\tau), \quad \tau = -\ln(T-t), \tag{3.66}$$

satisfying a typical singular (exponentially) perturbed equation,

$$w_\tau = -\tfrac{1}{2}\, w_\xi \xi + w - 1 + e^{-2\tau} F_4[w]. \tag{3.67}$$

Such PDEs occurred before for a number of the TFEs with absorption. Therefore, we expect that, as $\tau \to \infty$, bounded orbits $\{w(\cdot,\tau)\}$ converge to stationary profiles g, solving the linear ODE

$$-\tfrac{1}{2}\, g'\xi + g - 1 = 0,$$

that gives precisely the parabolic profiles $g(\xi)$ in (3.65). In this sense, solutions on the invariant subspace can detect the correct generic asymptotic behavior of extinction. As usual, a rigorous passage to the limit in (3.67) represents a difficult OPEN PROBLEM.

Case $n = 6$ ($\mu = \frac{1}{2}$). In this case, the diffusion-absorption equation takes the form

$$\boxed{u_t = -\left[u^3 u_{xxxx} + u^2 u_x u_{xxx} - \tfrac{3}{2} u^2 (u_{xx})^2 + \tfrac{3}{8}(u_x)^4\right] - 1.}$$

We have the following exact symmetric solutions on W_2:

$$u(x,t) = \left[C_1(t) + C_3(t)x^2\right]_+, \tag{3.68}$$

$$\begin{cases} C_1' = 6C_1^2 C_3^2 - 1, \\ C_3' = 12 C_1 C_3^3. \end{cases}$$

We cannot solve the system explicitly and will compute the generic asymptotic behavior of orbits near the extinction time, as $t \to T^-$. This yields $C_1(t) = T - t + \ldots$ and $C_3(t) \to -d < 0$, so that the behavior (3.65) remains valid asymptotically. Using the same rescaled variables (3.66) makes it possible to formulate the corresponding singular perturbed problem (3.67) for which the exact solutions on W_3 are expected to describe the generic asymptotic behavior. This is an OPEN PROBLEM.

Example 3.9 **(Blow-up)** *Case* $n = -2$ ($\mu = -\frac{3}{2}$). In (3.59), we take

$$g(v) = -\tfrac{3}{2}\, a v^{\frac{5}{3}} - \tfrac{3}{2}\, b v,$$

so, again choosing $a = -1$ and $b = 0$ yields the *TFE with source*

$$v_t = -\left(\tfrac{1}{v^2}\, v_{xxx}\right)_x + \tfrac{3}{2}\, v^{\frac{5}{3}}. \tag{3.69}$$

The reaction term $\sim v^{5/3}$ is superlinear for $v \gg 1$, so that blow-up is guaranteed for solutions of (3.69) with sufficiently large positive initial data $v_0(x)$.

Setting $v = u^{-3/2}$ yields quartic nonlinearities in the PDE

$$\boxed{\begin{aligned} u_t = -\big[&u^3 u_{xxxx} - 7u^2 u_x u_{xxx} - \tfrac{15}{2}\, u^2 (u_{xx})^2 \\ &+ 30\, u (u_x)^2 u_{xx} - \tfrac{105}{8} (u_x)^4\big] - 1. \end{aligned}} \tag{3.70}$$

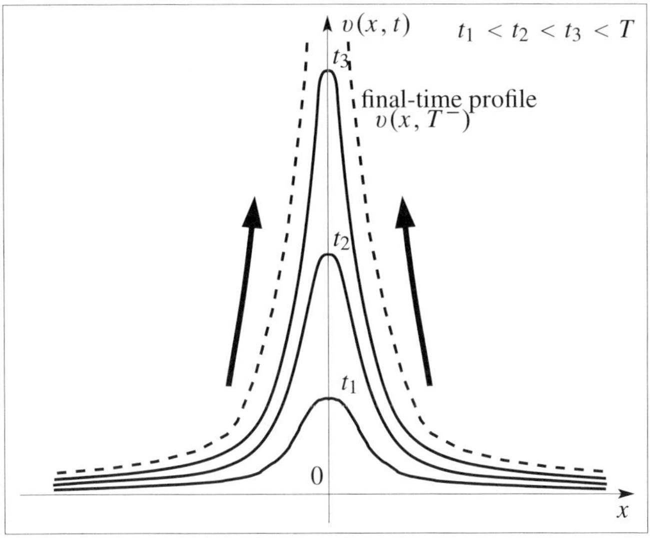

Figure 3.4 Single point blow-up described by positive exact solutions of (3.69) on W_3.

The blow-up behavior for (3.69) is equivalent to the quenching for (3.70). We look for nonnegative solutions (3.68) and obtain the DS for expansion coefficients

$$\begin{cases} C_1' = 30C_1^2C_3^2 - 1, \\ C_3' = -180C_1C_3^3. \end{cases}$$

Let $T > 0$ be the finite quenching time of a fixed bounded orbit. Then the DS gives as $t \to T^-$ the behavior $C_1(t) = T - t + \ldots$ and $C_3(t) \to d > 0$. This yields the blow-up behavior on W_3, as $t \to T^-$,

$$v(x,t) = u^{-\frac{3}{2}}(x,t) \approx (T-t)^{-\frac{3}{2}}\left(1 + d\,\xi^2\right)^{-\frac{3}{2}}, \quad \xi = \tfrac{x}{\sqrt{T-t}}. \tag{3.71}$$

These exact solutions are strictly positive, so that these are classical smooth solutions of the TFE (3.69) on $I\!R \times (0, T)$.

The blow-up evolution is shown in Figure 3.4. By (3.71), there occurs the *single point blow-up* as $x = 0$ only. Letting $t \to T^-$ yields the final-time profile

$$v(x, T^-) = d^{-\frac{3}{2}}|x|^{-3}(1 + o(1)) \quad \text{for } x \approx 0.$$

As shown in the previous example, these exact solutions asymptotically converge as $t \to T^-$ to the sufficiently smooth profiles given by the ODE $u_t = -1$. This implies that the blow-up patterns (3.71) are described as $t \to T^-$ by the ODE

$$v_t = \tfrac{3}{2}\, v^{\frac{5}{3}}.$$

This is easily seen from the equations (3.69), written for the rescaled function $v(x,t) = (T-t)^{-3/2}w(\xi, \tau)$ with the same spatial rescaling $\xi = x/\sqrt{T-t}$. The equation for

w contains a singular exponentially small perturbation as $\tau \to \infty$,

$$w_\tau = -\tfrac{1}{2}\, w_{\xi}\xi - \tfrac{3}{2}\, w + \tfrac{3}{2}\, w^{\frac{5}{3}} - \mathrm{e}^{-2\tau}\left(w^{-2} w_{\xi\xi\xi}\right)_\xi.$$

The phenomenon of such an asymptotic degeneracy of these TFEs near blow-up is an OPEN PROBLEM.

3.3 Exact solutions with zero contact angle

Let us return to exact solutions satisfying the more physically meaningful zero contact angle condition in (3.14). We again begin by studying the extinction phenomenon for the TFE with absorption, which is as a manifestation of an "evaporation" thin film phenomenon; see Section 3.2.

Example 3.10 **(Singularities and interfaces)** We consider the TFE with a strong non-Lipschitz absorption term,

$$v_t = -(v v_{xxx})_x - \sqrt{v}, \tag{3.72}$$

and now impose the zero contact angle condition, so, at each interface,

$$v = v_x = v v_{xxx} = 0 \quad \text{at} \quad x = s(t). \tag{3.73}$$

Setting $v = u^2$ yields an equation with the cubic operator,

$$\boxed{u_t = -\tfrac{1}{2}\, u (u^2)_{xxxx} - u_x (u^2)_{xxx} - \tfrac{1}{2} \equiv F_3[u] - \tfrac{1}{2},} \tag{3.74}$$

exhibiting the following invariant property.

Proposition 3.11 *Operator F_3 in (3.74) admits the subspace $W_3 = \mathcal{L}\{1, x, x^2\}$.*

Taking $u = C_1 + C_2 x + C_3 x^2$ yields

$$F_3[u] = -12(C_1 C_3 + C_2^2) C_3 - 60 C_2 C_3^2 x - 60 C_3^3 x^2 \in W_3. \tag{3.75}$$

For simplicity, setting $C_2(t) \equiv 0$ (meaning even and symmetric in x patterns) gives the following solutions of the original PDE (3.72) on $W_2 = \mathcal{L}\{1, x^2\}$:

$$v(x,t) = u^2(x,t) \equiv \left[C_1(t) + C_3(t) x^2\right]_+^2, \tag{3.76}$$

which, clearly, satisfy all three free-boundary conditions (3.73). In this case, (3.75) implies

$$\begin{cases} C_1' = -12 C_1 C_3^2 - \frac{1}{2}, \\ C_3' = -60 C_3^3. \end{cases} \tag{3.77}$$

The last ODE is solved,

$$C_3(t) = \pm \frac{1}{\sqrt{120}} \frac{1}{\sqrt{t}} \quad \text{for} \quad t > 0, \tag{3.78}$$

and, from the first,

$$C_1(t) = A_0 t^{-\frac{1}{10}} - \tfrac{5}{11}\, t, \quad \text{where } A_0 = \text{constant}. \tag{3.79}$$

Quenching and dynamic interface equation. Consider first the plus sign in (3.78), where, for convenience, setting $A_0 = \frac{5}{11}$ in (3.79) yields

$$u(x,t) = \left[\tfrac{5}{11}\, t^{-\frac{1}{10}} \left(1 - t^{\frac{11}{10}}\right) + \tfrac{1}{\sqrt{120}}\, \tfrac{1}{\sqrt{t}}\, x^2\right]_+. \tag{3.80}$$

This solution is strictly positive for all $t \in (0,1)$, and vanishes for the first time at $t = 1$ at the point $x = 0$ that describes the quenching phenomenon. After quenching, for $t > 1$, two interfaces $x = \pm s(t)$ appear, where

$$s(t) = R(t) \equiv \sqrt{\tfrac{5\sqrt{120}}{11}\, t^{\frac{2}{5}} \left(t^{\frac{11}{10}} - 1\right)}. \tag{3.81}$$

At the initial moment of time $t = 1^+$, the interface is not Lipschitz continuous,

$$s(t) = (30)^{\frac{1}{4}} \sqrt{t-1} + \dots,$$

and this poses a problem of a post quenching behavior of the extension of the solution after this evolution singularity occurred at $t = 1$.

In order to derive the regularity interface equation, we use the same elementary formulae (3.31) and calculate u_t from (3.74) for the solution (3.80). We obtain the following two-term expression:

$$s' = S[u] \equiv 6u_x u_{xx} + \tfrac{1}{2u_x} \quad \text{at } x = s(t) \text{ for } t > 1. \tag{3.82}$$

Recall that, by construction, our solution already satisfies the necessary three free-boundary conditions (3.73). Let us detect other interface equations by the time-parameterization $t = \frac{1}{30} \frac{1}{(u_{xx})^2}$ that follows from (3.80). Then (3.81) yields

$$s' = R'(t) \equiv R'\left(\tfrac{1}{30}\, \tfrac{1}{(u_{xx})^2}\right). \tag{3.83}$$

This equation is again of the second order, but is different from (3.82). Equating the right-hand sides of (3.82) and (3.83) yields a stationary second-order Neumann-type free-boundary condition

$$6u_x u_{xx} + \tfrac{1}{2u_x} = R'\left(\tfrac{1}{30}\, \tfrac{1}{(u_{xx})^2}\right) \quad \text{at } x = s(t) \text{ for } t > 1.$$

Towards local well-posedness of the FBP. Let us present some arguments justifying local existence/uniqueness of the FBP. We use the *von Mises transformation* by introducing the dependent variable

$$X = X(u,t), \quad \text{so} \quad X(u(x,t),t) \equiv x, \tag{3.84}$$

which is assumed to be well-defined in a neighborhood of the interface posed now at $u = 0$, at least for sufficiently small $t > 0$. Initial data at $t = 0$ are taken in the bell-shaped form $u_0(x) = [C_1(0) + C_3(0)x^2]_+$, where $C_1(0) > 0$ and $C_3(0) < 0$ are given constants, being initial data for the DS (3.77). This determines the unique solution on the invariant subspace with the known properties of extinction and singular interfaces. In terms of the new function (3.84), initial data are smooth,

$$X_0(u) = \sqrt{\tfrac{C_1(0) - u}{|C_3(0)|}} \quad \text{for small } u \geq 0. \tag{3.85}$$

Assuming that $X(u,t)$ is strictly monotone decreasing at least for small $u > 0$, and calculating derivatives yields

$$u_t = -\tfrac{X_t}{X_u}, \;\; u_x = \tfrac{1}{X_u}, \;\; u_{xx} = -\tfrac{X_{uu}}{(X_u)^3}, \;\; u_{xxx} = -\tfrac{X_{uuu}}{(X_u)^4} + \tfrac{3(X_{uu})^2}{(X_u)^5}, \; \dots .$$

On substitution into (3.74), the following PDE for X is derived:

$$\begin{aligned} X_t = H(X) \equiv -u^2\Big[\tfrac{X_{uuuu}}{(X_u)^4} - \tfrac{10X_{uu}X_{uuu}}{(X_u)^5} + \tfrac{15(X_{uu})^3}{(X_u)^6}\Big] \\ - u\Big[\tfrac{6X_{uuu}}{(X_u)^4} - \tfrac{21(X_{uu})^2}{(X_u)^5}\Big] - \tfrac{6X_{uu}}{(X_u)^4} + \tfrac{1}{2}\, X_u. \end{aligned} \tag{3.86}$$

Consider the principal fourth-order term in (3.86). Firstly, it follows that we need to deal with solutions satisfying $X_u(0,t) \neq 0$ (and finite) at the origin to exclude gradient singularities, so it we check that the initial function (3.85) satisfies this inequality of *transversality*. Secondly, (3.86) is degenerated at the boundary point $u = 0$, and the quadratic rate of degeneracy, $O(u^2)$, in the higher-order term makes it possible to pose some standard boundary conditions at $u = 0$.

The actual construction of the solution is as follows. The following decomposition of the solution is needed:

$$X = \hat{X} + \tilde{X}, \quad \text{where } \hat{X} \in \text{Span}\{1, u\} \text{ and } \tilde{X} = X - \hat{X}. \tag{3.87}$$

Projecting equation (3.86) onto $\hat{W}_2 = \text{Span}\{1, u\}$ yields the ODE–PDE system:

$$\begin{cases} \hat{X}_t = \hat{H}(X), \\ \tilde{X}_t = \tilde{H}(X), \end{cases} \tag{3.88}$$

where $\hat{H}$ and $\tilde{H}$ denote projections of H onto the corresponding subspaces. The first equation is a two-dimensional dynamical system, and the second equation is then a parabolic PDE with the principal degenerate operator

$$\mathbf{B} = -u^2 D_u^4. \tag{3.89}$$

This linear operator is symmetric in the weighted space $L^2_\rho((0,\delta))$, with a small $\delta > 0$ and $\rho = u^{-2}$. By classical theory of symmetric ordinary differential operators (see Naimark [432]), estimating the solutions of the eigenvalue equation

$$\mathbf{B}\psi = \pm \mathrm{i}\psi \quad \text{for } u \approx 0$$

gives four types of asymptotics $\psi_1(u) = O(1)$, $\psi_2(u) = O(u)$ (both contain logarithmic factors in higher-order terms), $\psi_3(u) = O(u^2)$, and $\psi_4(u) = O(u^3)$, where $\psi_{3,4} \in L^2_\rho$ and $\psi_{1,2} \notin L^2_\rho$. Therefore, the deficiency indices of $\mathbf{B}$ are (2,2), so that, any pair of self-adjoint Dirichlet boundary conditions at $u = 1$ gives a self-adjoint extension. We need to take the unique Friedrichs self-adjoint extension of $\mathbf{B}$ with a discrete spectrum, compact resolvent, and a complete, closed eigenfunction set. As usual, this extension is induced by the Dirichlet conditions at the singular endpoint $u = 0$, i.e., $w(0) = w'(0) = 0$ for functions from the domain $D(\mathbf{B})$. According to the representation (3.87), the parabolic problem for solutions $\tilde{X}$ is well-posed, provided that the transversality condition $X_u(0,t) \neq 0$, as well as suitable (free-boundary) conditions at $u = 0$, hold. Actually, the second equation in (3.88), after dividing by u^2, reduces to the PDE with the non-degenerate principal operator D_u^4, and the solution $\tilde{X} \in D(\mathbf{B})$ can be constructed by eigenfunction expansion, where the boundary conditions are key (see below). Note that, in view of analyticity of the coefficients of both equations, this gives a unique local analytic solution,

$$X(u,t) = \sum_{(k,j\geq 0)} c_{k,j} u^k t^j,$$

which can be constructed independently by substitution into (3.86) (convergence needs special involved majorant-type estimates).

Thus free-boundary conditions are crucial for local existence of smooth solutions. The dynamic interface equation (3.82) now reads

$$X_t = -\tfrac{6X_{uu}}{(X_u)^4} + \tfrac{1}{2} X_u \quad \text{at } u = 0. \tag{3.90}$$

Writing (3.86) in the form of

$$-u^2 \tfrac{X_{uuuu}}{(X_u)^4} + ... = X_t + \tfrac{6X_{uu}}{(X_u)^4} - \tfrac{1}{2} X_u,$$

in view of (3.90), the right-hand side is always zero at the boundary $u = 0$. It is precisely this that makes it possible to construct a unique local solution by semigroup theory based on the spectral properties of the linear degenerate operator (3.89).

Hence, the initial-boundary value problem for (3.86) in a small neighborhood of $u = 0$, $t = 0$ falls into the scope of the theory of higher-order parabolic PDEs; see [164, 205, 550]. For simplicity, we impose a pair of standard Dirichlet or Neumann boundary conditions at some fixed point $u = u_1 > 0$, small enough where there are no degeneracy and singularity. Once $\tilde{X}$ has been obtained, the DS for $\hat{X}$ in (3.88) gives the whole solution. The construction is local in u and t, and fails if extra degeneracy points appear in an arbitrarily small neighborhood of the origin $u = 0$. This would mean a new type of singularity which might affect the required regularity and the interface equation.

Extinction: singular perturbation problem. The extinction phenomenon corresponds to the minus sign in (3.78), so the explicit solution is

$$u(x,t) = \left[\tfrac{5}{11} t^{-\frac{1}{10}} \left(1 - t^{\frac{11}{10}}\right) - \tfrac{1}{\sqrt{120}} \tfrac{1}{\sqrt{t}} x^2\right]_+. \tag{3.91}$$

This solution has two interfaces for $t \in (0, 1)$ that coincide at the extinction time $t = 1$. The interface equations (3.82) or (3.83) remain the same. One can extract from (3.91) the asymptotic extinction pattern as $t \to 1^-$,

$$u(x,t) = \mathrm{e}^{-\tau} \left(\tfrac{1}{2} - \tfrac{1}{\sqrt{120}} y^2\right)_+ + O\left(\mathrm{e}^{-2\tau}\right), \tag{3.92}$$

with rescaled variables $y = x/\sqrt{1-t}$ and $\tau = -\ln(1-t) \to +\infty$. Therefore, asymptotic extinction theory uses the rescaled function $u(x,t) = \mathrm{e}^{-\tau} w(y,\tau)$, where w satisfies a singular perturbed first-order PDE,

$$w_\tau = -\tfrac{1}{2} w_y y + w - \tfrac{1}{2} + \mathrm{e}^{-\tau} F_3[w] \quad \text{for } \tau \gg 1. \tag{3.93}$$

For general solutions, the passage to the limit $\tau \to +\infty$ in (3.93) and stabilization to the stationary rescaled profile $g(y) = \left(\tfrac{1}{2} - \tfrac{1}{\sqrt{120}} y^2\right)_+$ given in (3.92) are difficult OPEN PROBLEMS. Translating (3.92) to the original solution $v(x,t)$ of the TFE (3.72), on the invariant subspace W_2, the following extinction behavior holds:

$$v(x,t) = (1-t)^2 \left[\left(\tfrac{1}{2} - \tfrac{1}{\sqrt{120}} \tfrac{1}{1-t} x^2\right)_+^2 + O(1-t)\right] \quad \text{as } t \to 1^-. \tag{3.94}$$

On self-similar extinction behavior. In addition, the TFE (3.72) admits standard similarity structures

$$v_S(x,t) = (1-t)^2 g(z), \quad z = \tfrac{x}{(1-t)^{3/4}}, \tag{3.95}$$

where, on substitution, $g \geq 0$ solves the ODE

$$\begin{aligned} &-(gg''')' - \tfrac{3}{4} g'z + 2g - \sqrt{g} = 0 \quad \text{for } z \in (0, z_0), \\ &g(z_0) = g'(z_0) = (gg''')(z_0) = 0, \quad g'(0) = g'''(0) = 0, \end{aligned} \tag{3.96}$$

with free-boundary conditions at some $z = z_0 > 0$ and symmetry ones at the origin $z = 0$. Then the ODE gives that, close to the interface,

$$g(z) = C(z_0 - z)^2 + C_1(z_0 - z)^3 + \dots, \ \text{ with } C > 0 \text{ and } 12C_1 = \tfrac{3}{2} z_0 - \tfrac{1}{\sqrt{C}}.$$

The space-time structure (3.95) is different from the behavior detected in (3.94) by exact solutions on the invariant subspace. Comparing the rescaled variables y in (3.92) and z in (3.95) yields that the self-similar extinction occurs on smaller sets $z = O(1)$, i.e.,

$$|x| \sim (1-t)^{\frac{3}{4}} \ll (1-t)^{\frac{1}{2}} \quad \text{as } t \to 1^-,$$

where the last sets are attributed to the rescaled variable $y = x/\sqrt{1-t} = O(1)$. Unlike the above case of explicit solutions, the ODE (3.96) for zero contact angle profiles $g(z)$ is difficult and existence/uniqueness (or nonexistence) are OPEN PROBLEMS.

We claim that a stable (generic) extinction behavior is given by the exact solution patterns, such as (3.94), so the asymptotic self-similar behavior given by (3.95) is expected to be unstable. This is an OPEN PROBLEM.

Oscillatory solutions of the Cauchy problem. The CP for the TFE with absorption (3.72) demands another setting for solutions of changing sign. It is known that, for $n \in \left(0, \frac{3}{2}\right)$, generalized strong (i.e., sufficiently regular) solutions $u(x,t)$ of the TFE

$$v_t = -\left(|v|^n v_{xxx}\right)_x \quad \text{in } I\!R \times I\!R_+$$

with bounded compactly supported initial data $v_0(x)$ should be oscillatory near interfaces in order to exhibit the *maximal regularity*. More precisely, such solutions behave near the left-hand interface $x = s(t)$ as

$$v(x,t) \sim (x - s(t))^{\frac{3}{n}} \varphi(\ln(x - s(t))) \quad \text{as } x \to s^+(t), \tag{3.97}$$

where $\varphi(s)$ denotes the *oscillatory component* and is typically a periodic changing sign solution of a fourth-order ODE. Explanations are postponed until Section 3.7, and here we discuss this phenomenon for the TFE with strong absorption (3.72).

First, in order to keep the parabolicity and absorption features, (3.72) is written as

$$v_t = -(|v| v_{xxx})_x - |v|^{-\frac{1}{2}} v \quad \text{in } I\!R \times I\!R_+. \tag{3.98}$$

Second, as usual, we first detect the oscillatory component for the TW solutions

$$\begin{aligned} &v(x,t) = f(y), \quad y = x - \lambda t, \quad \text{with the ODE} \\ &-\lambda f' = -(|f| f''')' - |f|^{-\frac{1}{2}} f. \end{aligned} \tag{3.99}$$

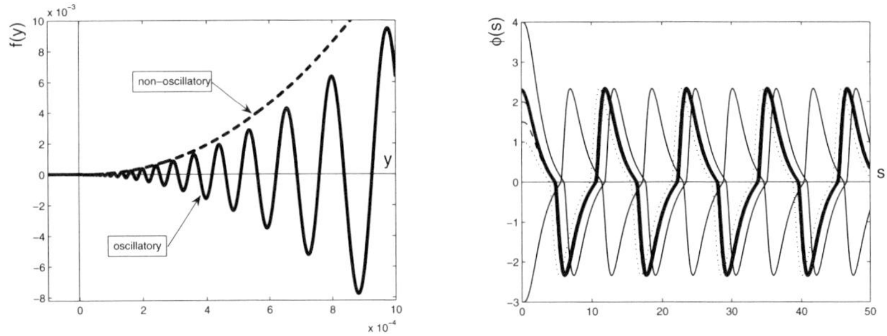

(a) two types of interfaces (b) stable periodic orbit

Figure 3.5 (a) Oscillatory and non-oscillatory behavior (3.101) close to the interface at $y = 0$, and (b) convergence to the stable periodic orbit of the ODE (3.102).

Assume that $f(y)$ has the interface at $y = 0$, with the trivial extension $f(y) \equiv 0$ for $y < 0$, and exhibits the "maximally regularity", i.e., the best which is admitted by this ODE. Then expressions like (3.97) with $n = 1$ are no longer true, i.e., the strong absorption term is seriously involved in the oscillatory behavior. The leading changing sign asymptotics are governed by the two-term ODE

$$(|f|f''')' + |f|^{-\frac{1}{2}} f = 0 \quad \text{for } y > 0, \quad f(0) = 0, \tag{3.100}$$

which is (3.99) with $\lambda = 0$. Hence, $v = f(x)$ is just a stationary solution of (3.98). The TW speed $\lambda \neq 0$ then enters the next expansion term, and this induces an interesting dynamic interface equation, which the interested reader may derive as usual (though the expansion is not easy). Instead of (3.97), with $n = 1$, we set

$$f(y) = y^{\frac{8}{3}} \varphi(s), \quad s = \ln y, \tag{3.101}$$

and, substituting into (3.100), derive the following ODE for $\varphi(s)$:

$$\begin{aligned} H[\varphi] \equiv \big[|\varphi|\big(\varphi''' + 5\varphi'' + \tfrac{22}{3}\,\varphi' + \tfrac{80}{27}\,\varphi\big)\big]' \\ + \tfrac{7}{3}\,|\varphi|\big(\varphi''' + 5\varphi'' + \tfrac{22}{3}\,\varphi' + \tfrac{80}{27}\,\varphi\big) + |\varphi|^{-\frac{1}{2}}\varphi = 0. \end{aligned} \tag{3.102}$$

We call $\varphi(s)$ in (3.101) the *oscillatory component* of $f(y)$. It turns out that (3.102) admits a periodic solution of changing sign, and (3.101) then yields the *oscillatory* behavior near the interface at $y = 0$; see Figure 3.5(a), the bold line. Figure 3.5(b) shows convergence to this unique asymptotically stable periodic solution of (3.102) for various initial data posed at $s = 0$. If the ODE (3.102) admits a positive equilibrium $\varphi(s) \equiv \varphi_0 > 0$ (actually, it does not), then (3.101) gives the *non-oscillatory* behavior close to the interface at $y = 0$; see Figure 3.5(a), the dashed line.

According to (3.101), the periodic $\varphi(s)$ gives the oscillatory behavior of solutions of the maximal regularity at interfaces for the TFE with absorption (3.98). A rigorous justification that the formulae (3.101) and (3.102) correctly describe a generic structure of multiple zeros of solutions $v(x,t)$ of the PDE is a hard OPEN PROBLEM.

Concerning similarity oscillatory extinction patterns satisfying the ODE (3.96) (with modified coefficients from (3.98)), numerical experiments show that such $g \neq$

0 is *nonexistent*, i.e., the generic extinction behavior in the CP is not self-similar, as in the FBP studied above.

Example 3.12 (Stabilization in the TFE) Consider the TFE (3.8) from lubrication theory with an unstable, backward parabolic diffusion term,

$$v_t = -(v v_{xxx})_x - (\sqrt{v}\, v_x)_x.$$

As shown in the previous example, we perform the change $v = u^2$ and obtain a PDE with the same cubic operator F_3 in (3.74) plus a quadratic term,

$$\boxed{u_t = -\tfrac{1}{2} u (u^2)_{xxxx} - u_x (u^2)_{xxx} - \tfrac{1}{2}(u^2)_{xx} - (u_x)^2.} \tag{3.103}$$

Clearly, the quadratic operator admits subspace W_3, and hence, by Proposition 3.11, there exist solutions (3.76) on W_2 driven by the DS

$$\begin{cases} C_1' = -12 C_1 C_3^2 - 2 C_1 C_3, \\ C_3' = -60 C_3^3 - 10 C_3^2. \end{cases}$$

Integrating the second ODE yields $-\frac{1}{C_3} + 6 \ln \left| 6 + \frac{1}{C_3} \right| = -10t \to -\infty$ as $t \to \infty$. Therefore, for any initial values $C_3(0) < 0$ and $C_1(0) > 0$, corresponding to bell-shaped compactly initial data $u_0(x)$, the following holds: $C_3(t) \to -\frac{1}{6}$ as $t \to +\infty$. Once $C_3(t)$ is known, one can find $C_1(t)$ from the first equation, showing that $C_1(t)$ stabilizes to a constant $B > 0$ as $t \to \infty$. Plugging these expansions into (3.76), we obtain the following asymptotic pattern:

$$u(x,t) = \left(B - \tfrac{1}{6} x^2\right)_+ + O\left(e^{-\frac{5}{3}t}\right) \quad \text{as } t \to +\infty,$$

with the uniform convergence on bounded intervals in x. These exact solutions describe exponentially fast stabilization to a single profile from the family of stationary solutions of (3.103) that are parameterized by a constant $B > 0$. Then, $\lambda_2 = -\frac{5}{3}$ in the exponential term is the first negative eigenvalue from the point (discrete) spectrum of the linear operator that appears in (3.103) after linearization about the stationary profile $(B - \frac{1}{6} x^2)_+$.

3.4 Extinction behavior for sixth-order thin film equations

Example 3.13 (Extinction, quenching, ...) A Stefan–Florin FBP and the CP can be posed for the sixth-order TFE with absorption

$$\boxed{u_t = (u u_{xxxxx})_x - 1,} \tag{3.104}$$

where the invariant subspace is $W_7 = \mathcal{L}\{1, x, x^2, ..., x^6\}$ on which the PDE is a seventh-order DS. Then, as shown in Section 3.2, we can describe the initial singularity via Dirac's mass, finite-time extinction, quenching and interface turning point patterns, and can develop local existence-uniqueness approach to the FBP based on the von Mises transformation (Example 3.10). For the sixth-order PDE (3.104), four free-boundary conditions should be posed. The Cauchy problem exhibits special oscillatory patterns to be briefly discussed.

Example 3.14 **(Zero contact angle conditions)** To get four conditions (3.15), another model is needed. Taking a sixth-order TFE with strong absorption,

$$v_t = (v v_{xxxxx})_x - v^{\frac{2}{3}},$$

we set $v = u^3$ and obtain a quartic operator F_4 in the PDE

$$\boxed{u_t = F_4[u] - \tfrac{1}{3}, \quad \text{where} \quad F_4[u] = \tfrac{1}{3}\big(u D_x^6 u^3 + 3u_x D_x^5 u^3\big).} \tag{3.105}$$

Similar to Proposition 3.11, F_4 admits $W_3 = \mathcal{L}\{1, x, x^2\}$, so there exists solution

$$v(x,t) = u^3(x,t) = \big[C_1(t) + C_3(t)x^2\big]_+^3 \quad (C_2 = 0), \tag{3.106}$$

satisfying all four free-boundary conditions

$$v = v_x = v_{xx} = v v_{xxxxx} = 0 \quad \text{at the interface } x = s(t).$$

The expansion coefficients solve the DS

$$\begin{cases} C_1' = \frac{1}{3}\, 6!\, C_1 C_3^3 - \frac{1}{3}, \\ C_3' = \frac{1}{3}\, 7!\, C_3^4. \end{cases}$$

The second ODE yields $C_3(t) = -\big(\frac{1}{7!\,t}\big)^{1/3} < 0$ for $t > 0$, which results in, e.g.,

$$u(x,t) = \big[\tfrac{7}{22}\, t^{-\frac{1}{21}}\big(1 - t^{\frac{22}{21}}\big) - \tfrac{1}{(7!t)^{1/3}}\, x^2\big]_+. \tag{3.107}$$

In rescaled variables, the extinction pattern as $t \to 1^-$ has the asymptotic structure

$$u(x,t) = \mathrm{e}^{-\tau}\big[\tfrac{1}{3} - \tfrac{1}{(7!)^{1/3}}\, y^2\big]_+ + O(\mathrm{e}^{-2\tau}), \ \ y = \tfrac{x}{\sqrt{1-t}}, \ \ \tau = -\ln(1-t).$$

This agrees with the singular perturbed problem for the original rescaled solution v given by (3.106), i.e., $v(x,t) = \mathrm{e}^{-3\tau} w(y,\tau)$, where w solves

$$w_\tau = -\tfrac{1}{2}\, w_y y + 3w - w^{\frac{2}{3}} + \mathrm{e}^{-\tau}(w w_{yyyyy})_y \quad \text{for} \quad \tau \gg 1.$$

The problem of passing to the limit $\tau \to +\infty$ in this PDE is OPEN.

Concerning the dynamic interface equation, formulae (3.31), together with the PDE (3.105), yield three interface operators on the right-hand side,

$$s' = S[u] \equiv -60(u_x)^2 u_{xxx} - 90 u_x (u_{xx})^2 + \tfrac{1}{3}\tfrac{1}{u_x} \tag{3.108}$$

(the first operator vanishes identically for $u \in W_3$). Other free-boundary conditions can be derived from the DS by using the parameterization via (3.107), $t = -\frac{8}{7!\,(u_{xx})^3}$. In the local von Mises variable $X = X(u,t)$, the free-boundary condition (3.108) (and others) is necessary for the well-posedness of the corresponding problem for a degenerate sixth-order symmetric linear differential operator. The analysis is similar to that shown in Example 3.10, but the mathematics and computations become technically more involved.

On oscillatory solutions of the Cauchy problem. Consider the PDE extended to $\{v < 0\}$ in the parabolic manner,

$$v_t = (|v| v_{xxxxx})_x - |v|^{-\frac{1}{3}} v \quad \text{in } I\!R \times I\!R_+. \tag{3.109}$$

TW analysis is the same, as in Example 3.10, and, instead of (3.99), the ODE is $-\lambda f' - (|f|f^{(5)})' - |f|^{-\frac{1}{3}} f$. Similarly, we neglect the left-hand side (or just set $\lambda = 0$),

$$(|f|f^{(5)})' - |f|^{-\frac{1}{3}} f = 0 \quad \text{for } y > 0, \quad f(0) = 0.$$

The oscillatory behavior exhibits a smoother envelope than in (3.101),

$$f(y) = y^{\frac{9}{2}} \varphi(s), \quad \text{with} \quad s = \ln y,$$

where the oscillatory component φ solves a sixth-order autonomous ODE which describes the changing sign character of solutions of (3.109). Later on, we will show a number of such ODEs that admit changing sign periodic solutions with oscillatory behavior, as shown in Figure 3.5(a). Such TW profiles satisfy at the interface at $y = 0$

$$f'(0) = f''(0) = f'''(0) = f^{(4)}(0) = 0,$$

and are smoother than $O(y^3)$ behavior given by (3.106) for the FBP.

Example 3.15 **(Stabilization)** Let us briefly consider stabilization in the lubrication equation with an unstable diffusion perturbation,

$$v_t = \left(v v_{xxxxx}\right)_x - \left(v^{\frac{1}{3}} v_x\right)_x.$$

As we have seen, setting $v = u^3$ yields the quartic operator F_4 in the PDE

$$\boxed{u_t = \tfrac{1}{3}\left(u D_x^6 u^3 + 3u_x D_x^5 u^3\right) - \left[u u_{xx} + 3(u_x)^2\right].} \tag{3.110}$$

We obtain solutions (3.106) on $W_2 = \mathcal{L}\{1, x^2\}$ with a slightly different DS

$$\begin{cases} C_1' = \frac{1}{3}\, 6!\, C_1 C_3^3 - 2C_1 C_3, \\ C_3' = \frac{1}{3}\, 7!\, C_3^4 - 14C_3^2. \end{cases}$$

Taking initial values $C_3(0) < 0$ and $C_1(0) > 0$ yields $C_3(t) \to -\frac{1}{2\sqrt{30}}$ and the following stabilization of solutions on W_2:

$$u(x, t) = \left(B - \tfrac{1}{2\sqrt{30}} x^2\right)_+ + O\left(\mathrm{e}^{-14t/\sqrt{30}}\right) \quad \text{as } t \to \infty.$$

This shows that exact solutions on W_2 belong to the stable manifold of unstable stationary solutions of (3.110), and that stabilization is exponentially fast.

3.5 Quadratic models: trigonometric and exponential subspaces

3.5.1 Blow-up and stability on W_3 for fourth-order TFEs

Consider thin film operators with an extra zero-order quadratic term,

$$\boxed{u_t = F[u] \equiv -u u_{xxxx} + \beta u_x u_{xxx} + \gamma\, (u_{xx})^2 + \delta u^2,} \tag{3.111}$$

where $\delta \neq 0$, so that F cannot preserve a nontrivial polynomial subspace.

Proposition 3.16 *Operator F in (3.111) preserves:*

(i) $W_3 = \mathcal{L}\{1, \cos x, \sin x\}$, *provided that*

$$\delta = 1 - \beta - \gamma; \tag{3.112}$$

(ii) $W_3 = \mathcal{L}\{1, \cosh x, \sinh x\}$, *provided that the same condition* (3.112) *holds;*
(iii) $W_5 = \mathcal{L}\{1, \cos x, \sin x, \cos 2x, \sin 2x\}$, *provided that*

$$\begin{cases} 10\beta + 8\gamma + 2\delta = 17, \\ 16\beta + 16\gamma + \delta = 16. \end{cases} \tag{3.113}$$

In the case of (i), which will be analyzed later on in greater detail, we find that the TFE (3.111) possesses solutions

$$u(x,t) = C_1(t) + C_2(t)\cos x + C_3(t)\sin x,$$

$$\begin{cases} C_1' = \delta C_1^2 - \beta\big(C_2^2 + C_3^2\big), \\ C_{2,3}' = (2\delta - 1)C_1 C_{2,3}. \end{cases} \tag{3.114}$$

In the case of (iii) where (3.113) holds, there exist even solutions

$$u(x,t) = C_1(t) + C_2(t)\cos x + C_3(t)\cos 2x,$$

$$\begin{cases} C_1' = \delta C_1^2 + \frac{1}{2}(\delta + \gamma - \beta - 1)C_2^2 - 16\beta C_3^2, \\ C_2' = (2\delta - 1)C_1 C_2 - 10\beta C_2 C_3, \\ C_3' = (2\delta - 16)C_1 C_3 + \frac{1}{2}(\delta + \gamma + \beta - 1)C_2^2. \end{cases}$$

It is not difficult to derive the DS on the full subspace W_5 in (iii). A typical example of the operator preserving W_5 is (set $\delta = 1$ in (3.113))

$$F[u] = -u u_{xxxx} + \tfrac{15}{4}\, u_x u_{xxx} - \tfrac{45}{16}\,(u_{xx})^2 + u^2.$$

Example 3.17 (**Regional blow-up: stability on** W_3) We next consider the TFE with source

$$\boxed{u_t = F[u] \equiv -(u u_{xxx})_x + 2u^2.} \tag{3.115}$$

Let us see how the source term $2u^2$ affects evolution properties of the solutions. As a physical motivation of the model, let us mention that such a reaction term in the TFE can be associated with a *condensation* of the film substance from the surrounding space (see the beginning of Section 3.2). The equation (3.115) is parabolic in the positivity domain $\{u > 0\}$, so, in general, we have to deal with nonnegative solutions (by taking the positive part, as usual) of the corresponding FBPs, with the functional setting as used in Sections 3.2 and 3.3. These well-posedness phenomena are not essential for the current stability analysis on the invariant subspace W_3.

Consider solutions on $W_3 = \mathcal{L}\{1, \cos x, \sin x\}$, i.e.,

$$u(x,t) = C_1(t) + C_2(t)\cos x + C_3(t)\sin x. \tag{3.116}$$

The corresponding DS (3.114) is

$$\begin{cases} C_1' = 2C_1^2 + C_2^2 + C_3^2, \\ C_2' = 3C_1 C_2, \\ C_3' = 3C_1 C_3. \end{cases} \tag{3.117}$$

First of all, the DS (3.117) admits solutions for which $C_1 = C_2$ and $C_3 = 0$. Hence, C_1 satisfies the single ODE

$$C_1' = 3C_1^2 \quad \Longrightarrow \quad C_1(t) = \tfrac{1}{3}\,\tfrac{1}{T-t}, \tag{3.118}$$

where $T > 0$ is the *blow-up time* of the explicit solution

$$u_S(x,t) = \tfrac{1}{3}\,\tfrac{1}{T-t}(1+\cos x) \equiv \tfrac{2}{3}\,\tfrac{1}{T-t}\cos^2(\tfrac{x}{2}). \tag{3.119}$$

Taking the central wave of $\cos^2(\frac{x}{2})$ and setting $u_S = 0$ for $|x| \ge \pi$ yields a localized blow-up solution satisfying, as $t \to T^-$,

$$u_S(x,t) \to \infty \quad \text{on the interval } \{|x| < \pi\} \quad \text{and} \quad u_S(\pm\pi, t) = 0.$$

The measure of this *localization domain* $L_S = 2\pi$ is called the *fundamental length* of the *regional blow-up*. Note that (3.119) is a smooth solution satisfying conditions (3.14) of zero contact angle and zero-flux on the stationary interfaces at $x = \pm\pi$, with $n = 1$. The local well-posedness of this FBP can be checked via the von Mises transformation, as done in Example 3.10, but we cannot guarantee that such a local unique construction can be extended up to the blow-up time, i.e., the solution will remain strictly monotone near the interfaces. Recall that the existence of the similarity solution (3.119) was interpreted as the result of invariance of the 1D subspace $W_1^+ = \mathcal{L}\{\frac{2}{3}\cos^2(\frac{x}{2})\}$ under the quadratic operator in (3.115). Another subspace is $W_1^- = \mathcal{L}\{\frac{2}{3}\sin^2(\frac{x}{2})\}$. In the CP, solutions are oscillatory; see Section 3.7.

Returning to the DS on W_3, let us emphasize another important aspect of evolution on W_3. We will show that all the blow-up orbits on W_3 asymptotically converge to the above similarity solution u_S on W_1; that means its asymptotic stability on W_3.

Proposition 3.18 *Let blow-up happen in the DS* (3.117)*, i.e.,* $C_1(t) \to +\infty$ *as* $t \to T^- < \infty$*. Then there exists a constant translational parameter* a *such that, uniformly in* $x \in I\!R$*,*

$$(T-t)u(x,t) \to \tfrac{2}{3}\cos^2(\tfrac{x+a}{2}) \quad \text{as } t \to T. \tag{3.120}$$

Proof. Multiplying the second and the third ODE in (3.117) by C_3 and C_2, respectively, and subtracting, one obtains

$$\left(\tfrac{C_3}{C_2}\right)' = 0 \quad \Longrightarrow \quad \exists\, B \in I\!R \text{ such that } C_3 = BC_2.$$

By (3.116), this gives a in (3.120), but we need to study the DS

$$C_1' = 2C_1^2 + (1+B^2)C_2^2, \quad C_2' = 3C_1C_2. \tag{3.121}$$

Setting $C_2 = C_1 P$ in the equivalent first-order ODE yields

$$\tfrac{\mathrm{d}C_2}{\mathrm{d}C_1} = \tfrac{3C_1C_2}{2C_1^2+(1+B^2)C_2^2} \quad \Longrightarrow \quad C_1\tfrac{\mathrm{d}P}{\mathrm{d}C_1} = \tfrac{P-(1+B^2)P^3}{2+(1+B^2)P^2}.$$

This is easily explicitly integrated, but, for our purpose, it suffices to note that

$$\int \tfrac{[2+(1+B^2)P^2]\,\mathrm{d}P}{P-(1+B^2)P^3} = \int \tfrac{\mathrm{d}C_1}{C_1} = \ln C_1 \to \infty$$

as $t \to T^-$. In this case, the denominator on the left-hand side tends to zero so

$P^2 \to \frac{1}{1+B^2}$ and $P \to \frac{1}{\sqrt{1+B^2}}$ (or, which is similar, $\to -\frac{1}{\sqrt{1+B^2}}$; both equilibria $\pm\frac{1}{\sqrt{1+B^2}}$ are *stable*). Summing up, we have that, as $t \to T$,

$$u(x,t) \approx C_1\Big(1 + \tfrac{1}{\sqrt{1+B^2}} \cos x + \tfrac{B}{\sqrt{1+B^2}} \sin x\Big). \tag{3.122}$$

Substituting $C_2 = C_1 P \approx \frac{C_1}{\sqrt{1+B^2}}$ into the first equation in (3.121), we conclude that $C_1' \approx 3C_1^2$, and hence,

$$C_1(t) = \tfrac{1}{3}\,\tfrac{1}{T-t}(1 + o(1))$$

as in (3.118). Finally, (3.122) implies that (3.120) holds with $\tan a = -B$. □

Thus, a symmetrization of the orbits of the DS occurs near the blow-up time, and solutions (3.116) converge as $t \to T^-$ to the separate variables pattern (3.119) up to translation in space. In the general treatment of such a stability, introducing the rescaled function

$$u(x,t) = \tfrac{1}{T-t}\, v(x,\tau), \quad \text{where } \tau = -\ln(T-t) \to +\infty,$$

yields the fourth-order rescaled PDE

$$v_\tau = F_*[v] \equiv -(v v_{xxx})_x + 2v^2 - v. \tag{3.123}$$

The asymptotic stability means that the rescaled solution tends to the similarity profile, i.e., setting for simplicity $a = 0$,

$$v(x,\tau) \to g(x) = \tfrac{2}{3}\cos^2\big(\tfrac{x}{2}\big) \quad \text{as } \tau \to \infty,$$

where $g(x)$ is indeed a stationary solution of (3.123) satisfying

$$-(g g''')' + 2g^2 - g = 0.$$

In rescaled variables, the stability of the blow-up solution reduces to the standard stabilization to stationary solutions in equation (3.123). For the fourth-order TFEs, this problem is OPEN. For the related second-order parabolic PME

$$v_\tau = (v v_x)_x + 2v^2 - v, \tag{3.124}$$

such stability of blow-up similarity solutions and stabilization results are well known since the 1980s; see [509, Ch. 4], where the Lyapunov and comparison techniques are key. (3.124) is a gradient system in L^2-metric (multiplication by $(v^2)_\tau$ and integration over $\mathbb{R}$ yields a Lyapunov function), while any potential and gradient properties for (3.123) are unknown. Also, it is easy to show that the linearized operator

$$F_*'[g] = -\tfrac{1}{3}(1+\cos x)\tfrac{\mathrm{d}^4}{\mathrm{d}x^4} + \tfrac{1}{3}\sin x\,\tfrac{\mathrm{d}^3}{\mathrm{d}x^3} - \tfrac{1}{3}\sin x\,\tfrac{\mathrm{d}}{\mathrm{d}x} + \big(\tfrac{1}{3} + \cos x\big) I$$

with the Dirichlet boundary conditions at $x = \pm\pi$ is not symmetric in $L^2_\rho(-\pi,\pi)$ for any weight $\rho \geq 0$, and hence, does not admit a self-adjoin extension. The point spectrum of $F_*'[g]$ is UNKNOWN. We expect that some useful estimates of the spectrum can be obtained, at least, by a hybrid analytic-numerical approach to guarantee linearized stability of the similarity profile g. As usual in blow-up stability problems, the positive eigenvalue $\lambda = 1$ with the eigenfunction g, which directly follows from the PDE (3.123), must be excluded, since it corresponds to the shifting of the blow-up time T that is fixed by scaling. Eigenvalue $\lambda = 0$ is related to shifting in x, and this unstable mode can be forbidden by, say, symmetry about the origin.

On single point blow-up patterns on exponential W_3. For the invariant subspace $W_3 = \mathcal{L}\{1, \cosh x, \sinh x\}$, the DS (3.117) remains the same, but the blow-up behavior is completely different. Namely, from the DS (3.121) with $B = 0$, we have

$$C_2 = -C_1 + 3^{\frac{1}{3}} A C_1^{\frac{1}{3}} + \dots \quad \text{as} \quad C_1 \to +\infty \quad (A > 0).$$

Estimating $C_{1,2}(t)$, this gives the blow-up behavior on W_3 as $t \to T^-$,

$$u(x,t) = (T-t)^{-\frac{1}{3}} \big(A - \tfrac{1}{6} \tfrac{x^2}{(T-t)^{2/3}} + \dots\big) = (T-t)^{-\frac{1}{3}} g(\xi) + \dots$$

for small $|x| > 0$, where $\xi = x/(T-t)^{1/3}$ is the rescaled spatial variable and $g(\xi) = (A - \frac{1}{6}\xi^2)$ is the limit rescaled profile. Taking the positive part for proper FBP setting, we observe two symmetric interfaces

$$s_{\pm}(t) = \pm\sqrt{6A}(T-t)^{\frac{1}{3}} + \dots$$

that collapse at the origin at $t = T^-$. Such blow-up patterns with a lower blow-up rate $O((T-t)^{-1/3})$ are expected to be evolutionary *unstable* (an OPEN PROBLEM); cf. the stable rate $\sim O((T-t)^{-1})$ in (3.120).

3.5.2 Sixth-order model with blow-up

It is easy to propose such a model, e.g.,

$$\boxed{u_t = (u u_{xxxxx})_x + 2u^2,}$$

with the same subspace $W_3 = \mathcal{L}\{1, \cos x, \sin x\}$ and the same DS (3.117). As in Example 3.17, there exist localized blow-up solutions on W_2, and the similarity separate variables solution is stable on W_3. Note that $W_3 = \mathcal{L}\{1, \cosh x, \sinh x\}$ is invariant for the operator with absorption $F[u] = (u u_{xxxxx})_x - 2u^2$.

3.5.3 Partially invariant subspaces (modules) for quadratic operators

These results have a counterpart in the class of quadratic second-order operators; see Example 1.43. Let us introduce the following one-parameter family of quadratic fourth-order operators:

$$F_\beta[u] = -u u_{xxxx} + \beta (u_{xx})^2 + (1-\beta) u_x u_{xxx} \quad (\beta \in I\!R), \tag{3.125}$$

which will be used later on in other applications.

Proposition 3.19 *For any constant $\gamma \neq 0$, operator* (3.125) *admits*

$$W_3^+ = \mathcal{L}\{1, \cosh \gamma x, \sinh \gamma x\} \quad \textit{and} \quad W_3^- = \mathcal{L}\{1, \cos \gamma x, \sin \gamma x\}. \tag{3.126}$$

Recall that, for second-order operators in Section 1.4, similar different trigonometric and exponential subspaces occurred for the unique remarkable operator

$$F_1[u] = uu_{xx} - (u_x)^2 \quad \big(= F_{\rm rem}[u]\big). \tag{3.127}$$

Its natural fourth-order analogy corresponds to $\beta = 1$ in (3.125),

$$F_1[u] = -uu_{xxxx} + (u_{xx})^2. \tag{3.128}$$

Using exponential functions. Continuing to detect other invariant properties, using Proposition 3.19, consider the following differential operator for functions $u = u(x,t)$ of two independent variables:

$$\hat{F}[u] = u_t - F_\beta[u].$$

We choose solutions in the form of

$$u(x,t) = C_1(t) + C_2(t)x + C_3(t){\rm e}^{\gamma(t)x}, \tag{3.129}$$

with a function $\gamma(t)$ to be specified. Our analysis here is similar to that shown in Example 1.37, so for functions (3.129) we find

$$\begin{aligned}\hat{F}[u] = C_1' + C_2'x + \big[C_3' + \gamma^4 C_1 C_3 - (1-\beta)\gamma^3 C_2 C_3\big]{\rm e}^{\gamma x} \\ + \big(\gamma' + \gamma^4 C_2\big)C_3 x{\rm e}^{\gamma x}.\end{aligned}$$

Then the ODE $\gamma' = -\gamma^4 C_2$ is a characterization of an invariant set M in the linear subspace (a partially invariant module) W_3, so that $\hat{F} : M \to W_3$.

Example 3.20 The fourth-order PDE

$$\boxed{u_t = -uu_{xxxx} + \beta(u_{xx})^2 + (1-\beta)u_x u_{xxx} + \mu u + \nu}$$

admits solutions (3.129), where the coefficients satisfy the DS

$$\begin{cases} C_1' = \mu C_1 + \nu, \\ C_2' = \mu C_2, \\ C_3' = -\gamma^4 C_1 C_3 + (1-\beta)\gamma^3 C_2 C_3 + \mu C_3, \\ \gamma' = -\gamma^4 C_2. \end{cases}$$

The second and the fourth ODEs give all possible functions $\gamma(t)$.

Using trigonometric functions. According to Proposition 3.19, we now deal with operator (3.128). Then, the fifth-order operator

$$\tilde{F}[u] = \tfrac{\rm d}{{\rm d}x}F_1[u] \equiv -uu_{xxxxx} - u_x u_{xxxx} + 2u_{xx}u_{xxx} \tag{3.130}$$

also preserves 3D trigonometric subspaces in (3.126) for arbitrary $\gamma \neq 0$. The extra differentiation in (3.130) is necessary to create a single ODE for $\gamma(t)$. Recall that, similarly, in Example 1.36 we used the quadratic operator of the RPJ equation as the derivative of (3.127). Such invariant properties of the corresponding operator

$$\hat{F}[u] = u_t - \tilde{F}[u]$$

are listed in the following example.

Example 3.21 The sixth-order semilinear parabolic equation

$$\boxed{u_t = \varepsilon u_{xxxxxx} - u u_{xxxxx} - u_x u_{xxxx} + 2u_{xx}u_{xxx} + \mu u + \nu}$$

possesses exact solutions

$$u(x,t) = C_1(t) + C_2(t)x + C_3(t)\cos(\gamma(t)x) + C_4(t)\sin(\gamma(t)x),$$

$$\begin{cases} C_1' = \mu C_1 + \nu, \\ C_2' = \mu C_2, \\ C_3' = -\varepsilon\gamma^6 C_3 - \gamma^5 C_1 C_4 - \gamma^4 C_2 C_3 + \mu C_3, \\ C_4' = -\varepsilon\gamma^6 C_4 + \gamma^5 C_1 C_3 - \gamma^4 C_2 C_4 + \mu C_4, \\ \gamma' = -\gamma^5 C_2. \end{cases}$$

3.6 $2m$th-order thin film operators and equations

We describe basic subspaces and extensions for $2m$th-order TFEs, such as (3.16).

3.6.1 Basic polynomial subspaces

Consider the standard higher-order quadratic thin film operator

$$F[u] = (-1)^{m+1} D_x\big(u D_x^{2m-1} u\big).$$

Since $M(l) = 2l - 2m$, i.e., $F[x^l] \sim x^{2l-2m}$, equating $2l - 2m = l$ yields the following basic subspace:

$$W_{2m+1} = \mathcal{L}\{1, x, x^2, ..., x^{2m}\}. \tag{3.131}$$

This makes it possible to study various singularity formation phenomena, such as finite time extinction, quenching, turning points, interface dynamics, etc., for the corresponding *TFE with absorption*

$$\boxed{u_t = (-1)^{m+1} D_x\big(u D_x^{2m-1} u\big) - 1.}$$

The corresponding exact solutions are given by

$$u(x,t) = \big[C_1(t) + C_2(t)x + C_3(t)x^2 + ... + C_{2m+1}(t)x^{2m}\big]_+,$$

where $\{C_i(t)\}$ solve a DS. These nonnegative functions are weak solutions of FBPs with Stefan–Florin free-boundary conditions, which can be detected in a manner similar to above lower-order models (cf. a simpler derivation below).

Example 3.22 **(Cubic TFE)** In the cubic equation

$$\boxed{u_t = F[u] \equiv (-1)^{m+1} D_x^{2m}\big(au^3 + bu^2 + cu\big)}$$

the operator admits the basic subspace $W_{m+1} = \mathcal{L}\{1, x, x^2, ..., x^m\}$. Free-boundary and asymptotic analysis of the extinction and quenching behavior in the corresponding absorption model

$$\boxed{u_t = F[u] - 1 \quad \text{on} \quad W_{m+1},}$$

are similar to the fourth-order models in Section 3.2.

3.6.2 Solutions with zero contact angle conditions

In order to obtain solutions $v(x,t)$ satisfying $m+1$ free-boundary conditions at the interface, including $m-1$ generalized zero contact angle conditions and a zero-flux condition, i.e.,

$$v = v_x = \ldots = D_x^{m-1} v = v D_x^{2m-1} v = 0 \quad \text{at} \quad x = s(t), \tag{3.132}$$

we need to introduce models with specific algebraic and invariant properties. As above, we consider two types of such equations.

Example 3.23 (**Strong absorption**) The first model is the $2m$th-order TFE with strong absorption

$$v_t = (-1)^{m+1} D_x\big(v D_x^{2m-1} v\big) - v^{\frac{m-1}{m}}. \tag{3.133}$$

Setting $v = u^m$ yields the operator F_{m+1} of the algebraic homogenuity $m+1$,

$$\boxed{u_t = F_{m+1}[u] - \tfrac{1}{m} \equiv (-1)^{m+1}\big(\tfrac{1}{m}\, u D_x^{2m} u^m + u_x D_x^{2m-1} u^m\big) - \tfrac{1}{m}.} \tag{3.134}$$

F_{m+1} admits $W_3 = \mathcal{L}\{1, x, x^2\}$, and hence, the FBP (3.133), (3.132) has solutions

$$v(x,t) = u^m(x,t) = \big[C_1(t) + C_3(t) x^2\big]_+^m, \tag{3.135}$$

$$\begin{cases} C_1' = (-1)^{m+1} \frac{(2m)!}{m}\, C_1 C_3^m - \frac{1}{m}, \\ C_3' = (-1)^{m+1} \frac{(2m+1)!}{m}\, C_3^{m+1}. \end{cases}$$

The dynamic interface equation is readily derived from (3.134),

$$s' = -\tfrac{1}{u_x}\, u_t = S[u] \equiv (-1)^m D_x^{2m-1} u^m + \tfrac{1}{m u_x} \quad \text{at} \;\; x = s(t).$$

This equation is the main regularity condition for solvability of the degenerate PDE, written in terms of the von Mises variable $X = X(u,t)$ near the interface. The manipulations become more technical for large m than those performed before for lower-order equations. Exact solutions (3.135) describe singular phenomena of quenching, extinction, and others.

On the Cauchy problem. For solutions of changing sign that exhibit the maximal regularity, we consider the TFE with absorption

$$v_t = (-1)^{m+1} D_x\big(|v| D_x^{2m-1} v\big) - |v|^{-\frac{1}{m}} v \quad \text{in} \;\; I\!R \times I\!R_+. \tag{3.136}$$

As in Example 3.10, we use the TWs $v(x,t) = f(x - \lambda t)$ satisfying the ODE

$$-\lambda f' = (-1)^{m+1}(|f| f^{(2m-1)})' - |f|^{-\frac{1}{m}} f,$$

or, neglecting the non-stationary, λ-dependent term for small f,

$$(-1)^{m+1}(|f| f^{(2m-1)})' - |f|^{-\frac{1}{m}} f = 0 \quad \text{for} \;\; y > 0, \quad f(0) = 0.$$

This gives the following oscillatory structure of solutions:

$$f(y) = y^{\gamma} \varphi(s), \quad s = \ln y, \quad \text{where} \;\; \gamma = \tfrac{2m^2}{m+1}, \tag{3.137}$$

which applies to many higher-order ODEs. Here the oscillatory component $\varphi(s)$

solves a $2m$th-order autonomous ODE. Such ODEs will be shown to admit oscillatory, periodic solutions that describe different types of multiple zeros of solutions of the PDE (3.136); see Figure 3.5(a) as an illustration. The exponent γ in the envelope in (3.137) is such that

$$\gamma > 2m-2, \quad \text{so that } f(y) = o(y^{2m-2}) \text{ as } y \to 0.$$

Hence, maximal regularity solutions satisfy on the interface the conditions

$$f'(0) = \ldots = f^{(2m-2)}(0) = 0,$$

and are smoother than those of the FBP, which are $O(y^m)$, as (3.132) suggests.

Example 3.24 **(Backward diffusion perturbation)** The second model includes a special divergent second-order operator generating an extra instability feature of the thin film flow,

$$v_t = (-1)^{m+1} D_x\big(v D_x^{2m-1} v\big) - \big(v^{\frac{1}{m}} v_x\big)_x, \quad \text{or}$$

$$\boxed{u_t = (-1)^{m+1}\big(\tfrac{1}{m}\, u D_x^{2m} u^m + u_x D_x^{2m-1} u^m\big) - \big[u u_{xx} + m(u_x)^2\big]}$$

for $v = u^m$. For solutions (3.135), the following DS is obtained:

$$\begin{cases} C_1' = (-1)^{m+1}\frac{(2m)!}{m}\, C_1 C_3^m - 2C_1C_3, \\ C_3' = (-1)^{m+1}\frac{(2m+1)!}{m}\, C_3^{m+1} - 2(2m+1)C_3^2. \end{cases}$$

It follows from the second ODE that, given an initial value $C_3(0) < 0$, the following holds: $C_3(t) \to -a = -[(2m-1)!]^{-1/(m-1)}$ as $t \to \infty$. This means the exponential stabilization to unstable stationary solutions,

$$u(x,t) = (B - ax^2)_+ + O(e^{-\gamma_m t}), \quad \text{with } \gamma_m = (m-1)(2m+1)a,$$

where $B > 0$ is a constant, depending on the initial data.

3.6.3 *Extensions of polynomial subspaces*

Consider a generalization of the thin film operator,

$$F[u] = (-1)^{m+1}\big(u D_x^{2m} u + \beta u_x D_x^{2m-1} u\big) \quad (\beta \in I\!R). \tag{3.138}$$

Here we keep two monomial operators of the same total differential order, where $i + j = k$, with $k = 2m$; cf. the general quadratic operator (1.86). The next result is proved directly.

Proposition 3.25 *Operator* (3.138) *preserves the following extensions of the subspace* (3.131)*:*

$$W_{2m+2} = \mathcal{L}\{1, x, \ldots, x^{2m}, x^{2m+1}\} \quad \textit{iff } \beta = \beta_1 = -\tfrac{2}{2m+1}; \tag{3.139}$$

$$W_{2m+3} = \mathcal{L}\{1, x, \ldots, x^{2m}, x^{2m+1}, x^{2m+2}\} \quad \textit{iff } \beta = \beta_2 = -\tfrac{3}{2m+2}. \tag{3.140}$$

Further extensions of polynomial subspaces are not possible for operator (3.138), but can be obtained for other similar operators containing extra monomials of the same exponent $M(l)$ with more free parameters β, γ, etc. (see (3.142) below). Constructing the corresponding exact solutions leads to higher-order DSs. In the case of (3.140), the DS is simplified if we take only even polynomials by introducing

$$W_{m+2}^{\text{even}} = \mathcal{L}\{1, x^2, ..., x^{2m}, x^{2m+2}\} \tag{3.141}$$

that is also invariant under operator (3.138) for $\beta = \beta_2$. Such subspaces are sufficient for studying most singularity, free boundary, and asymptotic phenomena for the TFE with symmetric initial data. Using subspaces of even functions, we formulate the following pretty general observation that is convenient for future computations and was observed before in particular cases.

Proposition 3.26 *If subspace* (3.141) *is invariant under a polynomial quadratic operator with constant coefficients* (*again, in each monomial,* $i + j = k = 2m$)

$$F[u] = (-1)^{m+1}\big(uD_x^{2m}u + \beta u_x D_x^{2m-1}u + \gamma u_{xx} D_x^{2m-2}u + ...\big) \tag{3.142}$$

for some $\beta = \tilde{\beta}_2$ *depending on* γ *and other parameters. Then subspace in* (3.140) *is also invariant under* F.

Proof. Indeed, taking

$$u = C_1 + C_3 x^2 + ... + C_{2m+1} x^{2m} + C_{2m+3} x^{2m+2} \tag{3.143}$$

and controlling the higher-degree term only yields

$$F[u] = (\beta - \tilde{\beta}_2) A_m (C_{2m+3})^2 x^{2m+4} + ..., \tag{3.144}$$

where $A_m \neq 0$ is a constant. The second (omitted) term contains $x^{2m+2} \in W_{m+2}^{\text{even}}$. Hence, always $F[u] \in W_{m+2}^{\text{even}}$ iff $\beta = \tilde{\beta}_2$. Taking now u from subspace (3.140),

$$u = C_{2m+3} x^{2m+2} + C_{2m+2} x^{2m+1} + ..., \quad \text{where} \quad C_{2m+3} \neq 0$$

($C_{2m+3} = 0$ is easy), and performing the translation $x \mapsto x - a$ gives

$$u = C_{2m+3} x^{2m+2} + \big[C_{2m+2} - (2m+2) a C_{2m+3}\big] x^{2m+1} + [\cdot] x^{2m} + ...$$

Choose now a such that $C_{2m+2} - (2m+2) a C_{2m+3} = 0$ to vanish the second coefficient. Then we get that the last two terms contain x^{2m+2} and x^{2m} as in (3.143). Hence, the argument for even degree polynomials applies, so the term with x^{2m+3} does not appear in $F[u]$. □

Similarly, a more general result holds, which applies to arbitrary quadratic operators with monomials of the same total order such as (1.86).

Lemma 3.27 *If operator* (3.142) *admits* $W_{2m+2}^{(1)} = \mathcal{L}\{1, x, ..., x^{2m-1}, x^{2m}, x^{2m+2}\}$ (*a subspace with "1-gap"*), *then it admits* $W_{2m+3} = \mathcal{L}\{1, x, ..., x^{2m}, x^{2m+1}, x^{2m+2}\}$ (*the full subspace*).

Remark 3.28 (**Open problem: gap completing**) The result is true for the "2-gap" subspace $W_{2m+1}^{(2)} = \mathcal{L}\{1, x, ..., x^{2m-3}, x^{2m-2}, x^{2m}, x^{2m+2}\}$. As an OPEN PROBLEM, taking the "$(m+1)$-gap" subspace of even polynomials (3.141), we conjecture that if $F : W_{m+2}^{\text{even}} \to W_{m+2}^{\text{even}}$, then $F : W_{2m+3} \to W_{2m+3}$. (For cubic operators, this is not true; see Proposition 3.5.) Invariant subspaces with more general distribution of gaps are also unclear.

3.6.4 Subspaces of irrational functions: dipole-type solutions

The origin of such subspaces and exact solutions is the quadratic PME

$$\boxed{u_t = F[u] \equiv (u^2)_{xx} \quad \text{in } I\!R \times I\!R_+,} \tag{3.145}$$

possessing the self-similar *dipole* Barenblatt–Zel'dovich solution [28]

$$u(x,t) = t^{-\frac{5}{8}} x^{\frac{1}{2}} \tfrac{1}{12}\big(A - t^{-\frac{3}{8}} x^{\frac{3}{2}}\big)_+ \quad \text{for } x \ge 0 \quad (A > 0), \tag{3.146}$$

which is extended for $x < 0$ as an odd function. The solution has the fixed point at $x = 0$, where $u(0,t) \equiv 0$, and the graph of $u(x,t)$ has a typical dipole form in x. The corresponding initial function at $t = 0$ is proportional to $\delta'(x)$, the weak derivative of Dirac's delta. Actually, this is a solution on the subspace

$$W_2 = \mathcal{L}\{x^{\frac{1}{2}}, x^2\}$$

that is invariant under the quadratic operator F in (3.145). At the same time, it is a standard similarity solution induced by a group of scaling transformations.

Let us show that such subspaces and solutions exist for higher-order parabolic PDEs. Consider the $2m$th-order quadratic operator

$$F[u] = (-1)^{m+1} D_x^{2m}\big(u^2\big). \tag{3.147}$$

Proposition 3.29 *Operator* (3.147) *preserves the* $(m+1)$*-dimensional subspace*

$$W_{m+1} = \mathcal{L}\big\{x^{\frac{1}{2}}, x^{\frac{3}{2}}, ..., x^{\frac{2m-3}{2}}, x^{\frac{2m-1}{2}}, x^{2m}\big\}.$$

Proof. It follows that, for any $u \in W_{m+1}$,

$$u^2 \in \mathcal{L}\big\{x, x^2, ..., x^{2m-1}, x^{2m+\frac{1}{2}}, ..., x^{2m+\frac{2m-1}{2}}, x^{4m}\big\},$$

which yields $D_x^{2m}(u^2) \in W_{m+1}$. □

Example 3.30 The fourth-order equation

$$\boxed{u_t = -(u^2)_{xxxx}}$$

is parabolic in the positivity domain $\{u > 0\}$. The corresponding solutions are

$$u(x,t) = C_1(t)x^{\frac{1}{2}} + C_2(t)x^{\frac{3}{2}} + C_3(t)x^4 \in W_3 \quad \text{for } t \ge 0,$$

$$\begin{cases} C_1' = -\frac{945}{8}\, C_1 C_3, \\ C_2' = -\frac{3465}{8}\, C_2 C_3, \\ C_3' = -1680\, C_3^2. \end{cases}$$

Solving the DS yields the explicit formula

$$u(x,t) = \big(A t^{-\frac{9}{128}} x^{\frac{1}{2}} + B t^{-\frac{33}{128}} x^{\frac{3}{2}} + \tfrac{1}{1680\,t}\, x^4\big)_+. \tag{3.148}$$

For different values of A and B, (3.148) describes a dipole-like singularity at $t = 0$, extinction/quenching phenomena, and the interface propagation. A proper setting of the corresponding Stefan–Florin FBP with positive solutions can be also revealed.

3.6.5 Trigonometric subspace

A typical example is the following TFE with source possessing blow-up solutions:

$$\boxed{u_t = F[u] \equiv (-1)^{m+1} D_x \big(u D_x^{2m-1} u\big) + 2u^2.} \tag{3.149}$$

Proposition 3.31 *Operator F in* (3.149) *admits $W_3 = \mathcal{L}\{1, \cos x, \sin x\}$.*

Equation (3.149) on W_3 is a third-order DS that describes blow-up of 2π-periodic solutions. As shown in Example 3.17, the 2D restrictions $\mathcal{L}\{1, \cos x\}$ and $\mathcal{L}\{1, \sin x\}$ are also invariant and generate simpler DSs on which blow-up similarity solutions are asymptotically stable.

3.7 Oscillatory, changing sign behavior in the Cauchy problem

For a moment we digress from invariant subspaces for TFEs and discuss some aspects concerning general solutions of changing sign of the Cauchy problem.

Source-type solutions of the TFE. Consider the *signed* $2m$th-order TFE with the higher-order term only,

$$u_t = (-1)^{m+1} D_x \big(|u|^n D_x^{2m-1} u\big) \quad \text{in } I\!R \times I\!R_+, \tag{3.150}$$

where $n > 0$ is a parameter. The fundamental source-type solution has the similarity form

$$b(x,t) = t^{-\beta} g(\xi), \quad \xi = \tfrac{x}{t^\beta}, \quad \text{where } \beta = \tfrac{1}{n+2m}$$

and, on integration once, g solves the $(2m-1)$th-order ODE

$$(-1)^{m+1} |g|^n g^{(2m-1)} + \beta g \xi = 0 \quad \text{in } I\!R. \tag{3.151}$$

The unit mass condition is also imposed

$$\textstyle\int g(\xi)\, \mathrm{d}\xi = 1 \tag{3.152}$$

and can always be achieved by scaling. Another usual normalization condition is $g(0) = 1$ that will be used sometimes later on. For the Cauchy problem, we are looking for solutions of the *maximal regularity*, which can be admitted by the ODE (3.151). We pose $m-1$ symmetry boundary conditions at the origin,

$$g'(0) = g'''(0) = \ldots = g^{(2m-3)}(0) = 0 \quad (g(0) > 0). \tag{3.153}$$

For $m = 2$, rigorous mathematical results on existence and uniqueness for the problem (3.151), (3.152) are known for $n \in (0, 1]$; see [174] where Bernis–McLeod approach [47] was used. For other n and $m \geq 3$, we rely on analytic-numerical evidence to be presented.

Let g be supported on the interval $[-\xi_0, \xi_0]$. We can also use normalization $\xi_0 = 1$ instead of (3.152). At the right-hand interface, for $\xi \approx \xi_0^-$, we then introduce the oscillatory component φ by

$$g(\xi) = (\xi_0 - \xi)^\gamma \varphi(s), \quad s = \ln(\xi_0 - \xi), \quad \text{where } \gamma = \tfrac{2m-1}{n}. \tag{3.154}$$

Setting

$$\xi = \xi_0 - \mathrm{e}^s \quad \text{for } s \ll -1$$

and omitting the exponentially small non-autonomous perturbation yields that $\varphi(s)$ satisfies

$$P_{2m-1}[\varphi] = c_0 |\varphi|^{-n} \varphi \quad \text{in } I\!R, \quad \text{where } c_0 = \beta \xi_0 (-1)^{m+1}. \tag{3.155}$$

Here, $\{P_k[\varphi],\ k \geq 0\}$ denote operators that are constructed by the iteration

$$P_{k+1}[\varphi] = (P_k[\varphi])' + (\gamma - k) P_k[\varphi] \ \text{ for } k = 0, 1, \ldots, \quad P_0[\varphi] = \varphi. \tag{3.156}$$

For instance, for $m = 2$ $\left(\gamma = \frac{3}{n}\right)$ and $m = 3$ $\left(\gamma = \frac{5}{n}\right)$, respectively,

$$\begin{aligned} P_3[\varphi] &= \varphi''' + 3(\gamma - 1)\varphi'' + (3\gamma^2 - 6\gamma + 2)\varphi' \\ &\quad + \gamma(\gamma - 1)(\gamma - 2)\varphi \quad \text{and} \end{aligned}$$

$$\begin{aligned} P_5[\varphi] &= \varphi^{(5)} + 5(\gamma - 2)\varphi^{(4)} + 5(2\gamma^2 - 8\gamma + 7)\varphi''' \\ &+ 5(\gamma - 2)(2\gamma^2 - 8\gamma + 5)\varphi'' + (5\gamma^4 - 40\gamma^3 + 105\gamma^2 \\ &- 100\gamma + 24)\varphi' + \gamma(\gamma - 1)(\gamma - 2)(\gamma - 3)(\gamma - 4)\varphi. \end{aligned} \tag{3.157}$$

Traveling waves. For TWs $u(x,t) = f(y)$, with $y = x - \lambda t$, the ODE is easier

$$(-1)^{m+1} |f|^n f^{(2m-1)} + \lambda f = 0 \quad \text{for } y > 0, \quad f(0) = 0, \tag{3.158}$$

where the left-hand interface is at $y = 0$. The oscillatory component is given by

$$f(y) = y^{\gamma} \varphi(s), \quad s = \ln y, \quad \text{where} \quad \gamma = \tfrac{2m-1}{n}, \tag{3.159}$$

where φ solves the ODE (3.155) with $c_0 = (-1)^m \lambda$.

As before, we are mainly interested in periodic solutions $\varphi(s)$ of ODEs (3.155), which, according to (3.154), determine typical (and sometimes generic) oscillatory behavior of solutions near interfaces, as shown in Figure 3.5(a). Existence and multiplicity of periodic solutions of dynamical systems in $I\!R^n$ is a classical area of applications of topological and geometric methods of nonlinear analysis, such as rotations of vector fields and index-degree theory; see Krasnosel'skii–Zabreiko [355, Sect. 13, 14]. Another mathematical direction is classical branching theory; see Vainberg–Trenogin [565, Ch. 6]. In our case, an n-branching approach may be effective, since for $n = 0$, the unique solution of the problem (3.151), (3.152) is indeed the rescaled kernel of the fundamental solution. A mathematical justification of such a branching is a hard OPEN PROBLEM. We also refer to [569, 399, 336] as a guide to modern theory of periodic solutions of higher-order nonlinear ODEs. In general, for large m, equations (3.155) are difficult to study, and especially as the main concern is the number of their different periodic solutions, as well as the identification of the most stable solution that describes a general structure and complexity of multiple zeros of solutions near interfaces. We will also rely on careful numerical evidence on existence, uniqueness and stability of periodic solutions. It is natural to begin recalling the properties of the linear PDE (3.150) with $n = 0$, which explain the oscillatory patterns in the quasilinear case for small $n > 0$.

Oscillations in the linear equations. For $n = 0$ in (3.150), the linear *polyharmonic* PDE is obtained,

$$u_t = (-1)^{m+1} D_x^{2m} u \quad \text{in } I\!R \times I\!R_+,$$

whose oscillatory properties are described by its *fundamental solution*

$$b(x,t) = t^{-\frac{1}{2m}} g(\xi), \quad \xi = \frac{x}{t^{1/2m}}, \tag{3.160}$$

where g is the unique exponentially decaying solution of the ODE

$$(-1)^{m+1} g^{(2m)} + \tfrac{1}{2m} g'\xi + \tfrac{1}{2m} g = 0 \quad \text{in } I\!R, \quad \int g = 1. \tag{3.161}$$

See Eidelman's classic book [164, Ch. 1] for the existence and sharp estimates of fundamental solutions of higher-order linear parabolic operators. By substituting, it is not hard to check that the behavior as $\xi \to +\infty$ of solutions of (3.161) is given by

$$g(\xi) \sim \xi^{-\mu} e^{a\xi^{\alpha}}, \quad \text{with } \alpha = \tfrac{2m}{2m-1} \text{ and } \mu = \tfrac{m-1}{2m-1}, \tag{3.162}$$

where a is the root with the maximal $\mathrm{Re}\, a < 0$ of the characteristic equation

$$(-1)^m (\alpha a)^{2m-1} = \tfrac{1}{2m}. \tag{3.163}$$

Estimating the number of complex conjugate pairs of roots, a, and their real parts yields the following useful conclusion.

Proposition 3.32 (i) *The asymptotic representation* (3.162) *of the rescaled kernel* $g(\xi)$ *of the fundamental solution* (3.160) *can be represented in terms of a quasi-periodic function containing not more than* ($[\cdot]$ *denotes the integer part*)

$$\left[\tfrac{m}{2}\right] \text{ fundamental frequencies.} \tag{3.164}$$

(ii) *There exists a constant* $D > 0$ *such that*

$$|g(\xi)| \le D e^{-d|\xi|^{\alpha}} \quad \text{in } I\!R, \quad \text{where } d = \tfrac{2m-1}{(2m)^{\alpha}} \left|\cos\left(\tfrac{m\pi}{2m-1}\right)\right|.$$

It follows from (3.164) that the total asymptotic linear bundle of exponentially decaying solutions (3.162) satisfies

$$\text{the bundle is } m\text{-dimensional.} \tag{3.165}$$

For m odd, this includes the 1D non-oscillatory bundle, corresponding to the real negative root of (3.163)

$$a_0 = -\tfrac{2m-1}{(2m)^{\alpha}}.$$

Oscillatory periodic patterns for m = 2 and a heteroclinic bifurcation. Let us begin with the simplest case, $m = 2$, where the third-order ODE (3.155) takes the form (here $c_0 = -1$ and $\gamma = \frac{3}{n}$)

$$\begin{aligned} P_3[\varphi] \equiv \varphi''' + 3(\gamma - 1)\varphi'' + (3\gamma^2 - 6\gamma + 2)\varphi' \\ + \gamma(\gamma - 1)(\gamma - 2)\varphi = -\tfrac{\varphi}{|\varphi|^n}. \end{aligned} \tag{3.166}$$

Numerical experiments (Figure 3.6) show that (3.166) admits a unique stable periodic solution for all $n \in (0, n_h)$, where

$$\boxed{n_h = 1.758665\ldots \quad (m = 2)}$$

is a critical heteroclinic bifurcation exponent.

The exponent n_h plays an important role and shows the precise parameter range

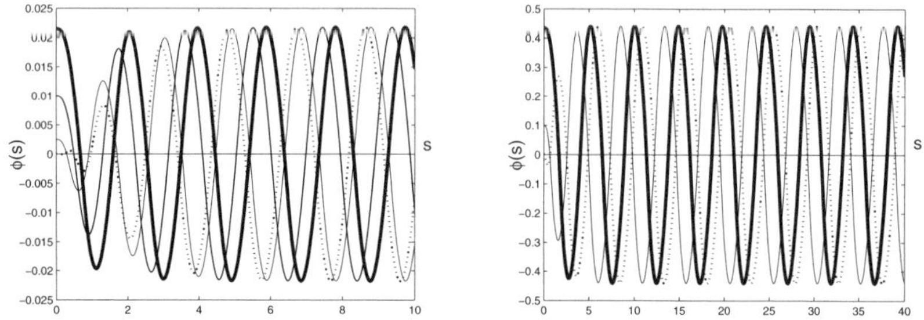

(a) $n = 1$ (b) $n = \frac{3}{2}$

Figure 3.6 Stable periodic behavior of (3.166), $m = 2$, for $n = 1$ (a) and $n = \frac{3}{2}$ (b).

of n for which all the ODE profiles near interfaces are *oscillatory* (this may also be key for the corresponding PDE). Numerical results reveal a non-local *heteroclinic* bifurcation at $n = n_{\rm h}$ associated with two unstable equilibria of (3.166)

$$\varphi_{\pm} = \pm\big[-\tfrac{1}{\gamma(\gamma-1)(\gamma-2)}\big]^{\frac{1}{n}} \quad \text{for} \quad n \in \big(\tfrac{3}{2}, 3\big).$$

Note that (3.159) with $\varphi(s) \equiv \varphi_+$ yields a non-oscillatory behavior; see the dashed line in Figure 3.5(a). Figure 3.7 shows a typical "heteroclinic" deformation of periodic patterns as $n \to n_{\rm h}^-$, where, in order to reveal the widest periodic pattern (the bold line), we need to take $n = 1.758664976837300$ (where not all the 15 decimals are reliable). This is a standard scenario of homoclinic-heteroclinic bifurcations of periodic solutions in ODEs; see Perko [460, Ch. 4]. A rigorous justification of such bifurcations is difficult and is an OPEN PROBLEM, especially for higher-order equations with $m = 4, 6, \ldots$ to be considered below.

For $n > n_{\rm h}$, the behavior of solutions of the ODE (3.166) becomes exponentially unstable and oscillatory or changing sign patterns are not observed. It is likely that precisely above $n = n_{\rm h}$, the ODE (and, to some extent, the corresponding PDE) loses its natural similarities with the linear equation for $n = 0$, i.e., some local properties of solutions dramatically change at $n = n_{\rm h}$.

$\mathrm{m} \geq 3$: unstable periodic behavior for odd m and stable for even. In numerical experiments, we have observed a single stable oscillatory behavior for even $m \geq 2$ and an unstable orbit for odd m. Figure 3.8(a) shows solutions of the ODE (3.155) for $m = 3$ and stability of equilibria $\varphi_{\pm}$. In this case, the unstable periodic solution lies in between those two stable flows of orbits tending to φ_+ and φ_- as (b) explains. It turns out that such an unstable periodic solution exists until the "heteroclinic" value

$$\boxed{\hat{n}_{\rm h} = 1.909\ldots \quad (m = 3).}$$

In view of its unstability, the corresponding analytical and numerical techniques become more involved; see [175] for extra details.

For $m = 4$, i.e., for the eighth-order TFE, the solutions $\varphi(s)$ approach a stable periodic pattern; see Figure 3.9. The periodic solution is easily detected above the

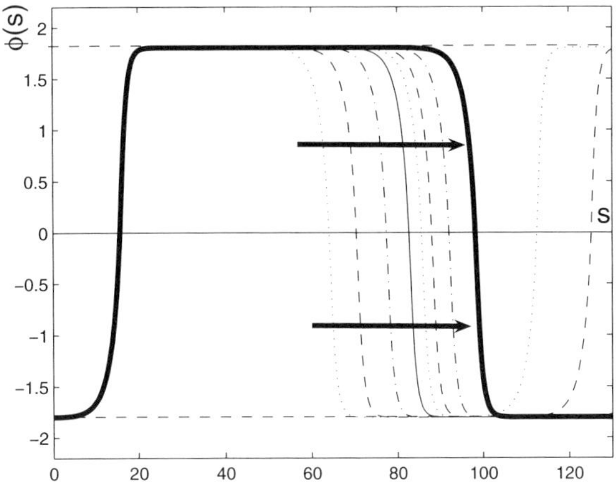

Figure 3.7 Formation of a heteroclinic connection $\varphi_+ \mapsto \varphi_-$ for (3.166) as $n \to n_h^-$.

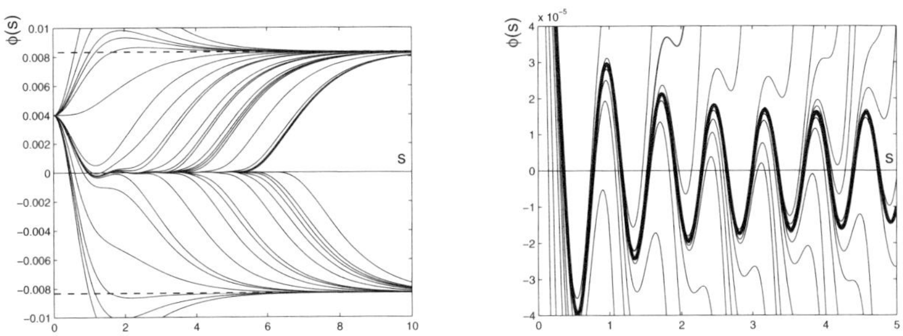

(a) stabilization to $\varphi_\pm$

(b) periodic behavior, enlarged

Figure 3.8 Solutions of (3.155) for $m = 3$ and $n = 1$: stability of constant equilibria $\varphi_\pm$ (a) and the unstable periodic behavior in between (b), $\varphi(0) = 0.04$, $\varphi''(0) = -0.0176896526588\ldots$.

first critical exponent $n_1 = \frac{7}{6}$ until the corresponding heteroclinic bifurcation at

$$\boxed{n_{\rm h} = 1.215053... \quad (m = 4);} \tag{3.167}$$

see Figure 3.10. For $n > n_{\rm h}$, the behavior becomes more and more unstable. Smaller values of n are discussed below.

For $m = 5$, the unstable periodic behavior is shown in Figure 3.11. Cauchy data for the periodic motion are: $\varphi(0) = 10^{-7}$, $\varphi''(0) = 4.838491256 \times 10^{-7}$ for $n = 1$ in Figure (a). In Figure (b), we have $\varphi''(0) = 1.942533642 \times 10^{-6}$ for $n = \frac{9}{8}$. The rest of the derivatives in both Figures (a) and (b) are equal to zero, $\varphi'(0) = \varphi'''(0) = \ldots = \varphi^{(8)}(0) = 0$.

According to these results checked numerically for several even and odd m, let us state the following:

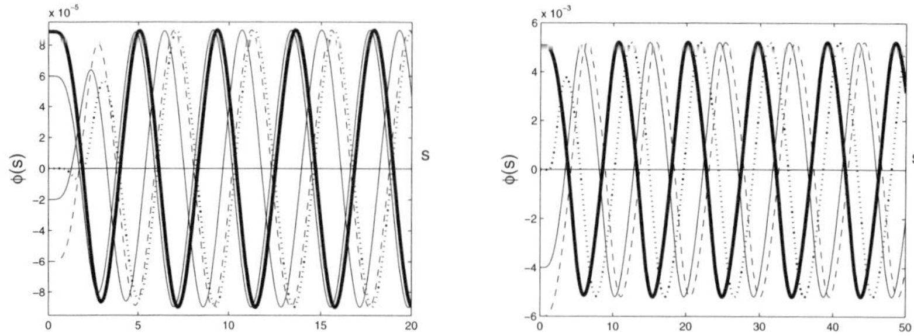

(a) $n = 1$ (b) $n = n_1 = \frac{7}{6} = 1.1666...$

Figure 3.9 Stable periodic behavior in the ODE (3.155), $m = 4$, for $n = 1$ (a) and $n = \frac{7}{6}$ (b).

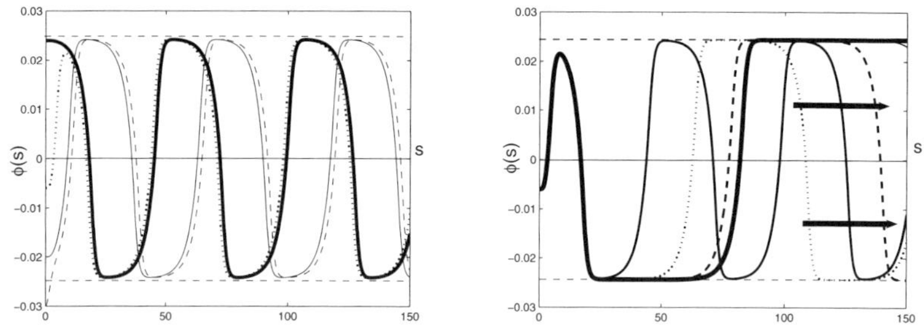

(a) $n = 1.215$ (b) $n \to n_{\rm h}^{-} = 1.2150534...$

Figure 3.10 (a) Periodic structures of (3.155), $m = 4$, for n close to $n_{\rm h}^{-}$ in (3.167); (b) describes formation of a heteroclinic connection (the wave moves to the right as n increases).

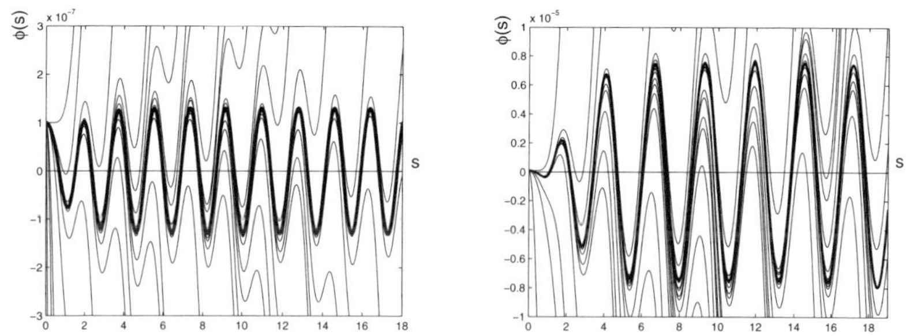

(a) $n = 1$ (b) $n = n_1 = \frac{9}{8}$

Figure 3.11 Unstable periodic behavior in (3.155), $m = 5$, for $n = 1$ (a) and $n = \frac{9}{8}$ (b).

Conjecture 3.1 ("On periodic solutions") *For any* $n \in (0, 1]$, *the ODE* (3.155) *admits a stable periodic solution for even* $m = 2, 4, \ldots$ *and an unstable one for for odd* $m = 3, 5, \ldots$.

Thus, for even m, stable periodic solutions, existence of which for small $n > 0$ can be associated with continuity in n, persist for all $n \in (0, 1]$. Moreover, based on numerical evidence, we also claim that they still exist for

$$n \in (1, n_{\rm h}), \quad \text{with some } n_{\rm h} \in \left(\tfrac{2m-1}{2m-2}, \tfrac{2m-1}{2m-3}\right) = (n_1, n_2),$$

where $n = n_{\rm h}$ is a heteroclinic bifurcation point for the corresponding $(2m$-$1)$th-order ODE, as explained above. Such a non-oscillatory character of interfaces for $n > n_{\rm h}$ suggests that, somewhere in this parameter range, sign-preserving properties of special classes of solutions may exist leading to nonnegative solutions of the Cauchy problem for sufficiently large n. This needs extra investigation and is an OPEN PROBLEM.

Extra scaling as n → 0. The above figures show that the oscillatory behavior is not directly detectable for small $n > 0$ and is numerically invisible if $n < \frac{1}{2}$. To revealing the limit oscillatory behavior as $n \to 0$, where solutions of changing sign are of the order

$$\max |\varphi(s)| \sim A(n) \equiv \left(\tfrac{n}{2m-1}\right)^{\frac{2m-1}{n}} \quad \text{for small } n > 0,$$

an extra rescaling in equation (3.155) is necessary,

$$\varphi(s) = A\psi(\eta), \quad \eta = \tfrac{s}{a}, \quad \text{where } a(n) = A^{\frac{n}{2m-1}} = \tfrac{n}{2m-1}. \tag{3.168}$$

Then, for small $n > 0$, function $\psi(\eta)$ solves the ODE with Euler's differential operator, which can be written in the form of

$$\mathrm{e}^{-\eta}(\psi \mathrm{e}^{\eta})^{(2m-1)} - c_0 \tfrac{\psi}{|\psi|^n} = 0 \tag{3.169}$$

(as usual, all higher-order perturbations have been omitted). This is an easier ODE than the original one (3.155), and contains the binomial coefficients. It turns out that (3.169) possesses similar periodic orbits.

Numerical analysis gives solid evidence of the existence of stable periodic motion for $m = 2$ in Figure 3.12, and for $m = 4$ in Figure 3.13, where the ODEs have the form

$$m = 2: \quad \psi''' + 3\psi'' + 3\psi' + \psi + \tfrac{\psi}{|\psi|^n} = 0, \quad \text{and}$$

$$\psi^{(7)} + 7\psi^{(6)} + 21\psi^{(5)} + 35\psi^{(4)} + 35\psi''' + 21\psi'' + 7\psi' + \psi + \tfrac{\psi}{|\psi|^n} = 0$$

for $m = 4$ (we take $c_0 = -1$). Already for $n = 0.1$ in Figure 3.13(a), oscillations are extremely small, since, by (3.168),

$$\max|\varphi| \sim 10^{-130} \quad (n = 0.1, \quad m = 4).$$

For $n = 0.03$ in (b), $\max|\varphi| \sim 10^{-460}$.

For $m = 3$, the corresponding rescaled ODE

$$m = 3: \quad \psi^{(5)} + 5\psi^{(4)} + 10\psi''' + 10\psi'' + 5\psi' + \psi - \tfrac{\psi}{|\psi|^n} = 0 \tag{3.170}$$

admits an unstable periodic solution for small $n > 0$; see Figure 3.14(a) for $n = \frac{1}{2}$

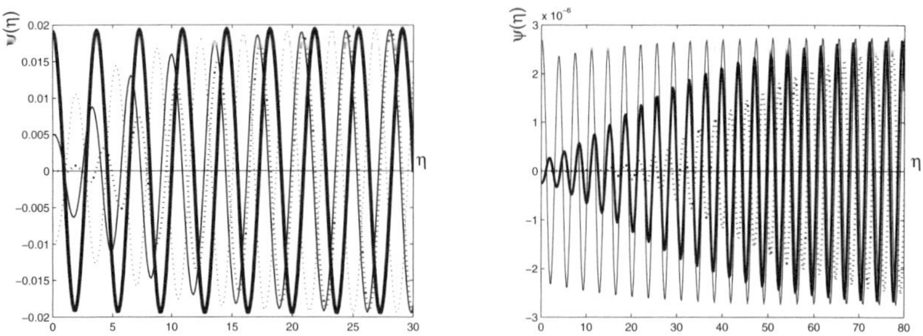

(a) $n = \frac{1}{2}$ (b) $n = 0.16$

Figure 3.12 Stable periodic behavior of (3.169), $m = 2$, for $n = \frac{1}{2}$ (a) and $n = 0.16$ (b).

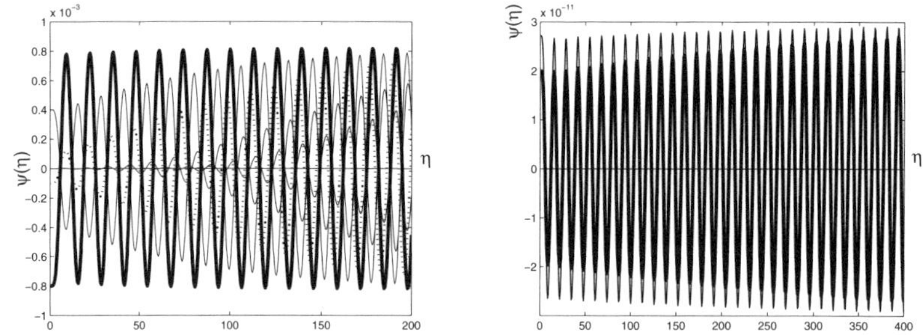

(a) $n = 0.1$ (b) $n = 0.03$

Figure 3.13 Oscillatory solutions of (3.169), $m = 4$, for $n = 0.1$ (a) and $n = 0.03$ (b).

(the Cauchy data to reveal the periodic motion are $\varphi(0) = 2 \times 10^{-5}$, $\varphi''(0) = -9.9850972244 \times 10^{-5}$, other derivatives are equal to zero). Figure 3.14(b) shows the case $n = 0.2$, where $\varphi(0) = 5 \times 10^{-10}$, $\varphi''(0) = -1.4296767 \times 10^{-9}$, and the periodic behavior is still not clearly seen. More details on such solutions and passage to the limit $n \to 0^+$ can be found in [175].

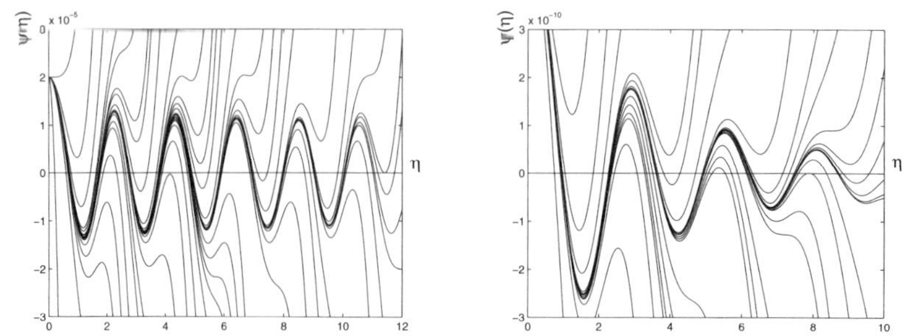

(a) $n = \frac{1}{2}$ (b) $n = 0.2$

Figure 3.14 Unstable periodic behavior for (3.170) for $n = \frac{1}{2}$ (a) and $n = 0.2$ (b).

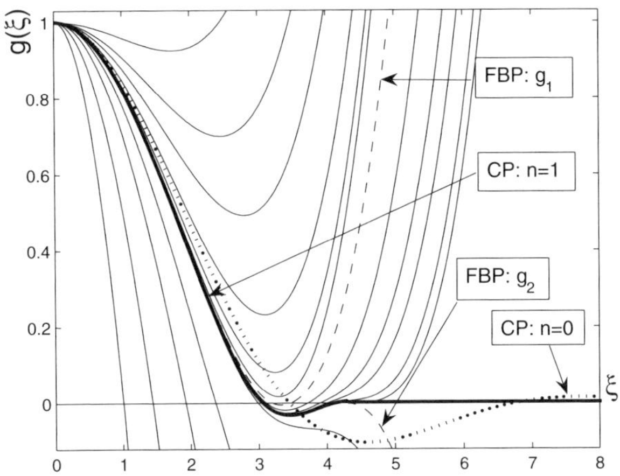

Figure 3.15 Similarity profiles $g(\xi)$ satisfying (3.151), $m = 2$, for $n = 1$ and $n = 0$. The dashed lines denote the first FBP profiles, $g_1(\zeta)$ and $g_2(\zeta)$.

On existence of the fundamental similarity solution: dimension of linear bundles. For $m = 2$, existence and uniqueness of the fundamental profile g satisfying (3.151) are proved for $n \in (0, 1)$, [174, Sect. 9]. For $m \geq 3$, such results are unknown. For using a shooting strategy from the interface point $\xi = \xi_0$ in (3.154) (where, e.g., $\xi_0 = 1$ is fixed by scaling invariance of the ODE) to the origin to match $m - 1$ symmetry conditions, one needs to have an $(m\text{-}1)$-parametric stability bundle about the periodic solution $\varphi(s)$. According to (3.165) for $n = 0$, by continuity in n, this bundle is expected to be precisely $(m\text{-}1)$-dimensional for small $n > 0$. Recall that $\xi_0 = 1$, i.e., one parameter is fixed. This shows existence of $g(\xi)$, in view of extremely good matching topology of the oscillatory bundle. Uniqueness is OPEN. For $n \in (0, n_{\rm h})$, we need to check the spectrum (the number of eigenvalues with positive real parts) of the singular, non self-adjoint linearized operator with periodic coefficients in equation (3.155),

$$P_{2m-1}[\varphi] - (1-n)c_0|\varphi|^{-n} I.$$

This is an OPEN PROBLEM.

In Figure 3.15, we present similarity profiles of the CP given by the ODE (3.151), $m = 2$, for $n = 1$ and, for comparison, the fundamental rescaled kernel for $n = 0$. Corresponding Cauchy data are $g(0) = 1$ and $g''(0) = -0.3696375...$ for $n = 1$, and $g''(0) = -0.337989123...$ for $n = 0$. Here, $g_{1,2}(\zeta)$ are the first similarity profiles of the zero contact angle, zero-flux FBP.

On negative exponents n. By continuity in $n \to 0$, it can be expected that oscillatory behavior of solutions of the Cauchy problem for TFEs persists for $n < 0$. The representation (3.159) suits this case with the interface at $y = +\infty$, where the oscillatory component φ solves the same ODE (3.155). Unlike in the case where $n > 0$, for $n < 0$, the changing sign character of such solutions is difficult to detect, since the ODEs involved exhibit extra blow-up singularities in finite y or s. This is easily seen from (3.158), written in the form of

$$f^{(2m-1)} = c_0|f|^{|n|}f \quad (c_0 = \lambda(-1)^m),$$

where the right-hand side is now superlinear for $f \gg 1$, and, hence, there exists monotone or oscillatory (always if $c_0 < 0$) blow-up in finite y. Moreover, typically an oscillatory behavior is not available, so that solutions of changing sign are constructed by asymptotic expansion as $n \to 0^-$ and extension in n; such WKBJ asymptotics are explained in [174, Sect. 7.6] for the TFEs with $n \to 0^+$.

3.8 Invariant subspaces in Kuramoto-Sivashinsky type models

The *Kuramoto–Sivashinsky* (KS) *equation*

$$\boxed{u_t = -u_{xxxx} - u_{xx} + (u_x)^2} \tag{3.171}$$

is a semilinear fourth-order parabolic PDE with a quadratic Hamilton–Jacobi term $(u_x)^2$. It was originally introduced as a model that describes flame front propagation in turbulent flows of gaseous combustible mixtures [526], and later it found other applications in many areas of physics, including 2D turbulence; see Remarks. The *modified Kuramoto–Sivashinsky* (mKS) *equation* contains an extra second-order quadratic operator,

$$\boxed{u_t = F[u] \equiv -u_{xxxx} - u_{xx} + (1-\lambda)(u_x)^2 + \lambda(u_{xx})^2,} \tag{3.172}$$

where $\lambda \in [0, 1]$ is a constant ($\lambda = 0$ leads to (3.171)), and describes dynamical properties of hyper-cooled melt.

Example 3.33 (**Instability**) Let us begin with simple instability phenomena for the mKS equation, which occur on the invariant subspace of periodic functions.

Proposition 3.34 *For any* $\lambda \in (0, 1)$, *the quadratic operator in* (3.172) *admits subspace* $W_3 = \mathcal{L}\{1, \cos\gamma x, \sin\gamma x\}$, *where* $\gamma = \sqrt{\frac{1-\lambda}{\lambda}}$.

Consider solutions in the subspace of cos-functions,

$$u(x,t) = C_1(t) + C_2(t)\cos\gamma x,$$
$$\begin{cases} C_1' = \frac{1}{\lambda}(1-\lambda)^2 C_2^2, \\ C_2' = \frac{1}{\lambda^2}(1-\lambda)(2\lambda-1)C_2. \end{cases}$$

It follows from the second ODE that $\lambda > \frac{1}{2}$ leads to the exponential instability, while for $\lambda = \frac{1}{2}$ it follows that $C_1(t) = O(t)$ for $t \gg 1$. On the other hand, for $\lambda < \frac{1}{2}$, $C_2(t)$ is exponentially small as $t \to \infty$, and $C_1(t)$ is bounded, thus describing the stability of the origin.

Let us introduce two models, exhibiting finite-time singularities.

Example 3.35 (**Blow-up**) Consider the *KS equation with source*

$$\boxed{u_t = -u_{xxxx} + (u_x)^2 + u^2,} \tag{3.173}$$

where the reaction term u^2 is added, and the linear second-order term $-u_{xx}$ (not important for blow-up solutions) is ignored. Invariant subspaces of first-order quadratic operators, such as $(u_x)^2 + u^2$ in (3.173), have been described in Section 1.4. Take $W_3 = \mathcal{L}\{1, \cos x, \sin x\}$ on which the solutions $u = C_1 + C_2 \cos x + C_3 \sin x$ are driven by the DS

$$\begin{cases} C_1' = C_1^2 + C_2^2 + C_3^2, \\ C_2' = 2C_1C_2 - C_2, \\ C_3' = 2C_1C_3 - C_3. \end{cases} \tag{3.174}$$

This quadratic system is similar to those studied for the TFEs in Example 3.17. Though the last two ODEs contain extra linear terms, the proof of Proposition 3.18 remains unchanged. This gives a typical stability result on the subspace W_3, where all the blow-up solutions converge to the similarity pattern belonging to the 1D subspaces $W_1^{\pm}$. It is curious that the whole linear operator $-u_{xxxx} - u_{xx}$ from (3.171) vanishes identically on W_3, the resulting DS does not contain any linear terms on the right-hand side, simplifying stability analysis. The precise asymptotic behavior as $t \to T$ for the DS (3.174) gives important properties of blow-up of 2π-periodic solutions of (3.173); see references in the Remarks.

Example 3.36 **(Extinction)** In order to exhibit extinction phenomena, let us choose the *KS equation with absorption*

$$\boxed{u_t = -u_{xxxx} + (u_{xx})^2 - 1.} \tag{3.175}$$

We restrict our attention to a lower-dimensional subspace $W_3 = \mathcal{L}\{1, x^2, x^4\}$ (which is extended to W_5 by adding extra basic functions x and x^3), and consider the evolution of the solutions $u(x,t) = C_1(t) + C_2(t)x^2 + C_3(t)x^4$ with the DS

$$\begin{cases} C_1' = 4C_2^2 - 24C_3 - 1, \\ C_2' = 48C_2C_3, \\ C_3' = 144C_3^2. \end{cases}$$

This system can be integrated and studied in a manner similar to that considered in Section 3.2. The positive solutions satisfy the corresponding FBP which can be studied by the von Mises transformation (see Example 3.10). The related singularity formation phenomena for the uniformly parabolic PDE (3.175) are not as interesting as those for the degenerate TFEs. On the other hand, the presence in the PDE of the non-Lipschitz absorption term -1, which is actually $|u|^{p-1}u$ for $p = 0$ (the Heaviside function), makes the problem rather consistent, especially in the Cauchy problem to be discussed next.

Finite propagation and oscillatory patterns in the Cauchy problem. Considering (3.175) in $I\!R \times I\!R_+$, we face the question of the maximal regularity of solutions at the interfaces. We present the *signed* version of this PDE for solutions of changing sign,

$$u_t = -u_{xxxx} + (u_{xx})^2 - \operatorname{sign} u.$$

As for TFEs with absorption, using the TWs $u(x,t) = f(y)$, $y = x - \lambda t$, yields

$$-\lambda f' = -f^{(4)} + (f'')^2 - \operatorname{sign} f \quad \text{for } y > 0, \quad f(0) = 0.$$

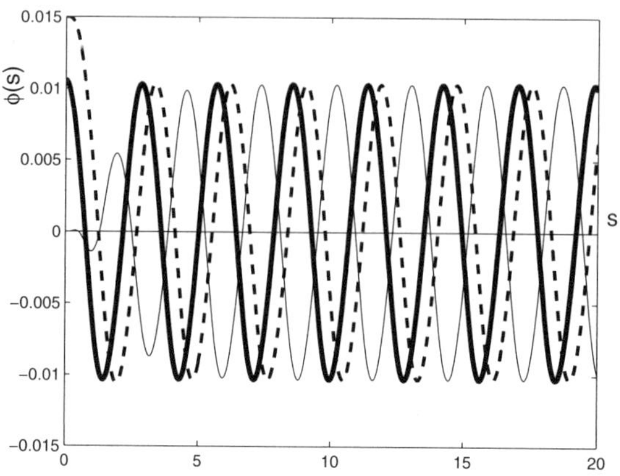

Figure 3.16 The stable periodic behavior in (3.177).

Using our TFE experience, we keep two leading terms,

$$f^{(4)} + \operatorname{sign} f = 0,$$

so the oscillatory component φ in the solution representation

$$f(y) = y^4 \varphi(s), \quad \text{where} \quad s = \ln y, \tag{3.176}$$

satisfies the autonomous fourth-order ODE

$$\varphi^{(4)} + 10\varphi''' + 35\varphi'' + 50\varphi' + 24\varphi + \operatorname{sign} \varphi = 0. \tag{3.177}$$

Existence and uniqueness of periodic solutions are OPEN.

Conjecture 3.2 *The ODE* (3.177) *has a unique nontrivial periodic solution* $\varphi(s)$ *which is asymptotically stable as* $s \to +\infty$.

Figure 3.16 shows this stable periodic motion obtained for different initial data. We expect that, according to (3.176), such TW patterns describe the generic behavior of solutions near interfaces in the CP.

Example 3.37 (Blow-up and interfaces in a quasilinear KS-type equation) Finally, let us present a simple *quasilinear* KS-type equation, which allows us to describe some crucial nonlinear phenomena by using simple mathematical tools. Consider the following quasilinear fourth-order parabolic equation:

$$\boxed{u_t = F[u] \equiv u(-u_{xxxx} + u),} \tag{3.178}$$

which is initially formulated for nonnegative solutions. Clearly, the quadratic operator F preserves

$$W_5 = \mathcal{L}\{1, \cos x, \sin x, \cosh x, \sinh x\},$$

with simple blow-up dynamics. We restrict our attention to blow-up solutions in separate variables

$$u(x,t) = \tfrac{1}{T-t}\, \theta(x), \tag{3.179}$$

where θ solves the linear ODE

$$-\theta^{(4)} + \theta = 1. \tag{3.180}$$

Zero contact angle FBP. Let us pose for (3.178) the FBP with the zero contact angle,

$$u = u_x = u_{xx} = 0 \quad \text{at the interface at} \quad x = s(t), \tag{3.181}$$

and bounded compactly supported initial data $u_0(x) \geq 0$. The well-posedness of this FBP can be checked locally by using the von Mises variables (see Example 3.10), or by exploiting a regularization in the degenerate term $-uu_{xxxx}$, e.g., by replacing

$$u \mapsto \varepsilon + |u| \quad (\varepsilon > 0),$$

in the PDE (3.178) and studying the convergence of the sequence $\{u_\varepsilon(x,t)\}$ of solutions of such uniformly parabolic equations. (It is likely that more "singular" as $\varepsilon \to 0^+$ approximation is necessary to catch this FBP; such hard questions are not discussed here and we refer to [44], where the dependence of solutions of TFEs on approximations of the degenerate coefficient was studied first.) Further references are given in the Remarks. The correspondence of the regularization to prescribed free-boundary conditions is a difficult asymptotic problem that, in general, is OPEN for most degenerate parabolic PDEs, including the present one. The analytic regularization

$$u \mapsto \sqrt{\varepsilon^2 + u^2}$$

is expected to induce, as $\varepsilon \to 0$, the solutions of the Cauchy problem; see below.

Thus, looking for even solutions of the ODE (3.180),

$$\theta(x) = 1 + C_1 \cos x + C_2 \cosh x \in W_5,$$

and assuming that $x_0 > 0$ is the corresponding interface point, where (3.181) are valid, we obtain the algebraic system

$$\begin{cases} 1 + C_1 \cos x_0 + C_2 \cosh x_0 = 0, \\ -C_1 \sin x_0 + C_2 \sinh x_0 = 0, \\ -C_1 \cos x_0 + C_2 \cosh x_0 = 0. \end{cases} \tag{3.182}$$

The last two equations give a single transcendental equation for x_0,

$$\tan x_0 = \tanh x_0,$$

that possesses a countable set of roots $\{x_{0k}\}$ such that

$$x_{0k} = \pi \left(k + \tfrac{1}{4}\right) + \ldots \quad \text{for} \quad k \gg 1.$$

Solving the rest of equations in (3.182) yields the sequence of solutions

$$\theta_k(x) = 1 - \tfrac{1}{2\cos x_{0k}} \cos x - \tfrac{1}{2\cosh x_{0k}} \cosh x \quad \text{for} \quad |x| < x_{0k}.$$

These functions are strictly positive on $(-x_{0k}, x_{0k})$, which is easily seen for $k \gg 1$, and the oscillatory patterns satisfy

$$\theta_k(x) \to \theta_\infty^\pm(x) \equiv 1 + \tfrac{(-1)^k}{\sqrt{2}} \cos x > 0 \quad \text{as} \quad k \to \infty \tag{3.183}$$

uniformly on compact subsets (θ^+ corresponds to even k and θ^- to odd). There exists a countable set $\{u_k(x,t)\}$ of standing wave blow-up solutions (3.179) of the FBP localized in bounded intervals $(-x_{0k}, x_{0k})$.

On solutions of the Cauchy problem. Consider now the signed PDE

$$u_t = |u|(-u_{xxxx} + u) \quad \text{in } I\!R \times I\!R_+, \tag{3.184}$$

with bounded compactly supported initial data $u_0(x)$ in $I\!R$. Then the separate-variable solutions (3.179) yield the ODE

$$-\theta^{(4)} + \theta = \operatorname{sign}\theta. \tag{3.185}$$

Evidently, there exist strictly positive solutions $\theta_\infty^\pm(x)$ given in (3.183), but we are interested in compactly supported patterns.

It follows from (3.185) that, in the positivity and negativity domains respectively, the profiles are

$$\theta^\pm(x) = \pm 1 + C_1^\pm \cos x + C_2^\pm \sin x + C_3^\pm \cosh x + C_4^\pm \sinh x.$$

Despite such a simple form of profiles, the matching assumes four conditions of continuity (here $[\cdot]$ denote the jumps of functions at zeros)

$$[\theta] = [\theta'] = [\theta''] = [\theta'''] = 0,$$

so that the construction of changing sign solutions of the FBP leads to algebraic problems for the eight expansion coefficients $\{C_k^\pm\}$ at each zero point $x = x_{0k}$. This is an OPEN PROBLEM that can be tackled numerically.

The existence of changing sign solutions of the ODE (3.185) for both the FBP and the CP problems is connected with general oscillatory properties of solutions of the PDE (3.184). It is convenient to describe the oscillatory character of solutions by TWs $u(x,t) = f(x - \lambda t)$, so

$$-\lambda f' = -|f| f^{(4)} \quad \text{in } I\!R,$$

where we omit $|f|f$ that is smaller near interfaces. Integrating once yields

$$f''' = \lambda \operatorname{sign} f \, \ln|f| \quad \text{for } y > 0, \quad f(0) = 0.$$

We next introduce the oscillatory component φ,

$$f(y) = y^3 \varphi(\ln y) \implies \varphi''' + 6\varphi'' + 11\varphi' + 6\varphi = 3\lambda s + \lambda \operatorname{sign}\varphi \, \ln|\varphi|. \tag{3.186}$$

For $\lambda < 0$, there exists the positive solution $\varphi(s) = \frac{1}{2}\lambda s + ...$ for $s \ll -1$. By transformation in (3.186), this gives non-oscillatory behavior at the interface,

$$f(y) = \tfrac{1}{2}\lambda y^3 \ln y + ... > 0 \quad \text{as } y \to 0, \tag{3.187}$$

which corresponds well to the free-boundary conditions (3.181). Oscillatory patterns in the CP need extra study.

Oscillatory component. As usual, we introduce the family of PDEs

$$u_t = -|u|^n u_{xxxx}, \quad \text{with parameter } n \in (0,1). \tag{3.188}$$

Dividing by $|u|^n$ and setting $v = |u|^{-n}u$ yields a divergent quasilinear equation,

$$\tfrac{1}{1-n} v_t = \Phi[v] \equiv -(|v|^\sigma v)_{xxxx}, \quad \text{with } \sigma = \tfrac{n}{1-n} > 0. \tag{3.189}$$

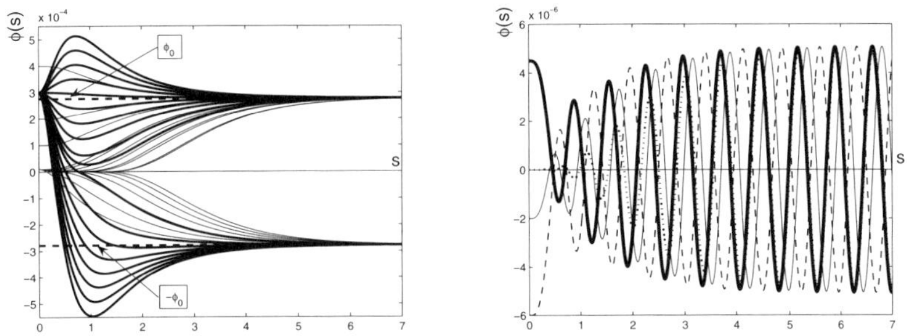

(a) $n = \frac{1}{2}$, $\lambda = 1$ (b) $n = \frac{1}{2}$, $\lambda = -1$

Figure 3.17 Solutions of the ODE (3.191) with $n = \frac{1}{2}$ in the stabilization case $\lambda = 1$ (a) and in the oscillatory case $\lambda = -1$ (b).

The operator $\Phi[v]$ is monotone in the topology of the Hilbert space $H^{-4}(I\!R)$. Existence and uniqueness of weak continuous solutions are based on general theory of monotone operators; see Lions [396, Ch. 2]. Writing (3.189) in the sense of distributions, it is not difficult to check that, for a special class of sufficiently regular (strong) weak solutions $\{u(x,t)\}$, there exists a limit as $n \to 0^+$, establishing a homotopic (non-singular) connection with the linear bi-harmonic PDE (3.2) in $I\!R \times I\!R_+$. The uniform convergence of strong solutions as $n \to 0^+$ is an important property that will be used later on. For (3.188), the TWs satisfy

$$|f|^n f^{(4)} = \lambda f' \quad \text{for } y > 0, \quad f(0) = 0,$$

so that, instead of the change (3.186),

$$f(y) = y^\gamma \varphi(s), \quad s = \ln y, \quad \text{with } \gamma = \tfrac{3}{n}, \quad \text{where} \tag{3.190}$$

$$\begin{aligned} &\varphi^{(4)} + 2(2\gamma - 3)\varphi''' + (6\gamma^2 - 18\gamma + 11)\varphi'' + 2(2\gamma^3 - 9\gamma^2 \\ &+ 11\gamma - 3)\varphi' + \gamma(\gamma - 1)(\gamma - 2)(\gamma - 3)\varphi = \tfrac{\lambda}{|\varphi|^n}(\varphi' + \gamma\varphi). \end{aligned} \tag{3.191}$$

This ODE has two equilibria

$$\pm\varphi_0 = \pm\left[\tfrac{\lambda}{(\gamma-1)(\gamma-2)(\gamma-3)}\right]^{\frac{1}{n}}. \tag{3.192}$$

They exist in the parameter ranges $n \in \left(1, \frac{3}{2}\right)$ and $n > 3$ if $\lambda < 0$, and for $n \in (0,1)$ and $n \in \left(\frac{3}{2}, 3\right)$ if $\lambda > 0$.

Figure 3.17 shows different behavior of solutions in the case of $n = \frac{1}{2}$ for $\lambda > 0$ (convergence to the positive constant equilibria (3.192); no traces of stable or unstable periodic patterns were observed) and for $\lambda < 0$ (oscillatory). Periodic patterns $\varphi(s)$ persist for larger n, including $n = 1.1$; see Figure 3.18. Further increasing n, a few oscillations appear followed by fast divergence of the solution $\varphi(s)$, according to the linear unstable exponential bundle.

Subcritical bifurcation. Numerically, the constant equilibrium φ_0 of (3.191) becomes asymptotically stable for $n > n_{\rm st}$, where

$$\boxed{n_{\rm st} = 1.287...}$$

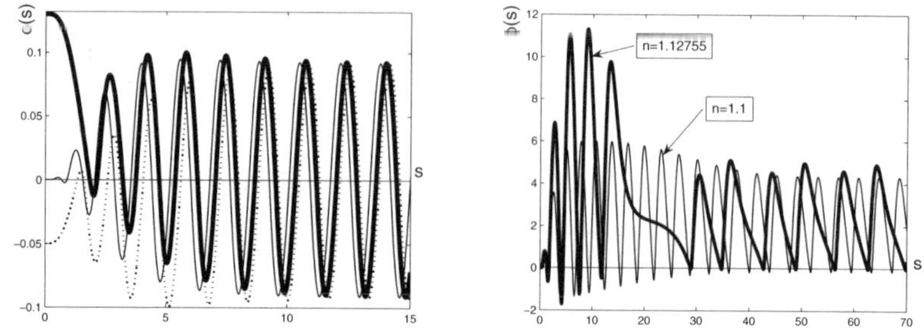

(a) $n = 0.9$ (b) $n = 1.1$ and 1.12755

Figure 3.18 Convergence to stable periodic solutions of the ODE (3.191), $\lambda = -1$, with various Cauchy data at $s = 0$ for $n = 0.9$ (a) and $n = 1.1$, $n = 1.12755$ (b).

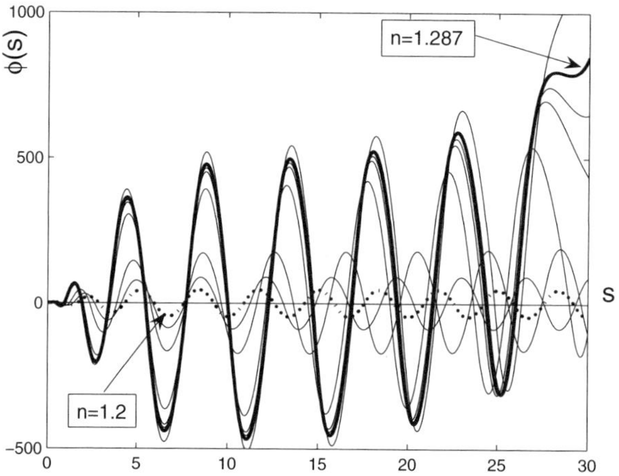

Figure 3.19 Behavior as $n \to n_{\rm st}^-$ of solutions of the ODE (3.191), with $\lambda = -1$.

and as $n \to n_{\rm st}^-$, the stable periodic solution becomes nonexistent, as shown in Figure 3.19. Therefore, for $n > n_{\rm st}$, the TW profiles are no longer oscillatory near such interfaces (only a finite number of zeros is available). Such non-oscillatory behavior becomes more and dominant for larger $n > \frac{3}{2}$, so that, possibly, the Cauchy problem can be posed in the class of nonnegative solutions; an OPEN PROBLEM.

Extra scaling for small n > 0. As usual, to see the oscillatory patterns for small $n > 0$, an extra scaling in (3.191) with $\lambda = -1$ is performed by setting

$$\varphi(s) = A\psi(\eta), \;\; \eta = \tfrac{s}{a}, \quad \text{where } a = \tfrac{n}{3} \text{ and } A = \left(\tfrac{n}{3}\right)^{\frac{3}{n}}.$$

This leads to the simplified equation with a linear binomial operator,

$$\psi^{(4)} + 4\psi''' + 6\psi'' + 4\psi' + \psi = -\tfrac{1}{|\psi|^n}\,(\psi' + \psi), \qquad (3.193)$$

where we omit terms of the order $O(n)$. In Figure 3.20, a stable periodic behavior

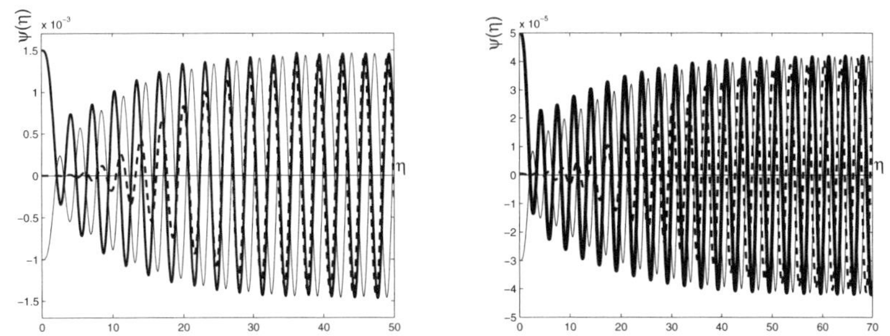

(a) $n = 0.3$ (b) $n = 0.2$

Figure 3.20 Stable periodic behavior for (3.193) for $n = 0.3$ (a) and $n = 0.2$ (b).

of the ODE (3.193) is shown. For $n = 0.2$, the periodic solution is very small, $\max|\varphi| \sim 10^{-22}$.

Comparison with the linear PDE. In view of the continuity at $n = 0^+$, consider the PDE (3.2) that admits TWs with the profiles satisfying

$$\lambda f = f''' \quad \Longrightarrow \quad f(y) = e^{\mu y}, \quad \text{where } \mu^3 = \lambda.$$

We are interested in the behavior as $y \to -\infty$ which is the left-hand "interface" for solutions of the linear equation. Then, taking $\lambda > 0$ gives the only possible non-oscillatory exponential behavior

$$f(y) \sim e^{\lambda^{1/3} y} \quad \text{as } y \to -\infty.$$

On the contrary, for $\lambda < 0$, oscillatory patterns are obtained,

$$f(y) \sim e^{\frac{1}{2}|\lambda|^{1/3} y} \cos\left(\tfrac{\sqrt{3}}{2}|\lambda|^{\frac{1}{3}} y + A_0\right), \quad \text{where } A_0 = \text{constant}. \tag{3.194}$$

This is in qualitative agreement with the properties of the nonlinear ODE (3.191) for $n \in (0, 1]$. It is curious that the period of linear oscillations in (3.194),

$$T\big|_{n=0} = \tfrac{4\pi}{\sqrt{3}} = 7.26... \quad (\lambda = -1),$$

is comparable with the nonlinear ones in Figure 3.17(b), for $n = \frac{1}{2}$, $T\big|_{n=\frac{1}{2}} \approx 1.3$.

The *fundamental solution* of the bi-harmonic equation (3.2),

$$b(x,t) = t^{-\frac{1}{4}} F(\zeta), \quad \zeta = \tfrac{x}{t^{1/4}},$$

corresponds to the case $\lambda > 0$ at $y = +\infty$. The rescaled kernel F is the unique radial solution of the ODE problem

$$-F^{(4)} + \tfrac{1}{4}\zeta F' + \tfrac{1}{4} F = 0 \quad \text{in } I\!R, \quad \int F = 1.$$

Then $F = F(|\zeta|)$ has exponential decay and oscillates as $|\zeta| \to \infty$; general estimates on fundamental solutions can be found in Eidelman [164, p. 115]. More precisely, using standard asymptotic expansion yields the following behavior of the kernel as $\zeta \to +\infty$:

$$F(\zeta) \sim \zeta^{-\frac{1}{3}} e^{a\zeta^{4/3}}, \quad \text{with complex } a \text{ satisfying } a^3 = \tfrac{1}{4}\left(\tfrac{3}{4}\right)^3, \ \operatorname{Re} a < 0.$$

There exist two complex conjugate roots for exponentially decaying profiles

$$a_{\pm} = -\tfrac{3}{8}\,4^{-\frac{1}{3}}(1 \pm \mathrm{i}\sqrt{3}) \equiv -c_1 \pm \mathrm{i}c_2.$$

This yields the following oscillatory behavior as $\zeta \to +\infty$:

$$F(\zeta) \sim \zeta^{-\frac{1}{3}} \mathrm{e}^{-c_1\zeta^{4/3}} \big[A_1 \cos\big(c_2\zeta^{\frac{4}{3}}\big) + A_2 \sin\big(c_2\zeta^{\frac{4}{3}}\big)\big], \tag{3.195}$$

where A_1 and A_2 are constants. The algebraic factor $\zeta^{-1/3}$ is obtained by the WKBJ-type technique. This is a periodic behavior with a *single* fundamental frequency. We have seen numerically that such a stable periodic structure is inherited in the behavior (3.190) for all $n \in (0, n_{\mathrm{h}})$.

Example 3.38 (Finite propagation and changing sign behavior in higher-order semilinear models) As we have seen, finite propagation and oscillatory solutions of the CP can be obtained in any semilinear parabolic models by adding a non-Lipschitz absorption-like term, e.g., (see [227] for properties and ε-regularization)

$$u_t = F[u] \equiv (-1)^{m+1} D_x^{2m} u - |u|^{p-1} u, \quad \text{with } p \in (-1, 1). \tag{3.196}$$

For TWs, there occurs the ODE

$$-\lambda f' = F[f] \quad \text{for } y > 0, \quad f(0) = 0,$$

which, close to the finite interface, becomes asymptotically stationary,

$$(-1)^{m+1} f^{(2m)} - |f|^{p-1} f = 0 \implies f(y) = y^{\gamma} \varphi(\ln y), \tag{3.197}$$

where $\gamma = \frac{2m}{1-p} > 0$. For any $m \geq 2$, such ODEs have oscillatory solutions; see Section 3.7. Note that (3.196) also describes the phenomenon of finite-time extinction proved by energy estimates based on Saint–Venant's principle; see [524] and survey in [240]. The range $p \leq -1$ corresponds to nonexistence, [227].

Similarly, equations of Cahn–Hilliard type with mass conservation

$$u_t = F[u] \equiv (-1)^{m+1} D_x^{2m} u - D_x^{2k} \big(|u|^{p-1} u\big) \quad \text{for} \quad 1 \leq k < m,$$

can also admit finite propagation. Solutions of changing sign (for $k \leq m-2$) are given by (3.197) with $2m \mapsto 2(m-k)$. Depending on m and k, the periodic behavior of $\varphi(s)$ can be stable or unstable, as in the examples discussed above.

3.9 Quasilinear pseudo-parabolic models: the magma equation

A typical pseudo-parabolic model to be studied is

$$u_t = \big(u^n u_t\big)_{xx} + \text{(lower-order terms)} \quad (u \geq 0), \tag{3.198}$$

where $n \in I\!R$ is a fixed exponent. Such PDEs arise in mathematical modeling of the segregation and migration of magma in the mantle of the Earth, [519, 420]. The full model, called the *magma equation*, is

$$u_t = \big[\big((u^n u_t)_x - 1\big) u^l\big]_x, \tag{3.199}$$

with two parameters n and l. Setting $l = 0$ gives the principal operator in (3.198). Such PDEs belong to the class of *pseudo-parabolic equations* that may enjoy some properties of standard parabolic flows. Examples include the linear analogy

$$u_t - u_{xxt} = u_{xx},$$

and the regularized PBBM equation (see Example 4.45)

$$u_t - u_{xxt} = uu_x + \varepsilon u_{xx} \quad (\varepsilon > 0).$$

The advantage of such pseudo-parabolic PDEs is that applying the inverse operator $(I - D_x^2)^{-1}$ with positive integrable kernel yields non-local equations with compact operators, which can be studied by nonlinear integral operators theory.

Standard theory does not apply to quasilinear PDEs, such as (3.198), which, in addition, are degenerate at the singular set $\{u = 0\}$. Mathematical theory of such degenerate PDEs with nonnegative solutions (or, possibly, solutions of changing sign) is unavailable. As usual, we will use solutions on invariant subspaces for formal detecting some evolution singularities.

Example 3.39 **(Quenching, interfaces, and the Cauchy problem)** Consider the PDE (3.198) for $n = 1$ with a constant absorption,

$$\boxed{u_t = F[u] \equiv (uu_t)_{xx} - 1,}$$

bearing in mind the phenomenon of finite-time extinction. F is associated with subspace (module) $W_2 = \mathcal{L}\{1, x^2\}$, so there exist exact solutions

$$u(x,t) = C_1(t) + C_2(t)x^2,$$

$$\begin{cases} C_1' = 2(C_1C_2)' - 1, \\ C_2' = 12C_2C_2'. \end{cases}$$

We take, e.g., $C_2 = \frac{1}{12}$ (any constant C_2 fits), and the first equation is then integrated,

$$u(x,t) = \tfrac{6}{5}(T - t) + \tfrac{1}{12}x^2, \tag{3.200}$$

where $T > 0$ is the finite quenching time. For $t < T$, the solution is strictly positive. At $t = T^-$, the solution touches the singularity level $\{u = 0\}$. We readily derive the asymptotic extinction behavior as $t \to T$,

$$u(x,t) = (T - t)g_0(\xi), \quad \xi = \tfrac{x}{\sqrt{T-t}}, \quad \text{where} \quad g_0(\xi) = \tfrac{6}{5} + \tfrac{1}{12}\xi^2. \tag{3.201}$$

Hence, (3.201) belongs to the class of self-similar solutions $u_s(x,t) = (T - t)g(\xi)$, where g (and g_0) solves the following third-order ODE:

$$\big[g\big(\tfrac{1}{2}g'\xi - g\big)\big]'' - \tfrac{1}{2}g'\xi + g - 1 = 0.$$

These solutions on W_2 explicitly describe the quenching behavior for this degenerate quasilinear PDE. A rigorous stability analysis of such quenching is OPEN (it is likely that it is not generic or robust at all). In the rescaled sense, this assumes, introducing the rescaled function $u(x,t) = (T-t)v(\xi,\tau)$, with $\tau = -\ln(T-t)$, to study the third-order rescaled PDE

$$v_\tau = \big[v\big(\tfrac{1}{2}v_\xi\xi - v + v_\tau\big)\big]_{\xi\xi} - \tfrac{1}{2}v_\xi\xi + v - 1.$$

This means, passing to the limit $\tau \to \infty$, that, for a class of positive, symmetric, and, say, convex initial data (recall the parabolic shapes, $\frac{6}{5}T + \frac{1}{12}x^2$, of initial functions in W_2), the convergence takes place,

$$v(\xi, \tau) \to g_0(\xi) \quad \text{as} \;\; \tau \to \infty \quad (\text{i.e., as } t \to T^-).$$

Interface propagation. For $t > T$, the problem admits a suitable setting in the class of nonnegative solutions in an FBP framework. It follows from (3.200) that the right-hand interface is located at

$$s(t) = 6\sqrt{\tfrac{2}{5}}\,\sqrt{t-T} \quad \text{for } t > T.$$

For such solutions $u_x = \frac{x}{6}$, so that, at $u = 0$ in (3.200), $-\frac{6}{5}(t-T) + 3(u_x)^2 = 0$. Plugging this into the expression for s' yields the dynamic interface equation

$$s' = \frac{6}{\sqrt{10}}\,\frac{1}{\sqrt{t-T}} \equiv \frac{6}{5u_x} \quad \text{at } x = s(t). \tag{3.202}$$

On the Cauchy problem: oscillatory and non-oscillatory patterns. Let us formally detect the maximal regularity that can be attributed to the CP. To keep pseudo-parabolicity and extinction properties, consider the *signed* equation for $n = 1$,

$$u_t = (|u|u_t)_{xx} - \operatorname{sign} u. \tag{3.203}$$

Substituting the TWs $u(x,t) = f(y)$, with $y = x - \lambda t$, yields the ODE $-\lambda f' = -\lambda(|f|f')'' - \operatorname{sign} f$. Studying the right-hand interface, and hence, using the reflection $y \mapsto -y$, we keep the leading two terms on the right-hand side,

$$\lambda(|f|f')'' - \operatorname{sign} f = 0 \quad \text{for } y > 0, \quad f(0) = 0. \tag{3.204}$$

It follows that the behavior near the interface at $y = 0$ is given by

$$f(y) = y^{\frac{3}{2}}\varphi(s), \quad \text{with } s = \ln y,$$

where the oscillatory component $\varphi(s)$ solves a third-order ODE. Such ODEs were studied earlier in the TFE applications; see (3.166). Therefore, $\varphi(s)$ is oscillatory and changes sign for $\lambda < 0$, and is non-oscillatory for $\lambda > 0$. Such TW properties are expected to make sense in the CP for (3.203). Its well-posedness is OPEN. Note that the same ODE (3.204) occurs for the *signed nonlinear dispersion equation* (see Section 4.3 for details)

$$u_t = \pm(|u|u_x)_{xx} - \operatorname{sign} u.$$

Secondly, for the original model (3.198) without strong absorption

$$u_t = \big(|u|^n u_t\big)_{xx} \quad (n > 0),$$

the same TW analysis leads to the ODE $(|f|^n f')' = f$ for $y > 0$, which always admits the strictly positive (non-oscillatory) solution

$$f(y) = \big[\tfrac{n^2}{2(n+2)}\big]^{\frac{1}{n}}\, y^{\frac{2}{n}}.$$

The regularity at the interface $y = 0$ increases without bound as $n \to 0^+$ when approaching the linear PDE.

Example 3.40 (Sixth-order PDE) Consider a higher-order pseudo-parabolic model, such as

$$u_t = \big(|u|^n u_t\big)_{xxxxxx}. \tag{3.205}$$

The TW profiles satisfy

$$(|f|^n f')^{(5)} = f \quad \text{for } y > 0, \quad f(0) = 0.$$

For $n \in (0, 1)$, there exists a strictly positive, non-oscillatory solution $f(y) = c_0 y^{\frac{6}{n}}$, where $c_0 > 0$ is a constant.

Example 3.41 (Single point blow-up) Consider next a pseudo-parabolic PDE with an extra reaction term,

$$v_t = \big(v^n v_t\big)_{xx} + v^{\frac{2-n}{2}}, \tag{3.206}$$

where, for $n < 0$, the source term is superlinear for $v \gg 1$ and may create finite-time blow-up singularities. In order to deal with polynomial subspaces, set

$$v = u^\mu, \quad \text{with } \mu = \tfrac{2}{n},$$

to get a cubic PDE of the form

$$\boxed{\begin{aligned} u_t &= u^2 u_{xxt} + 2(\mu+1) u u_x u_{xt} \\ &+ (\mu+1)\big[u u_{xx} + \mu (u_x)^2\big] u_t + \tfrac{1}{\mu} \equiv F_3[u] + \tfrac{1}{\mu}. \end{aligned}} \tag{3.207}$$

Proposition 3.42 *F_3 in (3.207) admits subspace $W_3 = \mathcal{L}\{1, x, x^2\}$ iff*

$$2\mu^2 + 7\mu + 6 = 0 \quad \Longleftrightarrow \quad \mu = -2, \quad \text{or } \mu = -\tfrac{3}{2}. \tag{3.208}$$

Substituting into (3.207) $u(x,t) = C_1(t) + C_2(t)x^2$ yields that (3.208) annuls the term on $\mathcal{L}\{x^4\}$. This gives

$$\begin{cases} C_1' = 2(\mu+1) C_1 C_2 C_1' + 2C_1^2 C_2' + \tfrac{1}{\mu}, \\ C_2' = 2(\mu+3) C_1 C_2 C_2' + 2(\mu+1)(2\mu+1) C_2^2 C_1'. \end{cases} \tag{3.209}$$

Thus, such solutions exist in two cases, where (3.206) is

$$v_t = \big(v^{-1} v_t\big)_{xx} + v^{\frac{3}{2}} \quad \text{for } n = -1 \ \ (\mu = -1);$$

$$v_t = \big(v^{-\frac{4}{3}} v_t\big)_{xx} + v^{\frac{5}{3}} \quad \text{for } n = -\tfrac{4}{3} \ \ (\mu = -\tfrac{3}{2}).$$

For both PDEs, we detect from the DS (3.209) the generic (on W_3) extinction behavior for the function $u(x,t)$, which is equivalent to blow-up for $v(x,t)$. For smooth bounded orbits of the DS, the first equation near extinction time T reads $C_1' = -\frac{1}{|\mu|} + \dots$, and the second equation yields $C_2(t) \to d > 0$, from which we obtain the asymptotic pattern

$$u(x,t) = (T-t)\tfrac{1}{|\mu|} + d\,x^2 + \dots = (T-t)\big(\tfrac{1}{|\mu|} + d\,\xi^2\big) + \dots \ \text{ as } t \to T^-,$$

with the spatial rescaled variable $\xi = x/\sqrt{T-t}$. In terms of the original solution $v = u^{2/n}$ of the initial PDE (3.206), this yields the following single point blow-up behavior as $t \to T^-$ for $n = -1$ or $n = -\frac{4}{3}$:

$$w(\xi, \tau) \equiv (T-t)^{-\frac{2}{n}} v(x,t) \to g(\xi) \equiv \big(\tfrac{1}{|\mu|} + d\,\xi^2\big)^{\frac{2}{n}} \tag{3.210}$$

uniformly in $\xi \in I\!R$. As happened before to other blow-up models, such behavior corresponds to a delicate case of a singular perturbed first-order Hamilton–Jacobi equation. To reveal this singular limit, we write down the perturbed PDE for the rescaled function $w(\xi, \tau)$ as follows:

$$w_\tau = -\tfrac{1}{2}\, w_\xi \xi + \tfrac{2}{n}\, w + w^{\frac{2-n}{2}} + \mathrm{e}^{-\tau}\big[w^2\big(w_\tau - \tfrac{2}{n}\, w + \tfrac{1}{2}\, w_\xi \xi\big)\big]_{\xi\xi}. \tag{3.211}$$

This is an exponential perturbation of the Hamilton–Jacobi equation $w_\tau = -\frac{1}{2}\, w_\xi \xi + \frac{2}{n}\, w + w^{\frac{2-n}{2}}$ that possesses $g(\xi)$ in (3.210) as a stationary solution. There is a large amount of mathematical literature devoted to infinite-dimensional singular perturbed DSs associated with blow-up and extinction phenomena for reaction-diffusion PDEs; see [245, Ch. 5, 9–11]. These well-developed techniques do not apply to the perturbed PDE (3.211) for which establishing uniform boundedness and compactness of the rescaled orbits in suitable metrics are also OPEN.

Remarks and comments on the literature

In many occasions we put references concerning specific models, equations, and applications alongside corresponding examples. Other references are given below.

§ 3.1. Earlier references on the derivation of the fourth-order TFE can be found in [263, 531], where the first analysis of some self-similar solutions was performed for $n = 1$. Source-type (ZKB) similarity solutions for arbitrary n were studied in [48] for $N = 1$ and [185] for the equation in $I\!R^N$. More information on similarity and other solutions can be found in [46, 45, 78]; see also a discussion of the TFE in the afterword of Barenblatt [25]. Thin film equations admit nonnegative solutions constructed by special parabolic approximations of the degenerate nonlinear coefficients; see the pioneering paper [44], various extensions in [264, 167, 376, 576], and the references therein. For estimates of not necessarily nonnegative solutions in $I\!R^N$, see [265] and the bibliography therein.

The family (3.7) of generalized TFEs was studied in [345], where further references and physical motivation can be found. The equation (3.3) was derived and studied in [155]; see [529] for a parallel development. Equation (3.8) with $m = n = 3$ has been used to describe bubble motion in a capillary tube and the Rayleigh–Taylor instability in a thin film [279]. For more information on the modeling and physics of thin liquid films, we refer to survey papers [430, 451, 34]; see also references in [431]. See [53]–[55], [293] for more general PDEs, such as (3.9). The doubly nonlinear equation

$$u_t = -(u^n |u_{xxx}|^l u_{xxx})_x$$

describes, for $n = l+3$, the surface tension-driven spreading of a power-law fluid and, for $n = 1$, a power-law fluid in a Hele–Shaw cell; see [345], survey [451], and mathematics in [14]. The exponent l is determined by rheological characteristics of the liquid, so $l = 0$ corresponds to a Newtonian liquid, while $l \neq 0$ appears for "power-law" (Ostwald–de Waele) liquids, called *shear-thinning* if $l > 0$. For such liquids, a typical sample relation of the viscosity η and the shear rate $\dot\gamma$ is of the form $\eta \sim \dot\gamma^{\,-l/(l+1)}$.

Concerning the Benney equation (3.10) [39] and models of falling liquid films, see [452]. Notice earlier experimental findings of P.L. and S.P. Kapitza [315] in 1949 related to traveling waves in such models. On Marangoni instability in thin film models (3.12), see [450]. Semilinear Cahn–Hilliard equations were introduced in [92]; see [438] and references therein.

The sixth-order thin film equation (3.13) was introduced in [338, 339] in the case of $n = 3$, and describes the spreading of a thin viscous fluid under the driving force of an elastica or light plate. In addition, [338, 339] treat a more general form of this equation (now allowing for a reaction at the usual solid interface), which is shown to arise in the industrial application of the isolation oxidation of silicon. Analysis for a finite length elastica or plate is given in [176], whilst a numerical scheme with subsequent parameter investigation of the more general system is derived in [177]. Delicate aspects of unusual similarity solutions and asymptotics can be found in [189]. Some particular features of the $2m$th-order PDEs (3.16) were already considered in [531], where the n-small and waiting-time solutions were noted. Doubly nonlinear equations, such as (3.22), which are relevant to capillary driven flows of thin film of power-law fluids, were derived in [345, 513]. Extra absorption terms in TFEs (see (3.25)) or the source terms are to model effects of evaporation (certain *permeability* of the surface may also be taken into account) or condensation, [451]. Actually, evaporation phenomena of thin films are well known for binary solutions (see references in [261]), which probably hardly apply to thin films on flat surfaces. The Florin problem for the fourth-order TFE with the constant non-zero angle $u_x = C$ at the interfaces, which is a lubrication model related to Darcy flow in a Hele–Shaw cell, was studied in [454].

Discussing other related higher-order models, note that similarity solutions of the equation with a monotone operator in H^{-2},

$$u_t = -(|u|^{m-1}u)_{xxxx},$$

were studied in [47], where, for $m > 1$, solutions were proved to be compactly supported and oscillatory (changing sign) near the interfaces. Other lubrication-type PDEs

$$u_t = -(u^n u_{xx})_{xx} \quad \text{and} \quad u_t = -|u|^n u_{xxxx} \;\; (n < 0)$$

were introduced in [52]. The fourth-order parabolic equation

$$u_t = -(u(\ln u)_{xx})_{xx}$$

arises in the context of interface fluctuations in spin systems and semiconductor theory; see mathematics and references in [64, 307]. The sixth-order PDEs

$$u_t = (u^n |u_{xxxxx}|^{m-1} u_{xxxxx})_x \quad \text{and} \quad u_t = \{u[\tfrac{1}{u}(u(\ln u)_{xx})_{xx} + \tfrac{1}{2}((\ln u)_{xx})^2]_x\}_x$$

are obtained, respectively, for power-law fluids spreading on a horizontal substrate [345] and as a generalized quantum drift-diffusion model for semiconductors, [146, 307]. Critical exponents and asymptotic and singularity phenomena for the eighth-order TFE

$$u_t = -(u^n u_{xxxxxxx})_x$$

are discussed in [189]. Various aspects concerning oscillatory solutions of the Cauchy problem for TFEs can be found in [174, 175], where further references are given.

Higher-order parabolic PDEs also occur in curve shortening flows for curves in two dimensions, whose normal velocity V_n is given by the Laplacian of its curvature; see [91]. The corresponding equation $V_n \equiv \mathbf{N}_t = -\kappa_{ss}$ (s is the arclength) for curves $y = u(x,t)$ on the $\{x, y\}$-plane, can be written as (see [82])

$$u_t = -\Big[\frac{1}{\sqrt{1+(u_x)^2}}\Big(\frac{u_{xx}}{\sqrt{1+(u_x)^2}}\Big)_x\Big]_x.$$

The hierarchy of such models of arbitrary order (the third-order PDE belongs to the KdV family) goes back to Mullins [428], who proposed classical theory of thermal grooving. Arbitrary order integrable models $V_n = D_s^l \kappa$ were analyzed in [82]. Other fourth-order parabolic PDEs appear [116] in blow-up analysis of the curve evolution of an immersed plane curve according to the H^{-1} gradient flow.

Exact solutions on invariant subspaces exist for some non-local parabolic equations related to the *b–l* model of the propagation of turbulent bursts from a horizontally uniform layer (see model statement in [26] and [25, p. 291]),

$$b_t = l(t)\Delta b^{\frac{3}{2}} - k\frac{b^{\frac{1}{2}}}{l^2(t)} \quad \text{in } I\!R^N \times I\!R_+,$$

where $k > 0$ is a fixed constant, and $l(t) > 0$ is the measure of the support of a nonnegative bounded solution, $l(t) = \text{meas supp}\, b(\cdot,t) \equiv \text{meas}\,\{x \in I\!R^N : b(x,t) > 0\}$. A class of such equations with invariant subspaces was studied in [224]. Another invariant subspace is available in the *b–ε* model of turbulence motion based on the ideas of A.N. Kolmogorov and L. Prandtl:

$$\begin{cases} b_t = \alpha(\frac{b^2}{\varepsilon}\, b_x)_x - \varepsilon, \\ \varepsilon_t = \beta(\frac{b^2}{\varepsilon}\, \varepsilon_x)_x - \gamma\, \frac{\varepsilon^2}{b}, \end{cases}$$

where α, β, and γ are positive constants. Here $b(x,t) \ge 0$ denotes the turbulent energy density, and $\varepsilon(x,t) \ge 0$ is the dissipation rate of turbulent energy. See [27, 26] and [56] for physical justification and further references. For $\alpha = \beta$, the finite-time extinction behavior can happen on a polynomial subspace invariant under quadratic operators [224].

§ 3.2. Some details concerning spectra, compact resolvent, and sectorial properties of linear operators like (3.52) can be found in [163]. Note that, as a rule, rescaled higher-order parabolic PDEs contain non-symmetric and non-potential operators. The spectral theory in [163, 87] of similar (but not precisely the same) operators appears in the study of asymptotic blow-up and global behavior for semilinear $2m$th-order parabolic equations

$$u_t = -(-\Delta)^m u \pm |u|^{p-1} u.$$

§ 3.3. The original von Mises transformation (see Example 3.10) was introduced in 1927, [424]. On oscillatory and other features of the CP for the fourth-order TFEs, see [174].

§ 3.4, 3.5. Proposition 3.7 is taken from [343] (transformation (3.60) was first used in [531] for formal perturbation analysis for small n), where Proposition 3.29 was used for $m = 2$. Various results on the FBP and the CP for the sixth-order TFEs are available in [175].

§ 3.6, 3.7. Basic properties of oscillatory solutions of the CP for TFEs of various orders with more detailed description are given in [174, 175]; see also [227].

§ 3.8. The KS-type equations are used as a description of the fluctuations of the position of flame front [202, 203], the motion of fluid on a vertical wall, and chemical reactions with spatially uniform oscillations on a homogeneous medium. Similar models also occur in solidification, [201]; see survey [411]. The mKS equation (3.172) is a model for the dynamics of a hyper-cooled melt [514]. A more general class of such models was introduced and discussed in [289]. Blow-up in the mKS equations was studied in [50].

§ 3.9. In the magma equation (3.199), $u(x,t) \ge 0$ measures the volume fraction of liquid phase [519]. The exponents n and l describe permeability and effective viscosity characterizing the rate of matrix compaction and distension on u. Similar pseudo-parabolic PDEs occur in modeling of thin film flows for poroviscous droplets over a planar substrate [347].

Open problems

- These are formulated throughout the chapter in Section 3.2, Examples 3.8–3.10, 3.14, 3.17, Remark 3.28, Section 3.7, and Examples 3.36, 3.37, 3.39, and 3.41.

CHAPTER 4

Odd-Order One-Dimensional Equations: Korteweg-de Vries, Compacton, Nonlinear Dispersion, and Harry Dym Models

In this chapter, we continue to describe applications of invariant subspaces to nonlinear PDEs in one dimension, and, unlike the previous chapter, concentrate on nonlinear evolution equations of odd orders. These models include the famous KdV, nonlinear dispersion, and Harry Dym-type equations, as well as their higher-order generalizations. Using exact solutions, we establish interesting similarities between classes of even and odd-order evolution PDEs and study singularity formation, interface propagation, and oscillatory, changing sign properties of solutions.

4.1 Blow-up and localization for KdV-type equations

In 1895, Korteweg and de Vries introduced the famous third-order *KdV equation* of shallow water waves

$$v_t = v_{xxx} + 2vv_x \quad \text{in } I\!R \times I\!R \tag{4.1}$$

and its explicit *soliton*

$$v(x,t) = \frac{3}{2\cosh^2[\frac{1}{2}(x+t)]}, \tag{4.2}$$

which is the *traveling wave* solution moving to the left with unit velocity.* This and other multi-soliton solutions play a determining role in general water waves theory and theory of integrable PDEs. By scaling $x \mapsto \lambda x$, $t \mapsto \lambda^3 t$, and $v \mapsto Cv$, (4.1) reduces to $v_t = v_{xxx} + 2C\lambda^2 vv_x$ and takes standard forms

$$v_t = v_{xxx} + vv_x \;\; \text{for } C = \tfrac{1}{2\lambda^2}, \quad \text{or}$$

$$v_t = v_{xxx} + 6vv_x \;\; \text{for } C = \tfrac{3}{\lambda^2}.$$

Setting $v = u_x$ in (4.1) and integrating once gives the *potential KdV equation*

$$u_t = u_{xxx} + (u_x)^2. \tag{4.3}$$

Then the soliton (4.2) is transformed into the front moving traveling wave, which is reconstructed from (4.2) by the inverse transformation

$$u(x,t) = \int_{-\infty}^{x} v(z,t)\,dz.$$

* It has long been recognized that both the KdV equation and its soliton solution were derived earlier by Boussinesq in 1872 [75], so the abbreviation *BKdV* can be used for (4.1).

4.1.1 Quadratic operators and invariant subspaces

Consider a third-order PDE with a more general quadratic operator associated with that in (4.3),

$$\boxed{u_t = F[u] = \alpha u_{xxx} + \beta (u_x)^2 + \gamma u^2.} \tag{4.4}$$

Let us begin with trigonometric subspaces for such simple quadratic operators already studied in Section 1.4.

Proposition 4.1 *Operator F in* (4.4) *admits $W_3 = \mathcal{L}\{1, \cos x, \sin x\}$ if $\beta = \gamma$.*

Looking for the solutions

$$u(x,t) = C_1(t) + C_2(t)\cos x + C_3(t)\sin x \in W_3 \quad \text{for } t \in I\!R$$

yields that the PDE (4.4) on W_3 is equivalent to the DS

$$\begin{cases} C_1' = \beta(C_1^2 + C_2^2 + C_3^2), \\ C_2' = -\alpha C_3 + 2\beta C_1 C_2, \\ C_3' = \alpha C_2 + 2\beta C_1 C_3. \end{cases}$$

These exact periodic solutions admit a *soliton-traveling wave* representation,

$$u(x,t) = C_1(t) + C_2(t)\cos(x \pm t) + C_3(t)\sin(x \pm t), \tag{4.5}$$

with a similar DS for the coefficients. Such moving 2π-periodic soliton-like solutions may blow-up in finite time. There exist 2D invariant reductions of such periodic moving waves.

Proposition 4.2 *If $\beta = \gamma$ and $\alpha = \pm 1$ in* (4.4)*, then there exist exact solutions*

$$u(x,t) = C_1(t) + C_2(t)\cos(x \mp t).$$

4.1.2 Applications to blow-up

Example 4.3 (Blow-up and localization) Consider the *KdV equation with source*

$$\boxed{u_t = u_{xxx} + (u_x)^2 + u^2.} \tag{4.6}$$

Here, $\alpha = \beta = \gamma = 1$, so by Proposition 4.2 there exist solutions on

$$W_2^- = \mathcal{L}\{1, \cos(x-t)\},$$

i.e.,

$$u(x,t) = C_1(t) + C_2(t)\cos(x-t), \tag{4.7}$$

$$\begin{cases} C_1' = C_1^2 + C_2^2, \\ C_2' = 2C_1C_2. \end{cases} \tag{4.8}$$

According to (4.7), the DS describes the time-deformation of a moving 2π-periodic soliton. An explicit self-similar solution in separate variables appears for $C_1 = C_2$, yielding the single ODE $C_1' = 2C_1^2$, and the solution on $\tilde{W}_1 = \mathcal{L}\{\cos^2[\frac{1}{2}(x-t)]\}$ given by

$$u(x,t) = \tfrac{1}{T-t}\cos^2[\tfrac{1}{2}(x-t)], \tag{4.9}$$

where $T > 0$ is the finite blow-up time. This is a *blow-up soliton solution* moving to the right with constant unit speed and blowing up as $t \to T^-$ everywhere, excluding the points $x_k = T + \pi(2k+1)$, $k = 0, \pm 1, ...$, where $u(x, T^-) = 0$. It is easy to detect the following *localization* property of these blow-up solutions:

$$u(x_k, t) = \tfrac{1}{4}(T-t) + ... \to 0 \quad \text{as } t \to T,$$

so that the solution remains bounded at all points $x = x_k$ and tends to infinity at any other $x \in I\!R$. Notice that (4.9) is a classical analytic periodic solution of the non-degenerate PDE (4.6) on $I\!R \times (0, T)$. To monitor a one-hump wave of such solutions on $\{|x-t| < \pi\}$, a proper FBP should be posed. This will be discussed later for degenerate third-order operators.

Returning to the quadratic DS (4.8) (cf. Proposition 3.18 for more general DSs), we easily solve it explicitly and obtain the following solutions:

$$u(x,t) = \tfrac{1}{T-t} \cos^2[\tfrac{1}{2}(x-t)] + \tfrac{1}{T_1-t} \sin^2[\tfrac{1}{2}(x-t)].$$

It follows that, for $T_1 > T > 0$, they asymptotically converge to the separate variables solution (4.9), in the sense that the rescaled function satisfies

$$w(x,t) \equiv (T-t)u(x,t) \to \cos^2[\tfrac{1}{2}(x-T)] \quad \text{as } t \to T. \tag{4.10}$$

The general asymptotic stability of this localized blow-up pattern leads to the study of the rescaled PDE for $w(x,t)$,

$$w_\tau = F_*[w] \equiv w_{xxx} + (w_x)^2 + w^2 - w, \quad \text{where } \tau = -\ln(T-t),$$

and proving convergence (4.10) (up to translation in x) for periodic initial data. Stability analysis demands a sharp upper bound for the rescaled orbit $\{w(\cdot, \tau)\}$, which remains an OPEN PROBLEM for general initial data. Similar conclusions apply to higher odd-order PDEs, such as

$$\boxed{u_t = D_x^{2m+1} u + (u_x)^2 + u^2, \quad m = 2, \dots,}$$

for which blow-up stability problems are OPEN.

4.2 Compactons and shocks waves in higher-order quadratic nonlinear dispersion models

In the next sections we study exact solutions of a number of quasilinear degenerate odd-order PDEs which have many applications and still rather poor mathematical understanding.

4.2.1 Compactons on 3D trigonometric subspace

As a typical simple example, consider the fifth-order quadratic operator,

$$F_5[u] = \alpha(u^2)_{xxxxx} + \beta(u^2)_{xxx} + \gamma(u^2)_x\,. \tag{4.11}$$

Proposition 4.4 *Operator* (4.11) *admits* $W_3 = \mathcal{L}\{1, \cos x, \sin x\}$ *iff*

$$16\alpha - 4\beta + \gamma = 0. \tag{4.12}$$

Example 4.5 (Dynamics on W_3 around compactons in quadratic models) Consider the *quintic nonlinear dispersion equation*

$$\boxed{u_t = \alpha(u^2)_{xxxxx} + \beta(u^2)_{xxx} + \gamma(u^2)_x \quad \text{in } I\!R \times I\!R.} \tag{4.13}$$

For $\alpha = 0$, $\beta = \gamma = 1$, this is the third-order *Rosenau–Hyman* (RH) *equation*

$$\boxed{u_t = (u^2)_{xxx} + (u^2)_x,} \tag{4.14}$$

which models the effect of nonlinear dispersion in the pattern formation in liquid drops [496]. It is the $K(2,2)$ equation from the general $K(m,n)$ family of *nonlinear dispersion equations*

$$u_t = (u^n)_{xxx} + (u^m)_x \quad (u \geq 0), \tag{4.15}$$

that also models phenomena of compact pattern formation, [491, 492]. Such PDEs appear in curve motion and shortening flows [494]. The $K(m,n)$ equation (4.15) with $n > 1$ is degenerated at $u = 0$, and therefore may exhibit finite speed of propagation and admit solutions with finite interfaces. Rigorously speaking, these questions, especially for degenerate higher-order models, lead to several OPEN PROBLEMS to be discussed.

The crucial advantage of the RH equation (4.14) is that it possesses *explicit* moving compactly supported soliton-type solutions, called *compactons* [496]:

$$u(x,t) = \begin{cases} -\frac{4\lambda}{3}\cos^2[\frac{1}{4}(x-\lambda t)], & \text{if } |x-\lambda t| \leq 2\pi, \\ 0, & \text{if } |x-\lambda t| > 2\pi, \end{cases} \tag{4.16}$$

where, for $\lambda < 0$, the solution is nonnegative; see Figure 4.1. These are the TW patterns with two interfaces moving to the left. Taking $\lambda > 0$ yields the negative compacton moving to the right.

For the fifth-order PDE (4.13), compacton solutions were first constructed in [147], where the more general $K(m,n,p)$ family of PDEs

$$u_t + \beta_1(u^m)_x + \beta_2(u^n)_{xxx} + \beta_3 D_x^5(u^p) = 0 \quad (m,n,p > 1),$$

was introduced. Some of these equations will be treated later on. Equation (4.13) is also associated with the family $Q(l,m,n)$ of more general quintic evolution PDEs with nonlinear dispersion,

$$u_t + a(u^{m+1})_x + \omega\big[u(u^n)_{xx}\big]_x + \delta\big[u(u^l)_{xxxx}\big]_x = 0, \tag{4.17}$$

possessing multi-hump, compact solitary solutions [499].

Using first the particular quadratic model (4.13), we will discuss the dynamic interface equations and a general mathematical meaning of compactons. The first important question is to establish which kind of FBPs or the Cauchy problem such compactons may be solutions of. In addition, as usual, our goal is to show that there exist extra explicit finite-dimensional dynamics on invariant subspaces around those compactons. These results can be extended to general odd-order quadratic dispersive PDEs, such as

$$\boxed{u_t = \textstyle\sum_{(k)} \alpha_k D_x^{2k+1}(u^2).}$$

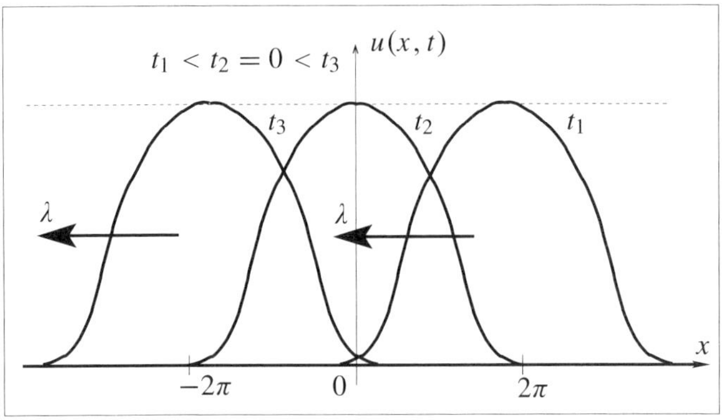

Figure 4.1 Moving compacton (4.16) with $\lambda < 0$ of the RH equation (4.14).

Let us return to the fifth-order PDE (4.13) and, assuming (4.12), look for exact solutions on W_3,

$$u(x,t) = C_1(t) + C_2(t)\cos x + C_3(t)\sin x, \tag{4.18}$$

$$\begin{cases} C_1' = 0, \\ C_2' = \mu C_1 C_3, \\ C_3' = -\mu C_1 C_2, \end{cases} \tag{4.19}$$

where $\mu = 2(\alpha - \beta + \gamma) \neq 0$. In this case, from the first ODE, $C_1(t) = A$, a constant, and the last two yield

$$C_2^2 + C_3^2 = B^2,$$

where $B \neq 0$ is a constant of integration. This gives the explicit solutions

$$u(x,t) = A + B\cos(x + \mu A t). \tag{4.20}$$

In the particular case $A = B$, denoting $\mu A = -\lambda$, and assuming that $\frac{\lambda}{\mu} < 0$, we obtain a nonnegative traveling wave solution

$$u(x,t) = -\tfrac{2\lambda}{\mu}\cos^2[\tfrac{1}{2}(x - \lambda t)]. \tag{4.21}$$

Choosing the one-hump profile on the moving interval $|x - \lambda t| \leq \pi$ and setting $u(x,t) = 0$ for $|x - \lambda t| > \pi$ yields a *compacton*, which is similar to (4.16) in Figure 4.1. Appropriate FBP setting for such solutions of the higher-order PDE will be studied in Section 4.2.3.

It turns out that there exists an intriguing similarity between *compacton* patterns in nonlinear dispersion media and *localized blow-up structures* in dissipative reaction-combustion models that were studied in the previous chapters.

Example 4.6 (Comparison of compactons and regional blow-up) Explicit TW compactons also exist for the nonlinear dispersion KdV-type equations with arbitrary power nonlinearities (formulae will be given shortly)

$$v_t = (v^{n+1})_{xxx} + \gamma\,(v^{n+1})_x, \quad \text{with } n > 0 \text{ and } \gamma = \tfrac{(n+1)^2}{n^2} \quad (v \geq 0), \tag{4.22}$$

where the parameter $\gamma > 0$ is chosen for convenience and can be scaled out. This is the $K(1+n, 1+n)$ model, [496]

Combustion model: regional blow-up. Firstly, we compare such compactons with blow-up solutions of the following parabolic *reaction-diffusion equation*:

$$v_t = (v^{n+1})_{xx} + \gamma v^{n+1}. \tag{4.23}$$

In (4.22), the right-hand side is the derivative D_x of that in (4.23). Mathematically, these equations belong to entirely different types of PDEs.

In the mid 1970s, Kurdyumov, with his former PhD students, Mikhailov and Zmitrenko, (see [510]) discovered the phenomenon of *heat* and *combustion localization* by studying properties of the following *Zmitrenko–Kurdyumov solution* of the equation (4.23):

$$v_S(x,t) = (T-t)^{-\frac{1}{n}} f(x), \tag{4.24}$$

where $T > 0$ is the blow-up time, and f satisfies the ODE

$$\tfrac{1}{n} f = (f^{n+1})'' + \gamma f^{n+1} \quad \text{for } x \in I\!R. \tag{4.25}$$

It turned out that (4.25) possesses the explicit compactly supported solution

$$f(x) = \begin{cases} \left[\frac{n}{(n+1)(n+2)} \cos^2\left(\frac{x}{2}\right)\right]^{\frac{1}{n}}, & \text{if } |x| \le \pi, \\ 0, & \text{if } |x| > \pi. \end{cases} \tag{4.26}$$

The striking *regional blow-up* (the so-called *S-regime of blow-up*, [510]) described by the solution (4.24), (4.26) is as follows: $v_S(x,t) \to \infty$ as $t \to T^-$ for all $|x| < \pi$ only and $v_S(x,t) \equiv 0$ otherwise. This is the localization phenomenon on the interval $\{|x| < \pi\}$ of the length 2π that is called the *fundamental length* of such a diffusive and combustion medium. Notice that (4.24) is a standard continuous weak solution of the PDE (4.23) on $I\!R \times (0, T)$.

Regional blow-up in a quasilinear wave equation. Secondly, a similar exact solution exists for the quasilinear hyperbolic equation

$$v_{tt} = (v^{n+1})_{xx} + \gamma v^{n+1}, \quad \text{where}$$

$$v_S(x,t) = (T-t)^{-\frac{2}{n}} \tilde{f}(x) \quad \Longrightarrow \quad \tfrac{2}{n}\left(\tfrac{2}{n}+1\right) \tilde{f} = (\tilde{f}^{n+1})'' + \gamma \tilde{f}^{n+1}. \tag{4.27}$$

Here f is given by a scaled function (4.26), $\tilde{f}(x) = \left[\frac{2(n+2)}{n}\right]^{1/n} f(x)$.

Compactons. Thirdly, returning to the compactons of (4.22) of the TW structure

$$v_c(x,t) = f(y), \quad y = x - \lambda t, \tag{4.28}$$

we find that f satisfies the ODE

$$-\lambda f' = (f^{n+1})''' + \gamma (f^{n+1})',$$

which gives, on integration once,

$$-\lambda f = (f^{n+1})'' + \gamma f^{n+1} + D, \tag{4.29}$$

where $D \in I\!R$ is the constant of integration. Setting $D = 0$ yields that the blow-up

ODE (4.25) and the compacton equation (4.29) *coincide*, provided that

$$-\lambda = \tfrac{1}{n} \quad \big(\text{or } -\lambda = \tfrac{2}{n}\big(\tfrac{2}{n}+1\big) \text{ to match (4.27)}\big).$$

This yields the compacton solution (4.28) with the same compactly supported profile (4.26) with translation $x \mapsto y = x - \lambda t$. Therefore, the blow-up solutions (4.24), (4.27) and the compacton solution (4.28) are essentially of a similar *mathematical* (*both the ODE and PDE*) *nature*, and, possibly, more than that. This reflects a certain *universality principle* of compact structure formation in nonlinear evolution PDEs.

On dynamics on invariant subspaces. The quasilinear heat equation (4.23) admits further restriction to the standard invariant subspace that we will briefly discuss. The pressure transformation $u = v^n$ in (4.23) yields

$$u_t = F[u] \equiv (n+1)\big[u u_{xx} + \tfrac{1}{n}(u_x)^2\big] + n\gamma\, u^2.$$

Operator F is known to preserve the 2D subspace $W_2 = \mathcal{L}\{1, \cos x\}$, so there exist exact solutions

$$u_S(x,t) = C_1(t) + C_2(t)\cos x, \tag{4.30}$$

$$\begin{cases} C_1' = \frac{(n+1)^2}{n}\, C_1^2 + \frac{n+1}{n}\, C_2^2, \\ C_2' = \frac{(n+1)(n+2)}{n}\, C_1 C_2. \end{cases}$$

This DS can be integrated in quadratures, so (4.30) describes some exceptional evolution and blow-up properties. Following [218] (or [509, p. 32]), where a detailed analysis and proofs can be found, we comment that, as $t \to 0^+$, the solution takes Dirac's delta as the initial function; see Figure 4.2. Next, the solution amplitude $u(0,t)$ decreases for some $t \in (0, t_3)$, and, after that, the solution starts to increase, and finally blows up and approaches as $t \to T^-$ the separate variables solution (4.24). Hence, the interfaces $s_\pm(t)$ of the compactly supported blow-up solution (4.30) converge to $\pm\pi$ respectively, as explained in Figure 4.2.

A similar invariant subspace analysis applies to the compacton equation (4.22), but only for $n = 1$, where the general solution (4.20) is not that consistent. It seems that, for $n \neq 1$, interesting invariant subspaces do not exist. Further interpretation is performed by using partially invariant subspaces. We explain this in Section 7.2.

Example 4.7 (**Higher-order signed PDEs: compactons and regional blow-up**) A similar, but not explicit, compacton–regional blow-up universality is available for the quintic *signed* nonlinear dispersion PDEs, such as

$$v_t = \alpha\big(|v|^n v\big)_{xxxxx} + \beta\big(|v|^n v\big)_{xxx} + \gamma\,\big(|v|^n v\big)_x \quad (n > 0), \tag{4.31}$$

and its parabolic reaction-diffusion counterpart ($\alpha < 0$ for parabolicity)

$$v_t = \alpha\big(|v|^n v\big)_{xxxx} + \beta\big(|v|^n v\big)_{xx} + \gamma\,|v|^n v.$$

Both equations are written for oscillatory solutions of changing sign; see more details in Section 4.3. Compactons (4.28) for (4.31) and the blow-up pattern (4.24) are then governed by the same quasilinear degenerate ODE

$$-\lambda f = \alpha\big(|f|^n f\big)^{(4)} + \beta\big(|f|^n f\big)'' + \gamma\,|f|^n f, \quad \text{with } \lambda = -\tfrac{1}{n} < 0.$$

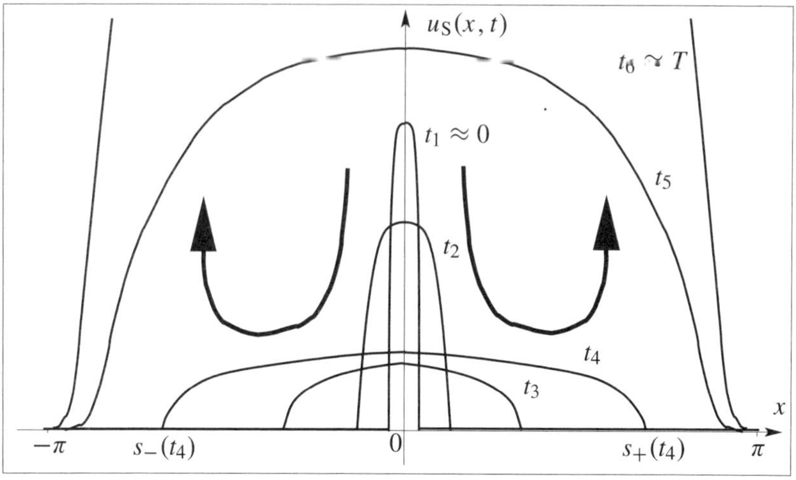

Figure 4.2 Non-monotone evolution of the blow-up solution (4.30); $0 < t_1 < ... < t_6 < T$.

For instance, consider two equations

$$\begin{aligned} u_t &= -\big(|u|^n u\big)_{xxxxx} + \big(|u|^n u\big)_x \ \text{(dispersive: compactons), and} \\ u_t &= -\big(|u|^n u\big)_{xxxx} + |u|^n u \ \text{(parabolic: blow-up).} \end{aligned} \tag{4.32}$$

Then the compacton for the first PDE is (4.28) and the localized blow-up pattern for the second is (4.24), where $f = f(y)$ solves

$$-(|f|^n f)^{(4)} + |f|^n f = \tfrac{1}{n} f \quad \text{in } I\!R.$$

Recall that, for the CP, similar to TFEs (Section 3.7), we are interested in solutions of maximal regularity with the following behavior:

$$f(y) \sim (y_0 - y)^{\frac{4}{n}} \varphi(\ln(y_0 - y))$$

near the interface at $y = y_0^-$, with, say, a bounded, periodic oscillatory component $\varphi(s)$. Figure 3.5(a) illustrates such oscillatory behavior. After the natural change, this gives the ODE with a non-Lipschitz nonlinearity,

$$F = |f|^n f \quad \Longrightarrow \quad F^{(4)} = F - \tfrac{1}{n} \left|F\right|^{-\frac{n}{n+1}} F \quad \text{in } I\!R. \tag{4.33}$$

Unlike the second-order case (4.26), for the CP, an explicit compactly supported solution f is not available. Proving existence and multiplicity results for such higher-order ODEs is a difficult OPEN PROBLEM to be tackled numerically. First, the simplest geometric patterns, $F_1(y)$ in 1D for various $n > 0$, are presented in Figure 4.3. Notice the clear oscillatory behavior of solutions close to interfaces. This is a key feature for the CP that was studied in Section 3.7 for TFEs, and will be continued and extended to odd-order PDEs.

As a rule, such localized profiles F (or f) are not unique. Three patterns denoted by F_1, F_2, and F_3 for $n = 1$, which exhibit a clear "approximate geometric order," are shown in Figure 4.4. Each F_k has precisely k dominant maxima (in view of

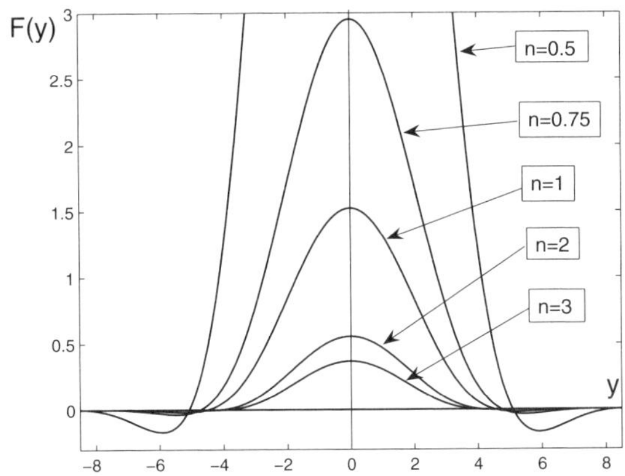

Figure 4.3 Compactly supported solutions of (4.33) for various $n > 0$.

the oscillatory behavior near interfaces, this Sturmian property is rigorously true for the second-order ODEs only). In general, we expect many different solutions F_σ with a multiindex σ of arbitrary length that characterizes oscillatory behavior of the pattern about equilibria $F_\pm = \pm n^{-(n+1)/n}$ and $F_0 = 0$. Some countable sequences of such blow-up-compacton patterns $\{F_k,\ k \geq 1\}$ can be attributed to Lusternik–Schnirel'man classic variational category (genus) theory from the 1930s; see Krasnosel'skii–Zabreiko [355, Ch. 8]). It is important that the ODE (4.33) is variational.

Figure 4.5 shows a complicated multi-hump pattern, F_σ, with the multiindex

$$\sigma = \{-8, +4, -10, +8, -2, 2, -8, 2, -2\}.$$

Here, the first number"-8" reflects the first eight intersections with $F_- = -1$, followed by "+4", i.e., four intersections with $F_+ = +1$, next "-10" means ten intersections with -1, etc. The number of intersections with $F_0 = 0$ is given without the sign, such as "2", that occurs two times in this multiindex. We omit mentioning the unique, transversal intersection with zero, i.e., "1", everywhere in this sequence. It seems that equation (4.33) admits compactly supported solutions $F_\sigma(y)$ of arbitrary "chaotic" complexity, corresponding to any suitable finite multiindex σ. A precise meaning of similar chaotic orbits for fourth-order ODEs (though of a different, non-oscillatory type close to interfaces as $F \to 0$; this destroys a standard homotopy-like approach to such orbits) is explained in [459, p. 198]. For the above nonlinear dispersion equation in (4.32), this is a compacton moving with the velocity $\lambda = -1$.

Similar standing blow-up patterns exist for the hyperbolic PDE

$$u_{tt} = -\big(|u|^n u\big)_{xxxx} + |u|^n u \quad \text{(hyperbolic: blow-up)},$$

where the only change in (4.33) is in the multiplier $\frac{1}{n} \mapsto \frac{2}{n}\big(\frac{2}{n} + 1\big)$. Therefore, as formerly, (4.28) and (4.24) give countable spectra of both moving *compactons* for

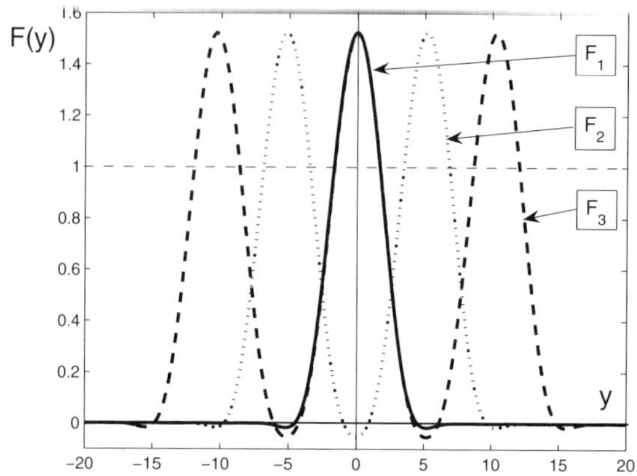

Figure 4.4 Three solutions of (4.33) for $n = 1$.

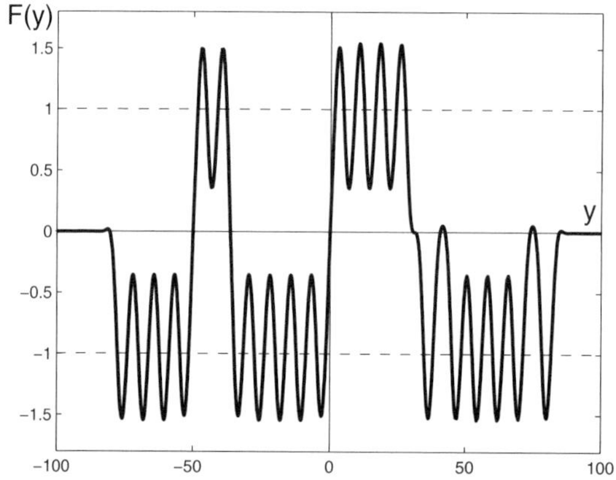

Figure 4.5 A complicated compacton pattern of the ODE (4.33) for $n = 1$.

nonlinear dispersion PDEs and standing *localized blow-up patterns* in parabolic and hyperbolic problems.

4.2.2 On shock and rarefaction waves in PDEs with nonlinear dispersion

As a key feature of quasilinear odd-order PDEs, it is important to note that sufficiently smooth profiles and evolution behavior exhibited by compactons and other solutions to be described are not generic for many such equations (excluding some special, "integrable" ones). This is in striking contrast with even-order quasilinear parabolic PDEs, which, according to classical theory, exhibit a strong internal regularity and the solutions are always, at least, continuous and sufficiently smooth at any regular point (cf. TFEs in Chapter 3).

For first-order PDEs that are known as *conservation laws*, such as the *Euler equation* originated from gas-dynamics

$$u_t + uu_x = 0, \tag{4.34}$$

discontinuous shocks have been recognized for more than a century. General theory of discontinuous *entropy solutions* of one-dimensional PDEs like (4.34) is due to Oleinik [441] developed in the 1950s; see Smoller [530, Part III] for names, results, references and amazing history of conservation laws. Among other important properties, one of the key features is that, in the most general case, the entropy solutions are obtained by regularization, i.e., at the limit as $\varepsilon \to 0^+$ of the family of smooth solutions $\{u_\varepsilon\}$ of uniformly parabolic Burgers' equation

$$u_t + uu_x = \varepsilon u_{xx}. \tag{4.35}$$

The first such ideas were due to Hopf (1950) and Burgers (1948).

Discontinuous solutions can occur for higher-order PDEs from compacton theory, though a suitable entropy-like approach is extremely difficult to develop along the lines of that for conservation laws. This is a principal OPEN PROBLEM. Due to highly oscillatory properties of solutions (see oscillatory asymptotics of the Airy function and other fundamental kernels in the next section), formation of shock waves cannot be described by exact solutions on simple invariant subspaces. We briefly discuss third or fifth-order PDEs with quadratic leading-order operators

$$u_t = (uu_x)_{xx}, \quad \text{or} \quad u_t + (uu_x)_{xxxx} = 0. \tag{4.36}$$

Consider two basic *Riemann's problems* for PDEs (4.36). First, this is the formation of the stationary *shock wave* $S_-(x) = -\operatorname{sign} x$ (it is entropy for (4.34)),

$$S_-(x) = \begin{cases} 1 & \text{for } x < 0, \\ -1 & \text{for } x > 0, \end{cases} \tag{4.37}$$

from smooth solutions in finite time, as $t \to T^-$. This phenomenon is described by the similarity solution

$$u_{\rm S}(x,t) = g(z), \quad \text{where } z = \tfrac{x}{(T-t)^{1/3}}, \quad \text{or} \quad z = \tfrac{x}{(T-t)^{1/5}}, \tag{4.38}$$

and g solves the following ODEs obtained on substitution into (4.36):

$$(gg')'' = \tfrac{1}{3}\, g'z, \quad \text{or} \quad (gg')^{(4)} = -\tfrac{1}{5}\, g'z, \quad \text{with } f(\pm\infty) = \mp 1. \tag{4.39}$$

For these higher-order ODEs, existence and uniqueness problems are not easily studied analytically and are OPEN. Numerically, we have evidence that, in each case, such a smooth odd profile g is unique. Figure 4.6 shows the profiles $G = g^2(z)$ for $z < 0$. For $z > 0$, $g(z)$ is extended anti-symmetrically to get the odd function. Such similarity profiles $g(z)$ describe formation of shocks, i.e.,

$$u_{\rm S}(x,t) \to S_-(x) \quad \text{as } t \to T^-$$

for any $x \in I\!R$, uniformly in $I\!R \setminus (\delta, \delta)$, with a $\delta > 0$ small, and in $L^1_{\rm loc}(I\!R)$. It is

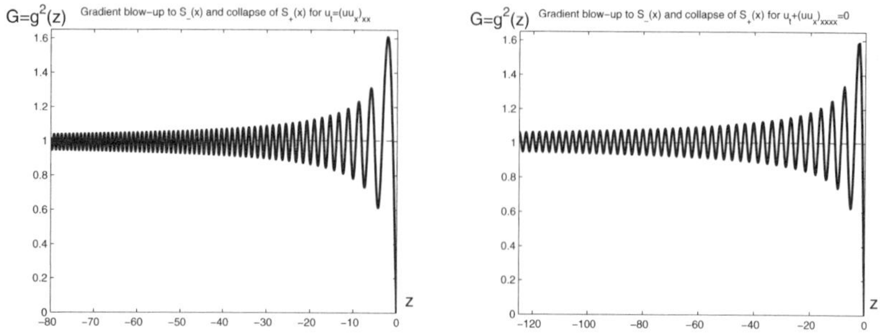

(a) third-order (b) fifth-order

Figure 4.6 The shock wave similarity profile $g(z)$ satisfying ODEs (4.39).

curious that, for the third-order case, in view of asymptotics of Airy functions given in (4.139), the total variation of $u_S(x,t)$ for any $t < T$ is *infinite*. The same is true for the fifth-order case. This strongly differs from the finite variation approach for first-order PDEs (4.34) that is key in scalar conservation laws theory, [441].

Using the reflection symmetry $u \mapsto -u$, $t \mapsto -t$ of PDEs (4.36) implies that the same similarity solutions defined for $t > 0$,

$$u_S(x,t) = g(z), \quad \text{with } z = \tfrac{x}{t^{1/3}}, \quad \text{or } z = \tfrac{x}{t^{1/5}}, \tag{4.40}$$

describe the collapse of the non-entropy shock $S_+(x) = \operatorname{sign} x$, posed as initial data. Then (4.40) plays the role of the *rarefaction wave* that, for the conservation law (4.34), has the simpler similarity piece-wise continuous form

$$u_S(x,t) = g(\tfrac{x}{t}) = \begin{cases} -1 & \text{for } x < -t, \\ \frac{x}{t} & \text{for } |x| < t, \\ 1 & \text{for } x > t. \end{cases}$$

This means that $S_+(x)$ is not an entropy shock. The same classification of stationary shocks $S_\pm(x)$ as solutions of two Riemann's problems applies to similar PDEs of arbitrary $(2m+1)$th order,

$$u_t = (-1)^{m+1} D_x^{2m}(uu_x) \quad \text{for } m = 1, 2, \dots . \tag{4.41}$$

For instance, consider parabolic ODE ε-approximations $\{u_\varepsilon(x)\}$ of the stationary shock $S_-(x)$ for (4.41),

$$(-1)^{m+1} D_x^{2m}(uu_x) + (-1)^m \varepsilon D_x^{2m+2} u = 0.$$

Integrating $2m$ times with zero constants, we obtain the problem

$$uu_x = \varepsilon u_{xx}, \quad \text{with } u(\pm\infty) = \mp 1.$$

This is precisely the correct entropy approximation (4.35) (with $u_t = 0$) for the first-order conservation law, and the approximating sequence is as follows:

$$u_\varepsilon(x) = \tfrac{1-e^{x/\varepsilon}}{1+e^{x/\varepsilon}} = \tanh \tfrac{x}{2\varepsilon} \to S_-(x) \quad \text{as } \varepsilon \to 0^+,$$

with pointwise and $L^1(I\!R)$ convergence. Notice that, unlike the above similarity solutions, such approximating profiles are strictly monotone and are not oscillatory about ∓ 1 as $x \to \pm\infty$ (this is related to the chosen special type of parabolic approximation). Therefore, for the corresponding $(2m+1)$th-order ODEs, the shock $S_-(x)$ is admissible in Gel'fand's sense (1963), or *G-admissible*, while $S_+(x)$ is not.

Later on, using invariant subspace techniques for odd-order PDEs, we will not pay the attention to the possible appearance of proper entropy shocks as solutions that are obtained via regular parabolic approximations. As we have seen, other similarity solutions are needed to revealing such singular phenomena.

4.2.3 On interface equations for compactons

Due to the degeneracy of the higher-order operators at $u = 0$, compacton (4.21) has finite interfaces. Similar to parabolic problems for TFEs (cf. Example 3.10), these explicit solutions help to identify the interface equation.

Example 4.8 (**Interface equation**) Let us begin with a slightly rescaled RH equation (4.14),

$$\boxed{u_t = (u^2)_{xxx} + 4(u^2)_x \equiv \left[(u^2)_{xx} + 4\,u^2\right]_x.} \tag{4.42}$$

Then (4.12) holds and the explicit traveling wave solution is obtained,

$$u(x,t) = f(y) \equiv -\tfrac{\lambda}{3}\,\sin^2(\tfrac{y}{2}), \quad \text{where} \quad y = x - \lambda t, \tag{4.43}$$

where, in order to have a nonnegative solution, it is assumed that $\lambda < 0$, i.e., the TW moves to the left. The compacton consists of the single hump for $y \in (0, 2\pi)$. For the $\sin^2$-wave in (4.43), the left-hand interface is fixed at the origin $y = 0$. We then naturally pose the free-boundary condition of a zero contact angle type from thin film theory (Section 3.1),

$$u = u_x = 0 \quad \text{at the interface } x = s(t). \tag{4.44}$$

For regular solutions, (4.44) implies the zero-flux condition for PDE (4.42), i.e.,

$$(u^2)_{xx} + 4u^2 = 0 \quad \text{at } x = s(t).$$

In a standard manner, a formal dynamic interface equation is derived by differentiating $u(s(t),t) = 0$ and using the PDE (4.42), so that

$$s' = S[u] \equiv -\tfrac{1}{u_x}\,u_t = -6u_{xx} \quad \text{at } x = s(t). \tag{4.45}$$

As usual, this is not an independent free-boundary condition, and is just a manifestation of the regularity, so (4.45) is true for any smooth solutions (not necessarily with the zero contact angle condition, i.e., remains valid for Stefan–Florin FBPs with $u_x = S[u] \neq 0$).

Let us detect other conditions for such sufficiently regular solutions. Now using either the explicit solution (4.43) or the ODE for f (this is necessary to do if explicit solutions are not available), that is

$$-\lambda f' = (f^2)''' + 4(f^2)', \tag{4.46}$$

we obtain the following expansion for small $y > 0$:

$$u(x,t) = f(y) = By^2 + Cy^4 + \dots, \tag{4.47}$$

with $B = \frac{1}{2} u_{xx}(0,t)$ and $C = \frac{1}{4!} u_{xxxx}(0,t)$. The first coefficient is then given by $B = -\frac{1}{12}\lambda = 6u_{xx}$ coinciding with (4.45). Recall that $s' = \lambda$ for the TWs. The second coefficient satisfies

$$-\lambda C = 60BC + 4B^2,$$

which yields the desired second interface equation

$$s' = S[u] \equiv -30\, u_{xx} - \frac{4!}{u_{xxxx}} (u_{xx})^2 \quad \text{at } x = s(t). \tag{4.48}$$

Using (4.45) reduces the dynamic condition (4.48) to a "stationary" higher-order Neumann-type condition

$$u_{xx} = -u_{xxxx} \quad \text{at } x = s(t).$$

These are interface free-boundary conditions which should be satisfied in order to generate a (unique) sufficiently regular solution. A rigorous justification needs the von Mises transformation for the new function $u = v^2$ with the transversal interface slope, $v_x \neq 0$ at $x = s(t)$. This leads to a third-order degenerate PDE for $X = X(v,t)$ with the boundary condition (4.45) at the origin $v = 0$, which is necessary for the correct functional setting of the corresponding degenerate operator. This problem is locally well-posed, provided that $v_x \neq 0$ at $v = 0$. Some features are similar to those in parabolic Example 3.10, though the proof is not easy. There are several OPEN PROBLEMS in such an approach.

4.2.4 On proper solutions by parabolic approximations

Compactons initiate further intriguing aspects of nonlinear PDE theory. Here, we face another principal question which remains OPEN for such weak solutions of wide classes of degenerate nonlinear dispersion models. Namely, it is key to identify the problem for solutions (4.43). If these are solutions of the Cauchy problem (so that the free-boundary conditions do not need to be posed explicitly), it is expected that (4.43) can be obtained by smooth approximations,

$$u(x,t) = \lim_{\varepsilon \to 0^+} u_\varepsilon(x,t), \tag{4.49}$$

via, say, regular analytic parabolic flows. For instance, using the family $\{u_\varepsilon,\ \varepsilon > 0\}$ of analytic solutions of the uniformly parabolic PDEs

$$u_t = -\varepsilon u_{xxxx} + (u^2)_{xxx} + 4(u^2)_x, \tag{4.50}$$

with the same compactly supported initial data $u_0(x)$. On the other hand, a sixth-order regularization

$$u_t = \varepsilon u_{xxxxxx} + (u^2)_{xxx} + 4(u^2)_x \tag{4.51}$$

may be applied. Do both ε-approximations lead to the same solution defined by (4.49)? [See a partial answer for TWs below.] For some good solutions of (4.50), e.g., those having a finite number of zeros that are uniformly transversal for all small $\varepsilon > 0$, the passage $\varepsilon \to 0$ is rather straightforward (each isolated transversal zero is localized, stable in ε and hence, cannot spoil the limit $\varepsilon \to 0$). The principal question is how to deal with generic highly oscillatory solutions $u_\varepsilon(x,t)$. This leads to the general problem of multiple zeros structure for solutions of the degenerate PDE (4.42) and other models. This is a part of the *Sturmian zero set analysis*, which was initiated by Sturm in 1836 [538] for the 1D second-order parabolic equations (e.g., for the heat equation); see history, modern developments, and many applications in [226, Ch. 1]. Notice that, for the semilinear KdV equation (4.1), the parabolic regularization, as in (4.50), has been recognized since the 1960s (Temam) to be an effective approach to nonlinear PDEs; see [396, Ch. 3].

Passing to the limit $\varepsilon \to 0^+$ in the regularized PDEs (4.50), (4.51) and similar equations represent a hard OPEN PROBLEM. Existence of a (unique) solution of the CP now becomes a delicate *ε-asymptotic problem*. This is a common unavoidable feature of many higher-order nonlinear degenerate and singular PDEs considered in this and other chapters. As an illustration, consider this asymptotic problem for the PDE (4.50) understood in the weak form

$$\int u_0 \chi \, \mathrm{d}x - \iint u_\varepsilon \chi_t = -\varepsilon \iint u_\varepsilon \chi_{xxxxx} + \iint (u_\varepsilon)^2 \chi_{xx} - 4 \iint (u_\varepsilon)^2 \chi_x, \tag{4.52}$$

where $\chi \in C_0^\infty$ is a test (cut-off) function. In general, the weak form of equations is not necessarily the best way for proper setting of many nonlinear problems. For instance, the weak approach fails for not fully divergent operators as for the THEs in Chapter 3 (though they are divergent in (4.50)). In these cases, we need to study directly the limit of the smooth family $\{u_\varepsilon\}$ as $\varepsilon \to 0$, which gives a number of OPEN PROBLEMS, especially for higher-order quasilinear nonlinear dispersion (or elliptic) operators; see Remarks for further comments.

Passage to the limit $\varepsilon \to 0$ in the integral identity (4.52) assumes the study of a couple of singular integrals. For compactly supported $u_0(x)$ given by (4.43),

$$u_0(x) = -\tfrac{\lambda}{3} \sin^2\!\left(\tfrac{x}{2}\right) \quad \text{for } x \in [0, 2\pi], \tag{4.53}$$

there are far field integrals for $|x| \gg 1$, which are extremely small via the exponential tails of the fundamental solution of the parabolic operator $\frac{\partial}{\partial t} + \varepsilon D_x^4$,

$$u_\varepsilon(x,t) \sim \exp\bigl\{a|x|^{\frac{4}{3}}/(\varepsilon t)^{\frac{1}{3}}\bigr\},$$

with some complex constant a such that $\operatorname{Re} a < 0$. Other harder integrals describe the behavior in domains with the resonance interaction between two higher-order terms in (4.50). For $x \approx 0$, we use scaling

$$u(x,t) = \varepsilon^{\frac{2}{3}} v(y,\tau), \quad y = \tfrac{x}{\varepsilon^{1/3}}, \quad \tau = \tfrac{t}{\varepsilon^{1/3}},$$

where $v(y,\tau)$ now solves the uniformly parabolic equation

$$v_\tau = -v_{yyyy} + (v^2)_{yyy} + 4\varepsilon^{\frac{2}{3}} (v^2)_y. \tag{4.54}$$

The initial function is calculated from (4.53) as follows:

$$v_0(y) \equiv -\tfrac{\lambda}{3}\, \varepsilon^{-\frac{2}{3}} \sin^2\!\left(\tfrac{1}{2}\, \varepsilon^{\frac{1}{3}} y\right) \to -\tfrac{\lambda}{12}\, y^2 \quad \text{as } y \to 0^+. \tag{4.55}$$

Since $\tau \to +\infty$ as $\varepsilon \to 0^+$, this fixes the asymptotic problem for the parabolic PDE (4.54) with $O(y^2)$ initial data on bounded intervals. It is not very difficult to justify that, by parabolic theory, this CP for (4.54) is well-posed and admits a unique global solution. This is the principal fact testifying that the sequence $\{u_\varepsilon\}$ cannot have non-small, $O(1)$-oscillations as $\varepsilon \to 0$. In this case, the limit (4.49) gives a unique (proper) solution of the Cauchy problem.

Let us demonstrate that the $O(y^2)$ behavior in (4.55) does not generate large, unbounded asymptotics as $\tau \to +\infty$, which can affect the corresponding singular integral in (4.52) and the limit (4.49). To this end, as a formal estimate, let us compare $v(x, \tau)$ with the standard similarity solution of (4.54) (without the last term negligible for $\varepsilon \ll 1$)

$$v_*(y, \tau) = \tau^{-\frac{1}{4}} g(z), \quad \text{with} \quad z = \tfrac{y}{\tau^{1/4}},$$

where g satisfies the ODE

$$-g''' + \tfrac{1}{4}\, gz + (g^2)'' = 0, \quad g(+\infty) = 0.$$

Then $g(z) \sim e^{az^{4/3}}$, with $\operatorname{Re} a = -\frac{3}{8}4^{-1/3} < 0$, has exponential decay as $z \to +\infty$, while for $z \ll -1$, it has a cubic growth of the form

$$g(z) \sim \tfrac{1}{120}(-z)^3 + \ldots \quad \text{as} \quad z \to -\infty.$$

It then follows that such a nontrivial solution occurs from more singular initial data

$$v_*(y, \tau) \sim \tfrac{1}{\tau}\, y^3 \quad \text{as} \quad \tau \to 0^+.$$

For $O(y^2)$ data, the asymptotic behavior is less singular and the influence in the integral identity of such an internal layer near the origin is negligible as $\varepsilon \to 0$.

This analysis indicates how ε-asymptotic theory penetrates into the existence and uniqueness construction of proper solutions. For more general, non-inverse bell-shaped initial data, we can have many internal singular layers (possibly an uncountable set?), which should be taken into account. In general settings, the passage $\varepsilon \to 0$ is very difficult and remains OPEN for most regularized PDEs considered later on.

On the other hand, for TWs, this asymptotic analysis is not hard and deals with standard matched asymptotic expansions. Figure 4.7 shows the (non-monotone) convergence as $\varepsilon \to 0$ of the TW profiles satisfying the third-order ODE

$$-\lambda f = -\varepsilon f''' + (f^2)'' + 4f^2. \tag{4.56}$$

Returning to our particular ODE approach, in order to detect maximal regularity solutions of the Cauchy problem that may be inherited from the analytic ε-regularization, it follows that expansion (4.47) yields the most smooth solutions of the ODE (4.46) at the interface point. This is easily proved for these second-order equations. Therefore, we claim (which may be obvious) that compactons (4.43) are *smooth solutions of the Cauchy problem* for (4.42) and can be obtained without *a priori* specified free-boundary conditions. A rigorous proof is not easy. The dynamic interface equation then follows from the PDE for $X = X(v, t)$ as explained above.

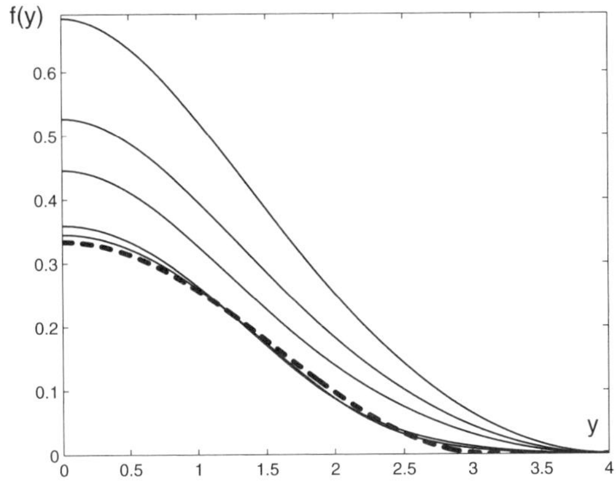

Figure 4.7 Convergence as $\varepsilon \to 0$ of solutions of (4.56), with $\lambda = -1$, to the compacton profile $\frac{1}{2}\cos^2(\frac{y}{2})$; $\varepsilon = 2, 1, 0.5, 0.05, 0.01$, and 0.001 (the dashed line).

4.2.5 Local behavior near interfaces for the $K(2,2)$ equation

Consider the quadratic PDE, keeping the leading differential term,

$$u_t = (u^2)_{xxx} \quad \text{in } I\!R \times I\!R_+, \tag{4.57}$$

with given compactly supported initial data u_0. Let us first study the TW solutions of the form (4.43) satisfying, after integrating once, a simple ODE

$$-\lambda f' = (f^2)''' \quad \Longrightarrow \quad (f^2)'' + \lambda f = 0. \tag{4.58}$$

These second-order nonlinear *Emden–Fowler-type equations*, introduced and studied by Emden (1907) [168] and Fowler (1914) [200], seem to be the most famous and well-studied ODEs in the twentieth century. We use in (4.58) the following standard change that follows from a group of scalings,

$$f(y) = y^2\varphi(s), \quad s = \ln y,$$

where φ solves the autonomous ODE

$$(\varphi^2)'' + 7(\varphi^2)' + 12\varphi^2 + \lambda\varphi = 0 \quad \text{in } I\!R. \tag{4.59}$$

For $\lambda \neq 0$, there exists the nontrivial constant profile

$$\varphi(s) \equiv \varphi_0 = -\tfrac{\lambda}{12}. \tag{4.60}$$

For $\lambda < 0$, this solution is positive and actually leads to the existence of the explicit nonnegative compacton (4.43). For $\lambda > 0$, (4.60) is negative and the ODE (4.58) has the obvious negative solution that is obtained by reflection $f \mapsto -f$, $\lambda \mapsto -\lambda$. By linearization, the constant solution (4.60) of (4.59) turns out to be asymptotically exponentially ($O(\mathrm{e}^{-s})$) stable as $s \to +\infty$, and is unstable as $s \to -\infty$; see Figure 4.8. The phase-plane of (4.59) shows that no other changing sign solutions exist. Thus the TW profiles are non-oscillatory for equation (4.57), though strictly positive solutions close to interfaces are possible for $\lambda < 0$ only.

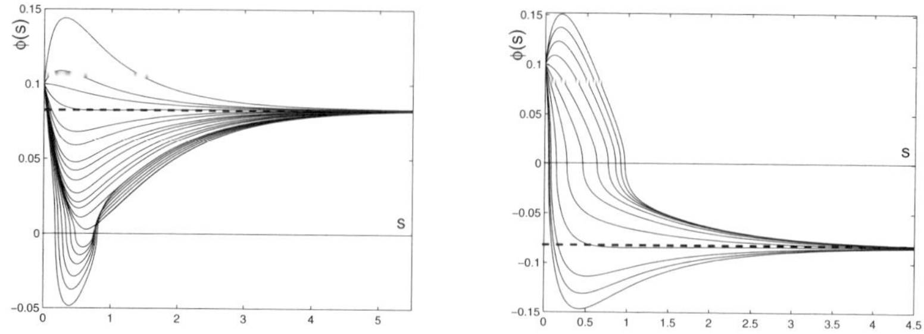

(a) $\lambda = -1$ (b) $\lambda = 1$

Figure 4.8 Asymptotic stability as $s \to +\infty$ of the constant solution (4.60) of (4.59) for $\lambda = -1$ (a) and $\lambda = 1$ (b).

This is an ODE analysis of the local structure of moving TWs near interfaces. The corresponding PDE result, after proving the existence of a proper solution $u(x,t)$ of the Cauchy problem for (4.57) (e.g., by the analytic regularization similar to that in (4.50)), assumes establishing that the generic behavior at the interfaces is governed by TWs. This is a difficult OPEN asymptotic problem that is typical for nonlinear rescaled evolution PDEs, which is briefly and formally discussed below. Namely, given a proper solution $u(x,t)$, the TW rescaling is performed as follows:

$$u(x,t) = v(y,t), \quad y = x - \lambda t,$$

so v solves the rescaled equation with the same ODE operator

$$v_t = (v^2)_{yyy} + \lambda v_y. \tag{4.61}$$

For convenience, we study the behavior of the solution at the initial moment $t = 0$, when the interface is at the origin, $s(0) = 0$, assuming that there exists the finite speed of propagation $\lambda = s'(0)$, i.e., the interface is a sufficiently smooth curve at $t = 0^-$. Proving such a regularity of interfaces is a hard OPEN PROBLEM. Then the solution of (4.61) is rescaled according to the invariant group of scalings by introducing the family of functions

$$w_\mu(z,\tau) = \tfrac{1}{\mu^2}\, v(\mu z, \mu\tau), \quad \text{with parameter } \mu > 0, \tag{4.62}$$

where w_μ solves the same equation (4.61),

$$(w_\mu)_\tau = [(w_\mu)^2]_{zzz} + \lambda (w_\mu)_z, \quad \text{with data } w(z,0) = \tfrac{1}{\mu^2} v_0(\mu z). \tag{4.63}$$

Then the limit $\mu \to 0$, describing, according to (4.62), the behavior of $v(y,t)$ for $(y,t) \approx (0,0^-)$, is equivalent to the passage to the limit $\tau \to +\infty$, $z \to \infty$ in (4.63), i.e., studying the asymptotic behavior of its solutions. Here, convergence of initial data $w_\mu(z,0)$ as $\mu \to 0$ determines the necessary speed λ. Had we proved the stabilization to a nontrivial stationary profile, we would have established the behavior of general solutions governed by the TWs given by (4.58). The equation (4.63) is not a gradient system, i.e., it does not admit a Lyapunov function, so passing to the limit as $\tau \to +\infty$ is an OPEN PROBLEM.

4.2.6 The signed $K(2,2)$ equation: TWs and oscillatory solutions

We now explain the origin of the *signed* versions of nonlinear dispersion PDEs that have already been used in Examples 4.6 and 4.7. Mathematically, this is related to another approach for checking the evolution properties of the compacton (4.43) and oscillatory properties of general solutions via the n-continuity (homotopy) construction. Instead of (4.57), consider the following PDE:

$$u_t = (|u|u)_{xxx}. \tag{4.64}$$

For smooth nonnegative solutions, these PDEs coincide. The mathematical advantage of (4.64) is that it admits a connection to the linear equation via the family on the *signed nonlinear dispersion equations*

$$u_t = (|u|^n u)_{xxx}, \quad \text{with parameter } n \geq 0. \tag{4.65}$$

Namely, at $n = 0$, the standard *linear dispersion equation* occurs,

$$u_t = u_{xxx}, \tag{4.66}$$

exhibiting the well-known local and global evolution properties. Its fundamental solution via Airy's function is described in Example 4.27. We say that the PDEs (4.64) and (4.66) belong to the same "homotopy class," if wide sets of solutions (with, say, stable transversal zeros only) of both can be continuously (in n) deformed to each other. Therefore, both quadratic (4.64) and linear (4.66) equations should exhibit similar local oscillatory properties of solutions.

This approach assumes the change of all the models, where in the PDEs and ODEs we replace

$$u^2 \mapsto |u|u, \quad \varphi^2 \mapsto |\varphi|\varphi, \quad u^m \mapsto |u|^{m-1}u, \quad \dots \,. \tag{4.67}$$

For nonnegative compactons this does not matter. Using such monotone nonlinearities in PDEs with nonlinear dispersion makes sense from a physical point of view, [495]. Furthermore, for the parabolic equations of any order, including the TFEs in Section 3.1, the only well-posed extension of quadratic models to solutions of changing sign assumes transformations (4.67), so the correct setting of PDEs is

$$u_t = (|u|u)_{xx}\,, \quad u_t = -(|u|u)_{xxxxx}\,, \quad u_t = (|u|u)_{xxxxxx}, \quad \text{etc.}$$

Indeed, equations with non-monotone nonlinearities, such as (cf. (4.57))

$$u_t = (u^2)_{xx}, \quad \text{or} \quad u_t = -(u^2)_{xxxx},$$

are backward parabolic in the negativity domain $\{u < 0\}$ and are not well-posed.

This motivates introducing the *signed* $K(m,n)$ ($sK(m,n)$) equation

$$u_t = (|u|^{n-1}u)_{xxx} + (|u|^{m-1}u)_x,$$

so that, for proper construction of solutions of the Cauchy problem, it is natural to use a homotopy connection as $n, m \to 1$ ($n \mapsto n+1$ later on for convenience) to the linear PDE

$$u_t = u_{xxx} + u_x \quad \text{in } I\!R \times I\!R,$$

with well-known evolution and oscillatory properties of solutions.

Traveling waves. Constructing TW solutions of (4.65) yields the second-order ODE

$$(|f|^n f)'' + \lambda f = 0 \quad \text{for } y > 0, \quad f(0) = 0, \tag{4.68}$$

which is easy to study by setting $F = |f|^n f$, so that

$$F'' = -\lambda |F|^{-\frac{n}{n+1}} F \quad \text{for } y > 0, \quad F(0) = 0.$$

Hence, the following behavior near interfaces occurs:

$$F(y) = \pm\left[-\tfrac{\lambda n^2}{2(n+1)(n+2)}\right]^{\frac{n+1}{n}} y^{\frac{2(n+1)}{n}} \quad \text{for } \lambda < 0.$$

For $\lambda > 0$, it follows that $F(y) \equiv 0$, meaning nonexistence of finite TW-interfaces. In this case, the ODE (4.68) admits arbitrarily small periodic solutions $f(y)$ in $I\!R$, which, formally, have interfaces at $y = \pm\infty$ and are not decaying as $y \to \infty$.

On regularization: convergence for the TWs. We will now slightly touch on the regularization problem, and, following (4.50), consider the parabolic PDE

$$u_t = -\varepsilon u_{xxxx} + \left(|u|^n u\right)_{xxx}.$$

Unlike (4.68), the TWs now solve the third-order singular perturbed ODE

$$-\varepsilon f''' + \left(|f|^n f\right)'' + \lambda f = 0. \tag{4.69}$$

Rigorous principles of singular perturbation methods for differential equations were established by Tikhonov in the 1940s and 50s. Boundary layer phenomena are explained in well-known monographs by Vasil'eva and Butuzov, O'Malley, Kevorkian and Cole, Lomov, and others.

Let us briefly comment on the passage to the limit $\varepsilon \to 0$ in (4.69). To study the behavior near the interface as $y \to 0^+$, the standard change in (4.68) is used,

$$f(y) = y^\gamma \varphi(s), \quad s = \ln y, \quad \text{where } \gamma = \tfrac{2}{n}. \tag{4.70}$$

Plugging (4.70) into (4.68) and denoting $\Phi = |\varphi|^n \varphi$ gives

$$G_n[\Phi] \equiv \Phi'' + \tfrac{3n+4}{n}\, \Phi' + \tfrac{2(n+1)(n+2)}{n^2}\, \Phi + \lambda |\Phi|^{-\frac{n}{n+1}} \Phi = 0. \tag{4.71}$$

Applying the same change (4.70) in (4.69) we obtain the non-autonomous ODE

$$\begin{aligned} &\varphi''' + 3(\gamma - 1)\varphi'' + (3\gamma^2 - 6\gamma + 2)\varphi' \\ &+ \gamma(\gamma - 1)(\gamma - 2)\varphi = \tfrac{1}{\varepsilon}\, e^{3s} G_n[|\varphi|^n \varphi]. \end{aligned} \tag{4.72}$$

Introducing the new independent variable

$$s \mapsto s + \tfrac{1}{3} \ln \varepsilon, \tag{4.73}$$

we get rid of the ε-dependence on the right-hand side of (4.72). The behavior as $\varepsilon \to 0$ can be studied by matching methods from ODE theory. It follows that, since the right-hand side in (4.72) becomes unbounded as $\varepsilon \to 0$, good solutions must approach the generic behavior of the ODE (4.71), formally corresponding to $\varepsilon = 0$.

Similarly, for the sixth-order regularization as in (4.51),

$$u_t = \varepsilon u_{xxxxxx} + (|u|^n u)_{xxx} \quad \Longrightarrow \quad \varepsilon f^{(5)} + \left(|f|^n f\right)'' + \lambda f = 0.$$

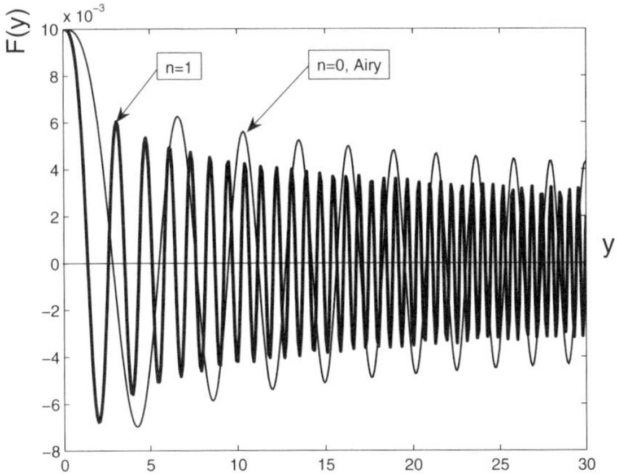

Figure 4.9 Oscillatory behavior of solutions of (4.75) for large $y > 0$.

Then the oscillatory component φ solves the ODE (operator P_5 is as in (3.157))

$$P_5[\varphi] \equiv \varphi^{(5)} + \ldots = \tfrac{1}{\varepsilon}\, \mathrm{e}^{5s} G_n[|\varphi|^n \varphi],$$

where, instead of (4.73), the translation $s \mapsto s + \frac{1}{5} \ln \varepsilon$ applies. As $\varepsilon \to 0$, the convergence of bounded solutions to those of the ODE (4.71) is observed.

Solutions of changing sign. We describe these by constructing the fundamental similarity solution of (4.65),

$$u(x,t) = t^{-\frac{1}{n+3}} f\big(x/t^{\frac{1}{n+3}}\big) \implies (|f|^n f)'' + \tfrac{1}{n+3}\, fy = 0, \quad \textstyle\int f = 1. \tag{4.74}$$

Setting $F = |f|^n f$ yields

$$F'' + \tfrac{1}{n+3} \left|F\right|^{-\frac{n}{n+1}} Fy = 0. \tag{4.75}$$

For $n = 0$, this is the ODE problem for the Airy function $\mathrm{Ai}(y)$ as the kernel of the fundamental solution of the linear operator in (4.66). For $n > 0$, assuming that $\mathrm{supp}\, f = [y_0, \infty)$ with some $y_0 < 0$, we have that, close to the finite left-hand interface, as $y \to y_0^+$, the behavior is non-oscillatory (see Figure 3.5(a))

$$f(y) = \big[\tfrac{n^2}{2(n+1)(n+2)(n+3)}\big]^{\frac{1}{n+1}} (y - y_0)^{\frac{2}{n}} (1 + o(1)).$$

For $y \gg 1$, the behavior is different and $f(y)$ does not have a finite interface. In Figure 4.9, we present a typical oscillatory behavior of the function $F(y) = (|f|^n f)(y)$ for $n = 1$ to be compared with similar oscillations of the Airy function for $n = 0$.

4.3 Higher-order PDEs: interface equations and oscillatory solutions

Example 4.9 (**Fifth-order PDE with nonlinear dispersion**) Let us return to higher-order degenerate PDEs, e.g.,

$$\boxed{u_t = (u^2)_{xxxxx} - 16(u^2)_x,} \tag{4.76}$$

with the corresponding compacton

$$u(x,t) = f(y) \equiv \tfrac{\lambda}{15} \sin^2\big(\tfrac{1}{2}\, y\big), \quad \text{where} \quad y = x - \lambda t. \tag{4.77}$$

Here, $\lambda > 0$, so that the TW moves to the right. These solutions satisfy the zero contact angle condition (4.44) (notice that the condition $u_{xxx} = 0$ holds for any even $f(y)$), but the flux in not zero,

$$(u^2)_{xxxx} - 16u^2 = 6(u_{xx})^2 = \tfrac{\lambda}{15} \quad \text{at} \quad x = s(t). \tag{4.78}$$

This is the first sign that compacton (4.77) *is not* a solution of the Cauchy problem. For this smooth case, the dynamic interface equation is standard,

$$s' = S[u] \equiv -\tfrac{1}{u_x}\, u_t = -10\, u_{xxxx} \quad \text{at} \quad x = s(t). \tag{4.79}$$

We expect that the zero contact angle conditions (4.44) plus the *Florin-type* condition (4.78) (with $\lambda = s'(t)$) comprise a locally well-posed FBP for (4.76) generating compactons with profiles (4.77). Hence, (4.79) will serve as a regularity solvability criterion for the degenerate equation that is obtained via the von Mises transformation $X = X(v,t)$ with $u = v^2$. The problem of the well-posedness is OPEN.

4.3.1 On the Cauchy problem

Concerning the correct setting for the Cauchy problem, we continue to study the ODE for TW profiles f,

$$-\lambda f' = (f^2)^{(5)} - 16(f^2)'. \tag{4.80}$$

Let us show that the quadratic behavior as $y \to 0$ as in (4.47), which remains true for the current compactons, does not provide us with the maximal regularity exhibited by the fifth-order ODE (4.80). To this end, consider the PDE with the leading higher-order term only,

$$u_t = (u^2)_{xxxxx} \quad \text{in} \quad I\!R \times I\!R_+. \tag{4.81}$$

For TWs, the following ODE is obtained on integration:

$$-\lambda f = (f^2)^{(4)} \quad \text{for} \quad y > 0, \quad f(0) = 0.$$

For $\lambda < 0$, it admits the positive solution

$$f(y) = -\tfrac{\lambda}{1680}\, y^4, \tag{4.82}$$

which is smoother at the interface $y = 0$ than (4.77) for the FBP. Set

$$f(y) = y^4 \varphi(s), \quad \text{where} \quad s = \ln y, \tag{4.83}$$

where φ solves the following fourth-order autonomous ODE:

$$(\varphi^2)^{(4)} + 26(\varphi^2)''' + 251(\varphi^2)'' + 1066(\varphi^2)' + 1680\varphi^2 + \lambda\varphi = 0. \tag{4.84}$$

For any $\lambda \neq 0$, there exists the constant equilibrium

$$\varphi(s) \equiv -\tfrac{\lambda}{1680}, \tag{4.85}$$

which for $\lambda < 0$ is positive and leads to (4.82). In Figure 4.10 we show the behavior of solutions of the ODE (4.84), so that (4.85) is asymptotically stable as $s \to +\infty$

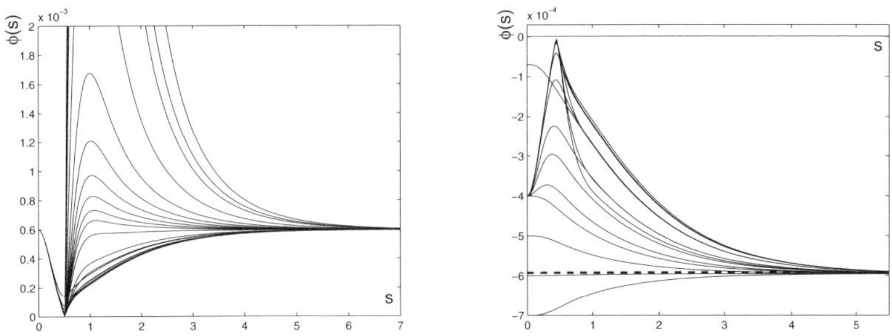

(a) $\lambda = -1$ (b) $\lambda = 1$

Figure 4.10 Non-oscillatory solutions of the ODE (4.84) for $\lambda = -1$ (a) and $\lambda = 1$ (b).

for both $\lambda > 0$ and $\lambda < 0$, with exponential convergence of the order $O(\mathrm{e}^{-s})$. The lower left-hand corner of Figure 4.10(a) and the upper one in (b) show that the trivial "equilibria" $\varphi = 0$ is highly unstable. Moreover, it seems that most of solutions of (4.84) cannot change sign at all, unlike a number of other higher-order TFEs; cf. Figure 3.8(a) and (b).

We expect that Figure 4.10 describes the generic non-oscillatory behavior near interfaces of some classes of solutions of the PDEs, such as (4.76) in $I\!R \times I\!R$, i.e., moving TWs exhibit the following behavior of the maximal regularity:

$$f(y) = O(y^4) \quad \text{as} \quad y \to 0. \tag{4.86}$$

Then the compactly supported function (4.77) is not a solution of the CP.

Example 4.10 **(Signed fifth-order PDE: solutions of changing sign)** We now describe the oscillatory interface behavior for the *signed* fifth-order nonlinear dispersion PDE with parameter $n > 0$,

$$u_t = (|u|^n u)_{xxxxx} \quad \text{in} \quad I\!R \times I\!R_+. \tag{4.87}$$

Oscillatory properties for the linear PDE. As usual, the advantage of the signed PDE (unlike (4.81)) is that it admits the formal passage to the limit $n \to 0$ as a connection to the *linear dispersion equation*

$$u_t = u_{xxxxx} \quad \text{in} \quad I\!R \times I\!R_+. \tag{4.88}$$

Using TW solutions of (4.88) gives

$$-\lambda f = f^{(4)},$$

so that setting $f(y) = \mathrm{e}^{\mu y}$ yields $\mu^4 = -\lambda < 0$ for $\lambda > 0$. The generic decaying behavior is *oscillatory* at the left-hand interface as $y \to -\infty$,

$$f(y) \sim \exp\{\lambda^{\frac{1}{4}} \tfrac{y}{\sqrt{2}}\}\big[A\cos\big(\lambda^{\frac{1}{4}} \tfrac{y}{\sqrt{2}}\big) + B\sin\big(\lambda^{\frac{1}{4}} \tfrac{y}{\sqrt{2}}\big)\big]. \tag{4.89}$$

For $\lambda < 0$, the only *decaying* (integrable at $y = -\infty$) solution is *non-oscillatory*,

$$f(y) \sim \exp\{|\lambda|^{\frac{1}{4}} y\} \quad \text{as} \quad y \to -\infty.$$

In addition, there are *bounded, non-integrable* solutions.

Alternatively, the fundamental solution of the corresponding linear operator $\frac{\partial}{\partial t} - D_x^5$ in (4.88) is

$$b(x,t) = t^{-\frac{1}{5}} g(\xi), \quad \text{with} \quad \xi = \frac{x}{t^{1/5}},$$

where g is a unique solution of the ODE problem

$$g^{(5)} + \tfrac{1}{5}(g\xi)' = 0 \quad \text{in } I\!R, \quad \int g = 1. \tag{4.90}$$

Then $g(\xi) \sim e^{a\xi^{5/4}}$ as $\xi \to +\infty$, with $a^4 = -\frac{1}{5}\left(\frac{4}{5}\right)^4$, so that the behavior is oscillatory of the type

$$g(\xi) \sim \xi^{-\frac{3}{8}} \exp\{-a_0 \xi^{\frac{5}{4}}\}\left[A \sin\left(a_0 \xi^{\frac{5}{4}}\right) + B \cos\left(a_0 \xi^{\frac{5}{4}}\right)\right],$$

where $a_0 = \frac{4}{5\sqrt{2}} 5^{-1/4}$. As $\xi \to -\infty$, $g(\xi)$ has stronger, not absolutely integrable on $(-\infty, 0)$, oscillations,

$$g(\xi) \sim |\xi|^{-\frac{3}{8}}\left[A \sin\left(a_0 |\xi|^{\frac{5}{4}}\right) + B \cos\left(a_0 |\xi|^{\frac{5}{4}}\right)\right].$$

As a rule, in order to compare such oscillatory patterns with those for the quasilinear model, we always formally mean that the left-hand interface for (4.88) is situated at $x = -\infty$, and not at a finite x, as for the degenerate PDE (4.87). Notice that, for (4.88), there exists a single fundamental frequency of the linear periodic motion and, most probably, this remains valid for (4.87) for small $n > 0$. In other words, we claim that equations (4.87) and (4.88) belong to the same homotopy class, and, in the CP, the solutions are then expected to be equally oscillatory to encourage their maximal regularity at interfaces.

Thus, by such a continuity in n, similar oscillatory properties are expected to be preserved in the quasilinear model (4.87), at least for sufficiently small $n > 0$. This helps to detect the maximal regularity of solutions and define *proper solutions* of the CP as those with the increasing regularity as $n \to 0$, i.e., approaching C^∞ (and analytic) regularity of the rescaled kernel in (4.90) for the linear PDE (4.88).

Oscillatory solutions for n > 0. The TW profiles for (4.87) solve the ODE

$$-\lambda f = \left(|f|^n f\right)^{(4)} \quad \text{for } y > 0, \quad f(0) = 0. \tag{4.91}$$

In Section 3.7, equations, such as (4.91), occurred in various aspects of thin film theory. In order to detect the character of sign changes of such solutions, we introduce the oscillatory components φ for (4.91) by setting

$$f(y) = y^\gamma \varphi(s), \quad s = \ln y, \quad \text{where } \gamma = \tfrac{4}{n}. \tag{4.92}$$

Then $F = |\varphi|^n \varphi$ satisfies the ODE with the operator $P_4[F]$ given by (3.156),

$$\begin{aligned} &F^{(4)} + 2(2\mu - 3)F''' + (6\mu^2 - 18\mu + 11)F'' + 2(2\mu^3 - 9\mu^2 \\ &+ 11\mu - 3)F' + \mu(\mu-1)(\mu-2)(\mu-3)F + \lambda\left|F\right|^{-\frac{n}{n+1}} F = 0, \end{aligned} \tag{4.93}$$

where $\mu = \frac{4(n+1)}{n} > 4$. According to (4.92), the oscillatory character of TW solutions near interfaces depends on the availability of periodic solutions of the ODE (4.93). Existence, nonexistence, multiplicity, and stability of periodic solutions of such higher-order equations are difficult OPEN questions of general ODE theory.

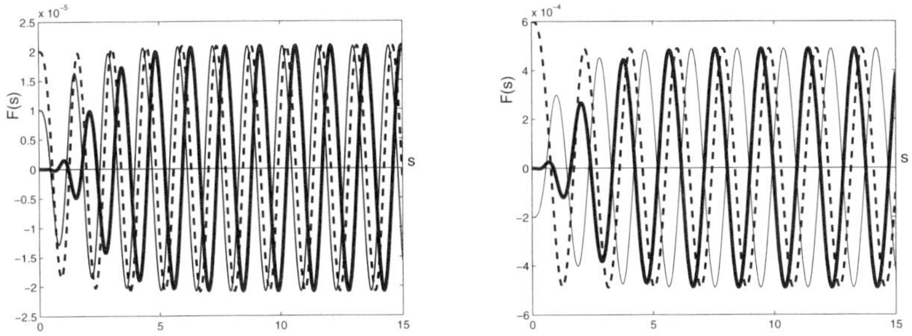

(a) $n = 2$ (b) $n = 4$

Figure 4.11 Convergence to stable periodic solutions of (4.93) with $\lambda = 1$ for $n = 2$ (a) and $n = 4$ (b).

Stable periodic solutions for positive TW speeds. For $\lambda > 0$, solutions of the ODE (4.93) are oscillatory; see Figure 4.11. A single stable periodic motion was always detected that agrees with a similar result for $n = 0$. As usual, the oscillation amplitude becomes extremely small as n approaches zero, so we need extra scaling.

Limit n → 0. This scaling is

$$F(s) = \left(\tfrac{n}{4}\right)^{\frac{4}{n}} \Phi(\eta), \quad \text{where} \quad \eta = \tfrac{4s}{n}, \tag{4.94}$$

where Φ solves a simpler limit ODE ($\lambda = 1$),

$$\Phi^{(4)} + 4\Phi''' + 6\Phi'' + 4\Phi' + \Phi + \left|\Phi\right|^{-\frac{n}{n+1}} \Phi = 0. \tag{4.95}$$

The stable oscillatory patterns for this equation are shown in Figure 4.12. For $n = 0.2$ in Figure 4.12(a), by scaling (4.94), the oscillatory component is estimated as follows:

$$\max|\varphi(s)| \sim 3 \cdot 10^{-4}\left(\tfrac{n}{4}\right)^{\frac{4}{n}} \sim 3 \cdot 10^{-30},$$

while

$$\max|\varphi(s)| \sim 10^{-93} \quad \text{for} \quad n = 0.08 \text{ in (b)}.$$

Limit n → ∞. Then $\mu \to 4$, so the original ODE (4.93) approaches the following equation with discontinuous nonlinearity:

$$F_\infty^{(4)} + 10F_\infty''' + 35F_\infty'' + 50F_\infty' + 24F_\infty + \operatorname{sign} F_\infty = 0, \tag{4.96}$$

which also admits a stable periodic solution, as shown in Figure 4.13. The same ODE (4.96) occurred for a KS-type equation in Example 3.36, see Figure 3.16.

Unstable non-periodic behavior for negative speeds. For $\lambda < 0$, two constant equilibria are asymptotically stable, and numerically we find an unstable oscillatory behavior in between; see Figure 4.14 for $n = 1.3$. This is a decaying behavior, which reminds us the asymptotics in the linear case $n = 0$, but cannot be extended to the interface at $s = -\infty$.

This regularity and oscillatory ODE analysis again confirms that compactons (4.77)

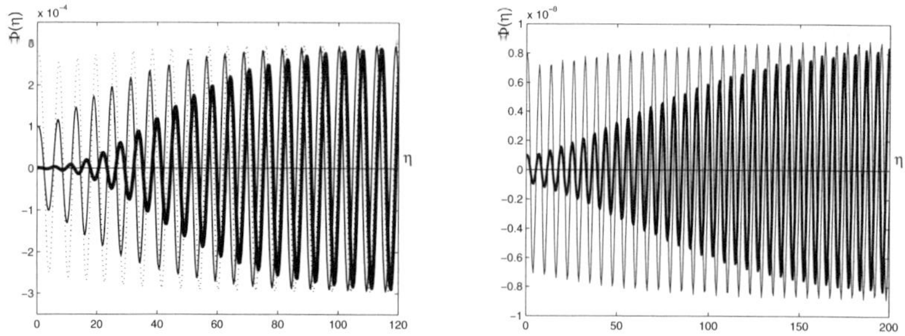

(a) $n = 0.2$ (b) $n = 0.08$

Figure 4.12 Stable periodic oscillations in the ODE (4.95) for $n = 0.2$ (a) and $n = 0.08$ (b).

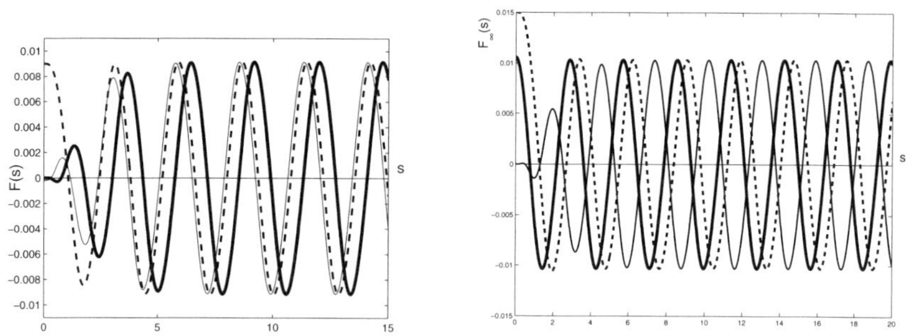

(a) (4.93): $n = 100$ (b) (4.96): $n = +\infty$

Figure 4.13 Stable periodic patterns of the ODE (4.93) ($\lambda = 1$) change slightly for $n = 100$ (a) and $n = +\infty$ (b), equation (4.96).

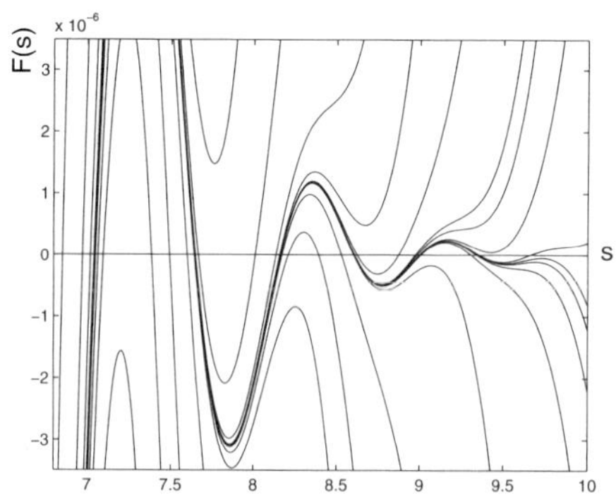

Figure 4.14 Unstable decaying structures of the ODE (4.93) with $\lambda = -1$ for $n = 1.3$. Cauchy data are $F(0) = 2$, $F'(0) = F'''(0) = 0$, $F''(0) = -2.67700981224690017\ldots$.

are not maximal regularity solutions of the Cauchy problem for (4.76) and solve the FBP specified above. This conclusion is true for similar $(2m+1)$th-order PDEs with $m \geq 2$.

4.3.2 Fast moving 2π-periodic solutions

Example 4.11 Consider a PDE similar to (4.13) with a linear perturbation on the right-hand side, where parameters satisfy $16\alpha - 4\beta + \gamma = 0$,

$$\boxed{u_t = \alpha(u^2)_{xxxxx} + \beta(u^2)_{xxx} + \gamma(u^2)_x + \delta u + \varepsilon \quad \text{in } I\!R \times I\!R.} \tag{4.97}$$

Consider solutions (4.18) on W_3. Then the DS (4.19) slightly changes,

$$\begin{cases} C_1' = \delta C_1 + \varepsilon, \\ C_2' = \delta C_2 + \mu C_1 C_3, \\ C_3' = \delta C_3 - \mu C_1 C_2, \end{cases} \tag{4.98}$$

where $\mu = 6(\beta - 5\alpha)$. This gives the following explicit 2π-periodic solutions:

(i) If $\delta = 0$, the solutions exhibit a quadratic propagation with time,

$$u(x,t) = \varepsilon t + A + B\cos\big[x + \mu\big(\tfrac{\varepsilon}{2}t^2 + At + D\big)\big] \quad (A, B, D \in I\!R); \tag{4.99}$$

(ii) If $\delta > 0$, the propagation is exponentially fast,

$$u(x,t) = e^{\delta t}\big\{-\tfrac{\varepsilon}{\delta}e^{-\delta t} + A + B\cos\big[x + \mu\big(\tfrac{A}{\delta}e^{\delta t} - \tfrac{\varepsilon}{\delta}t + D\big)\big]\big\}. \tag{4.100}$$

4.3.3 5D trigonometric subspaces

We begin with the fifth-order PDE (4.13) with $\alpha = 1$,

$$\boxed{u_t = F[u] \equiv (u^2)_{xxxxx} + \beta(u^2)_{xxx} + \gamma(u^2)_x.} \tag{4.101}$$

Proposition 4.12 *The only operator F in (4.101) preserving the 5D subspace*

$$W_5 = \mathcal{L}\{1, \cos x, \sin x, \cos 2x, \sin 2x\} \tag{4.102}$$

is as follows:

$$F[u] = (u^2)_{xxxxx} + 25(u^2)_{xxx} + 144(u^2)_x. \tag{4.103}$$

The algebraic manipulations yield the following invariance condition of W_5:

$$\begin{cases} 9\beta - \gamma = 81, \\ 16\beta - \gamma = 256, \end{cases}$$

from which $\beta = 25$, $\gamma = 144$, and (4.103) follows.

Example 4.13 (Quintic PDE on W$_5$) The quintic nonlinear dispersion equation (4.101), (4.103) possesses the solutions

$$u(x,t) = C_1(t) + C_2(t)\cos x + C_3(t)\sin x + C_4(t)\cos 2x + C_5(t)\sin 2x,$$

$$\begin{cases} C_1' = 0, \\ C_2' = 120(C_2C_5 - C_3C_4 + 2C_1C_3), \\ C_3' = -120(C_2C_4 + C_3C_5 + 2C_1C_2), \\ C_4' = 120(2C_1C_5 + C_2C_3), \\ C_5' = 60(C_3^2 - C_2^2 - 4C_1C_4). \end{cases} \tag{4.104}$$

The nonnegative compacton [147, p. 4734] exists for $\lambda < 0$,

$$u_c(x,t) = -\tfrac{\lambda}{105} \cos^4\big[\tfrac{1}{2}(x - \lambda t)\big]. \tag{4.105}$$

This corresponds to the following explicit solution of the DS (4.104):

$$C_1(t) = -\tfrac{\lambda}{280},\ C_2(t) = -\tfrac{\lambda}{210} \cos \lambda t,\ C_3(t) = -\tfrac{\lambda}{210} \sin \lambda t,$$
$$C_4(t) = -\tfrac{\lambda}{840} \cos 2\lambda t,\ C_5(t) = -\tfrac{\lambda}{840} \sin 2\lambda t.$$

The DS (4.104) describes a finite-dimensional evolution near the compacton, and possibly may detect its stability on the subspace W_5. The ODE analysis here is harder than that on W_3 in Example 3.17. According to (4.86), the compacton (4.105) satisfies the condition of the maximal regularity, so it is a solution of the CP.

It is easy to extend the above invariant analysis to the 7th-order PDE

$$\boxed{u_t = F[u] \equiv D_x^7(u^2) + \beta D_x^5(u^2) + \gamma\,(u^2)_{xxx} + \delta(u^2)_x,} \tag{4.106}$$

though such PDEs are still of no use in applications related to nonlinear dispersion phenomena, [495]. It follows from Proposition 4.12, that F admits W_5, if

$$\beta = 25,\ \ \gamma = 144,\ \ \text{and}\ \delta = 0.$$

This operator is unique up to a multiple of (4.103).

4.3.4 7D trigonometric subspace

Take

$$W_7 = \mathcal{L}\{1, \cos x, \sin x, \cos 2x, \sin 2x, \cos 3x, \sin 3x\}. \tag{4.107}$$

Proposition 4.14 *The only operator F in* (4.106) *preserving* (4.107) *is*

$$F[u] = D_x^7(u^2) + 77D_x^5(u^2) + 1876(u^2)_{xxx} + 14400(u^2)_x. \tag{4.108}$$

The invariance condition of W_7 is the linear system

$$\begin{cases} 256\beta - 16\gamma + \delta = 4096, \\ 1296\beta - 36\gamma + \delta = 46656, \\ 625\beta - 25\gamma + \delta = 15625, \end{cases}$$

which yields operator (4.108).

Example 4.15 (7th-order PDE on $\mathbf{W_7}$) The PDE (4.106), (4.108) on (4.107) is

$$u(x,t) = C_1 + C_2 \cos x + C_3 \sin x + C_4 \cos 2x + C_5 \sin 2x + C_6 \cos 3x + C_7 \sin 3x,$$

$$\begin{cases} C_1' = 0, \\ C_2' = 12600(C_4C_7 - C_5C_6 - C_3C_4 + C_2C_5 + 2C_1C_3), \\ C_3' = -12600(C_4C_6 + C_5C_7 + C_2C_4 + C_3C_5 + 2C_1C_2), \\ C_4' = 16128(C_2C_7 - C_3C_6 + 2C_1C_5 + C_2C_3), \\ C_5' = -8064(2C_2C_6 + 2C_3C_7 + 4C_1C_4 + C_2^2 - C_3^2), \\ C_6' = 9072(2C_1C_7 + C_3C_4 + C_2C_5), \\ C_7' = -9072(2C_1C_6 + C_2C_4 - C_3C_5). \end{cases}$$

Concerning compactons, the following result holds:

Proposition 4.16 *The only compacton admitted by* (4.106) *on* W_7 *occurs for*

$$u_t = D_x^7(u^2) + 56D_x^5(u^2) + 784(u^2)_{xxx} + 2304(u^2)_x, \tag{4.109}$$

and it is stationary ($\lambda = 0$),

$$u(x,t) = A \sin^3 x, \quad \textit{where} \quad A = \text{constant} > 0. \tag{4.110}$$

Namely, substituting $u(x,t) = A\sin^3(x - \lambda t)$ into (4.106) yields $\lambda = 0$ and the coefficients indicated in the PDE (4.109). Then W_7 is not invariant. Despite its sufficient regularity at the interfaces ($O(x^3)$ as $x \to 0^+$), (4.110) is governed by a special FBP with the zero contact angle conditions and interface equations, and is not a solution of the Cauchy problem. According to our conclusions in Example 4.9, the Cauchy problem needs another maximal regularity, which is given by the TWs $f(x - \lambda t)$, so that, on integration, keeping the leading term,

$$-\lambda f = (f^2)^{(6)}.$$

Then, instead of (4.83), the asymptotics behavior as $y \to 0$ is

$$f(y) = y^6 \varphi(s), \quad \text{where} \quad s = \ln y, \tag{4.111}$$

where the component φ satisfies a sixth-order ODE. As usual, for $\lambda < 0$, there exists the simple explicit solution

$$f(y) = -\lambda \, \tfrac{6!}{12!} \, y^6 > 0 \quad \text{for } y > 0.$$

Hence, the maximal regularity is $f(y) = O(y^6)$ as $y \to 0$, which is much smoother than $O(y^3)$ (here, $y = x$) given by (4.110).

On oscillatory patterns in the CP. The oscillatory behavior near interfaces occurs for the *signed* seventh-order PDE

$$u_t = D_x^7\big(|u|^n u\big) \quad \text{in } I\!R \times I\!R \quad (n > 0). \tag{4.112}$$

This has the natural connection as $n \to 0^+$ with the *linear dispersion equation*

$$u_t = u_{xxxxxxx},$$

whose fundamental solution is oscillatory at both interfaces $x = \pm\infty$.

For (4.112), the TW profiles $f(y)$ solve the ODE

$$\big(|f|^n f\big)^{(6)} = -\lambda f \quad \text{for } y > 0, \quad f(0) = 0.$$

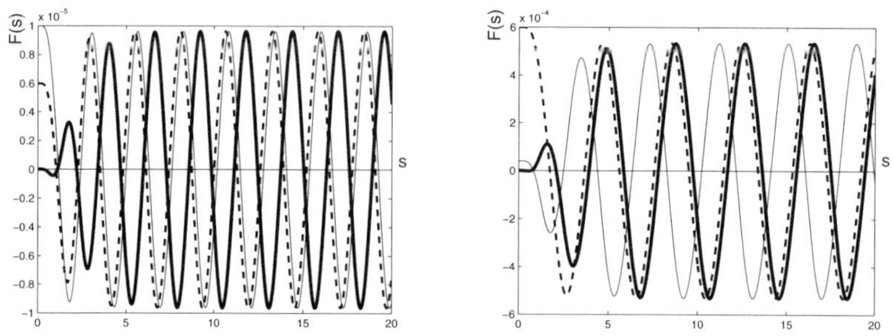

(a) $n = 5$ (b) $n = 500$

Figure 4.15 Stable periodic behavior for (4.114), $\lambda = 1, n = 5$ (a), and $n = 500$ (b).

Instead of (4.111), the oscillatory component φ is introduced as follows:

$$f(y) = y^{\mu}\varphi(s), \quad s = \ln y, \quad \text{with } \mu = \tfrac{6}{n}. \tag{4.113}$$

Setting $|\varphi|^n\varphi = F$ yields a sixth-order ODE (similar odd-order equations occurred in thin film analysis in Section 3.7)

$$P_6[F] + \lambda\big|F\big|^{-\frac{n}{n+1}}F = 0, \quad \text{with exponent } \gamma = \tfrac{6(n+1)}{n}, \tag{4.114}$$

where the linear operator P_6 is defined by the recursion (3.156). The stable periodic behavior for (4.114), $\lambda = 1$, which creates changing sign TW patterns by (4.113), is shown in Figure 4.15, where the part (b) corresponds to $n = 500$ and changes a little for larger values of n. Then $\gamma \to 6$, so the ODE admits the limit $n \to +\infty$, where the discontinuous nonlinearity sign F occurs.

In order to see periodic oscillations for smaller n (actually, there is a numerical difficulty already for $n \le 4$), we perform the scaling

$$F(s) = \left(\tfrac{n}{6}\right)^{\frac{6}{n}}\Phi(\eta), \quad \text{where} \quad \eta = \tfrac{6s}{n}, \tag{4.115}$$

to get in the limit the following simplified ODE with the binomial linear operator:

$$\begin{aligned} &\Phi^{(6)} + 6\Phi^{(5)} + 15\Phi^{(4)} + 20\Phi''' + 15\Phi'' + 6\Phi' + \Phi \\ &\equiv e^{-\eta}(e^{\eta}\Phi)^{(6)} = -\lambda\big|\Phi\big|^{-\frac{n}{n+1}}\Phi. \end{aligned} \tag{4.116}$$

Figure 4.16 shows the stable periodic behavior for (4.116) with $\lambda = 1$. According to scaling (4.115), the oscillatory component $\varphi(s)$ gets extremely small,

$$\max|\varphi| \sim 5 \times 10^{-10} \text{ for } n = 0.5, \text{ and } \max|\varphi| \sim 2 \times 10^{-111} \text{ for } n = 0.1.$$

For $\lambda < 0$, periodic solutions of (4.114) are unstable; see Figure 4.17 for $n = 15$ that is obtained by shooting from $s = 0$ with prescribed Cauchy data.

4.3.5 Cubic and higher-degree operators

Example 4.17 (**Cubic operators**) Consider the following cubic operator:

$$F[u] = D_x^5(u^3) + \beta(u^3)_{xxx} + \gamma(u^3)_x. \tag{4.117}$$

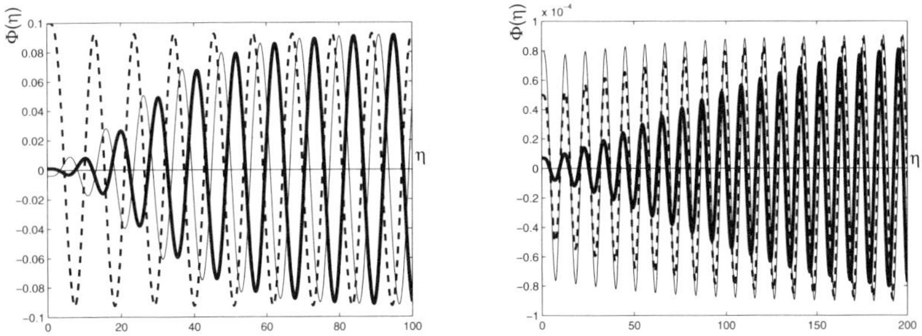

(a) $n = 0.5$ (b) $n = 0.1$

Figure 4.16 Periodic behavior for (4.116), $\lambda = 1$, $n = 0.5$ (a), and $n = 0.1$ (b).

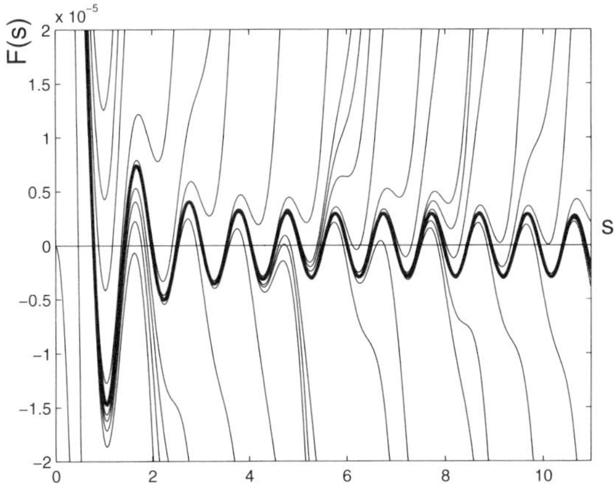

Figure 4.17 Unstable periodic behavior of the ODE (4.114), $\lambda = -1$, for $n = 15$. Cauchy data are $F(0) = 10^{-4}$, $F'(0) = F'''(0) = 0$, $F''(0) = -5.0680839826093907... \times 10^{-4}$.

Proposition 4.18 *Operator* (4.117) *admits* $W_3 = \mathcal{L}\{1, \cos x, \sin x\}$ *iff* $\beta = 13$ *and* $\gamma = 36$.

Hence, the PDE

$$u_t = D_x^5(u^3) + 13(u^3)_{xxx} + 36(u^3)_x \tag{4.118}$$

admits exact solutions (4.18) with the DS

$$\begin{cases} C_1' = 0, \\ C_2' = 18\big(4C_1^2 + C_2^2 + C_3^2\big)C_3, \\ C_3' = -18\big(4C_1^2 + C_2^2 + C_3^2\big)C_2. \end{cases}$$

There exist two first integrals

$$C_1 = A \quad \text{and} \quad C_2^2 + C_3^2 = B.$$

Then the DS provides us with explicit TW solutions

$$u(x,t) = D_1 + D_2 \cos(x - \lambda t) + D_3 \sin(x - \lambda t),$$

where D_k are arbitrary constants and

$$\lambda = -18\big(4D_1^2 + D_2^2 + D_3^2\big) < 0.$$

Choosing $D_3 = 0$ and $D_1 = D_2$ yields the compacton (obtained in [147] by a different approach)

$$u_c(x,t) = \sqrt{-\tfrac{2\lambda}{45}}\, \cos^2[\tfrac{1}{2}(x - \lambda t)] \quad \text{for } |x - \lambda t| < \pi. \tag{4.119}$$

According to the maximal regularity of the TW profiles satisfying

$$-\lambda f = (f^3)^{(4)} \quad \text{for } y > 0, \quad f(0) = 0,$$

we see that, for any uniformly bounded oscillatory component $\varphi(s)$,

$$f(y) = y^2 \varphi(\ln y) = O(y^2) \quad \text{as } y \to 0.$$

Since (4.119) exhibits precisely this maximal regularity at the interfaces, it can be considered as a solution of the Cauchy problem for (4.118). As usual, this does not mean positivity-like features of general solutions of (4.118). The function $F = \varphi^3(s)$ satisfies the ODE (4.93) with $n = 2$ (i.e., $\mu = 6$), so that the generic behavior is oscillatory for $\lambda > 0$ and is described by a stable periodic orbit $\varphi(s)$.

Example 4.19 (**Fifth-degree operators**) Consider operators of the algebraic homogenuity five,

$$F[u] = D_x^5(u^5) + \beta(u^5)_{xxx} + \gamma\,(u^5)_x. \tag{4.120}$$

Proposition 4.20 (i) *Operator* (4.120) *admits* $W_2 = \mathcal{L}\{\cos x, \sin x\}$ *iff* $\beta = 34$, $\gamma = 225$; *and* (ii) *does not preserve* $W_3 = \mathcal{L}\{1, \cos x, \sin x\}$.

In (ii), the invariance condition consists of fourteen nonlinear algebraic equations for β and γ which are not consistent. The corresponding PDE on W_2,

$$\boxed{u_t = D_x^5(u^5) + 34(u^5)_{xxx} + 225(u^5)_x,} \tag{4.121}$$

possesses solutions

$$u(x,t) = C_1(t)\cos x + C_2(t)\sin x,$$

$$\begin{cases} C_1' = 120\big(C_1^2 + C_2^2\big)^2 C_2, \\ C_2' = -120\big(C_1^2 + C_2^2\big)^2 C_1. \end{cases}$$

Using the first integral

$$C_1^2 + C_2^2 = A,$$

this leads to the TW solutions

$$u(x,t) = D_1\cos(x - \lambda t) + D_2\sin(x - \lambda t)$$

that depend on two arbitrary constants $D_{1,2}$, with

$$\lambda = -120\big(D_1^2 + D_2^2\big)^2 < 0.$$

Setting $D_2 = 0$ gives a formal compacton (cf. [147])

$$u_c(x,t) = \begin{cases} \left(-\frac{\lambda}{120}\right)^{\frac{1}{4}} \cos(x - \lambda t) & \text{for } |x - \lambda t| \le \frac{\pi}{2}, \\ 0 & \text{for } |x - \lambda t| > \frac{\pi}{2}. \end{cases} \tag{4.122}$$

It is curious that this simple Lipschitz continuous function is a solution of the maximal regularity and satisfies the Cauchy problem. Using TWs for the leading-order operator in (4.121), we have the ODE

$$-\lambda f = (f^5)^{(4)} \quad \text{for } y > 0, \quad f(0) = 0, \tag{4.123}$$

which has a linear envelope,

$$f(y) = y\,\varphi(s), \quad s = \ln y. \tag{4.124}$$

The oscillatory component $\varphi(s)$ is obtained from the fourth-order ODE (4.93) with $n = 4$ ($\mu = 5$). In general, solutions are oscillatory near interfaces. Hence, by (4.124), the Lipschitz continuity of solution (4.122) is the best regularity provided by the equation (4.123).

Example 4.21 (**Q(2, 2, 2)-family**) We now analyze a couple of operators from the family $Q(2,2,2)$ of equations (4.17) and study the corresponding PDEs on

$$W_3 = \mathcal{L}\{1, \cos x, \sin x\}.$$

First, it is easy to see that the third-order operator

$$F_3[u] = \left[u(u^2)_{xx}\right]_x + \beta(u^3)_x$$

does not admit W_3 (but it does $W_2 = \mathcal{L}\{\cos x, \sin x\}$ for $\beta = 4$). Consider next the fifth-order cubic operator

$$F_5[u] = \left[u(u^2)_{xxxx}\right]_x + \beta\left[u(u^2)_{xx}\right]_x + \gamma(u^3)_x. \tag{4.125}$$

Proposition 4.22 (i) *Operator* (4.125) *admits* W_3 *iff* $\beta = 5$ *and* $\gamma = 4$, *and* (ii) *does not admit* W_5 *given in* (4.102).

The corresponding evolution PDE

$$u_t = \left[u(u^2)_{xxxx}\right]_x + 5\left[u(u^2)_{xx}\right]_x + 4(u^3)_x$$

possesses exact solutions (4.18), with the DS

$$\begin{cases} C_1' = 0, \\ C_2' = 2(2C_1^2 + C_2^2 + C_3^2)C_3, \\ C_3' = -2(2C_1^2 + C_2^2 + C_3^2)C_2. \end{cases}$$

It follows that $C_1 = A$ and $C_2^2 + C_3^2 = B$, and the general solution is given by the TWs

$$u(x,t) = D_1 + D_2 \cos(x - \lambda t) + D_3 \sin(x - \lambda t),$$

where $D_{1,2,3} =$ constant, and

$$\lambda = -2(2D_1^2 + D_2^2 + D_3^2) < 0.$$

The compacton is obtained for $D_3 = 0$ and $D_1 = D_2$,

$$u_c(x,t) = \sqrt{-\tfrac{2\lambda}{3}}\,\cos^2[\tfrac{1}{2}(x-\lambda t)] \quad \text{for } |x-\lambda t| \le \pi. \tag{4.126}$$

By checking the maximal regularity of TWs via the leading term of the PDE, we obtain, on integration

$$-\lambda = (f^2)^{(4)} \quad \text{for } y > 0, \quad f(0) = 0.$$

The similarity structure of multiple zeros at $y = 0$ is then given by

$$f(y) = y^2\varphi(s), \quad \text{where } s = \ln y, \tag{4.127}$$

and this confirms that the non-oscillatory ($\lambda < 0$) compacton (4.126), exhibiting the same regularity at the interfaces, solves the Cauchy problem. As usual, the generic structure of multiple zeros for $\lambda > 0$ at interfaces depends on the behavior of the oscillatory component $\varphi(s)$ for $s = \ln y \ll -1$. Numerically, we did not see reliable periodic oscillations. For the *signed* $Q(2,2,2)$ equation,

$$u_t = [u(|u|u)_{xxxx}]_x \quad \Longrightarrow \quad -\lambda = (|f|f)^{(4)},$$

oscillatory solutions (4.127) do exist (see Example 4.10).

4.3.6 *Exponential subspaces*

For odd-order PDEs (4.97), dealing with exponential subspaces is easier.

Proposition 4.23 *The quadratic operator with constant coefficients*

$$F[u] = \textstyle\sum_{(i\ge 0)} \alpha_i D_x^i(u^2) \tag{4.128}$$

preserves the following subspaces:

$$W_2 = \mathcal{L}\{1, e^x\}, \quad \textit{if } \textstyle\sum 2^i\alpha_i = 0; \tag{4.129}$$

$$W_3 = \mathcal{L}\{1, e^x, e^{-x}\}, \quad \textit{if } \textstyle\sum 2^i\alpha_i = 0 \textit{ and } \sum(-2)^i\alpha_i = 0. \tag{4.130}$$

The result is straightforward by taking

$$u = C_1 + C_2 e^x + C_3 e^{-x} \tag{4.131}$$

($C_3 = 0$ for the subspace in (4.129)) and differentiating the equality

$$u^2 = C_1^2 + 2C_2C_3 + 2C_1C_2e^x + 2C_1C_3e^{-x} + C_2^2e^{2x} + C_3^2e^{-2x}.$$

Example 4.24 Equation (4.97), where

$$16\alpha + 4\beta + \gamma = 0$$

(two conditions in (4.130) coincide for all odd or even derivatives) admits solutions (4.131) with an easily derived DS for the expansion coefficients.

The order of the operator (4.128) can be arbitrary, so infinite-order equations can be considered. Such PDEs are well known in the mathematical literature; see Dubinskii [156].

Example 4.25 (**"Hyperbolic" PDE of infinite order**) Consider the PDE,

$$\boxed{u_{tt} = \sin(\alpha D_x)u^2 + u - 1,}$$

where α is a constant, and the linear operator $\sin(\alpha D_x)$ on the right-hand side is formally defined by

$$\sin(\alpha D_x) = \textstyle\sum_{(i \ge 0)} \frac{(-1)^i}{(2i+1)!}(\alpha D_x)^{2i+1}.$$

Looking for solutions on W_2 with

$$u(x,t) = C_1(t) + C_2(t)\mathrm{e}^x, \tag{4.132}$$

the invariance condition (4.129) reads

$$\sin 2\alpha = 0,$$

so we take $\alpha = \frac{\pi}{2}$. Then, substituting into the equation yields

$$C_1'' + C_2''\mathrm{e}^x = 2C_1C_2\big[\textstyle\sum \frac{(-1)^i}{(2i+1)!}\big(\frac{\pi}{2}\big)^{2i+1}\big]\mathrm{e}^x + C_1(t) + C_2(t)\mathrm{e}^x - 1,$$

and, since the sum in square brackets equals $\sin\big(\frac{\pi}{2}\big) = 1$, this gives the DS

$$\begin{cases} C_1'' = C_1 - 1, \\ C_2'' = 2C_1C_2 + C_2. \end{cases}$$

Then $C_1(t) = 1 + A\cosh t$ and C_2 solves a hyperbolic *Mathieu equation*,

$$C_2'' - (3 + 2A\cosh t)C_2 = 0.$$

Example 4.26 (**"Reaction-diffusion" equation of infinite order**) The PDE

$$\boxed{u_t = \sin(\alpha D_x)u^2 + \beta u^2,}$$

where $|\beta| \le 1$, admits solutions (4.132) if

$$\sin 2\alpha + \beta = 0.$$

4.4 Compactons and interfaces for singular mKdV-type equations

4.4.1 Preliminaries: Airy function, integral equation, and smooth solutions

Example 4.27 (**FBP-compactons for mKdV-type equations**) We now study some compacton-like solutions that are not associated with invariant subspaces or sets, but are simple and important for a general understanding of FBPs, the Cauchy problem, and finite propagation. Consider the mKdV-type PDE

$$u_t + u^m u_x + u_{xxx} = 0, \quad \text{with parameter } m \ge 0, \tag{4.133}$$

which at this time is formulated for nonnegative solutions. For $m = 1$, (4.133) is the KdV equation, and $m = 2$ yields the *modified KdV* (mKdV) *equation*. For $m = \frac{1}{2}$, (4.133) describes ion-acoustic waves in a cold-ion plasma with non-isothermal electrons, [517]; see Remarks for further applications and references. For such PDEs, basic computations are simple and sometimes explicit. To be precise, looking for standard TWs

$$u(x,t) = f(y), \quad \text{with } y = x - \lambda t, \;\; \lambda > 0, \tag{4.134}$$

yields the ODE

$$-\lambda f' + f^m f' + f''' = 0 \quad (f \geq 0),$$

so, on integration two times, bearing in mind the compacton with $f' = 0$ at $f = 0$,

$$(f')^2 = Af + \lambda f^2 - \tfrac{2}{(m+1)(m+2)}\, f^{m+2}, \tag{4.135}$$

where $A > 0$ is an arbitrary constant. In this case, for any $m > 0$, there exists a periodic solution $f(y)$. In particular, for $m = 2$, choosing $A = \frac{1}{3}\sqrt{2}\,\lambda^{3/2}$ yields the case of explicit integration (see [360] and [494])

$$u(x,t) = \tfrac{2}{3}\sqrt{2\lambda}\; \frac{\cos^2[\frac{1}{2}\sqrt{\lambda}(x-\lambda t)]}{1-\frac{2}{3}\cos^2[\frac{1}{2}\sqrt{\lambda}(x-\lambda t)]}.$$

Setting $u = 0$ for $\frac{1}{2}\sqrt{\lambda}|x - \lambda t| \geq \frac{\pi}{2}$, one obtains a (formal) compacton-like solution, which is localized in the interval of length $2\pi/\sqrt{\lambda}$. This and similar solutions given by the ODE (4.135) for any $m \geq 0$ are solutions of the FBP, so zero contact angle free-boundary conditions (4.44) are necessary to support such a compacton evolution. The dynamic interface equation is determined as in Example 4.8.

Let us present a further comment on this important issue to be dealt with later on: Such compacton-like solutions of *semilinear* PDEs with a *regular* lower-order nonlinear term $u^m u_x$ for $m > 0$ are not solutions of the Cauchy problem. The equation (4.133) in $I\!\!R \times I\!\!R_+$ describes processes with infinite propagation. To see this, set $m = 0$ (then the above compacton persists to exist if $\lambda < 1$) and consider the linear equation

$$u_t + u_{xxx} = 0 \quad \text{in } I\!\!R \times I\!\!R_+, \tag{4.136}$$

where the convection term u_x is eliminated by using the moving frame $x \mapsto x - t$. For initial data $u_0(x)$ with exponential decay at infinity, the unique solution of the Cauchy problem for (4.136) is given by the convolution

$$u(x,t) = b(\cdot,t) * u_0, \tag{4.137}$$

where $b(x,t)$ is the fundamental solution of the operator $\frac{\partial}{\partial t} + D_x^3$,

$$b(x,t) = t^{-\frac{1}{3}} g(\xi), \quad \xi = \tfrac{x}{t^{1/3}}, \tag{4.138}$$

and g satisfying $\int g = 1$ solves the linear ODE

$$g''' - \tfrac{1}{3}(g\xi)' = 0 \quad \Longrightarrow \quad g'' - \tfrac{1}{3}\, g\xi = 0.$$

The unique solution g is given by the Airy function $\mathrm{Ai}(\xi)$ and has the following behavior (see [4, p. 363] for details; recall the reflection $x \mapsto -x$ for PDE (4.66)):

$$g(\xi) \sim \begin{cases} \xi^{-\frac{1}{4}} e^{-a_0 \xi^{3/2}} & \text{as } \xi \to +\infty, \\ |\xi|^{-\frac{1}{4}} \cos\big(a_0|\xi|^{\frac{3}{2}} + A\big) & \text{as } \xi \to -\infty, \end{cases} \tag{4.139}$$

where $a_0 = \frac{2}{9}\sqrt{3}$ and A is a constant. Then (4.137) means that, for any compactly supported data $u_0(x) \not\equiv 0$, the solution $u(x,t)$ is not compactly supported for arbitrarily small $t > 0$.

Similarly, (4.137) implies that PDEs, such as (4.133) in $I\!R \times I\!R_+$ with the regular nonlinear term for $m \geq 0$, cannot admit nontrivial compactly supported solutions of the Cauchy problem. For sufficiently smooth solutions at $t = 0$ (i.e., for good initial data decaying fast enough at infinity), (4.133) can be written in the equivalent integral form

$$\begin{aligned} u(x,t) = M[u] &\equiv b(t) * u_0 - \textstyle\int_0^t b(t-s) * (|u|^m u_x)(s)\,\mathrm{d}s, \\ &= b(t) * u_0 - \tfrac{1}{m+1} \textstyle\int_0^t (t-s)^{-\frac{2}{3}} \int_{I\!R} g'\big(\tfrac{x-y}{(t-s)^{1/3}}\big)(|u|^m u)(y,s)\,\mathrm{d}y. \end{aligned} \tag{4.140}$$

Here, for solutions of changing sign, u^m is replaced by $|u|^m$. Integration by parts on the right-hand side of (4.140) needs extra estimates on the behavior of $u(x,t)$ as $x \to \pm\infty$. For compactly supported $u_0(x)$, we may expect that such behavior is similar to that shown in (4.139). A unique solution of (4.140) is constructed via the simple iteration

$$u_{n+1} = M[u_n] \quad \text{for } n = 0, 1, \ldots, \qquad u_0 \text{ is given}, \tag{4.141}$$

by using the fact that the integral operator M is a contraction in suitable functional spaces. In view of the slow decay of the Airy function in (4.139) as $\xi \to -\infty$, functional settings are rather tricky; details can be found in [179]. Semigroup approaches and Banach's Contraction Principle are effective tools to prove existence and uniqueness for the Cauchy problem; see [403, Ch. 7] for several advanced applications. Concerning the behavior of small enough solutions as $x \to \pm\infty$, we observe that, in iteration (4.141), the regular term $|u|^m u_x$ for any $m \geq 0$ does not affect the "essence" of asymptotics in the fundamental kernel (4.139). For compactly supported u_0, solution $u(x,t)$ of (4.140) will exhibit similar asymptotic decay as $x \to \infty$ for arbitrarily small $t > 0$.

Explicit FBP-compactons via simple ODEs can be constructed for other semilinear PDEs of KdV-type. For instance, the fifth-order KdV-type equation

$$u_t + \tfrac{5}{3}(2u + u^4) + 5u^2 u_{xxx} + u_{xxxxx} = 0$$

possesses the explicit solution with $\lambda = 1$ [360]

$$u(x,t) = B\, \frac{\cos^2[\frac{1}{2}(x-t)]}{1 - \frac{2}{3}\cos^2[\frac{1}{2}(x-t)]}, \tag{4.142}$$

where $B = \frac{4}{3}$. Another KdV-type equation

$$u_t + \tfrac{1}{10}\, u^4 u_x + (u_x)^3 + u^2 u_{xxx} + u_{xxxxx} = 0$$

admits the solution (4.142) with [570]

$$B = \tfrac{2}{3}\sqrt{10}.$$

All these and other similar solutions need a proper FBP setting. We agree with arguments of Rosenau's critics of such compactons [494, p. 202], which were sometimes wrongly treated as solutions of the Cauchy problem in $I\!R \times I\!R_+$.

4.4.2 Finite propagation and oscillatory solutions

Example 4.28 (Finite propagation for $m \in (-1, 0)$) Thus, in order to have finite propagation, we need a *singular* lower-order term with exponents

$$m < 0.$$

For solutions of changing sign, we take the *signed* mKdV-type equation,

$$u_t - |u|^m u_x + u_{xxx} = 0 \quad \text{in } I\!R \times I\!R_+ \tag{4.143}$$

(recall the sign change in the second term to ensure a suitable finite propagation).

In general, *existence* of a solution for $m \in (-1, 0)$ can be seen from the integral equation (4.140) analyzed by Schauder's Theorem in a proper functional setting (M is assumed to be compact and map a convex set into itself). It is principal that *uniqueness* cannot follow from the integral equation, since M is not a contraction for the non-Lipschitz nonlinearity $|u|^m u$, where $m + 1 \in (0, 1)$. This is an OPEN PROBLEM. As usual, *uniqueness* is associated with the approximation (ε-regularization) approach as in Section 4.2.4. For second-order parabolic PDEs with singular nonlinear coefficients, this leads to notions of maximal or minimal proper solutions (see [226, Ch. 7]).

As above, we reveal the interface behavior by using TWs (4.134) satisfying

$$\lambda f = f'' - \tfrac{1}{m+1} |f|^m f \quad \text{for } y > 0, \quad f(0) = 0 \tag{4.144}$$

(for simplicity, $A = 0$ in (4.135)). Let us begin by studying the crucial stationary case $\lambda = 0$, where the ODE is simpler,

$$f'' - \tfrac{1}{m+1} |f|^m f = 0 \quad \text{for } y > 0, \quad f(0) = 0.$$

This possesses the positive non-oscillatory solution

$$f(y) = \varphi_0 y^\gamma, \quad \text{where} \quad \gamma = \tfrac{2}{|m|} > 0 \quad \text{and} \qquad \varphi_0 = \Big[\tfrac{2(m+1)(m+2)}{m^2}\Big]^{\frac{1}{m}}, \tag{4.145}$$

which describes the behavior also near interfaces for $\lambda \neq 0$; see further asymptotic expansions below. The function (4.145) being extended by 0 for $y \le 0$ is at least $C^{[\gamma]-1}$ at $y = 0$, and the smoothness increases as $m \to 0^-$, since $\gamma = \frac{2}{|m|} \to +\infty$. Notice also that such solutions satisfy $u \in C_x^?$ for all $m \in (-1, 0)$, which seems difficult to prove for general weak solutions by the integral equation (4.140) or otherwise. These regularity problems are OPEN.

It is interesting to estimate the rate of divergence as $m \to 0^-$ of the interface $x = s(t)$ of solutions $u(x, t)$. For compactly supported u_0, using estimates in (4.139) yields, for small fixed $t > 0$, that

$$u(x, t) \sim b(x, t) \sim t^{-\frac{1}{3}} \big(\tfrac{|x|}{t^{1/3}}\big)^{-\frac{1}{4}} \quad \text{as } x \to -\infty.$$

Consider the expansion for $m \approx 0^-$,

$$|u|^m = 1 + m \ln |u| + \ldots ,$$

which is violated for the solution $u(x,t)$ almost everywhere (a.e.) in domains, where

$$|m \ln |u|| \sim |m| \, |\ln |b(x,t)|| \gg 1.$$

This characterizes the approximate position of the interface of $u(x,t)$ (indeed, if this is not true a.e., then u can be again approximated by b). Thus, for the interface, the following approximate equation holds:

$$\ln |b(s(t),t)| \equiv -\tfrac{1}{3} \ln t - \tfrac{1}{4} \ln\big(\tfrac{|s(t)|}{t^{1/3}}\big) \sim -\tfrac{1}{|m|}.$$

Resolving gives the exponential estimate of the interface position

$$s(t) \sim -\mathrm{e}^{\frac{4}{|m|}} \quad \text{as} \quad m \to 0^-.$$

Example 4.29 (Finite propagation and interface equation for $m \in (-2,-1]$**)** Let us now return to the original mKdV equation (we take $A = 0$)

$$u_t + |u|^m u_x + u_{xxx} = 0 \quad \Longrightarrow \quad \lambda f = f'' + \tfrac{1}{m+1} |f|^m f. \tag{4.146}$$

Then, for $m \in (-2,-1)$, there exists the positive solution

$$f(y) = \big[\tfrac{2|m+1|(2+m)}{m^2}\big]^{\frac{1}{m}} y^{\frac{2}{|m|}} (1 + o(1)) \quad \text{as} \quad y \to 0,$$

for which the interface equation is computed. For

$$m \le -2,$$

such a local TW profile does not exist. For the corresponding second-order diffusion-convection equations

$$u_t + u^m u_x - u_{xx} = 0 \quad (u \ge 0), \tag{4.147}$$

with $m \le -2$, this means *nonexistence* of a nontrivial solution $u(x,t) \not\equiv 0$ for any compactly supported data $u_0 \ge 0$; see [226, Sect. 7.5]. For the third and higher-order evolution PDEs with singular coefficients, similar nonexistence conclusions are unknown and represent an OPEN PROBLEM.

Taking into consideration the λf term in (4.146) yields that, close to the interface as $y \to 0^+$,

$$f(y) = C y^\gamma + \lambda \, \tfrac{C}{(\gamma+1)(\gamma+2)} \, y^{\gamma+2} + \ldots, \tag{4.148}$$

where

$$\gamma = \tfrac{2}{|m|} \quad \text{and} \quad C^m = -\tfrac{2(m+1)(m+2)}{m^2} > 0.$$

The expansion (4.148) determines the corresponding *pressure-like variable*

$$v = u^{|m|},$$

which, for TW profiles, exhibits the analytic-looking expansion

$$f^{|m|}(y) = C^{|m|} y^2 + \lambda \, \tfrac{|m| C^{|m|}}{(\gamma+1)(\gamma+2)} \, y^4 + \ldots. \tag{4.149}$$

Recalling the TW structure (4.134), this gives the following system of interface equations at $x = s(t)$, consisting of stationary and dynamic ones:

$$\begin{cases} (u^{|m|})_{xx} = 2C^{|m|}, \\ s' = \lambda = \frac{(\gamma+1)(\gamma+2)}{24|m|C^{|m|}} \, D_x^4 (u^{|m|}). \end{cases} \tag{4.150}$$

For singular second-order parabolic equations, including (4.147), such interface systems are rigorously justified by the Sturmian intersection comparison with TW solutions; see examples in [226, Sect. 7.11]. For $A \neq 0$ in (4.135), the expansion like (4.149) contains more terms, so an interface system, such as (4.150), may also include more equations for $m \approx -1^-$; cf. parabolic examples in [226, Sect. 8.5].

In the critical case $m = -1$, the expansion takes the form (for $A = 0$)

$$f(y) = -y^2 \ln y - \tfrac{1}{2}\, y^2 \ln(-\ln y) - \tfrac{1}{12}\, \lambda y^4 \ln y + ..., \tag{4.151}$$

so that the interface system is formulated in terms of another pressure variable

$$v = Q(u) \equiv \sqrt{\tfrac{2u}{|\ln u|}} + ... \quad \text{as } u \to 0.$$

Then $Q(f) = y^2 + ...$ and this implies the first stationary interface equation

$$[Q(u)]_{xx} = 2 \quad \text{at } x = s(t).$$

Expansion (4.151) then gives the second stationary equation, and the third equation containing λ that is dynamic. The justification of the interface system (4.150) for general solutions and the identification of the problem (an FBP, or the Cauchy problem) remain OPEN. The non-oscillatory property of the ODE (4.146) at interfaces suggests that solutions with the behavior (4.148) and (4.151) solve the CP.

Example 4.30 (**Solutions of changing sign**) Finite interface oscillatory solutions can be obtained in a KdV-type equation with a lower-order absorption-like term,

$$u_t + uu_x + u_{xxx} + |u|^{-n} u = 0 \quad \text{in } I\!R \times I\!R_+, \quad \text{with } n > 0.$$

The TWs solve

$$-\lambda f' + ff' + f''' + |f|^{-n} f = 0.$$

Near the interface at $y = 0$, keeping two leading terms yields

$$f''' + |f|^{-n} f = 0 \quad \text{for } y > 0, \quad f(0) = 0.$$

Introducing the oscillatory component

$$f(y) = y^\gamma \varphi(s), \quad \text{where} \quad s = \ln y, \tag{4.152}$$

where $\gamma = \frac{3}{n}$, gives the following ODE:

$$P_3[\varphi] = -\tfrac{\varphi}{|\varphi|^n}. \tag{4.153}$$

This is precisely equation (3.166) that occurred in thin film study in Section 3.7, where a stable periodic orbit of changing sign was detected for $n \in (0, n_{\rm h})$.

Example 4.31 (**Self-similar blow-up in (2k+1)th-order PDEs**) Neglecting the convective term uu_x, consider a general $(2k{+}1)$th-order equation with a source,

$$u_t = (-1)^{k+1} D_x^{2k+1} u + |u|^{p-1} u \quad \text{in } I\!R \times (0, T) \quad (k = 1, 2, ...),$$

where $p > 1$. Blow-up similarity solutions are given by

$$\begin{aligned} &u(x,t) = (T-t)^{-\frac{1}{p-1}} f\big(\tfrac{x}{(T-t)^{1/(2k+1)}}\big), \text{ where } f \text{ solves} \\ &(-1)^{k+1} f^{(2k+1)} - \tfrac{1}{2k+1}\, f' y - \tfrac{1}{p-1}\, f + |f|^{p-1} f = 0. \end{aligned} \tag{4.154}$$

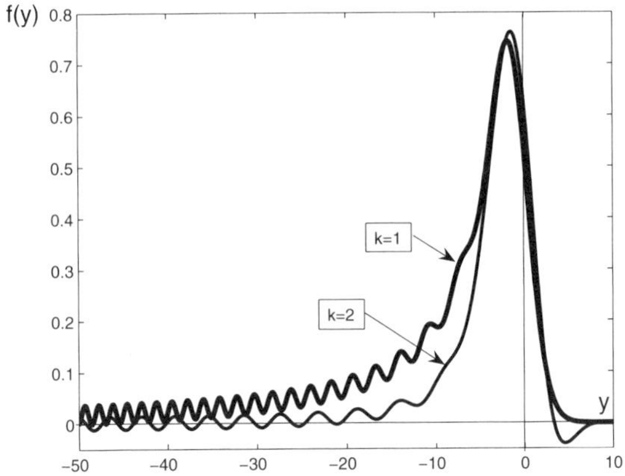

Figure 4.18 The first blow-up patterns of (4.154), $p = 3$, for $k = 1$ and $k = 2$.

The first blow-up patterns are calculated numerically and are shown in Figure 4.18 for $p = 3$ in the third, $k = 1$, and fifth-order, $k = 2$, cases. Notice the oscillatory tail as $y \to -\infty$, which for $k = 1$ corresponds to the Airy function (4.139). The decay as $y \to +\infty$ is exponentially fast, non-oscillatory for $k = 1$, and oscillatory for any $k \geq 2$. Existence and multiplicity of solutions for (4.154) in $I\!R$ are OPEN PROBLEMS and are more difficult in higher-order cases ($k \geq 2$). It is curious that blow-up is not single point. For instance, for $k = 1$, using expansions in (4.139), we find that, roughly speaking,

$$u(x, T^-) = \begin{cases} 0 & \text{for } x > 0, \\ \pm\infty & \text{for } x < 0, \end{cases}$$

where "$\pm\infty$" means unbounded oscillatory behavior. Namely, at every fixed point $x_0 < 0$, the final-time profile is unbounded for $p < 13$ (we expect that such solutions exist for $p > 1$ below the bifurcation point $p_0 = 4$),

$$u(x,t) \sim (T-t)^{-\frac{1}{p-1}} \left| \frac{x_0}{(T-t)^{1/3}} \right|^{-\frac{1}{4}} \cos(\ldots) = (T-t)^{\frac{p-13}{12(p-1)}} |x_0|^{-\frac{1}{4}} \cos(\ldots).$$

Example 4.32 (**Oscillatory solutions in higher-order mKdV equations**) Oscillatory, changing sign solutions at finite interfaces are achieved in higher-order mKdV-type models, such as the fifth-order one

$$u_t + \left(|u|^{-n} u\right)_x + u_{xxxxx} = 0 \quad (n > 0).$$

Then TWs satisfy $-\lambda f' + (|f|^{-n} f)' + f^{(5)} = 0$, so that, close to interfaces,

$$f^{(4)} + |f|^{-n} f = 0 \quad \text{for } y > 0, \quad f(0) = 0,$$

using representation (4.152) with $\gamma = \frac{4}{n}$. A periodic oscillatory behavior for ODEs like (4.153), with $P_3 \mapsto P_4$, was studied in Example 4.10.

4.5 On compactons in $I\!R^N$ for nonlinear dispersion equations

Invariant subspaces for operators in $I\!R^N$ will be systematically studied in Chapter 6, including compact moving structures. Here we present a few examples of compactons in $I\!R^N$, whose construction is based on the known 1D analysis. For compactons, this study is continued in Section 6.7. Many problems of existence, uniqueness, and asymptotics for the PDEs in this section are OPEN.

Example 4.33 (**KP equation with nonlinear dispersion**) Consider the PDE

$$\boxed{\left[u_t + uu_x + (u^2)_{xxx}\right]_x + \Delta_\perp u = 0,} \tag{4.155}$$

which is the *Kadomtsev–Petviashvili equation* in $I\!R^N \times I\!R$ *with nonlinear dispersion*. Here $X = (x, x')$ is the spatial variable, with $x \in I\!R$ and $x' = (x_2, ..., x_N) \in I\!R^{N-1}$. The Laplacian $\Delta_\perp$ takes into account the variable x'. This equation in $I\!R^3 \times I\!R$ was introduced by Rosenau and Hyman [496, p. 567] to demonstrate multi-dimensional compactons of special and unusual structure. These are treated below by using invariant subspaces. We continue to call such solutions *compactons*, though their supports are not bounded or of finite measure, but their solutions exhibit special types of finite propagation.

Consider solutions on the trigonometric subspace $W_3 = \mathcal{L}\{1, \cos\gamma x, \sin\gamma x\}$,

$$u(X,t) = C_1(x',t) + C_2(x',t)\cos\gamma x + C_3(x',t)\sin\gamma x, \tag{4.156}$$

with a constant γ to be determined. Plugging (4.156) into the PDE (4.155) yields the following expansion on $W_5 = W_3 \oplus \mathcal{L}\{\cos 2\gamma x, \sin 2\gamma x\}$:

$$\begin{aligned} \Delta_\perp C_1 + \left[\gamma\left(C_{3t} + \gamma(2\gamma^2-1)C_1C_2\right) + \Delta_\perp C_2\right]\cos\gamma x \\ - \left[\gamma\left(C_{2t} - \gamma(2\gamma^2-1)C_1C_3\right) + \Delta_\perp C_3\right]\sin\gamma x \\ + \gamma^2(8\gamma^2-1)\left(C_2^2 - C_3^2\right)\cos 2\gamma x + 2\gamma^2(8\gamma^2-1)C_2C_3\sin 2\gamma x = 0. \end{aligned}$$

The last two terms (projections onto $\cos 2\gamma x$ and $\sin 2\gamma x$) simultaneously vanish if

$$\gamma = \tfrac{1}{2\sqrt{2}}. \tag{4.157}$$

Then the restriction of (4.155) to W_3 is a system of three equations for expansion coefficients $\{C_1, C_2, C_3\}$,

$$\begin{cases} \Delta_\perp C_1 = 0, \\ C_{2t} + \frac{3}{8\sqrt{2}}\, C_1C_3 - 2\sqrt{2}\,\Delta_\perp C_3 = 0, \\ C_{3t} - \frac{3}{8\sqrt{2}}\, C_1C_2 + 2\sqrt{2}\,\Delta_\perp C_2 = 0. \end{cases} \tag{4.158}$$

Hence, $C_1(x', t)$ is an arbitrary solution of the Laplace equation in the variable x', and the last two PDEs give a linear system for $\{C_2, C_3\}$ that is not studied in detail. Note though that choosing a constant function $C_1(x', t) \equiv A$ yields a linear fourth-order hyperbolic PDE for C_3,

$$C_{3tt} + 8\Delta_\perp^2 C_3 - \tfrac{3A}{2}\,\Delta_\perp C_3 + \tfrac{9A^2}{128}\, C_3 = 0,$$

which e.g., possesses explicit solutions on simple polynomial subspaces.

We restrict ourselves to self-similar solutions of the system (4.158) by using the rescaled variables

$$C_1 = \tfrac{1}{t}\, U, \quad C_2 = \tfrac{1}{t}\, V, \quad C_3 = \tfrac{1}{t}\, W, \quad \xi = \tfrac{x'}{\sqrt{t}}, \quad \tau = \ln t. \tag{4.159}$$

Then (4.158) reduces to

$$\begin{cases} V_\tau - \frac{1}{2}\, \nabla_\xi V \cdot \xi - V + \frac{3}{8\sqrt{2}}\, UW - 2\sqrt{2}\, \Delta_\xi W = 0, \\ W_\tau - \frac{1}{2}\, \nabla_\xi W \cdot \xi - W - \frac{3}{8\sqrt{2}}\, UV + 2\sqrt{2}\, \Delta_\xi V = 0, \end{cases} \tag{4.160}$$

with an arbitrary function $U = U(\xi)$ satisfying Laplace's equation. Taking again the constant function $U = A$, and looking for solutions as radial TWs,

$$V = f(s) \text{ and } W = g(s), \quad \text{where } s = \tfrac{1}{2}\, |\xi|^2 - \lambda\tau, \tag{4.161}$$

where $\lambda > 0$ is a constant, gives the following system for $\{f, g\}$:

$$\begin{cases} \lambda f' + \frac{1}{2}\, f'|\xi|^2 + f - \frac{3}{8\sqrt{2}}\, Ag + 2\sqrt{2}\left[g''|\xi|^2 + (N-1)g'\right] = 0, \\ \lambda g' + \frac{1}{2}\, g'|\xi|^2 + g + \frac{3}{8\sqrt{2}}\, Af - 2\sqrt{2}\left[f''|\xi|^2 + (N-1)f'\right] = 0. \end{cases} \tag{4.162}$$

Projecting equations of (4.162) onto $|\xi|^2$ and 1, we obtain two standard systems

$$\begin{cases} \frac{1}{2}\, f' + 2\sqrt{2}\, g'' = 0, \\ \frac{1}{2}\, g' - 2\sqrt{2}\, f'' = 0, \end{cases} \quad \begin{cases} \lambda f' + f - \frac{3}{8\sqrt{2}}\, Ag + 2\sqrt{2}(N-1)g' = 0, \\ \lambda g' + g + \frac{3}{8\sqrt{2}}\, Af - 2\sqrt{2}(N-1)f' = 0. \end{cases} \tag{4.163}$$

The first system yields a single ODE for f, $f''' + \frac{1}{32}\, f' = 0$, which results in

$$f(s) = B\cos(as) \text{ and } g(s) = C\sin(as), \quad \text{where } a = \tfrac{1}{4\sqrt{2}}, \tag{4.164}$$

and $B, C \in I\!R$. Substituting these functions into the second system in (4.163) and projecting each equation onto $\cos as$ and $\sin as$ yields two linear algebraic equalities

$$\begin{cases} B + 2\sqrt{2}(N-1)Ca = 0, \\ \lambda Ba + \frac{3}{8\sqrt{2}}\, AC = 0, \end{cases} \quad \begin{cases} \lambda Ca + \frac{3}{8\sqrt{2}}\, AB = 0, \\ C + 2\sqrt{2}(N-1)Ba = 0. \end{cases} \tag{4.165}$$

From the first system, we find $A = \frac{1}{3}(N-1)\lambda$ and $C = -\frac{2}{N-1}\, B$. Substituting into the second equation of the second system yields

$$-\tfrac{2}{N-1} + \tfrac{N-1}{2} = 0 \quad \Longrightarrow \quad (N-1)^2 = 4, \quad \text{i.e., } N = 3.$$

Then the first equation implies $C = -B$. Notice that there is another formal hypothetical dimension $N = -1$ corresponding to the "unstable" radial Laplacian

$$\Delta_r = r^2 D_r\left(\tfrac{1}{r^2}\, D_r\right) \equiv D_r^2 - \tfrac{2}{r}\, D_r,$$

which occurs in some problems of plasma physics.

Thus, the overdetermined system (4.162) is consistent in the 3D geometry only, $N = 3$, and the solution is

$$C_1 = A = \tfrac{2\lambda}{3}, \quad f(s) = B\cos\left(\tfrac{s}{4\sqrt{2}}\right), \quad \text{and} \quad g(s) = -B\sin\left(\tfrac{s}{4\sqrt{2}}\right), \tag{4.166}$$

where $B \in I\!R$ is arbitrary. Substituting into (4.156) gives solutions on W_3,

$$u(X, t) = \tfrac{2\lambda}{3t} + \tfrac{B}{t}\, \cos\left[\gamma x + \tfrac{\gamma}{2}\left(\tfrac{1}{2t}\, |x'|^2 - \lambda \ln t\right)\right].$$

The compacton is achieved at $B = \frac{2\lambda}{3}$ and takes the form

$$u_c(X,t) = \frac{4\lambda}{3t} \cos^2\big[\tfrac{1}{4\sqrt{2}}\big(x + \tfrac{1}{4t}\,|x'|^2 - \tfrac{\lambda}{2}\,\ln t\big)\big], \tag{4.167}$$

where $u_c = 0$ if the absolute value of the argument of cos is larger than $\frac{\pi}{2}$.

In order to illustrate this type of finite propagation for the compacton (4.167), consider the corresponding linear PDE (for simplicity, with $x' = y \in I\!R$)

$$u_{tx} + u_{yy} = 0 \quad \text{in } I\!R^2 \times I\!R_+. \tag{4.168}$$

In particular, let us study its similarity solutions

$$u(x,y,t) = \tfrac{1}{\sqrt{1+t}}\, f(x,\eta), \quad \eta = \tfrac{y}{\sqrt{1+t}} \quad \Longrightarrow \quad f_{\eta\eta} = \tfrac{1}{2}(f_x \eta)_\eta.$$

This PDE is easily integrated (the fundamental solution belongs to the same family). For instance, there are solutions

$$u(x,y,t) = \tfrac{1}{\sqrt{1+t}}\, \psi\big(x + \tfrac{1}{4(1+t)}\, y^2\big), \tag{4.169}$$

where $\psi(s)$ is an arbitrary C^∞ function supported on the interval $[-1,1]$. Solution (4.169) has the support localized between two paraboloids; cf. (4.167). In other words, linear equation (4.168) supports this kind of finite propagation, and such typical support shapes are reflected by the nonlinear PDE (4.155). As in Example 4.8, we can use the TWs to check that such compactons are solutions of the Cauchy problem and do not need an FBP setting.

Example 4.34 (Sixth-order PDE with nonlinear dispersion) For higher-order extensions, we will use our 1D analysis in Example 4.5. Consider the sixth-order PDE composed in a similar manner,

$$\boxed{[u_t - \alpha(u^2)_{xxxxx} - \beta(u^2)_{xxx} - \gamma(u^2)_x]_x + \Delta_\perp u = 0} \tag{4.170}$$

in $I\!R^N \times I\!R$. The quadratic fifth-order operator admits $W_3 = \mathcal{L}\{1, \cos x, \sin x\}$ if $16\alpha - 4\beta + \gamma = 0$, and there exist solutions

$$u(X,t) = C_1(x',t) + C_2(x',t)\cos x + C_3(x',t)\sin x. \tag{4.171}$$

Substituting into (4.170) yields a similar system (here $\mu = 6(\beta - 5\alpha)$)

$$\begin{cases} \Delta_\perp C_1 = 0, \\ C_{2t} - \mu\, C_1 C_3 - \Delta_\perp C_3 = 0, \\ C_{3t} + \mu\, C_1 C_2 + \Delta_\perp C_2 = 0. \end{cases} \tag{4.172}$$

There exist similarity solutions (4.159), (4.161). For $N = 3$, the explicit solutions are given, instead of (4.166), by

$$C_1 = A = -\tfrac{\lambda}{2\mu}, \quad f(s) = B\cos\big(\tfrac{s}{2}\big), \quad \text{and} \quad g(s) = -B\sin\big(\tfrac{s}{2}\big).$$

The compacton solution takes place for $B = A = -\frac{\lambda}{2\mu}$,

$$u_c(X,t) = -\tfrac{\lambda}{\mu t} \cos^2\big[\tfrac{1}{2}\big(x + \tfrac{1}{4t}\,|x'|^2 - \tfrac{\lambda}{2}\,\ln t\big)\big]. \tag{4.173}$$

Its support is unbounded and lies between two paraboloids in $I\!R^3$,

$$\operatorname{supp} u_c(X,t) = \big\{4t\big(\tfrac{\lambda}{2}\,\ln t - \pi - x\big) \le |x'|^2 \le 4t\big(\tfrac{\lambda}{2}\,\ln t + \pi - x\big)\big\}.$$

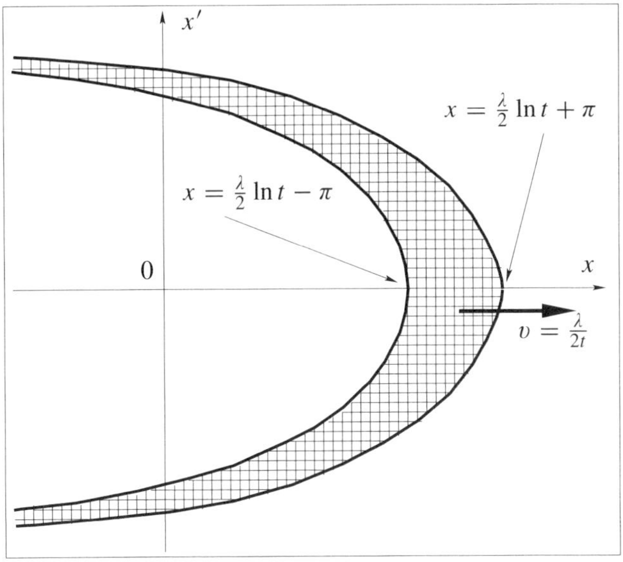

Figure 4.19 The support of compacton (4.173) moves to the right with the speed $v = \frac{\lambda}{2t}$.

This is shown in Figure 4.19, where, for convenience, we treat x' as a single variable. As in Example 4.8, we show that C^1-compactons (4.173) are not sufficiently smooth at $\{u = 0\}$ to satisfy the Cauchy problem, and need a standard FBP zero contact angle setting. In order to obtain C^3-compactons, as in (4.105), we need a PDE with the operator (4.103) for which manipulations with the expansion coefficients include five PDEs in the resulting difficult system, such as (4.172).

Example 4.35 (Zakharov–Kuznetsov PDE with nonlinear dispersion) In plasma physics, in the presence of strong magnetic fields, the evolution of the ion density in a strongly magnetized ion-acoustic plasma is described by the *Zakharov–Kuznetsov* (ZaK) *equation* [591]

$$u_t + uu_x + u_{xxx} + u_{xyy} = 0 \quad \text{in } I\!R^2 \times I\!R, \tag{4.174}$$

where a magnetic field is directed along the x-axis. This, together with the KP equation, is one of the best known 2D generalization of the KdV equation that describes two-dimensional modulations of solitons. Note that the KP equation is limited by the assumption of weak two-dimensionality; see details and references in [527]. But, unlike the KP equation, (4.174) is not integrable by the inverse scattering transform method.

Consider the higher-order ZaK-type equation with nonlinear dispersion,

$$\boxed{u_t - \left[\alpha(u^2)_{xxxxx} + \beta(u^2)_{xxx} + \gamma(u^2)_x\right] + \Delta_\perp u_x = 0} \tag{4.175}$$

in $I\!R^N \times I\!R$, and study it on $W_3 = \mathcal{L}\{1, \cos x, \sin x\}$ in the case where

$$16\alpha - 4\beta + \gamma = 0.$$

The corresponding solutions have the same form as (4.171) and are driven by the DS (4.172), where the first equation is replaced by the stationary one,

$$C_{1t} = 0.$$

Taking the constant function $C_1(x', t) \equiv A$ gives the linear system

$$\begin{cases} C_{2t} - \rho\, C_3 + \Delta_\perp C_3 = 0, \\ C_{3t} + \rho\, C_2 - \Delta_\perp C_2 = 0, \end{cases}$$

where $\rho = \mu A = 6(\beta - 5\alpha)A$. This is reduced to a linear fourth-order hyperbolic equation,

$$C_{3tt} = -\Delta_\perp^2 C_3 + 2\rho \Delta_\perp C_3 - \rho^2 C_3.$$

In particular, for $N = 2$, denoting $x' = y$ yields the solution

$$C_3(y, t) = B \sin[\kappa(y - \lambda t)],$$

where $\kappa > 0$ depends on the TW speed $\lambda \neq 0$ as follows: $\kappa^2 - \kappa\lambda + \rho = 0$. Determining a similar coefficient $C_2(y, t)$, we find the solution

$$u(x, y, t) = A + B \cos[\kappa(y - \lambda t) - x],$$

where setting $A = B$ yields the TW compacton

$$u_c(x, y, t) = 2A \cos^2[\tfrac{1}{2}(\kappa y - x - \kappa\lambda t)].$$

Its support is a moving strip in the $\{x, y\}$-plane,

$$|\kappa y - x - \kappa\lambda t| \leq \pi.$$

For the third-order PDE (4.175) with $\alpha = 0$, such a compacton seems to be a solution of the Cauchy problem by the same reasons detailed in Example 4.33, while, for the fifth-order case with $\alpha \neq 0$, a standard FBP setting is necessary to support it.

Example 4.36 **(N = 2)** Let us now demonstrate slightly different type of compacton solutions for quadratic PDEs. Without loss of generality, consider the simplest KP-type equation

$$\big[u_t - (uu_{xx})_x - (u^2)_x\big]_x + \Delta_\perp u = 0 \quad \text{in } I\!R^N \times I\!R_+. \tag{4.176}$$

Manipulations remain the same and lead to similar systems also for the fifth-order operators introduced above. Since the quadratic operator in (4.176) admits subspace W_3, looking for solutions (4.171) yields the PDE system (4.172) with $\mu = 1$. We now do not use the similarity scaling, as in (4.159), and set

$$C_1 = \tfrac{1}{\sqrt{t}}\, U, \quad C_2 = \tfrac{1}{\sqrt{t}}\, V, \quad C_3 = \tfrac{1}{\sqrt{t}}\, W, \quad \xi = \tfrac{1}{\sqrt{t}}\, x', \quad \tau = \ln t,$$

to get a non-autonomous PDE system,

$$\begin{cases} V_\tau - \tfrac{1}{2}\, \nabla_\xi V \cdot \xi - \tfrac{1}{2}\, V - \mathrm{e}^{\frac{\tau}{2}} UW - \Delta_\perp W = 0, \\ W_\tau - \tfrac{1}{2}\, \nabla_\xi W \cdot \xi - \tfrac{1}{2}\, W + \mathrm{e}^{\frac{\tau}{2}} UV + \Delta_\perp V = 0, \end{cases}$$

where, as usual, $U = A$. Looking for solutions (4.161) with

$$s = \tfrac{1}{2}\, |\xi|^2 - \lambda \mathrm{e}^{\frac{\tau}{2}},$$

we obtain the following ODE system:

$$\begin{cases} f'(-\frac{\lambda}{2})e^{\frac{\tau}{2}} - \frac{1}{2} f'|\xi|^2 - \frac{1}{2} f - e^{\frac{\tau}{2}} Ag - \left[g''|\xi|^2 + (N-1)g'\right] = 0, \\ g'(-\frac{\lambda}{2})e^{\frac{\tau}{2}} - \frac{1}{2} g'|\xi|^2 - \frac{1}{2} g + e^{\frac{\tau}{2}} Af + \left[f''|\xi|^2 + (N-1)f'\right] = 0. \end{cases} \tag{4.177}$$

Unlike in all the previous examples, this consists of three independent systems as "projections" onto $e^{\frac{\tau}{2}}$, $|\xi|^2$, and 1:

$$\begin{cases} \frac{\lambda}{2} f' + Ag = 0, \\ -\frac{\lambda}{2} g' + Af = 0, \end{cases} \quad \begin{cases} \frac{1}{2} f' + g'' = 0, \\ -\frac{1}{2} g' + f'' = 0, \end{cases} \quad \begin{cases} -\frac{1}{2} f - (N-1)g' = 0, \\ -\frac{1}{2} g' + (N-1)f' = 0. \end{cases}$$

From the second system, $f''' + \frac{1}{4} f' = 0$, so that (4.164) holds with $a = \frac{1}{2}$. The first system yields $A = -\frac{\lambda}{4}$ and $C^2 = B^2$, i.e., $C = \pm B$. Taking as usual $C = -B$ and substituting into the third one, one obtains

$$-\tfrac{1}{2} + \tfrac{N-1}{2} = 0 \quad \Longrightarrow \quad N = 2.$$

Similarly, for $C = B$ we get $-\frac{1}{2} - \frac{N-1}{2} = 0$, i.e., another hypothetical dimension $N = 0$ for which the radial Laplacian is (again appears in plasma physics)

$$\Delta_r = r D_r\left(\tfrac{1}{r} D_r\right) \equiv D_r^2 - \tfrac{1}{r} D_r.$$

Finally, the system (4.177) is consistent for $N = 2$ only with the solution

$$U = A = -\tfrac{\lambda}{4}, \quad f(s) = B\cos\left(\tfrac{s}{2}\right), \quad \text{and} \quad g(s) = -B\sin\left(\tfrac{s}{2}\right).$$

This yields a family of solutions of equation (4.176),

$$u(X,t) = \tfrac{1}{\sqrt{t}}\left[-\tfrac{\lambda}{4} + B\cos\left(x + \tfrac{1}{4t}|x'|^2 - \tfrac{\lambda}{2}\sqrt{t}\right)\right],$$

where B is arbitrary. Setting $B = -\frac{\lambda}{4}$ leads to the compacton

$$u_c(X,t) = -\tfrac{\lambda}{2\sqrt{t}} \cos^2\left[\tfrac{1}{2}\left(x + \tfrac{1}{4t}|x'|^2 - \tfrac{\lambda}{2}\sqrt{t}\right)\right] \ge 0 \quad \text{for } \lambda < 0, \tag{4.178}$$

where, as usual, we set $u_c = 0$ for arguments of $\cos^2$, $|\frac{1}{2}(\cdot)| \ge \frac{\pi}{2}$. The regularity of u_c is sufficient to satisfy the Cauchy problem for equation (4.176). For higher-order PDEs of the type (4.170), such C^1-compactons u_c need an FBP setting. The C^3 compactons as in (4.105) demand 5D subspaces for which the explicit integration of the corresponding PDE systems is unknown.

Example 4.37 (**Cubic PDE**) Consider a cubic PDE of the KP-type (cf. (4.155))

$$\left[u_t - (u(u^2)_{xx})_x - 4(u^3)_x\right]_x + u_{yy} = 0 \quad \text{in } I\!R^2 \times I\!R_+, \tag{4.179}$$

where the operator is chosen to admit $W_2 = \mathcal{L}\{\cos x, \sin x\}$. On W_2, i.e., for

$$u(x,y,t) = C_1(y,t)\cos x + C_2(y,t)\sin x,$$

equation (4.179) reduces to the PDE system in $I\!R \times I\!R_+$,

$$\begin{cases} C_{1t} - 2C_2\left(C_1^2 + C_2^2\right) - C_{2yy} = 0, \\ C_{2t} + 2C_1\left(C_1^2 + C_2^2\right) + C_{1yy} = 0. \end{cases} \tag{4.180}$$

In an analogous way, looking for similarity solutions of (4.180) in the form of

$$C_1 = \tfrac{1}{\sqrt{t}} f(s), \quad C_2 = \tfrac{1}{\sqrt{t}} g(s), \quad \text{where } s = \tfrac{1}{4t} y^2 + \lambda\tau \text{ and } \tau = \ln t,$$

yields the compacton solution for $\lambda < 0$, with interfaces as moving parabolas,

$$u_c(x, y, t) = \sqrt{-\tfrac{\lambda}{2t}}\, \cos\big(x + \tfrac{1}{4t}\, y^2 - \lambda \ln t\big),$$

where $u_c = 0$ in $\{|x + \frac{1}{4t}\, y^2 - \lambda \ln t| \ge \frac{\pi}{2}\}$. Comparing with the TWs shows that this Lipschitz continuous function can be a solution of the Cauchy problem.

4.6 "Tautological" equations and peakons

4.6.1 Trigonometric subspaces

We next consider another class of third and higher odd-order PDEs.

Example 4.38 (**Rosenau equation**) Consider the quasilinear degenerate *Rosenau equation* [491]

$$\boxed{u_t + u_{xxt} = 3uu_x + \left[uu_{xx} + \tfrac{1}{2}(u_x)^2\right]_x \equiv F_3[u] \quad \text{in } I\!R \times I\!R;} \tag{4.181}$$

see references, applications, and results in [449] and Remarks. We begin with some evolution aspects of the dynamics on the invariant subspace for a more general PDE with two parameters

$$\boxed{u_t + u_{xxt} = 3uu_x + \alpha uu_{xxx} + \beta u_x u_{xx} \equiv F[u] \quad \text{in } I\!R \times I\!R.} \tag{4.182}$$

In (4.181), $\alpha = 1$ and $\beta = 2$. This family was also introduced in [491].

Proposition 4.39 *Operator F in* (4.182) *admits $W_3 = \mathcal{L}\{1, \cos\gamma x, \sin\gamma x\}$ if*

$$\gamma^2 = \tfrac{3}{\alpha+\beta} > 0. \tag{4.183}$$

As usual, for the opposite sign in (4.183), the subspaces of exponential or hyperbolic functions are used. Describing the dynamics on W_3, and assuming that (4.183) holds, we obtain for (4.182) solutions

$$u(x,t) = C_1(t) + C_2(t)\cos\gamma x + C_3(t)\sin\gamma x,$$
$$\begin{cases} C_1' = 0, \\ (1-\gamma^2)C_2' = (3\gamma - \alpha\gamma^3)C_1C_3, \\ (1-\gamma^2)C_3' = -(3\gamma - \alpha\gamma^3)C_1C_2. \end{cases}$$

There exist three different cases:

Case 1: $\gamma^2 \ne 1$. Denoting

$$\rho = \tfrac{\gamma(\alpha\gamma^2-3)}{1-\gamma^2}$$

and integrating the DS yields

$$u(x,t) = A + B\cos(x - \rho A t) \quad (A, B \in I\!R).$$

Setting $A = B$ and $\rho A = \lambda$ determines the compacton solution

$$u_c(x,t) = \tfrac{2\lambda}{\rho}\, \cos^2\!\left[\tfrac{1}{2}(x - \lambda t)\right], \tag{4.184}$$

that is localized in the support $|x - \lambda t| \le \pi$ moving with constant speed. The analogy with the results in Example 4.8 suggests that these compactons are solutions of the Cauchy problem.

Case 2: $\gamma^2 = 1$. Hence,

$$\alpha + \beta = 3,$$

and if $\alpha \neq 3$ (this is true for (4.181)), then the DS takes a simpler form $C_1' = 0$, $C_1 C_3 = C_1 C_2 = 0$. Therefore, $C_1(t) \equiv 0$, so we can choose *arbitrary* functions $C_2(t)$ and $C_3(t)$ to obtain the following dynamics on such invariant subspaces.

Proposition 4.40 *For any $\alpha \in I\!R$, the equation*

$$u_t + u_{xxt} = 3uu_x + \alpha uu_{xxx} + (3-\alpha)u_x u_{xx} \equiv F_\alpha[u] \tag{4.185}$$

holds true on the subspace $W_2 = \mathcal{L}\{\cos x, \sin x\}$ that is invariant under F_α.

The result is obvious if (4.185) is written in the form

$$(u + u_{xx})_t = \alpha u(u + u_{xx})_x + (3-\alpha)u_x(u + u_{xx}).$$

Observe that the linear operator

$$L = I + \tfrac{\mathrm{d}^2}{\mathrm{d}x^2} \tag{4.186}$$

is the *annihilating operator* of the subspace W_2, i.e., $L : W_2 \to \{0\}$. We call such PDEs, whose right and left-hand sides are composed of annihilating operators of a given subspace, the *tautological equations* (on prescribed subspaces). As far as evolution properties are concerned, this implies that (4.185) admits an infinite-dimensional set of 2π-periodic solutions

$$u(x,t) = C_2(t)\cos x + C_3(t)\sin x, \tag{4.187}$$

where $C_{2,3}(t)$ are arbitrary C^1-smooth functions. So, given initial data

$$u_0(x) = a\cos x + b\sin x \in W_2,$$

the Cauchy problem, or the 2π-periodic initial-boundary value problem, for (4.185) has an infinite-dimensional set of smooth solutions (4.187), where $C_2(0) = a$ and $C_3(0) = b$. Therefore, the equation on W_2 (or in some neighborhood of W_2) is not a well-posed evolution PDE in the space of bounded periodic functions.

Let us next consider non-smooth solutions on W_2, which, possibly, make sense for such third-order PDEs. For instance, consider the following traveling wave solution:

$$u_1(x,t) = \lambda \sin|x - \lambda t|,$$

where $\lambda \in I\!R$ is the speed of propagation of the weak shock wave at $x = \lambda t$ being the point of discontinuity of the derivative u_x. Extending $u_1(x,t)$ from $\{|x - \lambda t| \leq \pi\}$ symmetrically yields a π-periodic structure. More general π-periodic solutions of this type have the form

$$u_N(x,t) = \textstyle\sum_{i=1}^N p_i(t)\sin|x - q_i(t)|,$$

where $\{p_i, q_i\}$ are smooth functions still remaining arbitrary.

According to PDE theory, evolution properties of such solutions (periodic multipeakons; see further interpretations below) should be checked in the π-periodic setting in the domain $(-\frac{\pi}{2}, \frac{\pi}{2}) \times I\!R$ with periodic boundary conditions by using the integral form of the PDE. Writing (4.181) as

$$L[u_t] = L[uu_x] + \big[u^2 - \tfrac{1}{2}(u_x)^2\big]_x$$

and applying L^{-1} with the kernel

$$\omega(s) = \tfrac{1}{2} \sin|s| \quad \text{in} \ \big(-\tfrac{\pi}{2}, \tfrac{\pi}{2}\big),$$

we formally obtain the integro-differential equation

$$u_t = uu_x + \big[\omega * \big(u^2 - \tfrac{1}{2}(u_x)^2\big)\big]_x. \tag{4.188}$$

Integration by parts, which is used for deriving (4.188), assumes certain "regularity" of solutions. This includes Rankine–Hugoniot, as well as some other "entropy-type", conditions at weak shocks. It would be interesting (an OPEN PROBLEM) to detect if such solutions can be obtained via a fourth-order parabolic regularization,

$$u_t + u_{xxt} = F[u] - \varepsilon u_{xxxx} \quad (\varepsilon > 0), \tag{4.189}$$

by passing to the limit $\varepsilon \to 0$. What if the sixth-order regularization $... + \varepsilon u_{xxxxxx}$ on the right-hand side of (4.189) applies?

Evolution properties of singularities of periodic multipeakons for (4.188) or (4.189) were less well studied in the literature. Another famous model with true multipeakons is considered in the next subsection.

Case 3: $\alpha = 3$. Then, if $\beta = 0$, the DS reduces to $C_1' = 0$, so the general solution of (4.182) is

$$u(x,t) = A + C_2(t)\cos x + C_3(t)\sin x,$$

with arbitrary functions $C_{2,3}(t)$ and any constant $A \in I\!R$.

Example 4.41 (**Tautological PDEs**) Using the annihilating operator (4.186), it is easy to construct many equations that are tautological on W_2. For instance, any fully nonlinear equations

$$\textstyle\sum_{(i)} a_i(u, u_x, ...) D_t^i L[u] = \sum_{(j)} b_j(u, u_x, ...) D_x^j L[u],$$

with arbitrary coefficients $\{a_i(\cdot)\}$ and $\{b_j(\cdot)\}$, are tautological. In particular, the following fourth-order PDE:

$$\boxed{u_t + u_{xxt} = 3uu_x + uu_{xxx} + 2u_x u_{xx} + (I + D_x^2)[|u|(u + u_{xx})]} \tag{4.190}$$

is tautological on W_2. In view of the analysis of higher-order degenerate TFEs (Example 3.10) and KdV equations (Example 4.8), correct evolution setting of compactly supported solutions is difficult to justify. For instance, in the integral representation for (4.190) via convolution with the kernel ω, we obtain the degenerate parabolic PDE (here it remains parabolic in $\{u < 0\}$)

$$u_t = |u|\, u_{xx} + |u|\, u + uu_x + \big[\omega * \big(u^2 - \tfrac{1}{2}(u_x)^2\big)\big]_x, \tag{4.191}$$

containing an extra non-local (compact) perturbation. Such parabolic PDEs naturally admit solutions that are non-smooth at the singularity level $\{u = 0\}$, though standard peakons are unlikely. Solutions of (4.191) may blow-up in finite time. Questions of existence and uniqueness are OPEN, as well as the regularization by adding $-\varepsilon u_{xxxx}$.

Example 4.42 Returning to the original models, recall that, for non-tautological PDEs with the annihilating operator (4.185), e.g.,

$$u_t + \varepsilon u_{xxt} = F_\alpha[u], \quad \text{with} \quad \varepsilon \neq 1,$$

looking for solutions (4.187) on W_2 yields the DS $C_1' = C_2' = 0$, so a stationary peakon solution

$$u(x,t) = A \sin |x|$$

can still be constructed. The fact that it is a proper weak solution should be checked by using the corresponding non-local integral equation or by parabolic ε-regularization. These are is OPEN PROBLEMS for such PDEs.

4.6.2 *Exponential subspaces: the FFCH equation*

Example 4.43 (The FFCH equation) We begin with another famous completely integrable shallow water model of a similar form,

$$\boxed{u_t - u_{xxt} = -3uu_x + 2u_x u_{xx} + uu_{xxx} \equiv F_3[u] \quad \text{in } I\!R \times I\!R.} \tag{4.192}$$

This is the *Fuchssteiner–Fokas–Camassa–Holm* (FFCH) *equation*, which arises as an asymptotic model that describes the wave dynamics at the free surface of fluids under gravity. It is derived from Euler equations for inviscid fluids under the long wave asymptotics of shallow water behavior (where the function u is the height of the water above a flat bottom); see Remarks.

We comment on some evolution aspects of such PDEs restricted to invariant subspaces of the operator F_3, and consider a three-parameter family of such equations

$$\boxed{u_t - u_{xxt} = \alpha uu_x + \beta u_x u_{xx} + \gamma uu_{xxx} \equiv F[u] \quad \text{in } I\!R \times I\!R.} \tag{4.193}$$

Proposition 4.44 *Equation* (4.193) *is tautological on* $W_2 = \mathcal{L}\{e^x, e^{-x}\}$ *if*

$$\alpha + \beta + \gamma = 0. \tag{4.194}$$

The annihilating operator of W_2 is now $L = I - \frac{d^2}{dx^2}$, and

$$F_3[u] = (\alpha u_x + \gamma u_{xxx})L[u] + (\alpha + \beta)u_{xx}(L[u])_x + (\alpha + \beta + \gamma)u_{xx}u_{xxx}.$$

Hence, any function

$$u(x,t) = C_1(t)e^x + C_2(t)e^{-x} \tag{4.195}$$

is a solution of (4.193) for arbitrary smooth coefficients $C_{1,2}(t)$, i.e., for initial data $u_0 \in W_2$, the Cauchy problem admits an infinite-dimensional set of solutions (4.195) satisfying $u(x,0) = u_0(x)$. These solutions are unbounded in x with exponential growth as $x \to \pm\infty$, so this non-uniqueness happens in the class of exponentially growing functions.

Another feature of equation (4.193), (4.194) is that the subspace W_2 makes it possible to construct, by gluing two opposite exponents, Lipschitz continuous solutions admitting the discontinuous derivatives u_x, and having exponential decay as $x \to \infty$. Such an elementary solitary wave, called *peakon* for the FFCH equation, is

$$u(x,t) = \lambda \mathrm{e}^{-|x-\lambda t|}, \tag{4.196}$$

where $\lambda \in I\!R$ is the traveling wave speed. The N-solitons, which are *multipeakons* discovered for (4.192) for any $N \geq 1$ in [94], adopt the form

$$u(x,t) = \textstyle\sum_{(1 \leq i \leq N)} p_i(t) \mathrm{e}^{-|x-q_i(t)|} \tag{4.197}$$

with $2N$ functions $\{q_i(t), p_i(t)\}$, which are the canonical coordinates $\{q_i\}$ and moments $\{p_i\}$, satisfying a DS to be presented. Function (4.197) has a cusp (a peak) at each $x = q_i(t)$, and $u(\cdot,t) \in W_2$ on any x-interval of C^1-regularity. As usual, in nonlinear PDE theory, dealing with odd-order equations admitting shock wave or weaker singularities, special *Rankine-Hugoniot*-type and *entropy* conditions (see mentioning *Oleinik's E-condition* for scalar conservation laws in Remarks) should be specified to detect proper unique (weak) solutions. This is achieved by writing the equation in the conservative integral form

$$u_t = -uu_x - \left[\omega * \left(u^2 + \tfrac{1}{2}(u_x)^2\right)\right]_x, \tag{4.198}$$

where $\omega(s) = \frac{1}{2}\mathrm{e}^{-|s|} > 0$ in $I\!R$ is the kernel of the linear operator $\left(I - \frac{\mathrm{d}^2}{\mathrm{d}x^2}\right)^{-1}$ in $L^2(I\!R)$. The integral representation (4.198) of the original third-order PDE correctly describes the propagation of weak shocks (peaks) and makes it possible to establish the global existence of a unique weak solution; see Remarks.

In particular, using the integro-differential equation (4.198) for the N-soliton solutions (4.197) yields that these are governed by the Hamiltonian ODEs

$$\begin{cases} \dot{q}_i = \frac{\partial H_A}{\partial p_i}, \\ \dot{p}_i = -\frac{\partial H_A}{\partial q_i}, \end{cases} \tag{4.199}$$

with the Hamiltonian

$$H_A(p_i, q_i) = \tfrac{1}{2} \textstyle\sum_{i,j=1}^{N} p_i p_j \mathrm{e}^{-|q_i - q_j|}. \tag{4.200}$$

This makes it possible to describe general evolution properties of such Lipschitz continuous multipeakons [94] which turn out to be generic in view of their orbital stability (see key references in the Remarks).

Similar results on the invariant subspace W_2 apply to the *generalized FFCH equation* [211]

$$\boxed{\begin{aligned} u_t - u_{xxt} = \alpha u_{xxx} - \beta(3uu_x - 2u_x u_{xx} - uu_{xxx}) \\ + \gamma\left[(u - u_{xx})(u^2 - (u_x)^2)\right]_x. \end{aligned}}$$

The set of tautological PDEs on W_2 is wide. Fixing an arbitrary linear differential operator M with constant coefficients and a nonlinear operator $A(u, u_x, ...)$, consider

$$u_t - u_{xxt} = M[u] - \beta(3uu_x - 2u_x u_{xx} - uu_{xxx}) + \gamma\{L[u]A(u, u_x, ...)\}_x.$$

For higher-order PDEs, a suitable definition of proper weak solutions with weak shocks (constructed either via suitable integral equations inheriting correct Rankine–

Hugoniot and entropy conditions, or by regular parabolic approximations) becomes much more delicate and generates many mathematical OPEN PROBLEMS.

4.6.3 Other related models

Example 4.45 (The PBBM equation) Consider the famous *Peregrine–Benjamin–Bona–Mahoney* (PBBM) *equation* [461, 38]

$$u_t - u_{xxt} = uu_x \quad \text{in } I\!R \times I\!R, \tag{4.201}$$

also known as the *regularized long-wave equation*. This is derived from the KdV equation written in the form

$$u_t + u_{xxx} = uu_x$$

by formally justifying that $u_t + u_x \approx 0$, and hence, replacing u_{xxx} by $-u_{xxt}$. Equation (4.201) also admits the integral representation

$$u_t = \tfrac{1}{2}\big(\omega * u^2\big)_x \quad \big(\omega(s) = \tfrac{1}{2}\,\mathrm{e}^{-|s|}\big), \tag{4.202}$$

where the derivation imposes the continuity condition of $u(x,t)$. Since $\omega'(s)$ is discontinuous at $s = 0$, (4.202) may admit solutions that are only Lipschitz continuous in x, similar to peakons. The *regularized PBBM equation*

$$u_t - u_{xxt} = uu_x + \varepsilon(I - D_x^2)u_{xx} \quad (\varepsilon > 0), \tag{4.203}$$

reduces to a uniformly parabolic equation with non-local perturbation,

$$u_t = \varepsilon u_{xx} + \tfrac{1}{2}\big(\omega * u^2\big)_x, \tag{4.204}$$

and admits smooth classical solutions. It would be interesting to describe the passage to the limit $\varepsilon \to 0$ in both PDEs (4.203) and (4.204); these are OPEN PROBLEMS.

Example 4.46 The non-tautological PDE on W_2,

$$u_t - \varepsilon u_{xxt} = F_3[u], \quad \text{with} \quad \varepsilon \neq 1,$$

with the general operator (4.193), (4.194) still admits a stationary 1-peakon

$$u(x,t) = A\,\mathrm{e}^{-|x|},$$

with unknown evolution properties; an OPEN PROBLEM.

Example 4.47 (Higher-order tautological PDEs) Consider, for instance,

$$u_t - D_x^6 u_t = F[u] \equiv \textstyle\sum_{(i,j)} a_{i,j}\, D_x^i u\, D_x^j u, \tag{4.205}$$

with the quadratic operator and polynomial $P(X,Y)$ defined as in (1.112), e.g.,

$$F[u] = \alpha u D_x^7 u + \beta u_x D_x^6 u + \gamma\, u_{xx} D_x^5 u + \dots .$$

The subspace $W_2 = \mathcal{L}\{\mathrm{e}^x, \mathrm{e}^{-x}\}$ is tautological for the operator in (4.205) iff

$$P(1,1) = P(-1,-1) = P(1,-1) = 0.$$

Therefore, one can define the peakon (4.196) and corresponding multipeakon solutions. Determining the kernel of $(I - D_x^6)^{-1}$ by the inverse Fourier transform,

$$\omega(s) = \mathcal{F}^{-1}\big(\tfrac{1}{1+\xi^6}\big) \equiv \tfrac{1}{\pi} \int_0^\infty \tfrac{\cos \xi s\, \mathrm{d}\xi}{1+\xi^6},$$

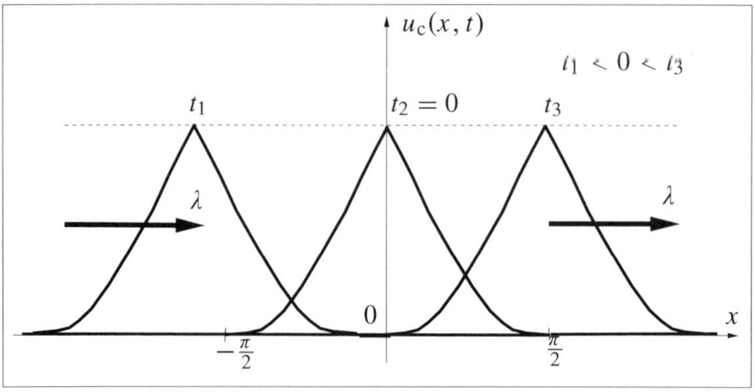

Figure 4.20 Compacton-peakon solution (4.206).

which is an exponentially decaying oscillatory function, we write down (4.205) for solutions $u(\cdot, t) \in H^6(I\!R)$ in the integral form

$$u_t = \omega * F[u] \quad \text{for } t > 0.$$

The evolution consistency of peakon solutions will depend on special divergent properties of the operator F and leads to many difficult OPEN PROBLEMS.

Example 4.48 (Compacton-peakon) We return to the PDE (4.182) and perform the shifting in x by $\pm\frac{\pi}{2}$ of the increasing and decreasing branches of the compacton solution(4.184) to create a formal *compacton-peakon* solution

$$u_c(x,t) = \frac{2\lambda}{\rho} \cos^2 \begin{cases} \left[\frac{1}{2}(x-\lambda t) + \frac{\pi}{4}\right] & \text{for } \lambda t \le x \le \lambda t + \frac{\pi}{2}, \\ \left[\frac{1}{2}(x-\lambda t) - \frac{\pi}{4}\right] & \text{for } \lambda t - \frac{\pi}{2} \le x < \lambda t, \end{cases} \tag{4.206}$$

which is shown in Figure 4.20. The evolution consistency is OPEN. It would be interesting to create other, possibly higher-order, models with guaranteed patterns like that in the CP, or FBP settings.

Example 4.49 (Nonlinear dispersion, finite propagation, and oscillatory solutions) We now check a possible character of finite interfaces for the fifth-order PDE. As a formal extension of our previous study of solutions of changing sign, consider the following nonlinear dispersion PDE:

$$u_t - u_{xxt} = \left(|u|^n u\right)_{xxxxxxx} \quad (n > 0). \tag{4.207}$$

Similar to the compacton analysis, we describe finite interfaces in the CP by using the TWs $u(x,t) = f(x - \lambda t)$, where f satisfies

$$-\lambda f' + \lambda f''' = \left(|f|^n f\right)^{(7)}.$$

Keeping the leading terms near the interface at $y = 0$, and integrating three times, yields the fourth-order ODE

$$\left(|f|^n f\right)^{(4)} = \lambda f \quad \text{for } y > 0, \quad f(0) = 0.$$

The oscillatory behavior close to the interface for $\lambda < 0$ was studied before (cf.

(4.91)) with the oscillatory component satisfying (4.93). Since the exponent there was $\gamma = \frac{4}{n} \to +\infty$ as $n \to 0^+$, such $f(y)$ can be arbitrarily smooth at the interface at $y = 0$ for small $n > 0$. We expect that such behavior close to interfaces plays a role for more general solutions of the PDE (4.207).

Example 4.50 (**Hunter–Saxton equation**) Neglecting the first term on the right-hand side of (4.192) gives

$$\boxed{u_t - u_{xxt} = 2u_x u_{xx} + u u_{xxx} \equiv F_3[u] \quad \text{in } I\!R \times I\!R.} \tag{4.208}$$

This is a modification of the *Hunter–Saxton equation*

$$u_{xxt} + 2u_x u_{xx} + u u_{xxx} = 0, \tag{4.209}$$

which was proposed as a model for the asymptotic behavior of neumatic fluids crystals [294] and belongs to the *Dym hierarchy* of integrable PDEs, [358]. Equation (4.209) is tautological on the subspace of linear functions

$$W_2 = \mathcal{L}\{1, x\}.$$

Instead of (4.197), the N-soliton solution takes the form

$$u(x,t) = \textstyle\sum_{(1 \le i \le N)} p_i(t)|x - q_i(t)|, \quad \text{with} \quad \textstyle\sum_{(1 \le i \le N)} p_i(t) = 0, \tag{4.210}$$

where the constraint is imposed to guarantee that the function (4.210) is uniformly bounded. The DS is (4.199) with the Hamiltonian

$$H_A(p_i, q_i) = \tfrac{1}{2} \textstyle\sum_{(i,j)} p_i p_j |q_i - q_j|.$$

A unique solution is obtained by using an integro-differential equation imposing a correct propagation mechanism of weak singularities; see Remarks.

Concerning more general invariant settings, let us point out that the quadratic operator F_3 in (4.208) is now associated with the subspace $W_4 = \mathcal{L}\{1, x, x^2, x^3\}$ of cubic polynomials. This gives the exact solutions

$$u(x,t) = C_1(t) + C_2(t)x + C_3(t)x^2 + C_4(t)x^3,$$

$$\begin{cases} C_1' = 4C_2C_3 + 6C_1C_4 + 84C_3C_4, \\ C_2' = 18C_2C_4 + 8C_3^2 + 252C_4^2, \\ C_3' = 42C_3C_4, \\ C_4' = 42C_4^2. \end{cases}$$

It follows from the last ODE that $C_4(t) = \frac{1}{42}(T - t)^{-1}$ blows up. This determines the same blow-up rate as $t \to T^-$ of all the other coefficients. The multipeakons on W_4 can be taken, e.g., in the form of

$$u(x,t) = \textstyle\sum_{(1 \le i \le N)} \big(p_i|x - q_i|^3 + r_i|x - q_i|^2 + s_i|x - q_i|\big),$$

with necessary constraints for functions $\{p_i, q_i, r_i, s_i\}$ that guarantee the uniform boundedness of these solutions. Existence of multipeakons as proper solutions of the Cauchy problem for (4.208) is an OPEN PROBLEM.

Example 4.51 (**Schwarzian KdV equation**) Some special annihilating properties are exhibited by the operator of the (2+1)-dimensional integrable generalization of the *Schwarzian KdV* (SKdV) *equation* [557]

$$u_t = F_*[u] \equiv -\tfrac{1}{4}\, u_{xxy} + \tfrac{u_x u_{xy}}{2u} + \tfrac{u_{xx} u_y}{4u} - \tfrac{(u_x)^2 u_y}{2u^2} + \tfrac{u_x}{8}\, D_x^{-1}\big[\tfrac{(u_x)^2}{u^2}\big]_y, \tag{4.211}$$

where $D_x^{-1} f$ denotes $\int f\,\mathrm{d}x$. F_* annihilates the 1D subspace $\mathcal{L}\{f(x)\}$ for *arbitrary* C^2-functions f, i.e., $F_*[Cf] = 0$ for any constant C. This gives a wide family of separable stationary solutions $u(x,y,t) = g(y)f(x)$ for any C^1-function g, including localized compactons, see [484]; justification is OPEN.

Example 4.52 **(Green–Naghdi equations)** Continuing a review of water wave models with invariant subspaces, consider the *Green–Naghdi* (GN) *equations*

$$\boxed{\begin{cases} \eta_t + (u\eta)_x = 0, \\ u_t + uu_x + g\eta_x = \frac{1}{3\eta}\left[\eta^2((\eta u_x)_t + u(\eta u_x)_x)\right]_x, \end{cases}} \tag{4.212}$$

which determine an approximate system of the full water problem, modeling surface wave propagation on an inviscid and incompressible gravity flow, [262]. Here u is the mean horizontal velocity, η is the surface disturbance, and g is the gravity acceleration. Invariant solutions of (4.212) are discussed in [21]. The last operator in (4.212) is quartic, so higher-dimensional subspaces are unlikely. It is curious that nontrivial dynamics for the GN equations already occur on the elementary subspace of linear functions $W_2 = \mathcal{L}\{1, x\}$, where the solutions are

$$\eta(x,t) = C_1(t) + C_2(t)x, \quad u(x,t) = D_1(t) + D_2(t)x,$$

$$\begin{cases} C_1' = -C_1 D_2 - C_2 D_1, \\ C_2' = -2C_2 D_2, \\ D_1' = -D_1 D_2 - gC_2 - \frac{2}{1-C_2^2}\, C_1 C_2 D_2^2, \\ D_2' = -\frac{1+C_2^2}{1-C_2^2}\, D_2^2. \end{cases}$$

On W_2, peakon solutions,

$$\eta(x,t) = p(t)|x - q(t)| \quad \text{and} \quad u(x,t) = r(t)|x - s(t)|,$$

can be formally defined with unclear evolution properties; an OPEN PROBLEM.

Example 4.53 **(The CDDD equation)** Longitudinal strain waves in isotropic cylindrical compressible elastic rods embedded in a viscoelastic medium can be described by higher-order models containing nonlinear dispersive operators. Consider a generalization of the *combined dissipative double-dispersive* (CDDD) *equation* (see, e.g., [469])

$$u_{tt} = \alpha u_{xxxx} + \beta u_{xxtt} + \gamma\,(u^2)_{xxxxt} + \delta(u^2)_{xxt} + \varepsilon(u^2)_t. \tag{4.213}$$

The quadratic operator $F[u]$ in this PDE preserves $W_3 = \mathcal{L}\{1, \cos x, \sin x\}$ of 2π-periodic solutions iff

$$16\gamma - 4\delta + \varepsilon = 0,$$

and (4.213) on W_3 is then a sixth-order DS possessing blow-up solutions.

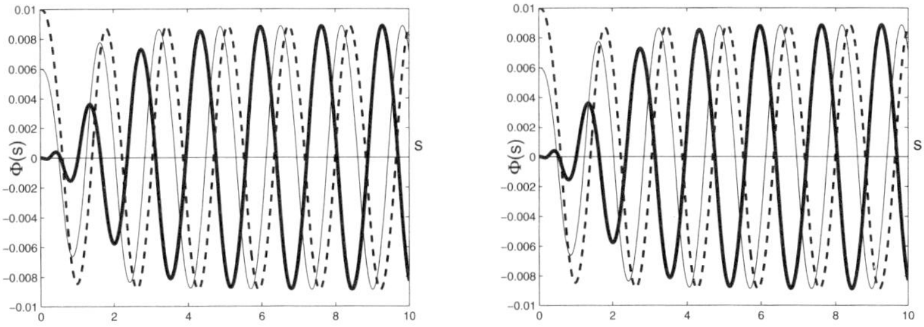

(a) $n = 1$ (b) $n = 10$

Figure 4.21 Stability of periodic solutions of (4.215) with $\lambda = -1$.

Example 4.54 (**On a model with finite interfaces**) The phenomenon of finite propagation in such models demands a proper monotone extension of nonlinearity u^2 for negative $u < 0$, as in the following *signed* PDE (here, $\beta = \delta = \varepsilon = 0$ and $\gamma = -1$ for well-posedness):

$$u_{tt} = -\big(|u|^n u\big)_{xxxxt} \quad (n > 0).$$

TW solutions $u(x, t) = f(x - \lambda t)$ give the ODE

$$\lambda^2 f'' = \lambda(|f|^n f)^{(5)}.$$

Integrating twice yields the third-order problem

$$\big(|f|^n f\big)''' - \lambda f = 0 \quad \text{for } y > 0, \quad f(0) = 0,$$

which appeared for TFEs; see Section 3.7. Namely, the behavior near interface at $y = 0$ needs introduction of the oscillatory component φ by

$$f(y) = y^{\frac{3}{n}} \varphi(s), \quad \text{where} \quad s = \ln y, \tag{4.214}$$

where $\Phi = |\varphi|^n \varphi$ solves the third-order ODE

$$\begin{aligned} &\Phi''' + 3(\mu - 1)\Phi'' + (3\mu^2 - 6\mu + 2)\Phi' \\ &+ \mu(\mu - 1)(\mu - 2)\Phi - \lambda \left|\Phi\right|^{-\frac{n}{n+1}} \Phi = 0, \quad \text{with} \quad \mu = \tfrac{3(n+1)}{n}. \end{aligned} \tag{4.215}$$

Transformation (4.214) shows that the regularity at the interface $y = 0$ improves as $n \to 0^+$. The ODE for $\Phi(s)$ with $\lambda < 0$ is similar to that given in (3.166) with known oscillatory properties (recall the difference between $\gamma = \frac{3}{n}$ therein, and μ above). Therefore, solutions of (4.215), with $\lambda = -1$, are oscillatory for all $n \in (0, \infty)$; see Figure 4.21.

For $\lambda > 0$, numerical unstable periodic solutions of (4.215) were not detected. To confirm this negative conclusion, using the idea of continuity (homotopy) at $n = 0^+$, we take the linear equation

$$u_{tt} = -u_{xxxxt},$$

which is the well-posed bi-harmonic PDE in terms of u_t (see (3.2)). For TWs, the ODE is

$$f''' - \lambda f = 0 \quad \Longrightarrow \quad f(y) = e^{\mu y}, \ \text{where} \ \mu^3 = \lambda.$$

As above, the solutions are oscillatory at the left-hand interface $y = -\infty$ for $\lambda = -1$

(roots are $\mu_\pm = \frac{1}{2} \pm \mathrm{i}\,\frac{\sqrt{3}}{2}$) and are not for $\lambda = 1$ (the only root is $\mu_0 = 1$). The fundamental solution is always oscillatory, since $\lambda < 0$ for it by symmetry.

4.6.4 *Quasilinear third-order remarkable operator*

Recall the definition of the remarkable second-order operator

$$F_{\rm rem}[u] = uu_{xx} - (u_x)^2,$$

which exhibits special invariant properties and has been used in the study of the RPJ equation in Example 1.36. In order to obtain a similar third-order remarkable operator, we take the derivative of $F_{\rm rem}$,

$$F[u] = \tfrac{\mathrm{d}}{\mathrm{d}x}\left[uu_{xx} - (u_x)^2\right] = uu_{xxx} - u_x u_{xx}\,,$$

which occurred in several models studied before.

Example 4.55 The quasilinear third-order PDE

$$\boxed{u_t = uu_{xxx} - u_x u_{xx} + \mu u + \nu}$$

possesses solutions

$$u(x,t) = C_1(t) + C_2(t)x + C_3(t)\cos(\gamma(t)x) + C_4(t)\sin(\gamma(t)x), \tag{4.216}$$

$$\begin{cases} C_1' = \mu C_1 + \nu, \;\; C_2' = \mu C_2, \\ C_3' = -\gamma^3 C_1 C_4 + \gamma^2 C_2 C_3 + \mu C_3, \\ C_4' = \gamma^3 C_1 C_3 + \gamma^2 C_2 C_4 + \mu C_4, \\ \gamma' = -\gamma^3 C_2. \end{cases}$$

4.7 Subspaces, singularities, and oscillatory solutions of Harry Dym-type equations

4.7.1 *PDEs and invariant subspaces*

Consider the following third-order PDE:

$$v_t = v^n v_{xxx}\,, \tag{4.217}$$

where $n \in I\!R$ is a parameter. This equation is posed for nonnegative solutions and, as usual, v^n will be replaced by $|v|^n$ for oscillatory solutions of changing sign. For $n = 3$, (4.217) gives the *Harry Dym* (HD) *equation*

$$v_t = v^3 v_{xxx}\,,$$

which is one of the most exotic soliton equations. It is associated with the classical *string problem* and is linearizable by the inverse spectral transform method; see details and references in [128]. Setting $v = -\frac{1}{\sqrt{1+q}}$ yields another form of the Harry Dym equation

$$q_t = \left(\tfrac{2}{\sqrt{1+q}}\right)_{xxx}\,,$$

which, on the complex plane, is relevant to several physical problems, such as the *Hele–Shaw problem*, the *Saffmane–Taylor problem* of the motion of the interface of two fluids with different viscosities, and the chiral dynamics of closed curves on the plane.

Let us find those PDEs (4.217) that admit exact solutions on polynomial subspaces. The case $n = 1$ is excluded, where the quadratic operator

$$F[v] = vv_{xxx}$$

preserves the obvious 4D subspace

$$W_4 = \mathcal{L}\{1, x, x^2, x^3\}.$$

This is suitable for PDEs, such as

$$\boxed{v_t = vv_{xxx} + \beta v_x v_{xx} + \mu v + \nu,}$$

which, on W_4, reduces to a 4D DS that can describe, e.g., finite-time extinction and interface propagation in various FBPs.

Looking for other cases, we introduce the "pressure" for the PDE (4.217),

$$v = u^\mu, \quad \text{with exponent } \mu = \tfrac{2}{n}.$$

This yields the following equation with a homogeneous cubic operator:

$$u_t = F[u] \equiv u^2 u_{xxx} + 3(\mu - 1)uu_x u_{xx} + (\mu - 1)(\mu - 2)(u_x)^3. \tag{4.218}$$

The basic subspace for F is trivial, $W_2 = \mathcal{L}\{1, x\}$. Substituting $u = x^2$ yields that F in (4.218) preserves the extended 3D subspace

$$W_3 = \mathcal{L}\{1, x, x^2\}, \quad \text{if } \ 12(\mu - 1) + 8(\mu - 1)(\mu - 2) = 0, \tag{4.219}$$

i.e., for $\mu = 1$ ($n = 2$, the trivial case: $u = v$ and $v_{xxx} = 0$ on W_3) and for $\mu = \frac{1}{2}$ ($n = 4$) that gives some applications and extensions.

Example 4.56 (Extinction and interfaces) Consider the Harry Dym-type equation with absorption

$$v_t = v^4 v_{xxx} - \tfrac{1}{v} \quad (v \geq 0). \tag{4.220}$$

The absorption term is unbounded and *singular* at $v \to 0^+$, so that the first question of PDE theory is to check if (4.220) can admit any *nontrivial* compactly supported solution, i.e., a solution $v(x, t) \not\equiv 0$. This is not an easy question, even for the second-order *PME with absorption*

$$v_t = (v^n)_{xx} - \tfrac{1}{v^p}, \quad \text{with } n > 0, \tag{4.221}$$

though the criterion for the existence is known: $p < n$. It is proved that, for $p \geq n$, any FBP with $v = 0$ on the interface has the trivial *unique proper* solution $v(x, t) = \lim v_\varepsilon(x, t) \equiv 0$ (i.e., v is the limit of a family $\{v_\varepsilon\}$ of smooth global solutions of the regularized non-singular equations), regardless of any nontrivial initial or regular boundary data. In particular, the *heat equation with absorption*

$$v_t = v_{xx} - \tfrac{1}{v}$$

belongs to the nonexistence range, so, for compactly supported initial data $v_0 \geq 0$, the unique *maximal* solution is trivial, $v(x, t) \equiv 0$. Hence, the same is true for all other non-maximal solutions of any FBPs. For parabolic PDEs, such as (4.221), the existence nonexistence criterion is proved by the Sturmian intersection comparison with TWs, [226, Ch. 7-9]. For higher-order equations, these are OPEN PROBLEMS.

Let us show that, regardless of such a singular absorption term $-\frac{1}{v}$, equation (4.220) admits nontrivial exact solutions and we will pose the corresponding FBP. Setting $v = \sqrt{u}$ $(\mu = \frac{1}{2})$ yields the PDE

$$\boxed{u_t = u^2 u_{xxx} - \tfrac{3}{2}\, u u_x u_{xx} + \tfrac{3}{4}(u_x)^3 - 2} \tag{4.222}$$

that possesses solutions $u \in W_3$,

$$u(x,t) = v^2(x,t) = C_1(t) + C_2(t)x + C_3(t)x^2, \tag{4.223}$$

$$\begin{cases} C_1' = -3C_1C_2C_3 + \frac{3}{4}C_2^3 - 2, \\ C_2' = -6C_1C_3^2 + \frac{3}{2}C_2^2C_3, \\ C_3' = 0. \end{cases}$$

Setting $C_3(t) \equiv 1$ gives the opportunity to study the *quenching* phenomenon, where the classical analytic strictly positive solution vanishes as $t \to T^-$ at some $x_0 \in I\!R$. Vice versa, $C_3(t) \equiv -1$ describes the single point extinction, when the solution vanishes identically as $t \to T^-$.

We consider the extinction phenomenon with $C_3 = -1$ using typical asymptotic arguments from thin film analysis (Section 3.2 and 3.10). We need an FBP setting where the quenching description is similar. We do not solve the DS in (4.223) explicitly and take orbits such that $C_1(t) = 2(T-t) + ...$ and $C_2(t) = b(T-t) + ...$ $(b \in I\!R)$ as $t \to T^-$. This gives the pattern

$$u(x,t) = \left[2(T-t) + b(T-t)x - x^2\right]_+ + ... \equiv \mathrm{e}^{-\tau}\left(2 + b\mathrm{e}^{-\frac{\tau}{2}}\xi - \xi^2\right)_+ + ...,$$

where $\xi = x/\sqrt{T-t}$ and $\tau = -\ln(T-t)$ are the standard blow-up rescaled variables, as in (3.39). In view of (4.223), this shows that, for the original equation (4.220), the rescaled solution satisfies

$$v(x,t) = \sqrt{T-t}\, w(\xi,\tau) \to g(\xi) = \sqrt{(2-\xi^2)_+} \quad \text{as } \tau \to \infty, \tag{4.224}$$

where w solves a singular perturbed PDE of the form

$$w_\tau = -\tfrac{1}{2}\, w_\xi \xi + \tfrac{1}{2}\, w - \tfrac{1}{w} + \mathrm{e}^{-\frac{3}{2}\tau} w^4 w_{\xi\xi\xi} \quad \text{for } \tau \gg 1. \tag{4.225}$$

Then $g(\xi)$ is a stationary solution of the *limit equation*

$$-\tfrac{1}{2}\, g'\xi + \tfrac{1}{2}\, g - \tfrac{1}{g} = 0,$$

which is (4.225) at $\tau = +\infty$. A rigorous passage to the limit $\tau \to +\infty$ in (4.225) to prove convergence (4.224) for a class of the FBP solutions with extinction at $t = T$ is a difficult OPEN PROBLEM. Notice that (4.220) admits similarity solutions (existence or nonexistence is OPEN)

$$v_s(x,t) = \sqrt{T-t}\, g(z), \;\; z = \tfrac{x}{T-t} \quad \Longrightarrow \quad g^4 g''' - g'z + \tfrac{1}{2}\, z - \tfrac{1}{g} = 0.$$

As usual, using the positive part $(\cdot)_+$ in the solutions means an FBP. The dynamic interface equation is obtained from (4.222) for any smooth solutions,

$$s' = -\tfrac{1}{u_x}\, u_t = S[u] \equiv -\tfrac{3}{4}(u_x)^2 + \tfrac{2}{u_x} \quad \text{at} \;\; x = s(t).$$

The governing interface equation for (4.223) is stationary, $u_{xx} = -2$ at $x = s(t)$.

New HD-type models appear by adding other operators to the right-hand side, preserving the invariant subspace. For instance, consider

$$v_t = v^4 v_{xxx} + (v^3)_{xx} - \tfrac{1}{v},$$

where setting $v = \sqrt{u}$ yields the PDE with an extra quadratic operator,

$$\boxed{u_t = u^2 u_{xxx} - \tfrac{3}{2}\, u u_x u_{xx} + \tfrac{3}{4}\,(u_x)^3 + 3\big[u u_{xx} + \tfrac{1}{2}(u_x)^2\big] - 2,}$$

that preserves W_3 in (4.219). This has similar exact solutions and interfaces, and exhibits typical singularity phenomena of extinction, quenching, and finite interfaces.

4.7.2 On the maximal regularity, oscillatory behavior, and the Cauchy problem

We use the TWs to check the maximal regularity for PDE (4.220), written now in the *signed* form for solutions of changing sign,

$$v_t = |v|^4 v_{xxx} - \tfrac{v}{|v|^2}. \tag{4.226}$$

Substituting the TW solution $v(x,t) = f(x - \lambda t)$ yields

$$-\lambda f' = |f|^4 f''' - \tfrac{f}{|f|^2} \quad \text{for} \;\; y > 0, \quad f(0) = 0,$$

so that

$$f(y) = \sqrt{y}\, \varphi(s), \quad \text{where} \quad s = \ln y, \tag{4.227}$$

with the following ODE for the oscillatory component φ:

$$\varphi''' - \tfrac{3}{2}\,\varphi'' + \big(\tfrac{\lambda}{|\varphi|^4} - \tfrac{1}{4}\big)\varphi' + \big(\tfrac{3}{8}\,\varphi + \tfrac{\lambda}{2|\varphi|^4} - \tfrac{1}{|\varphi|^6}\big)\varphi = 0. \tag{4.228}$$

For any $\lambda \in I\!R$, there exist constant solutions $\pm\varphi_0$, where

$$3\varphi_0^6 + 4\lambda\varphi_0^2 - 8 = 0.$$

It is not difficult to show that these constant equilibria are unstable for both $\lambda = \pm 1$. Numerically, solutions of (4.228) are non-oscillatory and of constant sign (this is associated with the strong singularity as $\varphi \to 0$ of the coefficients in (4.228)).

The TW solutions (4.227) have the Hölder continuity exponent $\frac{1}{2}$ attained at the interface $y = 0$ which coincides with that given by the square root $v = \sqrt{u}$ in (4.223). These solutions are expected to exhibit the maximal regularity for PDE (4.226), so (4.223) is assumed to correspond to the Cauchy problem. The non-changing sign properties of $\varphi(s)$ imply a possibility of constructing more general nonnegative solutions of the Cauchy problem for PDE (4.220).

Let us briefly discuss a curious interface equation for the critical case $n = 2$,

$$v_t = v^2 v_{xxx}.$$

We again use the TWs $v(x,t) = f(x - \lambda t)$ for deriving the interface equation, so

$$-\lambda f' = f^2 f''',$$

and, on integration,

$$\tfrac{\lambda}{f} = f''.$$

Hence, for $\lambda < 0$, the behavior close to the interface at $y = 0$ is not oscillatory,

$$f(y) = y\sqrt{2|\lambda||\ln y|} + ... \quad \text{as } \ y \to 0^+.$$

This asymptotic behavior can be used for deriving the dynamic interface equation, as in Example 4.29. The formal analysis is easier, since the speed, λ, enters the first expansion term; the proof is difficult and is an OPEN PROBLEM.

4.7.3 On the fundamental solution

As usual, the existence of nonnegative solutions does not mean positivity preserving property in the Cauchy problem. For FBPs, the positivity can be enforced by free-boundary conditions, as for the TFEs in Section 3.2.

Consider now the Cauchy problem for the *signed HD-type* equation (4.217) for $n \in (0, 1)$,

$$v_t = |v|^n v_{xxx} \quad \text{in } I\!R \times I\!R_+. \tag{4.229}$$

We study finite propagation and oscillatory properties of its fundamental solution

$$b(x,t) = t^\alpha g(\xi), \ \xi = \tfrac{x}{t^\beta}, \ \text{with } \alpha = -\tfrac{1}{3-2n} < 0, \ \beta = \tfrac{1-n}{3-2n} > 0, \tag{4.230}$$

where g solves the following ODE problem:

$$\mathbf{A}[g] \equiv |g|^n g''' + \beta g' \xi - \alpha g = 0 \ \text{in } \ I\!R, \quad \tfrac{1}{1-n} \int_{-\infty}^{\infty} |g|^{-n} g \, d\xi = 1. \tag{4.231}$$

Solutions (4.230) are induced by the conservation law

$$\tfrac{d}{dt} \int_{-\infty}^{\infty} |v|^{-n} v \, d\xi = 0, \tag{4.232}$$

which follows from the PDE (4.229), provided that the integral converges. For $n = 0$, (4.230) is the fundamental solution (4.138) of the operator in the linear PDE

$$v_t = v_{xxx} \quad \text{in } I\!R \times I\!R_+. \tag{4.233}$$

The normalization in (4.231) then becomes the standard condition of convergence to the δ-function as $t \to 0^+$,

$$\int_{-\infty}^{\infty} g \, d\xi = 1.$$

By (4.139) (recall the reflection $x \mapsto -x$), the kernel $g(\xi)$ is oscillatory as $\xi \to +\infty$ and is not as $\xi \to -\infty$, with different asymptotics at these infinite interfaces.

We are going to detect similar oscillatory properties in the nonlinear case $n \in (0, \frac{3}{2})$. The ODE (4.231) is integrated once by the conservation law, so

$$g'' + \tfrac{1}{3-2n} |g|^{-n} g \xi = 0. \tag{4.234}$$

Similar to solutions (4.74) of the signed $K(2,2)$ equation in Section 4.2.6, the fun-

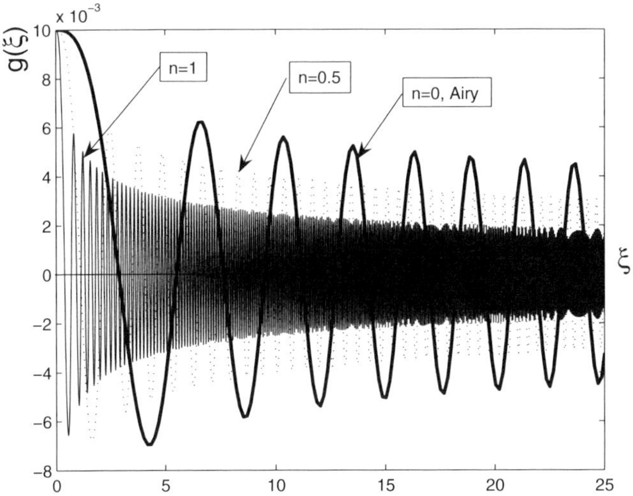

Figure 4.22 Oscillations for $\xi \gg 1$ of solutions of (4.234) with $n = 0, \frac{1}{2}$, and 1.

damental kernel $g(\xi)$ is compactly supported on $[\xi_0, \infty)$, $\xi_0 < 0$, with positive expansion at the left-end interface,

$$g(\xi) = \left[\frac{n^2|\xi_0|}{2(2-n)(3-2n)}\right]^{\frac{1}{n}} (\xi - \xi_0)^{\frac{2}{n}} + \ldots \quad \text{as} \quad \xi \to \xi_0^+.$$

The behavior as $\xi \to +\infty$ is oscillatory and can be compared with that for the Airy function in (4.139); see Figure 4.22.

4.7.4 *Oscillatory solutions of higher-order Harry Dym-type equations*

Similar phenomena take place for higher-order Harry Dym-type equations, such as

$$u_t = |u|^n u_{xxxxx}.$$

For instance, introducing the oscillatory component into the TW-equation yields

$$-\lambda f' = |f|^n f^{(5)}, \quad f(y) = y^\gamma \varphi(\ln y) \implies P_5[\varphi] = -\lambda \frac{\varphi' + \gamma\varphi}{|\varphi|^n}, \tag{4.235}$$

where $\gamma = \frac{4}{n}$ and the linear operator P_5 is as defined in Section 3.7, (3.157). There exists a stable periodic solution $\varphi(s)$ for $\lambda > 0$, at least for $n \in (0, 1]$; see Figure 4.23(a). We observed numerically the periodic behavior for all $n \in (0, n_{\rm h})$, where

$$\boxed{n_{\rm h} = 1.08...}$$

is a homoclinic bifurcation of periodic solutions, as shown in Figure 4.24. For $\lambda < 0$, a standard 1D shooting does not detect an unstable periodic solution. Figure 4.23(b) shows a non-monotone separatrix behavior between flows of stabilization to two constant equilibria $\pm\varphi_0$ (this needs extra study).

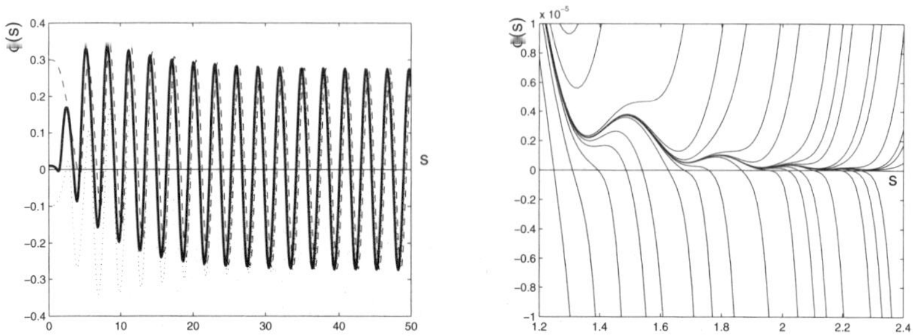

(a) $n = 1, \lambda = 1$ (b) $n = 0.9, \lambda = -1$

Figure 4.23 Solutions of (4.235) for $n = 1$, $\lambda = 1$ (a) and $n = 0.9$, $\lambda = -1$ (b).

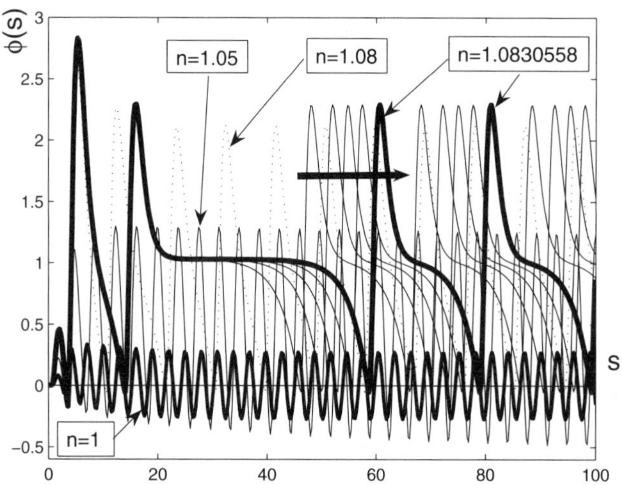

Figure 4.24 Deformation as $n \to n_{\mathrm{h}}^{-}$ of periodic solutions of the ODE in (4.235) with $\lambda = 1$. Cauchy data are $\varphi(0) = 10^{-3}$ and $\varphi'(0) = \varphi^{(4)}(0) = 0$.

Remarks and comments on the literature

There are many excellent books on fundamental mathematical techniques and discoveries related to the KdV equation and other integrable PDEs; see e.g., Newell [436], Remoissenet [487], and Ablowitz–Clarkson [2] for further references. We present below many classic and recent references, including surveys on these subjects. Concerning other questions, which are not directly related to integrability issues, such as existence, uniqueness, and regularity theory for semilinear odd-order PDEs, these have been developed for at least forty years. We refer to Lions [396, Ch. 3] for first results and key papers, and to Faminskii [179] for further references. Semigroup approaches to the Cauchy problem for the KdV equation (4.1) were first developed in papers by Kato and Faminskii–Kruzhkov in 1979 and 1980.

Soliton and KdV equation theory is one of most amazing scientific subjects relative to their physical origin, applications, history, and discoveries during its mathematical development.

The importance of solitary waves was emphasized by John Scott Russell's experimental observations of waves in August of 1834 in the Union (Edinburgh–Glasgow) canal, [503]. Further study were due to Airy (1845) [5] and Stokes (1847) [536]. See [511] and [2] for historical details. The KdV equation (4.1) appeared in 1895, [354], though Boussinesq studied it earlier in 1872 [75] and found the explicit $\cosh^{-2}$ formula for its solitary-wave solution. The KdV equation describes the evolution of weakly nonlinear and weakly dispersive waves in such physical contexts as plasma physics, ion-acoustic waves, stratified internal and atmospheric waves, etc. The term *soliton* is due to Zabusky and Kruskal [590], who, solving the KdV equation numerically, made the discovery on the elasticity of interaction of its solutions. The explicit form of N-soliton solutions for the KdV equation was obtained by the Baker–Hirota bilinear method [22, 284]; see historical details in surveys in [2, 95, 250], in papers [3, 387] and comments below. The integrable *modified KdV* (mKdV) *equation*

$$u_t = u_{xxx} + 2u^2 u_x,$$

is connected with the KdV equation (4.1) by the *Miura transformation* [426]

$$v = u^2 + \sqrt{-3}\, u_x.$$

The history of various soliton-type solutions for the KdV and other integrable PDEs is amazing. In 1903, Baker [22] derived the KdV hierarchy, including the fifth-order KdV equation

$$u_t + u_{xxxxx} + 30\,u^2 u_x + 20\,u_x u_{xx} + 10\,u u_{xxx} = 0 \quad \text{in} \quad I\!R \times I\!R,$$

as well as the *Kadomtsev–Petviashvili* (KP) *equation* [308]

$$(u_t + 6uu_x + u_{xxx})_x = u_{yy} \quad \text{in} \quad I\!R^2 \times I\!R.$$

The latter describes asymptotically weakly nonlinear and weakly dispersive long waves and is obtained, in the weakly 2D limit, from the full water wave equations, where the surface tension is large. It also occurs for weak amplitude ion acoustic waves in an unmagnetized plasma. Among Baker's other results, there are the *bilinear differential operator D* (see (7.30)), maps and transformations, which are referred to as *Baker–Hirota* transformations [20, p. 275], associated differential transformations (including, what we used to call, Cole–Hopf's transformation), giving the bilinear form of the equations and hence explicit forms of hyperelliptic, periodic multi-soliton solutions for a variety of integrable PDEs, etc.† A detailed survey on re-evaluation of the role of Baker's hyperelliptic sigma function and other results in modern soliton theory is available in [415]; see also comments in [165] and [86] for a review of the earlier part of Baker's theory. Actually, Baker derived the key differential identity of the hyperelliptic functions of arbitrary genus g (for odd, $2g$+1, or even $2g$+2, degree of the polynomial $f(x)$ of the corresponding hyperelliptic curve $y^2 = f(x)$) [22], which led to KdV hierarchy and the KP equations of higher orders, but, explicitly, Baker presented these for the genus $g = 3$ only. The curves of ($2g$+1) degree correspond to the KdV hierarchy, and the ones of ($2g$+2) are associated with the KP equation. The list of PDEs for Baker's $\wp$ function also includes the Boussinesq equation; see [19]. In the 19th century, the development of theory of hyperelliptic functions as generalizations of elliptic functions, as well as general algebraic and Abelian functions, was due to Weierstrass, Riemann, Abel, Klein, Jacobi, Poincaré, Burkhardt, Krazer, Königsberger, Kovalevskaya, Hermite, Goursat, Appel, Tikhomandritskii,

† "Surprisingly, even in the 19th century, there appeared most of the tools and objects in soliton theories," [417, p. 4322]. "It is not generally known that Baker solved a number of nonlinear integrable partial differential equations... ," [165].

Brioschi, Frobenius, Stickelberger, Picard, Burnside, Kiepert, Bolza, and many other famous mathematicians.

A direct relation of hyperelliptic functions to the KdV-type equations is seen from the definition of the classical Weierstrass $\wp$-function of genus $g = 1$, satisfying the ODE

$$(\wp'(z))^2 = 4\wp^3(z) - g_2\wp(z) - g_3, \tag{4.236}$$

where the constants $g_{2,3}$ are called the *invariants* of $\wp$. One can derive from (4.236) other ODEs for $\wp$, e.g.,

$$\wp''(z) = 6\wp^2(z) - \tfrac{1}{2}\, g_2 \quad \text{and} \quad \wp'''(z) = 12\wp'(z)\wp(z),$$

where, after scaling, the latter is the stationary KdV equation. The first deep addition theorems for such elliptic functions date back to 1849, [573].

Applications of Baker's (or Baker–Akhiezer's) functions for algebro-geometric, finite-gap and elliptic solutions of fully discretized KP and 2D Toda equations are given in [357]. Several facts from the modern finite-gap integration method of completely integrable PDEs, where Baker–Akhiezer functions are key (see e.g., [35]), were discovered by Drach as early as 1918–1919, [153, 154]. Drach also derived the stationary KdV hierarchy;‡ see [35, p. 84], a detailed survey in [250], and [311]. Earlier, in 1897, Drach was known for his general classification of PDE systems by reducing these to a first or second-order systems in one independent variable. The integrability of the associated linear spectral problem for the fundamental Ψ-function

$$\Psi'' - u\Psi = \lambda\Psi$$

goes back to Ermakov (1880) [170], and is equivalent to the integrability of Ermakov–Drach's equation

$$\psi'' - (u+\lambda)\psi = -\frac{\mu^2}{\psi^3}.$$

Concerning elliptic solutions of the stationary KdV hierarchy, or, equivalently, elliptic finite-band potentials, q, for the linear Schrödinger operator $L = \frac{\mathrm{d}^2}{\mathrm{d}x^2} + q(x)$, the famous finite-gap example of the *Lamé potential*

$$q(x) = -s(s+1)\wp(x+\omega_3), \quad s = 1, 2, \ldots,$$

with fundamental half periods $\omega_1 \in I\!R$ and $\omega_3 \in \mathrm{i}I\!R$, found by Hermite and Halphen in the 1870s and 1880s, remained the only explicit potential for almost a century; see [35, p. 81, p. 259], [250, Sect. 2.8], and [36, Sect. 4.2]. For another well-known example of integrable PDEs, which can be solved in terms of linear problems with spectral parameters on an elliptic curve, such as the *Krichever–Novikov* (KN) *equation* [356]

$$u_t = \tfrac{1}{4}\, u_{xxx} + \frac{3[1-(u_{xx})^2]}{8u_x} - \tfrac{3}{2}\, \wp(2u)(u_x)^3, \tag{4.237}$$

related linear spectral problems for third-order operators were studied by Halphen in 1884, [36, p. 302]. Particular degenerate cases of (4.237) are mapped to the *Schwarzian KdV* (SKdV) *equation*

$$\frac{z_t}{z_x} = \tfrac{1}{4}\,\{z, x\} \equiv \tfrac{1}{4}\left[\frac{z_{xxx}}{z_x} - \frac{3}{2}\,\frac{(z_{xx})^2}{(z_x)^2}\right],$$

with the *Schwarzian derivative* on the right-hand side. Notice that this equation is quadratic

‡ "It appears he was the first to make the explicit connection between completely integrable systems and spectral theory," [250, p. 288]. "It is amazing that this remarkable work containing the constructions rediscovered in connection with the study of the KdV equation by Dubrovin, Its, Matveev, Gel'fand, Dikii is referred to very early in the modern literature," [35, p. 85].

(bilinear). Baker's functions associated to the Lax equations, which determine the KP hierarchy of several variables, were studied in [377], where earlier references are given.

Exact solutions of the mKdV equation via Weierstrass σ and $\wp$-functions for the case of genus one and via Baker's hyperelliptic sigma functions for genus two were described in [417], where hyperelliptic solutions for arbitrary genus were also constructed by using Weierstrass al-function. Papers [86, 36] (here Kleinian functions are used as logarithmic derivatives of the hyperelliptic σ-ones) and [415]–[417] contain detailed explanations concerning the significance of hyperelliptic function theory from the nineteenth century, and especially of Baker's functions, for modern soliton theory. "...Miura transformation is a connection between the worlds of $\wp$ and the al. I think that the researchers in the 19th century might have implicitly already recognized these facts," [417, p. 4332].

A detailed overview of elliptic algebro-geometric solutions of the KdV and Ablowitz–Kaup–Newell–Segur (AKNS) hierarchies of integrable PDEs is available in [250], where special attention is paid to classical Floquet, Hermite, and Picard theories. In particular, it was emphasized there that the questions of stationary Lax equations $[P, L] = 0$ on commuting ordinary differential expressions (P, L) (the Lax pair) were raised by Floquet in 1879, and again considered by Wallenberg in 1903 and Schur in 1905. The criterion for commutativity of differential operators was established by Burchnall and Chaundy in the 1920s, and got further development later, including the work by Baker in 1928, simplifying these results; see also [109, p. 87]. "Theta solutions of the Sin-Gordon equation ... can be seen in Ch.11 of the book by H.F. Baker "*Abelian Functions–Abel's Theorem and the Allied Theory Including the Theory of the Theta Functions*", Cambridge, 1897... ," [109, p. 112]. This book was republished by Cambridge Univ. Press in 1995.

§ 4.1. Many exact solutions of various related nonlinear PDEs can be found by singularity analysis; see [137]. Moving periodic soliton-like solutions (4.5) were introduced in [232, 217]. Exact solutions on invariant subspaces and sets for semilinear third-order PDEs are given in [593]; see also [268] for a more recent work on the Lie point symmetries classification of the KdV-type equations

$$u_t = u_{xxx} + F(x, t, u, u_x, u_{xx}).$$

§ 4.2. Various families of quasilinear third-order KdV-type equations can be found in [128], where further references concerning such PDEs and their exact solutions can be found. Higher-order generalized KdV equations are of increasing interest; see e.g., the quintic KdV equation in [296] and [583], where the seventh-order PDEs are studied. For the $K(2, 2)$ equation (4.14), the compacton solutions were constructed in [491]. More general $B(m, k)$ equations

$$u_t + a(u^m)_x = \mu(u^k)_{xxx}$$

also admit simple semi-compacton solutions [498], as well as the $Kq(m, \omega)$ nonlinear dispersion equation (another nonlinear extension of the KdV) [491]

$$u_t + (u^m)_x + [u^{1-\omega}(u^{\omega}u_x)_x]_x = 0.$$

Setting $m = 2$ and $\omega = \frac{1}{2}$ yields a typical quadratic PDE

$$u_t + (u^2)_x + uu_{xxx} + 2u_x u_{xx} = 0$$

possessing solutions on standard trigonometric-exponential subspaces. Combining the $K(m, n)$ and $B(m, k)$ equations gives the dispersive-dissipativity entity $DD(k, m, n)$ [493]

$$u_t + a(u^m)_x + (u^n)_{xxx} = \mu(u^k)_{xx}$$

that can also admit solutions on invariant subspaces for some values of parameters.

Concerning the interface and approximation problems posed in Example 4.8 and others, for a class of degenerate third-order PDEs, a detailed approach concerning ODE *analytic* approximations of various non-smooth patterns (e.g., compactons or peakons) was performed in [388], establishing that some of these solutions can be approached by approximating analytic profiles. Among other things, this analysis reveals that, for higher-order PDEs and ODEs, such an approach can be practically intractable. Ideas of maximal regularity and homotopy families of PDEs help to detect correct classes of solutions of the Cauchy problem. For higher-order degenerate PDEs, such solutions are often of changing sign, which sometimes contradicts physical meaning and motivations of the models, but is a non-avoidable feature. Various FBPs can be dealt by von Mises transformations locally, near singularities, though the complexity of computations and analysis increase dramatically with the order of PDEs involved.

Concerning the ε-regularization aspects of proper (weak) solutions in Section 4.2.4, note that such difficulty occurs for semilinear higher-order parabolic PDEs with non-smooth coefficients, e.g.,

$$u_t = -u_{xxxx} - |u|^{p-1}u, \quad \text{where } p \in (-1, 1). \tag{4.238}$$

Since the strong absorption term $-|u|^{p-1}u$ is not Lipschitz continuous at $u = 0$ for $p < 1$ and even singular or discontinuous for $p \leq 0$, for construction of proper solutions, the following analytic ε-regularization is natural [227]

$$|u|^{p-1}u \mapsto (\varepsilon^2 + u^2)^{\frac{p-1}{2}} u \quad (\varepsilon > 0).$$

This gives the necessary smooth family $\{u_\varepsilon(x,t)\}$ of classical solutions. Equation (4.238) admits compactly supported solutions of changing sign. For $p \leq -1$, the limit $\varepsilon \to 0$ implies nonexistence of a proper solution, which is actually $u \equiv 0$, meaning that the limit proper semigroup is discontinuous at $t = 0$; see further examples in [226, Ch. 7 and 10].

§ 4.3. The *Florin FBP* for the heat equation in $I\!R^N$ (introduced by Florin in 1951) includes the free-boundary conditions of the form

$$u = 0 \quad \text{and} \quad \tfrac{\partial u}{\partial \mathbf{n}} = -1;$$

see [190] and references in [226, Ch. 8]. The FBP setting for the fifth-order equation (4.76) uses a second-order Florin-type condition. Third and higher odd-order evolution PDEs also appear in the hierarchy of *surface evolution equations* [428, 82]

$$V_n \equiv \mathbf{N}_t = BD_s^l \kappa,$$

where V_n is the local normal velocity of the surface, κ is the curvature of the surface, and s is the arclength. The case $l = 3$ governs long frontal waves in the quasigeostrophic approximation [471]. Concerning the relation between many integrable PDEs and the motion of curves on the plane in the various Klein geometries, see [120] and references therein.

§ 4.4. There is a large amount of literature devoted to existence, uniqueness, and asymptotic behavior for the KdV and KP-type equations; see [381, 69, 179]. Delicate structures of blow-up similarity solutions of the mKdV equation (4.133) with $m \geq 4$ are described in [70].

§ 4.5. This study will be continued in Section 6.7, where further references are given. Compactons of the type (4.178) was first detected by Rosenau [494, p. 197].

§ 4.6. Equation (4.181) can be considered as an integrable modification of the PBBM equation (4.201) (reduced to it by omitting the last two terms on the right-hand side). It was shown that (4.181) admits solitary-wave solutions with compact support. On the other hand, (4.181) is the nonlinear dispersive counterpart of the KdV equation (4.1) and forms the first member of the bi-Hamiltonian hierarchy; see an explicit algorithm in [449] based on the bi-Hamiltonian representation of the classically integrable systems.

Equation (4.192) was originally obtained by Fuchssteiner [209] by the method of recursion operators. As was first noticed by Fuchssteiner and Fokas in [212], it is a bi-Hamiltonian generalization of the KdV equation with infinitely many conservation laws and is formally integrable. It was subsequently derived via physical principles in [94], where its further new remarkable properties were established (see details on the derivation of the FFCH equation as a shallow water model in [306]). With the omission of the last two terms (formally, of higher quadratic asymptotic order for small solutions), (4.192) turns into the PBBM equation (4.201) which is not known to be integrable. This PDE also happens to be another (together with (4.181)) first member of the bi-Hamiltonian hierarchy of the KdV equation [449]. A principal feature of (4.192), observed first in [94], is that it admits soliton solutions with sharp corners. The resulting solutions, consisting of a collection of peaked waves, were called *multipeakons* in [94]. An algorithm for constructing solutions of the shallow water equation (4.192) by the inverse scattering technique and a Liouville transformation using the link with the KdV equation is explained in [96] (construction of N-soliton, cuspon, and soliton-cuspon solutions leads to difficult algebraic equations that can be studied numerically, [184]). A full classification of weak discontinuities (including various types of *cuspons* and *stumpons*) admitted by the traveling wave solutions $u(x,t) = \varphi(x-ct)$ of the FFCH equations is performed in [379]. Conservation laws are studied in [380].

Concerning uniqueness for PDEs, the most well-known example is the heat equation

$$u_t = u_{xx} ,$$

where the famous Tikhonov–Täklind uniqueness class was obtained in the 1930s [552, 548]. Uniqueness of the solution of the Cauchy problem takes place in the class

$$\{|u(x,t)| \le \mathrm{e}^{|x|\, h(|x|)}\},$$

with any positive increasing function $h(s)$ satisfying Osgood's criterion

$$\int^{\infty} \frac{\mathrm{d}s}{h(s)} = \infty.$$

We refer to results and the literature in [443], where a detailed analysis of uniqueness classes is performed for the second and higher-order parabolic PDEs and systems by energy estimates based on Saint-Venant's principle.

As far as singularity formation phenomena in integrable PDEs (e.g., formation and existence of soliton solutions with non-analytic singularities) are concerned, it seems that the first example was proposed in [300] (see also [568]), where the following equation:

$$u_{xt} + \left[\frac{u_{xx}}{(1+(u_x)^2)^{3/2}}\right]_{xx} = 0 \tag{4.239}$$

was derived as a model of nonlinear transverse oscillations of elastic beams. Due to the gradient-dependent nonlinearity, this equation can admit *cuspons*, i.e., weak solutions having cusps, where $u_x = \infty$, which propagate with finite speed. In particular, it was shown in [300] that there exist such traveling wave solutions

$$u(x,t) = g(\eta), \quad \text{with} \quad \eta = x \pm \lambda t,$$

where g solves a nonlinear ODE with the following behavior at the cusp at $\eta = 0$:

$$g(\eta) = \sqrt{\tfrac{2}{\lambda}} - \left(\tfrac{2}{\lambda}\right)^{\frac{1}{4}} \sqrt{|\eta|}\,(1 + o(1)) \quad \text{as } \eta \to 0 \quad (\lambda > 0).$$

These solutions are Hölder continuous with exponent $\frac{1}{2}$. We refer also to cusp structures revealed by N-soliton solutions of (4.192), corresponding to completely integrable Hamiltonian ODEs, in [94, 6, 31] and [449]. This model and the singularity formation study answer

Whitham's suggestion [575] to find mathematical PDEs for shallow water waves that include the wave breaking phenomena.

Analysis of the bi-Hamiltonian structure of the FFCH and Hunter–Saxton equations can be found in [211, 335].

An advanced theory of local solvability of the FFCH equation can be found in [387] (the regularization term εu_{xxxxt} is used, local existence and blow-up in Sobolev spaces H^s for $s > \frac{3}{2}$, and some results for $s \le \frac{3}{2}$), [134] and references in [136] (using Kato's semigroup approach to (4.198), local and global H^1-theory, and uniqueness), [579] (parabolic regularization εu_{xx} in (4.198), global H^1-theory for arbitrary data, with no uniqueness, recovering an analogy of Oleinik's entropy "E-condition" for shocks [441]), [364] (existence, uniqueness, and blow-up for initial-boundary value problems). See details on the propagation of weak singularities in [135] and on the orbital stability of multipeakons in [136]. In general, the FFCH (4.192) and similar PDEs belong to the class of the third-order quasilinear *pseudo-KdV*-type equations and their "pseudo" nature, via the integral representations with positive compact kernels, such as in (4.198), ensures solvability and uniqueness (cf. the PBBM equation (4.201) which has smoother solutions than the conservation law $u_t = uu_x$).

The PBBM equation, written in the form of

$$u_t + u_x + 6uu_x - u_{xxt} = 0$$

possesses a family of solitary wave solutions

$$u(x,t) = \frac{2\alpha^2}{1-4\alpha^2}\,\mathrm{sech}^2[\alpha(x-\lambda t)],$$

where $\lambda = \frac{1}{1-4\alpha^2}$. The PDE is not integrable, so their interaction is not elastic.

Well-developed mathematical theory of the FFCH equation (4.192) shows that it is precisely the PDE exhibiting all the desired properties of integrability, breaking of waves, and existence of solitons. The simpler *Whitham equation* (1967) [575] with a linear non-local term

$$u_t + uu_x + \int_{I\!R} k_0(x-\xi)u_x(\xi,t)\,\mathrm{d}\xi = 0,,$$

where the kernel is given by

$$k_0(x) = \frac{1}{2\pi}\int \sqrt{\frac{\tan\xi}{\xi}}\,\mathrm{e}^{i\xi x}\,\mathrm{d}\xi,$$

though admitting the effect of breaking of waves, does not have a soliton interaction of its traveling waves; see references in [136]. The corresponding generalized PDE with the extra fifth-order term u_{xxxxx} is globally well-posed for any H^1-data in the appropriate Bourgain function space X^1, [283]. Higher-dimensional versions of various shallow water equations, further extensions, history, and references on more abstract frameworks are available in [290].

There exists another integrable PDE from the family (4.193), (4.194), the *Degasperis–Procesi equation*

$$u_t - u_{xxt} = -4uu_x + 3u_xu_{xx} + uu_{xxx}\,;$$

see [145] and early references therein (no other integrable equations exist in this family, which was introduced in [491]). On existence, uniqueness (of entropy weak solutions in $L^1 \cap BV$), parabolic ε-regularization, Oleinik's entropy estimate and generalized PDEs, see [131].

It is curious that the whole three-parameter family of such PDEs indicated in Proposition 4.44 admits the single peakon solution (4.196). Restricting our attention to the one-parameter family of equations introduced in [145],

$$u_t - u_{xxt} = F_\beta[u] \equiv -(1+\beta)uu_x + \beta u_xu_{xx} + uu_{xxx}\,, \tag{4.240}$$

there exist the multipeakon solutions (4.197), where, for $\beta \neq 3, 2$, the functions $\{q_i, p_i\}$ are not canonical variables, and the corresponding DS takes a similar form

$$\begin{cases} \dot{q}_i = \frac{\partial H_A}{\partial p_i}, \\ \dot{p}_i = -(\beta - 1)\frac{\partial H_A}{\partial q_i}, \end{cases}$$

with the same "Hamiltonian" (4.200) (the canonical Hamiltonian form exists only in the special cases $\beta = 3$ and 2). This example shows the principal fact that integrability is not a necessary condition for the existence of countable families of N-solitons for arbitrary N, though, clearly, for $\beta = 3, 2$, explicit formulae for $\{q_i(t), p_i(t)\}$ can be found by inverse scattering techniques that describe the peakon interactions more clearly and in greater detail.

The *Fornberg–Whitham* (FW) *equation*

$$u_t - u_{xxt} = uu_{xxx} - uu_x + 3u_x u_{xx} - u_x$$

that describes qualitative behavior of wave-breaking [574], contains the quadratic operator from (4.240) with $\beta = 3$, so scaling $x \mapsto 2x$ leads to

$$\boxed{8u_t - 2u_{xxt} = F_3[u] - 4u_x.} \tag{4.241}$$

Looking for the 1-peakon solution

$$u(x,t) = C_1(t)\mathrm{e}^x + C_2(t)\mathrm{e}^{-x} \in W_2,$$

and bearing in mind that $F_3 = 0$ on W_2, we obtain linear ODEs for the coefficients and the general solution of (4.241) $u(x,t) = A\mathrm{e}^{x-\frac{2}{3}t} + B\mathrm{e}^{-(x-\frac{2}{3}t)}$, where A and B are arbitrary constants. This gives the well-known 1-peakon pattern [195]

$$u(x,t) = A\mathrm{e}^{-|x-\frac{2}{3}t|},$$

which, unlike the above tautological PDEs on W_2, has the fixed wave speed $\lambda = \frac{2}{3}$.

For the Hunter–Saxton equation (4.209), its reduction to a finite-dimensional completely integrable Hamiltonian system with phase space, consisting of piecewise linear solutions (4.210), was first discussed in [295]; see also references given in [32]. Another more general water wave equation takes the form [575]

$$\eta_t + \eta_{xxx} + 6\eta\eta_x + \varepsilon(\tfrac{19}{10}\,\eta_{xxxxx} + 10\eta\eta_{xxx} + 23\eta_x\eta_{xx} - 6\eta^2\eta_x) = 0 \tag{4.242}$$

(terms of the order $O(\varepsilon^2)$ are omitted), where $y = \eta(x,t)$ denotes the position of the free surface of a body of water, considered as an inviscid incompressible fluid, lying above a horizontal flat bottom. This PDE is associated with the generalized FFCH equation (*generalized integrable KdV equation*)

$$u_t + u_{xxx} + 6uu_x - \tfrac{19}{10}\,\varepsilon(u_{xxt} + 2uu_{xxx} + 4u_x u_{xx}) = 0, \tag{4.243}$$

in the sense that the function

$$\eta = u + \varepsilon(\tfrac{7}{5}\,u^2 + \tfrac{1}{5}\,u_{xx} - \tfrac{4}{5}\,u_x\, D_x^{-1}u)$$

(D_x^{-1} is integration in x) solves (4.242); see [193] for other examples and related references therein. The generalized SKdV equation (4.211) admits a wide range of moving blow-up and soliton-like solutions, depending on two arbitrary functions [247].

Further references and some mathematics on the Green–Naghdi equations (4.212) can be found in [386]. Concerning higher-order PDEs, we mention the *Kawahara equation* (1972)

$$u_t + uu_x + \alpha u_{xx} - u_{xxxxx} = 0$$

that describes propagation of long waves under ice cover in finite depth liquids and gravity waves on liquid surfaces with surface tension. The *Kawamoto equation* [330]

$$u_t = u^5 u_{xxxxx} + 5\,u^4 u_x u_{xxxx} + 10\,u^5 u_{xx} u_{xxx}$$

has higher-degree algebraic terms, as well as *Lax's seventh-order KdV equation*

$$u_t + [35u^4 + 70(u^2 u_{xx} + u(u_x)^2) + 7(2uu_{xxxx} + 3(u_{xx})^2 + 4u_x u_{xxx}) + u_{xxxxxx}]_x = 0,$$

and the *seventh-order Sawada–Kotara equation*

$$u_t + [63u^4 + 63(2u^2 u_{xx} + u(u_x)^2) + 21(uu_{xxxx} + (u_{xx})^2 + u_x u_{xxx}) + u_{xxxxxx}]_x = 0.$$

One and two-soliton solutions of the standard form, which can be also derived by Baker–Hirota-type methods of the multi-parameter family of equations

$$\begin{aligned} u_t + r_1 u + r_2 u_{xx} + r_3 u_{xxx} + r_4 (u_x)^2 + r_5 u u_x \\ + r_6 u u_{xx} + r_7 u^2 + r_8 u^2 u_x + r_9 u^3 + r_{10} u^4 = 0 \end{aligned} \tag{4.244}$$

were constructed in [595] by using a modification of the dressing method that was originally developed for application to completely integrable nonlinear PDEs.

In connection with solutions (4.216) in Example 4.55, notice that similar mixed subspaces appear for the *positon* and *negaton* solutions of the KdV equation, $u_t + 6uu_x + u_{xxx} = 0$, which, in terms of the function $v = 2(\ln u)_{xx}$ for the bilinear representation, take the form

$$v(x,t) = \sin(px + p^3 t) - p(x + 3p^2 t) \in \mathcal{L}\{1, x, \cos px, \sin px\},$$

$$v(x,t) = \sinh(px - p^3 t) + p(x - 3p^2 t) \in \mathcal{L}\{1, x, \cosh px, \sinh px\},$$

respectively, where $p \neq 0$. Positon solutions of the KdV equation (soliton-positon interaction) have been recognized since the 1970s as solutions with inverse square singularities and slow decay at infinity. See first results on *rational solutions* induced by polynomials obtained in 1978 by Ablowitz and Satsuma, Adler and Moser (see survey [407]) and Bordag and Matveev [72]; [16] (one-positon solution was studied by a variant of the inverse scattering method), and [418]. On two and higher-order positons, see [419, 407]. Negatons were obtained in [485]. Both types of solutions belong to the class of generalized Wronskian solutions of the bilinear KdV equation (i.e., (0.31) of the Introduction) introduced in [204, 418]. This approach is a generalization of the Wronskian representation of multi-solitons invented by Satsuma [515]. According to Matveev's Wronskian formula, positons correspond to choosing eigenfunctions with positive eigenvalues of the Schrödinger spectral problem. For the sine–Gordon equation

$$u_{xx} - u_{tt} = \sin u,$$

positons were constructed by Beutler [60]. These solutions, belonging to a combination of polynomial and trigonometric/exponential subspaces, exist for other integrable 1D and 2D PDEs; see [125].

§ **4.7.** Further references on the Dym hierarchy of integrable PDEs are available in [6] and [30], where the acoustic scattering theory is developed.

Open problems

- These have been formulated throughout the chapter in Examples 4.3, 4.7, Section 4.2.2, Example 4.8, Sections 4.2.4 and 4.2.5, Examples 4.9, 4.28, 4.29, 4.31, Section 4.5, Examples 4.38, 4.41, 4.43, 4.45–4.48, 4.50–4.52, 4.56, and in Section 4.7.2.

CHAPTER 5

Quasilinear Wave and Boussinesq Models in One Dimension. Systems of Nonlinear Equations

This chapter completes the description of exact solutions on invariant subspaces of 1D nonlinear evolution equations. We consider quasilinear wave and Boussinesq models. In the last section we study systems of evolution PDEs of various types.

5.1 Blow-up in nonlinear wave equations on invariant subspaces

5.1.1 Basic quasilinear wave models

We consider quasilinear PDEs which are second-order in the time variable, i.e., contain the derivative u_{tt}. This class includes the well-known second and higher-order PDEs of hyperbolic type, which provide us with interesting new examples of formation of evolution singularities. There are many applications of quasilinear hyperbolic equations possessing exact and explicit solutions. We present a few of equations below and include more references in the Remarks. In particular, Zabusky [589] proposed to use the hyperbolic equation

$$u_{tt} = k(u_x)u_{xx}$$

as a model for the dynamics of nonlinear strings. In this context, it is worth mentioning the standard derivation of the vibrating string equation by *Newton's Second Law* for a homogeneous thin string. Under the assumption that *Hook's Law* applies, this yields, in the dimensionless form, the quasilinear PDE

$$u_{tt} = \left[\frac{u_x}{\sqrt{1+(u_x)^2}}\right]_x.$$

For small deviations from equilibrium, where $(u_x)^2 \ll 1$, we arrive at the canonical linear *wave equation*

$$u_{tt} = u_{xx}.$$

As another example, wave operators appear from equations for steady transonic gas flow (here $t = y$ is the vertical coordinate)

$$\begin{cases} u_t = v_x, \\ v_t + uu_x = 0, \end{cases}$$

so, replacing $u \mapsto -u$, and excluding v yields the *quadratic wave equation*

$$u_{tt} = (uu_x)_x . \tag{5.1}$$

It is of hyperbolic type in the positivity domain of the solution $\{u > 0\}$. More general quasilinear wave models come from the 1D gas dynamics equations

$$\begin{cases} \rho_t + (\rho v)_x = 0, \\ v_t + v v_x + \frac{1}{\rho}\, p_x = 0, \end{cases}$$

where the pressure $p = p(\rho)$ is a given monotone increasing function of the density. Introducing the stream function $\psi(x,t)$ such that

$$\rho = \psi_x, \quad \rho v = -\psi_t,$$

and using the independent variables $\{\psi, t\}$, where $u(\psi,t) = \frac{1}{\rho}$, we find that $u(\psi,t)$ satisfies the hyperbolic PDE

$$u_{tt} = \left[\tfrac{1}{u^2}\, p'\left(\tfrac{1}{u}\right) u_\psi\right]_\psi.$$

In applications, we will study more general quasilinear wave equations, such as

$$u_{tt} = (\psi(u)u_x)_x + \text{(lower-order terms)},$$
$$u_{tt} = (\psi(u_x))_x + \text{(lower-order terms)},$$

with different types of coefficients $\psi(s)$, arising in nonlinear wave theory. Similar to other evolution models, the hyperbolic PDEs can create evolution singularities, exhibiting different asymptotic patterns.

5.1.2 Stability of blow-up on invariant subspaces

Example 5.1 **(Blow-up, localization, stability)** Let us begin with the following *quadratic wave equation with source* (a force term):

$$\boxed{u_{tt} = F[u] \equiv (uu_x)_x + 2u^2 \equiv uu_{xx} + (u_x)^2 + 2u^2,} \tag{5.2}$$

where F is a typical second-order operator from Section 1.4. In the positivity domain $\{(x,t) \in I\!R^2 : u > 0\}$, equation (5.2) is of *hyperbolic* type, where the Cauchy problem can be posed and, by the Cauchy–Kovalevskaya Theorem, there exists a unique local analytic solution in some neighborhood of any point of strict hyperbolicity. On the other hand, in the negativity domain, $\{(x,t) \in I\!R^2 : u < 0\}$, (5.2) is of *elliptic* type, where a boundary-value problem is natural. In general, (5.2) is a quasilinear *elliptic-hyperbolic* equation for which existence-uniqueness theory is not well developed. Exact solutions may help us to understand its singularities and main difficulties of the analysis.

We begin by describing blow-up properties of solutions

$$u(x,t) = C_1(t) + C_2(t)\cos x \tag{5.3}$$

of (5.2), which belong to the subspace $W_2 = \mathcal{L}\{1, \cos x\}$ that is invariant under the quadratic operator F. The corresponding DS takes the form

$$\begin{cases} C_1'' = 2C_1^2 + C_2^2, \\ C_2'' = 3C_1 C_2. \end{cases} \tag{5.4}$$

Consider 1D invariant subspaces from W_2. First of all, it is $W_1 = \mathcal{L}\{1\}$ on which

$u(t) = C_1(t)$ is obtained from (5.3) for $C_2(t) \equiv 0$, and the DS (5.4) reduces to

$$C_1'' = 2C_1^2. \tag{5.5}$$

This is the simplest solution of the PDE, which does not depend on the space variable x. Integrating (5.5) once via multiplying by C_1' yields that all the blow-up solutions have the following behavior close to the blow-up time T:

$$C_1(t) = \tfrac{3}{(T-t)^2}\,(1+o(1)) \quad \text{as} \quad t \to T^-.$$

Other invariant subspaces appear in symmetric cases $C_2 = \pm C_1$ in (5.3) that give

$$W_1^+ = \mathcal{L}\{2\cos^2(\tfrac{x}{2})\} \quad \text{and} \quad W_1^- = \mathcal{L}\{2\sin^2(\tfrac{x}{2})\}.$$

In both cases, setting $C_2 = \pm C_1$ in the DS (5.4) yields another ODE (cf. (5.5))

$$C_1'' = 3C_1^2 \quad \Longrightarrow \quad C_1(t) = \tfrac{2}{(T-t)^2}\,(1+o(1)) \quad \text{as} \quad t \to T^-.$$

Choosing the explicit solution $C_1(t) = 2(T-t)^{-2}$ leads to the 2π-periodic separate variables similarity solution of the PDE on W_1^+,

$$u_S(x,t) = \tfrac{1}{(T-t)^2}\,4\cos^2(\tfrac{x}{2}).$$

Next, noting that $u_S(x_k, t) \equiv 0$ at points $x_k = (2k+1)\pi$, we take the one-hump

$$u_S(x,t) = \begin{cases} \frac{1}{(T-t)^2}\,4\cos^2(\frac{x}{2}) & \text{for } |x| \le \pi, \\ 0 & \text{for } |x| > \pi, \end{cases} \tag{5.6}$$

which is a localized standing wave (a kind of blow-up standing compacton); see Figure 5.1. It exhibits *regional blow-up* in the interval $(-\pi, \pi)$. Then $L = 2\pi$ is the *fundamental length.*

The compactly supported function (5.6) is a standard weak solution of the Cauchy problem for (5.2) in $I\!R \times (0, T)$ with corresponding initial data $u_S(x, 0)$, $(u_S)_t(x, 0)$. This is easy to check via multiplying the equation by a smooth test function and integrating by parts. Of course, this does not mean that, for general initial data, the quasilinear hyperbolic equation (5.2) admits sufficiently smooth nonnegative solutions (actually, solutions may change sign and discontinuous shock waves may appear). In fact, we claim that the standing wave solution (5.6) with zero-flux and zero contact angle conditions is an exceptional one, so there are many other moving patterns which propagate with shock waves (for instance, via TWs).

Nevertheless, the separate variables solution (5.6) is expected to describe a stable generic blow-up behavior of this model. In a general setting, taking an arbitrary solution $u(x,t)$ (say, symmetric in x) blowing up at $t = T$, stability analysis assumes showing stabilization of the rescaled function as $\tau = -\ln(T-t) \to +\infty$,

$$w(x,\tau) = (T-t)^2 u(x,t) \to g(x) = 4\cos^2(\tfrac{x}{2}), \tag{5.7}$$

where g is the similarity profile given in (5.6). The rescaled equation for w is

$$w_{\tau\tau} + 5w_\tau = \mathbf{A}[w] \equiv F[w] - 6w = (ww_x)_x + 2w^2 - 6w. \tag{5.8}$$

The passage to the limit $\tau \to \infty$ to establish (5.7) for a certain class of solutions is a hard OPEN PROBLEM.

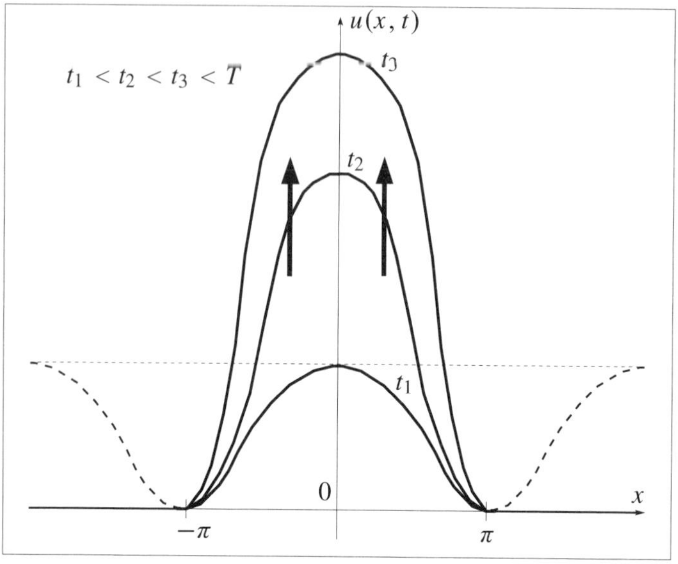

Figure 5.1 Regional blow-up exhibited by the solution (5.6) on the interval $(-\pi, \pi)$.

A comment on linear stability. Notice that even the linearized setting is not easy for such hyperbolic PDEs and leads to OPEN PROBLEMS of spectral theory of quadratic pencils of linear operators. We perform some preliminary computations which will help us later on to perform stability analysis on W_2. Using the linearization in (5.8) about the stationary similarity profile, we set $w(x,\tau) = g(x) + Y(x,\tau)$. This yields the following linear hyperbolic equation:

$$Y_{\tau\tau} + 5Y_\tau = \mathbf{A}'[g]Y, \tag{5.9}$$

where the second-order linearized operator $\mathbf{A}'[g]$ has the symmetric form

$$\mathbf{A}'[g]Y = (gY)'' + (4g-6)Y \equiv \tfrac{1}{g}(g^2Y')' + [g'' + 2(2g-3)]Y. \tag{5.10}$$

As usual, looking for separate variables solutions of (5.9),

$$Y_k(x,\tau) = \mathrm{e}^{\lambda_k \tau}\,\psi_k(x)$$

yields the spectral problem

$$\mathbf{A}'[g]\psi_k = \mu_k\psi_k, \quad \text{where} \;\; \mu_k = \lambda_k^2 + 5\lambda_k. \tag{5.11}$$

Here $\{\mu_k\}$ and $\{\psi_k\}$ are eigenvalues and eigenfunctions of the linear operator $\mathbf{A}'[g]$, but actually we deal with the simple quadratic pencil $(\lambda^2 + 5\lambda)I - \mathbf{A}'[g]$ of self-adjoint operators.

According to classical theory of linear singular ordinary differential operators [432], $\mathbf{A}'[g]$ admits a self-adjoint extension in the weighted space $L^2_\rho((-\pi,\pi))$ with the positive weight $\rho = g(x) \equiv 4\cos^2(\frac{x}{2})$. Let us take the unique minimal Friedrichs self-adjoint extension of the symmetric operator (5.10) that corresponds to zero Dirichlet boundary conditions at regular end-points; see Birman–Solomjak

[62] for details. We then need to know its real point spectrum. Comparing nonlinear and linearized operators in (5.8) and (5.10), one can find two first successive eigenvalues and eigenfunctions:

$$\mu_0 = 6 \;\; \text{with} \;\; \psi_0 = g(x), \quad \text{and} \quad \mu_1 = 0 \;\; \text{with} \;\; \psi_1 = g'(x),$$

where $g(x) > 0$ and $g'(x)$ has precisely a single zero at $x = 0$ in $(-\pi, \pi)$. Using the relation between μ_k and λ_k, we obtain the following eigenvalues of the pencil:

$$\lambda_1 = 1, \;\; \lambda_2 = 0, \;\; \lambda_3 = -5, \;\; \text{and} \;\; \lambda_4 = -6. \tag{5.12}$$

By Sturm's Theorem, other eigenfunctions (if any) $\psi_k(x)$ with $k \geq 2$ have k zeros and must correspond to eigenvalues $\mu_k < \mu_1 = 0$. Solving the quadratic equations $\lambda_k^2 + 5\lambda_k - \mu_k = 0$ yields, for any $k \geq 2$, eigenvalues λ_k with negative real parts. It is worth mentioning that the linear ODE (5.11) for $\mu_k < -3$ exhibits oscillatory behavior near the singular endpoint $x = \pm\pi$, so the spectrum of $\mathbf{A}'[g]$ is not discrete and contains a continuous counterpart belonging to the stable half of the complex plane. Indeed, this makes the linearized problem more difficult, as it includes integral terms in eigenfunction expansions over the continuous spectrum, which reflects the strong nonlinear degeneracy of the original PDE.

We have detected a single eigenvalue $\lambda_1 = 1$, corresponding to an unstable mode. It should be excluded from stability analysis, since it corresponds to the shifting of the blow-up time T (see computation below), which is fixed via rescaling (5.7). Therefore, g is exponentially asymptotically stable in the linear approximation. For many sufficiently smooth nonlinear evolution PDEs, and especially for parabolic ones, it is known that linear stability for the linearized equations implies that the nonlinear stability is true for the full PDE. This is called the *principle of linearized stability*; see Lunardi [403, Ch. 9]. For the degenerate quasilinear hyperbolic equation (5.8), such questions are OPEN and, bearing in mind complicated spectral properties of the corresponding quadratic pencil, are difficult to prove.

Linear stability on W_2 for the DS (5.4). Let us perform blow-up stability analysis on the subspace W_2. There exists a direct sum decomposition of W_2 into the two 1D invariant subspaces,

$$W_2 = W_1 \oplus W_1^+, \quad \text{or} \quad W_2 = W_1^+ \oplus W_1^-, \tag{5.13}$$

with clear and simple blow-up behavior on each of them. What kind of stable blow-up evolution can be detected on the wider subspace W_2? For simpler quasilinear parabolic PDEs, we managed to prove stability of self-similar blow-up evolution on $W_1^\pm$ (see Example 3.17). Unlike the parabolic case, for the hyperbolic PDE, the quadratic DS (5.4) is of fourth order and we cannot perform such a complete global stability analysis. We again restrict ourselves to a linearized stability study.

For linear stability analysis, we introduce the rescaled blow-up variables

$$C_{1,2}(t) = \tfrac{1}{(T-t)^2}\, \varphi_{1,2}(\tau), \quad \text{with} \quad \tau = -\ln(T-t) \to +\infty, \tag{5.14}$$

and obtain the DS

$$\begin{cases} \varphi_1'' = -5\varphi_1' + 2\varphi_1^2 + \varphi_2^2 - 6\varphi_1, \\ \varphi_2'' = -5\varphi_2' + 3\varphi_1\varphi_2 - 6\varphi_2. \end{cases} \tag{5.15}$$

The blow-up solution on W_1^+ given by $C_1(t) = C_2(t) = 2(T-t)^2$ corresponds to the equilibrium $(\varphi_1, \varphi_2) = (2, 2)$. Linearizing (5.15) about $(2, 2)$ by setting

$$\varphi_{1,2} = 2 + Y_{1,2}$$

yields the linearized system

$$\begin{cases} Y_1'' = -5Y_1' + 2Y_1 + 4Y_2, \\ Y_2'' = -5Y_2' + 6Y_1. \end{cases} \tag{5.16}$$

In variables $Z = (Y_1, Y_1', Y_2, Y_2')^T$, it is written as the DS

$$Z' = AZ, \quad \text{with the } 4 \times 4 \text{ matrix} \quad A = \begin{bmatrix} 0 & 1 & 0 & 0 \\ 2 & -5 & 4 & 0 \\ 0 & 0 & 0 & 1 \\ 6 & 0 & 0 & -5 \end{bmatrix}.$$

It is not surprising that the non-symmetric matrix A has four real eigenvalues $\lambda_1 = 1$, $\lambda_2 = -1$, $\lambda_3 = -4$, and $\lambda_4 = -6$, where the first and the last one are the same, as in the point spectrum (5.12) of the quadratic pencil. Hence, $(2, 2)$ is a hyperbolic equilibrium of the nonlinear system (5.15) and we can apply the Hartman–Grobman Theorem [460, p. 118] to classify this stationary point. Of course, it is a saddle and has a 1D unstable manifold which is tangent to the unstable subspace for $\lambda = 1$ of the linearized DS.

This unstable manifold should not be taken into account if the blow-up time T is fixed by scaling (5.14). Indeed, we perform a small change in T, setting $T' = T + \varepsilon$, to obtain that $C_1(t) \sim (T-t)^{-2}$. In terms of the time-variable $\tau = -\ln(T-t)$, this is transformed into

$$\begin{aligned} \tfrac{1}{(T'-t)^2} &= \tfrac{1}{(T-t)^2}\left(1 + \varepsilon\,\tfrac{1}{T-t}\right)^{-2} \\ &= \tfrac{1}{(T-t)^2}\left(1 - 2\varepsilon\,\tfrac{1}{T-t} + ...\right) = \tfrac{1}{(T-t)^2}\left(1 - 2\varepsilon e^{\tau} + ...\right) \end{aligned}$$

(here $t \approx T^-$ is fixed and we expand relative to the small parameter ε). Then the factor $\frac{1}{(T-t)^2}$ is scaled out by transformation (5.14), so that the remaining term $1 - 2\varepsilon e^{\tau} + ...$ describes a typical unstable behavior according to the mode with the eigenvalue $\lambda = 1$. Excluding the local 1D unstable manifold yields the following rescaled stability result.

Proposition 5.2 *Let the blow-up time T be fixed in rescaling* (5.14). *Then, for such orbits, the equilibrium* $(2, 2)$ *for the DS* (5.15) *is a stable node and, hence, is asymptotically stable.*

This implies that, in the rescaled sense for the full DS (5.4), the evolution on subspaces $W_1^{\pm}$ is locally asymptotically stable. Thus, any solution on W_2 being in a sufficiently small neighborhood of $W_1^{\pm}$ takes the form (5.6) of the similarity solution near the blow-up time.

Global analysis of orbits of the fourth-order DS (5.15) is a difficult OPEN PROBLEM. We expect that a certain stability result remains true for the sixth-order DS that describes blow-up evolution for the hyperbolic PDE (5.2) on the subspace $W_3 = \mathcal{L}\{1, \cos x, \sin x\}$.

Example 5.3 (**Generalized Boussinesq equation with source**) The blow-up dynamics (including local stability) do not essentially change if we add some extra linear terms to the quadratic operator F, leaving W_2 invariant. For instance, the stability conclusions remain the same for a generalized *Boussinesq equation with source* of the form

$$\boxed{u_{tt} = -u_{xxxx} + \beta u_{xx} + (uu_x)_x + \gamma u^2.}$$

Example 5.4 (**Improved Boussinesq equation**) Consider a quadratic perturbation of the *improved Boussinesq equation*

$$\boxed{u_{tt} - \alpha u_{ttxx} = \beta u_{xx} + (uu_x)_x + 2u^2 \quad (\alpha \neq -1),} \tag{5.17}$$

which admits solutions (5.3), with the DS

$$\begin{cases} C_1'' = 2C_1^2 + C_2^2, \\ C_2'' = \frac{3}{1+\alpha}\, C_1 C_2 - \frac{\beta}{1+\alpha}\, C_2. \end{cases}$$

Asymptotic and stability analysis is performed in a similar fashion. The extra linear term in the second ODE does not affect the asymptotics of blow-up and is scaled out in stability study.

5.2 Breathers in quasilinear wave equations and blow-up models

Example 5.5 (**Breathers**) In typical applications, breathers are periodic solutions of nonlinear hyperbolic models.

Classical breather. It has been recognized since the 1950s that the integrable *sine-Gordon* (sG) *equation* in 1D

$$u_{tt} = u_{xx} - \sin u \tag{5.18}$$

admits [520] explicit periodic solutions, called *breathers* (two-soliton solutions)

$$u(x,t) = 4\tan^{-1}\big[\tfrac{1}{\omega}\sqrt{1-\omega^2}\,\mathrm{sech}\big(\sqrt{1-\omega^2}\, x \sin \omega t\big)\big] \quad \big(\omega \in (0,1)\big).$$

In the differential geometry of pseudo-spherical surfaces of constant Gaussian curvature $K = -\rho^{-2}$, the study of (5.18) goes back to Edmond Bour (1862), Bonnet (1867), and Enneper (1868), [169], and the PDE is sometimes called the *Enneper equation*; see historical aspects in [521]. Detailed investigations of various forms of (5.18), including superposition behavior of its solutions, were performed by Bäcklund, Bianchi, Darboux and others, and "this work ... was essentially complete by the turn of the century...," [521, p. 1535]. The name *sine-Gordon* is associated with further exploitation of this equation as a 1D model of meson theory of nuclear forces developed in the 1960s, when this name has become customary (in 1967, G.L. Lamb used (5.18) for the study of propagation of ultrashort light pulses).

Existence and nonexistence of periodic solutions of the general *Klein–Gordon* (KG) *equation*

$$u_{tt} = u_{xx} - g(u),$$

with an arbitrary nonlinearity $g(u)$ on the right-hand side, is an important problem, in view of various physical applications in electromagnetizm, nonlinear optics, and quantum field theory. We refer to Segur–Kruskal [522], where an asymptotic approach to *nonexistence* of periodic solutions of the ϕ^4 *model*

$$u_{tt} = u_{xx} - 2u + 3u^2 - u^3$$

was proposed. Existence, nonexistence, and multiplicity of periodic solutions of nonlinear hyperbolic equations is an important direction of general PDE theory. We refer to Mitidieri–Pohozaev [425, Ch. 8], where a large amount of related existence-nonexistence results and further references are available.

More recent applications of breathers are associated with lattice theory that leads to discrete models. These models can involve many unit cells on the microscopic level and occur in the mathematical modeling of many physical processes, from chemical reaction theory and optics, to biology and acoustics. The *discrete sine-Gordon* (dsG) *equation* (or the *Frenkel–Kontorova* (FK) *model* from dislocation theory of plastic deformation in crystals, 1938) is an infinite-dimensional DS

$$\phi_n'' = \phi_{n+1} - 2\phi_n + \phi_{n-1} - \sin\phi_n, \quad n \in \mathbf{Z},$$

which, unlike its continuum counterpart (5.18), is not integrable, but is known to admit periodic solutions. In Section 9.5, we present further discussion of lattices and exact solutions on invariant subspaces for such discrete operators.

Compact breathers and localized blow-up patterns. Various lattices for the sG and more general KG equations are widely studied nowadays. Rosenau and Schochet [500] introduced *anharmonic lattices*, corresponding to the quasilinear *anharmonic KG equation*

$$u_{tt} + u = 3(u_x)^2 u_{xx} + u^3 \equiv \left[(u_x)^3\right]_x + u^3. \tag{5.19}$$

The operator on the right-hand side is variational and is a Frechet derivative of the following (Lagrangian) potential:

$$\Phi(u) = -\tfrac{1}{4}\int (u_x)^4\,\mathrm{d}x + \tfrac{1}{4}\int u^4\,\mathrm{d}x \quad \text{for} \quad u \in W^{1,4}(I\!R) \cap L^4(I\!R).$$

For us, equation (5.19) has particular interest, since it admits a *compact breather solution* in separable variables [500]

$$u_c(x,t) = \varphi(t) f(x), \tag{5.20}$$

where these two functions solve the ODEs

$$\varphi'' = \varphi^3 - \varphi \quad \text{and} \quad 3(f')^2 f'' + f^3 - f = 0.$$

The first ODE for $\varphi(t)$ admits a periodic solution, while the second equation for $f(x)$ has a compactly supported weak solution with finite interfaces yielding the compact breather. Furthermore, the ODE for $\varphi(t)$ admits blow-up solutions, so (5.20) then presents *localized* patterns of *regional blow-up* (an S-regime).

It is curious that the spatial part $f(x)$ of the compact breather (5.20) is the same as in blow-up analysis of the quasilinear parabolic p-Laplacian equation

$$u_t = 3(u_x)^2 u_{xx} + u^3. \tag{5.21}$$

These heat-type models are intensively studied in Section 6.4. The similarity blow-up solution of the S-regime takes the form

$$u_S(x,t) = \frac{1}{\sqrt{T-t}}\, f(x), \tag{5.22}$$

where $T > 0$ is blow-up time and $f(x)$ satisfies the same ODE

$$3(f')^2 f'' + f^3 - \tfrac{1}{2}\, f = 0 \tag{5.23}$$

(the constant $\frac{1}{2}$ is replaced by 1 via scaling in f). Such localized blow-up structures for the gradient-dependent diffusion have been studied from the beginning of the 1980s; see details and references in [509, Ch. 4]. The compactly supported solution of (5.23) is expressed in terms of the incomplete Euler Beta function B, and, in particular, the measure of the support is [216]

$$L_S = 2^{\frac{1}{2}} 3^{\frac{1}{4}} \pi, \tag{5.24}$$

which is called the *fundamental length* of this nonlinear medium with diffusion and reaction mechanisms. A fundamental character of the length (5.24) is supported mathematically: if $x = s(t)$ is the right-hand interface of an *arbitrary* blow-up solution $u(x,t) \geq 0$ of (5.21) with compactly supported initial data $u_0(x)$ having the right-hand interface at $s(0)$, then the localization holds, i.e.,

$$s(t) \leq s(0) + L_S \quad \text{for all} \quad t \in [0, T).$$

The proof is based on the Sturmian intersection comparison argument with the standing wave solution (5.22); details are given in [509, p. 245].

For the corresponding wave equation (5.19), any estimates of the blow-up interface propagation are OPEN PROBLEMS. There is also another fundamental difference between the above parabolic (5.21) and hyperbolic (5.19) PDEs with the same cubic operator. While the Cauchy problem for the parabolic equation is well-posed and there exists a unique weak local-in-time solution for any integrable compactly supported data u_0 (see [148, 309]), the hyperbolic PDE (5.19), admitting, possibly, an infinite number of shock wave discontinuities, needs a delicate adaptation to such nonlinearities of extensions of nonlinear semigroups (e.g., along the lines of Bressan's approach to 1D systems of conservation laws [81]).

The anharmonic lattices are obtained by a discretization of the PDE and admit a Hamiltonian representation. In Section 9.5, we study some lattices with breather solutions, which are "almost" compact (solutions cannot be compactly supported on a lattice, but the rate of spatial decay is *super-exponential*, which is typical for implicit difference schemes for degenerate quasilinear operators of the PME or p-Laplacian type). In the continuum limit, such discrete breathers correspond to compact ones for the PDE (5.19), which are periodic solutions in separable variables.

Example 5.6 (Blow-up patterns and compact breathers for quadratic operators) Consider the following quadratic p-Laplacian operator in $I\!R^N$:

$$F_2[u] = \nabla \cdot (|\nabla u| \nabla u) + u^2, \tag{5.25}$$

which is also variational with the potential

$$\Phi(u) = -\tfrac{1}{3}\int |\nabla u|^3\,dx + \tfrac{1}{3}\int u^3\,dx \quad \text{for} \quad u \in W^{1,3}(I\!R^N) \cap L^3(I\!R^N).$$

Invariant subspaces for operators in $I\!R^N$ are systematically studied in Chapter 6. Here, we borrow a simple result from Section 6.1.3 (Proposition 6.9): the operator (5.25) is associated with the 2D subspace

$$W_2 = \mathcal{L}\{1, f(x)\}, \tag{5.26}$$

where f is a solution of the following elliptic equation:

$$F_2[f] - f \equiv \nabla \cdot (|\nabla f| \nabla f) + f^2 - f = 0 \quad \text{in} \quad I\!R^N. \tag{5.27}$$

Namely, for any $C_1 \in I\!R$ and $C_2 \geq 0$,

$$\begin{aligned} F_2[C_1 + C_2 f] &= C_1^2 + 2C_1C_2 f + C_2^2 F_2[f] \\ &\equiv C_1^2 + (C_2^2 + 2C_1C_2) f \in W_2. \end{aligned} \tag{5.28}$$

In particular, this means that F_2 admits an *invariant cone*, $K_+ = \{C_{1,2} \geq 0\}$.

Compactly supported continuous weak (i.e., understood in the sense of distributions) solutions of (5.27) have been recognized since the beginning of the 1980s [216]; see details in Example 6.52. The equation (5.27) admits a nonnegative radially symmetric compactly supported solution $f(x)$ in any dimension $N \geq 1$, [229].

Using this invariant subspace, consider first the blow-up behavior in the corresponding parabolic equation

$$u_t = F_2[u] \quad \text{in} \quad I\!R^N \times I\!R_+. \tag{5.29}$$

Then, substituting

$$u_S(x,t) = C_1(t) + C_2(t) f(x) \tag{5.30}$$

into the PDE, in view of (5.28), yields the DS

$$\begin{cases} C_1' = C_1^2, \\ C_2' = C_2^2 + 2C_1C_2, \end{cases}$$

where $C_2(t)$ is assumed to be nonnegative. The first ODE gives $C_1(t) = -\frac{1}{t}$, and, integrating the second equation with $C_2 > 0$ yields the explicit blow-up pattern

$$u(x,t) = -\tfrac{1}{t} + \tfrac{T}{(T-t)t} f(x), \tag{5.31}$$

where $T > 0$ is the blow-up time. Turning a blind eye to the behavior of this solution at the initial moment of time $t = 0$ (see Example 6.52 for explanations), we observe the *regional blow-up* as $t \to T^-$, where the solution (5.31) blows up only inside the support of $f(x)$, i.e., for any x in the positivity domain $\{f > 0\}$.

Consider next the corresponding hyperbolic PDE with an extra linear term on the right-hand side necessary to create a breather solution,

$$u_{tt} = F_2[u] - u \quad \text{in} \quad I\!R^N \times I\!R. \tag{5.32}$$

Using solutions (5.30) yields a more difficult 4D DS (with $C_2 \geq 0$)

$$\begin{cases} C_1'' = C_1^2 - C_1, \\ C_2'' = C_2^2 + (2C_1 - 1)C_2. \end{cases} \tag{5.33}$$

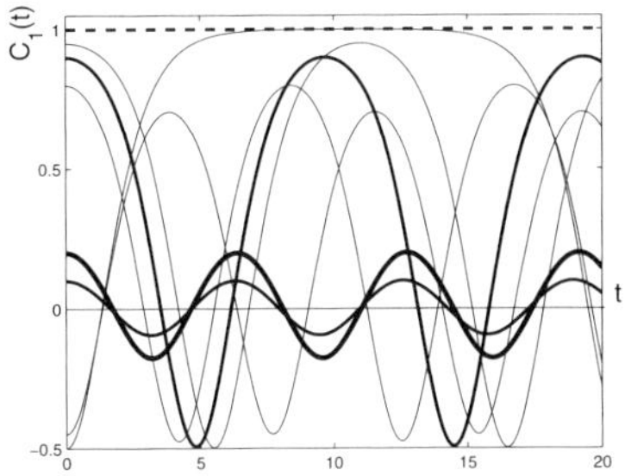

(a) (5.33): $C_1'' = C_1^2 - C_1$

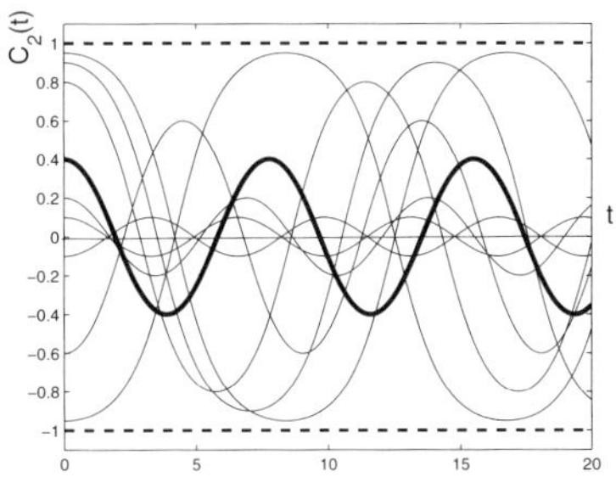

(b) (5.34): $C_2'' = |C_2|C_2 - C_2$

Figure 5.2 Periodic patterns in breather ODEs (5.33) and (5.34).

The standard breather solution is obtained for $C_1(t) \equiv 0$, where a slightly different PDE appears,

$$u_{tt} = \nabla \cdot (|\nabla u| \nabla u) + |u|u - u.$$

Here C_2 may change sign and satisfies a single ODE

$$C_2'' = |C_2|C_2 - C_2 \tag{5.34}$$

that possesses periodic solutions via Jacobi's elliptic functions; see Figure 5.2.

In the full DS (5.33), we initially find a solution $\tilde{C}_1(t)$, and next consider the second ODE with a given linear force $(2\tilde{C}_1(t) - 1)C_2$. Recall the hypothesis $C_2 \geq 0$ which may essentially affect dynamics on this invariant subspace.

Example 5.7 (**Breathers and blow-up in higher-order p-Laplacian PDEs**) Consider the fourth-order p-Laplacian operator with source

$$F_2[u] = -\Delta(|\Delta u| \Delta u) + u^2 \quad \big(\text{or with } u^2 \mapsto |u|u\big),$$

which has the potential

$$\Phi(u) = -\tfrac{1}{3} \int |\Delta u|^3 \, \mathrm{d}x + \tfrac{1}{3} \int u^3 \, \mathrm{d}x \quad \text{for} \quad u \in W^{2,3}(I\!R^N) \cap L^3(I\!R^N).$$

The above analysis is similar, where f solves a more complicated (variational) elliptic equation

$$F_2[f] - f \equiv -\Delta(|\Delta f| \Delta f) + f^2 - f = 0 \quad \text{in} \quad I\!R^N \quad (\text{or } f^2 \mapsto |f|f). \tag{5.35}$$

In particular, such compactly supported $f(x)$ generate compact breather solutions $u_c(x,t) = C_2(t) f(x)$ of the corresponding hyperbolic PDE (5.32), as well as localized blow-up patters $u_S(x,t) = (T-t)^{-1} f(x)$ of the parabolic equation (5.29). The same DSs occur on the invariant cone in subspace (5.26).

Restricted to radial solutions, (5.35) is a difficult fourth-order nonlinear ODE that admit oscillatory solutions near finite interfaces for the CP; see Example 5.9. For this ODE, existence is checked numerically. The profile (a) in Figure 5.3 is the first solution $f_1(x)$ of the ODE (5.35) in 1D. To underline a universality character of formation of compact and localized structures, for comparison, we include curves (b)

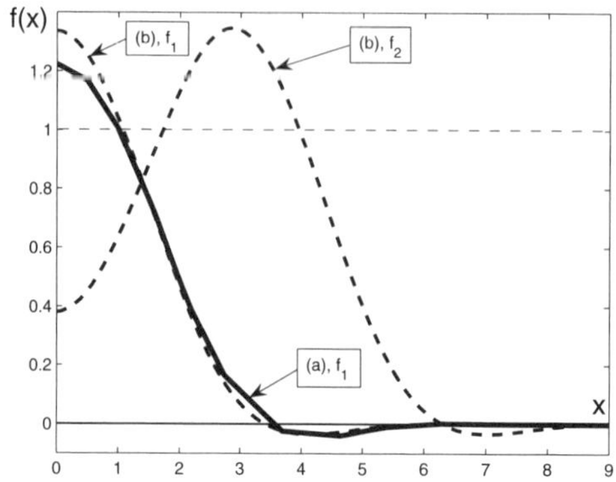

Figure 5.3 Compactly supported solutions of (5.35), $N = 1$ (a) and (5.36) (b).

that are first two compactly supported profiles $f_1(x)$ and $f_2(x)$ of the *non-variational* ODE

$$-(|f'''|f''')' + f^2 - f = 0 \quad \text{in} \quad I\!R. \tag{5.36}$$

This appears in constructing blow-up patterns for the reaction-diffusion PDE with a *non-potential* p-Laplacian,

$$u_t = -(|u_{xxx}|u_{xxx})_x + u^2 \quad \big(\text{or breathers for} \ \ u_t = -(|u_{xxx}|u_{xxx})_x + u^2 - u\big).$$

In both ODEs, replacing $f^2 \mapsto |f|f$ does not make any essential change in the solutions, in view of smallness of their oscillatory tail. Proof of existence, uniqueness of a bell-shaped solution, and overall multiplicity (typically countable sets for variational equations, as in the next example) are OPEN PROBLEMS.

Example 5.8 (**Higher-order porous medium-type operators**) Consider now equations containing PME-type operators with the parameter $n > 0$,

$$u_{tt} + u = -\Delta^2(|u|^n u) + |u|^n u \quad \text{(hyperbolic: breathers)},$$

$$u_t = -\Delta^2(|u|^n u) + |u|^n u \quad \text{(parabolic: blow-up)}.$$

The solutions in separate variables have the same form,

$$u(x,t) = C_2(t) f(x), \quad C_2'' = \tfrac{1}{n}\, |C_2|^n C_2 - C_2, \quad \text{or} \quad C_2(t) = (T-t)^{-\frac{1}{n}}, \tag{5.37}$$

where a compactly supported f solves the following elliptic (variational) equation:

$$-\Delta^2(|f|^n f) + |f|^n f - \tfrac{1}{n}\, f = 0 \quad \text{in} \quad I\!R^N. \tag{5.38}$$

For $N = 1$, such ODEs occurred earlier in compacton theory; see Example 4.7, where some profiles $\{f_k(x)\}$ were constructed numerically (such countable sets are associated with the variational setting).

Thus, formulae (5.37) and (5.38) give countable spectra of both compact breathers and localized blow-up patterns.

Example 5.9 (**p-Laplacian: oscillatory solutions in parabolic models**) As in Example 3.37, we take the 1D parabolic equation with the parameter $n > 0$,

$$u_t = F[u] \equiv -\big(|u_{xx}|^n u_{xx}\big)_{xx} \quad \text{in} \quad I\!R \times I\!R_+, \tag{5.39}$$

where F is a monotone monotone in $L^2(\mathbb{R})$, i.e., integrating by parts yields

$$\int (F[u] - F[v])(u - v)\,\mathrm{d}x = -\int \big(|u_{xx}|^n u_{xx} - |v_{xx}|^n v_{xx})(u_{xx} - v_{xx}\big)\,\mathrm{d}x \le 0$$

for all smooth compactly supported function $u, v \in C_0^\infty(\mathbb{R})$. For parabolic PDEs with monotone operators, there exists powerful existence-uniqueness theory of weak solutions; see [396, Ch. 2]. In order to understand the oscillatory nature of such solutions, as in Example 3.37 (see equation (3.189)), we write (5.39) in the sense of distributions, and consider sufficiently regular weak solutions that have a limit as $n \to 0^+$, meaning a connection with the linear bi-harmonic PDE

$$u_t = -u_{xxxx} \quad \text{in} \quad \mathbb{R} \times \mathbb{R}_+.$$

This has the oscillatory fundamental solution with the asymptotics (3.195).

Let us describe oscillatory properties of solutions of the quasilinear equation (5.39) by studying its fundamental solution

$$b(x,t) = t^{-\beta} F(\zeta), \quad \zeta = \tfrac{x}{t^\beta}, \quad \text{where} \quad \beta = \tfrac{1}{3n+4}$$

and F satisfies the ODE that is obtained after integration,

$$\big(|F''|^n F''\big)' = \beta \zeta F \quad \text{in} \quad \mathbb{R}. \tag{5.40}$$

In Figure 5.4(a), we present a compactly supported similarity profile $F(\zeta)$ for $n = 1$, normalized so that $F(0) = 1$ and $F'(0) = 0$ by symmetry. It is constructed by shooting from $\zeta = 0$ and corresponds to $F''(0) = -0.3938136507879....$ In (b), the oscillatory character of F near the interface at $\zeta = \zeta_0$ is shown and will be studied in detail next. We also specify in (b) a few zero contact angle FBP profiles F_1, ..., F_5, corresponding to smooth touching the ζ-axis (a correct setting of this FBP is not straightforward). In general, in view of changing sign behavior of $F(\zeta)$ as $\zeta \to \zeta_0^-$, there exists a sequence of FBP profiles $\{F_k(\zeta)\}$ such that the solution of the CP satisfies

$$F(\zeta) = \lim\nolimits_{k\to\infty} F_k(\zeta)$$

uniformly. The proof is straightforward for $n = 0$, i.e., for the linear equation (5.40), and is OPEN and difficult for $n > 0$. Notice that, by construction, each $F_k(\zeta)$ has precisely $k - 1$ zeros inside the support for $\zeta > 0$, which is a kind of Sturm's property for higher-order ODEs that is not associated with the Maximum Principle.

Assuming that F is compactly supported on some interval $[-\zeta_0, \zeta_0]$, let us introduce the oscillatory component by setting

$$F(\zeta) = (\zeta_0 - \zeta)^\gamma \varphi(s), \quad s = \ln(\zeta_0 - \zeta), \quad \text{where} \quad \gamma = \tfrac{3+2n}{n}, \tag{5.41}$$

so that we are looking for oscillatory behavior, as in Figure 3.5(a). Omitting exponentially small perturbations, we obtain the ODE for $\varphi(s)$,

$$\begin{aligned}(n+1)\big|\varphi'' + (2\gamma - 1)\varphi' + \gamma(\gamma - 1)\varphi\big|^n \big[\varphi''' + 3(\gamma - 1)\varphi''\\ + (3\gamma^2 - 6\gamma + 2)\varphi' + \gamma(\gamma - 1)(\gamma - 2)\varphi\big] = -\beta\zeta_0\varphi.\end{aligned} \tag{5.42}$$

The oscillatory character of solutions is shown in Figure 5.5 for $\zeta_0 = (3n+4)(n+1)$. The stable periodic solution gets smaller if n continues to decrease. For $n = 0.8$, the periodic orbit is already of the order 10^{-6}, while for $n = 0.5$, the oscillations are of

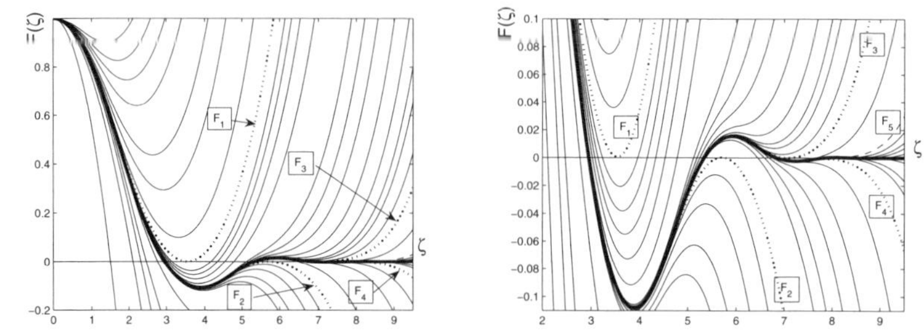

(a) $F(\zeta)$, the bold line (b) $F(\zeta)$, oscillations enlarged

Figure 5.4 The CP similarity profile satisfying (5.40), $n = 1$; F_k denote FBP profiles.

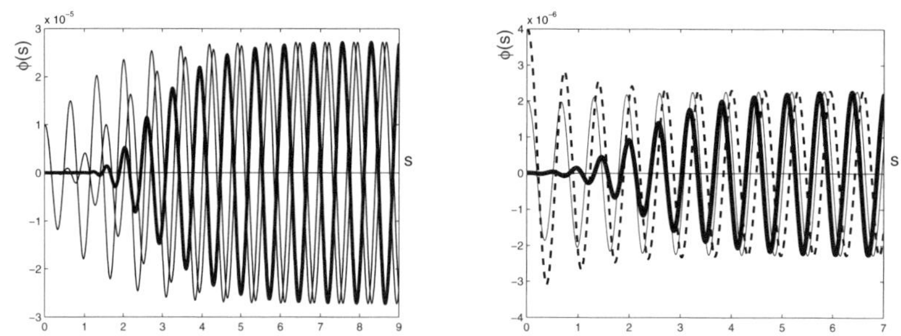

(a) $n = 1$ (b) $n = 0.8$

Figure 5.5 Periodic behavior for (5.42) with $n = 1$ (a) and $n = 0.8$ (b).

the order 10^{-9}. For larger $n \geq 2$, the stable oscillatory periodic patterns are shown in Figure 5.6.

On the other hand, using TWs $u(x, t) = f(y)$, with $y = x - \lambda t$ in the PDE (5.39)

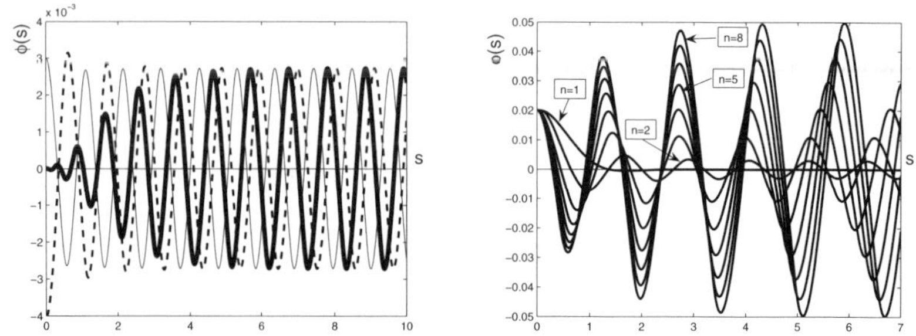

(a) $n = 2$ (b) $2 \leq n \leq 8$

Figure 5.6 Stable periodic behavior for (5.42) with $n = 2$ (a) and $n = 2, 3, 4, 5, 6, 7, 8$ (b) (the amplitude is monotone increasing with n).

yields the ODE

$$\lambda f' = \big(|f''|^n f''\big)'' \quad \text{for} \quad y > 0, \quad f(0) = 0,$$

where the interface is now at $y = 0^+$. Setting as in (5.41)

$$f(y) = y^\gamma \varphi(s), \quad \text{where} \quad s = \ln y,$$

yields precisely the ODE (5.42), with $\lambda = -\beta\zeta_0$. Therefore, Figures 5.5 and 5.6 also show the oscillatory character of TW solutions for any $\lambda < 0$. For $\lambda > 0$, (5.42) admits a positive constant solution, e.g., for $n = 1$, it is

$$\varphi(s) \equiv \tfrac{\lambda}{2400}. \tag{5.43}$$

The constant solutions for $\lambda > 0$, such as (5.43), are stable. As shown in Example 4.54, on the basis of the linear PDE for $n = 0$, we claimed that a periodic solution $\varphi(s)$ for $\lambda > 0$ does not exist.

Example 5.10 (**On degenerate hyperbolic models: finite propagation and oscillatory behavior**) Here, we review some hyperbolic PDEs with possible oscillatory behavior near finite interfaces.

Fourth and sixth-order PDEs. The TWs for the fourth-order wave equation

$$u_{tt} = -\big(|u_{xx}|^n u_{xx}\big)_{xx}, \quad \text{with} \quad n > 0, \tag{5.44}$$

are governed by the second-order Hamiltonian ODE $\lambda^2 f = -|f''|^n f''$ that does not admit solutions decaying to zero. This indicates that the propagation via smooth TWs is infinite and solutions are oscillatory at infinity (as for $n = 0$).

Consider next a similar sixth-order hyperbolic PDE,

$$u_{tt} = \big(|u_{xx}|^n u_{xx}\big)_{xxxx}, \quad \text{with parameter} \quad n > 0, \tag{5.45}$$

for which sufficiently smooth TW profiles solve the fourth-order ODE

$$\lambda^2 f = \big(|f''|^n f''\big)'' \quad \text{for} \quad y > 0, \quad f(0) = 0. \tag{5.46}$$

Solutions of the maximal regularity with a finite interface at $y = 0$ are given by

$$f(y) = y^\gamma \varphi(s), \quad s = \ln y, \quad \text{where} \quad \gamma = \tfrac{2(n+2)}{n}, \tag{5.47}$$

with e.g., $\varphi(s) \equiv \varphi_0$ (for $n = 1$, we have $f \in C^5$ and $f^{(5)}$ is Lipschitz continuous at $y = 0$). The oscillatory component φ solves the same fourth-order ODE, as appeared in Example 4.10 (with $\varphi \mapsto \varphi''$), where no periodic changing sign solutions were shown to exist.

Linear PDEs: fundamental solutions and TWs. To confirm the non-oscillatory character of solutions of (5.44) and (5.45), it is useful to apply the continuous connection as $n \to 0$ with the corresponding linear hyperbolic equations (cf. Example 3.37 and equation (3.188)). Concerning (5.44), we take

$$u_{tt} = -u_{xxxx} \quad \text{in} \quad \mathbb{R} \times \mathbb{R}_+, \tag{5.48}$$

with the fundamental solution $b(x,t) = \sqrt{t}\, g(y)$, with $y = x/\sqrt{t}$, where

$$g^{(4)} + \tfrac{1}{4} g'' y^2 + \tfrac{1}{4} g' y - \tfrac{1}{4} g = 0 \implies g(y) = \tfrac{1}{2\pi} \int_0^\infty \tfrac{\sin z \cos(\sqrt{z}y)}{z^{3/2}}\, dz.$$

It follows that $g(y)$ is not compactly supported (the oscillatory part is given by $g(y) \sim \cos(\frac{1}{2}y^2)$ as $y \to \infty$), so there is not finite propagation for (5.48) and solutions are oscillatory (cf. similar conclusions for (5.44)). Analogously, for (5.45) with $n = 0$ describing flexural oscillations of an elastic beam,

$$u_{tt} = u_{xxxxxx}\,, \tag{5.49}$$

the oscillatory and infinite propagation properties are seen from its fundamental solution

$$b(x,t) = t^{\frac{2}{3}} g(y), \quad y = x/t^{\frac{1}{3}}, \quad \text{where} \quad g(y) = \tfrac{1}{3\pi}\int_0^\infty \tfrac{\sin z \cos(z^{1/3}y)}{z^{5/3}}\,\mathrm{d}z.$$

Let us next study TWs, solving

$$f^{(4)} - \lambda^2 f = 0, \quad \text{with the characteristic equation } \mu^4 = \lambda^2 > 0,$$

so the decaying solutions are not oscillatory near the left-hand infinite interface at $y = -\infty$, which become the linear counterparts of those in (5.47) with the constant φ_0. In addition, there exist non-decaying oscillatory solutions like $f(y) = \cos(\sqrt{|\lambda|}y)$ (similar ones are admitted by (5.46)). The TW analysis also confirms that linear hyperbolic PDEs, such as (5.48), (5.49), and others do not allow finite propagation, unlike the canonical second-order model $u_{tt} = u_{xx}$.

Remark 5.11 Finite propagation with $\lambda = \pm 1$ exists in higher-order hyperbolic PDEs, such as

$$u_{tttt} = u_{xxxx}, \quad \text{etc.}$$

(strong estimates are obtained by multiplication by u_t in L^2). Odd-order linear dispersion equations $u_{ttt} = u_{xxx}$, etc. admit TW propagation with $\lambda = -1$ only.

More higher-order models. Consider TWs for a similar sixth-order PDE

$$u_{tt} = \big(|u_{xxx}|^n u_{xxx}\big)_{xxx}, \quad \text{where}$$

$$\big(|f'''|^n f'''\big)' - \lambda^2 f = 0 \quad \text{for} \quad y > 0, \quad f(0) = 0.$$

This ODE admits the strictly positive solution

$$f(y) = \varphi_0 y^\gamma, \quad \text{with a} \quad \varphi_0 > 0 \quad \text{and} \quad \gamma = \tfrac{3n+4}{n},$$

so there exists a class of non-oscillatory TW and other solutions. The ODE for the component $\varphi(s)$, with $s = \ln y$, is

$$f(y) = y^\gamma \varphi(s) \quad \Longrightarrow \quad P_4[\varphi] = \tfrac{1}{n+1}\,\tfrac{\varphi}{|P_3[\varphi]|^n}, \tag{5.50}$$

where the operators P_4 and P_3 are given in (4.93) and (3.166) respectively. Figure 5.7 shows a decaying unstable behavior for (5.50), $n = 1$, which is not periodic, so cannot be extended to the interface at $s = -\infty$ ($y = 0$).

Thus, oscillatory interfaces occur for, at least, eighth-order hyperbolic models,

$$u_{tt} = -\big(|u_{xxxx}|^n u_{xxxx}\big)_{xxxx}, \quad \big(\text{or } u_{tt} = -\big(|u_{xx}|^n u_{xx}\big)_{xxxxxx}\big).$$

Recall the fruitful change $v = u_{xxxx}$. Then the linear PDE for $n = 0$ has the TW equation of sixth order,

$$u_{tt} = -u_{xxxxxxxx} \quad \Longrightarrow \quad f^{(6)} + \lambda^2 f = 0,$$

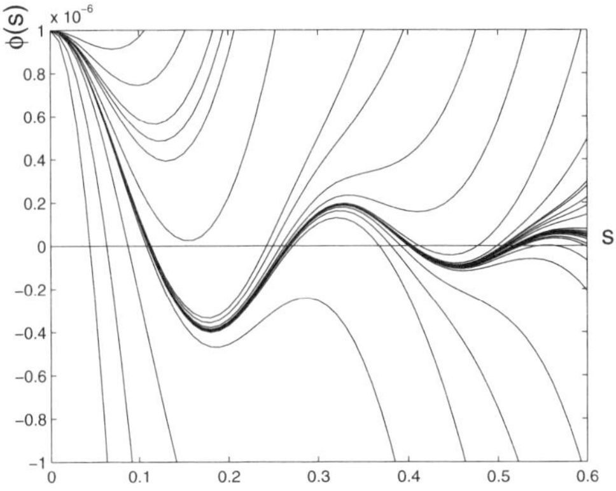

Figure 5.7 Unstable decaying behavior for ODE (5.50) for $n = 1$. Cauchy data are $\varphi(0) = 10^{-6}$, $\varphi'(0) = \varphi'''(0) = 0$, $\varphi''(0) = 0.000252638746\ldots$.

with the characteristic polynomial $\mu^6 + \lambda^2 = 0$, confirming that the interface at $y = -\infty$ is oscillatory, and remains oscillatory for small $n > 0$ by continuity; on passing to the limit $n \to 0$, see [174, Sect. 7.6].

Example 5.12 **(Finite propagation in singular dispersive Boussinesq equations)** Let us next briefly discuss finite propagation in semilinear hyperbolic models. First, consider the semilinear wave equation with a strong absorption (force) term,

$$u_{tt} = (-1)^{m+1} D_x^{2m} u - |u|^{p-1} u, \quad \text{with} \quad p < 1.$$

Studying TWs near finite interfaces leads to the ODE

$$(-1)^{m+1} f^{(2m)} - |f|^{p-1} f = 0 \implies f(y) = y^\gamma \varphi(\ln y), \ \gamma = \tfrac{2m}{1-p}. \tag{5.51}$$

Stable and unstable periodic behavior of $\varphi(s)$ appears in TFE theory; see Section 3.7. As $p \to 1^-$, i.e., approaching the linear equation, the smoothness of TWs at the interface point $y = 0$ increases (to C^∞ at $p = 1$). This mimics analytic TWs for $p = 1$, where no finite propagation is available.

For conservative Boussinesq-type models

$$u_{tt} = (-1)^{m+1} D_x^{2m} u - D_x^{2k}(|u|^{p-1} u), \quad 1 \le k < m,$$

(for $m = 2$, $k = 1$, it is the *signed* $B(1, p)$ dispersive Boussinesq equation, [496]), we have finite interfaces for TWs for $k < m$ (see explicit nonnegative compactons in [581] for $m = 2$, $k = 1$) and solutions of changing sign for $k \le m - 2$ with the behavior (5.51), where $2m$ is replaced by $2(m - k)$.

Example 5.13 **(A breather on invariant subspace for a cubic operator)** Consider now another cubic wave equation

$$u_{tt} = F[u] \equiv 2u^2 u_{xx} - u(u_x)^2 + u^3. \tag{5.52}$$

Such hyperbolic PDEs describe short-wave excitations of a nonlinear model, where each atom in a 1D lattice interacts with its neighbors by anharmonic forces, [349].

Note that operator F is not potential in L^2. The 2π-periodic breather is given by

$$u_*(x,t) = C(t)\cos x, \quad \text{where} \quad C'' = -C^3. \tag{5.53}$$

Clearly, the ODE for $C(t)$ has periodic solutions. Taking a single hump of $\cos x$ in (5.53) gives a solution of the Dirichlet problem for (5.52), where

$$u\left(\pm\tfrac{\pi}{2}, t\right) = 0 \quad \text{for} \quad t \geq 0,$$

so, unlike the compact breather (5.20), this is not a solution of the Cauchy problem. The operator in (5.52) has the advantage to admit the 3D subspace $W_3 = \mathcal{L}\{1, \cos x, \sin x\}$; see Proposition 1.30, operator F_6. Therefore, there exist exact solutions on its 2D restriction $W_2 = \mathcal{L}\{1, \cos x\}$,

$$u(x,t) = C_1(t) + C_2(t)\cos x, \tag{5.54}$$
$$\begin{cases} C_1'' = (C_1^2 - C_2^2)C_1, \\ C_2'' = (C_1^2 - C_2^2)C_2. \end{cases}$$

This DS exhibits a finite-dimensional evolution around the breather (5.53) that describes the periodic motion on the 1D invariant subspace $W_1 = \mathcal{L}\{\cos x\}$, with $C_1(t) \equiv 0$ in (5.54).

5.3 Quenching and interface phenomena, compactons

5.3.1 Basic singularity phenomena and stability

Example 5.14 (**Quenching, stability, and interfaces**) In order to describe singular quenching phenomena in quasilinear hyperbolic models, we introduce a simple equation combining the wave operator from (5.1) and a constant absorption term. This leads to a *quadratic wave equation with absorption*

$$\boxed{u_{tt} = (uu_x)_x - 1.} \tag{5.55}$$

We take smooth bounded initial functions $u(x,0) \geq a_0 > 0$ and $u_t(x,0)$, and, by classical theory of hyperbolic PDEs [550, Ch. 16], we obtain a local-in-time smooth positive solution $u(x,t)$ of the Cauchy problem for (5.55). In view of the constant negative absorption term -1, we may expect that there exists a finite time T such that $u(x,t)$ first touches the singular zero level $\{u = 0\}$ and ceases to exist as a classical solution. Without loss of generality, we assume that this happens the first time at the origin $x = 0$, i.e., $u(0,T) = 0$. Finally, the main (actually rather restrictive) assumption is that the classical solution without shock waves exists on $(0,T)$ (or, at least, shocks waves stay away from the extinction point $x = 0$).

As usual, we are interested in describing the formation of the quenching singularity as $x \to 0$ and $t \to T^-$ by using solutions on a polynomial subspace of the quadratic operator in (5.55). We choose the simplest subspace $W_2 = \mathcal{L}\{1, x^2\}$, so

$$u(x,t) = C_1(t) + C_2(t)x^2 \in W_2,$$

$$\begin{cases} C_1'' = 2C_1C_2 - 1, \\ C_2'' = 6C_2^2. \end{cases}$$

Choosing the positive function $C_2(t) = t^{-2}$ from the second ODE, and integrating the first equation yields the following solution:

$$u(x,t) = \tfrac{1}{t} - \tfrac{1}{3}\, t^2 \ln t + \tfrac{1}{t^2}\, x^2. \tag{5.56}$$

It is interesting first to analyze the *initial singularity* as $t \to 0^+$, where

$$u(x,t) = \tfrac{1}{t} + \eta^2 + \ldots, \quad \text{with the rescaled variable} \quad \eta = \tfrac{x}{t}.$$

This shows that initial functions for such solutions are entirely singular:

$$u(x,0) \equiv +\infty \quad \text{and} \quad u_t(x,0) \equiv -\infty \quad \text{in } I\!R.$$

Nevertheless, such initial data give rise to a local positive analytic solution with the given quadratic growth as $x \to \infty$.

Consider next the *quenching phenomenon* at the moment $t = T > 1$ such that

$$\tfrac{1}{T} = \tfrac{1}{3}\, T^2 \ln T.$$

Then, for all $t < T$, (5.56) is a smooth strictly positive solution of (5.55), and, as $t \to T^-$, the solution exhibits the following asymptotics:

$$u(x,t) = a_1(T-t) + a_2 x^2 + \ldots \equiv (T-t)\big(a_1 + a_2 \xi^2\big) + \ldots, \tag{5.57}$$

where a_1 and a_2 are positive constants, depending on T. Here ξ is the quenching rescaled spatial variable

$$\xi = \tfrac{x}{\sqrt{T-t}}. \tag{5.58}$$

Asymptotics (5.57) of the solutions shows a regular approach to the singularity point. In rescaled variables, there exists the limit as $t \to T^-$ of the rescaled function,

$$v(\xi,\tau) \equiv (T-t)^{-1} u(x,t) \to g(\xi) \equiv a_1 + a_2 \xi^2. \tag{5.59}$$

A general asymptotic stability problem is OPEN. Let us present some comments. By differentiating of $u = \mathrm{e}^{-\tau} v(\xi,\tau)$, $u_t = v_\tau - v + \frac{1}{2}\, v_\xi \xi$, $(uu_x)_x = \mathrm{e}^{-\tau}(vv_\xi)_\xi$, etc., we derive a singular perturbed PDE,

$$v_{\tau\tau} + v_{\tau\xi}\xi - v_\tau + \tfrac{1}{4}\, v_{\xi\xi}\xi^2 - \tfrac{1}{4}\, v_\xi \xi = \mathrm{e}^{-2\tau}(vv_\xi)_\xi - \mathrm{e}^{-\tau}, \tag{5.60}$$

where the second-order quadratic operator gives, on the right-hand side, an exponentially small perturbation as $\tau \to \infty$. The profile g given in (5.59) satisfies the limit (i.e., at $\tau = +\infty$) stationary ODE

$$\mathbf{C}[g] \equiv \tfrac{1}{4}\, g''\xi^2 - \tfrac{1}{4}\, g'\xi = 0.$$

On linear stability. The problem of the passage to the limit $\tau \to \infty$ in (5.60) remains OPEN. We briefly discuss the linear stability setting using similarities with that in Example 5.1. Linearization in (5.60) about the stationary profile by setting

$$v(\xi,\tau) = g(\xi) + Y(\xi,\tau)$$

leads to the following equation:

$$\begin{aligned} &Y_{\tau\tau} + Y_{\tau\xi}\xi - Y_\tau + \tfrac{1}{4}\, Y_{\xi\xi}\xi^2 - \tfrac{1}{4}\, Y_\xi \xi \\ &= \mathrm{e}^{-2\tau}\big[(gY_\xi)_\xi + (Yg')_\xi\big] - \mathrm{e}^{-\tau} + \mathrm{e}^{-2\tau}(YY_\xi)_\xi. \end{aligned} \tag{5.61}$$

In the first step of linearized analysis, we neglect both linear and nonlinear exponentially small perturbations on the right-hand side. Then, using the standard method of separation of variables for linear homogeneous PDEs and looking for

$$Y(\xi, \tau) = e^{\lambda \tau} \psi(\xi) \tag{5.62}$$

yields the following eigenvalue problem for the quadratic pencil of linear operators:

$$\{\lambda^2 I + \lambda \mathbf{B} + \mathbf{C}\}[\psi] \equiv \lambda^2 \psi + \lambda(\psi'\xi - \psi) + \tfrac{1}{4}\, \psi''\xi^2 - \tfrac{1}{4}\, \psi'\xi = 0. \tag{5.63}$$

Spectral theory of linear polynomial pencils is an important and classical area of differential operators theory; see Markus [412].

In general, for pencils of non-self-adjoint operators, even in the ordinary differential setting, the problem of discreteness of the spectrum, as well as completeness and closure of the eigenfunction sets, are often difficult. In the present case, we use a special advantage associated with the blow-up character of scaling variables. Since the coefficients of operators in (5.63) are unbounded as $\xi \to \infty$, the functional setting plays a key role. For such operators, a right setting is available in the weighted $L^2_\rho(I\!R)$ space with the exponential weight $\rho(\xi) = e^{-\xi^2}$. Since the operators are not self-adjoint, we are not obliged to be very determined in choosing a particular weight. As often happens in blow-up rescaled problems, our pencil generates some polynomial eigenfunctions.

Proposition 5.15 *The quadratic pencil* (5.63) *in* $L^2_\rho(I\!R)$ *has the discrete spectrum, consisting of two series of real eigenvalues*

$$\lambda_k^+ = -\tfrac{1}{2}(k-2) \quad \text{and} \quad \lambda_k^- = -\tfrac{1}{2}k \quad \text{for} \quad k = 0, 1, 2, \ldots, \tag{5.64}$$

with eigenfunctions $\psi_k(\xi)$ *being* k*th-order polynomials.*

The eigenvalues $\lambda_k^\pm$ in (5.64) are obtained by plugging $\psi_k(\xi) = \xi^k + \ldots$ into (5.63). We thus observe from (5.64) that, in the symmetric setting, there exist just two bad modes:
(i) $k = 0$ with $\lambda_0^+ = 1$, corresponding to the instability via perturbations of the blow-up time, and
(ii) $k = 1$ with $\lambda_1^+ = \frac{1}{2}$, corresponding to the shifting in x of the quenching point.
Both modes are excluded by fixing the time T and the point $x = 0$ of quenching in rescaled variables (5.58), (5.59). Since the polynomials are complete in any suitable weighted L^p-spaces, [352, p. 431], this gives certain evidence about the linear stability of the stationary profile g; though we should take into account that the constants a_1 and a_2 in (5.59) are arbitrary and depend on the initial data (this corresponds to the centre subspace behavior with $\lambda_k^\pm = 0$ for $k = 2$ and 0). The extension of the linear stability to the nonlinear one is a difficult OPEN PROBLEM.

On related blow-up problems on multiple zeros. It is curious that polynomial eigenfunctions of such pencils describe various pattern formation of multiple zeros in x of solution $u(x, t)$ of the linear wave equation

$$u_{tt} = u_{xx} \quad \text{in} \quad I\!R \times I\!R_- . \tag{5.65}$$

Namely, if a multiple zero occurs at the point (0, 0), we perform the blow-up scaling near this point (see [231])

$$u(x,t) = Y(\xi,\tau), \quad \text{with} \quad \xi = \tfrac{x}{(-t)}, \ \tau = -\ln(-t),$$

to get the PDE (cf. (5.61))

$$Y_{\tau\tau} + Y_\tau + 2Y_{\xi\tau}\xi = \mathbf{A}[Y] \equiv (1-\xi^2)Y_{\xi\xi} - 2Y_\xi \xi.$$

Looking for solutions in separate variables (5.62) yields the eigenvalue problem for a quadratic pencil,

$$\{(\lambda^2+\lambda)I + 2\lambda\xi D_\xi - \mathbf{A}\}[\psi] = 0, \quad \text{with eigenvalues}$$

$$\lambda_k^+ = -k \quad \text{and} \quad \lambda_k^- = -k-1 \quad \text{for} \quad k \ge 0,$$

and eigenfunctions $\psi_k(\xi)$ being kth-order polynomials. Again, in view of completeness and closure of the eigenfunction set $\Phi = \{\psi_k\}$ in weighted L^2 spaces, these eigenfunctions give a full countable set of different patterns of multiple zeros at $(0,0)$ for the wave equation (5.65),

$$u_k(x,t) = (-t)^{-\lambda_k^\pm}\psi_k\big(\tfrac{x}{(-t)}\big) \quad \big(\lambda_k^\pm < 0\big).$$

Therefore, each multiple zero of kth order is formed as $t \to 0^-$ by k zero curves focusing at $x=0$ with the behavior

$$x_j(t) = \xi_j(-t), \quad \text{where} \quad \psi_k(\xi_j) = 0.$$

For the linear parabolic *heat equation*

$$u_t = u_{xx},$$

a similar classification of multiple zeros with the heat kernel rescaled variable $\xi = x/\sqrt{-t}$ leads to the classic eigenvalue problem, where $\Phi = \{\psi_k\}$ consists of Hermite polynomials. These computations were first performed by Sturm (1836), [538], and initiated deep mathematical theory in relation to both PDEs and ODEs; see survey and references in [226, Ch. 1].

For hyperbolic equations, zero set theory is less developed. Similar quadratic pencils with the same spectrum occur [231] in the study of a different blow-up phenomenon for the semilinear wave equation with a nonlinear force term,

$$u_{tt} = \Delta u + |u|^{p-1}u \quad (p>1).$$

Interface propagation. For $t > T$, solution (5.56) takes negative values. To continue to deal with nonnegative solutions, one needs another FBP framework by introducing free-boundary conditions at the interfaces. Take now a simpler solution

$$u(x,t) = -\tfrac{1}{3}t^2\ln t + \tfrac{1}{t^2}x^2, \tag{5.66}$$

with the positive interface at

$$s(t) = \tfrac{1}{\sqrt{3}}t^2\sqrt{\ln t} \quad \text{for} \quad t > T = 1. \tag{5.67}$$

The dynamic interface equation is derived by using TWs $u(x,t) = f(y)$, with $y = x - \lambda t$, satisfying the ODE $\lambda^2 f'' = (ff')' - 1$, so that, for regular solutions,

$$(s')^2 = \lambda^2 = \tfrac{1}{f''}[(f')^2 - 1] \equiv \tfrac{1}{u_{xx}}[(u_x)^2 - 1] \quad \text{at} \ x = s(t). \tag{5.68}$$

Noticing that (5.68) holds for any regular solution (so it is a manifestation of regularity), we derive the actual dynamic equation by calculating $u_{xx} = \frac{2}{t^2}$ from (5.66), hence, $t = \sqrt{\frac{2}{u_{xx}}}$. Differentiating (5.67) yields

$$s' = \frac{2}{\sqrt{3\,u_{xx}}} \left[\left(\ln \frac{2}{u_{xx}}\right)^{1/2} + \frac{1}{2}\left(\ln \frac{2}{u_{xx}}\right)^{-1/2}\right].$$

This interface equation seems to be sufficient for the second-order PDE (5.55) to admit a local unique solution via the von Mises transformation (the analysis can be affected by possible formations of shock waves nearby). This is an OPEN PROBLEM. We will give further details of such a construction in the next model dealing with zero contact angle free-boundary conditions.

5.3.2 Zero contact angle and oscillatory solutions

Example 5.16 (Zero contact angle solutions) In order to create exact solutions with the zero contact angle at the interfaces, another modification of quasilinear wave equations is necessary. Consider the PDE with special left and right-hand sides

$$v\left(\sqrt{v}\right)_{tt} = (vv_x)_x + \alpha v + \beta v^{\frac{3}{2}} \quad \text{in} \quad I\!R \times I\!R_+ \tag{5.69}$$

for nonnegative solutions $v \geq 0$. This equation looks awkward, but by setting $v = u^2$ we obtain a simpler quadratic equation,

$$\boxed{u_{tt} = (u^2)_{xx} + 4(u_x)^2 + \alpha + \beta u}$$

that possesses solutions $u = C_1 + C_2 x + C_3 x^2$ on the subspace $\mathcal{L}\{1, x, x^2\}$. Obviously, then $v(x,t) = u^2(x,t)$ satisfies the zero contact angle condition

$$v = v_x = 0 \quad \text{at interfaces} \quad x = s(t). \tag{5.70}$$

To illustrate the behavior of such interfaces, set $\alpha = 0$ and $\beta = 1$, to get the PDE

$$\boxed{u_{tt} = (u^2)_{xx} + 4(u_x)^2 + u}$$

possessing solutions

$$u(x,t) = C_1(t) + C_3(t)x^2, \tag{5.71}$$

$$\begin{cases} C_1'' = 4C_1C_3 + C_1, \\ C_3'' = 28C_3^2 + C_3. \end{cases}$$

Choosing the equilibrium $C_3 = -\frac{1}{28}$ from the second ODE and substituting into the first equation yields $C_1'' = \frac{6}{7}\,C_1$, which gives, e.g., $C_1(t) = \cosh\sqrt{\frac{6}{7}}\,t$. The exact solution of the original equation (5.69) with $\alpha = 0$ and $\beta = 1$ takes the form

$$v(x,t) = \left(\cosh\sqrt{\tfrac{6}{7}}\,t - \tfrac{1}{28}\,x^2\right)_+^2, \tag{5.72}$$

which has exponentially expanding interfaces with the zero contact angle

$$s_\pm(t) = \pm R(t) \equiv \pm\sqrt{28\cosh\sqrt{\tfrac{6}{7}}\,t} \quad \text{for} \quad t > 0. \tag{5.73}$$

The regularity criterion at interfaces is obtained from the identity $\frac{d^2}{dt^2} u(s(t), t) = 0$, yielding

$$u_x s'' + 2u_{xt} s' + u_{xx}(s')^2 + u_{tt} = 0 \quad \text{at} \quad x = s(t). \tag{5.74}$$

The interface equation for the particular solutions (5.71), (5.72) is of stationary, Florin type,

$$u_{xx} = \tfrac{1}{28} \quad \text{at} \quad x = s(t).$$

In view of two already available free-boundary conditions (5.70), other dynamic interface equations look like regularity conditions. Introducing the von Mises transformation near the interface

$$X = X(u, t),$$

and assuming that it applies locally, by computations similar to those in Example 3.10, we arrive at the following second-order hyperbolic equation:

$$X_{tt} - \tfrac{1}{(X_u)^2}\big(2X_u X_{tu} - X_{uu} X_t\big) = 2u \, \tfrac{X_{uu}}{(X_u)^2} - \tfrac{6}{X_u} - u \, X_u. \tag{5.75}$$

We next impose the transversality assumption at the interface, $X_u \neq 0$ at $u = 0$, and then the local solvability and uniqueness of a sufficiently smooth solution depend on the spectral properties of the linear operator $\mathbf{B} = u \, \frac{d^2}{du^2}$. It is self-adjoint in the weighted space $L^2_\rho(0,1)$, $\rho = \frac{1}{u}$, has a compact resolvent, a discrete spectrum, and a complete, closed set of eigenfunctions. We refer to Naimark's monograph [432]. For the solvability of (5.75) by the semigroup approach and existence of a local regular solution, one needs to impose the conditions that this PDE is valid at $u = 0$, i.e.,

$$(X_u)^2 X_{tt} - (2X_u X_{tu} - X_{uu} X_t) = -6X_u,$$

which is the identity (5.74), written in terms of the von Mises variable X. As a consequence and an OPEN PROBLEM, it is believed that the compactly supported function (5.72) is a unique solution of the CP for the original equation (5.69).

Principles of formation of quenching patterns remain similar for higher-order wave equations with extra linear operators, e.g., for the Boussinesq-type equation

$$\boxed{u_{tt} = -u_{xxxx} + (uu_x)_x - 1.} \tag{5.76}$$

Questions of interface propagation becomes more delicate, as is usual for higher-order equations; cf. examples in Section 3.2. The fourth-order linear operator u_{xxxx} vanishes on W_2, and hence, will not affect the asymptotics of the singular extinction patterns.

On maximal regularity of oscillatory solutions. For solutions of changing sign in the CP, we consider the signed PDE with absorption

$$u_{tt} = -u_{xxxx} - \operatorname{sign} u \quad \text{in} \quad I\!R \times I\!R,$$

where we omit the quadratic term that is negligible near interfaces. The TWs satisfy

$$\lambda^2 f'' = -f^{(4)} - \operatorname{sign} f, \tag{5.77}$$

or, neglecting the left-hand side, which is smaller as $f \to 0$, we have a simple ODE

$$f^{(4)} + \operatorname{sign} f = 0 \quad \text{for} \quad y > 0, \quad f(0) = 0, \quad \text{so}$$

$$f(y) = y^4 \varphi(s), \quad \text{with} \quad s = \ln y, \tag{5.78}$$

where the oscillatory component φ solves

$$\varphi^{(4)} + 10\varphi''' + 35\varphi'' + 50\varphi' + 24\varphi + \operatorname{sign} \varphi = 0.$$

This is precisely the ODE (4.96) for a KS-type equation in Example 3.36, and for a fifth-order nonlinear dispersion equation in Example 4.10. This indicates a universality feature of formation of oscillatory patterns for nonlinear PDEs of different types. The stable periodic solution $\varphi(s)$ (the bold line) is shown in Figure 3.16 (see also Figure 4.13(b)). Therefore, the TWs (5.78) are oscillatory near interfaces. The λ-dependent right-hand side in (5.77) determines the next expansion term in (5.78), which states the dynamic interface equations for the CP (cf. Example 4.29).

5.3.3 Compactons in higher-order nonlinear dispersive Boussinesq equations

We now describe the dynamics of 2π-periodic solutions of the sixth-order dispersive Boussinesq equation $B(m, n, k)$

$$u_{tt} - u_{xx} + \alpha(u^m)_{xx} + \beta(u^n)_{xxxx} + \gamma (u^k)_{xxxxxx} = 0. \tag{5.79}$$

Taking $m = n = k = 2$ yields

$$\boxed{u_{tt} - u_{xx} + \alpha(u^2)_{xx} + \beta(u^2)_{xxxx} + \gamma (u^2)_{xxxxxx} = 0,} \tag{5.80}$$

where the quadratic operator admits $W_3 = \mathcal{L}\{1, \cos x, \sin x\}$ iff $-\alpha + 4\beta - 16\gamma = 0$. Hence, (5.80) restricted to W_2 possesses solutions

$$u(x, t) = C_1(t) + C_2(t) \cos x + C_3(t) \sin x,$$

$$\begin{cases} C_1'' = 0, \\ C_2'' = -(\mu C_1 + 1)C_2, \\ C_3'' = -(\mu C_1 + 1)C_3, \end{cases}$$

where $\mu = 2(\beta - \alpha - \gamma)$. Compactons are obtained for constant $C_1(t) \equiv A$, so denoting $\lambda^2 = \mu A + 1 > 0$, we obtain $u(x, t) = A + B \cos(x - \lambda t)$. Setting $A = B$ yields the compacton

$$u_c(x, t) = \tfrac{2(\lambda^2 - 1)}{2(\beta - \alpha - \gamma)} \cos^2[\tfrac{1}{2}(x - \lambda t)], \tag{5.81}$$

which is localized on $\{|x - \lambda t| < \pi\}$ and moves with constant speed $\lambda > 0$, as in Figure 4.1. The dynamic equation of the interfaces at $s(t) = \lambda t \pm \pi$ is derived from the PDE, as above, showing that compactons need an FBP setting.

Including the term δu^2 into (5.80) leads to blow-up behavior on W_3. If $C_1(t)$ is not constant, the dynamics of such periodic solutions are more complicated.

Oscillatory solutions of maximal regularity. The actual maximal regularity of solutions differs from the C^1 presented by (5.81). Consider the corresponding family of hyperbolic PDEs of the type (5.80), keeping only the main term

$$u_{tt} - u_{xx} = \big(|u|^n u\big)_{xxxxxx} \quad (n > 0).$$

The TWs satisfy, after integrating twice,

$$(\lambda^2 - 1) f = (|f|^n f)^{(4)}.$$

Therefore, for $\lambda \neq \pm 1$, the behavior near the interface at $y = 0$ is given by

$$f(y) = y^{\frac{4}{n}} \varphi(s), \quad \text{with} \quad s = \ln y, \tag{5.82}$$

where $F = |\varphi|^n \varphi$ satisfies an ODE of the form (4.93) from nonlinear dispersion KdV analysis (Example 4.10). Hence, there exists a periodic orbit $\varphi(s)$ if $|\lambda| < 1$ (TWs are oscillatory) and no such φ for $|\lambda| > 1$ (non-oscillatory TW interfaces). The same is true for $n = 0$.

For $n = 1$ in (5.82), we have the behavior $f(y) = O(y^4)$ as $y \to 0$, so the one-hump compacton (5.81), exhibiting less regularity $\sim O(y^2)$, is not a solution of the Cauchy problem in $I\!R \times I\!R$. The formula (5.82) gives a first approximation of multiple zeros behavior for such quasilinear hyperbolic equations and can be treated as a nonlinear counterpart of the linear operator pencil properties detected in Example 5.14. A classification of multiple zeros assumes solving a difficult nonlinear eigenvalue OPEN PROBLEM, which was studied for the second-order quasilinear parabolic equations and linear $2m$th-order PDEs, [226, pp. 29–34].

5.3.4 On the modified Zabolotskaya–Khokhlov equation

The dissipative *Zabolotskaya–Khokhlov* (ZK) *equation*

$$u_{tt} = -u_{xxx} - (uu_x)_x \tag{5.83}$$

arises in the theory of acoustic signals propagating through stratified media and has other applications, [587, 588, 363]. Actually, (5.83) is a *stationary* version of the following full PDE which is also known as the *3D Burgers equation*:

$$u_{xt} + (uu_x)_x + \nu u_{xxx} + \Delta_{\perp} u = 0, \tag{5.84}$$

where ν is a constant and $\Delta_{\perp} u$ denotes u_{yy}. Therefore, in (5.83), $u = u(x, y)$ with y replaced by t. For $\nu = 0$, (5.84) gives the *dispersionless Kadomtsev–Petviashvili equation*

$$u_{xt} + (uu_x)_x + u_{yy} = 0.$$

The ZK equation (5.83) contains a single quadratic operator of the PME type, with known polynomial or other subspaces that do not produce interesting singularity phenomena. Consider a modification of the ZK equation by adding an extra lower-order term

$$\boxed{u_{tt} = -u_{xxx} - (uu_x)_x - 2u^2.}$$

The quadratic operator $F[u] = (uu_x)_x + 2u^2$ admits the basic trigonometric subspace $W_3 = \mathcal{L}\{1, \cos, \sin x\}$, on which the solutions can be written as

$$u(x, t) = C_1(t) + C_2(t) \cos(x + \gamma(t)), \tag{5.85}$$

$$\begin{cases} C_1'' = -2C_1^2 - C_2^2, \\ C_2'' = \big[(\gamma')^2 - 3C_1\big] C_2, \\ \gamma'' = -\frac{2C_2'}{C_2} \gamma' - 1. \end{cases} \tag{5.86}$$

According to representation (5.85), the speed of propagation of such 2π-periodic waves is not constant (as in Example 4.3 for some extended KdV equations), and depends on the solutions under consideration. If solutions blow-up, the speed remains bounded, which follows from the last ODE in (5.86). Finite interfaces and oscillatory solutions can occur for PDEs with singular lower-order terms such as

$$u_{tt} = -u_{xxx} - (|u|^{p-1}u_x)_x, \quad \text{or} \quad u_{tt} = -u_{xxx} - |u|^{p-1}u,$$

where $p < 1$. These give OPEN PROBLEMS on existence, uniqueness and regularity.

5.4 Invariant subspaces in systems of nonlinear evolution equations

5.4.1 Main Theorem on invariant subspaces

Here we extend main results on linear subspaces (Section 2.1) to systems of evolution PDEs with vector-valued solutions $\mathbf{U} = (u_1, ..., u_n)^T$. In this case, we deal with symmetries of systems of linear ODEs

$$\frac{\mathrm{d}\mathbf{y}}{\mathrm{d}x} = P(x)\mathbf{y}, \tag{5.87}$$

where $x \in I\!R$ is the independent variable, $\mathbf{y} = (y_1, \dots, y_n)^T$ is a vector function, and $P(x)$ is a given $n \times n$ square matrix. Denote

$$\mathbf{F}(x, \mathbf{y}) = \big(F_1(x, \mathbf{y}), \dots, F_n(x, \mathbf{y})\big)^T \quad \text{and} \quad \frac{\partial}{\partial \mathbf{y}} = \big(\frac{\partial}{\partial y_1}, \dots, \frac{\partial}{\partial y_n}\big)^T.$$

All symmetries of system (5.87) are point and given by the operators

$$\begin{aligned} X &= (\mathbf{F}(x, \mathbf{y}))^T \frac{\partial}{\partial \mathbf{y}} \\ &\equiv F_1(x, y_1, \dots, y_n)\frac{\partial}{\partial y_1} + \dots + F_n(x, y_1, \dots, y_n)\frac{\partial}{\partial y_n}, \end{aligned} \tag{5.88}$$

with coefficients that are obtained from the invariance criterion

$$D\mathbf{F}(x, \mathbf{y}) = P(x)\mathbf{F}(x, \mathbf{y}).$$

Here, D is the operator of differentiation via system (5.87), i.e.,

$$\begin{aligned} D &\equiv \frac{\mathrm{d}}{\mathrm{d}x}\Big|_{(5.87)} = \frac{\partial}{\partial x} + (P(x)\mathbf{y})^T \frac{\partial}{\partial \mathbf{y}} \\ &\equiv \frac{\partial}{\partial x} + (P_1(x)\mathbf{y})\frac{\partial}{\partial y_1} + \dots + (P_n(x)\mathbf{y})\frac{\partial}{\partial y_n}, \end{aligned}$$

where $P_1(x), \dots, P_n(x)$ denote rows of the matrix $P(x)$.

Let Φ be an $n \times n$ *fundamental matrix of solutions* of system (5.87), i.e., the matrix, whose n columns are linearly independent solutions of this system. Let Ψ be the fundamental matrix of the adjoint system

$$\frac{\mathrm{d}\mathbf{z}}{\mathrm{d}x} = -(P(x))^T\mathbf{z},$$

where $\mathbf{z} = (z_1, \dots, z_n)^T$. According to ODE theory [132, Ch. 3], the product $\Psi^T\Phi$ is a constant matrix and the components $I_1, \dots, I_n$ of the column

$$\mathbf{I} = \Psi^T\mathbf{y} \tag{5.89}$$

give a system of independent integrals of system (5.87). Similarly to the study of single equations in Section 2.1, the following result is obtained.

Theorem 5.17 **("Main Theorem")** *The full algebra of symmetries of* (5.87) *is given by Lie-Bäcklund operators* (5.88) *with coefficients*

$$\mathbf{F}(x, \mathbf{y}) = \Phi\ A(\mathbf{I}),$$

where $A(\mathbf{I}) = (A^1(I_1, ..., I_n), ..., A^n(I_1, ..., I_n))^T$ *is a column of* n *arbitrary smooth functions.*

The fundamental matrix of the adjoint equation can be chosen as follows:

$$\Psi = (\Phi^{-1})^T.$$

Then the product $(\Psi)^T\Phi = E$ is the identity matrix and integrals (5.89) are

$$\mathbf{I} = \Phi^{-1}\mathbf{y}. \tag{5.90}$$

5.4.2 Examples: applications of the Main Theorem

We begin with simple examples illustrating Theorem 5.17.

Example 5.18 Consider the following system of ODEs:

$$\boxed{\frac{dy_1}{dx} = y_2, \quad \frac{dy_2}{dx} = -y_1.} \tag{5.91}$$

Its fundamental matrix takes the form

$$\Phi = \begin{bmatrix} \cos x & \sin x \\ -\sin x & \cos x \end{bmatrix} \quad \Longrightarrow \quad \Phi^{-1} = \Phi^T = \begin{bmatrix} \cos x & -\sin x \\ \sin x & \cos x \end{bmatrix},$$

and we find the integrals by formula (5.90),

$$I_1 = y_1 \cos x - y_2 \sin x \quad \text{and} \quad I_2 = y_1 \sin x + y_2 \cos x. \tag{5.92}$$

According to Theorem 5.17, the full algebra of symmetries of system (5.91) is given by operators

$$\begin{gathered} X = F_1(x, y_1, y_2)\frac{\partial}{\partial y_1} + F_2(x, y_1, y_2)\frac{\partial}{\partial y_2}, \ \text{where} \\ F_1 = A^1(I_1, I_2)\cos x + A^2(I_1, I_2)\sin x, \\ F_2 = -A^1(I_1, I_2)\sin x + A^2(I_1, I_2)\cos x, \end{gathered} \tag{5.93}$$

and $A^{(1)}$ and $A^{(2)}$ are arbitrary smooth functions of integrals (5.92).

Let us determine the symmetries that are independent of x, i.e., satisfying

$$\frac{\partial F_1}{\partial x} = \frac{\partial F_2}{\partial x} = 0.$$

This yields the system

$$\begin{cases} I_2 A^1_1 - I_1 A^1_2 = A^2, \\ I_2 A^2_1 - I_1 A^2_2 = -A^1, \end{cases}$$

which determines A^1 and A^2. Plugging these into (5.93), we obtain

$$\begin{bmatrix} F_1 \\ F_2 \end{bmatrix} = \begin{bmatrix} y_1 \\ y_2 \end{bmatrix} B^1(I) + \begin{bmatrix} y_2 \\ -y_1 \end{bmatrix} B^2(I),$$

where $I = (y_1)^2 + (y_2)^2$ and $B^1(I),\ B^2(I)$ are arbitrary smooth functions.

Example 5.19 **(Reaction-diffusion system)** Consider the system

$$\boxed{\begin{cases} u_t = u_{xx} + vB(u^2+v^2), \\ v_t = v_{xx} - uB(u^2+v^2), \end{cases}} \tag{5.94}$$

where B is an arbitrary function. Using the result of the previous example yields that the operator on the right-hand side admits $W_2 = \mathcal{L}\{\mathbf{f}_1, \mathbf{f}_2\}$, where

$$\mathbf{f}_1 = \begin{bmatrix} \cos x \\ -\sin x \end{bmatrix} \quad \text{and} \quad \mathbf{f}_2 = \begin{bmatrix} \sin x \\ \cos x \end{bmatrix}.$$

This makes it possible to look for solutions

$$\begin{bmatrix} u \\ v \end{bmatrix} = C_1(t) \begin{bmatrix} \cos x \\ -\sin x \end{bmatrix} + C_2(t) \begin{bmatrix} \sin x \\ \cos x \end{bmatrix}. \tag{5.95}$$

Substituting (5.95) into (5.94) yields

$$C_1'\mathbf{f}_1 + C_2'\mathbf{f}_2 = -C_1\mathbf{f}_1 - C_2\mathbf{f}_2 + (-C_1\mathbf{f}_2 + C_2\mathbf{f}_1)B(C_1^2+C_2^2), \quad \text{or}$$

$$\begin{cases} C_1' = -C_1 + C_2 B\big(C_1^2+C_2^2\big), \\ C_2' = -C_2 - C_1 B\big(C_1^2+C_2^2\big). \end{cases}$$

Solving this DS and using (5.95) gives

$$\begin{bmatrix} u \\ v \end{bmatrix} = D_1 \mathrm{e}^{-t} \begin{bmatrix} \cos(x+\bar{t}+D_2) \\ -\sin(x+\bar{t}+D_2) \end{bmatrix}, \quad \text{where} \quad \bar{t} = \int_{t_0}^{t} B(D_1^2 \mathrm{e}^{-2s})\, \mathrm{d}s,$$

where D_1 and D_2 are arbitrary constants. This solution is invariant under the operator $X = \frac{\partial}{\partial x} - u\frac{\partial}{\partial v} + v\frac{\partial}{\partial u}$.

Example 5.20 **(Nonlinear Schrödinger equation)** The following system of second-order PDEs:

$$\boxed{\begin{cases} u_t = -v_{xx} - \nu\, v(u^2+v^2), \\ v_t = u_{xx} + \nu\, u(u^2+v^2), \end{cases}}$$

is the real form of the cubic *nonlinear Schrödinger* (NLS) *equation**

$$\mathrm{i}z_t + z_{xx} + \nu\,|z|z = 0, \tag{5.96}$$

where $z = u + \mathrm{i}v$ and u, v are real-valued functions. It is straightforward that the operator on the right-hand side admits $W_2 = \mathcal{L}\{\mathbf{f}_1,\ \mathbf{f}_2\}$, where

$$\mathbf{f}_1 = \begin{bmatrix} \cos x \\ \sin x \end{bmatrix} \quad \text{and} \quad \mathbf{f}_2 = \begin{bmatrix} -\sin x \\ \cos x \end{bmatrix}.$$

Therefore, we look for solutions

$$\begin{bmatrix} u \\ v \end{bmatrix} = C_1(t) \begin{bmatrix} \cos x \\ \sin x \end{bmatrix} + C_2(t) \begin{bmatrix} -\sin x \\ \cos x \end{bmatrix}. \tag{5.97}$$

* Its derivation goes back to Da Rios, 1906.

Plugging this expression into (5.96) gives

$$C_1'\mathrm{f}_1 + C_2'\mathrm{f}_2 = -C_1\mathrm{f}_2 - C_2\mathrm{f}_1 + \nu(C_1\mathrm{f}_2 - C_2\mathrm{f}_1)\big(C_1^2 + C_2^2\big), \quad \text{or}$$

$$\begin{cases} C_1' = C_2 - \nu C_2\big(C_1^2 + C_2^2\big), \\ C_2' = -C_1 + \nu C_1\big(C_1^2 + C_2^2\big). \end{cases}$$

Solving the DS and substituting into (5.97) yields the explicit solution

$$\begin{bmatrix} u \\ v \end{bmatrix} = C_1 \begin{bmatrix} \cos(x + (D_1^2\nu - 1)t + D_2) \\ \sin(x + (D_1^2\nu - 1)t + D_2) \end{bmatrix},$$

where D_1 and D_2 are arbitrary constants. Indeed, this is the *traveling wave*

$$z = u + \mathrm{i}v = D_1 \mathrm{e}^{\mathrm{i}[x + (D_1^2\nu - 1)t + D_2]}.$$

The *fourth-order NLS equation* from nonlinear optics and quantum mechanics $\mathrm{i}z_t + z_{xxxx} + 2z_{xx} + \nu\,|z|^2 z = 0$ (see physics and references in Ablowitz–Segur [4]) possesses the same solution since $z_{xxxx} + z_{xx} = 0$ on W_2.

5.4.3 *Invariant subspaces for quadratic systems*

Further examples use simple invariant subspaces introduced in Sections 1.3–1.5 for the case of real-valued operators. In general, systems of nonlinear PDEs do not enjoy such a variety of subspaces of higher dimensions, but some results admit straightforward extensions. In what follows, we do not concentrate on establishing the optimal results on maximal dimensions of subspaces, a full classification of quadratic translation-invariant operators and other related delicate topics, but mainly just explain how some of the subspaces and solutions can be constructed.

On the one hand, in order to use our previous results for vector-valued operators $\mathbf{F}$, we may deal with vector sets defined in terms of the "span" of given linearly independent real-valued functions

$$W_{n,m} = \mathcal{L}\{f_1, ..., f_n\} \equiv \{\mathbf{C}_1 f_1 + ... + \mathbf{C}_n f_n : \ \mathbf{C}_1, ..., \mathbf{C}_n \in I\!R^m\},$$

which is calculated over $I\!R^m$. Then we will look for solutions

$$\mathbf{U}(x,t) = \mathbf{C}_1(t) f_1(x) + ... + \mathbf{C}_n(t) f_n(x) \tag{5.98}$$

for some vector-functions $\mathbf{C}_1(t)$, ..., $\mathbf{C}_n(t) \in I\!R^m$ for any $t \geq 0$. $W_{n,m}$ has dimension mn and (5.98) means that each component belongs to the scalar subspace $W_n = W_{n,1}$. We will use basic subspaces from Sections 1.3–1.5. We begin with some introductory examples of semilinear systems, and, as above, will discuss systems of two PDEs taking $m = 2$. Extensions of most of the results to larger m are straightforward. Here and later on, $a, b, c, ..., a_{1,2}, b_{1,2}, c_{1,2}, ...,$ and various Greek letters α, β, γ, ... with or without subscripts denote different real constants or constant vectors in $I\!R^m$ when necessary.

On the other hand, the majority of the results can be obtained by seeking invariant subspaces for standard real valued operators. We will use both approaches.

Example 5.21 **(Semilinear systems)** Consider the following system of semilinear heat equations:

$$\begin{cases} u_t = F_1[u,v] \equiv a_1 u_{xx} + b_1(u_x)^2 + c_1(v_x)^2 + d_1 u + e_1 v + f_1, \\ v_t = F_2[u,v] \equiv a_2 v_{xx} + b_2(u_x)^2 + c_2(v_x)^2 + d_2 u + e_2 v + f_2. \end{cases}$$

Invariant analysis is obvious, since each equation contains the quadratic operators preserving the polynomial subspace

$$W_3 = \mathcal{L}\{1, x, x^2\}, \tag{5.99}$$

in the sense that $F_{1,2} : W_3 \times W_3 \to W_3$. Looking for both components on W_3,

$$\begin{aligned} u(x,t) &= C_1(t) + C_2(t)x + C_3(t)x^2, \\ v(x,t) &= D_1(t) + D_2(t)x + D_3(t)x^2, \end{aligned} \tag{5.100}$$

yields a nonlinear DS for these six expansion coefficients,

$$\begin{cases} C_1' = 2a_1C_3 + b_1C_2^2 + c_1D_2^2 + d_1C_1 + e_1D_1 + f_1, \\ C_2' = 4b_1C_2C_3 + 4c_1D_2D_3 + d_1C_2 + e_1D_2, \\ C_3' = 4b_1C_3^2 + 4c_1D_3^2 + d_1C_3 + e_1D_3, \\ D_1' = 2a_2D_2 + b_2C_2^2 + c_2D_2^2 + d_2C_1 + e_2D_1 + f_2, \\ D_2' = 4b_2C_2C_3 + 4c_2D_2D_3 + d_2C_2 + e_2D_2, \\ D_3' = 4b_2C_3^2 + 4c_2D_3^2 + d_2C_3 + e_2D_3. \end{cases}$$

Subspace (5.99) remains invariant for 1D operators in a system with extra quadratic terms $\gamma_{1,2}u_xv_x$ added to both equations. Other linear operators with the subspace (5.99) can be put into equations. In further examples we will omit linear terms and operators and mainly deal with quadratic nonlinearities. For the corresponding system of hyperbolic PDEs

$$\begin{cases} u_{tt} = F_1[u,v], \\ v_{tt} = F_2[u,v], \end{cases}$$

the DS for solutions (5.100) takes the same form with $\frac{\mathrm{d}}{\mathrm{d}t}$ replaced by $\frac{\mathrm{d}^2}{\mathrm{d}t^2}$ on the left-hand side, so the DS is now of twelfth order.

Example 5.22 **(Trigonometric subspaces)** In order to use trigonometric and exponential subspaces, we need other semilinear quadratic models, e.g.,

$$\begin{cases} u_t = F_1[u,v] = u_{xx} + b_1(v_x)^2 + c_1 v^2, \\ v_t = F_2[u,v] \equiv v_{xx} + b_2(u_x)^2 + c_2 u^2. \end{cases} \tag{5.101}$$

The quadratic operators on the right-hand side are not coupled, and therefore, we can use the subspace

$$W_2 = \mathcal{L}\{1, \cos x\}, \tag{5.102}$$

or its 3D extension $W_3 = \mathcal{L}\{1, \cos x, \sin x\}$. Plugging

$$u(x,t) = C_1(t) + C_2(t)\cos x, \;\; v(x,t) = D_1(t) + D_2(t)\cos x \tag{5.103}$$

into (5.101) yields the following invariance conditions and the DS:

$$b_1 = c_1 \quad \text{and} \quad b_2 = c_2,$$

$$\begin{cases} C_1' = b_1 C_2 D_2 + b_1 D_1^2, \\ C_2' = -C_2 + 2b_1 D_1 D_2, \\ D_1' = b_2 C_2 D_2 + b_2 C_1^2, \\ D_2' = -D_2 + 2b_2 C_1 C_2. \end{cases} \tag{5.104}$$

In the opposite case,

$$b_1 = -c_1 \quad \text{and} \quad b_2 = -c_2,$$

we use the subspaces composed of exponential functions, $W_2 = \mathcal{L}\{1, \cosh x\}$, with some obvious changes in the DS. Second-order derivatives D_x^2 in these PDEs can be replaced by any $2m$th-order ones. Hyperbolic systems can also be treated similarly.

Example 5.23 (**KdV-type systems**) Integrable systems of coupled KdV-type equations have been intensively studied from the beginning of the 1980s. First examples include the *Hirota–Satsuma equations* [288]

$$\begin{cases} u_t = u_{xxx} + uu_x + vv_x, \\ v_t = -2v_{xxx} - uv_x, \end{cases}$$

possessing multi-soliton solutions (constructed by reduction to the bilinear form by setting $u = 2(\ln f)_{xx}$ and $v = \frac{g}{f}$) and an infinite number of conservation laws; and the *Ito equations* [302]

$$\begin{cases} u_t = u_{xxx} + 3uu_x + 3vv_x, \\ v_t = vu_x + uv_x, \end{cases}$$

again possessing infinitely many conservation laws and a recursion operator.

Consider a system of the third-order equations of the KdV-type with quadratic operators from the previous example,

$$\boxed{\begin{cases} u_t = F_1[u,v] \equiv u_{xxx} + b_1(v_x)^2 + c_1 v^2, \\ v_t = F_2[u,v] \equiv v_{xxx} + b_2(u_x)^2 + c_2 u^2. \end{cases}}$$

Then, unlike (5.103), the solutions are periodic moving with the constant speed 1,

$$u(x,t) = C_1 + C_2 \cos(x-t), \;\; v(x,t) = D_1 + D_2 \cos(x-t).$$

The DS is (5.104), where $-C_2$ is excluded from the second equation, and $-D_2$ from the last one. Solutions exist for systems of semilinear $(2m+1)$th-order PDEs.

Example 5.24 (**Quasilinear systems**) The system with homogeneous quadratic operators

$$\boxed{\begin{cases} u_t = F_1[u,v] \equiv a_1(vu_x)_x + b_1(v_x)^2, \\ v_t = F_2[u,v] \equiv a_2(uv_x)_x + b_2(u_x)^2, \end{cases}}$$

possesses solutions on $W_2 = \mathcal{L}\{1, x^2\}$,

$$u(x,t) = C_1(t) + C_2(t)x^2, \;\; v(x,t) = D_1(t) + D_2(t)x^2,$$

$$\begin{cases} C_1' = 2a_1 C_2 D_1, \\ C_2' = 6a_1 C_2 D_2 + 4b_1^2 D_2^2, \\ D_1' = 2a_2 C_1 D_2, \\ D_2' = 6a_2 C_2 D_2 + 4b_2 C_2^2. \end{cases}$$

One can construct more general solutions on the subspace (5.99).

Example 5.25 For some higher-order quasilinear equations, other polynomial subspaces can be used. For instance, the subspace $W_4 = \mathcal{L}\{1, x, x^2, x^3\}$ is suitable for the third-order PDEs

$$\boxed{\begin{cases} u_t = a_1(vu_{xx})_x + b_1(uu_{xx})_x + \text{(lower-order terms)}, \\ v_t = a_2(uv_{xx})_x + b_2(vv_{xx})_x + \text{(lower-order terms)}, \end{cases}}$$

where the lower-order (differential) terms are assumed to admit W_4.

Using systems of that type, we introduce an example of a higher-dimensional subspace. A more general analysis will be presented later on.

Example 5.26 (Extended polynomial subspace) Consider

$$\boxed{\begin{cases} u_t = F_1[u, v] \equiv a_1(vu_x)_x + b_1u_xv_x, \\ v_t = F_2[u, v] \equiv a_2(uv_x)_x + b_2u_xv_x. \end{cases}} \tag{5.105}$$

Let us look for components on $W_4 = \mathcal{L}\{1, x, x^2, x^3\}$ and set

$$\begin{aligned} u(x,t) &= C_1(t) + C_2(t)x + C_3(t)x^2 + C_4(t)x^3, \\ v(x,t) &= D_1(t) + D_2(t)x + D_3(t)x^2 + D_4(t)x^3. \end{aligned} \tag{5.106}$$

In general, quadratic operators in (5.105) do not leave W_4 invariant and map it onto W_5 containing the extra vector x^4. Substituting (5.106) into (5.105) yields that the coefficient of x^4 vanishes identically in both equations, provided that

$$\tfrac{a_1}{b_1} = \tfrac{a_2}{b_2} = -\tfrac{3}{5}.$$

This is the invariance condition of the subspace W_4, under which (5.105) on W_4 reduces to an eighth-order DS for the coefficients $\{C_i, D_i\}$.

Let us show that further extensions of the polynomial subspace W_4 is not possible. Unlike in the case of a single equation with quadratic nonlinearities (see Example 1.14 and Section 2.4), for system (5.105), the subspace

$$W_5 = \mathcal{L}\{1, x, x^2, x^3, x^4\} = W_4 \oplus \mathcal{L}\{x^4\}$$

cannot be invariant. Indeed, the quadratic operators map W_5 onto the extended subspace $W_7 = W_5 \oplus \mathcal{L}\{x^5, x^6\}$. Therefore, for functions

$$\begin{aligned} u &= C_1 + C_2x + C_3x^2 + C_4x^3 + C_5x^4, \\ v &= D_1 + D_2x + D_3x^2 + D_4x^3 + D_5x^4, \end{aligned}$$

too many invariant conditions are obtained. The first equation in (5.105) gives

$$6\big[2(2a_1 + b_1)C_5D_4 + (3a_1 + 2b_1)C_4D_5\big]x^5 + 4(7a_1 + 4b_1)C_5D_5x^6 \equiv 0,$$

from which we get three conditions $2a_1 + b_1 = 0$, $3a_1 + 2b_1 = 0$, and $7a_1 + 4b_1 = 0$, so $a_1 = b_1 = 0$. Similarly, a_2 and b_2 are annulled by the second equation,

$$6\big[2(2a_2 + b_2)C_4D_5 + (3a_2 + 2b_2)C_5D_4\big]x^5 + 4(7a_2 + 4b_2)C_5D_5x^6 \equiv 0.$$

There is another version of extending invariant subspaces for systems, where the components are taken from different subspaces; see King's solutions in Example

1.12. These systems are supposed to contain different operators in PDEs for each component, unlike a completely symmetric equations (5.105).

Example 5.27 Take $u(x,t)$ and $v(x,t)$ from different subspaces,

$$u(x,t) = C_1(t) + C_2(t)x + C_3(t)x^2 + C_4(t)x^3 + C_5(t)x^4 \in W_5,$$
$$v(x,t) = D_1(t) + D_2(t)x + D_3(t)x^2 \in W_3.$$

It is easy to reconstruct a number of operators supporting these subspaces in the corresponding component equations,

$$\begin{cases} u_t = a_1 v u_{xxxx} + b_1 v u_{xx} + c_1 u v_{xx} + d_1 u_{xx}(v_x)^2 + e_1(v^3)_{xx} + \dots, \\ v_t = a_2(vu)_{xxxx} + b_2 v_{xx} + c_2 u_{xx} + d_2(v_x)^2 + (v^3)_{xxxx} + \dots . \end{cases}$$

Example 5.28 More-dimensional subspaces appear for higher-order systems. For instance, the coupled quasilinear third-order equations

$$\boxed{\begin{cases} u_t = a_1(v u_{xx})_x + b_1 u_{xx} v_x, \\ v_t = a_2(u v_{xx})_x + b_2 u_{xx} v_x, \end{cases}}$$

possess solutions on $W_5 = \mathcal{L}\{1, x, x^2, x^3, x^4\}$, provided that

$$\tfrac{a_1}{b_1} = \tfrac{a_2}{b_2} = -\tfrac{2}{3}.$$

Example 5.29 (**Trigonometric subspaces**) We will look for solutions of system

$$\boxed{\begin{cases} u_t = F_1[u,v] \equiv a_1(v u_x)_x + b_1 u_x v_x + c_1 uv, \\ v_t = F_2[u,v] \equiv a_2(u v_x)_x + b_2 u_x v_x + c_2 uv, \end{cases}}$$

on the subspace (5.102). Substituting solutions (5.103) into the system yields the invariance condition $2a_1 + b_1 = c_1$, $2a_2 + b_2 = c_2$, and the DS

$$\begin{cases} C_1' = (a_1 + b_1)C_2 D_2 + c_1 C_1 D_1, \\ C_2' = -a_1 C_2 D_1 + c_1(C_2 D_1 + C_1 D_2), \\ D_1' = (a_2 + b_2)C_2 D_2 + c_2 C_1 D_1, \\ D_2' = -a_2 C_1 D_2 + c_2(C_2 D_1 + C_1 D_2). \end{cases}$$

Extensions of this example to systems of third and higher-order PDEs are similar to those for polynomial subspaces.

Example 5.30 Combining semilinear and quasilinear equations, a special invariance condition appears. The first PDE in the parabolic system

$$\boxed{\begin{cases} u_t = u_{xx} + b_1(u_x v_x + uv), \\ v_t = \alpha v v_{xx} + \beta(v_x)^2 + \gamma v^2 + \delta u, \end{cases}}$$

suggests using the subspace $W_2 = \mathcal{L}\{1, \cos x\}$, i.e., solutions (5.103). The quadratic operator in the second equation admits this subspace if $\gamma - \alpha - \beta = 0$, and the coefficients of solutions (5.103) satisfy the DS

$$\begin{cases} C_1' = b_1(C_1 D_1 + C_2 D_2), \\ C_2' = -C_2 + b_1(C_1 D_2 + D_1 C_2), \\ D_1' = \beta C_2 D_2 + \gamma D_1^2 + b_2(C_1 D_1 + C_2 D_2) + \delta C_1, \\ D_2' = -\alpha D_1 C_2 + 2\gamma D_1 D_2 + b_2(D_1 C_2 + C_1 D_2) + \delta C_2. \end{cases}$$

5.4.4 On restricted invariance: reductions to real-valued operators

For systems, consisting of similar quadratic or cubic PDEs for u and v, there is also the possibility to use some *restricted* invariant properties, assuming that the components are algebraically related to each other. In this case, we can directly use the results of classification of real-valued operators from Chapters 1 and 2. The following simple computations illustrate such an approach.

Example 5.31 Consider the following system $\big(m = 2,\, \mathbf{U} = (u, v)^T\big)$ of hyperbolic PDEs with general quadratic and linear operators:

$$\begin{aligned} \mathbf{U}_{tt} = \mathbf{F}[\mathbf{U}] \equiv \mathbf{a}_1 u u_{xx} + \mathbf{a}_2 (u_x)^2 + \mathbf{a}_3 u^2 + \mathbf{b}_1 v u_{xx} + \mathbf{b}_2 v_x u_x \\ + \mathbf{b}_3 u v + \mathbf{c}_1 v v_{xx} + \mathbf{c}_2 (v_x)^2 + \mathbf{c}_3 v^2 + \mathbf{a} u_{xx} + \mathbf{b} u + \mathbf{c} v_{xx} + \mathbf{d} v + \mathbf{e}, \end{aligned} \tag{5.107}$$

containing fourteen parameters from $I\!R^2$ denoted, as usual, by the boldface Latin letters, i.e., 28 arbitrary scalar parameters altogether. Looking for solutions of the PDEs (5.107) in the form

$$v = Au + B, \tag{5.108}$$

where A and B are two extra scalar unknowns, yields two quadratic PDEs for u

$$u_{tt} = P_{i1} u u_{xx} + P_{i2} (u_x)^2 + P_{i3} u^2 + P_{i4} u_{xx} + P_{i5} u + P_{i6}, \quad i = 1, 2, \tag{5.109}$$

with twelve constants $\{P_{ij}\}$, depending on all the parameters. For instance,

$$P_{11} = a_{11} + b_{11} A + c_{11} A^2, \quad P_{21} = \tfrac{1}{A}\big(a_{12} + b_{12} A + c_{12} A^2\big), \ \text{ etc.}$$

Solutions (5.108) exist, provided that the following six conditions hold:

$$P_{1j} = P_{2j} \quad \text{for} \quad j = 1, 2, ..., 6.$$

This system is consistent, since the total number of parameters including A and B is

$$28 + 2 = 30.$$

Thus, we can use the usual subspaces W_n from Chapters 1 and 2 for a single quadratic equation, such as (5.109), to obtain exact solutions for the system of PDEs (5.107). This corresponds to restricted invariance analysis, since the invariance conditions are checked for $u \in W_n$ on the affine manifold (5.108) in terms of v.

5.4.5 On extensions of basic invariant subspaces

Extensions of invariant subspaces are restricted by the following result, which is formulated for $m = 2$:

Theorem 5.32 **("Theorem on maximal dimension")** *Let $F[u, v]$ be a nonlinear ordinary differential operator of the order k that admits the invariant subspace W_n, i.e., $F : W_n \times W_n \to W_n$. Then $n \le 2k + 1$.*

The proof is contained in the proof of Theorem 2.44.

Extending polynomial subspaces. We now use the scalar form of operators for a more systematic extension analysis concerning the following quadratic operator defined for the vector function $\mathbf{U} = (u, v)^T$:

$$F[u, v] = \alpha u_{xx} v + \beta u_x v_x + \gamma u v_{xx}\,, \tag{5.110}$$

where α, β, and γ are scalar parameters. Clearly, $W_3 = \mathcal{L}\{1, x, x^2\}$ is invariant, and next we formulate the conditions of its extensions.

Proposition 5.33 *Operator* (5.110) *preserves:*

$$\text{(i)} \quad W_4 = \mathcal{L}\{1, x, x^2, x^3\} \quad \textit{iff} \;\; 2\alpha + 3\beta + 2\gamma = 0; \;\; \textit{and}$$
$$\text{(ii)} \quad W_5 = \mathcal{L}\{1, x, x^2, x^3, x^4\} \quad \textit{iff} \;\; 3\alpha + 4\beta + 3\gamma = 0.$$

(iii) *$F \ne 0$ does not admit polynomial subspaces of dimension six or more.*

Proof. By easy computations of $F[u]$, the above hypotheses annul all terms containing: (i) x^4, and (ii) x^5 and x^6. (iii) For $n = 6$, there is a direct proof by taking the expansion on W_6,

$$\mathbf{U} = \mathbf{C}_1 + \mathbf{C}_2 x + \mathbf{C}_3 x^2 + \mathbf{C}_4 x^3 + \mathbf{C}_5 x^4 + \mathbf{C}_6 x^5$$

with six coefficients $\mathbf{C}_i = (C_{1i}, C_{2i})^T \in I\!R^2$, the terms with x^6, x^7, and x^8 are annulled iff $\{\alpha, \beta, \gamma\}$ satisfy the linear system

$$\begin{cases} 20\alpha + 15\beta + 6\gamma = 0, \;\; 3\alpha + 4\beta + 3\gamma = 0, \;\; 6\alpha + 15\beta + 20\gamma = 0, \\ 3\alpha + 5\beta + 5\gamma = 0, \;\; 5\alpha + 5\beta + 3\gamma = 0, \;\; 4\alpha + 5\beta + 4\gamma = 0. \end{cases}$$

The first three equations annul the terms with $C_{16}C_{24}x^6$, $C_{15}C_{25}x^6$, and $C_{14}C_{26}x^6$, while the last three annul the terms with $C_{15}C_{26}x^7$, $C_{16}C_{25}x^7$, and $C_{16}C_{26}x^8$, respectively. The first three equations have the trivial solution only.

For arbitrary $n \ge 6$, the negative result follows from Theorem 5.32. □

Extending trigonometric subspaces. For this type of subspaces, consider operators with an extra reaction term,

$$F[u, v] = \alpha u_{xx} v + \beta u_x v_x + \gamma u v_{xx} + \delta u v. \tag{5.111}$$

Proposition 5.34 *Operator* (5.111) *preserves:*

$$\text{(i)} \quad W_3 = \mathcal{L}\{1, \cos x, \sin x\} \quad \textit{iff} \;\; \alpha + \beta + \gamma - \delta = 0;$$

$$\text{(ii)} \quad W_5 = \mathcal{L}\{1, \cos x, \sin x, \cos 2x, \sin 2x\}$$

for $\alpha = 2$, $\beta = -3$, $\gamma = 2$, and $\delta = 4$, where the operator is

$$F[u, v] = 2u_{xx}v - 3u_x v_x + 2u v_{xx} + 4uv.$$

(iii) *$F \ne 0$ does not admit trigonometric subspaces of more than five dimensions.*

Proof. (i) is straightforward by substituting

$$\mathbf{U} = \mathbf{C}_1 + \mathbf{C}_2 \cos x + \mathbf{C}_3 \sin x \in W_3$$

and observing that the invariance condition annuls the terms with $\cos 2x$ and $\sin 2x$ that contain four members having $C_{21}C_{22}$, $C_{21}C_{32}$, $C_{22}C_{31}$, and $C_{31}C_{32}$.

(ii) Taking the cos-expansion $\mathbf{U} = \mathbf{C}_1 + \mathbf{C}_2 \cos x + \mathbf{C}_3 \cos 2x$ yields the following homogeneous linear system:

$$\begin{cases} 4\alpha + 4\beta + 4\gamma = \delta, \\ \alpha + 2\beta + 4\gamma = \delta, \\ 4\alpha + 2\beta + \gamma = \delta, \end{cases}$$

where the first equation annuls the terms, containing $C_{13}C_{23} \cos 4x$. Two other linear equations, written in the form of $-\alpha - 4\gamma + \delta = 2\beta$ and $-4\alpha - \gamma + \delta = 2\beta$, guarantee that the terms

$$\begin{aligned} &\big[(-\alpha - 4\gamma + \delta) \cos x \cos 2x + 2\beta \sin x \sin 2x\big] C_{12}C_{23} \\ &+ \big[(-4\alpha - \gamma + \delta) \cos x \cos 2x + 2\beta \sin x \sin 2x\big] C_{22}C_{13} \end{aligned}$$

belong to $\mathcal{L}\{\cos x\}$. Using Reduce shows that the whole subspace W_5 is invariant.

(iii) See Theorem 5.32. □

Remark 5.35 **(Open problem: gap completing)** In the proof of (ii), the cos-expansion correctly determines the operator. We conjecture that this is the case for any suitable quadratic operator F, such as (1.86) with, say, even $k \equiv i + j$. Namely, if F preserves the invariant cos-subspace, it does the full cos/sin-subspace. (For polynomial subspaces, see a related conjecture in Remark 3.28.) Most of problems of "non-uniform" distributions of gaps are also OPEN.

Remark 5.36 **(Symmetric restriction)** Another approach to nonexistence in (iii) is based on using the *symmetric restriction* of F (cf. Section 2.7.3),

$$\hat{F}[u] = F[u,u] = (\alpha + \gamma) u u_{xx} + \beta (u_x)^2 + \delta u^2.$$

Hence, if F admits W_n, then $\hat{F} \neq 0$ (and is not linear) must also admit W_n. Hence $n > 2k + 1$ is impossible by Theorem 2.8 on maximal dimension. This proof does not cover the case of the null projection, $\alpha = -\gamma$ and $\beta = \delta = 0$.

Example 5.37 The following system of Boussinesq-type equations:

$$\mathbf{U}_{tt} = -\mathbf{U}_{xxxx} + a\mathbf{U}_{xx} + \big(2v u_{xx} - 3v_x u_x + 2u v_{xx} + 4uv\big) \begin{bmatrix} \alpha \\ \beta \end{bmatrix}$$

admits exacts solutions

$$\mathbf{U}(x,t) = \mathbf{C}_1(t) + \mathbf{C}_2(t) \cos x + \mathbf{C}_3(t) \sin x + \mathbf{C}_4(t) \cos 2x + \mathbf{C}_5(t) \sin 2x,$$

where ten coefficients $\{C_{ij},\ i = 1,2,\ j = 1,2,3,4,5\}$ solve a 20D DS.

Extending exponential subspaces. As in Sections 1.3–1.5, the *exponential* subspace for (5.111) is $W_3 = \mathcal{L}\{1, e^x, e^{-x}\}$ with the straightforward invariance condition $\alpha + \beta + \gamma + \delta = 0$. The analysis of the corresponding extended 5D subspace $W_5 = \mathcal{L}\{1, e^x, e^{-x}, e^{2x}, e^{-2x}\}$ is performed in a similar fashion. Some other examples of systems with invariant subspaces will be presented in the next chapter, where operators in $\mathbb{R}^N$ are studied.

Remarks and comments on the literature

§ 5.1–5.3. Another area of applications, where various quasilinear wave PDEs appear, is theory of relativity and gravitational *instantons* dealing with models

$$v_{xx} + v_{yy} = (\psi(v))_{tt},$$

where $\psi(v)$ is a nonlinear coefficient. Setting $u = \psi(v)$, in the case where $\psi' \geq 0$, yields a quasilinear wave equation in $I\!R^2 \times I\!R$. In particular, the exponential coefficient $\psi(v) = e^v$ leads to the *"heavenly" equation* (a continuous version $u_{tt} = (e^u)_{xx}$ of the *Toda lattice*). This plays a role in the theory of gravitational instantons and describes self-dual Einstein spaces with Euclidean signature and a rotational Killing vector, area preserving diffeomorphisms, and is a completely integrable system; see references in [413] and new exact analytical solutions therein. The integrable *Plebański second heavenly equation*

$$v_{tx} + v_{zy} + v_{xx}v_{yy} - (v_{xy})^2 = 0 \quad \text{in} \quad I\!R^3 \times I\!R \tag{5.112}$$

is descriptive of self-dual Einstein spaces with Ricci-flat metrics, [463]. More precisely, the Einstein field equations that govern self-dual gravitational fields are known to be reduced to single scalar-valued equations which are the first and second heavenly equations. Equation (5.112) belongs to the M-A-type and admits solutions on polynomial subspaces (see Example 6.55), $v = C_1 + C_2x + C_3x^2$, where the coefficients $\{C_k(y, z, t)\}$ satisfy a PDE system which can be simplified on other subspaces.

The classical *Boussinesq equation*

$$u_{tt} = -u_{xxxx} + u_{xx} - (u^2)_{xx} \tag{5.113}$$

first appeared in 1871 [74, p. 258]. It arises in many physical applications, such as propagation of long waves in shallow water, 1D nonlinear lattice-waves, vibration in nonlinear strings, and ion sound waves in plasma; see a survey in [127], where various exact solutions and related references are presented. The improved Boussinesq equation (5.17) (without the quadratic perturbation $2u^2$) is a model of vibrations in elastic rods and of DNA dynamics; see references in [127, Sect. 6]. N-soliton solutions of (5.113) were given by Hirota† [285]; see also [320] for other solutions that describe, for instance, an elastic soliton-breather interaction. Boussinesq equations, such as (5.113), admit interacting soliton solutions creating finite-time singularities; see [68], where earlier references are traced back to the beginning of the 1980s.

For the quasilinear hyperbolic-elliptic PDEs similar to those in Section 5.1, some evolution properties of solutions on various invariant subspaces and the corresponding DSs were studied in [162, 161]. The results of the group classification of such PDEs can be found in [10, Ch. 12], [299] for PDEs

$$u_{tt} = f(x, u_x)u_{xx} + g(x, u_x),$$

and in [366] for $u_{tt} = u_{xx} + F(x, t, u, u_x)$, where a detailed list of references can be found. The quadratic wave equation (5.1),

$$u_{tt} = (uu_x)_x,$$

which arises in different other physical contexts (e.g., longitudinal wave propagation on a moving threadline and electromagnetic transmission), possesses interesting exact solutions. In [199], where conditional symmetries of (5.1) were studied, the following solution was derived:

$$u(x, t) = f(x - \tfrac{1}{2}t^2) + t^2,$$

† The first exact soliton-type solutions of the Boussinesq equation were constructed by Baker in 1903 [22] via hyperelliptic $\wp$ function of genus two; see [19].

with f solving the ODE $(ff')' = 2 - f'$. This ansatz was originally proposed in 1949, [558], for the nonlinear wave equation $u_{tt} = u_{xx} + \beta(u^2)_{xx}$ that arises in transonic gas flows. A general Klein–Gordon equation, $u_{tt} = u_{xx} + g(u)$, admits the solution

$$u(x,t) = f(y), \quad y = x^2 - t^2, \quad \text{with the ODE} \quad (f'y)' + \tfrac{1}{4}\, g(f) = 0.$$

In the canonical form $u_{xt} = g(u)$, the solution is $u(x,t) = f(xt)$, where $(f'y)' = g(f)$. In [89], potential symmetries were used to show that (5.1) possesses implicit solutions

$$x^2 = ut^2 + C(\tfrac{t}{u})^{4/5}, \quad t^2 = \tfrac{x^2}{u} + C(\tfrac{x}{u^2})^{8/7} \quad (C = \text{constant}).$$

There is a large amount of mathematical literature devoted to *blow-up* singularities in semilinear and quasilinear wave equations. Let us mention the classical papers [305, 328, 332] and more recent, general results and surveys in [8, 90, 255, 382].

Compactons and compact breathers for the *modified improved Boussinesq equation*

$$u_{tt} = hu_{xxtt} + \sigma'(u_x)u_{xx} + V'(u)$$

(and more realistic PDEs with damping mechanism, $+\nu u_{xxt}$) are studied in [506], where further references are given. Here, $h = 0$ leads to a nonlinear Klein–Gordon equation also admitting compact structures.

In Section 5.3.3, sixth-order Boussinesq-type equations are considered, and the results are easily extended to similar quadratic models of arbitrary order $2m$. Dispersive fourth-order Boussinesq equations, called $B(m,n)$, i.e., (5.79) with $\gamma = 0$, were considered in [496] and studied in [581], where explicit TW compactons were constructed.

Various results on exact solutions, classical and nonclassical symmetry reductions of the ZK equation (5.83), and earlier references are given in [129], where, on p. 390 and p. 405, solutions on the subspace $\mathcal{L}\{1, x, x^2\}$ are considered.

§ 5.4. Parabolic systems of quasilinear PDEs occur in combustion theory [594, 33], in chemical reaction theory [187], in fluid mechanics [528], and many other areas of application. Exact solutions on invariant subspaces for a number of systems of nonlinear PDEs were considered in [217, 232, 110, 113], where further references can be found. A systematic analysis of invariant subspaces for general systems of two second-order parabolic PDEs

$$\begin{cases} u_t = f(u, v, u_x, v_x)u_{xx} + g(u, v, u_x, v_x), \\ v_t = p(u, v, u_x, v_x)v_{xx} + q(u, v, u_x, v_x), \end{cases}$$

was performed in [477]. A Lie group classification for systems of reaction-diffusion equations is available in [10, p. 171]. Explicit solutions for a particular quadratic system from the class in Example 5.31 were constructed in [111], where the standard trigonometric and exponential subspaces were used. For systems with first-order convection terms (not considered here) some new non-Lie group-invariant exact solutions were obtained in [114].

Concerning integrable KdV-type systems (i.e., possessing infinitely many generalized symmetries), notice earlier examples introduced by Fuchssteiner [210]; see a general classification in [197, 198, 562] and references therein. Some of the KdV-type systems and various extensions possess exact solutions on standard basic invariant subspaces. Soliton solutions of more general vector KdV and Ito equations were constructed in [287] by reduction to bilinear forms. Explicit multi-soliton solutions exist for a number of *sypersymmetric* integrable equations, which are systems of coupled equations for $u(x,t)$ (a *bosonic* field) and $\xi(x,t)$ (a *fermionic* field). One of the supersymmetric KdV equations is the *Manin–Radul super KdV equation*

$$\begin{cases} u_t = -u_{xxx} - 6uu_x + 3\xi\xi_{xx}\,, \\ \xi_t = -\xi_{xxx} - 3\xi_x u - 3\xi u_x\,. \end{cases}$$

This system, which was introduced by Manin and Radul in 1985 [410] (who also proposed a supersymmetric extension to the whole KP hierarchy), is integrable and, for $\xi = 0$, yields the KdV equation. Earlier Kupershmidt's super KdV-version [361] is not invariant under a space supersymmetric transformation. For several supersymmetric systems, the Baker–Hirota bilinear method applies and soliton solutions can be constructed; see references and results in [101, 251], though existence of higher-order N-solitons is not guaranteed. For systems of two semilinear equations with rather general cubic nonlinearities, the system

$$\begin{cases} u_t = u_{xx} - 2u^2 v + 2ku, \\ v_t = -v_{xx} + 2uv^2 - 2kv, \end{cases} \tag{5.114}$$

where k is a constant, is the only one admitting infinite-dimensional prolongation Lie algebras [7]. See [512] and references therein for a complete classification of two-component homogeneous polynomial symmetry-integrable systems (i.e., admitting a generalized symmetry of infinitely many orders). System (5.114), also emerging in the gauge gravity formulation, admits a Lax pair and Bäcklund transformation, and affords soliton-like solutions that are called *dissipatons*; see references in [7]. Note also the *Thirring equations*, [551] representing the most famous solvable field theory model,

$$\begin{cases} -\mathrm{i}u_x + 2v + 2|v|^2 u = 0, \\ -\mathrm{i}v_t + 2u + 2|u|^2 v = 0, \end{cases}$$

possess exact solitons constructed by using Baker's hyperelliptic functions generated by even hyperelliptic curves (similar to the KP equation; recall that the KdV one needs odd curves, [22, 415]); see results and earlier references in [166]. Baker's hyperelliptic functions associated with algebraic curves of genus two were used in [126] to construct traveling wave periodic and quasi-periodic solutions of two coupled nonlinear Schrödinger equations

$$\begin{cases} \mathrm{i}u_t + u_{xx} + (\kappa uu^* + \chi vv^*)u = 0, \\ \mathrm{i}v_t + v_{xx} + (\chi uu^* + \rho vv^*)v = 0, \end{cases}$$

in the case where $\kappa = \chi = \rho$, corresponding to the integrable *Manakov system* [409].

A large number of parabolic systems admitting second and third-order differential constraints were introduced in [324], where some of the examples deal with linear invariant subspaces. A classification of first-order differential constraints and substitutions was performed in an earlier paper by Kaptsov [318]. For systems, determining exact solutions via known differential constraints is often a difficult problem. Consider [318] the reaction-diffusion system

$$\begin{cases} u_t = u_{xx} + \mu(u-v) + \nu(u-v)\mathrm{e}^{u+v} - \mathrm{e}^{2(u+v)}, \\ v_t = d\, v_{xx} + \nu(u-v)\mathrm{e}^{u+v} - \mu(u-v) - 2d\mathrm{e}^{2(u+v)}, \end{cases} \tag{5.115}$$

where d is constant (d = Le, the *Lewis number* in reaction-diffusion theory, [383]). The functions $\mu(u-v)$ and $\nu(u-v)$ are arbitrary. The system is compatible with two differential constraints $u_x = v_x = \mathrm{e}^{u+v}$, which on integration imply the following solutions:

$$u(x,t) = \tfrac{1}{2}\left[C_1(t) - \ln(2x + C_2(t))\right], \;\; v(x,t) = \tfrac{1}{2}\left[-C_1(t) - \ln(2x + C_2(t))\right],$$

so $u - v = C_1(t)$, as the structure of the system suggests. Plugging into (5.115) yields the DS $C_1' = 2\mu(C_1)$, $C_2' = -2\nu(C_1)$. One may observe a direct similarity of nonlinearities in (5.115) and that of the hyperbolic *Tzitzéica equation* (sometimes also called the *Mikhailov–Dodd–Boullough equation*) $u_{tt} - u_{xx} = \mathrm{e}^u + \mathrm{e}^{2u}$, or of the corresponding elliptic *nonlinear Poisson equation*, $u_{tt} + u_{xx} = \mathrm{e}^u + \mathrm{e}^{2u}$, describing vortical structure in inviscid fluid, where u is the stream function; see [29]. Both PDEs admit Lie–Bäcklund symmetries [10, p. 205] and possess extra solutions on 1D subspaces; see the generalized separation of variables (GSV) method in [316] and [13, Ch. 5]. Application of the GSV to the corresponding parabolic PDE $u_t = u_{xx} + e^u - e^{2u}$ gives a simple traveling wave solution $u(x,t) = \frac{1}{2} \ln[\sqrt{2}(x - \sqrt{2}\,t)]$.

A classical example of quadratic systems is the completely anisotropic *XYZ Landau–Lifshitz* (LL) *equations* [369], [371, p. 417]

$$S_t = S \times S_{xx} + S \times JS, \tag{5.116}$$

where $\times$ denotes the vector product in $I\!R^3$, $S = (S_1, S_2, S_3)^T(x,t)$ is a vector-function, and $J = \text{diag}\,\{j_1, j_2, j_3\}$ is a constant diagonal 3×3 matrix. The scalar form of (5.116) contains quadratic operators,

$$\begin{cases} S_{1t} = S_2 S_{3xx} - S_3 S_{2xx} + (j_3 - j_2) S_2 S_3, \\ S_{2t} = S_3 S_{1xx} - S_1 S_{3xx} + (j_1 - j_3) S_1 S_3, \\ S_{3t} = S_1 S_{2xx} - S_2 S_{1xx} + (j_2 - j_1) S_1 S_2. \end{cases}$$

This model was derived for magnetic crystals as a modification of the Hamiltonian *Heisenberg spin equation* $S_t = S \times S_{ss}$, $S^2 = 1$ (t is time, s is the arc length) from solid state physics. The LL equations describe perturbations propagating in a direction orthogonal to the anisotropy axis in a ferromagnet (the anisotropy parameters satisfy $j_1 < j_2 < j_3$), and has since been identified as a soliton-bearing system possessing 1, 2, and N-soliton solutions constructed by the Baker–Hirota method; see references in [539, 507] and [35, p. 218] for finite-gap periodic solutions. Existence, uniqueness, and regularity results (solutions are smooth, except, at most, finitely many points) for equations such as (5.116) have been studied in a number of papers; see [106] and [105], where the Maximum Principle and comparison aspects for the LL equations are pointed out. Some exceptional symmetries exist for the radially symmetric LL equations for the N-dimensional magnetic spin system with an external magnetic field

$$Z_t = Z \times \Delta_N Z + Z \times H,$$

where Δ_N denotes the radial Laplacian in $I\!R^N$, and $H = (0,0,h)^T$ is a constant vector. For instance, there exist exact solutions

$$Z = c(r)(k_2 \cos k \sin m, k_2 \sin k \sin m, k_1 \cos m)^T,$$

where $k_{1,2}$ are constants, $c(r)$ is an arbitrary function, and $k = k(r,t)$, $m = m(r,t)$ satisfy

$$\begin{cases} \frac{\sin m}{c} k_t = \Delta_N m + \frac{2c'}{c} m_r - \sin m \cos m \,(k_r)^2 - h \frac{\sin m}{c}, \\ \frac{\sin m}{c} m_t = -\Delta_N k - \frac{2c'}{c} k_r - \frac{2 \cos m}{\sin m} m_r k_r. \end{cases} \tag{5.117}$$

This system generates a number of explicit solutions, including those with blow-up, [398]. In the case where $N = 2$ and $h = 0$, there exist blow-up similarity solutions [269]

$$Z(r,t) = \frac{1}{\sqrt{T-t}} f(y), \quad y = \frac{x}{(T-t)^{1/4}}, \quad f(y) = -\frac{1}{y^2}\left(\frac{1}{4}, \cos\frac{y^4}{4}, \sin\frac{y^4}{4}\right)^T$$

(similar solutions can be derived from system (5.117)), and the following global non-scaling invariant solutions:

$$Z(r,t) = \frac{1}{C\varphi_C(t)}\left(\cos\frac{r^2}{4\varphi_C(t)}, \sin\frac{r^2}{4\varphi_C(t)}, Ct\right)^T, \quad \text{with } \varphi_C(t) = \sqrt{\frac{1}{C^2} + t^2}$$

($C = \infty$ yields similarity solutions), [270]. Some of these solutions can be associated with invariant subspaces or sets.

Open problems

- In particular, these appeared in Examples 5.1, 5.5, 5.7, 5.9, 5.14, 5.16, Sections 5.3.3, 5.3.4, and in Remark 5.35.

CHAPTER 6

Applications to Nonlinear Partial Differential Equations in $I\!R^N$

We present invariant subspace methods and results that are applied to nonlinear second and higher-order differential operators in $I\!R^N$. The analysis of invariant conditions is now associated with systems of nonlinear elliptic PDEs, and results of the generality and completeness achieved earlier for $N = 1$ in the three previous chapters are difficult to obtain. Nevertheless, we give a number of examples of invariant subspaces in radial and non-radial geometry in $I\!R^N$, especially when dealing with polynomial and trigonometric functions.

6.1 Second-order operators and some higher-order extensions

6.1.1 Basic polynomial subspaces

As in 1D geometry, we begin with a class of quadratic second-order operators admitting polynomial subspaces. Consider a family of operators given by

$$F[u] = \alpha(\Delta u)^2 + \beta u \Delta u + \gamma |\nabla u|^2 \quad \text{in } I\!R^N, \tag{6.1}$$

where, as usual, α, β, and γ are arbitrary parameters. Here $x = (x_1, x_2, ..., x_N)^T$ is a vector in $I\!R^N$. Obviously, F preserves the subspace of linear functions

$$W_N^{\rm lin} = \mathcal{L}\{x_1, ..., x_N\}. \tag{6.2}$$

This defines a simple map $F : W_N^{\rm lin} \to \mathcal{L}\{1\}$. The next observation is also easy.

Proposition 6.1 *Operator* (6.1) *preserves:*
(i) *the 2D subspace of radial functions*

$$W_2^{\rm r} = \mathcal{L}\{1, |x|^2\}; \tag{6.3}$$

(ii) *the* $(N+1)$*-dimensional subspace of diagonal quadratic forms*

$$W_{N+1}^{\rm q} = \mathcal{L}\{1, x_1^2, ..., x_N^2\}; \tag{6.4}$$

(iii) *the subspace of arbitrary quadratic forms*

$$W_n^{\rm q} = \mathcal{L}\{1, x_i x_j,\ 1 \le i, j \le N\} \quad \left(n = 1 + \tfrac{N(N+1)}{2}\right); \ \textit{and} \tag{6.5}$$

(vi) *the direct sum of subspaces* (6.2) *and* (6.5)*,*

$$W_n^{\rm q} \oplus W_N^{\rm lin}. \tag{6.6}$$

Let us perform basic computations for the general case (6.6).

Example 6.2 Consider the quasilinear PDE with the operator (6.1),

$$\boxed{u_{tt} = F[u] \equiv \alpha(\Delta u)^2 + \beta u \Delta u + \gamma |\nabla u|^2 \quad \text{in } I\!R^N \times I\!R_+,} \tag{6.7}$$

which is formally hyperbolic in the domain $\{\alpha \Delta u + \beta u > 0\}$. Using subspace (6.6), let us look for solutions

$$u(x,t) = C(t) + \mathbf{D}(t)x + x^T A(t)x, \tag{6.8}$$

where $C(t)$ is a function, $\mathbf{D}^T(t) = (d_1(t), ..., d_N(t))^T \in I\!R^N$, and $A(t) = \|a_{i,j}(t)\|$ is an $N \times N$ symmetric matrix with elements $a_{i,j}(t) = a_{j,i}(t)$ for all i, j. Then,

$$\Delta u = 2 \textstyle\sum_{(i)} a_{i,i} \equiv 2\,(\text{tr}\, A),$$

and substituting (6.8) into (6.7), after some manipulations, we obtain the DS

$$\begin{cases} C'' = 4\alpha\,(\text{tr}A)^2 + 2\beta\,(\text{tr}\,A)\,C + \gamma\,|\mathbf{D}|^2, \\ \mathbf{D}'' = 2\beta\,(\text{tr}\,A)\,\mathbf{D} + 4\gamma\,\mathbf{D}A, \\ A'' = 2\beta\,(\text{tr}\,A)\,A + 4\gamma\,A^2. \end{cases} \tag{6.9}$$

In general, (6.9) is a difficult DS even in lower-dimensional cases $N = 3$ or 2; see references and related results in Remarks.

Example 6.3 (**Quadratic wave equation**) The PDE

$$\boxed{u_{tt} = \nabla \cdot (u \nabla u) + a + bu \quad \text{in } I\!R^2 \times I\!R}$$

is hyperbolic in $\{u > 0\}$ and admits solutions

$$u(x,t) = C(t) + a_{1,1}(t)x_1^2 + 2a_{1,2}(t)x_1x_2 + a_{2,2}(t)x_2^2 \in W_4^{\rm q}. \tag{6.10}$$

The corresponding DS is eighth-order,

$$\begin{cases} C'' = 2(a_{1,1} + a_{2,2})C + a + bC, \\ a_{1,1}'' = 6a_{1,1}^2 + 2a_{1,1}a_{2,2} + 4a_{1,2}^2 + ba_{1,1}, \\ a_{1,2}'' = 6(a_{1,1} + a_{2,2})a_{1,2} + ba_{1,2}, \\ a_{2,2}'' = 4a_{1,2}^2 + 2a_{1,1}a_{2,2} + 6a_{2,2}^2 + ba_{2,2}, \end{cases} \tag{6.11}$$

so the set of solutions is 8D. Setting $a_{1,2}(t) \equiv 0$ gives a sixth-order DS for $\{C, a_{1,1}, a_{2,2}\}$ and a 6D manifold of solutions

$$u(x,t) = C(t) + a_{1,1}(t)x_1^2 + a_{2,2}(t)x_2^2 \in W_3^{\rm q} = \mathcal{L}\{1, x_1^2, x_2^2\}.$$

Therefore, $W_4^{\rm q}$ contains new solutions, which cannot be obtained from those on $W_3^{\rm q}$ despite the fact that the quadratic form in (6.10) is diagonalizable by an orthogonal transformation in $I\!R^2$. Clearly, this reflects the fact that, for hyperbolic (second-order in t) PDEs, non-diagonal terms in solutions can occur evolutionarily, even if initial data $u(x,0)$ was given by a diagonal quadratic form. For parabolic (first-order in t) PDEs, initially diagonal forms remain diagonal for all times.

Example 6.4 (**PME-type equations**) Next, consider solutions on $W_{N+1}^{\rm q}$ of a quasilinear parabolic equation. The pressure transformation $u = v^\sigma$, $\sigma \neq 0$, in the PME with extra reaction-absorption terms

$$v_t = \nabla \cdot (v^\sigma \nabla v) + \alpha v^{1-\sigma} + \beta v \quad (v > 0)$$

yields the PDE with quadratic operators

$$\boxed{u_t = F[u] \equiv u\Delta u + \tfrac{1}{\sigma}\,|\nabla u|^2 + \sigma\alpha + \sigma\beta u.}$$

We use the subspace (6.4) and take

$$u(x,t) = C(t) + \textstyle\sum_{(i)} a_{i,i}(t)\,x_i^2,$$

where $a_{i,i}(t)$ for $i = 1, 2, ..., N$ are the only non-zero diagonal terms of the matrix $A(t)$. This gives the following $(N{+}1)$-dimensional DS:

$$\begin{cases} C' = 2\,(\mathrm{tr}\,A)\,C + \sigma\alpha + \sigma\beta C, \\ a_{i,i}' = 2\,(\mathrm{tr}\,A)\,a_{i,i} + \frac{4}{\sigma}\,a_{i,i}^2 + \sigma\beta a_{i,i}. \end{cases} \tag{6.12}$$

For equal coefficients $a_{1,1} = ... = a_{N,N} = a$, i.e., for radially symmetric solutions

$$u(r,t) = C(t) + a(t)r^2 \quad (r = |x|)$$

on the subspace (6.3), the DS (6.12) reduces to a simple explicitly solvable system

$$\begin{cases} C' = 2NaC + \sigma\alpha + \sigma\beta C, \\ a' = 2(N + \frac{2}{\sigma})a^2 + \sigma\beta a. \end{cases}$$

Example 6.5 (**Ernst equation from general relativity**) To illustrate application to systems, consider the nonstationary version of the *Ernst equation*

$$\boxed{\begin{cases} f_t = f\Delta f - |\nabla f|^2 + |\nabla g|^2, \\ g_t = f\Delta g - 2\nabla f \cdot \nabla g. \end{cases}} \tag{6.13}$$

For time-independent functions in $I\!R^3$, setting $\mathcal{E} = f + \mathrm{i}g$ yields the stationary Ernst equation

$$(\mathrm{Re}\,\mathcal{E})\Delta\mathcal{E} = \nabla\mathcal{E}\cdot\nabla\mathcal{E}$$

that describe stationary axisymmetric space-times in general relativity, [171]. Using the polynomial subspace (6.4) yields solutions of (6.13)

$$f(x,t) = C_0(t) + \textstyle\sum_{i=1}^N C_i(t)x_i^2, \quad g(x,t) = D_0(t) + \sum_{i=1}^N D_i(t)x_i^2,$$

where the expansion coefficients solve the DS

$$\begin{cases} C_0' = 2\big(\sum C_k\big)C_0, \;\; C_i' = 2\big(\sum C_k\big)C_i + 4\big(D_i^2 - C_i^2\big), \\ D_0' = 2\big(\sum D_k\big)C_0, \;\; D_i' = 2\big(\sum D_k\big)C_i - 8C_iD_i. \end{cases}$$

In radial setting for solutions

$$f(x,t) = C_0(t) + C(t)|x|^2, \quad g(x,t) = D_0(t) + D(t)|x|^2,$$

using $\Delta|x|^2 = 2N$, we obtain a simpler DS,

$$\begin{cases} C_0' = 2NC_0C, \quad C' = 4D^2 + 2(N-2)C^2, \\ D_0' = 2NC_0D, \quad D' = 2(N-4)CD. \end{cases}$$

Example 6.6 These polynomial subspaces can be used for another class of second-order parabolic equations. Consider the following homogeneous quadratic PDE:

$$vv_t = \alpha v\Delta v + \beta|\nabla v|^2 + \gamma v^2 \quad \text{in} \;\; I\!R^N \times I\!R_+,$$

where, unlike many previous examples, the left-hand side is also quadratic. By the exponential change $v = e^u$, this PDE reduces to the standard form

$$\boxed{u_t = \alpha \Delta u + (\alpha + \beta)|\nabla u|^2 + \gamma,}$$

containing the quadratic Hamilton–Jacobi operator from the class (6.1). By Proposition 6.1, we can construct various solutions on the subspaces listed therein.

6.1.2 *Two quadratic water wave models*

Example 6.7 **(Benney–Luke equation for water waves)** The isotropic *Benney–Luke* (BL) *equation* in $I\!R^3 \times I\!R$

$$\boxed{u_{tt} - \Delta u + \mu(a\Delta^2 u - b\Delta u_{tt}) + \varepsilon[u_t \Delta u + (|\nabla u|^2)_t] = 0,} \tag{6.14}$$

which describes the evolution of long water waves with small amplitude, was derived in [40] as an approximation of the full water wave problem. The original model includes $\varepsilon = \mu$, $a = \frac{1}{6}$, and $b = \frac{1}{2}$ in the absence of surface tension (in general, a and b are positive constants satisfying $a - b = \sigma - \frac{1}{3}$, where σ is the *Bond number* proportional to the coefficient of surface tension). In appropriate limits, the BL equation (6.14) can be reduced to the KdV and the KP equation; see details and references in [458]. Both the quadratic and the linear operators in (6.14) preserve the polynomial subspace (6.4) for $N = 3$ (or the more general one (6.5)), so the restriction to $W_4^{\rm q}$ yields an eighth-order DS (or higher-order for $W_n^{\rm q}$).

Let us point out another reduction of (6.14) describing interesting mathematical features. In 1D, neglecting the linear terms gives the quadratic model

$$\boxed{u_{tt} + \varepsilon(u_t u_{xx} + 2u_x u_{xt}) = 0.} \tag{6.15}$$

In the domain $\{\varepsilon u_t < 0\}$, it is formally hyperbolic, for which an FBP can be posed, exhibiting the free boundaries on the set $\{u_t = 0\}$, where the PDE is degenerate. We discuss the interface propagation by using the elementary exact solutions

$$u(x,t) = C_1(t) + C_2(t)x^2,$$

$$\begin{cases} C_1'' = -2\varepsilon C_2 C_1', \\ C_2'' = -10\varepsilon C_2 C_2'. \end{cases}$$

Taking $\varepsilon = -1$, consider a particular explicit solution of the form

$$u(x,t) = -t^{\frac{3}{5}} - \tfrac{1}{5t}x^2 \quad \text{for } t > 0. \tag{6.16}$$

It follows that $u_t = 0$ on the curve $s(t) = \sqrt{3}\,t^{4/5}$, so that (6.16) is a solution of an FBP in $\{x > s(t)\}$ with the following free-boundary conditions:

$$u_t = 0 \quad \text{and} \quad u = \tfrac{8}{5}\big(\tfrac{2}{5u_{xx}}\big)^{\frac{3}{5}} \quad \text{at } x = s(t).$$

These are easily derived from (6.16) by differentiating, with the t-parameterization $t = -\frac{2}{5}\frac{1}{u_{xx}}$. Similar to FBPs for the TFEs in Example 3.10, one way to justify the well-posedness of such FBPs is by using the von Mises transformation $X = X(z,t)$,

$z = u_t > 0$, in a neighborhood of the interface. This is a hard OPEN PROBLEM for a singular non-local-in-time PDE with a non-local Neumann-type condition at the origin $z = 0$.

Example 6.8 **(Boussinesq system)** Consider the *Boussinesq system* that generates the BL equation in Example 6.7,

$$\begin{cases} \left(I - \frac{\mu}{2}\,\Delta\right)\eta_t = -\Delta\Phi - \varepsilon\nabla\cdot(\eta\nabla\Phi) + \frac{2\mu}{3}\,\Delta^2\Phi, \\ -\left(I - \frac{\mu}{2}\,\Delta\right)\Phi_t = \eta - \mu\sigma\,\Delta\eta + \frac{\varepsilon}{2}\,|\nabla\Phi|^2, \end{cases} \tag{6.17}$$

where η is the surface elevation with the velocity potential on the free surface

$$\xi = \phi\big|_{z=h_0+\eta} = \Phi - \tfrac{\mu}{2}\,\Delta\Phi + O(\mu^2, \mu\varepsilon);$$

see details in [458, pp. 481–483]. The system (6.17) is close to that derived by Boussinesq in 1877 [76, p. 314]. It is not difficult to find the DS reduction of (6.17), e.g., on the subspace (6.4). Both PDEs in (6.17) are degenerate at the interface surface $\{\nabla\Phi = 0\}$ on which certain free-boundary conditions should be posed. Solvability analysis, even in 1D, becomes more involved than that for (6.15); cf. the FFCH equation in Example 4.43.

6.1.3 Invariant subspaces driven by elliptic equations

Linear elliptic equations. Consider other types of subspaces governed by linear elliptic PDEs for fully nonlinear second-order operators

$$F[u] = \alpha(\Delta u)^2 + \beta u\,\Delta u + \varepsilon u^2 \quad \text{in } I\!R^N. \tag{6.18}$$

Proposition 6.9 *Operator* (6.18) *admits* $W_2 = \mathcal{L}\{1, f(x)\}$, *where* f *satisfies the linear elliptic equation*

$$\Delta f + \mu f = 0 \quad \text{in } I\!R^N, \quad \text{with } \alpha\mu^2 - \beta\mu + \varepsilon = 0. \tag{6.19}$$

Proof. Setting $u = C_1 + C_2 f \in W_2$ yields

$$F[u] = \varepsilon C_1^2 + (2\varepsilon f + \beta\Delta f)C_1C_2 + F[f]C_2^2,$$

where $\Delta f = -\mu f \in W_2$ and $F[f] = (\alpha\mu^2 - \beta\mu + \varepsilon)f^2 = 0$ by (6.19). □

Here, Δ can be replaced by any homogeneous higher-order operator,

$$L = \textstyle\sum_{(|\sigma|\ge 1)} a_\sigma(x)D^\sigma \quad (e.g.,\ L = \Delta^m \text{ for } m \ge 2).$$

Example 6.10 **(Semilinear wave equation)** Consider a fourth-order semilinear wave equation with nonlinearities from the Boussinesq PDEs,

$$\boxed{u_{tt} = a\Delta^2 u + b\Delta u + \beta u\,\Delta u + \varepsilon u^2 + cu + d.} \tag{6.20}$$

It admits solutions

$$u(x,t) = C_1(t) + C_2(t)f(x), \tag{6.21}$$

where f solves the linear elliptic PDE

$$\beta\Delta f + \varepsilon f = 0 \quad \text{in } I\!R^N.$$

So $\Delta f = -\frac{\varepsilon}{\beta} f$ and $\Delta^2 f = \left(\frac{\varepsilon}{\beta}\right)^2 f$, and expansion coefficients satisfy the DS

$$\begin{cases} C_1'' = \varepsilon C_1^2 + cC_1 + d, \\ C_2'' = \varepsilon C_1 C_2 + \left[a\left(\frac{\varepsilon}{\beta}\right)^2 - b\,\frac{\varepsilon}{\gamma} + c\right] C_2. \end{cases}$$

Example 6.11 (**Quasilinear heat equation**) Consider the following quasilinear parabolic equation with exponential nonlinearities:

$$v_t = \beta \nabla \cdot (\mathrm{e}^v \nabla v) + \mathrm{e}^{-v}(\varepsilon \mathrm{e}^{2v} + c\mathrm{e}^v + d).$$

By the pressure-type change $u = \mathrm{e}^v$, it reduces to the quadratic PDE

$$\boxed{u_t = \beta u \Delta u + \varepsilon u^2 + cu + d.}$$

There exist solutions (6.21), where f satisfies (6.19) with $\mu = \frac{\varepsilon}{\beta}$, and the DS is

$$\begin{cases} C_1' = \varepsilon C_1^2 + cC_1 + d, \\ C_2' = \varepsilon C_1 C_2 + cC_2. \end{cases}$$

Example 6.12 (**Fourth-order diffusion equation**) Keeping the same exponential nonlinearities, let us introduce a fourth-order parabolic PDE (we write principal operators only)

$$v_t = \alpha \Delta \nabla \cdot (\mathrm{e}^v \nabla v) + \beta \nabla \cdot (\mathrm{e}^v \nabla v) + \varepsilon \mathrm{e}^v + \ldots,$$

which on substitution $u = \mathrm{e}^v$ yields

$$\boxed{u_t = \alpha u \Delta^2 u + \beta u \Delta u + \varepsilon u^2 + \ldots.}$$

For solutions (6.21), f solves (6.19).

Nonlinear elliptic systems. We consider families of quadratic evolution PDEs for which it is convenient to deal with subspaces defined on the functions of the time-variable t. This leads to systems of elliptic PDEs. Let us study two main cases, corresponding to parabolic and hyperbolic PDEs. The proofs are elementary and have applications to a number of classical models. We begin by describing the polynomial subspaces.

Proposition 6.13 *The following holds:*

$$\begin{aligned} &\text{(i)}\quad F[u] = uu_t \quad \text{admits} \quad W_2 = \mathcal{L}\{1, t\}; \\ &\text{(ii)}\quad F[u] = uu_{tt} \quad \text{admits} \quad W_3 = \mathcal{L}\{1, t, t^2\}; \\ &\text{(iii)}\quad F[u] = u_t u_{tt} \quad \text{admits} \quad W_4 = \mathcal{L}\{1, t, t^2, t^3\}. \end{aligned}$$

Example 6.14 (**2mth-order parabolic equation**) This construction is associated with the Oron–Rosenau solutions in Example 1.9. Consider a quasilinear parabolic equation,

$$v_t = L[\sqrt{v}] + p(x) \quad (v \geq 0), \tag{6.22}$$

where L is a linear $2m$th-order elliptic operator with coefficients independent of t, and $p(x)$ is a given function. Setting $v = u^2$ yields

$$\boxed{uu_t = \tfrac{1}{2}\, L[u] + \tfrac{1}{2}\, p.} \tag{6.23}$$

By Proposition 6.13(i), there exist solutions

$$u(x,t) = C_1(x) + C_2(x)t \in W_2. \tag{6.24}$$

Substituting this into (6.23) yields a system of elliptic PDEs in $I\!R^N$,

$$\begin{cases} C_1C_2 = \frac{1}{2}\,L[C_1] + \frac{1}{2}\,p, \\ C_2^2 = \frac{1}{2}\,L[C_2]. \end{cases}$$

For instance, the fast diffusion equation with reaction or absorption terms

$$v_t = \Delta(\sqrt{v}) + \delta\sqrt{v} + p$$

possesses solutions $v = u^2$, where u is given by (6.24) with the elliptic system

$$\begin{cases} C_1C_2 = \frac{1}{2}(\Delta C_1 + \delta C_1 + p), \\ C_2^2 = \frac{1}{2}(\Delta C_2 + \delta C_2). \end{cases}$$

Example 6.15 (**Hyperbolic equation**) It follows from Proposition 6.13(ii) that the hyperbolic PDE

$$\boxed{uu_{tt} + \mu u_t = L[u] + p(x)}$$

has solutions

$$u(x,t) = C_1(x) + C_2(x)t + C_3(x)t^2 \in W_3,$$

$$\begin{cases} 2C_1C_3 + \mu C_2 = L[C_1] + p, \\ 2C_2C_3 + 2\mu C_3 = L[C_2], \\ 2C_3^2 = L[C_3]. \end{cases} \tag{6.25}$$

For instance, such solutions exist for the quasilinear hyperbolic (in $\{u > 0\}$) equation with damping,

$$\boxed{uu_{tt} + \mu u_t = \Delta u + p(x) \quad (L = \Delta).}$$

Note also that W_3 is admitted by the operator $(uu_t)_t$. Therefore, setting $v = u^2$ ($v \ge 0$) in the quasilinear hyperbolic PDE yields

$$v_{tt} = L[\sqrt{v}] + p(x) \quad \Longrightarrow \quad \boxed{2(uu_t)_t = L[\sqrt{u}] + p,}$$

possessing solutions on W_3.

Example 6.16 Using Proposition 6.13(iii), consider the PDE

$$\boxed{u_tu_{tt} = L[u] + p(x),}$$

possessing solutions on W_4,

$$u(x,t) = C_1(x) + C_2(x)t + C_3(x)t^2 + C_4(x)t^3, \tag{6.26}$$

$$\begin{cases} 2C_2C_3 = L[C_1] + p, \\ 4C_3^2 + 6C_2C_4 = L[C_2], \\ 18C_3C_4 = L[C_2], \\ 18C_4^2 = L[C_4]. \end{cases}$$

Example 6.17 **(Forced quasilinear wave equation)** It is not difficult to extend this approach to other classes of PDEs. For instance, setting $v = u^3$ in the forced quasilinear wave equation yields a cubic equation,

$$v_{tt} = \nabla \cdot \left(v^{-\frac{2}{3}} \nabla v\right) + \mu v^{\frac{1}{3}} + \nu \implies \boxed{(u^2 u_t)_t = \Delta u + \tfrac{\mu}{3}\, u + \tfrac{\nu}{3}.}$$

Since the operator on the left-hand side admits $W_2 = \mathcal{L}\{1, t\}$, there exist solutions (6.24) with the coefficients satisfying the elliptic system

$$\begin{cases} 2C_1 C_2^2 = \Delta C_1 + \frac{\mu}{3} C_1 + \frac{\nu}{3}, \\ 2C_2^3 = \Delta C_2 + \frac{\mu}{3} C_2. \end{cases}$$

We next consider trigonometric, exponential, and other subspaces.

Proposition 6.18 *The following holds:*

(i) $F[u] = uu_t - u^2$ *admits* $W_2 = \mathcal{L}\{1, \mathrm{e}^t\}$;

(ii) $F[u] = uu_{tt} - u^2$ *admits* $W_3 = \mathcal{L}\{1, \mathrm{e}^t, \mathrm{e}^{-t}\}$;

(iii) $F[u] = uu_{tt} + u^2$ *admits* $W_3 = \mathcal{L}\{1, \cos t, \sin t\}$;

(iv) $F[u] = u_t u_{tt} - u^2$ *admits* $W_4 = \mathcal{L}\{1, \varphi(t)\}$, *where* $\varphi' \varphi'' = \varphi^2$.

Concerning case (ii), see more general Proposition 4.23.

Example 6.19 Consider the PDE (6.22) with an extra linear term,

$$v_t = L[\sqrt{v}] + p(x) + 2v,$$

so that $v = u^2$ yields

$$\boxed{F[u] \equiv uu_t - u^2 = \tfrac{1}{2}\, L[u] + \tfrac{1}{2}\, p.}$$

Using Proposition 6.18(i) yields the exact solutions

$$u(x,t) = C_1(x) + C_2(x)\mathrm{e}^t,$$

$$\begin{cases} -C_1^2 = \frac{1}{2}\, L[C_1] + \frac{1}{2}\, p, \\ -C_1 C_2 = \frac{1}{2}\, L[C_2]. \end{cases}$$

Example 6.20 As a new second-order PDE, consider

$$v_{tt} = L[\sqrt{v}] + p(x) + 4v.$$

Setting $v = u^2$ again yields

$$\boxed{F[u] \equiv (uu_t)_t - 2u^2 = \tfrac{1}{2}\, L[u] + \tfrac{1}{2}\, p.}$$

This operator F admits the exponential subspace W_3 in (ii), so there exists

$$u(x,t) = C_1(x) + C_2(x)\mathrm{e}^t + C_3(x)\mathrm{e}^{-t},$$

$$\begin{cases} -2C_1^2 - 4C_2 C_3 = \frac{1}{2}\, L[C_1] + \frac{1}{2}\, p, \\ -3C_1 C_2 = \frac{1}{2}\, L[C_2], \\ -3C_1 C_3 = \frac{1}{2}\, L[C_3]. \end{cases}$$

6.1.4 Models from transonic gas dynamics

Quadratic operators are common in this area.

Example 6.21 **(KFG equation)** The *Kármán–Fal'kovich–Guderley* (KFG) *equation* of transonic gas flows, [395, 504, 267],

$$\boxed{u_t u_{tt} = \Delta u \quad \text{in } I\!R^N \times I\!R,} \tag{6.27}$$

possesses exact solutions (6.26). Then $p = 0$ and $L = \Delta$ in the system (6.25). The invariance of W_4 remains valid for linear perturbations of the PDE. For instance, solutions (6.26) also satisfy the PDE

$$\boxed{\mu u_{ttt} + u_t u_{tt} + \alpha u_t = \Delta u \quad \text{in } I\!R^N \times I\!R}$$

that describe transonic flows around a thin body with effects of viscosity and heat conductivity, for which the velocity of the gas is close to the local speed of sound, [504]. The expansion coefficients then satisfy

$$\begin{cases} 6\mu C_4 + 2C_2C_3 + \alpha C_2 = \Delta C_1, \; 4C_3^2 + 6C_2C_4 + 2\alpha C_3 = \Delta C_2, \\ 18C_3C_4 + 3\alpha C_4 = \Delta C_3, \; 18C_4^2 = \Delta C_4. \end{cases}$$

Example 6.22 **(LRT equation)** Next, consider the *Lin–Reissner–Tsien* (LRT) *equation* [395]

$$\boxed{2u_{xt} + u_x u_{xx} = u_{yy} \quad \text{in } I\!R^2 \times I\!R,} \tag{6.28}$$

which plays an important role as a simplified equation of gas motion at sonic velocities in a channel (nozzle) with the plane geometry. It is derived by the so-called transonic expansion using small deviations from unity of the actual Mach number of the flow; see also [504, 133] and further comments below.

First of all, using the invariant subspace $W_4 = \mathcal{L}\{1, x, x^2, x^3\}$ of $u_x u_{xx}$ (a module), we find

$$u(x, y, t) = C_1(y, t) + C_2(y, t)x + C_3(y, t)x^2 + C_4(y, t)x^3, \tag{6.29}$$

where the coefficients satisfy the PDE system

$$\begin{cases} C_{2t} = \frac{1}{2}\, C_{1yy} - C_2C_3, \\ C_{3t} = \frac{1}{4}\, C_{2yy} - C_3^2 - \frac{3}{2}\, C_2C_4, \\ C_{4t} = \frac{1}{6}\, C_{3yy} - 3C_3C_4, \\ 0 = C_{4yy} - 18C_4^2. \end{cases}$$

As happened a few times before, the last ODE is independent of the others and represents a classical semilinear elliptic equation from combustion theory (t is a parameter). Once $C_4(y, t)$ is known, the rest of the equations can be solved step by step, as with usual ODEs for C_3, C_2, and C_1.

Secondly, concerning simpler reductions of the LRT equation, there exist solutions of (6.28) on $W_7 = \mathcal{L}\{1, x, x^2, y, y^2, xy, xy^2\}$,

$$u(x, y, t) = C_1 + C_2x + C_3x^2 + C_4y + C_5y^2 + C_6xy + C_7xy^2, \tag{6.30}$$

$$\begin{cases} C_2' = -C_2C_3 + C_5, \\ C_3' = -C_3^2 + \frac{1}{2}\,C_7, \\ C_6' = -C_3C_6, \\ C_7' = -C_3C_7. \end{cases}$$

The same low-dimensional reductions apply to the *complete* version of the LRT equation

$$\boxed{2u_{xt} - \big[K - (\gamma + 1)u_x\big]u_{xx} = u_{yy},} \tag{6.31}$$

where γ is the *adiabatic exponent* of the gas and $K \sim 1 - M_\infty^2$, with the Mach number $M_\infty = \frac{U}{a_\infty}$, a_∞ and U being the characteristic velocity of sound and the velocity of main flow in the channel respectively. The viscosity version of the LRT equation [504]

$$\boxed{2u_{xt} - \mu u_{xxx} + u_x u_{xx} = \Delta u} \tag{6.32}$$

also possesses such exact solutions.

The *Ryzhov–Khristianovitch equations* (or *"short waves" equations*) [504] belong to the same class of models,

$$\begin{cases} u_y - 2v_t - 2(v - x)v_x - 2kv = 0, \\ v_y + u_x = 0, \end{cases}$$

where (u, v) is the velocity field, and $k = 0$ or 1 for plane or axisymmetric waves respectively. In the potential form with $u = w_y$, $v = -w_x$, the system reduces to the single equation

$$\boxed{2w_{tx} - 2(x + w_x)w_{xx} + 2kw_x + w_{yy} = 0 \quad \text{in } I\!R^2 \times I\!R,} \tag{6.33}$$

possessing solutions (6.29) and (6.30) with similar DSs.

Example 6.23 (Full model of potential flows) Finally, we treat the *full model of potential transonic flow*, where the shock waves (if any) remain weak, the flow vorticity is low, and hence, the velocity potential Φ can be introduced by

$$\begin{cases} u = \Phi_x, \\ v = \Phi_y. \end{cases}$$

The PDE for Φ follows from the equations of motion and continuity (see Guderley [267, p. 7] and [133])

$$\boxed{\begin{aligned} \Phi_{tt} + 2\Phi_x\Phi_{xt} + 2\Phi_y\Phi_{yt} = \big[a^2 - (\Phi_x)^2\big]\Phi_{xx} \\ -2\,\Phi_x\Phi_y\Phi_{xy} + \big[a^2 - (\Phi_y)^2\big]\Phi_{yy}, \end{aligned}} \tag{6.34}$$

where the velocity of sound a is determined by the equation of the conservation of energy

$$\Phi_t + \tfrac{1}{2}\,|\nabla\Phi|^2 + \tfrac{1}{\gamma-1}\,a^2 = \tfrac{1}{2}\,U^2 + \tfrac{1}{\gamma-1}\,a_\infty^2. \tag{6.35}$$

In particular, the complete LRT equation (6.31) is derived from this system by expansion of the potential Φ under the assumptions that $M_\infty \approx 1$ in a special $\{x, y\}$-geometry corresponding, e.g., to the plane or circular (where u_{yy} is replaced by $u_{yy} + u_{zz}$) Laval nozzle, [133, 267].

Consider the full system and substitute a^2 from (6.35),

$$a^2 = a_*^2 - (\gamma - 1)\Phi_t - \tfrac{\gamma-1}{2}\,|\nabla\Phi|^2, \quad \text{with} \quad a_*^2 = a_\infty^2 + \tfrac{\gamma-1}{2}\,U^2,$$

into (6.34). In this case, the resulting quadratic and cubic operators on the left- and right-hand sides admit $W_4 = \mathcal{L}\{1, x^2, xy, y^2\}$ (for simplicity, we do not include the subspace of linear functions $\mathcal{L}\{x, y\}$ that gives new solutions),

$$\Phi(x, y, t) = C_1(t) + C_2(t)x^2 + C_3(t)xy + C_4(t)y^2,$$

$$\begin{cases} C_1'' + 2(\gamma - 1)(C_2 + C_4)C_1' = 2a_*^2(C_2 + C_4), \\ C_2'' + 2[(\gamma + 3)C_2 + (\gamma - 1)C_4]C_2' + 2C_3C_3' = -4(\gamma + 1)C_2^3 \\ \quad - (\gamma + 3)C_2C_3^2 - 4(\gamma - 1)C_2^2C_4 - (\gamma + 1)C_3^2C_4, \\ C_3'' + 4C_3C_2' + 2(\gamma + 1)(C_2 + C_4)C_3' + 4C_3C_4' \\ \quad = -4(\gamma + 1)\big(C_2^2 + C_4^2\big)C_3 - 8\gamma\, C_2C_3C_4 - 2C_3^3, \\ C_4'' + 2C_3C_3' + 2[(\gamma - 1)C_2 + (\gamma + 3)C_4]C_4' = -(\gamma + 1)C_2C_3^2 \\ \quad - 4(\gamma - 1)C_2C_4^2 - (\gamma + 3)C_3^2C_4 - 4(\gamma + 1)C_4^3. \end{cases}$$

6.1.5 *Maxwell equations in nonlinear optics*

We consider the model [254], consisting of *Maxwell equations* coupled to a single Lorentz oscillator governing the polarization field $\mathbf{P}$, in which the oscillator is driven by the electric field $\mathbf{E}$. Assuming a transverse plane wave propagating along the z-axis with $\mathbf{E} = (E(z,t), 0, 0)^T$, displacement current $\mathbf{D} = (D(z,t), 0, 0)^T$, polarization $\mathbf{P} = (P(z,t), 0, 0)^T$ all along the x-axis, and the magnetic field induction $\mathbf{B} = (0, B(z,t), 0)^T$ along the y-axis, *Faraday's law* and the *Ampère equation*, together with the Lorentz oscillator equation, can be written as

$$\begin{cases} B_t + E_z = 0, \\ D_t + B_z = 0, \\ P_{tt} + P - \alpha E = 0. \end{cases} \tag{6.36}$$

Here the displacement current

$$D = E + \tfrac{1}{2k+1}\, E^{2k+1}, \quad \text{with} \;\; k = 1 \;\; \text{or} \;\; 2,$$

depends in a nonlinear manner on the electric field. The coupling parameter α in the Lorentz equation ensures that the polarization oscillations are driven by the electric field E. In view of the first PDE in (6.36), introducing the potential $u = u(z,t)$ by

$$\begin{cases} B = u_z, \\ E = -u_t, \end{cases}$$

reduces the system to two PDEs

$$\begin{cases} \big[-u_t - \tfrac{1}{2k+1}\,(u_t)^{2k+1}\big]_t + u_{zz} = 0, \\ P_{tt} + P + \alpha u_t = 0. \end{cases}$$

Finally, applying operator $\frac{d^2}{dt^2} + I$ to the first equation and using the second yields the following quasilinear wave equation in $I\!R \times I\!R_+$ [572]:

$$\boxed{F_k[u] \equiv \left(\tfrac{d^2}{dt^2} + I\right)\left[\left(1 + (u_t)^{2k}\right)u_{tt}\right] = \left(\tfrac{d^2}{dt^2} + I\right)u_{zz} - \alpha u_{tt}.} \tag{6.37}$$

For $k = 1$, the quadratic operator $F_1[u]$ admits $\mathcal{L}\{1, t, t^2\}$, so (6.37) possesses

$$u(z,t) = C_1(z) + C_2(z)t + C_3(z)t^2,$$

$$\begin{cases} C_1'' = 2(1 + C_2^2)C_3 + 2\alpha C_3, \\ C_2'' = 8C_2C_3^2, \\ C_3'' = 8C_3^3. \end{cases}$$

6.2 Extended invariant subspaces for second-order operators

6.2.1 Fourth-degree radial polynomial subspaces

Let us return to general quadratic operators (6.1) and first look for an extension of the standard polynomial subspaces listed in Proposition 6.1. Consider the following simple extension result in the class of radially symmetric functions.

Proposition 6.24 *Operator* (6.1) *preserves the 3D subspace*

$$W_3^r = \mathcal{L}\{1, |x|^2, |x|^4\} \quad \textit{iff} \;\; \gamma = -\tfrac{N+2}{4}\,\beta. \tag{6.38}$$

Proof. Setting

$$u = C_1 + C_2|x|^2 + C_3|x|^4 \tag{6.39}$$

and using equalities $\Delta|x|^2 = 2N$, $\Delta|x|^4 = 4(N+2)|x|^2$, we find that

$$F[u] = \Psi_1 + \Psi_2|x|^2 + \Psi_3|x|^4 + 4C_3^2[(N+2)\beta + 4\gamma]|x|^6, \tag{6.40}$$

so always $F[u] \in W_3^r$ iff $(N+2)\beta + 4\gamma = 0$. The expansion coefficients are

$$\begin{aligned} \Psi_1 &= 2\beta N C_1 C_2 + 4\alpha N^2 C_2^2, \\ \Psi_2 &= 4\beta(N+2)C_1C_3 + 16\alpha N(N+2)C_2C_3 + \beta(N-2)C_2^2, \\ \Psi_3 &= 2\beta N C_2 C_3 + 16\alpha(N+2)^2 C_3^2. \end{aligned}$$

□

Example 6.25 (**Blow-up for fast diffusion equation with source**) Consider positive solutions of the following fast diffusion equation with source:

$$v_t = \nabla \cdot \left(v^{-\frac{4}{N+2}} \nabla v\right) + b v^{\frac{N+6}{N+2}} + cv, \tag{6.41}$$

where $b > 0$ and $c > 0$ are constants. Using the pressure transformation $u = v^{-\frac{4}{N+2}}$ yields the quasilinear heat equation with absorption

$$\boxed{u_t = u\Delta u - \tfrac{N+2}{4}|\nabla u|^2 - \tfrac{4}{N+2}(b + cu),} \tag{6.42}$$

where the quadratic operator admits the subspace in (6.38). Hence, there exist solutions (6.39), where the expansion coefficients solve the DS

$$\begin{cases} C_1' = \Psi_1 - \frac{4}{N+2}(b + c\, C_1), \\ C_2' = \Psi_2 - \frac{4}{N+2} c\, C_2, \\ C_3' = \Psi_3 - \frac{4}{N+2} c\, C_3. \end{cases}$$

The *quenching* behavior for (6.42), where $u \to 0$ as $t \to T^-$, is equivalent to blow-up for the original PDE (6.41). The asymptotic setting is similar to that for the TFE in Section 3.2 and is easier. Let $b = 1$ and $c = 0$ in (6.41). Estimating

$$C_1(t) = \tfrac{4}{N+2}(T - t) + ... \quad \text{as} \quad t \to T^-$$

from the first ODE, and taking any orbit such that $C_2(t) \to a_0 > 0$ and $C_3(t) \to a_1$ as $t \to T$, we obtain the following blow-up pattern for (6.41):

$$v(x,t) = (T-t)^{-\frac{N+2}{4}} \left[\tfrac{4}{N+2} \left(1 + a_0 |\xi|^2 + a_1 \mathrm{e}^{-\tau} |\xi|^4 + ...\right)\right]^{-\frac{N+2}{4}}, \tag{6.43}$$

where $\xi = r/\sqrt{T-t}$ and $\tau = -\ln(T-t)$ are the new variables. According to (6.43), let us introduce the rescaled function

$$v(x,t) = (T-t)^{-\frac{N+2}{4}} w(\xi, \tau)$$

satisfying the singular perturbed first-order PDE

$$w_\tau = -\tfrac{1}{2} w_\xi \xi - \tfrac{N+2}{4} w + w^{\frac{N+6}{N+2}} + \mathrm{e}^{-\tau} \nabla \cdot \left(w^{-\frac{4}{N+2}} \nabla w\right). \tag{6.44}$$

The passage to the limit as $\tau \to \infty$ in (6.44) uses specific stability techniques; see [245, Ch. 5]. Recalling that $\{u = 0\} = \{v = +\infty\}$ for nonnegative solutions $u(x,t)$, the *extinction* behavior as $t \to T^-$ of compactly supported solutions of (6.42) describes the evolution and finite-time focusing of the *burnt zone* $\{v = +\infty\}$ of blow-up solutions $v(x,t)$, which are now the *minimal* proper solutions of (6.41) uniquely constructed by a monotone approximation (a truncation) of the PDE. FBP theory for such blow-up solutions is developed for the second-order parabolic problems, [226, Ch. 5]. Notice an interesting feature: the blow-up interfaces $r = s(t)$ are not C^2 smooth for $t > 0$ and $s'(t)$ is not more than Lipschitz continuous, [226, p. 140].

We next perform an extension of the invariant subspace (6.38).

Proposition 6.26 *If the condition in* (6.38) *holds, operator* (6.1) *admits*

$$\mathcal{L}\{W_n^{\mathrm{q}}, |x|^4\}, \tag{6.45}$$

where W_n^{q} is the subspace (6.5) *with general quadratic forms in $I\!R^N$.*

Proof. Take an arbitrary function on the subspace (6.45),

$$u(x,t) = C(t) + x^T A(t)\, x + G(t)|x|^4, \quad \text{so} \tag{6.46}$$

$$\begin{gathered} \Delta u = 2(\operatorname{tr} A) + 4(N+2)G|x|^2, \;\; |\nabla u|^2 = 4 \textstyle\sum_{(l)} \left(\sum_{(j)} a_{l,j} x_j + 2G|x|^2 x_l\right)^2, \\ F[u] = 4\alpha (\operatorname{tr} A)^2 + 2\beta\, (\operatorname{tr} A)\, C + 4(N+2)\big[4\alpha\, (\operatorname{tr} A) + \beta C\big] G|x|^2 \\ + 2\beta (\operatorname{tr} A) x^T A\, x + 4\gamma\, x^T A^2 x + 2\big[8\alpha (N+2)^2 G + \beta\, (\operatorname{tr} A)\big] G|x|^4 + S. \end{gathered}$$

Here, S denotes the fourth and sixth-degree polynomials,

$$S = 4[\beta(N+2) + 4\gamma]\big(|x|^2 x^T A\, x + G|x|^6\big)G,$$

which, in general, do not belong to the subspace. S vanishes due to the condition in (6.38). □

Example 6.27 **(Hyperbolic equation)** The formally hyperbolic equation

$$\boxed{u_{tt} = F[u] \equiv \alpha(\Delta u)^2 + \beta u \Delta u - \tfrac{N+2}{4}\beta|\nabla u|^2 \quad \text{in} \quad I\!R^N \times I\!R}$$

on the subspace (6.45) reduces to the DS for the coefficients of expansion (6.46),

$$\begin{cases} C' = 4\alpha(\mathrm{tr}\, A)^2 + 2\beta(\mathrm{tr}\, A)C, \\ A' = 2\beta(\mathrm{tr}\, A)A - (N+2)A^2 + 4(N+2)\big[4\alpha(\mathrm{tr}\, A) + \beta C\big]I_N, \\ G' = 16\alpha(N+2)^2G^2 + 2\beta G(\mathrm{tr}\, A). \end{cases}$$

In the case of diagonal quadratic forms as in (6.4), the computations are easier and subspace $\mathcal{L}\{W^{q}_{N+1}, |x|^4\}$ remains invariant. In fact, this is the generic case, since, by an orthogonal transformation $x \mapsto Px$ (Δ and $|\nabla(\cdot)|^2$ are invariant), $x^T A\, x$ is reduced to a diagonal form, while the canonical form $|x|^2$, and hence $|x|^4$, remain unchanged. On the other hand, it seems that, in this case, we cannot add linear functions (6.2) to the invariant subspace. Indeed, translations in $I\!R^N$ split $|x|^4$ into a number of terms, including a cubic form, such as $\sum b_{i,j}x_i^2x_j$, which is not invariant. More general quartic forms, such as $\sum b_{i,j}x_i^2x_j^2$, do not also lead to invariant subspaces (unless they reduce to radial function $|x|^4$ for $b_{i,j} = 1$). Recall in this context that the actual dimension of the spaces of exact solutions is determined by the order of the corresponding DS for the expansion coefficients, i.e., depends on the evolution order (the highest time-derivative) of the PDE.

Example 6.28 **(On non-symmetric blow-up)** Using solutions (6.46) makes it possible to describe non-symmetric blow-up and non-radial blow-up interfaces for (6.41) with $b = 1$ and $c = 0$. The simpler case $G(t) \equiv 0$ has been studied in detail, [226, p. 152]. As a result, it was shown that blow-up analytic interfaces in general are not the proper ones for minimal solutions that have finite regularity. For general blow-up solutions, a fully developed mathematical regularity theory of singular interfaces is still absent. It would be interesting to extend some of the results on non-symmetric blow up interfaces in the corresponding wave PDEs, such as

$$\boxed{u_{tt} = u\Delta u - \tfrac{N+2}{4}|\nabla u|^2 - \tfrac{4}{N+2}(b + cu),}$$

using similar exact solutions. These give several OPEN PROBLEMS.

6.2.2 *Logarithmic perturbation of quadratic polynomials*

Here, a special example of non-autonomous quadratic operators and ODEs governing the invariant subspaces is considered.

Critical fast diffusion equation and invariant subspace. We begin with a parabolic

PDE in a radially symmetric case,

$$\boxed{u_t = F[u] + \Phi[u] \quad \text{in } I\!R^N \times I\!R_+,} \tag{6.47}$$

where F is the quadratic operator

$$F[u] = u\,\Delta u - \tfrac{N}{2}\,|\nabla u|^2 \equiv u\big(u_{rr} + \tfrac{N-1}{r}\,u_r\big) - \tfrac{N}{2}(u_r)^2, \quad \text{with } r = |x| > 0,$$

and Φ is a linear lower-order operator, including a convection-like term,

$$\Phi[v] = \beta(x \cdot \nabla v) + \gamma\, v \equiv \beta r v_r + \gamma\, v.$$

The quasilinear operator F occurs in the *fast diffusion equation* with another critical exponent

$$v_t = \nabla \cdot \big(v^{-\frac{2}{N}}\,\nabla v\big) \quad \text{in } I\!R^N \times I\!R_+ \quad (N \geq 3). \tag{6.48}$$

This is a famous equation in diffusion theory. The asymptotic behavior of its positive L^1-solutions is complicated and inherits certain features of similarity solutions for the diffusivity coefficient v^σ of the *first* ($\sigma > -\frac{2}{N}$) and the *second kind* ($\sigma < -\frac{2}{N}$). This represents an important example of the *matched* asymptotics; see details, references, and discussion in [245, Ch. 5]. The pressure transformation $v^{-2/N} = u$ reduces (6.48) to

$$\boxed{u_t = F[u] \equiv u\,\Delta u - \tfrac{N}{2}\,|\nabla u|^2.}$$

It turns out that, in the so-called *outer region* (i.e., sufficiently remote from the origin $x = 0$), the asymptotic behavior of finite mass solutions can be described by special orbits belonging to a peculiar subspace $W_2 = \mathcal{L}\{|x|^2, |x|^2 \ln |x|\}$ invariant under F. Such radial self-similar solutions are

$$v(x,t) = \tfrac{1}{N-2}\,\tfrac{|x|^2}{t}\,\ln\big(\tfrac{|x|}{at^\gamma}\big) \in W_2, \quad \text{with } \gamma = \tfrac{N}{2(N-2)},$$

where $a > 0$ is arbitrary. The solutions are positively defined in $\{|x| > at^\gamma\}$ and correctly describe the *outer* asymptotic behavior governed by a singular perturbation of a first-order conservation law. This is a key feature of such a matched asymptotic structure.

The outer behavior shows that the evolution on the invariant subspace W_2 is stable, in the sense that it attracts a wide set of other general non-invariant orbits, [245, Theorem 6.3]. The internal structure of W_2 plays a role. In particular, it is important that

$$W_2 = \mathcal{L}\{|x|^2\} \oplus \mathcal{L}\{|x|^2 \ln |x|\},$$

where the first 1D subspace is also invariant under F, while the second one is not. Moreover, in rigorous mathematical analysis of the asymptotic behavior, it is proved that, in the outer region, evolution orbits move towards the second (non-invariant) subspace.

Let us now state a simple invariant property of operators F and Φ.

Proposition 6.29 *Both operators F and Φ admit subspace $W_2 = \mathcal{L}\{r^2, r^2 \ln r\}$ given by the linear homogeneous elliptic PDE with non-constant coefficients*

$$\Delta u - (N+2)\,\tfrac{x \cdot \nabla u}{|x|^2} + 4\,\tfrac{u}{|x|^2} \equiv u_{rr} - 3\,\tfrac{u_r}{r} + 4\,\tfrac{u}{r^2} = 0.$$

We derive the corresponding exact solutions of the general evolution PDE

$$\boxed{P(\tfrac{\partial}{\partial t})[u] = F[u] + \Phi[u],} \tag{6.49}$$

where $P(\frac{\partial}{\partial t})$ is a linear polynomial operator with coefficients, depending on t only.

Corollary 6.30 *The PDE* (6.49) *on* W_2 *with* $u(x,t) = C_1(t)|x|^2 + C_2(t)|x|^2 \ln|x|$ *is the DS*

$$\begin{cases} P[C_1] = (2-N)C_1C_2 - \frac{N}{2}C_2^2 + (2\beta+\gamma)C_1 + \beta C_2, \\ P[C_2] = (2-N)C_2^2 + (2\beta+\gamma)C_2. \end{cases}$$

In particular, for (6.47), the following holds:

$$\begin{cases} C_1' = (2-N)C_1C_2 - \frac{N}{2}C_2^2 + (2\beta+\gamma)C_1 + \beta C_2, \\ C_2' = (2-N)C_2^2 + (2\beta+\gamma)C_2. \end{cases}$$

In the case of the hyperbolic Boussinesq-type equation

$$\boxed{u_{tt} = F[u] + \Phi[u],}$$

the DS is of fourth order with $C_{1,2}'$ replaced by $C_{1,2}''$.

Reduction to the autonomous operator and ODE. For convenience, replacing r by x and u by y as in Section 2.1, yields the quadratic operator

$$F[y] = y\big(y'' + \tfrac{N-1}{x}\,y'\big) - \tfrac{N}{2}(y')^2,$$

while the subspace W_2 is given by Euler's ODE

$$y'' - \tfrac{3y'}{x} + \tfrac{4y}{x^2} = 0.$$

Both are not autonomous relative to the space variable x. Observing that, on W_2, the operator F is equivalent to the first-order one

$$F[y] \equiv -\tfrac{N}{2}(y')^2 + \tfrac{3}{x}\,yy' - \tfrac{4}{x^2}\,y^2,$$

we use the transformation

$$y = \mathrm{e}^{2\tilde{x}}\tilde{y} \equiv x^2\tilde{y}, \quad x = \mathrm{e}^{\tilde{x}},$$

yielding the autonomous quadratic operator

$$\tilde{F}[\tilde{y}] = \tfrac{1}{x^2}F[y] = \tilde{y}\tilde{y}'' - \tfrac{N}{2}(\tilde{y}')^2 + (2-N)\tilde{y}\tilde{y}'.$$

Then the invariant subspace consists of linear functions,

$$\tilde{y} \in \tilde{W}_2 = \mathrm{e}^{-2\tilde{x}}\mathcal{L}\{x^2, x^2\ln x\} = \mathcal{L}\{1, \tilde{x}\}.$$

This operator and the corresponding subspace are described in Section 2.5.2.

It is easy to present a more general set of fully nonlinear second-order operators preserving subspace W_2, e.g.,

$$\tilde{F}[\tilde{y}] = G(\tilde{y}, \tilde{y}', \tilde{y}'') + \alpha(\tilde{y}')^2 + \beta\tilde{y}\tilde{y}',$$

where G is an arbitrary function satisfying $G(a,b,0) \equiv 0$.

6.2.3 Polynomial-exponential subspaces

Example 6.31 (**Quasilinear heat equation**) Consider the quasilinear heat equation with absorption or source

$$v_t = (v^\sigma v_x)_x - \gamma\, v^{\sigma+1}, \quad \text{where} \;\; \gamma = \tfrac{4(\sigma+1)}{\sigma^2} \quad (\sigma \neq 0). \tag{6.50}$$

Let $\sigma > -1$, i.e., $\gamma > 0$. In this case, as shown in Section 1.5, there exist solutions on the 3D exponential subspace

$$v^\sigma \in W_3 = \mathcal{L}\{1, \cosh 2x, \sinh 2x\} \tag{6.51}$$

(for $\gamma < 0$, a subspace of trigonometric functions occurs). By the transformation

$$\begin{cases} x = \ln r, \\ v^\sigma = \frac{1}{r^2}\, u^\sigma, \end{cases}$$

(6.50) reduces to the quasilinear purely diffusive equation in radial geometry,

$$u_t = \nabla_r \cdot \left(u^\sigma \nabla_r u\right) \equiv \tfrac{1}{r^{N-1}} \left(r^{N-1} u^\sigma u_r\right)_r \quad \text{in } I\!R^N \times I\!R_+, \tag{6.52}$$

with dimension $N = -\frac{2(\sigma+2)}{\sigma}$ which is not necessarily an integer and is negative for $\sigma > 0$. It follows from (6.51) that (6.52) admits exact solutions on the following subspace:

$$u^\sigma \equiv r^2 v^\sigma \in \hat{W}_3 = \mathcal{L}\big\{r^2, r^2\cosh(2\ln r), r^2 \sinh(2\ln r)\big\}.$$

Similar to Example 1.15, for $\sigma = -\frac{4}{3}$, there exists the 5D subspace

$$\hat{W}_5 = \mathcal{L}\big\{r^2, r^2\cosh(2\ln r), r^2\sinh(2\ln r), r^2\cosh(4\ln r), r^2\sinh(4\ln r)\big\}.$$

6.2.4 Singular subspaces for the critical diffusion exponent

Proposition 6.32 *The quadratic diffusion operator*

$$F_*[u] = u\Delta u + \gamma_* |\nabla u|^2, \quad \textit{with} \;\; \gamma_* = -\tfrac{N-4}{2(N-3)}, \;\; N \neq 3,$$

admits the subspace

$$W_s = \mathcal{L}\big\{|x|^2, |x|^{4-N}, |x|^{6-2N}\big\}. \tag{6.53}$$

The subspace W_s is 3D, except in the case of $N = 2$, where it is 1D with $\gamma_* = -1$, corresponding to the *remarkable* operator to be studied in detail in the next section. For $N = 1$, we have $\gamma_* = -\frac{3}{4}$, and $W_s = \mathcal{L}\{x, x^2, x^3\}$ is indeed a subspace of the more general 5D one, $W_5 = \mathcal{L}\{1, x, x^2, x^3, x^4\}$, described in Example 1.14. For $N \geq 4$, the subspace contains a singular function at the origin, e.g.,

$$W_s = \mathcal{L}\big\{1, |x|^2, |x|^{-2}\big\} \quad \text{for } N = 4. \tag{6.54}$$

Example 6.33 (**Fast diffusion equation**) Consider the PDE

$$\boxed{v_t = \nabla \cdot \left(v^{-\frac{2(N-3)}{N-4}} \nabla v\right) \;\; \text{for } N > 4.}$$

The pressure $u = v^{-\frac{2(N-3)}{N-4}}$ satisfies the equation $u_t = F_*[u]$. Therefore, there exist solutions on the subspace (6.53),

$$u(x,t) = C_1(t)r^2 + C_2(t)r^{4-N} + C_3(t)r^{6-2N},$$

$$\begin{cases} C_1' = \frac{2(N-2)^2}{N-3}\, C_1^2, \\ C_2' = \frac{2(N-2)^2}{N-3}\, C_1 C_2, \\ C_3' = (N-2)^2\big[2C_1C_3 - \frac{N-4}{2(N-3)}\, C_2^2\big]. \end{cases}$$

Similarly, exact solutions $u = v^{\sigma_*} \in W_3$ exist for the hyperbolic PDE

$$\boxed{v^{1-\sigma_*}(v^{\sigma_*})_{tt} = \nabla \cdot (v^{\sigma_*} \nabla v), \quad \text{with } \sigma_* = \tfrac{1}{\gamma_*}.}$$

Example 6.34 **(Quenching on a sphere)** The quasilinear degenerate heat equation with constant absorption

$$\boxed{u_t = u\Delta u - 1 \quad \text{in } I\!R^4 \times I\!R_+}$$

has solutions on the subspace (6.54),

$$u(x,t) = C_1(t)r^2 + C_2(t) + C_3(t)r^{-2}, \tag{6.55}$$

$$\begin{cases} C_1' = 8C_1^2, \\ C_2' = 8C_1C_2 - 1, \\ C_3' = 8C_1C_3. \end{cases}$$

Integrating yields the explicit solution

$$u(x,t) = -\tfrac{1}{8t}\,|x|^2 + \tfrac{B}{t} - \tfrac{1}{2}\,t + \tfrac{A}{8t}\,|x|^{-2} \quad (A, B \in I\!R).$$

Here, $u(x,t)$ is always unbounded at the origin $x = 0$, showing that such singular solutions can be global in time. Taking positive initial data $C_k(0) > 0$, due to the absorption term -1 in the second ODE, the solution may vanish in finite time, as $t \to T^-$, at some point $r = r_0 > 0$. This is *quenching on a sphere* $\{|x| = r_0\}$ (not a standard single point quenching), which is a subject of asymptotic singularity theory for the PME-type equations, [245, Ch. 5]. Here we detect such behavior via explicit formulae; see Figure 6.1. For $t > T$, an FBP appears with two moving interfaces at $r = s_\pm(t)$.

In a similar manner, extinction phenomenon on W_3 can be described for the degenerate quadratic wave equation with absorption

$$\boxed{u_{tt} = u\Delta u - 1 \quad \text{in } I\!R^4 \times I\!R_+.}$$

6.2.5 *On a cubic operator in $I\!R^N$*

Proposition 6.35 *The cubic operator in $I\!R^N$*

$$F[u] = u^2\Delta u - \tfrac{N}{2}\, u|\nabla u|^2 \quad \text{admits } W_2^{\rm r} = \mathcal{L}\{1, |x|^2\}.$$

Example 6.36 **(Fast diffusion equation in $I\!R^N$)** Setting $v = u^{-\frac{N+2}{2}}$ (v is not the pressure!) in the fast diffusion equation with critical exponents

$$v_t = \nabla \cdot \big(v^{-\frac{4}{N+2}}\nabla v\big) - \tfrac{N+2}{2}\big(av + bv^{\frac{N+4}{N+2}}\big) \tag{6.56}$$

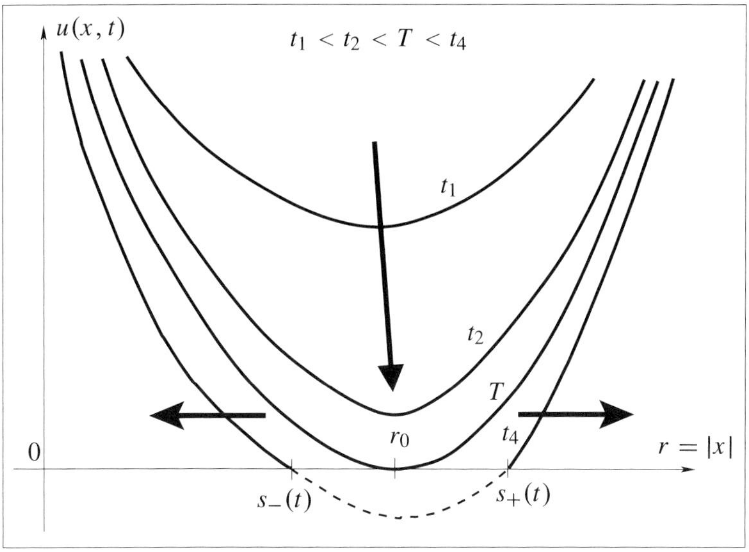

Figure 6.1 Quenching at $t = T$ of positive solutions (6.55). For $t > T$, there appear two interfaces and an FBP should be properly posed.

yields the PDE

$$\boxed{u_t = F[u] + au + b.}$$

There exist solutions $v(x,t) = \big[C_1(t) + C_2(t)|x|^2\big]^{-\frac{N+2}{2}}$ of (6.56), where

$$\begin{cases} C_1' = 2NC_1^2C_2 + aC_1 + b, \\ C_2' = 2NC_1C_2^2 + aC_2, \end{cases}$$

which are not new. By the true pressure transformation $v = u^{-\frac{N+2}{4}}$, (6.56) reduces to a quadratic PDE, which, in general, possesses a wider class of exact solutions on $W_3^{\rm r} = \mathcal{L}\{1, |x|^2, |x|^4\}$, where Proposition 6.24 applies.

6.3 On the remarkable operator in $I\!R^2$

Consider the following *remarkable* quadratic operator:

$$F_{\rm rem}[u] = u\,\Delta u - |\nabla u|^2 \quad \text{on the plane } (x, y) \in I\!R^2, \tag{6.57}$$

which, according to (6.38), has the critical parameter $\gamma = -\frac{N+2}{4} = -1$. The corresponding quadratic parabolic equation

$$\boxed{u_t = u\,\Delta u - |\nabla u|^2} \tag{6.58}$$

is transformed into the fast diffusion PDE by the pressure change $v = \frac{1}{u}$,

$$v_t = \nabla \cdot \big(\tfrac{1}{v}\,\nabla v\big) \equiv \Delta \ln |v|. \tag{6.59}$$

This PDE arises as a model for long van der Waals interactions in thin films of a fluid spreading on a solid surface if fourth order effects are neglected [249, 54] (cf. Section 3.1). It also represents (see [278] and, e.g., [577]) the evolution of the conformally equivalent metric $g_{i,j}$ $\big(\mathrm{d}s^2 = v(\mathrm{d}x^2 + \mathrm{d}y^2)\big)$ under the *Ricci flow* which evolves a general metric $\mathrm{d}s^2 = g_{i,j}\mathrm{d}x^i\mathrm{d}y^j$ by its Ricci curvature $R_{i,j}$ by the PDE

$$\tfrac{\partial}{\partial t}\, g_{i,j} = -2R_{i,j}. \tag{6.60}$$

In the present case, the conformal metric $g_{i,j} = vI_{i,j}$ has scalar curvature $R = -\frac{1}{v}\,\Delta \ln v$ and $R_{i,j} = \frac{1}{2}\,Rg_{i,j}$, where $R(x,t)$ satisfies the semilinear heat equation

$$R_t = \Delta R + R^2, \tag{6.61}$$

which admit solutions blowing-up as $t \to T^- < \infty$. As a principal feature, note that (6.60) is a system of second-order nonlinear parabolic equations which obey the Maximum Principle (similar to (6.61)). In Hamilton [278], it was established that, for a given compact 3D manifold with initially positive Ricci curvature, after scaling, the Ricci flow (6.60) evolves to a metric of positive *constant* curvature (so the manifold is diffeomorphic to the sphere $\mathbf{S}^3$). A crucial part of Hamilton's analysis is proving that solutions $R(x,t) > 0$ of (6.61), after scaling, form a symmetric in x blow-up singularity as $t \to T$. The phenomenon of symmetrization close to blow-up time is a fundamental property of many nonlinear evolution PDEs. In the previous chapters, this has been checked for a number of parabolic equations admitting exact solution on invariant subspaces. New refined estimates of solutions of equations (6.60) and (6.61) (and, implicitly, (6.58) and (6.59)), including a *monotonicity formula* [97, p. 254], are a core of Perel'man's approach to the Poincaré Conjecture (a closed connected 3D manifold is homeomorphic to $\mathbf{S}^3$; see [97] for history, references, and recent development).

It is known [10, Sect. 10.7] that the symmetry Lie algebra of equation (6.59) contains an infinite-dimensional subalgebra generated by operators

$$X = \xi_1 \tfrac{\partial}{\partial x} + \xi_2 \tfrac{\partial}{\partial y} - 2\xi_{1x}\, v \tfrac{\partial}{\partial v}, \tag{6.62}$$

where $\xi_1(x,y)$ and $\xi_2(x,y)$ are arbitrary *harmonic conjugate* functions, i.e., satisfying the *Cauchy–Riemann conditions*

$$\xi_{1x} = \xi_{2y}, \quad \xi_{1y} = -\xi_{2x}.$$

The existence of this algebra is connected with the invariance of equation (6.59) with respect to the transformation

$$\bar{x} = \eta(x,y), \quad \bar{y} = \delta(x,y), \quad \bar{v} = \tfrac{v}{(\eta_x)^2 + (\eta_y)^2}, \tag{6.63}$$

where η and δ are arbitrary independent harmonic conjugate functions,

$$\delta_x = \eta_y, \quad \delta_y = -\eta_x \quad (\delta_x \eta_y - \delta_y \eta_x \neq 0). \tag{6.64}$$

The first two formulae (6.63) define a group of conformal transformations of the (x,y)-plane. Imposing in addition the conditions

$$\delta_x = \tfrac{\xi_1}{\xi_1^2 + \xi_2^2}, \quad \delta_y = \tfrac{\xi_2}{\xi_1^2 + \xi_2^2}$$

yields that, in the variables (6.63), the operator (6.62) takes the form $\bar{X} = \partial/\partial \bar{y}$ which corresponds to translations in $\bar{y}$.

Hence, for equation (6.58), we obtain the transformation

$$\bar{x} = \eta(x, y), \quad \bar{y} = \delta(x, y), \quad \bar{u} = \big[(\eta_x)^2 + (\eta_y)^2\big]u, \tag{6.65}$$

leaving this equation invariant. Applying the transformation (6.65) to a solution $u = f(x, y, t)$ of (6.58) yields an infinite family of solutions,

$$u(x, y, t) = \tfrac{1}{(\eta_x)^2+(\eta_y)^2}\, f(\eta(x, y), \delta(x, y), t), \tag{6.66}$$

containing a pair of harmonic conjugate functions. In particular, taking solution $u = f(x, t)$ independent of y, i.e., satisfying the 1D equation

$$u_t = uu_{xx} - (u_x)^2,$$

yields the solution

$$u(x, y, t) = \tfrac{1}{(\eta_x)^2+(\eta_y)^2}\, f(\eta(x, y), t) \tag{6.67}$$

of the 2D equation (6.58).

Thus, operator (6.57) is "doubly critical", since, in addition to the property in Proposition 6.24, the associated parabolic (and also elliptic) PDE admits an infinite-dimensional Lie algebra of symmetries. This is a rare situation for nonlinear equations that allows us to discuss new aspects concerning the structure of invariant subspaces and nonlinear separation of variables.

6.3.1 Polynomial subspaces

Proposition 6.37 *Operator* (6.57) *preserves the 9D subspace*

$$W_9 = \mathcal{L}\big\{1, x, y, x^2, xy, y^2, xr^2, yr^2, r^4\big\} \quad \big(r^2 = x^2 + y^2\big). \tag{6.68}$$

This subspace has larger dimension than most other quadratic operators studied so far. Subspace (6.68) is an extension for $N = 2$ of that given in Proposition 6.26.

Example 6.38 (**Fast diffusion equation and remarkable operators**) The fast diffusion equation with a quadratic reaction-absorption term

$$v_t = \nabla \cdot \big(\tfrac{1}{v}\, \nabla v\big) - av^2 - bv \tag{6.69}$$

by the pressure change $v = \frac{1}{u}$ reduces to

$$\boxed{u_t = F_{\rm rem}[u] + a + bu \quad \text{in } I\!R^2 \times I\!R_+.} \tag{6.70}$$

There exist solutions, which are more general than (6.46),

$$u(x, t) = C + d_1 x + d_2 y + a_1 x^2 + a_2 xy + a_3 y^2 + e_1 xr^2 + e_2 yr^2 + Gr^4, \tag{6.71}$$

where the expansion coefficients satisfy the DS

$$\begin{cases} C' = [2(a_1 + a_3) + b]C - d_1^2 - d_2^2 + a, \\ d_1' = [2(a_3 - a_1) + b]d_1 - 2a_2 d_2 + 8Ce_1, \\ d_2' = [2(a_1 - a_3) + b]d_2 - 2a_2 d_1 + 8Ce_2, \\ a_1' = [2(a_3 - a_1) + b]a_1 - a_2^2 + 2d_1 e_1 - 2d_2 e_2 + 16CG, \\ a_2' = [-2(a_1 + a_3) + b]a_2 + 4d_2 e_1 + 4d_1 e_2, \\ a_3' = [2(a_1 - a_3) + b]a_3 - a_2^2 - 2d_1 e_1 + 2d_2 e_2 + 16CG, \\ e_1' = [2(a_3 - a_1) + b]e_1 - 2a_2 e_2 + 8Gd_1, \\ e_2' = [2(a_1 - a_3) + b]e_2 - 2a_2 e_1 + 8Gd_2, \\ G' = [2(a_1 + a_3) + b]G - e_1^2 - e_2^2. \end{cases} \tag{6.72}$$

Let us return to equation (6.58). Note that the following three types of transformations, which are particular cases of (6.65), do not change the form of the solution under consideration affecting only the coefficients of the expansion (6.71):
(i) translations in x and in y;
(ii) rotations in the (x, y)-plane; and
(iii) the inversion

$$\bar{x} = \tfrac{1}{r^2}\, x, \quad \bar{y} = \tfrac{1}{r^2}\, y, \quad \bar{u} = \tfrac{1}{r^4}\, u. \tag{6.73}$$

The full subspace W_9 contains the invariant subspace $W_4 = \mathcal{L}\{1, x^2, y^2, r^2\}$ on which the DS (6.72) is simplified (we set $a = b = 0$ and also $d_1 = d_2 = a_2 = e_1 = e_2 = 0$), so the general solution

$$u(x, y, t) = C(t) + a_1(t)x^2 + a_3(t)y^2 + G(t)r^4$$

will contain four arbitrary constants. Applying to this solution the above transformations in the (x, y)-plane, inversion and translations in $\bar{x}$ and $\bar{y}$, we find, at most, a nine-parametric family of solutions (6.71) on the subspace W_9.

Remark 6.39 Equation (6.70) with non-zero parameters a or b is not invariant under the inversion (6.73) and is mapped into

$$\bar{u}_t = F_{\rm rem}[\bar{u}] + a\bar{r}^4 + b\bar{u},$$

where the "inhomogeneous" term $a\bar{r}^4$ belongs to the subspace (6.68), with the new variables $\{\bar{x}, \bar{y}\}$.

Example 6.40 (**Quasilinear wave equation**) Consider the corresponding quasilinear wave equation

$$\boxed{u_{tt} = F_{\rm rem}[u] + a + bu.} \tag{6.74}$$

Looking for solutions (6.71) on W_9, we obtain the DS (6.72) with the second-order derivatives $\frac{\mathrm{d}^2}{\mathrm{d}t^2}$ on the left-hand side. Similarly, (6.73) maps this equation into that with the right-hand side in W_9,

$$\bar{u}_{tt} = F_{\rm rem}[\bar{u}] + a\bar{r}^4 + b\bar{u}.$$

Example 6.41 (**Pseudo-hyperbolic PDEs**) The following equation which is formally pseudo-hyperbolic in $\{u > 0\}$:

$$\boxed{(I - \Delta)u_{tt} \equiv u_{tt} - \Delta u_{tt} = u\Delta u - |\nabla u|^2} \tag{6.75}$$

is a dispersion model that describes the propagation of long 2D waves [337]. Being restricted to (6.68), this PDE reduces to an 18D DS. A similar reduction exists for higher-order pseudo-hyperbolic PDEs, such as

$$\boxed{(I-\Delta)u_{tt} = -\alpha\Delta^2 u + u\Delta u - |\nabla u|^2 + \beta\Delta u + \gamma u + \delta.} \tag{6.76}$$

6.3.2 The GSV and other invariant subspaces

The remarkable operator (6.57) makes it possible to clarify interesting relations between the extended polynomial subspace and others. In general, determining invariant subspaces is equivalent to a hard problem of the *generalized separation of variables* (GSV), which in the present remarkable case admits a special treatment. We consider the parabolic equation (6.70). The same analysis applies to the hyperbolic PDE (6.74).

Let us apply transformation (6.65) to the equation (6.58) to obtain another equation

$$\bar u_t = F_{\rm rem}[\bar u] + a\big[(\eta_x)^2 + (\eta_y)^2\big] + b\bar u. \tag{6.77}$$

We then impose the condition

$$(\eta_x)^2 + (\eta_y)^2 \in \bar W_9 = \mathcal{L}\big\{1, \bar x, \bar y, \bar x^2, \bar x\bar y, \bar y^2, \bar x\bar r^2, \bar y\bar r^2, \bar r^4\big\} \quad \big(\bar r^2 = \bar x^2 + \bar y^2\big).$$

Therefore, for $\eta(x,y)$, we obtain the following multi-parameter first-order PDE (a *nonlinear eigenvalue problem*): to find parameters $\{p_1, ..., p_9\}$ and $\eta \not\equiv 0$ such that

$$\begin{aligned}(\eta_x)^2 + (\eta_y)^2 = {} & p_1 + p_2\bar x + p_3\bar y + p_4\bar x^2 + p_5\bar x\bar y + p_6\bar y^2 \\ & + p_7\bar x\bar r^2 + p_8\bar y\bar r^2 + p_9\bar r^4.\end{aligned} \tag{6.78}$$

Then (6.77) will have solutions on W_9. For instance, take

$$\eta(x,y) = e^x\cos y = \bar x, \quad \text{so that } \delta(x,y) = -e^x \sin y = \bar y \tag{6.79}$$

by (6.64). Then $(\eta_x)^2 + (\eta_y)^2 = e^{2x} = \bar x^2 + \bar y^2 = \bar r^2$ and $\bar u = e^{2x}u$. The solutions of (6.77) on $\bar W_9$ are

$$\bar u = C_1 + C_2\bar x + C_3\bar y + C_4\bar x^2 + C_5\bar x\bar y + C_6\bar y^2 + C_7\bar x\bar r^2 + C_8\bar y\bar r^2 + C_9\bar r^4.$$

In the original variables, this gives

$$\begin{aligned}e^{2x}u = {} & C_1 + C_2e^x\cos y - C_3e^x\sin y \\ & + C_4e^{2x}\cos^2 y - C_5e^{2x}\cos y\sin y + C_6e^{2x}\sin^2 y \\ & + C_7e^{3x}\cos y - C_8e^{3x}\sin y + C_9e^{4x}.\end{aligned}$$

Combining similar terms, we finally obtain

$$\begin{aligned}u = {} & \tfrac12(C_4 + C_6) + C_1e^{-2x} + C_9e^{2x} + (C_2e^{-x} + C_7e^x)\cos y \\ & - (C_3e^{-x} + C_8e^x)\sin y + \tfrac12(C_4 - C_6)\cos 2y + \tfrac12 C_5\sin 2y,\end{aligned}$$

or, which is the same, passing to the hyperbolic functions and renaming the expansion coefficients,

$$\begin{aligned}u = {} & \tilde C_1 + \tilde C_2\cosh 2x + \tilde C_3\sinh 2x + \tilde C_8\cos 2y + \tilde C_9\sin 2y \\ & + (\tilde C_4\cosh x + \tilde C_5\sinh x)\cos y + (\tilde C_6\cosh x + \tilde C_7\sinh x)\sin y.\end{aligned}$$

Therefore, the polynomial subspace is transformed into the new exponential-trigonometric invariant subspace.

Proposition 6.42 *Operator* (6.57) *preserves the following subspaces:*

$$W_4 = \mathcal{L}\{1,\ \cosh 2x,\ \cos 2y,\ \cosh x \cos y\} \quad \text{and} \tag{6.80}$$

$$\begin{aligned} W_9 = \mathcal{L}\big\{&1,\ \cosh 2x,\ \sinh 2x,\ \cos 2y,\ \sin 2y, \\ &\cosh x \cos y,\ \sinh x \cos y,\ \cosh x \sin y,\ \sin x \sin y\big\}. \end{aligned} \tag{6.81}$$

Of course, (6.80) is a subspace of (6.81) to functions that are even in both x and y. Note that $F_{\rm rem}$ maps even functions into even, so that the subspace of even function from W_9 is also invariant. The above calculus of invariant transformations illustrate how the lower-dimensional invariant subspace W_4 can be extended to W_9.

Are there other interesting solutions of the nonlinear eigenvalue problem (6.78) rather than (6.79)? This is an OPEN PROBLEM.

Example 6.43 (**Fast diffusion equation**) The equation (6.70) admits exact solutions on the subspace (6.80),

$$u(x,t) = C_1 + C_2 \cosh 2x + C_3 \cos 2y + C_4 \cosh x \cos y, \tag{6.82}$$

$$\begin{cases} C_1' = 4\big(C_2^2 - C_3^2\big) + a + bC_1, \\ C_2' = 4\big(C_1C_2 - \frac{1}{8}\, C_4^2\big) + bC_2, \\ C_3' = 4\big(\frac{1}{8}\, C_4^2 - C_1C_3\big) + bC_3, \\ C_4' = 4C_4(C_2 - C_3) + bC_4. \end{cases} \tag{6.83}$$

For $a < 0$, (6.69) admits positive blowing up patterns, so that solutions (6.82), (6.83) can be used for detecting a fine structure of blow-up singularities for the case of this critical fast diffusion operator.

Example 6.44 The fourth-order pseudo-hyperbolic equation (6.76) can be restricted to subspaces (6.80), or (6.81).

6.3.3 *An application to systems*

As in Section 5.4, the above results can be used for generating various systems of PDEs reduced to DSs on invariant subspaces.

Example 6.45 Let us first present a formal, but exceptional application of the operator (6.57) in the following hyperbolic system in $I\!R^2 \times I\!R$:

$$\boxed{\mathbf{U}_{tt} = -A\Delta^2\mathbf{U} + B\Delta\mathbf{U} + C\mathbf{F}[\mathbf{U}] + D\mathbf{U} + \mathbf{E},} \tag{6.84}$$

where $\mathbf{U} = (u_1, ..., u_m)^T$, $A > 0$, B, C, D are $m \times m$ matrices, $\mathbf{E} \in I\!R^m$, and

$$\mathbf{F}[\mathbf{U}] = (F_{\rm rem}(u_1), ..., F_{\rm rem}(u_m))^T.$$

The right-hand side in (6.84) preserves subspace (6.81), provided that $\mathbf{E} \in W_9$, so system (6.84) restricted to W_9 is an $18m$th-order DS.

Example 6.46 (**Parabolic system**) The parabolic system for $m = 2$ with remarkable operators,

$$\begin{cases} u_t = a_1 F_{\rm rem}[u] + b_1 \Delta u + c_1 \Delta v + d_1 u + e_1 v + f_1, \\ v_t = a_2 F_{\rm rem}[v] + b_2 \Delta u + c_2 \Delta v + d_2 u + e_2 v + f_2, \end{cases}$$

where $f_{1,2} \in W_9$, being restricted to W_9, is an 18th-order DS. Setting $u = \frac{1}{U}$ and $v = \frac{1}{V}$ for positive solutions recovers a system with remarkable elliptic operators given in (6.59), admitting an infinite-dimensional Lie group of symmetries,

$$\begin{cases} U_t = a_1 \Delta \ln U + b_1 \Delta U - 2b_1 \frac{|\nabla U|^2}{U} - c_1 U^2 \Delta\big(\frac{1}{V}\big) - d_1 U - e_1 \frac{U^2}{V} - f_1 U^2, \\ V_t = a_2 \Delta \ln V - b_2 V^2 \Delta\big(\frac{1}{U}\big) + c_2 \Delta V - 2c_2 \frac{|\nabla V|^2}{V} - d_2 \frac{V^2}{U} - e_2 V - f_2 V^2. \end{cases}$$

6.3.4 On partial invariant modules

Let us briefly describe invariant modules related to $F_{\rm rem}$ that, in the 1D case, were studied in Examples 1.37 and 1.43. Consider the PDE operator

$$\hat{F}[u] = u_t - (u \Delta u - |\nabla u|^2) \quad \text{in } I\!R^2 \times I\!R_+,$$

and solutions of the form

$$u(x,t) = C_1(t) + C_2(t)x + C_3(t) e^{\gamma(t)x} e^y, \tag{6.85}$$

where $\gamma(t)$ is a function. For a constant γ, the module with given basic functions is not invariant under $F_{\rm rem}$, though it is composed of the invariant modules of linear functions $\mathcal{L}\{1, x, y\}$ and the exponential function $W_2 = \mathcal{L}\{1, e^{\gamma x} e^y\}$. For any function (6.85), the following holds:

$$\hat{F}[u] = C_1' + C_2^2 + C_2' x + \big[C_3' - (\gamma^2+1) C_1 C_3 + 2\gamma C_2 C_3\big] e^{\gamma x} e^y \\ + \big[\gamma' - (\gamma^2+1) C_2\big] C_3 x e^{\gamma x} e^y,$$

so that we need the condition

$$\gamma' = (\gamma^2 + 1) C_2,$$

which deletes the last term and keeps the subspace invariant (see more details in Example 1.37).

Example 6.47 (**Parabolic equation**) The fourth-order semilinear parabolic PDE

$$\boxed{u_t = -\Delta^2 u + (u + \alpha)\Delta u - |\nabla u|^2 + \mu u + \nu}$$

admits solutions (6.85) with the DS

$$\begin{cases} C_1' = -C_2^2 + \mu C_1 + \nu, \\ C_2' = \mu C_2, \\ C_3' = (\gamma^2+1) C_1 C_3 - 2\gamma C_2 C_3 + \big[-(\gamma^2+1)^2 + \alpha(\gamma^2+1) + \mu\big] C_3, \\ \gamma' = (\gamma^2+1) C_2. \end{cases}$$

If $\mu \neq 0$, the second ODE yields $C_2(t) = e^{\mu t}$, and the last implies

$$\gamma(t) = \tan\big(\tfrac{1}{\mu} e^{\mu t} + \text{constant}\big).$$

For $\mu = 0$, it follows that $C_2 = 1$ and $\gamma_2(t) = \tan t$.

Similar more general solutions can be taken in the form of

$$u(x, y, t) = C_1(t) + C_2(t)x + C_3(t)y + C_4(t)e^{\gamma_1(t)x}e^{\gamma_2(t)y} \tag{6.86}$$

with two unknown functions $\gamma_{1,2}(t)$. The invariance property is checked similarly. Let us present final computations for a slightly modified model.

Example 6.48 **(2mth-order parabolic equation)** A higher-order PDE

$$\boxed{u_t = (-1)^{m+1}\Delta^m u + (u + \alpha)\Delta u - |\nabla u|^2 + \mu u + \nu}$$

admits solutions (6.86), where the following DS occurs:

$$\begin{cases} C_1' = -C_2^2 - C_3^2 + \mu C_1 + \nu, \\ C_2' = \mu C_2, \\ C_3' = \mu C_2, \\ C_4' = (\gamma_1^2 + \gamma_2^2)C_1C_4 - 2\gamma_1 C_2 C_4 - 2\gamma_2 C_3 C_4 \\ \qquad +\big[(-1)^{m+1}(\gamma_1^2 + \gamma_2^2)^m + \alpha(\gamma_1^2 + \gamma_2^2) + \mu\big]C_4, \\ \gamma_1' = (\gamma_1^2 + \gamma_2^2)C_2, \\ \gamma_2' = (\gamma_1^2 + \gamma_2^2)C_3. \end{cases}$$

Solving independently the ODEs for coefficients $\{C_2, C_3, \gamma_1, \gamma_2\}$ yields expressions for $\gamma_{1,2}(t)$ similar to those given above.

6.4 On second-order p-Laplacian operators

Here, we describe specific subspaces that are admitted by operators in $I\!R^N$ with gradient-dependent nonlinearities.

Example 6.49 **(p-Laplacian operator)** Consider the following quasilinear parabolic p-Laplacian equation with extra nonlinear and linear terms:

$$v_t = \nabla \cdot (|\nabla v|^\sigma \nabla v) + a v^{\frac{1}{\sigma+1}} + bv \quad (v \geq 0), \tag{6.87}$$

where $\sigma > -1$, $\sigma \neq 0$, is a fixed exponent. Equations with such nonlinearities describe filtration in non-Newtonian *dilatable* (for $\sigma > 0$) and *pseudo-plastic* (for $\sigma \in (-1, 0)$) fluids, and also occur in solid fuel combustion theory (see details in [444] and in the Remarks to Section 8.5).

In radial geometry, for solutions $v = v(r, t)$ with $r = |x| \geq 0$, (6.87) becomes

$$v_t = |v_r|^\sigma \big[(\sigma + 1)v_{rr} + \tfrac{N-1}{r}\, v_r\big] + a v^{\frac{1}{\sigma+1}} + bv.$$

By the transformation $v = u^{\frac{\sigma+1}{\sigma}}$ (similar to the PMEs, u is called the *pressure* in filtration theory), we obtain

$$\boxed{u_t = F[u] + A + Bu,} \tag{6.88}$$

where F is the quasilinear operator

$$F[u] = \mu |u_r|^\sigma \big\{(\sigma + 1)\big[u u_{rr} + \tfrac{1}{\sigma}(u_r)^2\big] + \tfrac{N-1}{r}\, u u_r\big\}, \tag{6.89}$$

with constants $\mu = (\frac{\sigma+1}{\sigma})^\sigma$, $A = \frac{a\sigma}{\sigma+1}$, and $B = \frac{b\sigma}{\sigma+1}$. Though (6.89) is not a quadratic or a cubic operator, it has the following invariant property.

Proposition 6.50 *Operator* (6.89) *preserves the 2D subspace*

$$W_2^{\rm r} = \mathcal{L}\{1, r^\gamma\}, \quad \textit{where} \;\; \gamma = \tfrac{\sigma+2}{\sigma+1}.$$

Proof. Taking an arbitrary function from $W_2^{\rm r}$,

$$u(x,t) = C_1(t) + C_2(t) r^\gamma, \tag{6.90}$$

yields

$$F[u] = \alpha C_1 C_2 |C_2|^\sigma + \beta |C_2|^{\sigma+2} r^\gamma \in W_2^{\rm r},$$

where $\alpha = \mu\gamma^{\sigma+1} N$ and $\beta = \mu\gamma^{\sigma+1}\big(N + \gamma + \frac{\gamma}{\sigma}\big)$. □

The regularity properties of solutions given by the pressure (6.90) well correspond to the known typical smoothness of weak solutions of the p-Laplacian equations. In particular, the Hölder continuity in r of the function r^γ in (6.90) is optimal, at least for $N = 1$, and in radial geometry for bell-shaped solutions (i.e., for those without holes in the support). In the theory of quasilinear degenerate PDEs, this is proved by *Bernstein's method*; see DiBenedetto [148] and Kalashnikov [309].

Equation (6.88) restricted to $W_2^{\rm r}$ is equivalent to the DS

$$\begin{cases} C_1' = \alpha C_1 C_2 |C_2|^\sigma + B C_1 + A, \\ C_2' = \beta |C_2|^{\sigma+2} + B C_2. \end{cases}$$

Let us integrate this system. Assuming for definiteness that $C_2 \le 0$ (this corresponds to finite-time extinction to be considered below) and setting $C_2 \equiv -C$ with $C \ge 0$, we arrive at

$$\begin{cases} C_1' = -\alpha C_1 C^{\sigma+1} + B C_1 + A, \\ C' = -\beta C^{\sigma+2} + B C. \end{cases} \tag{6.91}$$

The second ODE is independent and is integrated explicitly,

$$C^{\sigma+1}(t) = B\big[\beta + H_0 \mathrm{e}^{-B(\sigma+1)t}\big]^{-1},$$

where H_0 is an arbitrary constant. In this case, from the first ODE in (6.91),

$$C_1(t) = -\tfrac{A}{B(\sigma+1)}\, \mathrm{e}^{Bt}\big(\beta \mathrm{e}^{B(\sigma+1)} + H_0\big)^{-\lambda} E(t), \tag{6.92}$$

where the function E is given by the integral related to Euler's Beta function

$$E(t) = \textstyle\int z^\delta (\beta + H_0 z)^\lambda \, \mathrm{d}z, \tag{6.93}$$

with $z = \mathrm{e}^{-B(\sigma+1)t}$, $\lambda = \frac{\alpha}{\beta(\sigma+1)} \equiv \frac{N}{(\sigma+1)(N\sigma+\sigma+2)}$, and $\delta = -\frac{\sigma[N(\sigma+1)+\sigma+2]}{(\sigma+1)(N\sigma+\sigma+2)}$. The special case $H_0 = 0$ yields the constant function

$$C_2 = -C = -\big(\tfrac{B}{\beta}\big)^{\frac{1}{\sigma+1}}, \quad \text{with} \;\; B > 0.$$

Then (6.92) implies

$$C_1(t) = \big[H_1 \mathrm{e}^{B(1-\frac{\alpha}{\beta})t} - A\big]\big[B\big(1 - \tfrac{\alpha}{\beta}\big)\big]^{-1},$$

where H_1 is a constant of integration. Another explicit elementary solution exists for $A = B = 0$. This solution on W_2^r yields precisely the fundamental, instantaneous point-source (or source-type) solution of the p-Laplacian equation $u_t = F[u]$, which is similar to the ZKB solution for the PME; see [509, p. 84] and history therein.

Example 6.51 **(Interface equation and finite-time extinction)** Consider nonnegative solutions of the p-Laplacian equation with absorption

$$\boxed{v_t = \nabla \cdot \left(|\nabla v|^\sigma \nabla v\right) - v^{\frac{1}{\sigma+1}} \quad \text{in} \quad I\!R^N \times I\!R_+,} \tag{6.94}$$

where $\sigma > 0$. This PDE describes processes with the finite propagation, i.e., with interfaces (free boundaries) similar to the PME. In addition, due to the presence of the *strong* absorption term $-v^p$ with the exponent $p = p_* = \frac{1}{\sigma+1} < 1$, any bounded solution has finite extinction time T and $v(x,t) \to 0$ as $t \to T^-$ for any $x \in I\!R^N$. Furthermore, the critical absorption exponent $p_* = \frac{1}{\sigma+1}$ in (6.94) is known to generate the following phenomena:
(i) the finite interface propagation no longer obeys the standard Darcy law (see [226, Ch. 7]), and
(ii) the singular extinction behavior of the solution as $t \to T^-$ becomes special and differs from those which occur for p above and below the critical exponent p_* (see details in [245, Ch. 4]).

Let us show that solutions on W_2^r correctly describe both critical phenomena. Using (6.90) leads to the following explicit solutions:

$$v(x,t) \equiv u^{\frac{\sigma+1}{\sigma}}(x,t) = \left[C_1(t) - C(t)|x|^\gamma\right]_+^{\frac{\sigma+1}{\sigma}}, \tag{6.95}$$

where the coefficients are given by

$$C_1(t) = \tfrac{\sigma}{(\sigma+1)(1+\lambda)}(T^{1+\lambda} - t^{1+\lambda}) \quad \text{and} \quad C(t) = [\beta(\sigma+1)t]^{-\frac{\lambda}{\sigma+1}}. \tag{6.96}$$

Here, $T > 0$ is the finite extinction time and λ is as in (6.93).

Interface equation. It follows from (6.95) that

$$r = s(t) = \left[\tfrac{C_1(t)}{C(t)}\right]^{\frac{\sigma+1}{\sigma+2}}$$

is the interfaces of the solution. Let us derive the corresponding dynamic interface equation. As usual, differentiating the identity for the pressure $u(s(t),t) \equiv 0$ implies $u_r(s,t)s' + u_t = 0$, which yields

$$s'(t) = -\tfrac{1}{u_r(s(t),t)}\, u_t(s(t),t), \tag{6.97}$$

provided that both derivatives involved exist and the limit as $r \to s^-(t)$ makes sense. For the explicit solution (6.95), formula (6.97) is obviously valid. In order to simplify the right-hand side, we calculate u_t from the pressure equation (6.88), (6.89) with $A = -\frac{\sigma}{\sigma+1}$, $B = 0$, so passing to the limit as $r \to s^-$ yields

$$u_t = \left(\tfrac{\sigma+1}{\sigma}\right)^{\sigma+1}|u_r|^{\sigma+2} - \tfrac{\sigma}{\sigma+1} \quad \text{at} \quad r = s(t).$$

Substituting into (6.97) determines the following *dynamic interface equation*:

$$s'(t) = S[u] \equiv -\left(\tfrac{\sigma+1}{\sigma}\right)^{\sigma+1}|u_r|^{\sigma+1} + \tfrac{\sigma}{\sigma+1}\,\tfrac{1}{u_r} \quad \text{at} \quad r = s(t), \tag{6.98}$$

consisting of two terms. The standard Darcy law for the pure p-Laplacian equation contains the first term only, so, in this case, the strong absorption essentially changes the local interface propagation. Using the whole family of exact solutions on $W_2^{\rm r}$ makes it possible to extend (6.98) to general radial monotone solutions of (6.94) by using the geometric Sturmian argument of intersection comparison, [226, Ch. 7].

Extinction behavior. This is also easily detected from the explicit solution (6.95). Calculating the asymptotics of $C_1(t)$ in (6.96) yields, as $t \to T^-$,

$$v(x,t) = \big[\tfrac{\sigma}{\sigma+1}(T-t)\big(1 - \tfrac{|\xi|^\gamma}{a_0^\gamma}\big)\big]_+^{\frac{\sigma+1}{\sigma}} + \dots, \quad \xi = x/(T-t)^{\frac{\sigma+1}{\sigma+2}}, \tag{6.99}$$

where $a_0 > 0$ is a constant and ξ is the rescaled spatial variable. It is important that the extinction behavior (6.99) is not self-similar, since the similarity rescaled variable for (6.94) is different, $\eta = x/(T-t)$. By the Sturmian intersection comparison, the asymptotic extinction behavior (6.99) holds true for general solutions, [245, Ch. 4].

Example 6.52 (**Regional blow-up**) We next consider a combustion problem and describe a blow-up behavior for the p-Laplacian equation with source

$$\boxed{u_t = \nabla \cdot (|\nabla u| \nabla u) + u^2 \quad \text{in} \;\; I\!R^N \times I\!R_+.} \tag{6.100}$$

The operator on the right-hand side is quadratic and we use Proposition 6.9 to derive exact solutions on a 2D subspace,

$$u(x,t) = \tfrac{T}{(T-t)t}\big[-\tfrac{T-t}{T} + f(x)\big] \in \mathcal{L}\{1, f(x)\}, \tag{6.101}$$

where $T > 0$ is the blow-up time and f is a solution of the following quasilinear elliptic PDE (the same as in Example 5.6):

$$\nabla \cdot (|\nabla f| \nabla f) + f^2 - f = 0 \quad \text{in} \;\; I\!R^N \tag{6.102}$$

that admits weak compactly supported solutions. For $N = 1$, there exists a nonnegative profile $f(x)$ given by the incomplete Euler Beta function such that [216]

$$f(x) > 0 \quad \text{on the interval} \;\; \big\{|x| < \tfrac{1}{2} L_{\rm S} = 2^{\frac{1}{3}} 3^{-\frac{1}{2}} \pi\big\}.$$

Existence of a radial weak solution $f \geq 0$ in any dimension $N > 1$ was proved in [229] (the proof, in the case of regional blow-up, is the same as for the porous medium operator in [509, p. 183]; the structure of single point blow-up similarity patterns is completely different). It is reasonable to expect that any nonnegative compactly supported weak solution of (6.102) is radially symmetric relative to a point in $I\!R^N$. Such a result should involve the moving plane method via *Aleksandrov's Reflection Principle*; see [245, p. 51].

Given a suitable nonnegative solution (6.101), we have the phenomenon of *localization* of blow-up (regional blow-up), where, as $t \to T^-$,

$$u(x,t) \to +\infty \quad \text{for any } x \text{ inside } \operatorname{supp} f,$$

and $u(x,t) \to -\frac{1}{T}$ for all $x \in I\!R^N \setminus \operatorname{supp} f$, as shown in Figure 6.2.

It is curious that the solution (6.101) exhibits an extremely singular behavior at the initial moment of time $t = 0$. Passing to the limit $t \to 0^+$ yields the following

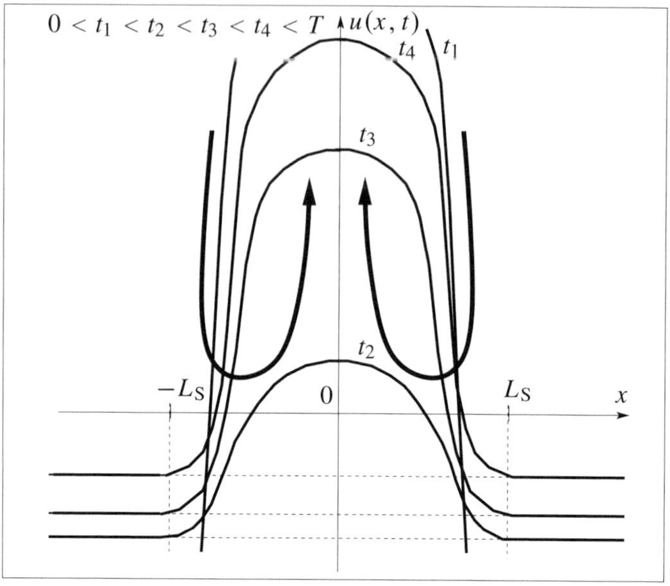

Figure 6.2 Regional blow-up of solutions (6.101) on the interval $[-L_S, L_S]$.

initial function for such solutions:

$$u(x,0^+) = \begin{cases} -\infty, & \text{if } f(x) < 1, \\ +\infty, & \text{if } f(x) > 1, \\ \frac{1}{T}, & \text{if } f(x) = 1. \end{cases}$$

These initial data are unbounded almost everywhere in $I\!R^N$, but generate a bounded weak solution that exists in $I\!R^N \times (0, T)$. Such a phenomenon is portrayed in Figure 6.2 (see the steep profile for $t = t_1 \approx 0$). This is an exceptional situation. Obviously, such a local solvability of (6.100) for initial data $u_0 \notin L^\infty_{\rm loc}(\Omega)$ for any arbitrarily small domain $\Omega \subset I\!R^N$ is not a generic property of this PDE.

Using functions $f(x)$ from Example 5.7, similar blow-up patterns can be constructed for higher-order reaction-diffusion PDEs, such as

$$u_t = -(|u_{xx}|u_{xx})_{xx} + u^2 \quad \text{and} \quad u_t = -(|u_{xxx}|u_{xxx})_x + u^2.$$

6.5 Invariant subspaces for operators of Monge–Ampère type

6.5.1 Second-order equations

For a given function $u \in C^2(I\!R^N)$, let D^2u be the corresponding $N \times N$ *Hessian* matrix $\|u_{x_ix_j}\|$. Parabolic *Monge–Ampère* (M-A) *equations*

$$u_t = g\big(\det D^2 u\big) + h(x, u, D_x u) \quad \text{in } I\!R^N \times I\!R_+ \tag{6.103}$$

play an important role in various geometric problems and applications, such as logarithmic Gauss and Hessian curvature flows, the Minkowski problem (1897), the

Weyl problem, the Calabi conjecture in complex geometry, and many others. See references and basic mathematical results in [550, Ch. 14,15], [253, Ch. 17], and [272]. For increasing functions $g(s)$, equation (6.103) is parabolic if $D^2u(\cdot,t)$ remains positively definite for $t > 0$, assuming that $D^2u_0 > 0$ for initial data u_0. For a class of lower-order operators $h(\cdot)$ satisfying necessary growth estimates, typical monotone increasing, concave nonlinearities g in the principal operator are

$$g(s) = \ln s, \quad g(s) = -\tfrac{1}{s}, \quad \text{and} \quad g(s) = s^{\frac{1}{N}} \quad \text{for } s > 0,$$

for which the global-in-time solvability is known. For other $g(s)$ functions with a faster growth as $s \to \infty$, local-in-time solutions existing by standard parabolic theory (see e.g., the classic book [365, p. 320]) may blow-up in finite time. This gives special asymptotic patterns, which sometimes are also of interest for geometric applications.

We will verify a few types of singularity formation phenomena occurring on linear subspaces admitted by such Monge–Ampère operators. We discuss our basic examples in the 2D case $N = 2$ with independent variables $\{x, y\}$, so

$$F_2[u] = \det D^2 u \equiv u_{xx}u_{yy} - (u_{xy})^2. \tag{6.104}$$

Example 6.53 (**Motion with constant speed**) In the subspace of convex parabolic surfaces

$$u(x,y,t) = C_1(t) + C_2(t)x^2 + C_3(t)y^2 \in W_3 = \mathcal{L}\{1, x^2, y^2\}, \tag{6.105}$$

with $C_2 > 0$, $C_3 > 0$, the parabolic flow with an arbitrary g,

$$\boxed{u_t = g(F_2[u]),}$$

is globally well-defined by the DS

$$\begin{cases} C_1' = g(4C_2C_3), \\ C_2' = 0, \;\; C_3' = 0. \end{cases}$$

So $C_{2,3}(t) \equiv C_{2,3}(0)$, from which we obtain the solution

$$u(x,y,t) = C_1(0) + g(4C_2(0)C_3(0))t + C_2(0)x^2 + C_3(0)y^2$$

corresponding to the motion of the surface with constant speed. For the *hyperbolic M-A equation*

$$\boxed{u_{tt} = g(F_2[u]),}$$

the DS is of sixth order (it is of eighth order if $\mathcal{L}\{xy\}$ is included into the subspace)

$$\begin{cases} C_1'' = g(4C_2C_3), \\ C_2'' = 0, \;\; C_3'' = 0. \end{cases}$$

Initial data $u_t(x,0)$ are important to keep $u(x,y,t)$ convex and the equation hyperbolic. Otherwise, if $D^2u(\cdot,t)$ loses its positivity, the PDE becomes of elliptic type and possibly loses evolution setting as the Cauchy problem.

Example 6.54 (**Blow-up via PME-term**) Adding a standard quasilinear quadratic operator from PME theory to the right-hand side,

$$\boxed{u_t = g\left(\det D^2 u\right) + u\Delta u,} \tag{6.106}$$

for the same polynomial solutions (6.105), we obtain the DS

$$\begin{cases} C_1' = g(4C_2C_3) + 2C_1(C_2 + C_3), \\ C_2' = 2C_2(C_2 + C_3), \\ C_3' = 2C_3(C_2 + C_3). \end{cases}$$

From the last two ODEs, it follows that $C_3 = AC_2$, with a constant $A > 0$, so $C_2' = 2(1 + A)C_2^2$, which yields finite-time blow-up of $C_{1,2,3}(t)$ with the rate

$$C_2(t) = \tfrac{1}{2(1+A)} \, \tfrac{1}{T-t}.$$

This blow-up is induced by the quadratic operator on the right-hand side of (6.106), so, in this sense, the M-A operator is weaker on such polynomial profiles.

Example 6.55 (Blow-up in quadratic equations) We next consider the parabolic equation with $g(s) = s$,

$$\boxed{u_t = F_2[u] \equiv u_{xx}u_{yy} - (u_{xy})^2.} \tag{6.107}$$

The basic subspace for F_2 consists of arbitrary fourth-degree polynomials,

$$W_{15} = \mathcal{L}\{x^\alpha y^\beta, \;\; 0 \le \alpha + \beta \le 4\}. \tag{6.108}$$

For simplicity, consider solutions on a smaller subspace W_6,

$$u(x, y, t) = C_1 + C_2x^2 + C_3y^2 + C_4x^2y^2 + C_5x^4 + C_6y^4, \tag{6.109}$$

$$\begin{cases} C_1' = 4C_2C_3, \\ C_2' = 24C_3C_5 + 4C_2C_4, \\ C_3' = 4C_3C_4 + 24C_2C_6, \\ C_4' = -12C_4^2 + 144C_5C_6, \\ C_5' = 24C_4C_5, \\ C_6' = 24C_4C_6. \end{cases} \tag{6.110}$$

The last two ODEs yield $C_6 = AC_5$, with $A > 0$. Then we obtain two independent equations for $\{C_4, C_5\}$,

$$\begin{cases} C_4' = -12C_4^2 + 144AC_5^2, \\ C_5' = 24C_4C_5, \end{cases}$$

which are integrated in quadratures. The phase-plane of the first-order ODE

$$\tfrac{\mathrm{d}C_5}{\mathrm{d}C_4} = \tfrac{2C_4C_5}{12AC_5^2 - C_4^2} \tag{6.111}$$

is given in Figure 6.2, which shows that all positive orbits blow-up in finite time and approach the explicit solution $C_5 = \frac{1}{2\sqrt{A}} C_4$, corresponding to coefficients

$$\begin{cases} C_4(t) = \tfrac{1}{24} \, \tfrac{1}{T-t}, \\ C_5(t) = \tfrac{1}{48\sqrt{A}} \, \tfrac{1}{T-t}, \\ C_6(t) = \tfrac{\sqrt{A}}{48} \, \tfrac{1}{T-t}. \end{cases} \tag{6.112}$$

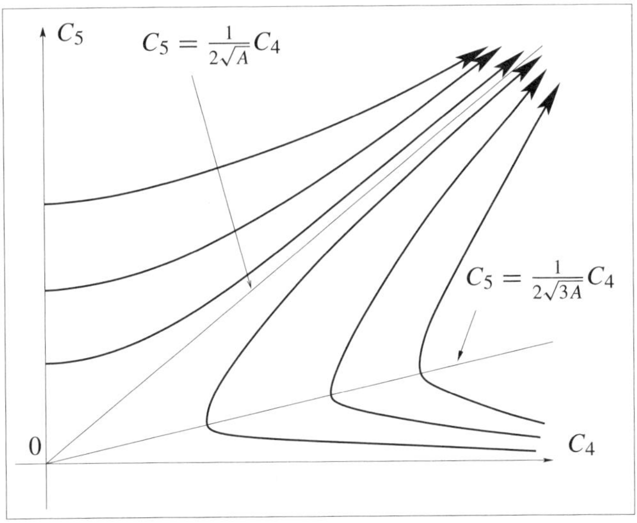

Figure 6.3 The phase-plane of (6.111): stability of the explicit solution (6.112).

The rest of the coefficients can also be calculated explicitly from the first three ODEs (6.110), and these are also singular as $t \to T^-$. In the class of fourth-degree polynomials (6.109), the solution of the parabolic M-A equation (6.107) is essentially local in time, $u(x, y, t) \to \infty$ as $t \to T^-$. This blow-up is connected with the exceeding growth of convex initial data at infinity.

The blow-up phenomenon can also be detected for the hyperbolic equation

$$\boxed{u_{tt} = u_{xx}u_{yy} - (u_{xy})^2,}$$

where the DS with the right-hand sides from (6.110) is of twelfth order and, as is usual for such quadratic systems, the blow-up rate is now $\sim (T-t)^{-2}$. For the whole subspace (6.108), the DS is of thirties order and cannot be reduced for arbitrary initial data, unlike the parabolic case.

Example 6.56 (**Higher-order regularization**) Consider a fourth-order parabolic regularization of the above M-A equation,

$$\boxed{u_t = u_{xx}u_{yy} - (u_{xy})^2 - u\Delta^2 u \quad (u > 0).}$$

Then the last three ODEs of the DS (6.110) take the form

$$\begin{cases} C_4' = -20C_4^2 + 144C_5C_6 - 24C_4(C_5 + C_6), \\ C_5' = 8(2C_4 - 3C_5 - 3C_6)C_5, \\ C_6' = 8(2C_4 - 3C_5 - 3C_6)C_6. \end{cases}$$

Again $C_6 = AC_5$ from the last two ODEs, so the blow-up behavior (or a global one due to the fourth-order regularization) can be studied on the $\{C_4, C_5\}$-plane.

Example 6.57 (**Extinction for a parabolic equation**) We next consider the extinction phenomenon for the M-A equation associated with the nonlinearity $g(s) = -\frac{1}{s}$, where an extra linear multiplier u is included,

$$\boxed{u_t = -\frac{u}{\det D^2 u}.} \tag{6.113}$$

For solutions (6.105), the DS takes the form

$$\begin{cases} C_1' = -\frac{C_1}{4C_2C_3}, \\ C_2' = -\frac{1}{4C_3}, \\ C_3' = -\frac{1}{4C_2}. \end{cases}$$

Integrating yields the following solutions:

$$u(x, y, t) = \sqrt{T-t}\,\big(B + \sqrt{\tfrac{1}{2A}}\, x^2 + \sqrt{\tfrac{A}{2}}\, y^2\big), \tag{6.114}$$

where A and B are arbitrary positive constants. These separate variables solutions vanish as $t \to T^-$, and represent a singular extinction pattern. Let us describe the corresponding asymptotic problem for more general solutions having extinction at the same moment $t = T$. Introducing the rescaled function according to (6.114),

$$u(x, t) = \sqrt{T-t}\; v(x, y, \tau), \quad \tau = -\ln(T-t) \to +\infty, \tag{6.115}$$

yields the rescaled PDE

$$v_\tau = -\frac{v}{\det D^2 v} + \tfrac{1}{2}\, v. \tag{6.116}$$

Typical second-order Hessian operators are known to be potential and the corresponding smooth parabolic flows are gradient systems. For M-A PDEs, these ideas go back to Bernstein (1910) [51]; see also Reilly [486], and [462] for a particular case. For (6.116), the Lyapunov functional is given by

$$\Phi(v) = -\tfrac{1}{3} \int v \det D^2 v + 2 \int v.$$

Here, we assume integration over a bounded domain $\Omega \subset I\!R^2$ with suitable boundary conditions on $\partial\Omega$, so the manipulations make sense. For the problem in $I\!R^N$, extra conditions at infinity are necessary, as well as suitable changing of the functional. Using (6.116) then yields

$$\tfrac{\mathrm{d}}{\mathrm{d}\tau} \Phi(v) = -\int v_\tau \big(\det D^2 v - 2\big) \equiv -\int \frac{v(\det D^2 v - 2)^2}{2 \det D^2 v} \le 0. \tag{6.117}$$

On the other hand, this is equivalent to an L^2-bound of the time-derivative v_τ,

$$\int^\infty \mathrm{d}\tau \int \frac{2 \det D^2 v}{v}\, (v_\tau)^2 < \infty, \tag{6.118}$$

which shows that the ω-limit set of the rescaled orbit satisfying (6.116) consists of stationary points only. A rigorous application of gradient systems theory for (6.116) also demands some delicate estimates of the rescaled orbits; e.g., to prove that $v(\cdot, \tau)$ is uniformly bounded and bounded from below for $\tau \gg 1$. This is not easy for scaling (6.115), which is singular as $t \to T^-$ ($\tau \to +\infty$). These problems are OPEN.

Example 6.58 (**Extinction for a hyperbolic equation**) The corresponding hyperbolic M-A equation

$$\boxed{u_{tt} = -\frac{u}{\det D^2 u}}$$

admits solutions

$$u(x, y, t) = C_1(t)\big(1 + Ax^2 + By^2\big),$$

where $A, B > 0$ are constants and C_1 satisfies the ODE $C_1'' = -\frac{1}{4AB}\frac{1}{C_1}$. It admits solutions vanishing in finite time with the generic behavior

$$C_1(t) = (T-t) - \tfrac{1}{4AB}(T-t)\ln(T-t) + \ldots \quad \text{as} \quad t \to T^-.$$

This represents an asymptotic separate variables extinction pattern that describes a stable extinction behavior on W_3. A stability theory is absent and OPEN.

Example 6.59 (**Blow-up**) For the parabolic M-A equation

$$\boxed{u_t = u\sqrt{\det D^2 u} \quad \text{in } I\!R^2 \times I\!R_+,} \tag{6.119}$$

the solutions (6.105) lead to the DS

$$\begin{cases} C_1' = 2C_1\sqrt{C_2C_3}, \\ C_2' = 2C_2\sqrt{C_2C_3}, \\ C_3' = 2C_3\sqrt{C_2C_3}. \end{cases}$$

This gives blow-up solutions in separate variables,

$$u(x, y, t) = \tfrac{1}{2\sqrt{AB}}\,\tfrac{1}{T-t}\,\big(1 + Ax^2 + By^2\big) \to \infty \quad \text{as } t \to T^-,$$

where $A, B > 0$. For the corresponding hyperbolic M-A equation

$$\boxed{u_{tt} = u\sqrt{\det D^2 u} \quad \text{in } I\!R^2 \times I\!R_+,}$$

there are the following blow-up solutions in separate variables:

$$u(x, y, t) = C_1(t)\big(1 + Ax^2 + By^2\big), \quad \text{with} \quad C_1'' = 2\sqrt{AB}\,C_1^2.$$

The DS for the coefficients $\{C_1, C_2, C_3\}$ shows that this pattern is stable on W_3 (the analysis is similar to that for hyperbolic equations in Example 5.1). The generic blow-up behavior on the subspace W_3 (in general, stability is OPEN) is given by

$$C_1(t) \sim \tfrac{3}{\sqrt{AB}}\,\tfrac{1}{(T-t)^2} \quad \text{as } t \to T^-.$$

Example 6.60 The logarithmic Gauss curvature flow in terms of the support function (see [122]) is described by the M-A-type equation

$$u_t = \sqrt{1 + |x|^2}\,\ln\big(\det D^2 u\big) + h \quad \text{in } I\!R^N \times I\!R_+,$$

where h is given. Depending on the initial uniformly convex hypersurfaces, this PDE is known to admit either a local solution, corresponding to shrinking to a point in finite time or a global solution that describes the uniform convergence to an expanding sphere.

Consider a simpler related version of this M-A-type equation

$$\boxed{u_t = (1 + |x|^2)g\big(\det D^2 u\big) \quad \text{in } I\!R^N \times I\!R_+}$$

and pose the same question of the blow-up or global existence of solutions. Taking solutions on the subspace of second-degree polynomials,

$$u(x,t) = C_0(t) + C_1(t)x_1^2 + ... + C_N(t)x_N^2,$$

their dynamics are governed by the DS

$$C_k' = g\big(2^N C_1 ... C_N\big) \quad \text{for } \ k = 0, 1, ..., N.$$

It follows that blow-up solutions appear iff *Osgood's criterion* holds

$$\int^{\infty} \mathrm{d}s \big/ g(s^N) < \infty.$$

If the integral diverges, the flow is globally defined in t. For the hyperbolic PDE

$$\boxed{u_{tt} = (1+|x|^2) g\big(\det D^2 u\big) \quad \text{in } \mathbb{R}^N \times \mathbb{R}_+,}$$

taking the solutions in separate variables

$$u(x,t) = C_1(t)\big(1 + A_1 x_1^2 + ... + A_N x_N^2\big)$$

yields the ODE

$$C_1'' = g\big(2^N A_1 ... A_N C_1^N\big),$$

so blow-up occurs via another Osgood criterion

$$\int^{\infty} \mathrm{d}s \Big/ \sqrt{\textstyle\int g(s^N)\mathrm{d}s} < \infty.$$

Example 6.61 Finally, consider the following PDE (see Remarks for a motivation for such models):

$$u_t = F_2[u] \equiv \det D^2 u - u^2 \quad \text{in } \mathbb{R}^2 \times \mathbb{R}_+. \tag{6.120}$$

In order to describe main aspects of blow-up for such M-A equations, we present an invariant analysis of (6.120) in radial geometry with the spatial variable $r = |x|$, so

$$\boxed{u_t = F_2[u] \equiv \tfrac{1}{r}\, u_r u_{rr} - u^2 \quad \text{in } \mathbb{R}_+ \times \mathbb{R}_+.} \tag{6.121}$$

The origin, $r = 0$, is a regular point of the operator on the right-hand side. Indeed, by L'Hospital's rule, for strictly convex C^2-solutions, $\frac{1}{r}\, u_r u_{rr} \to (u_{rr})^2 > 0$ as $r \to 0$. Consider solutions satisfying

$$u < 0, \quad u_r > 0, \quad \text{and} \quad u_{rr} > 0, \tag{6.122}$$

on which the equation is uniformly parabolic. Let us construct exact solutions

$$u(r,t) = C_1(t) + C_2(t) f(r) \in \mathcal{L}\{1, f\}, \tag{6.123}$$

with some unknown function f. Plugging (6.123) into (6.121), and assuming that f solves the ODE problem

$$\tfrac{1}{r}\, f' f'' = f^2, \;\; f < 0 \;\text{ for } r \in (0, r_0), \quad f'(0) = 0, \;\; f(r_0) = 0, \tag{6.124}$$

(where $\mathcal{L}\{1, f\}$ is invariant under F_2) yields the DS

$$\begin{cases} C_1' = -C_1^2, \\ C_2' = -2C_1 C_2. \end{cases}$$

This gives blow-up solutions

$$u(r,t) = -\tfrac{1}{T-t} + \tfrac{A}{(T-t)^2} f(r), \quad \text{with } A > 0. \tag{6.125}$$

These solutions have singularity at the degeneracy point $r = r_0$, where $u_{rr} = 0$ by (6.124), so solutions are not strictly convex. The PDE is degenerate and loses its uniform parabolicity. It follows that, for solutions (6.123) defined for all $r > 0$, the point $r = r_0$ is the only point of degeneracy ($u_{rr} > 0$ for all $r \neq r_0$). Nevertheless, the strong Maximum Principle for u_{rr} does not apply, and this singularity point persists until the blow-up time $t = T$. Introducing the rescaled function

$$u(r,t) = \tfrac{1}{(T-t)^2}\, v(r,\tau), \quad \text{where } \tau = \tfrac{1}{T-t},$$

gives the rescaled equation with an $O(\frac{1}{\tau})$-perturbation,

$$v_\tau = \tfrac{1}{r}\, v_r v_{rr} - v^2 - \tfrac{2}{\tau}\, v. \tag{6.126}$$

Notice that, for solutions $v(r,\tau)$ of (6.126), the "waiting time" of the singularity at $r = r_0$ is infinite. Exact solutions (6.125) show the convergence as $\tau \to \infty$,

$$v(r,\tau) \equiv -\tfrac{1}{\tau} + Af(r) \to Af(r).$$

The passage to the limit in the general equation (6.126) is an OPEN PROBLEM.

In addition, (6.121) admits separate variables solutions with a weaker blow-up rate

$$\bar{u}(r,t) = \tfrac{1}{T-t}\, \bar{f}(r),$$

where $\bar{f}$ solves the ODE

$$\tfrac{1}{r}\, \bar{f}'\bar{f}'' = \bar{f}^2 + \bar{f} \ \text{ for } r \in (0, r_0), \quad \bar{f}'(0) = 0, \ \bar{f}(r_0) = 0;$$

cf. (6.124). The corresponding exact solutions are

$$u(r,t) = C_1(t) + C_2(t)\bar{f}(r),$$

$$\begin{cases} C_1' = -C_1^2, \\ C_2' = -2C_1C_2 + C_2^2. \end{cases}$$

The regularity convexity assumption, $\bar{f}'' > 0$, determines another singularity point $r = \bar{r}_0 > 0$ such that $\bar{f}(\bar{r}_0) = -1$ with different boundary conditions that give such a blow-up solution. The rescaled solution now takes the form

$$u(r,t) = \tfrac{1}{T-t}\, w(r,s), \quad \text{where } s = -\ln(T-t),$$

and satisfies the autonomous, unperturbed PDE with a potential operator

$$w_s = \tfrac{1}{r}\, w_r w_{rr} - w^2 - w.$$

A Lyapunov function is obtained by multiplying by rw_s in L^2 with suitable boundary conditions. The passage to the limit as $s \to \infty$ demonstrates that, for general solutions, $w(r,s) \to \bar{g}(r)$.

For the M-A equation with absorption

$$\boxed{u_t = \tfrac{1}{r}\, u_r u_{rr} + u^2 \quad (u < 0)}$$

in the class (6.122), the generic decay behavior is expected to be described by the separable solution

$$u(r,t) = \tfrac{1}{T+t}\, \tilde{f}(r), \quad \text{where } \tfrac{1}{r}\, \tilde{f}'\tilde{f}'' + \tilde{f}^2 = -\tilde{f}$$

and $\tilde{f}(0) \in (-1,0)$. Hence, the equation is degenerate at some point r_0, at which $\tilde{f}(r_0) = 0$. It is easy to construct solutions on the subspace $\mathcal{L}\{1, \tilde{f}\}$.

The above subspaces can be used for studying singularity formation phenomena for hyperbolic M-A models in radial geometry,

$$\boxed{u_{tt} = \tfrac{1}{r}\, u_r u_{rr} \pm u^2.}$$

6.5.2 *On higher-order Monge–Ampère-type equations*

We discuss some types of *higher-order* M-A equations admitting invariant subspaces. Existence, uniqueness, and regularity theory of such PDEs is not well developed, excluding a few particular cases; see Remarks.

Example 6.62 (**Fourth-order Hessian equations**) In order to formulate fourth-order Hessian equations in $I\!R^2$, let us write down the fourth differential of a C^4-function $u = u(x,y)$ as a *quartic form*

$$\mathrm{d}^4u = u_{xxxx}\mathrm{d}x^4 + 4u_{xxxy}\mathrm{d}x^3\mathrm{d}y + 6u_{xxyy}\mathrm{d}x^2\mathrm{d}y^2 + 4u_{xyyy}\mathrm{d}x\mathrm{d}y^3 + u_{yyyy}\mathrm{d}y^4.$$

This gives the *catalecticant determinant*

$$F_4[u] = \det D^4u \equiv \det \begin{bmatrix} u_{xxxx} & u_{xxxy} & u_{xxyy} \\ u_{xxxy} & u_{xxyy} & u_{xyyy} \\ u_{xxyy} & u_{xyyy} & u_{yyyy} \end{bmatrix},$$

which plays an important role in the theory of quartic forms. For instance, each such form in two variables can be expressed via a sum of three fourth powers of linear forms and via two powers, provided that $\det D^4u = 0$; see [281].

We will use $F_4[u]$ for construction of some geometric flows. Clearly, F_4 preserves the subspace of fourth-degree polynomials $W_7 = \mathcal{L}\{1, x^2, xy, y^2, x^4, x^2y^2, y^4\}$ and $F_4 : W_7 \to \mathcal{L}\{1\}$. Therefore, the flow $u_t = F_4[u]$ is global on W_7. The basic invariant subspace of sixth-degree polynomials is

$$W = \mathcal{L}\{x^\alpha y^\beta,\ 0 \le \alpha + \beta \le 6\}, \tag{6.127}$$

on which blow-up may happen via a cubic DS. Singular patterns also exist for other fourth-order M-A-type models that are constructed in accordance to their second-order counterparts.

Example 6.63 (**Extinction and blow-up**) Equation

$$\boxed{u_t = -\frac{u}{\det D^4 u} \quad \text{in } I\!R^2 \times I\!R_+}$$

admits solutions on W_4,

$$u(x,y,t) = C_1(t) + C_2(t)x^4 + C_3(t)x^2y^2 + C_4(t)y^4.$$

Then $|D^4u| = 64(36C_2C_3C_4 - C_3^3)$. The corresponding fourth-order DS yields solutions in separate variables

$$u(x,y,t) = C_1(t)\big(1 + Ax^4 + Bx^2y^2 + Cy^4\big), \tag{6.128}$$

where A, B, and C are positive constants satisfying $\gamma = 64(36ABC - B^3) > 0$ by the convexity assumption on initial data. Here C_1 solves the ODE $C_1' = -\frac{1}{\gamma C_1^2}$. This gives finite-time extinction with the rate

$$C_1(t) = \big[\tfrac{3}{\gamma}(T-t)\big]^{\frac{1}{3}} \to 0 \quad \text{as } t \to T^-.$$

Vice versa,

$$\boxed{u_t = u\sqrt{\det D^4 u} \quad \text{in } I\!R^2 \times I\!R_+}$$

admits solutions (6.128) driven by the ODE $C_1' = \sqrt{\gamma}\, C_1^{5/2}$ with blow-up,

$$C_1(t) = \big[\tfrac{3\sqrt{\gamma}}{2}(T-t)\big]^{-\frac{2}{3}} \to +\infty \quad \text{as } t \to T^-.$$

Similar singularity phenomena are traced for the corresponding hyperbolic M-A flows on these subspaces. A number of typical conclusions for the second-order Hessian flows can be extended to this fourth-order, as well as higher-order, though the well-posedness of such parabolic or hyperbolic PDEs in classes of "convex" functions is a difficult OPEN PROBLEM.

Example 6.64 (**Sixth-order equations**) Similarly, one can define the sixth-order operator of the M-A-type by

$$\det D^6 u = \det \begin{bmatrix} u_{60} & u_{51} & u_{42} & u_{33} \\ u_{51} & u_{42} & u_{33} & u_{24} \\ u_{42} & u_{33} & u_{24} & u_{15} \\ u_{33} & u_{24} & u_{15} & u_{06} \end{bmatrix}, \tag{6.129}$$

where $u_{ij} = u_{x^i y^j}$. The basic subspace is $W = \mathcal{L}\{x^\alpha y^\beta,\ 0 \le \alpha+\beta \le 8\}$. The above similar "parabolic" and "hyperbolic" models admitting extinction and blow-up can be studied on this invariant subspace, though, as for $|D^4u|$, many basic questions concerning convexity and potential properties of such operators are OPEN and unclear (see Remarks).

Example 6.65 Consider the fourth-order PDE

$$F_4[u] \equiv \Delta\big(\det D^2 u\big)^{-\frac{1}{N+2}} = 0, \tag{6.130}$$

where Δ denotes the Laplacian with respect to the Blaschke metric

$$G = \big(\det D^2 u\big)^{-\frac{1}{N+2}} \sum u_{x_i x_j} \mathrm{d}x_i \mathrm{d}x_j.$$

For a locally strongly convex function $u(x)$ defining a hypersurface

$$M:\ x_{N+1} = f(x) \quad \text{for } x \in I\!R^N,$$

equation (6.130) occurs in the theory of affine maximal hypersurfaces; see [385] for basics of affine differential geometry and [384] for recent results. For many years, this theory was driven by various versions of so-called *affine Bernstein conjectures*, including Chern's and Calabi's conjecture. A typical result says that, in the class of complete hypersurfaces, $\mathcal{M}$ must be an elliptic paraboloid [384] (cf. the Jörgens–Calabi–Pogorelov results for the inhomogeneous M-A equation $\det D^2 u = 1$ in $I\!R^N$; see Remarks). The corresponding fourth-order parabolic flows

$$u_t = F_4[u] + \text{(lower-order terms)}$$

can be locally well-defined on the classes of convex functions. In particular, a related modified equation

$$\boxed{u_t = \Delta\big(\det D^2 u\big) + \text{(lower-order terms)} \quad \text{in } I\!R^2 \times I\!R_+}$$

(for simplicity, Δ is the Laplacian in $I\!R^N$) can be restricted to the subspace (6.127), and the DS determines the blow-up or global evolution of convex surfaces in $I\!R^2$.

Example 6.66 (**mth-order equations**) Consider a general mth-order fully nonlinear equation of the M-A-type in $I\!R^2 \times I\!R_+$ $(m \geq 2)$

$$\boxed{u_t = \textstyle\sum_{(|\mu|=|\nu|=m)} a_{\mu,\nu} D_x^{\mu} u D_x^{\nu} u + \text{(lower-order terms)},} \tag{6.131}$$

where μ and ν are multi-indices, and the matrix $\|a_{\mu,\nu}\|$ satisfies a positivity-type assumption for local existence (see Remarks). This equation admits a finite-dimensional restriction on the polynomial subspace $W = \mathcal{L}\{x^\alpha y^\beta,\ 0 \leq \alpha + \beta \leq 2m\}$, with a typical blow-up dynamics of convex solutions driven by a quadratic DS.

Example 6.67 (**Third-order equation**) We finish our list of M-A-type models with a related third-order PDE, written in the following evolution-looking form:

$$\boxed{u_{ttt} = F[u] \equiv (u_{txx})^2 - u_{ttx} u_{xxx}.} \tag{6.132}$$

This is known as one of the *associativity equations* in 2D field theory [157]. On the other hand, it is a parameterized form of the M-A-type equation

$$v_{xxx} v_{yyy} - v_{xxy} v_{xyy} = 1,$$

as the compatibility condition for a PDE system (a reduction of the Gauss–Codazzi equations) governing hypersurfaces $M^2 \subset A^3$ with a flat centroaffine metric, where x and y are the asymptotic coordinates on M^2, [182]. An abundance of explicit solutions of (6.132) is available in [157]; see also the table in [182, p. 41].

The quadratic operator F in (6.132) admits the 5D subspace (now a module) $W_5 = \mathcal{L}\{1, x, x^2, x^3, x^4\}$, so that the PDE restricted to W_5 with solutions

$$u(x,t) = C_1(t) + C_2(t)x + C_3(t)x^2 + C_4(t)x^3 + C_5(t)x^4 \tag{6.133}$$

is equivalent to the following fifteenth-order DS:

$$\begin{cases} C_1''' = -6C_4C_2'' + 4(C_3')^2, \\ C_2''' = -24C_5C_2'' - 12C_4C_3'' + 24C_3'C_4', \\ C_3''' = -48C_5C_3'' - 18C_4C_4'' + 36(C_4')^2 + 48C_3'C_5', \\ C_4''' = -72C_5C_4'' - 24C_4C_5'' + 144C_4'C_5', \\ C_5''' = -96C_5C_5'' + 144(C_5')^2. \end{cases}$$

This is a difficult DS in general, but some explicit solutions can be found on 2D and 3D partially invariant subspaces from W_5. The majority of the solutions in [157] belong to such classes. In addition, looking for solutions on another module

$$u(x,t) = C_1(x) + C_2(x)\mathrm{e}^t \in \mathcal{L}\{1, \mathrm{e}^t\} \tag{6.134}$$

yields a simpler sixth-order DS of the form

$$\begin{cases} C_2' C_1''' = -C_2, \\ C_2' C_2''' = (C_2'')^2. \end{cases}$$

The second ODE is integrated, providing us with two types of solutions, e.g., (i) $C_2(x) = x$ and $C_1(x) = -\frac{1}{24}x^4$, which is contained in (6.133), and (ii) $C_2(x) = \mathrm{e}^x + 1$, $C_1(x) = \mathrm{e}^{-x} - \frac{1}{6}x^3$ that is not available in (6.133).

6.6 Higher-order thin film operators

6.6.1 Basic polynomial subspaces

Let us return to the fourth-order quadratic thin film operator in $I\!R^N$

$$F[u] = -\nabla \cdot (u \nabla \Delta u) \equiv -u\Delta^2 u - \nabla u \cdot \nabla \Delta u. \tag{6.135}$$

In the statement below, i, j, k, l and $\alpha, \beta, \gamma, \delta$ are various nonnegative integers.

Proposition 6.68 *Operator* (6.135) *preserves the following subspaces:*

$$W_1^{\mathrm{r}} = \mathcal{L}\{|x|^4\}, \quad W_2^{\mathrm{r}} = \mathcal{L}\{1, |x|^4\}, \quad W_3^{\mathrm{r}} = \mathcal{L}\{1, |x|^2, |x|^4\}; \tag{6.136}$$

$$W_n^{\mathrm{f}} = \mathcal{L}\{1, x_i^2 x_j^2, \ 1 \le i, j \le N\}; \tag{6.137}$$

$$W^{\mathrm{q,f}} = \mathcal{L}\{1, x_i x_j, x_i^2 x_j^2, \ 1 \le i, j \le N\}; \tag{6.138}$$

$$\bar{W}^{\mathrm{f}} = \mathcal{L}\big\{1, x_i^{\alpha} x_j^{\beta}, \ 1 \le i, j \le N, \ 1 \le \alpha + \beta \le 4\big\}; \tag{6.139}$$

$$\tilde{W}^{\mathrm{f}} = \mathcal{L}\big\{1, x_i^{\alpha} x_j^{\beta} x_k^{\gamma} x_l^{\delta}, \ 1 \le i, j, k, l \le N, \ 1 \le \alpha + \beta + \gamma + \delta \le 4\big\}. \tag{6.140}$$

Proof. In the radial case, calculations are straightforward. Setting $u = C_1 + C_2|x|^2 + C_3|x|^4$ shows that W_3^{r} is invariant,

$$F[u] = -8(N+2)\big[NC_1C_3 + (N+2)C_2C_3|x|^2 + (N+4)C_3^2|x|^4\big].$$

Then the straight line W_1^{r}: $C_1 = C_2 = 0$ and the plane W_2^{r}: $C_2 = 0$ in W_3^{r} are also invariant. For the subspace (6.138), take

$$u = C + \textstyle\sum a_{i,j}\, x_i x_j + \sum b_{i,j}\, x_i^2 x_j^2 \equiv C + x^T A x + (x^2)^T B x^2,$$

where x^2 denotes $\big(x_1^2, ..., x_N^2\big)^T$. Writing the quartic form as

$$\textstyle\sum b_{i,j}\, x_i^2 x_j^2 = \sum_{(i)} b_{i,i}\, x_i^4 + \sum_{(i \ne j)} b_{i,j}\, x_i^2 x_j^2, \tag{6.141}$$

and differentiating, we obtain

$$\Delta u = 2(\mathrm{tr}\, A) + 4\textstyle\sum_{(l,i)} b_{i,l}(2\delta_{il} + 1)x_i^2, \quad \Delta^2 u = 8\big[2(\mathrm{tr}\, B) + \mathrm{sum}\, B\big],$$

where sum B denotes $\sum_{(i,j)} b_{i,j}$. Hence,

$$F[u] = -\big(C + \sum a_{i,j} x_i x_j + \sum b_{i,j}\, x_i^2 x_j^2\big)\,\Delta^2 u$$
$$- 16 \sum_{(m,j,l)} b_{m,l}(2\delta_{ml} + 1)\big[a_{m,j} x_j x_m + 2b_{m,j} x_m^2 x_j^2\big]$$

belongs to $W^{\rm q,f}$. Choosing $A = 0$ yields that (6.137) is invariant. For the subspace (6.139) of fourth-order polynomials, for arbitrary

$$u = \sum_{(i,j,\alpha,\beta)} C_{i,j,\alpha,\beta}\, x_i^\alpha x_j^\beta,$$

the operator (6.135) maps this polynomial into another polynomial of the degree not exceeding $\max\{2(\alpha+\beta) - 4\} = 4$, provided that $\alpha + \beta \le 4$. The same argument applies to the most general subspace (6.140). □

Example 6.69 (**Dispersive Boussinesq equations**) Another area of application of polynomial subspaces is connected with the family of 2D *dispersive Boussinesq equations* denoted by $B(m,n,k,p)$ (see [580]),

$$\boxed{(u^m)_{tt} + \alpha(u^n)_{xx} + \beta(u^k)_{xxxx} + \gamma\,(u^p)_{yyyy} = 0.}$$

Then the quadratic operators of $B(1,1,2,2)$ or $B(1,2,2,2)$ equations admit basic polynomial subspaces on which the PDE reduces to a DS.

Similar basic subspaces exist for higher-order thin film operators.

Proposition 6.70 *For any $m \ge 2$, the thin film operator*

$$F[u] = (-1)^{m+1} \nabla \cdot (u \nabla \Delta^{m-1} u) \quad in\ I\!R^N$$

admits the subspaces

$$W_{N+1} = \mathcal{L}\{1, x_1^{2m}, ..., x_N^{2m}\}; \tag{6.142}$$
$$W_n = \mathcal{L}\{1, x_i^m x_j^m\}; \tag{6.143}$$
$$\bar W = \mathcal{L}\big\{1, x_i^\alpha x_j^\beta,\ 1 \le \alpha{+}\beta \le 2m\big\};$$
$$\tilde W = \mathcal{L}\big\{1, x_{i_1}^{\alpha_1} x_{i_2}^{\alpha_2} ... x_{i_{2m}}^{\alpha_{2m}},\ 1 \le \alpha_1{+}...{+}\alpha_{2m} \le 2m\big\}.$$

Example 6.71 (**Non-symmetric extinction and interface equation**) The TFE with the constant absorption,

$$\boxed{u_t = (-1)^{m+1} \nabla \cdot (u \nabla \Delta^{m-1} u) - 1,} \tag{6.144}$$

admits solutions on the subspace (6.142),

$$u(x,t) = C(t) + a_{1,1}(t) x_1^{2m} + ... + a_{N,N}(t) x_N^{2m}, \tag{6.145}$$

where the expansion coefficients solve the DS

$$\begin{cases} C' = (-1)^{m+1}(2m)!\,(\operatorname{tr} A)\, C - 1, \\ a_{1,1}' = (-1)^{m+1}(2m)!\,\big[(\operatorname{tr} A) + 2m a_{1,1}\big] a_{1,1}, \\ \quad ... \qquad ... \qquad ... \\ a_{N,N}' = (-1)^{m+1}(2m)!\,\big[(\operatorname{tr} A) + 2m a_{N,N}\big] a_{N,N}. \end{cases} \tag{6.146}$$

Here, $\operatorname{tr} A = \sum a_{i,i}$ is the trace of the diagonal $N \times N$ matrix $A(t)$. These exact solutions correspond to a special FBP for equation (6.144) and describe the non-symmetric finite-time extinction. The asymptotic behavior near the extinction time, as $t \to T^-$, is easy to obtain from (6.146),

$$u(x,t) = (T-t)\big(1 - b_1\xi_1^{2m} - ... - b_N\xi_N^{2m}\big)_+ + ..., \quad \xi = \tfrac{x}{(T-t)^{1/2m}}, \tag{6.147}$$

where $b_i = -a_{i,i}(T) > 0$ for $i = 1, ..., N$. Rescaling the PDE (6.144) according to (6.147) by introducing the rescaled function

$$u(x,t) = (T-t)w(\xi,\tau), \quad \text{where } \tau = -\ln(T-t),$$

yields a difficult singular perturbed PDE

$$w_\tau = -\tfrac{1}{2m}\nabla w \cdot \xi + w - 1 + (-1)^{m+1}\mathrm{e}^{-\tau}\nabla \cdot (w \nabla \Delta^{m-1} w).$$

The pattern (6.147) is not expected to describe a stable generic extinction behavior for the TFE (6.144) (an OPEN PROBLEM), except in the porous medium case, $m = 1$, where it was proved to be stable, [245, Ch. 5].

The FBP generating extinction asymptotics (6.147) is obtained in the usual way. Differentiating $u(x,t) = 0$ on the interface yields the normal interface velocity $V^\perp$ in the outward direction

$$V^\perp = \tfrac{1}{|\nabla u|}u_t \equiv \tfrac{1}{|\nabla u|}\big(C' - a'_{1,1}x_1^2 - ... - a'_{N,N}x_N^2\big), \tag{6.148}$$

which is satisfied by any sufficiently smooth solution. In order to get the dynamic interface equation for solutions (6.145), we use the time-parameterization $\Delta^m u = (2m)!\,\mathrm{tr}\,A(t)$, and substitute this $t = t(\Delta^m u)$ into (6.148). The unique solvability of such FBPs is a hard OPEN PROBLEM to be discussed in a simpler example.

On the subspace (6.143) with solutions

$$u(x,t) = C(t) + (x^m)^T A(t)\, x^m \quad \big(x^m = (x_1^m, ..., x_N^m)^T\big),$$

the DS becomes more complicated. For instance, for $m = 2$, using calculations at the beginning of this section, we have

$$\begin{cases} C' = -[12(\mathrm{tr}\,A) + 8\,\mathrm{sum}\,A]C - 1, \\ A' = -[12(\mathrm{tr}\,A) + 8\,\mathrm{sum}\,A]A - 32A^2 - 64\,\mathrm{diag}\,A^2, \end{cases}$$

where $\mathrm{sum}\,A = \sum_{(i,j)} a_{i,j}$ and $\mathrm{diag}\,A^2$ is the diagonal matrix $\mathrm{diag}\{a_{1,1}^2, ..., a_{N,N}^2\}$.

6.6.2 On solutions with zero contact angle at interfaces

The above exact solutions do not have zero contact angle, and correspond to less physically motivated Stefan–Florin FBPs. As in the 1D problems in Section 3.3, we show how to construct zero contact angle solutions for two TFE problems.

Example 6.72 **(Extinction for TFE with absorption)** Consider the PDE

$$v_t = -\nabla \cdot (v \nabla \Delta v) - \sqrt{v} \quad \text{in } I\!R^N \times I\!R_+. \tag{6.149}$$

Setting $v = u^2$ yields the cubic equation

$$\boxed{u_t = -\tfrac{1}{2}u\Delta^2 u^2 - \nabla u \cdot \nabla \Delta u^2 - \tfrac{1}{2} \equiv F_3[u] - \tfrac{1}{2}.} \tag{6.150}$$

For the main application, it suffices to use the following result, which is not optimal, as is known from the general representation of such subspaces in Section 6.1.

Proposition 6.73 *The cubic operator F_3 admits $W_{N+1} = \mathcal{L}\{1, x_1^2, ..., x_N^2\}$.*

For solutions with the diagonal matrix $A(t)$, we perform as in (6.141),

$$\begin{aligned} u(x,t) &= C(t) + a_1(t)x_1^2 + ... + a_N(t)x_N^2 \in W_{N+1}, \quad \text{so} \\ u^2(x,t) &= C^2 + 2C \textstyle\sum_{(i)} a_i x_i^2 + \sum_{(i)} a_i^2 x_i^4 + \sum_{(i \neq j)} a_i a_j x_i^2 x_j^2, \end{aligned} \tag{6.151}$$

and the following holds:

$$\begin{aligned} \Delta u &= 2(\text{tr}\, A), \quad \Delta u^2 = 4(\text{tr A})\, C + 4 \textstyle\sum a_i a_j (2\delta_{ij} + 1) x_j^2, \\ \Delta^2 u^2 &= 8 \textstyle\sum a_i a_j (2\delta_{ij} + 1) \equiv 8(\text{tr}\, A)^2 + 16 \sum a_i^2, \end{aligned} \tag{6.152}$$

where δ_{ij} is Kronecker's delta. Substituting (6.151) into (6.150) yields the DS

$$\begin{cases} C' = -4[(\text{tr}\, A)^2 + 2 \sum a_i^2]\, C - \frac{1}{2}, \\ a_k' = -4[(\text{tr}\, A)^2 + 2 \sum a_i^2]\, a_k - 16(\text{tr}\, A) a_k^2 - 32 a_k^3, \end{cases} \tag{6.153}$$

for all $k = 1, 2, ..., N$. Since $v = u^2$, the corresponding solutions of the original equation (6.149)

$$v(x,t) = \big[C(t) + a_1(t)x_1^2 + ... + a_N(t)x_N^2\big]_+^2$$

satisfy the zero contact angle and the zero-flux conditions

$$v = \nabla v = v \frac{\partial}{\partial \mathbf{n}} \Delta v = 0 \quad \text{at the interface,} \tag{6.154}$$

where $\mathbf{n}$ denotes the unit outward normal to the given smooth interface. We next characterize the extinction phenomenon. We fix bounded initial values for the expansion coefficients $C(0) > 0$ and $a_k(0) < 0$ for all k. Hence, the initial function

$$v_0(x) = \big[C(0) + a_1(0)x_1^2 + ... + a_N(0)x_2^2\big]_+^2$$

is compactly supported. It follows from (6.153) that $C(t)$ vanishes in finite time with the asymptotics $C(t) = \frac{1}{2}(T - t) + O((T - t)^2)$ as $t \to T^-$, while all $a_k(t)$ are assumed to have finite negative limits, $a_k(t) \to -\nu_k < 0$. This gives the non-symmetric asymptotic extinction pattern

$$u(x,t) = \mathrm{e}^{-\tau} \big(\tfrac{1}{2} - \nu_1 \xi_1^2 - ... - \nu_N \xi_N^2\big)_+ + O\big(\mathrm{e}^{-2\tau}\big), \tag{6.155}$$

where $\zeta = x/\sqrt{T - t}$ and $\tau = -\ln(T - t)$. As usual, this suggests introducing the rescaled function $u(x,t) = \mathrm{e}^{-\tau} w(\xi, \tau)$, where w solves a singular perturbed first-order PDE of the form

$$w_\tau = -\tfrac{1}{2} \nabla w \cdot \xi + w - \tfrac{1}{2} + \mathrm{e}^{-\tau} F_3[w]. \tag{6.156}$$

The function $g(\xi) = \frac{1}{2} - \nu_1 \xi_1^2 - ... - \nu_N \xi_N^2$ derived in (6.155) is a non-symmetric stationary solution of the limit ($\tau = \infty$) equation (6.156). A rigorous passage to the limit $\tau \to \infty$ in this equation is a difficult OPEN PROBLEM. Our solutions on W_{N+1} show that existence of the limit is plausible. In a similar fashion, solutions (6.151) describe *non-symmetric quenching* of strictly positive analytic solutions as $t \to T^-$, with the rescaled equation (6.156) and the same FBP that appears for $t > T$.

The dynamic interface equation for this Stefan–Florin FBP is given by (6.148). Since v already satisfies three free-boundary conditions (6.154), we do not need any extra ones, so (6.148) plays the role of the regularity condition. Unlike the 1D case in Example 3.10, in non-radial geometry, the von Mises transformation with x_1 replaced by u (this applies to some typical geometries), $X = X(u, x_2, ..., x_N, t)$, still transforms the free boundary onto the hyperplane $\{u = 0\}$, but leads to more complicated higher-order Neumann-like boundary constraints for the PDE satisfied by X. Local existence and uniqueness for such problems are OPEN.

Example 6.74 (**Stabilization**) Consider the TFE with the unstable second-order diffusion operator

$$v_t = -\nabla \cdot (v \nabla \Delta v) - \nabla \cdot (\sqrt{v}\ \nabla v) \quad \text{in } I\!R^N \times I\!R_+. \tag{6.157}$$

Then $u = \sqrt{v}$ solves the following cubic equation:

$$\boxed{u_t = -\tfrac{1}{2}\, u \Delta^2 u^2 - \nabla u \cdot \nabla \Delta u^2 - u \Delta u - 2|\nabla u|^2.}$$

Apparently, the second-order quadratic operator therein admits the subspace in Proposition 6.73, so there exist exact solutions (6.151) with the governing DS

$$\begin{cases} C' = -\big(\tfrac{1}{2}\, \Delta^2 u^2 + \Delta u\big) C, \\ a_k' = -\big(\tfrac{1}{2}\, \Delta^2 u^2 + \Delta u\big) a_k - 8(1 + \Delta u) a_k^2 - 32 a_k^3. \end{cases} \tag{6.158}$$

This system describes stability (stabilization) or instability of stationary solutions of (6.157). It is easy to find the symmetric stationary solution for which

$$a_1 = ... = a_k = a = -\tfrac{1}{2(N+2)} < 0. \tag{6.159}$$

By the first ODE in (6.158), $C(t)$ stabilizes exponentially fast to a constant $B > 0$,

$$u(x,t) = \big[B - \tfrac{1}{2(N+2)}\, |x|^2\big]_+ + O(e^{-\gamma t}) \quad \text{as } t \to \infty.$$

For non-symmetric stationary solutions, a complicated algebraic system occurs that can have solutions other than the symmetric one (6.159). For instance, in $I\!R^2$, there exists the solution of changing sign with $a_1 = \frac{1}{18}$ and $a_2 = -\frac{2}{9}$, where $a_1 + a_2 = -\frac{1}{6}$. In this case, at the stationary point, $C(t)$ converges to zero, and the solution exhibits the following large time behavior:

$$u(x,t) = \big(e^{-\frac{8}{9}t} + \tfrac{1}{18}\, x_1^2 - \tfrac{2}{9} x_2^2 + ...\big)_+ .$$

This means stabilization to an unbounded stationary solution, with the free boundary composed of two intersecting straight lines. Stability problems are OPEN.

Example 6.75 (**2mth-order TFEs**) Consider two $2m$th-order TFEs

$$v_t = (-1)^{m+1} \nabla \cdot (v \nabla \Delta^{m-1} v) - v^{\frac{m-1}{m}},$$

$$v_t = (-1)^{m+1} \nabla \cdot (v \nabla \Delta^{m-1} v) - \nabla \cdot \big(v^{\frac{1}{m}} \nabla v z\big).$$

Transformation $v = u^m$ yields

$$\boxed{u_t = (-1)^{m+1}\big(\tfrac{1}{m}\, u \Delta^m u^m + \nabla u \cdot \nabla \Delta^{m-1} u^m\big) - \tfrac{1}{m} \equiv F_{m+1}[u] - \tfrac{1}{m},}$$

$$\boxed{u_t = F_{m+1}[u] - (u \Delta u + m|\nabla u|^2),}$$

where the operator F_{m+1} satisfies Proposition 6.73. Therefore, solutions (6.151) can be constructed, where, in deriving the DS, $\Delta^m u^m = (2m)! \sum u_i^m$. It follows that solutions of the original equations

$$v(x,t) = \big[C(t) + a_1(t)x_1^2 + ... + a_N(t)x_N^2\big]_+^m$$

satisfy the zero contact angle condition, as well as the zero-flux one,

$$\nabla v = ... = \tfrac{\partial^{m-1}}{\partial \mathbf{n}^{m-1}} v = v \tfrac{\partial}{\partial \mathbf{n}} \Delta^{m-1} v = 0 \quad \text{at the interface.}$$

The first model describes finite-time extinction or quenching with a singular perturbed rescaled first-order PDE (an OPEN PROBLEM), while the second PDE is suitable for studying stabilization to stationary solutions (also hard and OPEN).

Example 6.76 As a further generalization of the TFEs, for a fixed parameter $n \neq 0$, let us introduce another pair of $2m$th-order equations in $I\!R^N \times I\!R_+$,

$$v_t = (-1)^{m+1} \nabla \cdot \big(v \Delta^{m-1} v^n\big) - v^{\frac{m-n}{m}},$$

$$v_t = (-1)^{m+1} \nabla \cdot \big(v \Delta^{m-1} v^n\big) - \nabla \cdot \big(v^{\frac{n}{m}} \nabla v\big).$$

We use the transformation $v = u^\mu$ with $\mu = \frac{m}{n}$ to get

$$\boxed{u_t = (-1)^{m+1}\big(\tfrac{n}{m}\, u \Delta^m u^m + \nabla u \cdot \nabla \Delta^{m-1} u^m\big) - \tfrac{n}{m} \equiv F_{m+1}[u] - \tfrac{n}{m},}$$

$$\boxed{u_t = F_{m+1}[u] - \big(u \Delta u + \tfrac{m}{n} |\nabla u|^2\big),}$$

where the operator F_{m+1} admits the subspace from Proposition 6.73. Thus, we can proceed as usual, using the same formulae in describing singular phenomena.

6.6.3 *Blow-up and extinction for* $n = -\frac{6}{N+2}$

Interesting singularity phenomena can be described by using radially symmetric functions from the subspace

$$W_2 = \mathcal{L}\{1, |x|^2\}. \tag{6.160}$$

Our basic model is the TFE with source (for $n < 0$) or absorption (for $n > 0$),

$$v_t = -\nabla \cdot \big(v^n \nabla \Delta v\big) - \tfrac{3}{n}\, v^{\frac{3-n}{3}} \quad \text{in } I\!R^N \times I\!R_+. \tag{6.161}$$

Setting $v = u^\mu$, with $\mu = \frac{3}{n}$, yields a PDE with a quartic operator which can be analyzed in a manner similar to Proposition 3.7. Then the last reaction-absorption term in (6.161) reduces to -1, and we obtain the PDE

$$\boxed{u_t = -\tfrac{1}{\mu}\, u^{1-\mu} \nabla \cdot \big(u^3 \nabla \Delta u^\mu\big) - 1 \equiv F_4[u] - 1.} \tag{6.162}$$

Using the radial form of the operator,

$$F_4[u] = -\tfrac{1}{\mu}\, u^{1-\mu}\, \tfrac{1}{r^{N-1}} \big[r^{N-1} u^3 \big(\tfrac{1}{r^{N-1}} [r^{N-1} (u^\mu)_r']_r'\big)_r'\big]_r',$$

we arrive at the following conclusion:

Proposition 6.77 *Operator* F_4 *in* (6.162) *admits subspace* (6.160) *iff*

$$(\mu - 1)\big(4\mu^2 + 4N\mu + N^2 - 4\big) = 0,$$

i.e., in the following three cases:
(i) $n = 3$ $(\mu = 1)$;
(ii) $n = -\frac{6}{N-2}$ for $N \neq 2$ $\left(\mu = -\frac{N-2}{2}\right)$; and
(iii) $n = -\frac{6}{N+2}$ $\left(\mu = -\frac{N+2}{2}\right)$.

Example 6.78 **(Single point blow-up and extinction)** Take $n = -\frac{6}{N+2}$ and consider the TFE with source

$$v_t = -\nabla \cdot \left(v^{-\frac{6}{N+2}} \nabla \Delta v\right) + \tfrac{N+2}{2}\, v^{\frac{N+4}{N+2}}. \tag{6.163}$$

The corresponding exact solutions are

$$\begin{gathered} v(x,t) = u^{\frac{3}{n}}(x,t) = \left[C_1(t) + C_2(t)|x|^2\right]^{-\frac{N+2}{2}}, \\ \begin{cases} C_1' = aC_1^2C_2^2 - 1, \\ C_2' = -bC_1C_2^3, \end{cases} \\ \text{where } a = 2N(N+2)(N+4), \\ b = 12(N+2)(N+4), \quad \text{so that} \\ a - b = 2(N+2)(N+4)(N-6) > 0 \quad \text{for} \quad N > 6. \end{gathered} \tag{6.164}$$

Figure 6.4 shows the phase-plane of the equivalent first-order ODE

$$\frac{\mathrm{d}C_2}{\mathrm{d}C_1} = \frac{bC_1C_2^3}{1 - aC_1^2C_2^2} \tag{6.165}$$

in the case $N > 6$. We show there the infinity-cline (the dashed line) and the *separatrix* (the bold-face curve) with the equations

$$C_2 = \frac{1}{\sqrt{a}}\frac{1}{C\,1} \quad \text{and} \quad C_2 = \frac{1}{\sqrt{a-b}}\frac{1}{C_1}, \tag{6.166}$$

respectively. The separatrix separates two classes of orbits with different evolution. These are:
(I) The positive orbits $\{C_1(t), C_2(t)\}$ below the separatrix which end up the evolution at finite points on the C_2-axis. This means *blow-up* of the exact solutions.
(II) Neglecting 1 in the denominator in (6.165) yields for $C_1 \gg 1$,

$$\frac{\mathrm{d}C_2}{\mathrm{d}C_1} = -\frac{b}{a}\frac{C_2}{C_1} + \ldots \quad \Longrightarrow \quad C_2 = DC_1^{-\frac{b}{a}} + \ldots, \text{ where } \frac{b}{a} = \frac{6}{N}, \tag{6.167}$$

where $D > 0$ is a constant. Then

$$aC_1^2C_2^2 \sim C_1^{2(1-\frac{b}{a})} \gg 1 \quad \text{for} \quad C_1 \gg 1,$$

provided that $a > b$, i.e., $N > 6$. This gives monotone decreasing orbits above the separatrix, which actually describe the *extinction* of small solutions in the pure TFE (6.163) without the source term. Let us present some details.

(I) Single point blow-up. According to Figure 6.4, for all positive initial values $\{C_1(0), C_2(0)\}$ *below* the separatrix, $C_1(t)$ vanishes in finite time with the behavior

$$C_1(t) = T - t + \ldots \quad \text{as} \quad t \to T^-,$$

while $C_2(t)$ has a finite limit, $C_2(t) \to \nu > 0$ as $t \to T^-$. Then $u(x,t)$ is strictly

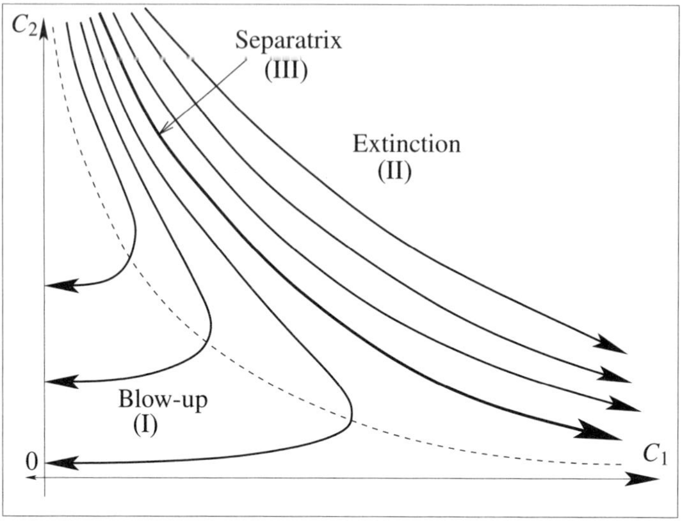

Figure 6.4 The phase-plane of (6.165) for $N > 6$: blow-up (I) and extinction (II) orbits.

positive in $I\!R^N \times (0, T)$, and hence, is a smooth solution of the TFE (and moreover is analytic by classical theory of uniformly parabolic PDEs with analytic coefficients). Translating this behavior to the original function $v(x, t)$ yields the following blow-up behavior of the exact solutions:

$$\begin{aligned} v(x,t) &= \big[C_1(t) + C_2(t)|x|^2\big]^{-\frac{N+2}{2}} \approx \big(T - t + \nu|x|^2\big)^{-\frac{N+2}{2}} \\ &= (T-t)^{-\frac{N+2}{2}}\big(1+\nu|\xi|^2\big)^{-\frac{N+2}{2}} \equiv (T-t)^{-\frac{N+2}{2}} g(\xi), \end{aligned} \tag{6.168}$$

where $\xi = x/\sqrt{T-t}$ is the corresponding spatial rescaled variable. As usual, according to these asymptotics, we introduce the rescaled function

$$v(x,t) = (T-t)^{-\frac{N+2}{2}} w(\xi,\tau), \quad \text{where } \tau = -\ln(T-t),$$

which satisfies a singular perturbed *Hamilton–Jacobi equation* of the form

$$w_\tau = -\tfrac{1}{2}\nabla w \cdot \xi - \tfrac{N+2}{2}\, w + \tfrac{N+2}{2}\, w^{\frac{N+4}{N+2}} - \mathrm{e}^{-2\tau}\nabla \cdot \big(w^{-\frac{6}{N+2}}\nabla \Delta w\big).$$

Passage to the limit $\tau \to \infty$ to get the rescaled stationary profile $g(\xi)$ given in (6.168) for a class of solutions is an OPEN PROBLEM.

It follows from (6.168) that the *final-time profile* for these *single point* blow-up solutions is as follows:

$$v(x,t) \to v(x,T^-) = \big(\nu|x|^2\big)^{-\frac{N+2}{2}} \quad \text{as } t \to T^-.$$

(II) Generic finite or infinite-time extinction. On the other hand, if positive initial values $\{C_1(0), C_2(0)\}$ lie *above* the separatrix in Figure 6.4, the ODE (6.165) yields (6.167) as $C_1 \to +\infty$. Then the first equation from the DS (6.164) implies

$$C_1' = aD^2C_1^\gamma + \dots, \quad \text{with } \gamma = 2\big(1 - \tfrac{6}{N}\big). \tag{6.169}$$

If $\gamma > 1$, i.e., $N > 12$, $C_1(t)$ blows up in finite time with the behavior

$$C_1(t) = \big[(\gamma - 1)aD^2(T - t)\big]^{-\frac{N}{N-12}}(1 + o(1)).$$

In terms of the original solution $v(x,t)$, we obtain the finite-time extinction pattern

$$v(x,t) \approx (T-t)^{\frac{N(N+2)}{2(N-12)}}\big(\nu_1 + \nu_2|\zeta|^2\big)^{-\frac{N+2}{2}}, \quad \zeta = x(T-t)^{\frac{N+6}{2(N-12)}}, \tag{6.170}$$

where $\nu_{1,2} > 0$ are some constants. This extinction is due to the singular character of the higher-order diffusion operator, where the negative exponent $n = -\frac{6}{N+2}$ mimics a very fast diffusion with non-zero flux at $x = \infty$. This phenomenon is common for the second-order quasilinear fast-diffusion equations (see Remarks), but, for higher-order TFEs, it is not well understood and creates many OPEN PROBLEMS.

If $\gamma = 1$ in (6.169), i.e., $N = 12$, we have

$$C_1(t) \sim e^{aD^2t} \quad \text{for} \quad t \gg 1,$$

and obtain exponentially decaying patterns $v(x,t) \sim e^{-7aD^2t}$ as $t \to +\infty$. For $\gamma < 1$, i.e., $N \in (6, 12)$,

$$C_1(t) \sim t^{\frac{N}{12-N}} \implies v(x,t) \sim t^{-\frac{N(N+2)}{2(12-N)}} \quad \text{as} \quad t \to \infty,$$

so the solution $v(x,t)$ has infinite-time extinction with a power-like decay.

(III) Separatrix behavior for N $>$ 6. This new, most probable, unstable behavior corresponds to the separatrix in (6.166). Then from the second ODE in (6.164), $C_2(t) = \frac{\sqrt{a-b}}{b}\frac{1}{t}$, and finally we find the following new explicit solution of the TFE (6.163):

$$v(x,t) = t^{-\frac{N+2}{2}}\Big(\tfrac{6}{N-6} + \tfrac{1}{6}\sqrt{\tfrac{N-6}{2(N+2)(N+4)}}\,|\xi|^2\Big)^{-\frac{N+2}{2}}, \quad \text{where} \quad \xi = \tfrac{x}{t}.$$

6.6.4 *On some extensions of polynomial subspaces*

Quadratic operators. Extended subspaces that are composed of higher-degree polynomials exist for more general operators of the thin film type, such as

$$F[u] = \alpha u\Delta^2 u + \beta\nabla u \cdot \nabla\Delta u + \gamma(\Delta u)^2, \tag{6.171}$$

which is considered in radial geometry in $I\!R^N$.

Proposition 6.79 *Operator* (6.171) *preserves the following subspaces:*

$$\begin{gathered}\text{(i)} \quad W_4^{\rm r} = \mathcal{L}\{1, |x|^2, |x|^4, |x|^6\} \quad \textit{iff} \\ 2(N+2)\alpha + 12\beta + 3(N+4)\gamma = 0;\end{gathered} \tag{6.172}$$

$$\text{(ii)} \quad W_5^{\rm r} = \mathcal{L}\{1, |x|^2, |x|^4, |x|^6, |x|^8\} \quad \textit{iff} \tag{6.173}$$

$$\begin{cases} (N+4)(3N+14)\alpha + 4(5N+26)\beta + 4(N+4)(N+6)\gamma = 0, \\ 3(N+4)\alpha + 24\beta + 4(N+6)\gamma = 0, \end{cases} \tag{6.174}$$

i.e., for $\alpha = \frac{2(N-2)}{N+4}\beta$ *and* $\gamma = -\frac{3(N+2)}{2(N+6)}\beta$.

Proof. (i) The condition in (6.172) annuls the terms containing $|x|^8$ in $F[u]$. (ii) (6.174) annuls all terms in $F[u]$ containing $|x|^{10}$ and $|x|^{12}$. □

It is easy to check that (6.171) does not preserve the 6D subspace

$$W_6^{\rm r} = \mathcal{L}\{1, |x|^2, |x|^4, |x|^6, |x|^8, |x|^{10}\}$$

unless $\alpha = \beta = \gamma = 0$. Using $W_6^{\rm r}$ yields the terms containing $|x|^{12}$, $|x|^{14}$, $|x|^{16}$ and, for annulling it, we obtain a homogeneous system of linear algebraic equations for α, β, and γ that has a trivial solution only.

Example 6.80 Consider the fourth-order quasilinear parabolic PDE

$$\boxed{u_t = \alpha u \Delta^2 u + \beta \nabla u \cdot \nabla \Delta u + \gamma (\Delta u)^2}$$

for which the condition in (6.172) holds. Using subspace (6.172) with

$$u(x,t) = C_1(t) + C_2(t)|x|^2 + C_3(t)|x|^4 + C_4(t)|x|^6$$

yields the following DS:

$$\begin{cases} C_1' = 8\alpha N(N+2)C_1C_3 + 4\gamma N^2 C_2^2, \\ C_2' = 8(N+2)(N\alpha + 2\beta + 2N\gamma)C_2C_3 + 24\alpha(N+2)(N+4)C_1C_4, \\ C_3' = 8(N+2)\big[N\alpha + 4\beta + 2(N+2)\gamma\big]C_3^2 \\ \qquad + 24(N+4)\big[(N+2)\alpha + 2\beta + N\gamma\big]C_2C_4, \\ C_4' = 8\big[4\alpha(N+2)(N+3) + 6\beta(3N+10) + 6\gamma(N+2)(N+4)\big]C_3C_4. \end{cases}$$

These solutions can be used for detecting interface equations for the corresponding FBPs, and for revealing various quenching and extinction patterns (OPEN PROBLEMS).

On a cubic operator in $I\!R^N$. As an example of another type, consider a cubic operator of the form

$$F[u] = -\Delta^{2l}(au^3 + bu^2 + cu) \quad \text{for} \quad l = 1, 2, \ldots . \tag{6.175}$$

Proposition 6.81 *Operator* (6.175) *admits* $W_{l+1}^{\rm r} = \mathcal{L}\{1, r^2, ..., r^{2l}\}$.

In particular, solutions on $W_{l+1}^{\rm r}$ describe the fine structure of singularity formation for the quasilinear parabolic equation with absorption

$$\boxed{u_t = -\Delta^{2l}(u^3) - 1,}$$

where a suitable FBP should be posed.

6.6.5 *On remarkable thin film operators in $I\!R^2$*

In the second-order case in Section 6.3, we have studied the remarkable operator (6.57), exhibiting a variety of invariant properties. Such related remarkable operators exist in thin film theory.

First remarkable operator. Let us begin with special properties of the following thin film operator:

$$F_1[u] = -u\Delta^2 u - (\Delta u)^2 + 2\nabla u \cdot \nabla \Delta u \quad \text{for } (x,y) \in I\!R^2. \tag{6.176}$$

This is a 2D version of the special operator (3.125) with $\beta = -1$. According to Proposition 3.19, we expect that (6.176) still inherits extended invariant properties of the 1D operator, so that invariant subspaces can be composed of both trigonometric and hyperbolic functions.

Proposition 6.82 *Operator* (6.176) *admits*

$$W_4 = \mathcal{L}\{1, \cosh 2x, \cos 2y, \cosh x \cos y\}.$$

Proof. The following holds:

$$u = C_1 + C_2 \cosh 2x + C_3 \cos 2y + C_4 \cosh x \cos y \implies \tag{6.177}$$

$$\begin{aligned} F_1[u] = -32\big(C_2^2 + C_3^2\big) - 16C_1C_2 \cosh 2x \\ -16C_1C_3 \cos 2y - 16(C_2 + C_3)C_4 \cosh x \cos y. \end{aligned} \tag{6.178}$$

□

Example 6.83 (**TFE with absorption**) Exact solutions (6.177) of the TFE

$$\boxed{u_t = -u\Delta^2 u - (\Delta u)^2 + 2\nabla u \cdot \nabla \Delta u - 1 \quad \text{in } I\!R^2 \times I\!R_+}$$

describe extinction or quenching phenomena. The DS follows from (6.178).

Second remarkable operator. The thin film operator

$$F_2[u] = -u\Delta^2 u + (\Delta u)^2$$

is associated with the same second-order one (6.57), but in a different manner. To be precise, there exist solutions (6.86) associated with invariant modules for

$$\hat{F}_2[u] = u_t - F_2[u].$$

Example 6.84 Consider the fourth-order parabolic PDE

$$\boxed{u_t = \alpha[-u\Delta^2 u + (\Delta u)^2] + \beta(u\Delta u - |\nabla u|^2) \quad \text{in } I\!R^2 \times I\!R_+.}$$

Then there exist solutions (6.86), where the DS takes the form

$$\begin{cases} C_1' = -\beta\big(C_2^2 + C_3^2\big), \\ C_2' = C_3' = 0, \\ C_4' = \big[-\alpha\big(\gamma_1^2 + \gamma_2^2\big)^2 + \beta\big(\gamma_1^2 + \gamma_2^2\big)\big]C_1C_4 - 2\beta(\gamma_1 C_2 + \gamma_2 C_3)C_4, \\ \gamma_1' = -\alpha\big(\gamma_1^2 + \gamma_2^2\big)^2 C_2 + \beta\big(\gamma_1^2 + \gamma_2^2\big)C_2, \\ \gamma_2' = -\alpha\big(\gamma_1^2 + \gamma_2^2\big)^2 C_3 + \beta\big(\gamma_1^2 + \gamma_2^2\big)C_3. \end{cases} \tag{6.179}$$

The DS for $\gamma_{1,2}$ is integrated in quadratures, but solutions are not explicit.

Example 6.85 (**2mth-order remarkable operators**) For higher-order operators, similar invariant properties in $I\!R^2$ are exhibited by

$$F_2[u] = -u\Delta^{2m} u + (\Delta^m u)^2 \quad \text{and} \quad F_2[u] = u\Delta^{2m-1} u - |\nabla \Delta^{m-1} u|^2,$$

which, for any $m = 1, 2, \ldots$, are elliptic in the positivity domain of u. Parabolic PDEs with such operators generate the DSs, such as (6.179). The resulting solutions are convenient for studying the interface propagation in FBPs, as well as for describing quenching and extinction singularity formation phenomena.

6.7 Moving compact structures in nonlinear dispersion equations

Continuing the study of compactons in $I\!R^N$ that was began in Section 4.5, we extend the above results to PDEs of nonlinear dispersion types. Let $F[u]$ be a higher-order operator admitting the subspace of quadratic and linear functions,

$$W_{2N+1} = \mathcal{L}\{1, x_1^2, ..., x_N^2, x_1, ..., x_N\}, \tag{6.180}$$

e.g., $F = F_3$ given in (6.150). Let $\hat{F}[u]$ be a simpler operator admitting W_{2N+1}, for instance,

$$\hat{F}[u] = \gamma u \quad \text{with a constant} \quad \gamma \neq 0.$$

Looking for compact structures moving in the x_1-directions, consider the PDE

$$\boxed{u_t = \big(F[u] + \hat{F}[u]\big)_{x_1},} \tag{6.181}$$

where the "inhomogeneous" lower-order term with $\hat{F}$ is necessary to initiate such a drift. For the operator in (6.150), this is a fifth-order nonlinear dispersion equation, while choosing the second-order operator (6.1) yields a third-order PDE similar to those studied in Section 4.2 in 1D. In general, F can be composed from various second and higher-order operators admitting the necessary subspace. Using the $2m$th-order operator from Proposition 6.70 needs another subspace

$$\hat{W} = \mathcal{L}\{1, x_1^{2m}, ..., x_N^{2m}, x_1^{2m-1}, ..., x_N^{2m-1}, ...\}.$$

For the subspace (6.180), we look for solutions

$$u(x,t) = C(t) + a_1(t)x_1^2 + ... + a_N(t)x_N^2 + d_1(t)x_1 + ... + d_N(t)x_N, \tag{6.182}$$

and obtain a standard DS for expansion coefficients. Note that the differentiation D_{x_1} on the right-hand side of (6.181) implies that

$$a_1' = a_2' = ... = a_N' = 0,$$
$$d_2' = ... = d_N' = 0.$$

This suggests that solutions (6.182) can describe dynamics around the radial TW compactons moving in the x_1-direction,

$$u_c(x,t) = f(y), \quad \text{where } y = \sqrt{(x_1 - \lambda t)^2 + x_2^2 + ... + x_N^2}. \tag{6.183}$$

Plugging (6.183) into (6.181), and integrating once in y with the zero constant of integration, yields the radial ODE for f,

$$-\lambda f = F[f] + \hat{F}[f], \tag{6.184}$$

where Δ, Δ^2 and other differential forms in F and $\hat{F}$ are calculated in radial y-geometry. Equation (6.184) possesses stationary, independent of t solutions (6.182). More complicated dynamics on W_{2N+1} or $\hat{W}$ (not necessarily associated with TWs) and compact patterns exist for the corresponding second-order PDE

$$\boxed{u_{tt} = \big(F[u] + \hat{F}[u]\big)_{x_1}.}$$

Example 6.86 (Compacton TWs in $I\!R^N$) For instance, take $F = F_3$ as in (6.150) and the simplest $\hat{F}[u] = \gamma u$, i.e., the PDE

$$\boxed{u_t = \left(-\tfrac{1}{2} u \Delta^2 u^2 - \nabla u \cdot \nabla \Delta u^2 + \gamma u\right)_{x_1}.} \tag{6.185}$$

Looking for solutions moving in the x_1-direction (cf. (6.151)),

$$u(x,t) = C + a_1(x_1 - \lambda t)^2 + a_2 x_2^2 + ... + a_N x_N^2,$$

substituting into (6.185), and using the right-hand side of the first equation in the DS (6.153), we obtain the following projection of the PDE onto $2(x_1 - \lambda t)$:

$$-\lambda a_1 = -\tfrac{1}{2}(\Delta^2 u^2) a_1 - 16(\text{tr}\, A) a_1^2 - 32 a_1^3 + \gamma a_1.$$

Assuming for simplicity that $a_1 = a_2 = ... = a_N = a$ (obviously, there exist other not that symmetric solutions) yields by (6.152)

$$\Delta^2 u^2 = 8(\text{tr}\, A)^2 + 16 \textstyle\sum a_i^2 = 8N(N+2)a^2, \quad \text{so}$$
$$\gamma + \lambda = 4N(N+2)a^2 + 16Na^2 + 32a^2 = 4(N+2)(N+4)a^2.$$

This gives the explicit moving compacton for $\lambda > -\gamma$,

$$u_c(x,t) = \left[B - \tfrac{1}{2}\sqrt{\tfrac{\gamma+\lambda}{(N+2)(N+4)}}\left((x_1 - \lambda t)^2 + x_2^2 + ... + x_N^2\right)\right]_+,$$

where $B = C(0) > 0$. Thus, solutions on the invariant subspace (6.182) can describe more complicated non-radial dynamics around the TW $u_c(x,t)$.

As in the TFE model (6.150) and in Example 6.74, the transformation of such nonnegative solutions on W_{2N+1} (or on $\hat{W}$) into solutions with a zero contact angle on interfaces is done by setting $v = u^2$, so that (6.181) reads

$$v_t = 2\sqrt{v}\left(F[\sqrt{v}] + \hat{F}[\sqrt{v}]\right)_{x_1}. \tag{6.186}$$

In view of our analysis of oscillatory behavior of compacton equations in Sections 4.2 and 4.3, such solutions of the $(2m+1)$th-order PDE (6.186) on polynomial invariant subspaces $\hat{W}$ satisfy the Cauchy problem for $m = 1$ only, while for $m \geq 2$, these need a suitable FBP setting.

For the $C_N(m, a+b)$ *equation* that describes the sedimentation of particles in a dilute dispersion (see [494] for further references and applications)

$$v_t + (v^m)_x + \tfrac{1}{b}\left[v^a(\Delta v^b)\right]_{x_1} = 0 \quad (v \geq 0),$$

there exist almost explicit TWs (6.183) in $I\!R \times I\!R^3$ for $m = a = 1 + b$ and $m = 2$, $a + b = 3$; see [497]. For such third-order PDEs, these solutions do not belong to polynomial invariant subspaces.

6.8 From invariant polynomial subspaces in $I\!R^N$ to invariant trigonometric subspaces in $I\!R^{N-1}$

In this section, we deal with trigonometric subspaces in the multi-dimensional case. We show that, using simple polynomial subspaces for quadratic operators in $I\!R^N$

in spherical coordinates, it is possible to generate various trigonometric subspaces for classes of nonlinear operators in $I\!R^{N-1}$. Actually, excluding a few examples, trigonometric subspaces in $I\!R^N$ for $N \geq 2$ have not been treated before in a systematic manner, and their existence was associated with rather technical algebraic manipulations. Now, we link such trigonometric subspaces with simpler polynomial ones for quadratic second and higher-order operators.

6.8.1 Second-order operators

Operators in $I\!R^2$. We begin with a simple example. Consider the quadratic hyperbolic PDE

$$\boxed{u_{tt} = F[u] \equiv u\Delta u + \gamma\,|\nabla u|^2 \quad \text{in } I\!R^2 \times I\!R.} \tag{6.187}$$

By Proposition 6.1(ii), there exist solutions given by a general quadratic form (the free coefficient $C(t) \equiv 0$)

$$u(x,t) = x^T A(t) x \equiv a_{1,1}(t)x_1^2 + 2a_{1,2}(t)x_1x_2 + a_{2,2}(t)x_2^2,$$

where the expansion coefficients satisfy a DS; see (6.11). Introducing the polar coordinates on the plane

$$\begin{cases} x_1 = r\cos\theta, \\ x_2 = r\sin\theta, \end{cases} \tag{6.188}$$

gives the finite trigonometric expansion

$$\begin{aligned} u(x,t) &= r^2\big(a_{1,1}\cos^2\theta + 2a_{1,2}\cos\theta\sin\theta + a_{2,2}\sin^2\theta\big) \\ &\equiv r^2(C_1 + C_2\cos 2\theta + C_3\sin 2\theta), \end{aligned}$$

where $C_1 = \frac{1}{2}(a_{1,1} + a_{2,2})$, $C_2 = \frac{1}{2}(a_{1,1} - a_{2,2})$, and $C_3 = a_{1,2}$. Therefore,

$$u(x,t) = r^2 U(\theta,t), \quad \text{where } U(\theta,t) \in W_3 = \mathcal{L}\{1, \cos 2\theta, \sin 2\theta\} \tag{6.189}$$

belongs to a standard trigonometric subspace for ordinary differential operators; cf. Proposition 1.29. Let us derive a PDE for the function U by using that

$$\Delta = \Delta_r + \tfrac{1}{r^2}\,\Delta_\sigma, \tag{6.190}$$

where Δ_r is the radial part of the Laplacian in $I\!R^2$, and Δ_σ is the Laplace–Beltrami operator on the circle S^1,

$$\Delta_r = \tfrac{\partial^2}{\partial r^2} + \tfrac{1}{r}\tfrac{\partial}{\partial r} \quad \text{and} \quad \Delta_\sigma = \tfrac{\partial^2}{\partial \theta^2}.$$

By the polar representation of the gradient

$$\begin{cases} u_{x_1} = u_r\cos\theta + u_\theta\,\frac{\sin\theta}{r}, \\ u_{x_2} = u_r\sin\theta - u_\theta\,\frac{\cos\theta}{r}, \end{cases} \tag{6.191}$$

it is easy to see that $|\nabla u|^2 = (u_r)^2 + \frac{1}{r^2}(u_\theta)^2$. Substituting (6.189) into the original equation (6.187) yields the following hyperbolic PDE:

$$\boxed{U_{tt} = \hat{F}[U] \equiv UU_{\theta\theta} + \gamma\,(U_\theta)^2 + 4(1+\gamma)U^2,}$$

so the standard operator $\hat{F}$ preserving the subspace W_3 in (6.189) is detected.

Operators in $I\!R^3$ and $I\!R^N$. Similarly, for such second-order parabolic or hyperbolic PDEs in $I\!R^N$, using the spherical coordinates in N-dimensional geometry explains how the polynomial subspaces, such as (6.5) or (6.4), generate the multi-dimensional trigonometric subspaces spanned by the cos, sin, and the products of such functions of all the N-1 spherical angles. An example of such subspaces has been considered in Proposition 6.42 for the remarkable quadratic operator. Let us present another example in $I\!R^3$, showing how the trigonometric subspaces in $I\!R^2$ can be constructed for more general operators.

Example 6.87 (**Deriving PME with source**) Consider the quadratic PME

$$\boxed{u_t = \Delta(u^2) \quad \text{in } I\!R^3 \times I\!R_+ ,} \tag{6.192}$$

for which the main computations are quite simple. On the 6D polynomial subspace (6.5), exact solutions are

$$u(x,t) = a_{1,1}x_1^2 + a_{2,2}x_2^2 + a_{3,3}x_3^2 + 2a_{1,2}x_1x_2 + 2a_{1,3}x_1x_3 + 2a_{2,3}x_2x_3. \tag{6.193}$$

Introducing the spherical coordinates in $I\!R^3$ with two angles $\sigma = \{\theta, \vartheta\}$,

$$\begin{cases} x_1 = r \sin\vartheta \cos\theta, \\ x_2 = r \sin\vartheta \sin\theta, \\ x_3 = r \cos\vartheta, \end{cases} \tag{6.194}$$

and plugging into (6.193) yields

$$\begin{aligned} u(x,t) = r^2 U(\sigma,t) &\equiv r^2(a_{1,1}\sin^2\vartheta\cos^2\theta \\ &+ a_{2,2}\sin^2\vartheta\sin^2\theta + a_{3,3}\cos^2\vartheta + 2a_{1,2}\sin^2\vartheta\sin\theta\cos\theta \\ &+ 2a_{1,3}\sin\vartheta\cos\vartheta\cos\theta + 2a_{2,3}\sin\vartheta\cos\vartheta\sin\theta). \end{aligned}$$

Here U belongs to the 8D trigonometric subspace

$$\begin{aligned} W_8^{\rm tr} = \mathcal{L}\{&1, \cos 2\theta, \sin 2\theta, \cos 2\vartheta, \sin 2\vartheta \cos\theta, \\ &\sin 2\vartheta \sin\theta, \cos 2\vartheta \cos 2\theta, \sin 2\theta \cos 2\vartheta\}. \end{aligned}$$

Let us next derive the corresponding equation with a quadratic operator preserving such a subspace. Using (6.190) with the Laplace–Beltrami operator on the unit sphere S^2 in $I\!R^3$

$$\Delta_\sigma = \tfrac{1}{\sin\vartheta}\tfrac{\partial}{\partial\vartheta}\big(\sin\vartheta\,\tfrac{\partial}{\partial\vartheta}\big) + \tfrac{1}{\sin^2\vartheta}\tfrac{\partial^2}{\partial\theta^2}. \tag{6.195}$$

Substituting $u = r^2 U$ into (6.192) gives

$$r^2 U_t = r^2 \Delta_\sigma U^2 + U^2 \Delta_r r^4,$$

which yields the following PME with source:

$$\boxed{U_t = \Delta_\sigma U^2 + 20\,U^2 \quad \text{in } S^2 \times I\!R_+.}$$

This PDE generates blow-up, so exact solutions describe interesting features of such localized patterns. The blow-up is regional here and these solutions explain properties of a non-symmetric, non-radial singular behavior.

6.8.2 Higher-order operators

Space $I\!R^2$. We begin with the quadratic thin film equation

$$\boxed{u_t = F[u] \equiv -\nabla \cdot (u \nabla \Delta u) \quad \text{in } I\!R^2 \times I\!R_+.}$$

Taking the polynomial subspace as in (6.137) yields solutions

$$u(x,t) = (x^2)^T A(t) x^2 \equiv a_{1,1}(t) x_1^4 + 2a_{1,2}(t) x_1^2 x_2^2 + a_{2,2}(t) x_2^4.$$

In polar coordinates (6.188), this gives the trigonometric expansion

$$\begin{aligned} u(x,t) &= r^4\big(a_{1,1}\cos^4\theta + 2a_{1,2}\cos^2\theta\sin^2\theta + a_{2,2}\sin^4\theta\big) \\ &\equiv r^4(C_1 + C_2\cos 2\theta + C_3\cos 4\theta) \equiv r^4 U(\theta,t), \end{aligned} \tag{6.196}$$

so $U \in W_3 = \mathcal{L}\{1, \cos 2\theta, \cos 4\theta\}$. Using formulae (6.190) and (6.191) yields the following quadratic PDE for the function U in (6.196):

$$\boxed{U_{tt} = -U U_{\theta\theta\theta\theta} - U_\theta U_{\theta\theta\theta} - 28\, U U_{\theta\theta} - 16\,(U_\theta)^2 - 192\, U^2,}$$

with a more general quadratic operator than those in Proposition 3.16. Using other polynomial subspaces in Proposition 6.68 leads to extended trigonometric ones.

Example 6.88 **(2mth-order equations)** This approach applies to higher-order operators. For instance, consider the $2m$th-order quadratic wave equation

$$\boxed{u_{tt} = (-1)^{m+1} \Delta^m (u^2) \quad \text{in } I\!R^2 \times I\!R,} \tag{6.197}$$

which is hyperbolic in the positivity domain $\{u > 0\}$. As in Proposition 6.70, we take the polynomial solutions

$$u(x,t) = \textstyle\sum_{(i+j=m)} a_{i,j}(t)\, x_1^{2i} x_2^{2j}.$$

In polar coordinates, this yields the trigonometric expansion

$$u = r^{2m} U \equiv r^{2m} \textstyle\sum a_{i,j}(t) \cos^{2i}\theta \sin^{2j}\theta,$$

so that U belongs to the trigonometric subspace

$$\mathcal{L}\{1, \cos 2\theta, \cos 4\theta, ..., \cos 2m\theta\}.$$

The operator preserving this subspace is obtained by plugging $u = r^{2m} U$ into the PDE (6.197),

$$U_{tt} = F_m[U] \equiv \tfrac{1}{r^{2m}} (-1)^{m+1} \big(\Delta_r + \tfrac{1}{r^2}\Delta_\sigma\big)^m (r^{4m} U^2).$$

For instance, for $m = 3$,

$$F_3[U] = \Delta_\sigma^3(U^2) + 308\, \Delta_\sigma^2(U^2) + 30016\, \Delta_\sigma(U^2) + 921600\, U^2,$$

with the trigonometric subspace

$$W_4 = \mathcal{L}\{1, \cos 2\theta, \cos 4\theta, \cos 6\theta\}.$$

This subspace can be extended to 7D by adding necessary sin functions.

Space $I\!R^N$. As in the second-order case, polynomial subspaces for operators in $I\!R^N$ detect the trigonometric ones in $I\!R^{N-1}$. Consider an example in $I\!R^3$.

Example 6.89 The fourth-order equation which is parabolic in $\{u > 0\}$,

$$\boxed{u_t = -\Delta^2(u^2) \quad \text{in } I\!R^3 \times I\!R_+,} \tag{6.198}$$

admits polynomial solutions

$$u(x,t) = \sum_{(i+j+k=2)} a_{i,j,k}(t)\, x_1^{2i} x_2^{2j} x_3^{2k} \tag{6.199}$$

(cf. subspace (6.142)). Using in (6.199) the spherical coordinates (6.194) yields

$$u = r^4 U \equiv r^4 \sum a_{i,j,k} (\sin\vartheta\, \cos\theta)^{2i} (\sin\vartheta\, \sin\theta)^{2j} \cos^{2k}\vartheta,$$

where U belongs to the subspace

$$\mathcal{L}\{\cos(2i\theta)\cos(2j\vartheta),\ 0 \le i, j \le 2\}.$$

Substituting $u = r^4 U$ into (6.198) gives a quadratic equation for $U(\sigma, t)$,

$$\boxed{U_t = -\Delta_\sigma^2(U^2) - 114\, \Delta_\sigma(U^2) - 3024\, U^2,}$$

where Δ_σ is the Laplace–Beltrami operator (6.195).

Remarks and comments on the literature

As has been mentioned, the general theory of annihilating operators and annihilators of finite-dimensional invariant modules in $I\!R^N$ was developed in [312]. "The formulae for the affine annihilators and annihilators are often extremely complicated, even for relatively simple subspaces" [312, p. 316], so general results are often difficult to apply in practice.

§ 6.1. Some of the results are taken from [220] and [239]. For the PME, polynomial exact solutions were obtained in [342], where the resulting DS was carefully studied. A general theorem on finite-dimensional subspaces of analytic functions of several variables was obtained in [312] (similar to the Main Theorem in Section 2.1 in 1D), where a complete description of the set of operators preserving a given subspace was obtained. The Ernst equation (6.13) in Example 6.5 is connected with the sdYM equation; see [1] and comments below. Non-symmetric blow-up and extinction on multi-dimensional polynomial subspaces for the quasilinear heat equation

$$u_t = \nabla \cdot (u^\sigma \nabla u) \pm u^{1-\sigma}$$

(as well as for (6.41)) were studied in [235]. On group classification of semilinear and quasilinear heat equations in $I\!R^N \times I\!R_+$, see [10, pp. 144-163], [121] for

$$u_t = u_{xx} + u_{yy} + F(t, x, y, u, u_x, u_y),$$

and references therein for more recent results.

Concerning general matrix equations, such as (6.9), note that the quadratic first-order and related second-order cubic equations

$$A' = A^2 + B \quad \text{and} \quad A'' = 2A^3 + BA + AB,$$

where B is a constant matrix, are integrable and indeed solvable [84, 301]. A related matrix equation

$$A' = -(\operatorname{tr} A)A + (\operatorname{adj} A)A + A^T A \quad (\operatorname{adj} A = (\det A)A^{-1}) \tag{6.200}$$

appears [102] in the process of reduction of the *self-dual Yang–Mills* (sdYM) *equations* (a system of equations for Lie algebra-valued functions on $\mathbb{C}^4$). In $I\!R^3$, introducing simple factorization, (6.200) can be transformed into a quadratic system [1]

$$\begin{cases} \omega_1' = \omega_2\omega_3 - \omega_1(\omega_2+\omega_3) + \tau^2, \\ \omega_2' = \omega_3\omega_1 - \omega_2(\omega_3+\omega_1) + \tau^2, \\ \omega_3' = \omega_1\omega_2 - \omega_3(\omega_1+\omega_2) + \tau^2, \end{cases} \tag{6.201}$$

where $\tau^2 = \tau_1^2 + \tau_2^2 + \tau_3^2$ and

$$\begin{cases} \tau_1' = -\tau_1(\omega_2+\omega_3), \\ \tau_2' = -\tau_2(\omega_3+\omega_1), \\ \tau_3' = -\tau_3(\omega_1+\omega_2). \end{cases}$$

For $\tau = 0$, (6.201) is the classical *Darboux–Halphen system* derived by Darboux (1878) in the analysis of triple orthogonal surfaces [140], and later solved by Halphen (1881) [277] using linearization in terms of Fuchsian differential equations with three regular singular points. In the case where $\tau = 0$, the function $y = -2(\omega_1+\omega_2+\omega_3)$ satisfies the *Chazy equation*

$$y''' = 2yy'' - 3(y')^2,$$

which is explicitly solved in terms of solutions of a linear hypergeometric equation [104]. Given a solution $y(t)$, the three distinct roots $\omega_1(t)$, $\omega_2(t)$, and $\omega_3(t)$ of the cubic equation

$$\omega^3 + \tfrac{1}{2}\,y\omega^2 + \tfrac{1}{2}\,y'\omega + \tfrac{1}{12}\,y'' = 0$$

solve the Darboux–Halphen system (6.201) with $\tau = 0$. See [1] for a survey, multiple references, and many reductions of the sdYM equations to integrable PDEs.

It seems that second-order quadratic equations for A in (6.9) occurring in the hyperbolic case are not solvable even in the class of diagonal matrices. The first-order equations, such as given in (6.12), are indeed solvable.

Proposition 6.13 and related examples were taken from [239]. Such solutions were first observed in [453] for diffusion-absorption equations in $I\!R$ (see also [435], where the pseudo-symmetries were used), and further extensions to parabolic and wave PDEs were developed in [341] (Example 6.17 deals with a slight modification of an equation from this paper). Example 6.19 explains the invariant subspace essence of particular exact solutions in 1D with $L = D_x^2 + \alpha I$ that were constructed in [325, p. 1409], by determining a third-order differential constraint.

Exact solutions of the LRT equation (6.28) were studied in a number of papers; see [504, 505] and further references in [133, 267]. Invariant and partially invariant solutions were considered in [567, 566]. Local and global (for small initial data) existence results can be proved for the viscosity LRT equation (6.32), [375]. A group classification and some exact solutions of (6.33) are presented in [10, p. 301].

Notice a quadratic system from *binormal flow*

$$\mathbf{X}_t = \mathbf{X}_s \times \mathbf{X}_{ss} \equiv \frac{\mathbf{X}_t \times \mathbf{X}_{tt}}{(s')^3}$$

for curves $\{\mathbf{X}(\cdot,t),\ t>0\}$ in $I\!R^3$, which are parameterized by the arclength s. This equation of the evolution of isolated vortexes in an inviscid liquids was proposed by Da Rios in 1906, whose study established the first geometric link between soliton theory and the motion of inextensible curves. Nowadays this equation is used as an approximate model for a vortex tube of infinitesimal cross section $\mathbf{X}(\cdot,t)$ described by Euler equations; see details and references in [275], where similarity solutions were constructed.

§ 6.2. Proposition 6.24 is taken from [220]; see also [343]. The subspace in Proposition 6.29

and the corresponding exact solutions play the key role in asymptotic analysis of the critical fast diffusion equation as explained in [245, Sect. 6]. The transformation in Example 6.31 is given in [343], where further extensions are available. Subspace (6.53) in Proposition 6.32 was obtained in [439] by using a nonclassical symmetry approach. In [439, p. 130], a 3D non-polynomial subspace (governed by an ODE with non-constant coefficients) was detected for the 1D operator $F[u] = uu_{xx} - \frac{3}{4}(u_x)^2$, which is studied and applied throughout this book.

§ 6.3. The results, transformations and reductions around (6.62)–(6.67) are taken from [540, 542]. In the representation of the results, we have used several ideas from [343]. The subspace (6.80) was constructed in [220, 232]. For equation

$$v_t = \Delta \ln v,$$

a solution similar to that in Example 6.47 was constructed in [323] by deriving an invariant manifold via a pair of vector fields not belonging to the algebra of symmetries of the equation. A similar technique was applied there to obtain some solutions on the 6D subspace $\mathcal{L}\{1, x, y, x^2, xy, y^2\}$. Solutions of the type (6.86) of equation (6.58) were obtained in [325] by compatible differential constraints.

Polynomial subspaces are also typical for quadratic operators in $I\!R^2$ that appeared in the Gibbons–Tsarev (1.63) and Prandtl (1.145) equations. A similar operator is available in the (2+1)-dimensional integrable *breaking soliton equation* [93]

$$u_{xt} = -u_{xxxy} + 4u_x u_{xy} + 2u_y u_{xx} \quad \text{in } I\!R^2 \times I\!R, \tag{6.202}$$

which describes the interaction of a Riemann wave along the y-axis and a long wave along the x-axis. See its N-solitons and algebro-geometric solutions in [248], where further details can be found. Clearly, (6.202) possesses low-dimensional solutions on the 2D subspace,

$$u(x, y, t) = C_1(x, t) + C_2(x, t)y \in W_2 = \mathcal{L}\{1, y\},$$

$$\begin{cases} C_{1xt} - 4C_{1x}C_{2x} - 2C_2C_{1xx} + C_{2xxx} = 0, \\ C_{2xt} - 4(C_{2x})^2 - 2C_2C_{2xx} = 0. \end{cases}$$

This is an alternative invariant treatment of solutions in [584], which correspond to the simple choice $C_2 = C_2(t)$ is arbitrary (the first ODE is then integrated). Similar solutions on W_2 exist for other PDEs, such as (6.202), and are not necessarily integrable.

These PDEs suggest a general quasilinear quadratic operator with the parameter β,

$$F[u] = u_x u_{xy} + \beta u_y u_{xx} \quad \text{in } I\!R^2.$$

Apparently, for any β, the basic subspace is $W_7 = \mathcal{L}\{1, x, y, x^2, xy, y^2, x^2y\}$, while, for $\beta = -2$, there exists the extended

$$W_8 = \mathcal{L}\{1, x, y, x^2, xy, y^2, x^2y, x^2y^2\}.$$

The value $\beta = \frac{1}{2}$ occurs in the *potential Calogero–Bogoyavlenskij–Schiff equation*

$$u_{xt} + \tfrac{1}{4}\, u_{xxxz} + u_x u_{xz} + \tfrac{1}{2}\, u_z u_{xx} = 0,$$

and in a two-directional generalization of the *potential KP equation*

$$u_{xt} + \tfrac{1}{4}\, u_{xxxz} + u_x u_{xz} + \tfrac{1}{2}\, u_z u_{xx} + \tfrac{1}{4}\, D_x^{-1}(u_{zzz}) = 0.$$

This is an integrable PDE possessing N-soliton solutions. See references and results in [586], where the technology of solving the related cubic *trilinear* equation for τ, given by $u = \frac{2\tau_x}{\tau}$ (unlike the KdV equation that is reduced to a quadratic bilinear one) by Baker–Hirota-type differential operators is explained.

For existence, uniqueness, and regularity for the odd-order PDEs, such as KdV, KP, and ZaK-type equations, see [381, 69, 179] and references therein.

§ **6.4.** We follow [234]. Example 6.52 was used in [220, 232].

§ **6.5.** The origin of fully nonlinear M-A equations dates back to Monge's paper [427] in 1781, where Monge proposed a civil-engineering problem of moving a mass of earth from one configuration to another in the most economical way. This problem has been further studied by Appel [15] and Kantorovich [313, 314]; see references and a survey in [186].

The parabolic M-A equation associated with (6.113),

$$-u_t \det D^2 u = f \quad \text{in } Q_T = \Omega \times (0, T),$$

where $\Omega \subset I\!R^N$ is a bounded smooth domain, was first introduced in [359]; see recent related references in [274] and more general models like that in [516]. Note the pioneering paper by Hamilton [278] on the evolution of a metric in direction of its Ricci curvature. Conditions of the global unique solvability of the M-A equation associated with (6.119),

$$u_t = (\det D^2 u)^{\frac{1}{N}} + g \quad \text{in } Q_T,$$

were obtained by Ivochkina and Ladyzhenskaya [303]. This model corresponds to special curvature flows. Potential properties of Hessian operators are described in [561]. Parabolic M-A equations as gradient flows, and the related questions of the asymptotic behavior of solutions, were studied, e.g., in [123, 516], where further references concerning various types of Gaussian flows can be found. The classical Gauss curvature flow describes the deformation of a convex compact surface $\Sigma : z = u(x, y, t)$ in $I\!R^3$ by its Gauss curvature and is governed by the PDE

$$u_t = \frac{\det D^2 u}{(1+|\nabla u|^2)^{3/2}}, \tag{6.203}$$

which is uniformly parabolic on strictly convex solutions. Singularity formation phenomena for (6.203) appear if the initial surface Σ has flat sides, where the curvature becomes zero and the equation degenerate. This leads to an FBP for (6.203) with the unknown domain of singularity, $\{(x, y) : u(x, y, t) = 0\}$, and specific regularity properties; see [143, 144] and references therein. Alternatively, finite-time formation of *non-smooth* free boundaries with flat parts is a typical phenomenon for blow-up solutions of the reaction-diffusion PDEs (see the first equation in (6.207) below) via extended semigroup theory. In 1D, optimal regularity of such $C^{1,1}$-interfaces is well understood [226, Ch. 5]. For $N > 1$, the regularity problem remains essentially OPEN; see some estimates and examples in [226, p. 151]. As a formal extension, note that a nontrivial ($u(x, t) \not\equiv 0$) proper convex solution exists for such PDEs with an arbitrarily strong (as $u \to 0$) absorption term, e.g.,

$$u_t = \frac{\det D^2 u}{(1+|\nabla u|^2)^{3/2}} - e^{1/u},$$

where a similar FBP occurs. Therefore, the parabolic operator of the Gauss curvature flow is extremely powerful, in the sense that it prevents a *complete extinction* (i.e., $u(x, t) \equiv 0$ for arbitrarily small $t > 0$; this can happen for many other parabolic PDEs). For any initial data with flat sides, $\{(x, y) : u_0(x, y) = 0\} \neq \emptyset$, this proper solution of the FBP can be constructed by regular approximations of the equations and initial data by replacing $e^{1/u} \mapsto \min\{\frac{u}{\varepsilon}, e^{1/u}\}$, with $\varepsilon > 0$, (a uniformly Lipschitz continuous approximation), and $u_0 \mapsto u_0 + \varepsilon$. Uniform *a priori* estimates for $\{u_\varepsilon\}$ are obtained by local (near the interface) comparison with 1D TW solutions or other radial sub- and super-solutions, [226, Ch. 7].

Concerning other nonlinearities, the elliptic M-A equation

$$(\det D^2 u)^{\frac{1}{N+2}} = -\frac{1}{u}, \quad u < 0, \quad D^2 u > 0 \ \text{in } \Omega, \quad u = 0 \text{ on } \partial\Omega, \tag{6.204}$$

where Ω is a bounded convex domain in $I\!R^N$, was derived by Loewner–Nirenberg [400] in the study of the metric of the form $-\frac{1}{u}\,D^2u$ (u is then treated as a section of a certain line bundle), and was proved to admit a unique C^∞ solution; see [108] and earlier references therein. The general Hessian equation has the form

$$S_k(D^2u) = (-u)^p, \;\; u < 0 \quad \text{in } \Omega, \qquad u = 0 \;\text{ on } \partial\Omega, \tag{6.205}$$

where Ω is a ball in $I\!R^N$, $N \geq 3$, and S_k is given by the elementary symmetric function

$$S_k(D^2u) = \textstyle\sum_{(1 \leq i_1 < \ldots < i_k \leq n)} \lambda_{i_1} \ldots \lambda_{i_k},$$

with $\{\lambda_i\}$ being the eigenvalues of the Hessian D^2u (so $k = 1$ and $k = n$ correspond to the Laplace and the M-A operators, respectively). (6.205) is known to exhibit the *critical exponents*

$$\gamma(k) = \tfrac{(N+2)k}{(N-2k)_+},$$

such that no smooth solution $u < 0$ exists for $p \geq \gamma(k)$, and a negative radial solution exists for $p \in (0, \gamma(k))$; see [561] (nonexistence is proved by a Pohozaev-type inequality) and [117] for extensions. For $k = 1$, $\gamma(1) = \frac{N+2}{N-2}$ is the critical Sobolev exponent. The nonexistence result for the elliptic equation

$$\Delta u + u^p = 0$$

is associated with Pohozaev's classic inequality [464]. The exponents $\gamma(k)$ above are to be compared with the critical ones

$$\gamma(k) = \tfrac{N+2k}{(N-2k)_+} \quad \text{for elliptic PDEs} \quad -(-\Delta)^k u + |u|^{p-1}u = 0,$$

where the existence-nonexistence results are proved by higher-order Pohozaev's inequalities [465] applied to general quasilinear $2k$th-order PDEs.

This suggests generalized second-order M-A parabolic flows (or others with the elliptic operator as in (6.203))

$$u_t = (\det D^2u)^m \pm (-u)^p, \tag{6.206}$$

with some exponents $m > 0$ and $p \in I\!R$, that generate OPEN PROBLEMS concerning local existence of convex solutions, free-boundary (degeneracy set) propagation, extinction, and blow-up singularity patterns, etc. In a radial setting, where (6.206) reduces to a 1D quasilinear parabolic PDE, the interface equations and their regularity, moduli of continuity of proper solutions, waiting time phenomena, etc., are characterized by Sturmian intersection comparison techniques, [226, Ch. 7]. For $N > 1$, the majority of the problems are OPEN, and particular exact solutions might be key. Such models are natural counterparts of the PME with reaction/absorption, and of thin film (or Cahn–Hilliard-type, $n = 0$) models,

$$u_t = \Delta u^m \pm u^p \quad \text{and} \quad u_t = -\nabla \cdot (|u|^n \nabla \Delta u) \pm \Delta |u|^{p-1}u, \tag{6.207}$$

that were studied throughout this chapter; cf. Example 6.61.

Basic properties of hyperbolic M-A equations are explained in [550, Ch. 16]. Quadratic polynomials $p(x)$ occur in the celebrated result of the theory of elliptic M-A PDEs, establishing that any convex solution of the elliptic M-A equation

$$\det D^2u = 1 \quad \text{in } I\!R^N$$

is $u(x) = p(x)$. This result is due to Jörgens (1954) for $N = 2$, Calabi (1958) for $N = 3$,4, and 5 and to Pogorelov (1978) for any $N \geq 2$ (see also [108] for a more general result). The same conclusion holds for the *Hessian quotient equation*

$$S_k(D^2u) = 1 \quad \text{for} \;\; u(x) \leq A(1 + |x|^2) \;\; \text{strictly convex},$$

with any $1 \le k < N$ [24]. Similarly, if $u(x,t)$ is a smooth solution of the parabolic PDE

$$-u_t \det D^2 u = 1 \quad \text{in } I\!\!R^N \times I\!\!R_+,$$

where u is convex in x, nonincreasing with t, and u_t is bounded away from 0 and $-\infty$, then $u(x,t) = Ct + p(x)$; see [273] for the results and a survey.

Fourth and higher-order M-A PDEs have been less well studied, though some of the equations correspond to classical geometric problems, and several general results have been established. We refer to [585] (existence for mth-order elliptic M-A equations with the principle operators in (6.131), where a Riemann–Hilbert factorization condition appears), [560] ($W^{1,p}$-regularity estimates for fourth-order M-A equations and establishing an analogy of the Jörgens–Calabi–Pogorelov result for such PDEs), and [384] (homogeneous fourth-order PDEs for affine maximal hypersurfaces), where further references are given.

The homogeneous equation $|D^4 u| = 0$ is a direct sum of two identical copies of the second-order M-A equation (see [181])

$$v_{xx} v_{yy} - (v_{xy})^2 = 0. \tag{6.208}$$

Similarly, the sixth-order equation $|D^6 u| = 0$ with the operator (6.129) decouples into three copies of (6.208), [181]. Possibly, this means that some problems with such higher-order M-A operators are associated with the second-order ones. In particular, the inhomogeneous equations $|D^4 u| = 1$, $|D^6 u| = 1$ might be handled by reduction to second-order equations, and a result associated with the Jörgens–Calabi–Pogorelov theorem might be expected (though some basics of such PDEs remain obscure).

§ 6.6, 6.7. Invariant subspaces (6.136) were obtained in [59], where it was shown that (i) the 4D extension with an extra singular component,

$$W_4^{\rm r} = W_4^{\rm r} \oplus \mathcal{L}\{r^{2-N}\},$$

is also invariant, and (ii) the multi-dimensional subspace (6.139) in $I\!\!R^2$ was found. This paper contains a detailed description of interesting evolution properties of such solutions on invariant subspaces, especially, in 2D. Example 6.76 is also based on the idea formulated in [59] for the operator with $m = 2$ in $I\!\!R^2$.

Concerning Example 6.78, extinction was studied for the *fast diffusion equation*

$$v_t = \nabla \cdot (v^{-\sigma} \nabla v) \quad \text{in} \quad I\!\!R^N \times I\!\!R_+ \text{ for } N \ge 3.$$

It is known that extinction happens for $\sigma > \frac{2}{N}$ [37], and, in the *critical Sobolev case* $\sigma = \frac{4}{N+2}$, the asymptotic behavior is driven by solutions in separate variables [230]; see [245, Ch. 6] for more details about critical exponents for such PDEs.

§ 6.8. For the second-order operators, the idea of using the polar coordinates on the plane for reducing the PME $u_t = \Delta u^m$ to the reaction-diffusion equation

$$u_t = (u^m)_{xx} + \gamma u^m$$

is due to King [342].

Open problems

- These are formulated in Examples 6.7 and 6.28, Section 6.3.2, Examples 6.57–6.59, 6.61, 6.63, 6.64, 6.69, 6.71, 6.72, 6.74, 6.75, and 6.78.

CHAPTER 7

Partially Invariant Subspaces, Invariant Sets, and Generalized Separation of Variables

Here we more systematically apply the concept of partially invariant subspaces (invariant sets). In general, partial invariance of W_n reduces the PDE to an overdetermined dynamical system that may be inconsistent. We study a few classes of nonlinear operators for which such DSs admit nontrivial solutions. Sometimes these exact solutions are easier to obtain by differential constraints approach, or by sign-invariants which are the subject of Chapter 8.

Recall that, given an operator F, a set $M \subseteq W_n$ is said to be *invariant on the linear subspace* W_n, and then W_n is *partially invariant* if

$$F[M] \subseteq W_n.$$

7.1 Partial invariance for polynomial operators

7.1.1 Basic ideas and examples

Let us begin with an extension of the results on invariant subspaces in Section 1.5.2, where we considered general quadratic operators

$$F[u] = \textstyle\sum_{(i,j)} a_{i,j}\, D_x^i u\, D_x^j u, \tag{7.1}$$

with the real symmetric matrix $\|a_{i,j}\|$ and the corresponding polynomial

$$P(X,Y) = \textstyle\sum_{(i,j)} a_{i,j}\, X^i Y^j.$$

We are looking for ODE reductions of PDEs with the operator (7.1) on the subspace

$$W_n = \mathcal{L}\{e^{p_k x},\ k = 1, ..., n\}, \quad \text{where, for convenience,} \quad p_1 = 0.$$

Let $\Gamma = \{p_1, ..., p_n\}$ denote the set of all the exponents. Such finite-dimensional subspaces are natural for N-soliton solutions of many integrable PDEs, including the KdV, Harry Dym, and Boussinesq equations (see Section 4.1). Here we construct exact solutions of non-integrable PDEs.

Consider the first-order (in time) quadratic PDE

$$\boxed{u_t = F[u] = \textstyle\sum_{(i,j)} u_{i,j}\, D_x^i u\, D_x^j u.} \tag{7.2}$$

Let us look for solutions

$$u(x,t) = \textstyle\sum_{(k)} C_k(t) e^{p_k x} \in W_n \quad \text{for any} \quad t \geq 0.$$

Plugging into (7.1) yields

$$\begin{aligned} F[u] &= \sum_{(i,j)}\Big(\sum_{(m,l)} a_{i,j}\, p_m^i p_l^j C_m C_l \mathrm{e}^{(p_m+p_l)x}\Big) \\ &\equiv \sum_{(m,l)} P(p_m, p_l) C_m C_l \mathrm{e}^{(p_m+p_l)x}. \end{aligned} \tag{7.3}$$

Introducing, as before, the set

$$\Gamma' = \{p_m + p_l : \ p_m,\ p_l \in \Gamma,\ \ p_m + p_l \notin \Gamma\},$$

and denoting by $n' \geq 1$ its cardinal number, we need n' conditions to guarantee that $F[u] \in W_n$, where W_n then becomes invariant. Here it is assumed that there exists at least one term $\mathrm{e}^{(p_m+p_l)x}$ in (7.3) with $p_m + p_l \in \Gamma'$ that is obtained by two or more multiplications of elementary exponential factors (a nontrivial *correlation* between exponential factors occurs). This gives the corresponding invariant conditions depending not only on P, but also on expansion coefficients $\{C_j\}$, unlike in the case of invariant subspaces in Section 1.5. So Γ satisfies the following property:

$$\exists \text{ pairs } \{p_m, p_l\} \neq \{p_{m'}, p_{l'}\} \text{ such that } p_m + p_l = p_{m'} + p_{l'} \in \Gamma'.$$

Let us begin by using simple examples.

Example 7.1 **(4D subspace)** Set $p_1 = 0 < p_2 < p_3$, where $2p_2 \neq p_3$, $p_4 = \frac{1}{2}(p_2 + p_3)$, and look for solutions

$$u(x,t) = C_1(t) + C_2(t)\mathrm{e}^{p_2 x} + C_3(t)\mathrm{e}^{p_3 x} + C_4(t)\mathrm{e}^{p_4 x}. \tag{7.4}$$

Then $\Gamma' = \{2p_2,\ p_2+p_4,\ p_2+p_3 = 2p_4,\ p_3+p_4,\ 2p_3\}$ and $n' = 5$. Four invariance conditions are the same as for invariant subspaces,

$$P(p_2, p_2) = P(p_2, p_4) = P(p_3, p_4) = P(p_3, p_3) = 0. \tag{7.5}$$

The $(p_2{+}p_3)$-one fails to be like that and contains two different terms

$$2P(p_2, p_3)C_2C_3 + P(p_4, p_4)C_4^2 = 0, \tag{7.6}$$

since the exponential $\mathrm{e}^{(p_2+p_3)x} = \mathrm{e}^{2p_4 x}$ occurs twice in bilinear products. We next add to (7.6) the usual ODEs on W_4,

$$\begin{cases} C_1' = P(0,0)C_1^2, \\ C_2' = 2P(0,p_2)C_1C_2, \\ C_3' = 2P(0,p_3)C_1C_3, \\ C_4' = 2P(0,p_4)C_1C_4. \end{cases} \tag{7.7}$$

The system (7.7), (7.6) is overdetermined and we need to check its consistency. The first ODE is independent and determines

$$C_1(t) = -\tfrac{1}{P(0,0)}\, t^{-1}, \quad \text{if } P(0,0) \neq 0, \tag{7.8}$$

and, from the other equations,

$$C_k(t) = A_k t^{-\rho_k}, \quad \text{with } \rho_k = \tfrac{2P(0,p_k)}{P(0,0)} \text{ for } k = 2,3,4,$$

where constants $\{A_k\}$ are initial data for the DS at $t = 1$. Plugging these functions into the algebraic equation (7.6) yields

$$2P(p_2,p_3)A_2A_3 t^{-(\rho_2+\rho_3)} + P(p_4,p_4)A_4^2 t^{-2\rho_4} = 0,$$

from which we get $\rho_2 + \rho_3 = 2\rho_4$, i.e., an extra condition on the operator follows,

$$P(0, p_2) + P(0, p_3) = 2P(0, p_4). \tag{7.9}$$

For initial data $\{A_k\}$, this gives

$$2P(p_2, p_3)A_2A_3 + P(p_4, p_4)A_4^2 = 0. \tag{7.10}$$

Thus, under the assumptions (7.5) and (7.9) on $F[u]$, the overdetermined DS admits a two-parameter (A_2 and A_3 are arbitrary) family of explicit solutions, which are not exponential functions in both variables x and t. If $P(0,0) = 0$, we take $C_1(t) = 1$, and the solutions are composed of purely exponential terms as for solitons.

Example 7.2 **(Second-order PDEs)** Keeping the same notation, for the second-order evolution equation

$$\boxed{u_{tt} = F[u] \equiv \textstyle\sum_{(i,j)} a_{i,j}\, D_x^i u\, D_x^j u,} \tag{7.11}$$

under the same invariance conditions on F, a harder higher-order DS occurs,

$$\begin{cases} C_1'' = P(0,0)C_1^2, \\ C_2'' = 2P(0,p_2)C_1C_2, \\ C_3'' = 2P(0,p_3)C_1C_3, \\ C_4'' = 2P(0,p_4)C_1C_4, \end{cases} \tag{7.12}$$

with the same algebraic relation (7.6). Let us again derive the necessary consistency conditions. We begin with the simpler case, $P(0,0) = 0$. Then, taking, e.g., $C_1(t) = 1$ yields, for all $P(0, p_k) > 0$,

$$C_k(t) = A_k \mathrm{e}^{\rho_k t}, \quad \text{where } \rho_k = \pm\sqrt{2P(0,p_k)} \ \text{ for } k = 2,3,4,$$

so substituting into (7.6) gives (cf. (7.10))

$$\rho_2 + \rho_3 = 2\rho_4 \quad \text{and} \quad 2P(p_2,p_3)A_2A_3 + P(p_4,p_4)A_4^2 = 0.$$

If $P(0,p_k) < 0$, we may choose $C_k(t) = A_k \sin \rho_k t$, with $\rho_k = \sqrt{2|P(0,p_k)|}$, and solutions exist for $\rho_2 = \rho_3 = \rho_4$. Similarly, for $C_1(t) = t$, the solutions of

$$C_k'' = 2P(0,p_k)tC_k$$

are given by Airy functions.

If $P(0,0) \neq 0$, the first ODE in (7.12) admits a particular solution

$$C_1(t) = \tfrac{6}{P(0,0)}\,(\pm t)^{-2},$$

where the factor $(-t)$ corresponds to blow-up as $t \to 0^-$. Then the rest of the functions are chosen as follows:

$$C_k(t) = A_k(\pm t)^{\rho_k}, \quad \text{where } \rho_k^2 - \rho_k = \tfrac{12P(0,p_k)}{P(0,0)} \ \text{ for } k = 2,3,4$$

are assumed to be real. Plugging into (7.6) yields (7.10) and

$$\rho_2 + \rho_3 = 2\rho_4.$$

Example 7.3 **(Another 4D subspace)** Let us return to the first-order PDE (7.2) and apply another expansion on W_4 with more correlations in the bilinear product,

$$u(x,t) = C_1(t) + C_2(t)e^x + C_3(t)e^{2x} + C_4(t)e^{\frac{3}{2}x}, \tag{7.13}$$

where $\Gamma' = \{\frac{5}{2}, 3, \frac{7}{2}, 4\}$ and $n' = 4$, so that there are three invariance conditions, $P(1, \frac{3}{2}) = P(2, \frac{3}{2}) = P(2,2) = 0$, plus the algebraic equation

$$2P(1,2)C_2C_3 + P(\tfrac{3}{2}, \tfrac{3}{2})C_4^2 = 0. \tag{7.14}$$

The DS on this W_4 is modified,

$$\begin{cases} C_1' = P(0,0)C_1^2, \\ C_2' = 2P(0,1)C_1C_2, \\ C_3' = 2P(0,2)C_1C_3 + P(1,1)C_2^2, \\ C_4' = 2P(0,\frac{3}{2})C_1C_4. \end{cases} \tag{7.15}$$

Let $P(0,0) \neq 0$. In this case, $C_1(t)$ is given by (7.8) and

$$C_k(t) = A_k t^{-\rho_k} \quad \text{for} \quad k = 2, 4, \quad \text{where } \rho_2 = \tfrac{2P(0,1)}{P(0,0)} \text{ and } \rho_4 = \tfrac{2P(0,\frac{3}{2})}{P(0,0)},$$

so the ODE for C_3 takes the form

$$C_3' = -\tfrac{2P(0,2)}{P(0,0)}\, t^{-1}\, C_3 + P(1,1)A_2^2 t^{-2\rho_2}.$$

Hence,

$$C_3(t) = A_3 t^{-\rho_3},$$

where

$$\rho_3 = 2\rho_2 - 1 \quad \text{and} \quad -\rho_3 A_3 = -\tfrac{2P(0,2)}{P(0,0)}\, A_3 + P(1,1)A_2^2.$$

Plugging into the algebraic equation (7.14), we obtain extra conditions on the operator and initial data,

$$\rho_2 + \rho_3 = 2\rho_4 \quad \text{and} \quad 2P(1,2)A_2A_3 + P(\tfrac{3}{2}, \tfrac{3}{2})A_4^2 = 0.$$

If $P(0,0) = 0$ and $C_1(t) = 1$, then

$$C_2(t) = A_2 e^{2P(0,1)t} \quad \text{and} \quad C_4(t) = A_4 e^{2P(0,\frac{3}{2})t}.$$

Therefore,

$$C_3(t) = \tfrac{P(1,1)}{2[2P(0,1)-P(0,2)]}\, A_2^2 e^{4P(0,1)t}.$$

Hence, (7.14) yields

$$3P(0,1) = 2P(0, \tfrac{3}{2}) \quad \text{and} \quad \tfrac{2P(1,1)P(1,2)}{2[2P(0,1)-P(0,2)]}\, A_2^3 + P(\tfrac{3}{2}, \tfrac{3}{2})A_4^2 = 0.$$

Example 7.4 **(5D subspace)** Consider another expansion on W_5,

$$u(x,t) = C_1(t) + C_2(t)e^x + C_3(t)e^{2x} + C_4(t)e^{\frac{5}{2}x} + C_5(t)e^{3x}, \tag{7.16}$$

where $\Gamma' = \{\frac{7}{2}, 4, \frac{9}{2}, 5, \frac{11}{2}, 6\}$ and $n' = 6$, so six invariance conditions are obtained. The first four are standard,

$$P(1, \tfrac{5}{2}) = P(2, \tfrac{5}{2}) = P(3, \tfrac{5}{2}) = P(3,3) = 0, \tag{7.17}$$

and the later two are equations of partial invariance

$$2P(1,3)C_2C_5 + P(2,2)C_3^2 = 0 \quad \text{and} \quad 2P(2,3)C_3C_5 + P(\tfrac{5}{2},\tfrac{5}{2})C_4^2 = 0. \quad (7.18)$$

For the first-order PDE (7.2), the DS takes the form

$$\begin{cases} C_1' = P(0,0)C_1^2, \\ C_2' = 2P(0,1)C_1C_2, \\ C_3' = P(1,1)C_2^2 + 2P(0,2)C_1C_3, \\ C_4' = 2P(0,\tfrac{5}{2})C_1C_4. \\ C_5' = 2P(0,3)C_1C_5 + 2P(1,2)C_2C_3, \end{cases} \quad (7.19)$$

There are two cases:

Case I: $P(0,0) = 0$, where $C_1(t) = 1$, and hence,

$$C_k(t) = A_k e^{\rho_k t} \quad \text{for } k = 2, 4, \quad \text{where } \rho_2 = 2P(0,1), \;\; \rho_4 = 2P(0,\tfrac{5}{2}). \quad (7.20)$$

Then the algebraic equations (7.18) take the form

$$\begin{cases} 2P(1,3)A_2 e^{\rho_2 t} C_5 + P(2,2)C_3^2 = 0, \\ 2P(1,3)C_3C_5 + P(\tfrac{5}{2},\tfrac{5}{2})A_4^2 e^{2\rho_4 t} = 0, \end{cases}$$

so that both C_3 and C_5 are exponential functions

$$C_{3,5}(t) = A_{3,5}\, e^{\rho_{3,5} t},$$

where $\rho_3 = \frac{1}{3}(2\rho_4 + \rho_2)$ and $\rho_5 = \frac{1}{3}(4\rho_4 - \rho_2)$, and

$$A_3^3 = \frac{P(\frac{5}{2},\frac{5}{2})P(1,3)}{P(2,3)P(2,2)} A_2 A_4^2 \quad \text{and} \quad A_5 = -\frac{P(2,2)A_3^2}{2P(1,3)A_2}.$$

The functions $C_{3,5}(t)$ satisfy the corresponding ODEs in (7.19), provided that $\rho_3 = 2\rho_2$ and $\rho_5 = \rho_2 + \rho_3$ plus two conditions on the initial data $\{A_k\}$. On the whole, we then obtain a linear system for the coefficients $\{\rho_k,\ k = 2, 3, 4, 5\}$,

$$\begin{cases} 2\rho_2 - \rho_3 = 0, \\ \rho_2 + \rho_3 - \rho_5 = 0, \\ \rho_2 - 3\rho_3 + 2\rho_4 = 0, \\ \rho_2 + 3\rho_5 - 4\rho_4 = 0, \end{cases} \quad (7.21)$$

which has a singular 4×4 matrix and hence nontrivial solutions

$$(\rho_2, \rho_3, \rho_4, \rho_5)^T = t\big(1, 2, \tfrac{5}{2}, 3\big)^T, \quad \text{where} \quad t \in I\!R.$$

Case II: $P(0,0) \neq 0$. Then, instead of the exponential functions,

$$C_k(t) = A_k t^{\rho_k} \quad \text{for } k = 1, ..., 5, \quad (7.22)$$

where $\rho_1 = -1$ (instead of $\rho_1 = 0$, as in the previous case), and the rest of the exponents are the same as above. Since the linear system (7.21) is consistent, solutions on W_5 exist.

Example 7.5 **(Quadratic thin film operators)** Consider the general fourth-order thin film operator with ten real parameters,

$$\begin{aligned} F[u] &= \alpha_1 u u_{xxxx} + \alpha_2 u_x u_{xxx} + \alpha_3 u_{xx} u_{xxx} + \alpha_4 (u_{xxx})^2 \\ &+ \alpha_5 u u_{xx} + \alpha_6 u_x u_{xx} + \alpha_7 (u_{xx})^2 + \alpha_8 u u_x + \alpha_9 (u_x)^2 + \alpha_{10} u^2. \end{aligned} \quad (7.23)$$

The polynomial is

$$\begin{aligned}P(X,Y) = {} & \tfrac{1}{2}\alpha_1(X^4+Y^4)+\tfrac{1}{2}\alpha_2(XY^3+X^3Y)+\tfrac{1}{2}\alpha_3(X^2Y^3\\ & +X^3Y^2)+\alpha_4X^3Y^3+\tfrac{1}{2}\alpha_5(X^2+Y^2)+\tfrac{1}{2}\alpha_6(XY^2+X^2Y)\\ & +\alpha_7X^2Y^2+\tfrac{1}{2}\alpha_8(X+Y)+\alpha_9XY+\alpha_{10}.\end{aligned} \tag{7.24}$$

For solutions (7.16), we have four conditions (7.17) and three linearly independent consistency relations from (7.21). Hence, there exists at least a three-parameter family of TFEs (7.2), (7.23) with such solutions.

Example 7.6 In a similar fashion, for the hyperbolic PDE (7.11), looking for solutions (7.16), in Case I we obtain from the second-order ODEs (7.19) (with $C_k'' = ...$) the exponential expressions (7.20) with

$$\rho_2 = [2P(0,1)]^{1/2} \quad \text{and} \quad \rho_4 = \left[2P(0,\tfrac{5}{2})\right]^{\frac{1}{2}}.$$

By (7.18), $C_{3,5}(t)$ are then the same exponential functions, the linear system (7.21) is also the same, and substituting into the DS gives the consistency conditions on the initial data. Calculations in Case II remain the same, and the only difference is that $\rho_1 = -2$ in (7.22).

Example 7.7 (**Another 5D subspace**) Consider equation (7.2) on another W_5,

$$u(x,t) = C_1(t) + C_2(t)e^x + C_3(t)e^{-x} + C_4(t)e^{\frac{1}{3}x} + C_5(t)e^{-\frac{1}{3}x}, \tag{7.25}$$

where $\Gamma' = \left\{-2, -\frac{4}{3}, -\frac{2}{3}, \frac{2}{3}, \frac{4}{3}, 2\right\}$ and $n' = 6$. In this case, we obtain

$$P(-1,-1) = P(-1,-\tfrac{1}{3}) = P(1,\tfrac{1}{3}) = P(1,1) = 0$$

and two algebraic equations

$$\begin{aligned}&P(\tfrac{1}{3},\tfrac{1}{3})C_4^2 + 2P(1,-\tfrac{1}{3})C_2C_5 = 0,\\ &P(-\tfrac{1}{3},-\tfrac{1}{3})C_5^2 + 2P(-1,\tfrac{1}{3})C_3C_4 = 0,\end{aligned} \tag{7.26}$$

with the DS

$$\begin{cases} C_1' = P(0,0)C_1^2 + 2P(1,-1)C_2C_3 + 2P(\tfrac{1}{3},-\tfrac{1}{3})C_4C_5,\\ C_2' = 2P(0,1)C_1C_2,\\ C_3' = 2P(0,-1)C_1C_3,\\ C_4' = 2P(0,\tfrac{1}{3})C_1C_4,\\ C_5' = 2P(0,-\tfrac{1}{3})C_1C_5. \end{cases} \tag{7.27}$$

From the last four ODEs,

$$C_k = A_k C_2^{\rho_k} \quad \text{for} \quad k = 3,4,5,$$

with ρ_k calculated as above,

$$\rho_3 = \tfrac{P(0,-1)}{P(0,1)}, \quad \rho_4 = \tfrac{P(0,\frac{1}{3})}{P(0,1)}, \quad \text{and} \quad \rho_5 = \tfrac{P(0,-\frac{1}{3})}{P(0,1)}.$$

Then the algebraic equations (7.26) are valid, provided that

$$\rho_4 = \tfrac{1}{3}(\rho_3+2) \quad \text{and} \quad \rho_5 = \tfrac{1}{3}(2\rho_3+1),$$

and a DS for the remaining two coefficients $C_{1,2}$ is obtained,

$$\begin{cases} C_1' = P(0,0)C_1^2 + \big[2P(1,-1)A_3 + 2P(\tfrac{1}{3},-\tfrac{1}{3})A_4A_5\big]C_2^{1+\rho_3}, \\ C_2' = 2P(0,1)C_1C_2. \end{cases}$$

This can be integrated in quadratures and possesses blow-up solutions. For the thin film operator (7.23), we obtain six extra conditions that imply existence of a four-parameter family of TFEs possessing such exact solutions.

For the second-order PDE (7.11), the DS (7.27) with $C_k'' = ...$ admit, in general, either exponential particular solutions $C_k(t) = A_k e^{\rho_k t}$ with $\rho_1 = 0$, or the algebraic ones (7.22), where $\rho_1 = -2$. If $P(0,1) = P(0,-1) = P(0,\frac{1}{3}) = P(0,-\frac{1}{3})$, there exist solutions $C_k(t) = A_k C_2(t)$ for $k = 3, 4, 5$, where $C_1(t)$ and $C_2(t)$ solve a DS, and (7.26) give two extra conditions on $\{A_k\}$.

7.1.2 On further extensions

1. $\frac{\partial}{\partial t}$-dependent operators. We considered such operators in Sections 1.5.2 (see (1.135)) and 2.7, where invariant modules were dealt with. For invariant sets (partially invariant modules), a similar computational analysis can be performed, though the overdetermined DSs become more cumbersome and difficult. We must admit that there is a higher probability to obtain purely, in both x and t variables, exponential solutions that sometimes can be more efficiently manipulated by using Baker–Hirota derivatives and bilinear operators, or by other methods from theory of integrable PDEs.

Example 7.8 (**PDE with nonstationary part from the KdV equation**) Let us begin with the following rather artificial PDE:

$$\boxed{u_x u_t - u u_{xt} = F[u],} \tag{7.28}$$

where the quadratic left-hand side has been borrowed from the bilinear form of the KdV equation (0.31) in the Introduction. Here $F[u]$ is the general operator (7.1). First, consider the expansion (7.4) with the same assumptions on the exponents. Not specifying the DS, we claim that the only possible solutions are exponential ones,

$$C_k(t) = A_k e^{\rho_k t} \quad \text{for all } k = 1, 2, 3, 4, \tag{7.29}$$

where $\{\rho_k\}$ satisfies a number of conditions associated with the values of the polynomial P at p_k. One can think this rigid exponential solution structure as being associated with the specific form of the left-hand side of (7.28), which is a full Baker–Hirota derivative $-2Du \cdot u_t$, defined as, [22, 284]:

$$Dv \cdot w = \tfrac{d}{dz}\, v(x+z)w(x-z)\big|_{z=0} = v_x w - v w_x. \tag{7.30}$$

Therefore, the Baker–Hirota method applies ensuring that the right solutions are supposed to be of exponential soliton-type only.

Furthermore, for the PDE with the first term only on the left-hand side

$$\boxed{u_x u_t = F[u],} \tag{7.31}$$

the conclusion remains the same: the overdetermined DS has the exponential solutions (7.29) only. The exponential character of solutions is true generic for sets on W_5. For instance, using expansion (7.16) in (7.31) yields an overdetermined DS, consisting of eleven equations,

$$\begin{cases} 3C_4C_4' = P(3,3)C_4^2, \\ 3C_4C_5' + \frac{5}{2}\,C_5C_4' = 2P(3,\frac{5}{2})C_4C_5, \\ 2C_3C_5' + \frac{5}{2}\,C_5C_3' = 2P(2,\frac{5}{2})C_3C_5, \\ C_2C_5' + \frac{5}{2}\,C_5C_2' = 2P(1,\frac{5}{2})C_1C_5, \\ 2C_3C_3' + 3C_4C_2' + C_2C_4' = 2P(1,3)C_2C_4 + P(2,2)C_3^2, \\ 2C_3C_4' + 3C_4C_3' + \frac{5}{2}\,C_5C_5' = 2P(2,3)C_3C_4 + P(\frac{5}{2},\frac{5}{2})C_5^2, \\ 0 = P(0,0)C_1^2, \\ C_2C_1' = 2P(0,1)C_1C_2, \\ 2C_3C_1' + C_2C_2' = P(1,1)C_2^2 + 2P(0,3)C_1C_3, \\ 3C_4C_1' + C_2C_3' = 2P(0,3)C_1C_4 + 2P(1,2)C_2C_3, \\ \frac{5}{2}\,C_5C_1' = 2P(0,\frac{5}{2})C_1C_5. \end{cases}$$

This system is solved, giving exponential solutions (7.29). The same holds for the related second-order PDE

$$\boxed{u_x u_{tt} = F[u].}$$

Example 7.9 Consider the expansion (7.13) for PDE (7.31). Then, using the same action formula for both quadratic operators therein, we derive the DS, consisting of eight equations, where the invariance conditions, such as (7.14), also become ODEs (cf. (7.15)),

$$\begin{cases} 0 = P(0,0)C_1^2, \;\; 2C_3C_3' = P(2,2)C_3^2, \\ C_2C_3' + 2C_3C_2' + \frac{3}{2}\,C_4C_4' = 2P(1,2)C_2C_3 + P(\frac{3}{2},\frac{3}{2})C_4^2, \\ C_2C_2' + 2C_3C_1' = 2P(0,2)C_1C_3 + P(1,1)C_2^2, \\ \frac{3}{2}\,C_4C_2' + C_2C_4' = 2P(1,\frac{3}{2})C_1C_4, \;\; C_2C_1' = 2P(0,1)C_1C_2, \\ \frac{3}{2}\,C_4C_3' + 2C_3C_4' = 2P(2,\frac{3}{2})C_3C_4, \;\; \frac{3}{2}\,C_4C_1' = 2P(0,\frac{3}{2})C_1C_4. \end{cases}$$

It is easy to check that this DS has exponential solutions.

2. Complex exponents. Taking $p_k = a_k + {\rm i}b_k$ with $b_k \neq 0$ yields quadratic systems of ODEs; see (1.134). Such systems can possess solutions on partially invariant subspaces.

3. Cubic operators. For cubic (see (1.136)) and higher-degree operators, the construction becomes more technical and can be performed for classes of operators and linear subspaces by using codes for computer-supported algebraic manipulations.

7.2 Quadratic Kuramoto–Sivashinsky equations

We now study partially invariant subspaces for another family of quadratic fourth-order operators

$$\begin{aligned} F[u] = {} & uu_t + \alpha uu_x + \beta uu_{xx} + \gamma\,(u_x)^2 \\ & + \delta uu_{xxxx} + \varepsilon u_x u_{xxx} + \mu\,(u_{xx})^2 + \nu uu_{xxx}. \end{aligned}$$

The corresponding PDE

$$F[u] = 0 \tag{7.32}$$

can be treated as a generalized Kuramoto–Sivashinsky equation; see Section 3.8. Our goal is to describe the partially invariant trigonometric subspace

$$W_3 = \mathcal{L}\{1, \cos x, \sin x\},$$

and find all consistent overdetermined DSs and the corresponding exact solutions. This can be done analytically, though it leads to technical difficulties. The approach applies to more general PDEs

$$F[u] = L[u] \equiv \textstyle\sum_{(k \le 2m)} a_k\, D_x^k u,$$

where L is a linear $2m$th-order elliptic operator with constant coefficients $\{a_k\}$.

We look for the following solutions on W_3 for any $t \ge 0$:

$$u(x,t) = C_1(t) + C_2(t)\cos x + C_3(t)\sin x.$$

Denote for convenience

$$s = \varepsilon - \gamma + \mu, \;\; p = \beta - \delta - \varepsilon + \gamma - \mu, \;\; q = \alpha - \nu, \;\; r = \varepsilon - \gamma.$$

Substituting into (7.32) yields the following overdetermined DS:

$$\begin{cases} C_1C_1' + C_2C_2' - (p+r)C_2^2 + qC_2C_3 - rC_3^2 = 0, \\ C_2C_1' + C_1C_2' - (p+s)C_1C_2 + qC_1C_3 = 0, \\ C_3C_1' + C_1C_3' - qC_1C_2 - (p+s)C_1C_3 = 0, \\ C_2C_2' - C_3C_3' - pC_2^2 + 2qC_2C_3 + pC_3^2 = 0, \\ C_3C_2' + C_2C_3' - qC_2^2 - 2pC_2C_3 + qC_3^2 = 0, \end{cases} \tag{7.33}$$

where the equations are projections of $F[u] = 0$ onto 1, $\cos x$, $\sin x$, $\frac{1}{2}\cos 2x$, and $\frac{1}{2}\sin 2x$ respectively. Assuming that

$$C_1^2 - C_2^2 \ne 0 \quad \text{and} \quad C_2^2 + C_3^2 \ne 0,$$

we take C_1' and C_2' from the first two equations of (7.33) and C_2' and C_3' from the last two, and obtain the system

$$\begin{cases} C_1' = sC_1, \\ C_2' = pC_2 - qC_3, \\ C_3' = qC_2 + pC_3, \\ sC_1^2 = r\left(C_2^2 + C_3^2\right). \end{cases} \tag{7.34}$$

Solving the first three equations of (7.34) yields

$$\begin{aligned} C_1(t) &= D_1 e^{st}, \\ C_2(t) &= e^{pt}(D_2 \cos qt - D_3 \sin qt), \\ C_3(t) &= e^{pt}(D_2 \sin qt + D_3 \cos qt), \end{aligned} \tag{7.35}$$

where D_1, D_2, and D_3 are arbitrary constants. Substituting these functions into the last algebraic equation in (7.34) gives the following equality:

$$sD_1^2 e^{2st} = re^{2pt}\left(D_2^2 + D_3^2\right). \tag{7.36}$$

Analyzing (7.36) yields the following:

Proposition 7.10 *There exist three cases:*
(i) $s = r = 0$, *i.e.,*

$$\varepsilon = \gamma \;\; \text{and} \;\; \mu = 0.$$

Then the explicit solution of (7.32) *is*

$$u(x,t) = D_1 + e^{pt}[D_2 \cos(x-qt) + D_3 \sin(x-qt)],$$

where D_1, D_2, *and* D_3 *are arbitrary.*
(ii) $r = 0$ *and* $s \neq 0$. *Then*

$$u(x,t) = e^{pt}[D_2 \cos(x-qt) + D_3 \sin(x-qt)],$$

where D_2 *and* D_3 *are arbitrary.*
(iii) $r \neq 0$ *and* $s \neq 0$. *Then* (7.36) *is equivalent to two conditions*

$$s = p \quad \text{and} \quad sD_1^2 = r\big(D_2^2 + D_3^2\big).$$

and the solution takes the form

$$u(x,t) = e^{pt}[D_1 + D_2 \cos(x-qt) + D_3 \sin(x-qt)].$$

The only reasonable compacton-like solution exists in (iii), where we set $D_1 = D_2 = \frac{1}{2}$ and $D_3 = 0$ to get

$$u_c(x,t) = e^{pt} \cos^2[\tfrac{1}{2}(x-qt)]. \tag{7.37}$$

This implies the extra condition $s = r$, so solution (7.37) exists for $s = p = r \neq 0$, or, in the original notation,

$$\beta = \delta + 2(\varepsilon - \gamma) \quad \text{and} \quad \mu = 0.$$

We do not check if, being extended by zero in $\{|x-qt| \geq \pi\}$, (7.37) will be a solution of the Cauchy problem (most probably not, for such PDEs). In any case, the compacton (7.37) localized in the domain $\{|x-qt| < \pi\}$ with the exponentially varying amplitude is a solution of an FBP with necessary free-boundary conditions, including the zero contact angle one. Usually, a correct setting of such FBPs is a difficult OPEN PROBLEM that was discussed in Section 3.2 for some TFEs.

7.3 Method of generalized separation of variables

In this section, another extension of notions and techniques related to (partially) invariant subspaces is presented.

7.3.1 The general scheme for GSV

Consider a class of evolution PDEs of the form

$$T[u] = F[u], \tag{7.38}$$

where $T = T(\frac{\partial}{\partial t})$ and $F = F(\frac{\partial}{\partial x})$ are some nonlinear differential operators, in most applications, of the polynomial type. The method of *generalized separation of variables* (GSV) consists of looking for solutions in the form of finite sums

$$u = U(x,t) \equiv \sum_{i=1}^{n} a_i(t) f_i(x) = \mathbf{a}(t)\mathbf{f}^T(x), \tag{7.39}$$

where unknown sufficiently smooth vector functions $\mathbf{a}(t) = (a_1(t), ..., a_n(t))$ and $\mathbf{f}(x) = (f_1(x), ..., f_n(x))$ are assumed to have linearly independent components. The mathematical basis of the GSV is as follows. By $A_i[\mathbf{a}(t)] = A_i(\mathbf{a}(t), \mathbf{a}'(t), ...)$ and $B_i[\mathbf{f}(x)] = B_i(\mathbf{f}(x), \mathbf{f}'(x), ...)$ we denote some differential operators.

Lemma 7.11 *Let for functions* (7.39)

$$\begin{aligned} T[U] &= \sum_{i=1}^{m} A_i[\mathbf{a}(t)]\tilde{f}_i(x), \\ F[U] &= \sum_{i=1}^{k} B_i[\mathbf{f}(x)]\tilde{a}_i(t), \end{aligned} \tag{7.40}$$

where the sets $\{\tilde{f}_i(x)\}$ *and* $\{\tilde{a}_i(t)\}$ *are linearly independent. Then, equation* (7.38) *on solutions* (7.39) *reduces to two systems*

$$\begin{aligned} &\text{(I)}\ \ (A_1[\mathbf{a}(t)], ..., A_m[\mathbf{a}(t)]) = (\tilde{a}_1(t), ..., \tilde{a}_k(t))C, \\ &\text{(II)}\ \ (B_1[\mathbf{f}(x)], ..., B_k[\mathbf{f}(x)]) = (\tilde{f}_1(x), ..., \tilde{f}_m(x))C^T, \end{aligned} \tag{7.41}$$

where $C = \|C_i^j\|$ *is a constant* $k \times m$ *matrix.*

Proof. Plugging (7.39) into (7.38) yields

$$\sum_{i=1}^{m} A_i[\mathbf{a}(t)]\tilde{f}_i(x) = \sum_{j=1}^{k} B_j[\mathbf{f}(x)]\tilde{a}_j(t).$$

Differentiating in t leads to the system

$$\sum_{i=1}^{m} (A_i[\mathbf{a}(t)])^{(p)}\tilde{f}_i(x) = \sum_{j=1}^{k} B_j[\mathbf{f}(x)]\tilde{a}_j^{(p)}(t) \ \text{ for } \ p = 0, ..., k-1,$$

or, in the matrix form,

$$A[\mathbf{a}(t)]\tilde{\mathbf{f}}^T(x) = \alpha(t)B^T[\mathbf{f}(x)], \tag{7.42}$$

where $A[\mathbf{a}(t)] = \|A_i^{(p)}[\mathbf{a}(t)]\|$ and $\alpha(t) = \|\tilde{a}_j^{(p)}(t)\|$ are matrices, and we denote $\tilde{\mathbf{f}} = (\tilde{f}_1, ..., \tilde{f}_m)$, $B = (\tilde{B}_1, ..., \tilde{B}_k)$. Since the set $(\tilde{a}_1(t), ..., \tilde{a}_k(t))$ is linearly independent, the corresponding Wronskian $\det\alpha(t)$ is non-zero, so there exists the inverse matrix $\alpha^{-1}(t)$, [132, Ch. III]. Multiplying (7.42) by $\alpha^{-1}(t)$ yields

$$\alpha^{-1}(t)A[\mathbf{a}(t)]\,\tilde{\mathbf{f}}^T(x) = B^T[\mathbf{f}(x)].$$

On differentiation in t, we find that

$$\tfrac{\mathrm{d}}{\mathrm{d}t}\left(\alpha^{-1}(t)A[\mathbf{a}(t)]\right)\,\tilde{\mathbf{f}}^T(x) = 0. \tag{7.43}$$

Since $(\tilde{f}_1(x), ..., \tilde{f}_m(x))$ is also a linearly independent set, it follows from (7.43) that $\alpha^{-1}A = C$ must be a constant matrix. Hence, $A[\mathbf{a}(t)] = \alpha(t)C$ and $B[\mathbf{f}(x)] = \tilde{\mathbf{f}}(x)C^T$. That completes the proof. □

It follows from this proof that the GSV does not assume any specific form of the solutions such as (7.39), and deals with operator equalities (7.40) only.

7.3.2 The GSV in application: systems (I) and (II)

Example 7.12 (GSV for a parabolic equation) Consider the operator

$$F[u] = \alpha u u_{xx} + \beta (u_x)^2 + \gamma u^2 + \varepsilon u u_x + \mu u_x + \nu u + \delta. \tag{7.44}$$

Let $T[u] = uu_t$, so that (7.38) is a quasilinear parabolic equation of the reaction-diffusion-absorption type. Fix $n = 2$ and look for solutions

$$u = U(x,t) = g(x) + f(x)a(t),$$

where $a(t) \not\equiv$ constant and $f(x)$ and $g(x)$ are linearly independent. In this case, (7.40) reads

$$\begin{gathered} T[U] = aa'f^2 + a'fg, \\ F[U] = a^2\big[\alpha f f'' + \beta (f')^2 + \gamma f^2 + \varepsilon f f'\big] \\ + a\big[\alpha(gf'' + fg'') + 2\beta f'g' + 2\gamma fg + \varepsilon (fg)' + \mu f' + \nu f\big] \\ + \big[\alpha g g'' + \beta (g')^2 + \gamma g^2 + \varepsilon g g' + \mu g' + \nu g + \delta\big]. \end{gathered}$$

According to Lemma 7.11, the PDE $T[U] = F[U]$ reduces to two systems

$$\text{(I)} \quad \begin{cases} aa' = c_1 a^2 + c_2 a + c_3, \\ a' = \tilde{c}_1 a^2 + \tilde{c}_2 a + \tilde{c}_3, \end{cases}$$

and

$$\text{(II)} \quad \begin{cases} \alpha f f'' + \beta (f')^2 + \gamma f^2 + \varepsilon f f' = c_1 f^2 + \tilde{c}_1 fg, \\ \alpha(gf'' + fg'') + 2\beta f'g' + 2\gamma fg + \varepsilon (fg)' + \mu f' + \nu f = c_2 f^2 + \tilde{c}_2 fg, \\ \alpha g g'' + \beta (g')^2 + \gamma g^2 + \varepsilon g g' + \mu g' + \nu g + \delta = c_3 f^2 + \tilde{c}_3 fg. \end{cases}$$

Here c_i and $\tilde{c}_i$ are some constants (elements of the 2×3 matrix C). It is worth mentioning that the GSV analysis can be naturally associated with a partially invariant module, $W_2 = \mathcal{L}\{1, a(t)\}$, for the given operators, that establishes links with the previous context.

Concerning the first system (I), it is not difficult to show that, up to translations and scalings, there exist just two essentially different cases

$$a(t) = t \quad \text{and} \quad a(t) = \mathrm{e}^t.$$

For other $T[u]$ operators to be studied, more sophisticated functions can appear. The second overdetermined system (II) is more difficult to handle, but it admits a complete classification. Later on, we study various overdetermined systems that are particular cases of (II).

Before introducing other parabolic and hyperbolic examples, it is worth discussing some common aspects of such systems. In the general case, where T is an arbitrary quadratic homogeneous differential operator such that $T[1] = 0$ (i.e., $b_{00} = 0$),

$$T[u] = \textstyle\sum_{(i,j,\, i+j \neq 0)} b_{i,j}\, D_t^i u\, D_t^j u, \tag{7.45}$$

the following holds:

$$T[g(x) + f(x)a(t)] = A_1[a(t)] f^2(x) + A_2[a(t)] f(x) g(x).$$

Here $A_1[a] = T[a]$ and $A_2[a] = T[a+1] - T[a]$. Then system (I) takes the form

$$\text{(I)} \quad \begin{cases} A_1[a] = c_1 a^2 + c_2 a + c_3, \\ A_2[a] = \tilde{c}_1 a^2 + \tilde{c}_2 a + \tilde{c}_3. \end{cases}$$

The following two cases are mainly treated later on:

$$1) \begin{cases} A_1[a] = A a^2, \\ A_2[a] = B a, \end{cases} \quad \text{and} \quad 2) \begin{cases} A_1[a] = \tilde{A} a, \\ A_2[a] = \tilde{B}, \end{cases}$$

with various constants A, B and $\tilde{A}$, $\tilde{B}$. The overdetermined system (II) remains practically the same in both cases and admits a unified treatment.

7.4 Generalized separation and partially invariant modules

Here a slightly different version of the GSV is used. We present another case of classification of overdetermined DSs occurring in the analysis of nonlinear inhomogeneous PDEs for which invariant subspaces or sets are not prescribed by polynomial, trigonometric, or exponential functions. For convenience, unlike (7.44), for $F[u]$ we now denote a purely quadratic second-order operator,

$$F[u] = \alpha u u_{xx} + \beta (u_x)^2 + \gamma u^2 + \varepsilon u u_x, \tag{7.46}$$

and introduce linear terms separately. Here $T[u]$ is still given by (7.45), and we set

$$\mathbf{A}[u] = T[u] - F[u]. \tag{7.47}$$

In fact, F can be taken in the general form (7.1), though, as will be shown, for $2m$th-order ordinary differential operators F with $m \geq 2$, the solvability and consistency analysis of overdetermined systems become illusive.

Again, for convenience of notation, we use symmetric bilinear forms of quadratic operators (7.45) and (7.46),

$$\begin{aligned} T[a,b] &= T[a+b] - T[a] - T[b], \\ F[f,g] &= F[f+g] - F[f] - F[g], \end{aligned} \tag{7.48}$$

and the corresponding polynomials, so that $F[u] = \frac{1}{2} F[u,u]$ and $T[u] = \frac{1}{2} T[u,u]$. Let us next introduce two linear operators (in the previous GSV analysis $Q[u]$ was included into F)

$$P(D_t)[u] = \sum b_k D_t^k u \quad \text{and} \quad Q(D_x)[u] = \sum d_l D_x^l u,$$

with real constant coefficients $\{b_k\}$ and $\{d_l\}$. Denote, for convenience,

$$R = Q(D_x) - P(D_t). \tag{7.49}$$

Consider a quadratic inhomogeneous PDE of the form

$$\mathbf{A}[u] = R[u] + p, \tag{7.50}$$

where $p = p(x,t)$ is a given function. As a new example, this includes the following hyperbolic equation:

$$\boxed{u u_{tt} = \alpha u u_{xx} + \beta (u_x)^2 + \gamma u^2 + \varepsilon u u_x + \mu u_x + \nu u + \delta.}$$

7.4.1 Partially invariant 2D modules

According to the GSV method, we study properties of the operator (7.47) on

$$W_2 = \mathcal{L}\{1, a(t)\}, \tag{7.51}$$

with a smooth function $a(t)$ to be determined.

1. Partial invariance. We first establish a formal existence of a set $M \subset W_2$, such that, for any $u(x,t) = g(x) + f(x)\,a(t)$ (i.e., $u(x,t) \in M$ for any $x \in I\!R$),

$$\mathbf{A}[g + fa] \in W_2. \tag{7.52}$$

This leads to the system 1) in Section 7.3.2.

Lemma 7.13 *Let $a(t)$ satisfy the system*

$$\begin{cases} T[a] = Aa^2, \\ T[a,1] = Ba, \end{cases} \tag{7.53}$$

where A and B are some constants. Then there exists the invariant set

$$M_A = \{u = g + fa \in W_2 : f \text{ satisfies } F[f] = Af^2\}. \tag{7.54}$$

Proof. By (7.48),

$$\begin{aligned} T[g + fa] &= f^2 T[a] + gf\,T[a,1], \\ F[g + fa] &= F[g] + F[g,f]a + F[f]\,a^2, \end{aligned} \tag{7.55}$$

so that

$$\mathbf{A}[g + fa] \equiv -F[g] + \{Bgf - F[g,f]\}a + \{Af^2 - F[f]\}a^2. \tag{7.56}$$

Hence, (7.52) is valid for all $v \in M_A$. □

It follows from (7.55) that (7.52) holds in a more general case, where

$$T[a] = A_2a^2 + A_1a + A_0 \quad \text{and} \quad T[a,1] = B_2a^2 + B_1a + B_0.$$

In Lemma 7.13, we set $A_1 = A_0 = B_2 = B_0 = 0$. It follows from (7.45) that $a(t)$ can be taken in the exponential form

$$a(t) = \mathrm{e}^{\lambda t}, \quad \text{with a constant } \lambda \in I\!R. \tag{7.57}$$

Then (7.53) holds with

$$A = \textstyle\sum_{(i,j,\,i\neq j)} b_{i,j}\,\lambda^i\lambda^j, \quad \text{where} \quad B = \sum_{(j\neq 0)} b_{0,j}\lambda^j + \sum_{(i\neq 0)} b_{i,0}\lambda^i.$$

2. PDEs on partially invariant modules. Lemma 7.13 gives a family $\{M_A,\ A \in I\!R\}$ of sets which are invariant on W_2 under the operator (7.47). Let us now show that the PDE (7.50) can be restricted to such sets M_A. We assume that

$$P(D_t)[a(t)] = 0, \quad \text{so} \quad a \in \ker P. \tag{7.58}$$

The case of an arbitrary eigenfunction, $P[a(t)] = \rho a(t)$ for some constant $\rho \in I\!R$, is treated in a similar fashion.

Proposition 7.14 *Let $a(t)$ satisfy (7.53) and (7.58). Then, equation (7.50) on M_A, with $p \in W_2$,*

$$\mathbf{A}[u] = R[u] + p \quad \text{for} \;\; u = g + fa \in M_A, \tag{7.59}$$

is equivalent to the system

$$\begin{cases} F[g] = -Q[g] - p, \\ F[g, f] - Bgf = -Q[f]. \end{cases} \tag{7.60}$$

Proof. It follows from (7.49) that

$$R[g + fa] = Q[g] + Q[f]a.$$

Hence, (7.60) is a consequence of (7.59) and (7.56). □

In the case of (7.57), denoting by $\bar{P}(\lambda)$ the polynomial of P, the following holds:

$$P[\mathrm{e}^{\lambda t}] = \rho \mathrm{e}^{\lambda t}, \quad \text{with} \;\; \rho = \bar{P}(\lambda).$$

7.4.2 Existence of solutions via partial invariance

Let us now formulate a class of problems (7.59) having a nontrivial solution on M_A. Consider the case where $Q(D_x)$ is a first-order linear operator,

$$Q(D_x)[u] = c_0 u + c_1 u_x, \;\; \text{where } c_0, \, c_1 \in I\!R, \;\; \text{and} \;\; p(x) \equiv \delta \in I\!R. \tag{7.61}$$

Let $\sigma \in I\!R$ be a real root of

$$\beta \sigma^2 - c_1 \sigma + \delta = 0 \quad (c_1^2 \ge 4\beta\delta). \tag{7.62}$$

Below D, D_1, D_2, ... denote different arbitrary constants. The type of the solution $u(x, t)$ of (7.59) on M_A differs in the cases where $A \ne 0$ and $A = 0$.

1. Case A ≠ 0.

Theorem 7.15 *Assume that* (7.53) *holds, where*

$$A = B \ne 0. \tag{7.63}$$

Then, if

$$\beta = -2\alpha \ne 0, \;\; \varepsilon = -\tfrac{\alpha c_0}{2\alpha\sigma + c_1}, \;\; A = \tfrac{\gamma(3\alpha\sigma + c_1)}{2\alpha\sigma + c_1} \ne 0, \tag{7.64}$$

where $2\alpha\sigma + c_1 \ne 0$, problem (7.59) has the nontrivial solution

$$u(x, t) = g(x) + f(x)\, a(t). \tag{7.65}$$

The functions $\{g, f\}$ in (7.65) are determined as follows:
(i) *If $c_0^2 > 4\sigma\gamma(2\alpha\sigma + c_1)$, then*

$$g(x) = -\tfrac{(2\alpha\sigma + c_1)}{\gamma} \tfrac{\mu_1 D_1 \mathrm{e}^{\mu_2 x} + \mu_2 D_2 \mathrm{e}^{\mu_1 x}}{D_1 \mathrm{e}^{\mu_2 x} + D_2 \mathrm{e}^{\mu_1 x}} - \tfrac{c_0}{\gamma}, \;\; f(x) = \tfrac{D \mathrm{e}^{(\mu_1 + \mu_2)x}}{D_1 \mathrm{e}^{\mu_2 x} + D_2 \mathrm{e}^{\mu_1 x}}, \tag{7.66}$$

where μ_1 and μ_2 are different (real) roots of

$$(2\alpha\sigma + c_1)\mu^2 + c_0\mu + \sigma\gamma = 0; \tag{7.67}$$

(ii) *If $c_0^2 = 4\sigma\gamma(2\alpha\sigma + c_1)$, then (7.67) has a unique root μ and*

$$g(x) = -\tfrac{(2\alpha\sigma + c_1)}{\gamma}\big(\mu - \tfrac{D_2}{D_1 + D_2 x}\big) - \tfrac{c_0}{\gamma}, \;\; f(x) = \tfrac{D \mathrm{e}^{\mu x}}{D_1 + D_2 x}; \tag{7.68}$$

(iii) *If* $c_0^2 < 4\sigma\gamma(2\alpha\sigma + c_1)$, *then*

$$g(x) = -\frac{(2\alpha\sigma + c_1)}{\gamma}[\nu_2 \tan(\nu_2(x + D_1)) + \nu_1] - \frac{c_0}{\gamma},\\ f(x) = \frac{De^{\nu_1 x}}{\cos(\nu_2(x+D_1))}, \tag{7.69}$$

where $\mu = \nu_1 \pm i\nu_2$ *are complex roots of* (7.67).

Proof. We rewrite the equation of M_A in (7.54) as

$$\alpha f f'' + \beta(f')^2 + \varepsilon f f' = (A - \gamma) f^2. \tag{7.70}$$

Equations (7.60) can be written as follows:

$$\alpha g g'' + \beta(g')^2 + \gamma g^2 + \varepsilon g g' + c_0 g + c_1 g' + \delta = 0, \tag{7.71}$$

$$\alpha(g f'' + f g'') + 2\beta g' f' + \varepsilon(g f)' + c_0 f + c_1 f' - (A - 2\gamma) g f = 0. \tag{7.72}$$

Here we use that, by (7.48),

$$F[g, f] \equiv \alpha(g f'' + f g'') + 2\beta g' f' + \varepsilon(g f)' + 2\gamma g f. \tag{7.73}$$

We now show that, under the above hypotheses, the overdetermined system (7.70)–(7.72) has a nontrivial solution $\{g(x), f(x)\}$. Substituting g'' and f'' from (7.70) and (7.71) into (7.72) yields

$$\beta(g f' - f g')^2 = c_1 f(g f' - f g') - \delta f^2, \quad \text{so} \tag{7.74}$$

$$g' = \frac{f'}{f} g - \sigma, \tag{7.75}$$

where $\sigma = \sigma_\pm = \frac{1}{2\beta}\left(c_1 \pm \sqrt{c_1^2 - 4\beta\delta}\right)$ from (7.62). By (7.75),

$$g'' = \frac{f''}{f} g - \sigma \frac{f'}{f}. \tag{7.76}$$

Substituting (7.75) and (7.76) into (7.71) and using (7.70) implies

$$A g = [(\alpha + 2\beta)\sigma - c_1]\frac{f'}{f} + (\varepsilon\sigma - c_0). \tag{7.77}$$

Thus, in general, problem (7.70)–(7.72) is equivalent to the overdetermined system (7.70), (7.75), and (7.77). Let us next derive the identity satisfied by $f(x)$, which is precisely a criterion for solvability of our equation on the invariant set.

Since $A \neq 0$, (7.75) can be rewritten as $(Ag)f' - f(Ag)' = \sigma A f$, so substituting Ag from (7.77) yields

$$[(\alpha + 2\beta)\sigma - c_1]\left(\frac{f'}{f}\right)^2 + (\varepsilon\sigma - c_0)\frac{f'}{f} - [(\alpha + 2\beta)\sigma - c_1]\left(\frac{f'}{f}\right)' = \sigma A. \tag{7.78}$$

Since $\alpha \neq 0$, it follows from (7.70) that

$$\left(\frac{f'}{f}\right)' = -\frac{\alpha+\beta}{\alpha}\left(\frac{f'}{f}\right)^2 - \frac{\varepsilon}{\alpha}\frac{f'}{f} + \frac{A-\gamma}{\alpha}, \tag{7.79}$$

and, therefore, (7.78) implies that $f(x)$ must satisfy the identity

$$[(\alpha + 2\beta)\sigma - c_1]\frac{\beta+2\alpha}{\alpha}\left(\frac{f'}{f}\right)^2 + (\varepsilon\sigma - c_0)\frac{f'}{f}\\ - [(\alpha + 2\beta)\sigma - c_1]\left[-\frac{\varepsilon}{\alpha}\left(\frac{f'}{f}\right) + \frac{A-\gamma}{\alpha}\right] \equiv \sigma A. \tag{7.80}$$

Assume that

$$\frac{f'}{f} \not\equiv \text{constant} \tag{7.81}$$

(exponential $f(x) = e^{\mu x}$ will be discussed last). Then, (7.80) is valid iff

$$\begin{aligned} &[(\alpha+2\beta)\sigma - c_1](\beta+2\alpha) = 0, \\ &\varepsilon\sigma - c_0 + \tfrac{\varepsilon}{\alpha}[(\alpha+2\beta)\sigma - c_1] = 0, \\ &[(\alpha+2\beta)\sigma - c_1]\tfrac{\gamma - A}{\alpha} = \sigma A. \end{aligned} \tag{7.82}$$

We may suppose that $(\alpha + 2\beta)\sigma - c_1 \neq 0$. (If not, this yields $c_1 = c_0 = \delta = 0$ and $g(x) \equiv 0$, and hence, $u(x,t) = f(x)a(t)$ is a simple solution in separate variables.) Hence, from the first condition in (7.82), we obtain (7.64), while the rest of conditions are equivalent to the last two hypotheses in (7.64). Here it is assumed that $2\alpha\sigma + c_1 \neq 0$; see comment below.

Finally, we need to determine the functions $f(x)$ and $g(x)$. Equation (7.70) with $\beta = -2\alpha$ is equivalent to (cf. (7.79))

$$\left(\tfrac{f'}{f}\right)' = \left(\tfrac{f'}{f}\right)^2 - \tfrac{\varepsilon}{\alpha}\tfrac{f'}{f} + \tfrac{A-\gamma}{\alpha},$$

that can easily be integrated. This yields the functions $f(x)$ in (7.66), (7.68), and (7.69), and $g(x)$ follows from (7.77). This completes the proof. □

Concerning exponential functions, if (7.81) is not valid and

$$f(x) = e^{\mu x}, \quad \text{with a } \mu \neq 0,$$

it follows from (7.75) and (7.77) that there exists the solution

$$u(x,t) = \tfrac{\sigma}{\mu} + D_1 e^{\mu x} a(t),$$

provided that μ satisfies the following two equations (cf. (7.80) and (7.70)):

$$\begin{aligned} &[(\alpha+2\beta)\sigma - c_1]\mu^2 + (\varepsilon\sigma - c_0)\mu - \sigma A = 0, \\ &(\alpha+\beta)\mu^2 + \varepsilon\mu + \gamma - A = 0. \end{aligned}$$

Here α, β, and A are arbitrary constants.

If $2\alpha\sigma + c_1 = 0$ $(c_1 \neq 0)$, it follows from (7.82) and (7.62) that explicit solutions exist in the case where

$$\beta = -2\alpha \neq 0, \quad \gamma = c_0 = \delta = 0 \quad \left(\sigma = \tfrac{c_1}{\beta}\right).$$

Here $A \neq 0$ is arbitrary. It follows from (7.70) that $Y = \frac{1}{f}$ solves a linear second-order ODE, $\alpha Y'' + \varepsilon Y' + AY = 0$, and (7.77) yields

$$g = \tfrac{c_1}{2A}\tfrac{f'}{f} + \tfrac{\varepsilon c_1}{\beta A}.$$

2. Case A = 0. Solutions $u(x,t) \in M_0$ with $A = 0$ have another form.

Theorem 7.16 *Assume that (7.53) with $A = B = 0$ holds. Then, there exists the nontrivial solution (7.65) in the following cases:*

$$\text{(i)}\ \alpha + 2\beta \neq 0,\ \beta \neq 0,\ c_1 \neq 0,\ \delta = \tfrac{(\alpha+\beta)c_1^2}{(\alpha+2\beta)^2},\ c_0 = \tfrac{\varepsilon c_1}{\alpha+2\beta}, \tag{7.83}$$

with functions $Y = \ln|f|$ *and* g *given by*

$$\begin{aligned} &\alpha Y'' + (\alpha + \beta)(Y')^2 + \varepsilon Y' + \gamma = 0, \\ &g(x) = -\sigma f(x) \int \tfrac{\mathrm{d}x}{f(x)} + D f(x); \end{aligned} \tag{7.84}$$

$$\text{(ii)}\;\; \alpha = -2\beta,\;\; \beta \neq 0,\;\; c_0 = \varepsilon\sigma,\;\; c_1 = 0,\;\; \delta \neq 0,$$

$Y = \ln|f|$ *solves the ODE* $\alpha Y'' + \frac{\alpha}{2}(Y')^2 + \varepsilon Y' + \gamma = 0$, *and* g *is defined by* (7.84);

$$\text{(iii)}\;\; \beta = 0,\;\; \alpha \neq 0,\;\; c_1^2 = \alpha\delta \neq 0,\;\; c_0 = \tfrac{\varepsilon\delta}{c_1},$$

and $f(x)$ *satisfies the linear ODE* $\alpha f'' + \varepsilon f' + \gamma f = 0$, g *being given by* (7.84).

Proof. Setting $A = 0$ in (7.77), and assuming that (7.81) holds, we find that a nontrivial solution exists iff $c_1 = (\alpha + 2\beta)\sigma$ and $c_0 = \varepsilon\sigma$, where σ is given by (7.62). A simple analysis of this algebraic system for the parameters and (7.62) leads to the above conclusions. ODEs for $f(x)$ follow from (7.70) with $A = 0$, while (7.84) is equivalent to (7.75). □

Let us now present a more detailed analysis of such solutions of particular PDEs.

Example 7.17 (**Quadratic parabolic PDE**) We begin with a parabolic model and first use Theorem 7.15. In view of (7.64), without loss of generality, set $\alpha = 1$ and $\beta = -2$. The equation (7.50), (7.61) with $P(D_t) = 0$ reads

$$\boxed{u u_t = u u_{xx} - 2(u_x)^2 + \gamma u^2 + \varepsilon u u_x + c_0 u + c_1 u_x + \delta.} \tag{7.85}$$

Transformation $v = \frac{1}{u}$ gives a semilinear heat equation from combustion theory,

$$v_t = v_{xx} + (\varepsilon + c_1 v)v_x - (\gamma v + c_0 v^2 + \delta v^3). \tag{7.86}$$

It follows from (7.53) that $a(t) \not\equiv 0$ satisfies

$$a' = Aa \quad \Longrightarrow \quad a(t) = \mathrm{e}^{At} \quad (A \neq 0).$$

Then $T[a, 1] \equiv a' = Aa$, and, therefore, (7.63) is valid. It follows from Theorem 7.15 that, under the last two hypotheses in (7.64), equation (7.85) possesses solutions

$$u(x, t) = g(x) + f(x)\mathrm{e}^{At}, \tag{7.87}$$

where the coefficients f and g are given in (7.66), (7.68), or (7.69). These solutions coincide with soliton-type solutions of a similar PDE constructed by Kawahara and Tanaka [329] and Carriello and Tabor [100] by Baker–Hirota-type techniques and related Penlevé-type analysis.

Example 7.18 (**Hyperbolic equation**) Consider now a quadratic hyperbolic PDE,

$$\boxed{u u_{tt} = u u_{xx} - 2(u_x)^2 + \gamma u^2 + \varepsilon u u_x + c_0 u + c_1 u_x + \delta.} \tag{7.88}$$

Setting $v = \frac{1}{u}$ yields

$$v_{tt} - \tfrac{2}{v}(v_t)^2 = v_{xx} + (\varepsilon + c_1 v)v_x - (\gamma v + c_0 v^2 + \delta v^3).$$

It follows from (7.53) that $a(t) \not\equiv$ constant satisfies

$$a'' = Aa. \tag{7.89}$$

Since, in this case, $T[a, 1] = a''$, we have by (7.89) that (7.63) holds. By Theorem 7.15, the type of the explicit solution (7.65) depends on the sign of the parameter A given in (7.64):

$$A = \tfrac{\gamma(3\sigma + c_1)}{2\sigma + c_1}, \quad \text{so that}$$

$$u(x,t) = g(x) + f(x)\big(D_3 \mathrm{e}^{\sqrt{A}t} + D_4 \mathrm{e}^{-\sqrt{A}t}\big) \;\; \text{for } A > 0,$$

$$u(x,t) = g(x) + f(x)\big[D_3 \sin\big(|A|^{\frac{1}{2}} t\big) + D_4 \cos\big(|A|^{\frac{1}{2}} t\big)\big] \;\; \text{for } A < 0.$$

If $A = 0$, by Theorem 7.16, we study the general PDE

$$\boxed{u u_{tt} = \alpha u u_{xx} + \beta (u_x)^2 + \gamma u^2 + \varepsilon u u_x + c_0 u + c_1 u_x + \delta.}$$

It follows from (7.89) with $A = 0$ that $a(t) = t$, and hence,

$$u(x,t) = g(x) + f(x)\, t \in M_0.$$

Then, in the case of (7.83), the ODE for f in (7.84) can easily be integrated. For instance, in the simplest case $\alpha + \beta = 0$ (where $\delta = 0$), we find

$$f(x) = \exp\big\{D_1 \mathrm{e}^{-\varepsilon x/\alpha} - \tfrac{\gamma}{\varepsilon} x + D_2\big\}.$$

For arbitrary D_1 and D_2, the integral in (7.84) cannot be calculated explicitly.

Example 7.19 Consider the operator

$$T[u] = u(u_{tt} + 2\omega u_t),$$

where $\omega \neq 0$. Then, the ODE for $a(t)$ (see (7.53)) has the form $a'' + 2\omega a' = Aa$, and, in particular,

$$a(t) = t\,\mathrm{e}^{-\omega t}, \quad \text{if } A = -\omega^2.$$

One can see that (7.63) is valid. It then follows from Theorem 7.15 that, under hypotheses (7.64) with $A = -\omega^2$, there exists a solution of the problem (7.59),

$$u(x,t) = g(x) + f(x)\, t\, \mathrm{e}^{-\omega t},$$

where $g(x)$ and $f(x)$ are given by (7.66), (7.68), or (7.69).

7.4.3 On quasilinear hyperbolic PDEs for $A \neq B$

In what follows, unlike Theorems 7.15 and 7.16, assume that

$$A \neq B. \tag{7.90}$$

This condition is valid for the following hyperbolic PDE with the left-hand side of the KFG equation (6.27):

$$\boxed{u_t u_{tt} = \alpha u u_{xx} + \beta (u_x)^2 + \gamma u^2 + \varepsilon u u_x + c_0 u + c_1 u_x + \delta,} \tag{7.91}$$

which will be used in future illustrations of the main results. Then $B = 0$. Suppose that $\alpha \neq 0$.

As in Theorem 7.15, consider a general equation. It follows from (7.54), (7.60), and (7.90) that functions g and f satisfy (7.70), (7.71) and

$$\alpha(gf'' + fg'') + 2\beta g' f' + \varepsilon (gf)' + c_0 f + c_1 f' - (B - 2\gamma) gf = 0. \tag{7.92}$$

The only difference with (7.72) is the last term, where $B \neq A$. In this case, using the same technique as in the proof of Theorem 7.15, we derive that (cf. (7.71))

$$\beta(gf' - fg')^2 = c_1 f(gf' - fg') - [\delta - (A - B)g^2]f^2. \tag{7.93}$$

Hence, instead of (7.75), for $\beta \neq 0$,

$$g' = \tfrac{f'}{f} g - \tfrac{1}{2\beta}(c_1 + \hat{R}), \text{ where } \hat{R} = \pm\sqrt{c_1^2 - 4\beta\delta + 4\beta(A - B)g^2}. \tag{7.94}$$

Let us next derive a condition of solvability of the equation on M_A. Substituting g' and g'' from (7.94) into (7.71) and using (7.70) yields

$$\begin{aligned} \tfrac{f'}{f} \equiv z = \big\{&(A - B)\alpha c_1 g + [A(\alpha + 2\beta) - B(\alpha + \beta)]g\hat{R} \\ &+ \big(c_0\beta - \tfrac{\varepsilon c_1}{2}\big)\hat{R} - \tfrac{\varepsilon}{2}\hat{R}^2\big\}\big[\tfrac{\alpha c_1}{2}\hat{R} + 2(A - B)\alpha\beta g^2 + \tfrac{\alpha+2\beta}{2}\hat{R}^2\big]^{-1}, \end{aligned} \tag{7.95}$$

where the expression in the last square bracket is assumed to be non-zero. Let

$$\big(\tfrac{f'}{f}\big)(x) \not\equiv \text{constant and } g(x) \not\equiv \text{constant}. \tag{7.96}$$

It follows from (7.70) that

$$z' \equiv \big(\tfrac{f'}{f}\big)' = \tfrac{1}{\alpha}\,[A - \gamma - (\alpha + \beta)z^2 - \varepsilon z]. \tag{7.97}$$

By substituting into (7.94) $g' = g_z z'$ with g_z and z from (7.95), we arrive at the following solvability criterion:

$$zg - \tfrac{1}{\alpha}[A - \gamma - (\alpha + \beta)z^2 - \varepsilon z](z_g')^{-1} \equiv \tfrac{c_1}{2\beta} + \tfrac{1}{2\beta}\,\hat{R}, \tag{7.98}$$

which, together with (7.95), must be the identity for all suitable $F \in I\!R$.

1. Existence for the rational case. Let us begin with a simple case where the irrational function in (7.94) does not appear, i.e.,

$$c_1^2 = 4\beta\delta \text{ and } \hat{R} = \sigma g, \ \sigma = \pm 2\sqrt{\beta(A - B)}, \ \beta \neq 0. \tag{7.99}$$

Then (7.95) is equivalent to

$$g = \big[\tfrac{\alpha c_1}{2\beta} z + n_2\big]\big[n_1 - \tfrac{\sigma(\alpha+\beta)}{\beta} z\big]^{-1}, \tag{7.100}$$

where $n_1 = A + \frac{\sigma^2(\alpha+\beta)}{4\beta^2} - \frac{\varepsilon\sigma}{2\beta}$ and $n_2 = \frac{\varepsilon c_1}{2\beta} - \frac{\alpha\sigma c_1}{4\beta^2} - c_0$. Plugging (7.100) into (7.98) must yield the identity. One can verify that it is true iff a certain cubic polynomial is trivial,

$$\big[\tfrac{\alpha c_1 \sigma(\alpha+\beta)}{2\beta^2}\big]z^3 + \ldots \equiv 0 \quad \text{for all } z \in I\!R.$$

Finally, we obtain from (7.95) that the identity holds, if

$$\alpha + \beta = c_0 = c_1 = \delta = 0 \text{ and } \varepsilon^2 = \tfrac{\beta A^2}{A-B}. \tag{7.101}$$

These are the existence conditions of a nontrivial solution $u(x, t)$ on M_A ((7.101) corresponds to the case for which both square brackets in (7.100) vanish).

Proposition 7.20 *Under hypotheses* (7.90) *and* (7.101), *system* (7.70), (7.71), *and* (7.92) *admits a nontrivial solution* $\{g, f\}$.

Example 7.21 If (7.101) holds, equation (7.91) with $\alpha = -\beta = 1$ has the form

$$\boxed{u_t u_{tt} = u u_{xx} - (u_x)^2 + \gamma u^2 + \varepsilon u u_x.} \tag{7.102}$$

Transformation $v = \ln |u|$ yields

$$v_t\big[v_{tt} + (v_t)^2\big] = v_{xx} + \gamma + \varepsilon v_x.$$

By Proposition 7.20, there exist solutions (7.65), where $a(t)$ satisfies

$$T[a] \equiv a'a'' = Aa^2, \quad \text{with} \quad A = -\varepsilon^2 \neq 0. \tag{7.103}$$

Recall that $B = 0$. In particular, the simplest function is $a(t) = \mathrm{e}^{-\varepsilon^{2/3}t}$. It then follows from (7.70) that $Y = \ln |f|$, and $g(x)$ are given by

$$Y'' + \varepsilon Y' + (\gamma + \varepsilon^2) = 0 \ \text{ and } \ g' = Y' \pm \varepsilon; \tag{7.104}$$

see (7.94) and (7.99). This yields the following solution:

$$u(x,t) = D \exp\{-\tfrac{\varepsilon^2+\gamma}{\varepsilon}\, x + D_1 \mathrm{e}^{-\varepsilon x}\}\big[D_2 \mathrm{e}^{\varepsilon x} + a(t)\big]. \tag{7.105}$$

2. Linear Case: $\beta = 0$. From (7.94), we get two subcases:

2.1. Nonexistence with $c_1 \neq 0$. Then, instead of (7.94),

$$g' = zg + \tfrac{1}{c_1}\,[(A-B)g^2 - \delta], \quad \text{with } z = \tfrac{f'}{f}.$$

Therefore, we derive similarly that (cf. (7.95))

$$\begin{aligned} z = &-\big[2(A-B)^2\alpha g^3 + \varepsilon(A-B)c_1 g^2 + (2A-B)c_1^2 g \\ &-2(A-B)\alpha\delta g + c_0 c_1^2 - \varepsilon\delta c_1\big]\big[c_1\big(3(A-B)\alpha g^2 - \delta\alpha + c_1^2\big)\big]^{-1}. \end{aligned} \tag{7.106}$$

Finally, we obtain the following solvability criterion (cf. (7.98)):

$$zg - \tfrac{1}{\alpha}\left(A - \gamma - \alpha z^2 - \varepsilon z\right)(z'_g)^{-1} \equiv -\tfrac{(A-B)}{c_1}\, g^2 + \tfrac{\delta}{c_1} \ \text{ for } g \in I\!R. \tag{7.107}$$

Substituting $z(g)$ and $z'_g(g)$ from (7.106), we infer that (7.107) is the identity if a certain eighth-order polynomial satisfies

$$\alpha^3(A-B)^5 g^8 + \ldots \equiv 0 \quad \text{for all } g \in I\!R.$$

Since $\alpha \neq 0$ by the assumption, a nontrivial solution exists in the case of $A = B$ only, which has been studied before.

2.2. Existence. Suppose now that

$$\beta = c_1 = 0. \tag{7.108}$$

Then (7.93) implies

$$g^2 \equiv g_0^2 = \tfrac{\delta}{A-B} > 0. \tag{7.109}$$

In this case, functions $\{g_0, f\}$ solve the following system:

$$\alpha f'' + \varepsilon f' + (\gamma - A)f = 0, \quad g_0^2 = \tfrac{\delta}{A-B}, \quad \gamma g_0^2 + c_0 g_0 + \delta = 0. \tag{7.110}$$

It is easily seen that such a constant solution g_0 of (7.110) exists if

$$c_0^2 \geq 4\gamma\delta \ \text{ and } \ \tfrac{\gamma\delta}{A-B} \pm c_0\sqrt{\tfrac{\delta}{A-B}} + \delta = 0. \tag{7.111}$$

Proposition 7.22 *Under hypotheses* (7.90), (7.108), (7.109), *and* (7.111), *system* (7.70), (7.71), *and* (7.92) *has a nontrivial solution* $\{g_0, f\}$.

Example 7.23 For parameters (7.108), the equation (7.91) takes the form

$$\boxed{u_t u_{tt} = \alpha u u_{xx} + \gamma u^2 + \varepsilon u u_x + c_0 u + \delta.}$$

In view of Proposition 7.22, it possesses the solution

$$u(x,t) = g_0 + f(x)a(t),$$

where $a(t)$ solves (7.103), the constant A is given by (7.111) (here $B = 0$), and $|g_0| = \sqrt{\frac{\delta}{A}}$. From (7.110), it follows that if $G \equiv \varepsilon^2 - 4\alpha(\gamma - A) > 0$, then $f(x) = D_1 e^{\mu_1 x} + D_2 e^{\mu_2 x}$, where μ_1 and μ_2 are different roots of $\alpha\mu^2 + \varepsilon\mu + (\gamma - A) = 0$. Hence,

$$f(x) = (D_1 x + D_2)e^{-\varepsilon x/2\alpha}, \quad \text{if } G = 0,$$
$$f(x) = D_1 e^{\nu_1 x}\cos[\nu_2(x + D_2)], \quad \text{if } G < 0,$$

where $\nu_1 = -\frac{\varepsilon}{2\alpha}$ and $\nu_2 = \frac{1}{2\alpha}\sqrt{4\alpha(\gamma - A) - \varepsilon^2}$.

3. General case. Let

$$\beta \neq 0 \quad \text{and} \quad c_1^2 \neq 4\beta\delta, \tag{7.112}$$

i.e., (7.94) contains an irrational function. Substituting (7.95) into (7.98) and making simplifications yields a linear combination of functions

$$1,\ g,\ g^2,\ \ldots,\ g^6,\ \hat{R},\ g\hat{R},\ \ldots,\ g^6\hat{R}, \tag{7.113}$$

which must be identically zero. This gives fourteen algebraic equations for the parameters. In particular, from this system the following result is obtained.

3.1. Existence. Two existence cases are derived.

Proposition 7.24 *Let* (7.90) *and* (7.112) *hold. Then, system* (7.70), (7.71), *and* (7.92) *possesses a nontrivial solution* $\{g, f\}$ *with constant* $f(x) \equiv f_0$ *if*

$$\beta = \tfrac{\alpha(\gamma - B)}{B - 2\gamma} \ (B \neq 2\gamma), \quad c_0 = c_1 = \varepsilon = 0, \quad \delta \neq 0, \quad A = \gamma \neq 0, \tag{7.114}$$

where $\gamma\delta > 0$ *if* $\beta < 0$.

Note that (7.114) is the sufficient condition for identity (7.98) to be valid with $f = f_0$. Using (7.95) gives $z = 0$, and plugging this into (7.97) yields $A = \gamma$. Other conditions in (7.114) follow from (7.95) with $z = 0$, if we look for a nontrivial function $g(x) \not\equiv$ constant, such that all rational and irrational terms on the right-hand side of (7.95) are linearly independent. This yields $c_0 = c_1 = \varepsilon = 0$ and $A(\alpha + 2\beta) - B(\alpha + \beta) = 0$. A more general solution $\{g, f_0\}$ is given in the example below.

Example 7.25 Under hypotheses (7.114) with $B = 0$ (where $\alpha = -2\beta$), (7.91) is

$$\boxed{u_t u_{tt} = -2\beta u u_{xx} + \beta(u_x)^2 + \gamma u^2 + \delta.} \tag{7.115}$$

Then, system (7.70), (7.71), (7.92) reduces to

$$\begin{cases} -2ff'' + (f^2)' = 0, \\ -2\beta g g'' + \beta(g^2)' + \gamma g^2 + \delta = 0, \\ -2\beta(gf'' + g''f) + 2\beta g'f' + 2\gamma gf = 0. \end{cases} \tag{7.116}$$

As has been shown above, this system is compatible iff $f(x) \equiv f_0$. In this case, (7.115) admits the solution $u(x,t) = g(x) + f_0 a(t)$, where $a'a'' = \gamma a^2$. It then follows from (7.116) with $f = f_0$ that $g(x)$ has the form ($D \neq 0$ is arbitrary):

$$g(x) = \sqrt{\tfrac{\delta}{\gamma}} \sin\big(\sqrt{-\tfrac{\gamma}{\beta}}\, x + D\big), \;\; \gamma\delta > 0, \;\; \gamma\beta < 0,$$
$$g(x) = D \exp\big\{\sqrt{\tfrac{\gamma}{\beta}}\, x\big\} + \tfrac{\delta}{4\gamma D} \exp\big\{-\sqrt{\tfrac{\gamma}{\beta}}\, x\big\}, \;\; \gamma\beta > 0.$$

Proposition 7.26 *Let* (7.90) *and* (7.112) *hold. Then, system* (7.70), (7.71), (7.92) *admits nontrivial solution* $\{g_0, \mathrm{e}^{\mu x}\}$ *with* $g_0 \neq 0,\;\; \mu \neq 0$, *provided that*

$$(\alpha+\beta)\mu^2 + \varepsilon\mu - (A-\gamma) = 0, \;\; \gamma g_0^2 + c_0 g_0 + \delta = 0,$$
$$[\alpha\mu^2 + \varepsilon\mu - (B - 2\gamma)]g_0 + c_0 + c_1\mu = 0.$$

Assume that (7.96) is not valid and

$$g(x) = \mathrm{e}^{mx}, \quad f(x) = \mathrm{e}^{nx}, \quad \text{where } m \neq n.$$

It follows from (7.70), (7.71), and (7.93) that this solution exists if $c_0 = c_1 = \delta = 0$, and

$$(\alpha+\beta)n^2 + \varepsilon n + \gamma - A = 0,$$
$$(\alpha+\beta)m^2 + \varepsilon m + \gamma = 0,$$
$$\beta(m-n)^2 = A - B.$$

For equation (7.102), from (7.53), we have $B = 0$, and the above solution is included into (7.105).

3.2. Nonexistence. We consider the last case where (7.96) holds. Then, as above, (7.98) is equivalent to the identity on the linear span of the functions (7.113). The final result (obtained via computer supported symbolic manipulations) can be stated as follows: under assumptions (7.90), (7.96) and (7.112), (7.98) is not the identity. This means nonexistence of solutions $\{g, f\}$ satisfying (7.96).

7.4.4 On quadratic operators with linear properties

This linear case is connected with the system 2) in Section 7.3.2. Previously, we have studied the case where the operator T in (7.45) satisfied

$$T[u] \in W_3 = \mathcal{L}\{1, a, a^2\} \quad \text{for all } \; u \in W_2 = \mathcal{L}\{1, a\}.$$

Let us now show that if $a(t)$ is such that

$$T[u] \in W_2 \quad \text{for all } \; u \in W_2, \tag{7.117}$$

then, in some particular cases, there exist explicit solutions that differ from those given above. Under hypothesis (7.117), the equation (7.50) on an invariant set has a different form than (7.60). In particular, it follows from (7.55) that (7.117) is valid, provided that (cf. (7.53))

$$T[a] = \tilde{A}a, \ \ T[a, 1] = \tilde{B}, \tag{7.118}$$

where $\tilde{A} \neq 0$ and $\tilde{B}$ are some constants. In this case, the first expression in (7.55) reads $T[g + fa] = \tilde{A}f^2a + \tilde{B}gf$. Hence, instead of (7.56),

$$\mathbf{A}[g + fa] = \tilde{B}gf - F[g] + (\tilde{A}f^2 - F[g, f])a - F[f]a^2.$$

Therefore, there exists the set M_0, given by (7.54), and the equation (7.59) has the form (cf. (7.60))

$$\begin{cases} F[g] - \tilde{B}gf = -Q[g] - p, \\ F[g, f] - \tilde{A}f^2 = -Q[f]. \end{cases} \tag{7.119}$$

Observe that (7.117) holds if (cf. (7.118))

$$T[a] = A_1 a + A_0 \quad \text{and} \quad T[a, 1] = B_1 a + B_0. \tag{7.120}$$

In fact, below the case $A_0 = B_1 = 0$ is considered.

Theorem 7.27 *Let* (7.61) *be valid. Let* (7.118) *with* $\tilde{A} = \tilde{B} \neq 0$ *hold. Then the problem* (7.59) *admits the following solutions:*

(i) *for* $2\alpha + \beta = \gamma = 0, \ \varepsilon\sigma - c_0 \neq 0, \ \alpha c_0 = \varepsilon(\beta\sigma - c_1)$, *with* σ *in* (7.62),

$$u(x,t) = \tfrac{D_1 e^{-x/n_1}}{1 - n_2 D_1 e^{-x/n_1}} \big[D_2 - \tfrac{\sigma n_1}{D_1} e^{x/n_1} + \sigma n_2 x + a(t)\big], \tag{7.121}$$

where $n_1 = \frac{\sigma(\alpha + 2\beta) - c_1}{\varepsilon\sigma - c_0}$ *and* $n_2 = \frac{\tilde{A}}{\varepsilon\sigma - c_0}$;

(ii) *for* $2\alpha + \beta = \varepsilon = \gamma = c_0 = 0$,

$$u(x,t) = \tfrac{1}{nx + D_1}\big[D_2 - \sigma\big(\tfrac{1}{2}nx^2 + D_1 x\big) + a(t)\big], \ \ n = \tfrac{\tilde{A}}{\sigma(\alpha + 2\beta) - c_1}. \tag{7.122}$$

Proof. By using the same technique as in the proof of Theorem 7.15, we obtain (7.75) and, instead of (7.77),

$$[(\alpha + 2\beta)\sigma - c_1]\tfrac{f'}{f} + (\varepsilon\sigma - c_0) + \tilde{A}f = 0. \tag{7.123}$$

The above solution follows from the solvability of (7.123), together with the equation of the set M_0, $F[f] = 0$ (see (7.54)), yielding

$$\alpha f f'' + \beta(f')^2 + \varepsilon f f' + \gamma f^2 = 0. \tag{7.124}$$

Having a solution $f(x)$ of (7.123), (7.124), the function $g(x)$ is calculated from (7.75). □

Example 7.28 **(Parabolic PDE)** Consider equation (7.85). Under the hypotheses of Theorem 7.27, there exist explicit solutions (7.121) or (7.122) with $a(t) = \tilde{A}t$. These solutions are different from those given in Example 7.17.

Example 7.29 **(Hyperbolic PDE)** Under the same hypotheses, equation (7.88) has explicit solutions (7.121) or (7.122) with $a(t) = \frac{1}{2}\tilde{A}t^2 + Dt$, where D is arbitrary.

In addition to the case of (7.117), consider $a(t)$ such that

$$T[u] \in W_1 = \mathcal{L}\{1\} \quad \text{on} \ W_2.$$

It follows from (7.55) that this is true if

$$T[a] = \bar{A}, \quad T[a,1] = \bar{B},$$

where $\bar{A}$ and $\bar{B}$ are constants. For instance, for the operator $T[u] = u_t u_{tt}$, there exists $a(t) = \frac{2}{3}(2\bar{A})^{\frac{1}{2}} t^{\frac{3}{2}}$, provided that $\bar{A} > 0$ and $\bar{B} = 0$. The corresponding equation on M_0 has the form (cf. (7.119))

$$\begin{cases} F[g] - \bar{A} f^2 - \bar{B} g f = -Q[g] - p, \\ F[g,f] = -Q[f]. \end{cases}$$

Then, as in the proof of Theorem 7.15, we derive the equation (cf. (7.74) and (7.93))

$$\beta (gf' - fg')^2 = c_1 f (gf' - fg') - f^2(\delta - \bar{B} g f - \bar{A} f^2). \tag{7.125}$$

In general, both existence and nonexistence are OPEN PROBLEMS. For $\beta = c_1 = 0$, where (7.125) takes the most simple form, a nontrivial solution does not exist.

Example 7.30 (On higher-order PDEs: dispersive Boussinesq equation) Consider the fourth-order hyperbolic $B(1, \frac{1}{2})$ equation [496]

$$v_{tt} - v_{xx} - \left(\sqrt{v}\right)_{xx} - v_{xxxx} = 0$$

that is formulated for nonnegative solutions. In general, such equations may admit solutions of changing sign; see Example 5.12. Setting $v = u^2$ yields

$$\begin{aligned} T[u] \equiv u u_{tt} + (u_t)^2 = & -u u_{xxxx} - 4 u_x u_{xxx} \\ & - 3(u_{xx})^2 + u u_{xx} + (u_x)^2 + u_{xx}. \end{aligned} \tag{7.126}$$

According to the GSV in Section 7.3.1, looking for solutions $u = g + f a(t)$ yields a standard system (I) for $a(t)$ that is easily solved, giving t or e^t as usual. In contrast, system (II) for f, g becomes essentially more involved than for the above second-order cases. A complete consistency analysis is not straightforward. Moreover, for such higher-order PDEs (7.126), looking for solutions on a trigonometric subspace,

$$u(x,t) = C_1(t) + C_2(t)\cos(\lambda x) + C_3(t)\sin(\lambda x) \quad (\lambda \in I\!R)$$

yields an overdetermined DS that allows only traveling wave compactons (similar to those in [581]), with no exact dynamics around, as was shown in several examples in Section 7.1. In other words, for higher-order operators $F[u]$ and multi-term $\frac{\partial}{\partial t}$-dependent operator $T[u]$, the GSV becomes complicated and can provide us with new exact solutions for some exceptional PDEs only.

7.4.5 On some other extensions

1. Partially invariant modules $\mathbf{W_3}$. In general, the operator (7.47) admits invariant sets on the 3D subspace $W_3 = \mathcal{L}\{1, a(t), b(t)\}$, where the functions 1, $a(t)$, and $b(t)$ are linearly independent. If $b^2(t) = a(t)$, for instance,

$$a(t) = e^{\lambda t} \quad \text{and} \quad b(t) = e^{\frac{\lambda t}{2}} \quad \text{for some } \lambda \neq 0 \tag{7.127}$$

(this is reminiscent of soliton-type solutions), we find a family of sets M on W_3 such that, for any $u = g(x) + f(x)a(t) + h(x)b(t) \in M$, $A[g + fa + hb] \in W_3$. In this case,

$$T[F + fa + hb] = f^2 T[a] + h^2 T[b] + gfT[a, 1] \\ + ghT[b, 1] + fhT[a, b], \; F[g + fa + hb] = F[g] + F[g, f]a + F[g, h]b \\ + F[f]a^2 + F[h]b^2 + F[f, h]ab.$$

Using (7.127) yields the following result.

Lemma 7.31 *Under hypothesis* (7.127), *there exists the set on* W_3

$$M_A = \{u = g + fa + hb : \; F[f] = A_1 f^2, \; F[f, h] = A_2 fh\},$$

where $A_1 = \sum a_{i,j}\, \lambda^{i+j}$ *and* $A_2 = \sum a_{i,j}(2^{-i} + 2^{-j})\lambda^{i+j}$.

Equation (7.50) can easily be stated on $M_A \subset W_3$ with $A = (A_1, A_2)$. The general problem of the existence and nonexistence of solutions on M_A is OPEN. Some examples given above can be treated as the existence result on the set $M_A \subset W_3$. For instance, Example 7.29 shows that the PDE under consideration admits the explicit solution which formally satisfies $u(x,t) \in M_A \subset W_3 = \mathcal{L}\{1, t, t^2\}$. By translation in time $t \to t+1$ in Example 7.19, we obtain a solution $u(x,t) \in M_A \subset \tilde{W}_3 = \mathcal{L}\{1, \mathrm{e}^{-\omega t}, t\, \mathrm{e}^{-\omega t}\}$.

2. N-dimensional operators. Such sets M can be constructed for elliptic operators in the N-dimensional space, e.g., for

$$F[u] = \alpha u \Delta u + \beta |\nabla u|^2 + \gamma u^2 + \varepsilon u(\nabla u \cdot d) \quad (d \in I\!R^N),$$

as well as for others from Chapter 6. The problem of the existence and nonexistence of solutions on M becomes more difficult.

3. Operators with cubic nonlinearities. In general, a similar invariant analysis can be done for operators of higher algebraic homogenuity. For instance, as in Section 1.5, we can consider cubic operators

$$\mathbf{A}[u] = T[u] - F[u] \equiv \sum b_{i,j,k}\, D_t^i u\, D_t^j u D_t^k u - \sum a_{i,j,k}\, D_x^i u\, D_x^j u\, D_x^k u$$

(or use a combination of quadratic and cubic operators). Then, e.g., for exponential functions $a(t) = \mathrm{e}^{\lambda t}$ with a $\lambda \neq 0$, $\mathbf{A}[u] \in \mathcal{L}\{1, a, a^2, a^3\}$ for $u = g + fa \in W_2$. Invariant sets $M \subset W_2$ are defined so that $\mathbf{A}[M] \subseteq \mathcal{L}\{1, a\}$. This yields two equations for coefficients $\{g(x), f(x)\}$, which determine M with a harder consistency analysis.

7.5 Evolutionary invariant sets for higher-order equations

In this section, we use the notion of *evolutionary* invariant sets which are associated with a prescribed nonlinear PDE. Using a particular class of such equations, the actual relation between evolutionary invariance and the standard concept of partial invariance (invariant sets on linear modules) is shown. More general and sophisticated classes of such problems will be introduced in Chapter 8.

Consider a 1D ((1+1)-dimensional) mth-order evolution PDE

$$u_t = F[u] \equiv F(x, u, u_1, ..., u_m) \quad \text{in} \quad Q = I\!R \times [0, 1], \tag{7.128}$$

where $u_i = D_x^i u$. Suppose that $F \in C^\infty(I\!R^{m+2})$, and solutions of (7.128) are assumed to be smooth, $u \in C^\infty(Q)$.

7.5.1 Scaling order, index and evolutionary invariant sets

Simple exact solutions presented here admit other treatments to be discussed later on. Currently, we apply a version explaining connection with the most well-known group-invariant solutions that are admitted by many scaling invariant PDEs.

1. Group of scalings, order, and index. Consider a simple Lie group of scaling transformations

$$x^* = e^{\varepsilon} x, \quad t^* = e^{\mu\varepsilon} t, \tag{7.129}$$

where $\mu \in I\!R \setminus \{0\}$ is a fixed constant, with the infinitesimal generator

$$X = x \tfrac{\partial}{\partial x} + \mu t \tfrac{\partial}{\partial t}. \tag{7.130}$$

Then, equation (7.128) is invariant under (7.129) if operator F satisfies the following homogenuity condition: for any $s > 0$,

$$F\big(sx, u, \tfrac{1}{s}u_1, ..., \tfrac{1}{s^m}u_m\big) \equiv \big(\tfrac{1}{s}\big)^{\mu} F(x, u, u_1, ..., u_m). \tag{7.131}$$

From the invariance equation $Xu = 0$, it follows that (7.128) admits self-similar solutions, depending on a single variable (the invariant of the group):

$$u(x,t) = \theta(\xi), \quad \xi = \tfrac{x}{t^{1/\mu}}. \tag{7.132}$$

On the set of invariant solutions (7.132), the PDE (7.128) reduces to the following ODE for $\theta(\xi)$:

$$F\big(\xi, \theta, \theta', ..., \theta^{(m)}\big) + \tfrac{1}{\mu}\theta'\xi = 0.$$

Then, $\mu = \mu(F)$ is said to be the *scaling order* of the operator F. For the operator with power-like nonlinearities in derivatives

$$F[u] = \phi(u)x^{\beta}\Pi_{j=1}^{m}(u_j)^{\alpha_j}, \tag{7.133}$$

where ϕ is a smooth function, the scaling order is $\mu(F) = \sum_{j=1}^{m} j\alpha_j - \beta$.

Consider a more general evolution PDE

$$u_t = F[u] \equiv \textstyle\sum_{i=1}^{I} F_i[u], \tag{7.134}$$

where each operator F_i is of the form (7.128) with the scaling orders $\mu_i \neq \mu_j$ for all $i \neq j$. We call $I = I(F)$ the *scaling index* of the operator F.

2. Evolution invariance and algebraic differentiation. If $I > 1$, (7.134) does not admit any Lie group (7.129). In this case, we begin our analysis by introducing the set of functions

$$M_0 = \big\{v \in C^{\infty}(I\!R): \; v_x = \tfrac{1}{x} H(v)\big\}, \tag{7.135}$$

where H is an unknown C^{∞}-function. As a standard practice, M_0 is said to be *evolutionary invariant* for PDE (7.134) if

$$u(\cdot,0) \in M_0 \quad \Longrightarrow \quad u(\cdot,t) \in M_0 \quad \text{for } \; t \in (0,1]. \tag{7.136}$$

The contact, first-order structure of the equality

$$u_x = \tfrac{1}{x} H(u) \tag{7.137}$$

(a differential constraint) includes the scaling invariant (7.132). This is seen by integration. The simple differential structure of M_0 makes it possible to introduce the following algebraic differentiation in the set, that reduces the PDE on M_0 to a system of ODEs.

Lemma 7.32 *Let $u \in M_0$. Then,*

$$u_k = \tfrac{1}{x^k}\, G_k[H], \quad k = 1, 2, \ldots, \tag{7.138}$$

where the ordinary differential operators G_k satisfy the recursion relation: for any $H(u) \in C^\infty$,

$$G_{k+1}[H] = H \tfrac{\mathrm{d}}{\mathrm{d}u}\, G_k[H] - kG_k[H] \;\; \text{for } k = 1, 2, \ldots, \quad G_1[H] = H.$$

Proof. Differentiating (7.138) yields

$$u_{k+1} = -\tfrac{k}{x^{k+1}}\, G_k[H] + \tfrac{1}{x^k}\, \tfrac{\mathrm{d}}{\mathrm{d}u}\, G_k[H] u_1 \equiv \tfrac{1}{x^{k+1}}\, G_{k+1}[H],$$

since $u_1 = \frac{1}{x}\, H(u)$ by the definition (7.135). □

We present below a few of the quasilinear polynomial differential operators G_k, which are sufficient to deal with PDEs up to the fifth order:

$$\begin{aligned} G_2[H] &= H(H' - 1), \\ G_3[H] &= HG_2' - 2G_2 = H[(HH')' - 3H' + 2], \\ G_4[H] &= HG_3' - 3G_3 = H[(H(HH')')' - 6(HH')' + 11H' - 6], \\ G_5[H] &= H[(H(H(HH')')')' - 10(H(HH')')' \\ &\quad + 35(HH')' - 50H' + 24]. \end{aligned} \tag{7.139}$$

4. Evolutionary invariant sets. Using the rule (7.138) and the scaling properties (7.131) with $\mu = \mu_i$ of operators F_i, we find that, on M_0,

$$\begin{aligned} F[u] &= \textstyle\sum_{i=1}^{I} F_i\big(x, u, \tfrac{1}{x} G_1, \ldots, \tfrac{1}{x^m} \mathbf{G}_m\big) \\ &\equiv \textstyle\sum_{i=1}^{I} x^{-\mu_i} F_i(1, u, G_1, \ldots, G_m) \equiv \sum_{i=1}^{I} x^{-\mu_i} HC_i[H], \end{aligned} \tag{7.140}$$

where, for convenience, we introduce the operators

$$C_i[H] = \tfrac{1}{H}\, F_i(1, u, G_1[H], \ldots, G_m[H]) \quad \text{for } i = 1, \ldots, I.$$

The main result is as follows:

Theorem 7.33 *The set M_0 is evolutionary invariant iff the function $H(u)$ satisfies the ODE system*

$$H \tfrac{\mathrm{d}}{\mathrm{d}u}\, C_i[H] = \mu_i C_i[H], \quad i = 1, \ldots, I, \tag{7.141}$$

and then solutions $u(x,t)$ on M_0 take the form

$$v \equiv \int_1^u \tfrac{\mathrm{d}z}{H(z)} = \ln x + D(t). \tag{7.142}$$

Here, $D(t)$ solves the ODE

$$D' = \textstyle\sum_{i=1}^{I} d_i \mathrm{e}^{\mu_i D}, \quad \text{where } d_i = C_i[H]\big|_{u=1}. \tag{7.143}$$

The ODE (7.143) is always integrated in quadratures and is the PDE (7.134) restricted to the set M_0.

Proof. In view of (7.142),

$$u_t = H(u)D'(t). \tag{7.144}$$

Substituting (7.144) into (7.134) and using (7.140) yields

$$D'(t) = \sum_{i=1}^{I} x^{-\mu_i} C_i[H(u)] \equiv R(x,t). \tag{7.145}$$

Therefore, $R(x,t)$ does not depend on x, which results in

$$R_x = \sum_{i=1}^{I} \{H \tfrac{\mathrm{d}}{\mathrm{d}u} C_i[H] - \mu_i C_i[H]\} x^{-(\mu_i+1)} \equiv 0.$$

Since functions $\{x^{-(\mu_i+1)}\}$ are linearly independent in $I\!R^I$, we obtain the system (7.141), meaning that

$$\tfrac{\mathrm{d}}{\mathrm{d}v} C_i[H] \equiv H \tfrac{\mathrm{d}}{\mathrm{d}u} C_i[H] = \mu_i C_i[H], \;\; \text{so} \;\; C_i[H] = d_i \mathrm{e}^{\mu_i v}. \tag{7.146}$$

Plugging the functions (7.146) with $v = \ln x + D$ given by (7.142) into (7.145) yields the ODE (7.143). □

5. Group of scalings for I = 1. In the case $I(F) = 1$, for which the group of scalings exists, one can solve (7.143) explicitly. Set $\mu_1 = \mu$ and $d_1 = d$, so

$$D' = d\mathrm{e}^{\mu D} \quad \Longrightarrow \quad D(t) = -\tfrac{1}{\mu} \ln t + \text{constant}.$$

Substituting this into (7.142) yields

$$\int_1^{u(x,t)} \tfrac{\mathrm{d}z}{H(z)} = \ln\big(\tfrac{x}{t^{1/\mu}}\big) + \text{constant},$$

i.e., $u(x,t)$ depends on the single scaling group invariant as in (7.132).

Remark 7.34 **(On extensions)** As an extension of the evolutionary invariant set (7.137), it makes sense to consider

$$u_x = \tfrac{1}{x} H(u) + G(u), \tag{7.147}$$

with two unknown functions H and G. The ODE conditions on all the coefficients (slightly harder than (7.138)) can be obtained. The main difficulty is that (7.147) cannot be integrated explicitly in general. A convenient integrable case (see Remarks) corresponds to the *Bernoulli equation* $v_x = \frac{1}{x} v + v^n$ for $v = H(u)$. Then again a single function H appears in (7.147). Are there other cases (an OPEN PROBLEM)?

7.5.2 Back to invariant subspaces: equivalent formulation

We now comment on other equivalent (and sometimes simpler) forms of such exact solutions. The first form is based on the representation (cf. (7.142))

$$u(x,t) = U(\xi), \quad \xi = \ln x + D(t), \tag{7.148}$$

for solutions of (7.134). Differentiating (7.148) yields, as in the previous case,

$$\begin{aligned} u_1 &= x^{-1} U'(\xi) \equiv x^{-1} P_1[U], \\ u_2 &= x^{-2}\{U''(\xi) - U'(\xi)\} \equiv x^{-1} P_2[U], \\ &\cdots \qquad \cdots \qquad \cdots \\ u_k &= x^{-k}\{(P_{k-1}[U])' - (k-1)P_{k-1}[U]\} \equiv x^{-k} P_k[U], \end{aligned}$$

where $P_k[U] = G_k[U']$ given in (7.139). Substituting into the PDE (7.134) yields

$$U'D' = \sum_{i=1}^{I} F_i(x, U, x^{-1}P_1[U], ..., x^{-m}P_m[U])$$
$$\equiv \sum x^{-\mu_i} F_i(1, U, P_1[U], ..., P_m[U]), \quad \text{or}$$
$$D' = \sum_{i=i}^{I} x^{-\mu_i} \tfrac{1}{U'} \tilde{F}_i[U], \quad \text{where } \tilde{F}_i[U] = F_i(1, U, P_1[U], ..., P_m[U]). \quad (7.149)$$

Differentiating the last equality in x leads to

$$0 = \sum x^{-\mu_i - 1}\big\{\big(\tfrac{1}{U'}\tilde{F}[U]\big)' - \mu_i \tfrac{1}{U'}\tilde{F}[U]\big\}.$$

Therefore, U satisfies I conditions

$$\big(\tfrac{1}{U'}\tilde{F}[U]\big)' = \mu_i \tfrac{1}{U'}\tilde{F}[U], \quad \text{so on integration } \tfrac{1}{U'}\tilde{F}[U] = d_i e^{\mu_i \xi}, \quad (7.150)$$

where d_i are arbitrary constants. Finally, the ODE (7.149) takes the form

$$D' = \sum_{i=1}^{I} d_i e^{\mu_i D}. \quad (7.151)$$

Determining $U(\xi)$ from (7.150) and $D(t)$ from (7.151) yields the exact solutions (7.148) of the equation (7.134).

Another representation of solutions (7.148) is as follows:

$$u(x,t) = V(\zeta), \quad \zeta = xs(t), \quad (7.152)$$

where $\zeta = e^{\xi}$ and $s(t) = e^{D(t)}$. Then, instead of (7.150) and (7.151), we find

$$\tfrac{\tilde{1}}{V'} \tilde{F}_i[V] = d_i \zeta^{\mu_i + 1}, \quad \text{where } \tilde{F}_i[V] = \zeta^{\mu_i} F_i(\zeta, V, V', ..., V^{(m)})$$

and

$$s' = \sum_{i=1}^{I} d_i s^{\mu_i + 1}.$$

Both the representations are equivalent. Notice that the second form (7.152) is connected with the invariant subspace $\mathcal{L}\{x\}$.

Example 7.35 (**General quasilinear heat equation**) As a generalization to be also treated later on, consider a parabolic PDE of reaction-diffusion type,

$$\boxed{u_t = F_2[u] \equiv \big[\phi(u)u_{xx} + \psi(u)(u_x)^2\big] + f(u)} \quad (7.153)$$

that contains three functions $\phi(u) \geq 0$, $\psi(u)$, and $f(u)$. Bearing in mind an arbitrary smooth change $u = R(v)$, which leaves the structure of the PDE (7.153) invariant, we are going to find solutions in the separable form

$$u = e(x)g(t) \quad (7.154)$$

that are associated with the 1D subspace $\mathcal{L}\{e\}$ with an unknown function e. Plugging (7.154) into (7.153) and dividing by eg yields

$$\tfrac{g'}{g} = \phi \tfrac{e''}{e} + u\psi\big(\tfrac{e'}{e}\big)^2 + \tfrac{e}{u}. \quad (7.155)$$

Denote

$$\tfrac{e'(x)}{e(x)} = h(x) \implies \tfrac{e''}{e} - \tfrac{(e')^2}{e^2} = h' \text{ and } \tfrac{e''}{e} = h' + h^2.$$

Then (7.155) reads

$$\tfrac{g'}{g} = \phi(h' + h^2) + u\psi h^2 + \tfrac{e}{u} = h' + (\phi + u\psi)h^2 + \tfrac{e}{u}. \quad (7.156)$$

Differentiating in x vanishes the left-hand side, so

$$\phi h'' + [\phi' u + 2(\phi + u\psi)]hh' + (k + u\psi)' uh^3 + \left(\tfrac{e}{u}\right)' uh = 0. \tag{7.157}$$

As in the GSV, it follows from (7.157) that $h(x)$ must satisfy, for some constants α_1, α_2, and α_3, the ODE

$$h'' = \alpha_1 h^3 + \alpha_2 h + \alpha_3 hh'. \tag{7.158}$$

Then (7.157) reduces to the following system:

$$\begin{cases} (\phi + u\psi) + \alpha_1 \phi = 0, \\ (\frac{e}{u})' u + \alpha_2 \phi = 0, \\ \phi' u + 2(\phi + u\psi) + \alpha_3 \phi = 0. \end{cases} \tag{7.159}$$

We will integrate it later on.

7.5.3 Second-order parabolic PDEs

Let us return to the original treatment of PDEs via evolutionary invariant sets (7.135).

Example 7.36 (**Quasilinear heat equation, continued**) Consider again equation (7.153). The scaling orders of two operators in (7.153) are $\mu_1 = 2$ and $\mu_2 = 0$, so that the scaling index is $I(F_2) = 2$. Substituting the differential rule (7.138), one can calculate that, on M_0,

$$F_2[u] = H\left\{\tfrac{1}{x^2}[\phi(H' - 1) + \psi H] + \tfrac{f}{H}\right\} \equiv H\left\{C_1[H] + \tfrac{1}{x^2} C_2[H]\right\}. \tag{7.160}$$

By Theorem 7.33, the invariant conditions (7.141) take the form

$$H\left(\tfrac{f}{H}\right)' = 0, \;\; H[\phi(H' - 1) + \psi h]' = 2[\phi(H' - 1) + \psi H], \;\; \text{or} \tag{7.161}$$

$$\begin{cases} f = \nu H, \\ HH'' - 2H' + 2 + \frac{\phi'}{\phi} H(H' - 1) + \frac{\psi}{\phi} H(H' - 2) + \frac{\psi'}{\phi} H^2 = 0, \end{cases} \tag{7.162}$$

where $\nu \in I\!R$ is a constant. The equation (7.153) on M_0 reduces to the following ODE for the function $D(t)$ in the exact solution (7.142):

$$D' = d_1 + d_2 e^{2D},$$

where $d_1 = (\frac{f}{H})(1) = \nu$ and $d_2 = [\phi(H' - 1) + \psi H](1)$. Notice that (7.161) is a system of two equations with four arbitrary unknown functions ϕ, ψ, f, and H.

Example 7.37 (**Radial N-dimensional equations**) Consider the quasilinear heat equation in $I\!R^N$

$$\boxed{u_t = [\phi(u)\Delta u + \psi(u)|\nabla u|^2] + f(u).}$$

In the radial case, where $x > 0$ is the radial spatial variable and $\Delta u = u_{xx} + \frac{N-1}{x} u_x$, $|\nabla u|^2 = (u_x)^2$, this PDE has the same scaling orders as (7.153). Then, in (7.160),

$$C_1[H] = \phi(H' + N - 2) + \psi H.$$

Therefore, the second equation in (7.162) is replaced by the following one:

$$HH'' - 2H' - 2(N - 2) + \frac{\phi'}{\phi} H(H' + N - 2) + \frac{\psi}{\phi} H(H' - 2) + \frac{\psi'}{\phi} H^2 = 0.$$

Example 7.38 **(Equations with gradient-dependent diffusivity)** Consider a generalization of the p Laplacian equations studied in Section 6.4,

$$\boxed{u_t = \phi(u)(u_x)^\alpha u_{xx} + f(u)(u_x)^\beta \equiv F_1[u] + F_2[u].} \tag{7.163}$$

Here, $\mu_1 = \alpha + 2$ and $\mu_2 = \beta$. If $\alpha + 2 = \beta$, then (7.163) is invariant under the scaling group (7.129). Let $\alpha + 2 \neq \beta$. In this case, on M_0,

$$F_1[u] + F_2[u] = H\big(\tfrac{1}{x^{\alpha+2}}\, \phi H^\alpha (H' - 1) + \tfrac{1}{x^\beta}\, H^{\beta-1} f\big),$$

and the invariant conditions are

$$H[\phi H^\alpha (H' - 1)]' = (\alpha + 2)\phi H^\alpha (H' - 1), \;\; H(H^{\beta-1} f)' = \beta H^{\beta-1} f, \tag{7.164}$$

or, which is the same,

$$HH'' + \alpha(H')^2 - 2(\alpha + 1)H' + (\alpha + 2) + \tfrac{\phi'}{\phi}\, H(H' - 1) = 0,$$
$$Hf' + [(\beta - 1)H' - \beta]f = 0.$$

The ODE for $D(t)$ is

$$D' = d_1 \mathrm{e}^{(\alpha+2)D} + d_2 \mathrm{e}^{\beta D},$$

with constants $d_1 = [\phi H^\alpha (H' - 1)](1)$ and $d_2 = (H^{\beta-1} f)(1)$.

Example 7.39 **(Fully nonlinear parabolic PDEs)** Consider the equation

$$\boxed{u_t = \phi(u)(u_x)^\alpha (u_{xx})^\gamma + f(u)(u_x)^\beta \equiv F_1[u] + F_2[u].}$$

The term $(u_{xx})^\gamma$ is typical for the *dual porous medium equation*, which was discussed in Example 1.18. Then $\mu_1 = \alpha + 2\gamma$ and $\mu_2 = \beta$, and we assume that $\mu_1 \neq \mu_2$, so that $I = 2$. On M_0,

$$F_1[u] + F_2[u] = H\big(\tfrac{1}{x^{\alpha+2}}\, \gamma \phi H^{\alpha+\gamma-1} (H' - 1)^\gamma + \tfrac{1}{x^\beta}\, H^{\beta-1} f\big).$$

Therefore, in the invariant conditions (7.164), the first equation is replaced by

$$H[\phi H^{\alpha+\gamma-1} (H' - 1)^\gamma]' = (\alpha + 2\gamma)\phi H^{\alpha+\gamma-1} (H' - 1)^\gamma,$$

and the ODE for $D(t)$ becomes

$$D' = d_1 \mathrm{e}^{(\alpha+2\gamma)D} + d_2 \mathrm{e}^{\beta H}, \;\; \text{with} \;\; d_1 = [\phi H^{\alpha+\gamma-1} (H' - 1)^\gamma](1).$$

7.5.4 Fourth-order generalized thin film equations

Example 7.40 Consider a general fourth-order PDE

$$\boxed{u_t = F_4[u] \equiv q(u)u_{xxx} + p(u)u_{xxxx} + F_2[u],} \tag{7.165}$$

where F_2 is the second-order operator given in (7.153). In Section 3.1, we dealt with a number of such nonlinear thin film models possessing various exact solutions. The analysis of (7.165) is similar to that for the second-order PDE (7.153). The third and fourth-order term in (7.165) have the scaling orders $\mu_3 = 3$ and $\mu_4 = 4$, respectively, and $I(F_4) = 3$. Hence, on M_0, we add two extra terms to the right-hand side of

(7.160),

$$F_4[u] = F_2[u] + H\big(\tfrac{1}{x^3}\, qC_3[H]\big) + H\big(\tfrac{1}{x^4}\, \rho C_4[H]\big),$$

$$C_3[H] = \tfrac{1}{H}\, G_3[H], \;\; C_4[H] = \tfrac{1}{H}\, G_4[H],$$

where $G_{3,4}$ are given in (7.139). In this case, we add two equations to the system (7.162),

$$H(qC_3)' = 3qC_3 \quad \text{and} \quad H(\rho C_4)' = 4\rho \mathbf{C}_4. \tag{7.166}$$

The ODE for $D(t)$ takes the form

$$H' = d_1 + d_2 e^{2H} + d_3 e^{3H} + d_4 e^{4H},$$

where $d_3 = (qC_3[H])(1)$ and $d_4 = (\rho C_4[H])(1)$. The TFE (7.165) contains five arbitrary functions $\{\phi, \psi, f, q, \rho\}$, so, together with H, four ODEs (7.162), (7.166) for six functions are obtained. Hence, such solutions exist for PDEs (7.165) with two arbitrary coefficients (say, $\rho(u)$ and $q(u)$ of the higher-order operators).

7.5.5 Extensions

1. Other evolutionary invariant sets. In Section 1.4, we revealed several evolution PDEs, in particular, of type (7.153) that possessed solutions on the trigonometric subspace $W = \mathcal{L}\{\cos x\}$, or on the exponential subspace $W = \mathcal{L}\{\cosh x\}$. These solutions can be obtained by constructing sets of a different structure,

$$M_1 = \big\{u : \;\; u_x = \tan x \; H(u)\big\}.$$

The algebraic differentiation in M_1 is different from that in M_0. Setting

$$w(x) = \tfrac{1}{\cos^2 x},$$

we obtain

$$u_2 \equiv u_{xx} = wH(H'+1) - HH' \equiv H(w\mathbf{P}_{21} + \mathbf{P}_{20}).$$

Denoting

$$\mathbf{Q}[H] = H[(HH')' + 3H' + 2]$$

yields

$$u_4 = H(w^2\mathbf{P}_{42} + w\mathbf{P}_{41} + \mathbf{P}_{40}), \quad \text{where}$$

$$\mathbf{P}_{42} = 3\mathbf{Q} + (H\mathbf{Q})' \quad \big((\cdot)' = \tfrac{\mathrm{d}}{\mathrm{d}u}\big),$$

$$\mathbf{P}_{41} = -2\mathbf{Q} - (H\mathbf{Q})' - (HH')' - (H(HH')')',$$

$$\mathbf{P}_{40} = (H(HH')')'.$$

In particular, the even derivatives u_{2k} belong to a $(k+1)$-dimensional subspace,

$$u_{2k} \in \mathcal{L}\{1, w, ..., w^k\}.$$

Example 7.41 (**Thin film-type equations**) Consider the fourth-order PDE

$$\boxed{u_t = F_4[u] \equiv \rho(u)u_{xxxx} + \phi(u)u_{xx} + \psi(u)(u_x)^2.} \tag{7.167}$$

Since for $u \in M_1$,

$$(u_x)^2 = wH^2 - H^2,$$

it follows that, on M_1,

$$\begin{aligned} F_4[u] - H\big[w^2(\rho\mathbf{P}_{42}) + w(\rho\mathbf{P}_{41} + \phi\mathbf{P}_{21} + \psi H) \\ + \rho\mathbf{P}_{40} + \phi\mathbf{P}_{20} - \psi H\big] \equiv H(w^2\Phi_2 + w\Phi_1 + \Phi_0) \in \mathcal{L}\{1, w, w^2\}. \end{aligned} \tag{7.168}$$

The index of F_4 (the number of linearly independent terms in (7.168)) is equal to 3. Integrating $u_x = \tan x\, H(u)$ yields

$$v \equiv \int_1^u \frac{\mathrm{d}z}{H(z)} = -\ln\cos x + D(t) \quad (\cos x > 0).$$

Substituting this expression into (7.167) and using (7.168), we obtain

$$D'(t) = w^2(x)\Phi_2 + w(x)\Phi_1 + \Phi_0. \tag{7.169}$$

In view of the linear independence, functions $\Phi_2(v)$, $\Phi_1(v)$, and $\Phi_0(v)$ must be exponential and equal to $d_2\mathrm{e}^{4v}$, $d_1\mathrm{e}^{2v}$, and d_0, respectively. This yields the system of three equations for H,

$$H\Phi_2' = 4\Phi_2, \quad H\Phi_1' = 2\Phi_1, \quad \text{and } H\Phi_0' = 0. \tag{7.170}$$

Then (7.169) becomes

$$H' = d_2\mathrm{e}^{4H} + d_1\mathrm{e}^{2H} + d_0,$$

with the coefficients $d_2 = \Phi_2|_{u=1}$, $d_1 = \Phi_1|_{u=1}$, and $d_0 = \Phi_0|_{u=1}$. Here (7.170) is a system of three ODEs for four arbitrary functions $\{\phi, \psi, \rho, H\}$. Hence, such solutions exist for a family of equations (7.167) with a single arbitrary coefficient, e.g., $\rho(u)$. Adding another thin film-type monomial, $\kappa(u)u_x u_{xxx}$, to the right-hand side does not change the scaling index, since on M_1,

$$u_1 u_3 \in \mathcal{L}\{1, w, w^2\}.$$

Hence, no extra ODE is added to the system (7.170). A similar analysis is performed for the sets given by the constraint

$$u_x = \tanh x\, H(u).$$

On more general approaches. We now clarify the origin of all the evolutionary invariant sets studied above; cf. Example 7.35. Namely, consider the set

$$M: \quad u_x = g(x)H(u), \tag{7.171}$$

and determine all possible functions g and H that guarantee the invariance in time. Of course, integrating this constraint yields the following exact solution:

$$u(x,t) = U(e(x) + D(t)),$$

with three unknown functions e, D, and U. Such solution structures, which we have revealed above in particular cases, in general, are related to Stäckel's form (1893) and generalized separation of variables; see Remarks to Section 8.4.

We now borrow the following result from Section 8.3 (cf. identity (8.50)): (7.171) is *evolutionary invariant* under the flow induced by the quasilinear heat equation

$$u_t = (k(u)u_x)_x + f(u), \quad \text{provided that} \tag{7.172}$$

$$\begin{aligned} \mathcal{F}_1 \equiv\, & g''(x)kH + g^3(x)\, H^2(kH)'' \\ & + g(x)(f'H - fH') + gg'(x)(3k'H^2 + 2kHH') \equiv 0. \end{aligned} \tag{7.173}$$

The right-hand side of (7.173) belongs to the subspace

$$W = \mathcal{L}\{g'', g^3, g, gg'\} \tag{7.174}$$

and so, if it is 4D, we obtain four ODEs for three functions $\{k, f, H\}$, which are not expected to admit any interesting solution.

3D reduction. The next step is to consider an extra reduction of the subspace (7.174) by choosing $g(x)$ such that, for some constants $\alpha_{1,2,3}$, (7.158) holds. Assuming that $\mathcal{L}\{g^3, g, gg'\}$ is 3D yields the following system for three functions:

$$\begin{cases} H(kH)'' + \alpha_1 k = 0, \\ f'H - fH' + \alpha_2 kH = 0, \\ 3k'H + 2kH' + \alpha_3 k = 0. \end{cases} \tag{7.175}$$

It is not difficult to characterize the whole family of such equations (7.172). Setting $kH = P$, we derive from the first two equations

$$HH' = -\tfrac{\alpha_1}{2}\left(\tfrac{P}{P''}\right)' \quad \text{and} \quad \left(\tfrac{f}{H}\right)' = \tfrac{\alpha_2}{\alpha_1} P''.$$

Integrating the first one,

$$H^2 = -\alpha_1\left(\tfrac{P}{P''}\right) + A \quad (A \in I\!R)$$

and substituting this $H = H(P)$ and $k = \frac{P}{H(p)}$ into the third equation in (7.175) yields an autonomous third-order ODE for the single function $P(u)$,

$$PP''' = 2\tfrac{\alpha_3}{\alpha_1}(P'')^2\sqrt{-\alpha_1\left(\tfrac{P}{P''}\right) + A} - 5P'P''$$

(for $A = 0$, this can be reduced to a first-order equation).

Thus, in general, we obtain a four-parametric family of PDEs (7.172) possessing such explicit solutions.

The algebraic differentiation on the set (7.171) depends on g and is not as perfect as in the case where $g(x) = \frac{1}{x}$. For instance,

$$\begin{aligned} u_{xx} &= g^2 HH' + g'H, \\ u_{xxx} &= g^3[\alpha_1 H + H(HH')'] + g(\alpha_2 H) + gg'(3HH' + \alpha_3 H), \end{aligned}$$

etc. It is not difficult to derive the ODE governing the evolution on M for an arbitrary admissible $g(x)$. It follows from (7.171) that the function

$$\begin{gathered} v(x,t) = G(u) \equiv \int \tfrac{du}{H(u)} = \int_0^x g(y)\,dy + D(t) \ \text{ solves} \\ v_t = k(u)v_{xx} + (kH)'(u)(v_x)^2 + \left(\tfrac{f}{H}\right)(u). \end{gathered} \tag{7.176}$$

Since $v_x = g(x)$ and $v_{xx} = g'(x)$, setting $x = 0$ yields $v(0,t) \equiv G(u(0,t)) = D(t)$, and so $u(0,t) = G^{-1}(v(0,t))$, from which comes the following ODE for $D(t)$:

$$D' = k(G^{-1}(D))g'(0) + (kH)'(G^{-1}(D))g^2(0) + \left(\tfrac{f}{H}\right)(G^{-1}(D)). \tag{7.177}$$

2D reductions. It is now easy to obtain from (7.158) further reductions leading to some specific functions g,

$$\begin{aligned} \alpha_1 = \alpha_2 = 0,\ \alpha_3 = -2 \quad &\Longrightarrow \quad g(x) = \tfrac{1}{x}, \\ \alpha_1 = \alpha_3 = 0,\ \alpha_2 = -1 \quad &\Longrightarrow \quad g(x) = \cos x, \\ \alpha_1 = \alpha_3 = 0,\ \alpha_2 = 1 \quad &\Longrightarrow \quad g(x) = \cosh x. \end{aligned} \tag{7.178}$$

The most transparent degenerate case is

$$\alpha_1 = \alpha_2 = \alpha_3 = 0 \quad \Longrightarrow \quad g(x) = x. \tag{7.179}$$

These cases will play a role for constructing sign-invariants for quasilinear parabolic PDEs which is the main subject of the next chapter.

7.5.6 Evolutionary invariant sets for hyperbolic PDEs

Consider the set (7.171) for the second-order hyperbolic equation

$$u_{tt} = (k(u)u_x)_x + f(u). \tag{7.180}$$

Equating u_{xtt} given by (7.171) and (7.180), we obtain the invariance condition (cf. (7.173))

$$\begin{aligned} \mathcal{F}_2 \equiv & -gH''(u_t)^2 + g''kH + g^3(k''H^3 + 2k'H^2H' + kH^2H'') \\ & + g(f'H - fH') + gg'(2kHH' + 3k'H^2) \equiv 0. \end{aligned} \tag{7.181}$$

If this identity is true, performing the same change (7.176) yields the PDE

$$v_{tt} + H'(u)(v_t)^2 = k(u)v_{xx} + (kH)'(u)(v_x)^2 + \left(\tfrac{f}{H}\right)(u),$$

so that, exactly as in (7.177), we obtain the ODE

$$\begin{aligned} D'' + H'(G^{-1}(D))(D')^2 = & \, k(G^{-1}(D))g'(0) \\ & + (kH)'(G^{-1}(D))g^2(0) + \left(\tfrac{f}{H}\right)(G^{-1}(D)). \end{aligned}$$

There is an essential difference with the parabolic case that is related to the first term on the right-hand side of (7.181), so that one needs H to be a linear function,

$$H'' = 0.$$

The rest of the analysis is equally based on the dimension of the linear space (7.174). In particular,

$$g(x) = \tfrac{1}{x} \quad \Longrightarrow \quad W = \mathcal{L}\{\tfrac{1}{x}, \tfrac{1}{x^3}\},$$

and (7.181) yields two ODEs for the coefficients

$$\begin{cases} f'H - fH' = 0, \\ k''H^2 + 2k'HH' - 2kH' - 3k'H + 2k = 0. \end{cases}$$

The first ODE implies

$$f(u) = \mu H(u).$$

In particular, choosing a linear function

$$H(u) = \alpha u$$

yields, for $k(u)$, second-order Euler's equation

$$\alpha^2u^2k'' + \alpha(2\alpha - 3)uk' + 2(1 - \alpha)k = 0.$$

For $\alpha = 1$, we have

$$u^2k'' - uk' = 0,$$

which yields the following hyperbolic equation:

$$\boxed{u_{tt} = F[u] \equiv \big[(a_1 + a_2u^2)u_x\big]_x + \mu u}$$

with the solution $u(x,t) = C(t)x$ on the invariant subspace $\mathcal{L}\{x\}$ of F, where

$$C'' = 2a_2C^3 + \mu C.$$

For $\alpha = 2$, k satisfies

$$2u^2k'' + uk' - k = 0,$$

giving the PDE

$$\boxed{u_{tt} = F[u] \equiv \big[(a_1u + \tfrac{a_2}{\sqrt{u}})u_x\big]_x + 2\mu u}$$

with exact solutions $u(x,t) = C(t)x^2$, where

$$C'' = 6a_1C^2 + 2\mu C.$$

In the case where

$$H(u) = 1,$$

we find $k'' - 3k' + 2k = 0$, and hence, the equation

$$\boxed{u_{tt} = F[u] \equiv \big[(a_1\mathrm{e}^u + a_2\mathrm{e}^{2u})u_x\big]_x + \mu,}$$

which admits the solution $u(x,t) = \ln x + D(t)$, where

$$D'' = 2a_2\mathrm{e}^{2D} + \mu.$$

Similarly, it is easy to consider other cases from (7.178). In the degenerate case (7.179), where $W = \mathcal{L}\{x, x^3\}$, there occur two ODEs

$$\begin{cases} 2kHH' + 3k'H^2 + f'H - fH' = 0, \\ k''H + 2k'H' = 0, \end{cases}$$

so taking $H(u) = u$ yields the PDE

$$\boxed{u_{tt} = \big[(a_2 - \tfrac{a_1}{u})u_x\big]_x - 2a_2u\ln u + a_1 + a_3u}$$

possessing the solution $u(x,t) = C(t)\mathrm{e}^{x^2/2}$, where

$$C'' = -2a_2C\ln C + (a_2 + a_3)C.$$

7.6 A separation technique for the porous medium equation in $I\!R^N$

In this last section, we discuss a specific version of separation of variables. We deal with the N-dimensional PME

$$v_t = \Delta v^m \quad \text{in } I\!R^N \times I\!R_+ \quad (v \geq 0), \tag{7.182}$$

which, by the pressure transformation $u = \frac{m}{m-1} v^{m-1}$, reduces to the quadratic form

$$\boxed{u_t = F_m[u] \equiv (m-1)u\Delta u + |\nabla u|^2.} \tag{7.183}$$

The PME has $m > 1$, but the fast diffusion range $m \in (0,1)$ is also included, as well as $m = 0$, corresponding to

$$v_t = F_0[v] \equiv \Delta \ln v.$$

The pressure equation (7.183) with $m = 0$ is obtained by setting $u = -\frac{1}{v}$. Recall also the PDE with the exponential nonlinearity

$$v_t = \Delta e^v,$$

which reduces to a similar quadratic pressure-like equation

$$\boxed{u_t = u\Delta u, \quad \text{where } v = \ln u.}$$

According to Section 6.1, basic subspaces for the operator F_m consist of either linear, $W^{\rm lin} = \mathcal{L}\{1, x_i, \; i = 1, ..., N\}$, or quadratic functions, $W^{\rm q} = \mathcal{L}\{1, x_i x_j\}$. On each of the subspaces or on its sum, the PDE (7.183) reduces to finite-dimensional DSs which can be integrated in quadratures (Section 6.1). We will use these subspaces in what follows.

7.6.1 A separation problem

Let us look for solutions in the following form:

$$u(x,t) = [f(x,t)]^p + g(x,t) \quad (f > 0), \tag{7.184}$$

where f and g are two unknown functions from the above spaces of linear or quadratic functions, and $p \neq 0, 1$ is a parameter. Plugging into (7.183) yields a three-term expansion

$$\begin{aligned} & f^{p-1}p\{-f_t + (m-1)\big[g\Delta g + \tfrac{1}{p} f\Delta g\big] + 2\nabla f \cdot \nabla g \\ & + (m-1)(p-1)g\tfrac{|\nabla f|^2}{f}\} + \{-g_t + (m-1)g\Delta g \\ & + |\nabla g|^2\} + f^{2p-1}p\{(m-1)\Delta f + (mp-m+1)\tfrac{|\nabla f|^2}{f}\} = 0. \end{aligned} \tag{7.185}$$

This will be treated as a formal linear combination (with coefficients, depending on x and t as parameters) of three vectors f^{p-1}, 1, and f^{2p-1}. Assuming that

$$f(x,t),\ g(x,t) \in \mathcal{L}\{1, x_i, x_i x_j\} \quad \text{for all } t \geq 0 \tag{7.186}$$

yields, in general, three PDEs on $\{f, g\}$ if $p \neq \frac{1}{2}$,

$$\begin{aligned} & -f_t + (m-1)\big[g\Delta g + \tfrac{1}{p} f\Delta g\big] + 2\nabla f \cdot \nabla g \\ & + (m-1)(p-1)g\tfrac{|\nabla f|^2}{f} = 0, \quad -g_t + (m-1)g\Delta g + |\nabla g|^2 = 0, \\ & (m-1)\Delta f + (mp-m+1)\tfrac{|\nabla f|^2}{f} = 0. \end{aligned} \tag{7.187}$$

The case $p = \frac{1}{2}$ (where $f^{2p-1} = 1$) is special when the system contains two equations,

$$\begin{cases} -f_t + (m-1)[g\Delta f + 2f\Delta g] + 2\nabla f \cdot \nabla g - \frac{1}{2}(m-1)g\frac{|\nabla f|^2}{f} = 0, \\ -g_t + (m-1)g\Delta g + |\nabla g|^2 + \frac{1}{2}\big[(m-1)\Delta f + \frac{1}{2}(2-m)\frac{|\nabla f|^2}{f}\big] = 0. \end{cases}$$

Roughly speaking, we are dealing with a kind of GSV on the set (a module) of linear combinations $W_2 = \mathcal{L}\{1, \sqrt{f(x,t)}\}$. In fact, we are already familiar with some other representations of solutions, such as (7.184), given by extended fourth-order polynomial subspaces studied in Section 6.2. By Proposition 6.24, for the operator (7.183), such a subspace exists if

$$m = \tfrac{N-2}{N+2}.$$

In terms of representation (7.184), this gives the integer $p = 2$. The same $p = 2$ applies to the case $m = 0$, corresponding to the remarkable operator

$$F_{\rm rem}[u] = u\Delta u - |\nabla u|^2 \quad \text{in } I\!R^2. \tag{7.188}$$

As was shown in Section 6.3, this admits subspaces of fourth-degree polynomials and many others. It will be shown that $p = \frac{1}{2}$ also makes sense for (7.188).

Nonlinear separation problem in the class of general quadratic forms. Consider the most important and promising case of $p = \frac{1}{2}$, where (7.185) can be written in the form of

$$\begin{aligned} &1\big[-f_t + (m-1)g\Delta f + 2(m-1)f\Delta g + 2\nabla f \cdot \nabla g\big] \\ &+ \sqrt{f}\,\big[-2g_t + (m-1)\Delta f + 2(m-1)g\Delta g + 2|\nabla g|^2\big] + J = 0. \end{aligned} \tag{7.189}$$

Here, J denotes two extra *singular* terms, which should satisfy the following problem of nonlinear separation (in other words, a "nonlinear eigenvalue problem"):

$$J \equiv -(m-1)g\,\frac{|\nabla f|^2}{2f} + (2-m)\,\frac{|\nabla f|^2}{2\sqrt{f}} = \lambda_1 + \lambda_2\sqrt{f}, \tag{7.190}$$

where $\lambda_{1,2}$ are some constants or suitable functions.

Consider the problem (7.190) separately and independently of the PDE and its symmetries. Let $f = f(x)$ be a general quadratic form in $I\!R^N$. Using an orthogonal transformation, we reduce it to the diagonal form,

$$f = \textstyle\sum a_i x_i^2.$$

Let us see if this can solve the eigenvalue problem (7.190). Thus

$$|\nabla f|^2 = 4\textstyle\sum a_i^2 x_i^2,$$

so that, to resolve the singularities in the main first term in (7.190), there must exist a constant a such that

$$a_i^2 = a a_i \quad \text{for all } i = 1, \ldots, N.$$

This gives two possibilities: either

Case (i): $a_i = 0$ for some i, or

Case (ii): $a_i = a \neq 0$ for all i.

We exclude case (i) that, as will be shown, leads to solutions belonging to lower-dimensional subspaces in $I\!R^N$, and corresponds to the same analysis in $I\!R^K$ with some $K < N$, where we take

$$f = a \sum_{i=1}^{K} x_i^2.$$

Case (ii) implies that $f(x)$ should be canonical, with the matrix being the multiple of the identity one, i.e., aI. In this case,

$$\frac{|\nabla f|^2}{2f} = 2a \quad \text{and} \quad \frac{|\nabla f|^2}{2\sqrt{f}} = 2a\sqrt{f}, \tag{7.191}$$

and this solves the separation problem (7.190).

Therefore, bearing in mind translations in x, we can take $f(x,t)$ in the form of

$$f(x,t) = a(t) \sum (x_i - \varphi_i(t))^2 \quad \Longrightarrow \quad \Delta f = 2Na, \tag{7.192}$$

and, in view of (7.191), we obtain from (7.189) the following system of PDEs:

$$\begin{cases} f_t = (m-1)g\Delta f - 2a(m-1)g + 2(m-1)f\Delta g + 2\nabla f \cdot \nabla g, \\ 2g_t = (m-1)\Delta f + 2a(2-m) + 2(m-1)g\Delta g + 2|\nabla g|^2. \end{cases} \tag{7.193}$$

The case where $\mathbf{p \neq \frac{1}{2}}$ for quadratic function $\mathbf{f(x,t)}$. Using formulae (7.191) and (7.192) in the last equation in (7.187) yields another critical value of p,

$$2a\big[(m-1)N + 2(mp - m + 1)\big] = 0,$$

i.e.,

$$p = \frac{(m-1)(2-N)}{2m}.$$

Hence, as for $p = \frac{1}{2}$, there occur two equations

$$\begin{cases} f_t = (m-1)[g\Delta g + 2f\Delta g] + 2\nabla f \cdot \nabla g - 2a(m-1)g, \\ g_t = (m-1)g\Delta g + |\nabla g|^2 + a[(m-1)N - m + 2]. \end{cases}$$

This system is analyzed similarly to that for $p = \frac{1}{2}$, which we now begin to investigate in detail.

7.6.2 Linear functions $g(x,t)$

Let us now return to $p = \frac{1}{2}$ and consider the first simpler case, assuming that $g \in W^{\rm lin}$, i.e.,

$$g(x,t) = A(t) + \mathbf{C}(t)x \equiv A(t) + \sum C_i(t)x_i.$$

Then $\Delta g = 0$, $|\nabla g|^2 = |\mathbf{C}|^2$, and system (7.193) takes a simpler form

$$\begin{cases} a' \sum (x_i - \varphi_i)^2 - 2a \sum \varphi_i'(x_i - \varphi_i) \\ = (m-1)(N-1)2a\big(A + \sum C_i x_i\big) + 4a \sum C_i(x_i - \varphi_i), \\ 2\big(A' + \sum C_i' x_i\big) = 2a(N-2)(m-1) + 2|\mathbf{C}|^2. \end{cases}$$

Equating the coefficients of monomials x_i^2, x_i, and 1 in both sides of the equations yields the following system of ODEs:

$$\begin{cases} a' = 0, \\ \varphi' = -\beta \mathbf{C}, \\ \varphi \cdot \varphi' = (\beta - 2)A - 2\mathbf{C} \cdot \varphi, \\ \mathbf{C}' = 0, \\ A' = (\beta - 1)a + |\mathbf{C}|^2, \end{cases} \tag{7.194}$$

where $\beta = 2 + (m-1)(N-1)$. From the first, fourth, and second ODEs,

$$a(t) \equiv a_0, \quad \mathbf{C}(t) \equiv \mathbf{C}_0, \quad \varphi(t) = \mathbf{V}t, \quad \text{with } \mathbf{V} = -\beta \mathbf{C}_0,$$

and the last equation gives

$$A(t) = \left[(\beta - 1)a_0 + |\mathbf{C}_0|^2\right]t. \tag{7.195}$$

Finally, the third ODE in (7.194) implies that either

$$a_0 = |\mathbf{C}_0|^2 \quad \Longrightarrow \quad A(t) = \beta |\mathbf{C}_0|^2 t, \tag{7.196}$$

or $\beta = 1$, i.e.,

$$m = \tfrac{N-2}{N-1}.$$

Assuming that $m > 1$, and denoting $|\mathbf{V}|^2 = \sum V_i^2$ yields the solution

$$u(x,t) = \tfrac{1}{\beta}\left[|\mathbf{V}|\sqrt{\textstyle\sum (x_i - V_i t)^2} - \left(\textstyle\sum V_i x_i - |\mathbf{V}|^2 t\right)\right]_+, \tag{7.197}$$

where, as is usual for the PME, the positive part gives the weak solution with free boundaries. This solution is strictly positive and, hence, is a smooth analytic function everywhere in $I\!R^N \times I\!R_+$, except the unbounded segment of the straight line in $I\!R^N$

$$\Gamma_t: \quad x = \mathbf{V}s, \quad \text{with } s \geq t$$

(we are assuming that $V_i > 0$), on which $u = 0$, and $u(x,t)$ is Lipschitz continuous in a neighborhood of the free boundary Γ_t. This interface has the cusp end-point moving linearly with time,

$$x_{\text{cusp}}(t) = \mathbf{V}t, \tag{7.198}$$

so the interface is not a smooth surface for all $t > 0$. In Figure 7.1 we present level sets of this exact solution in $I\!R^2$.

Let us next briefly consider two easy extensions of such solutions, exhibiting other evolution properties.

Example 7.42 (**Cusp localization**) Consider the PME with a linear absorption,

$$v_t = \Delta v^m - v, \quad m > 1. \tag{7.199}$$

Then, setting $v = \mathrm{e}^{-t} w$ yields

$$w_t = \mathrm{e}^{-(m-1)t} \Delta w^m.$$

Changing the time-variable

$$s = \tfrac{1}{m-1}\left[1 - \mathrm{e}^{-(m-1)t}\right] : I\!R_+ \to \left(0, \tfrac{1}{m-1}\right)$$

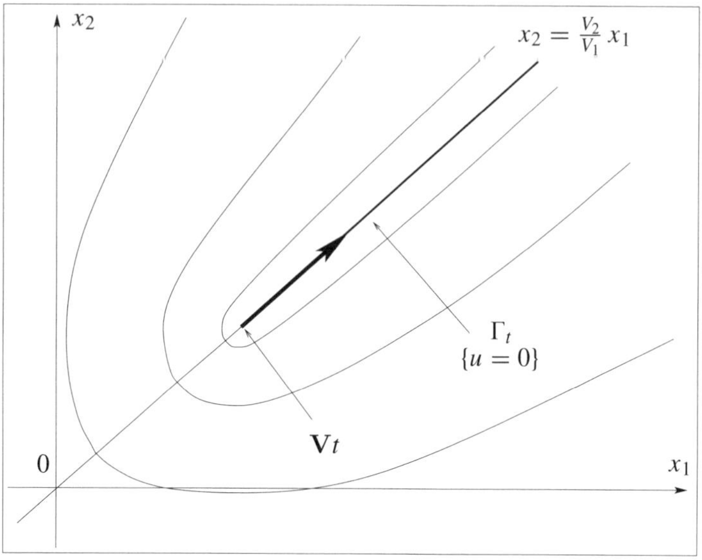

Figure 7.1 The "razor blade" described by the explicit solution (7.197) in $I\!R^2$.

returns us to the original PME

$$w_s = \Delta w^m$$

that possesses the above explicit cusp solution (7.197). Hence, its spatial structure remains the same, but the asymptotic behavior of this space-time pattern is different. Since $t = +\infty$ corresponds to the finite $s = \frac{1}{m-1}$, it follows that

$$v(x,t) \approx \mathrm{e}^{-t} w\big(x, \tfrac{1}{m-1}\big) \quad \text{for } t \gg 1,$$

so, for the model (7.199), the cusp is localized and propagates through the finite distance $\frac{1}{m-1}\,|\mathbf{V}|$ during the whole time of evolution $t \in I\!R_+$.

Example 7.43 (**Single point and conical singularities**) Let us add a nonlinear reaction or strong absorption term to the right-hand side of (7.182),

$$v_t = \Delta v^m + \mu\, v^{2-m}.$$

The pressure u will then solve the quadratic equation with an extra constant term,

$$\boxed{u_t = (m-1)u\,\Delta u + |\nabla u|^2 + \mu m.}$$

Let us see how this affects explicit solutions with $p = \frac{1}{2}$. Solving the separation problem (7.190), we then add an extra single term into the second equation in (7.193), so it has the form

$$2g_t = (\text{same terms}) + 2\mu m.$$

This affects the last ODE in (7.194) only,

$$A' = (\beta - 1)a + |\mathbf{C}|^2 + \mu m.$$

Therefore, instead of (7.195),

$$A(t) = \big[(\beta - 1)a_0 + |\mathbf{C}_0|^2 + \mu m\big]t,$$

so that the third ODE in (7.194) yields another relation between a_0 and $\mathbf{C}_0$,

$$a_0 = |\mathbf{C}_0|^2 - \tfrac{\mu m}{\beta - 1} > 0.$$

Finally, this gives the same function $A(t)$, as in (7.196).

Thus we obtain the explicit solution

$$u(x,t) = \tfrac{1}{\beta}\big[\sqrt{|\mathbf{V}|^2 - \varepsilon_0}\,\sqrt{\textstyle\sum (x_i - V_i t)^2} - \big(\textstyle\sum V_i x_i - |\mathbf{V}|^2 t\big)\big]_+, \qquad (7.200)$$

where

$$\varepsilon_0 = \tfrac{\mu m \beta^2}{\beta - 1},$$

and we have to assume that $|\mathbf{V}|^2 > \varepsilon_0$. Though this solution looks similar to (7.197) for the pure PME, it exhibits other evolution and interface properties.

Strong absorption: a single point singularity. Here, $\varepsilon_0 < 0$ in (7.200), so this explicit solution has the *unique* point of singularity $x_{\rm cusp}(t) = \mathbf{V}t$, at which $u(\mathbf{V}t, t) = 0$. As above, this cusp point moves linearly with time.

Reaction: a conical singularity. Here, $\varepsilon_0 > 0$, and hence, the support of the solution (7.200) is a conical surface K_t in $I\!R^N$ composed of straight lines with the parametric equations

$$x_i - V_i t = d_i s,$$

where $\mathbf{d} \in I\!R^N$ satisfies

$$\big(|\mathbf{V}|^2 - \varepsilon_0\big)|\mathbf{d}|^2 = \textstyle\sum V_i V_j d_i d_j.$$

K_t has the vertex at the moving point (7.198). In $I\!R^2$, the cone K_t is the interior of halves of two straight lines intersecting at $x_{\rm cusp}(t) = \mathbf{V}t$; see Figure 7.2.

7.6.3 Quadratic functions $g(x,t)$

Here, g is treated as a general quadratic polynomial. We have already fixed the canonical structure of f in (7.192) that is necessary for resolving the separation problem (7.190). From the first equation in (7.193), it is easy to see that the quadratic form of g is then also diagonal and canonical. This reminds us of the well-known fact from linear algebra saying that there exists an orthogonal transformation that simultaneously reduces a given quadratic form in $I\!R^N$ to the diagonal kind, and the second, positive one, to the canonical kind.

Thus, take a general quadratic polynomial, which it is convenient to write down in a form similar to (7.192),

$$g(x,t) = b(t)\textstyle\sum (x_i - \psi_i(t))^2 + A(t).$$

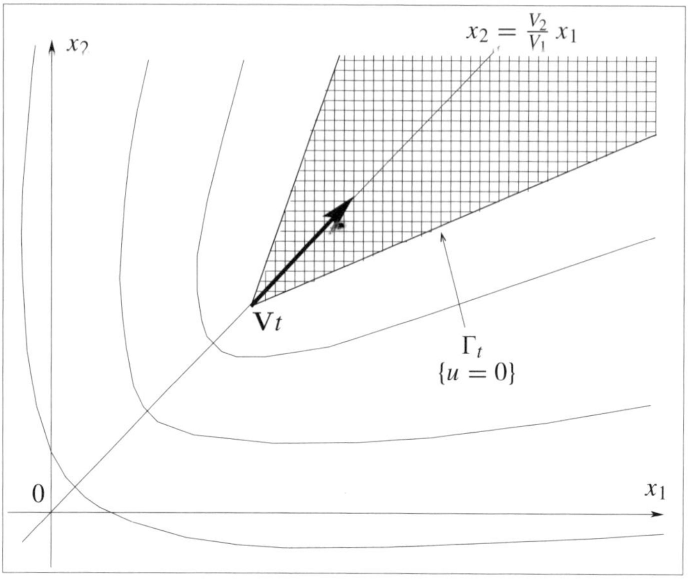

Figure 7.2 Moving triangular support described by the solution (7.200) of the PME with source in $I\!R^2$.

Substituting both expressions for f and g into (7.193) yields two equalities

$$
\begin{aligned}
-2a \textstyle\sum(x_i - \varphi_i)\varphi_i' + a' \sum(x_i - \varphi_i)^2 &= 2a(m-1)(N-1)\big[b \textstyle\sum(x_i - \psi_i)^2 + A\big] \\
&\quad + 4(m-1)Nab \textstyle\sum(x_i - \varphi_i)^2 + 8ab \sum(x_i - \varphi_i)(x_i - \psi_i), \\
- 2b \textstyle\sum(x_i - \psi_i)\psi_i' + b' \sum(x_i - \psi_i)^2 + A' & \\
= (\beta - 1)a + 2b^2(\beta + m - 1) \textstyle\sum(x_i - \psi_i)^2 + 2(m-1)NAb.
\end{aligned}
$$

Projecting both equations onto 1, x_i, and x_i^2 respectively, we obtain the overdetermined system

$$
\begin{aligned}
&2a\varphi \cdot \varphi' + a'|\varphi|^2 = 2a(m-1)(N-1)A \\
&\quad + 2ab(m-1)(N-1)a|\psi|^2 + 4(m-1)Nab|\varphi|^2 + 8ab\varphi \cdot \psi, \\
&\varphi' + \tfrac{a'}{a}\,\varphi = 2b(m-1)(N-1)\psi + 4(m-1)Nb\varphi + 4b(\varphi + \psi), \\
&a' = 2[(m-1)(N-1) + 2(\beta + m - 1)]ab, \\
&2b\psi \cdot \psi' + b'|\psi|^2 + A' = (\beta - 1)a \\
&\quad + 2b^2(\beta + m - 1)|\psi|^2 + 2(m-1)NAb, \\
&\psi' + \tfrac{b'}{b}\,\psi = 2b(\beta + m - 1)\psi, \\
&b' = 2b^2(\beta + m - 1).
\end{aligned}
\tag{7.201}
$$

The last ODE is independent of the others, so we need to consider two cases:

Degenerate exponential case: $\beta + m - 1 = 0$. Assume that

$$\beta + m - 1 \equiv (m-1)N + 2 = 0,$$

i.e.,

$$m = \tfrac{N-2}{N}.$$

Then $m \in (0,1)$ for $N \geq 3$ and $m = 0$ for $N = 2$, i.e., we deal with the fast diffusion equation (7.182) rather than the PME for $m > 1$. The value $m_* = \frac{N-2}{N}$ is a well-known important *critical exponent* in the theory of the fast diffusion equation (7.182). In particular, finite-time extinction of L^1-solutions occurs precisely in the range $m \in (0, m_*)$, $N \geq 3$, [37]. The critical $m = m_*$ corresponds to the unusual asymptotic behavior for $t \gg 1$ in the Cauchy problem for (7.182); see [245, Ch. 6] for further details concerning other critical issues of this exponent.

Then we take

$$b(t) \equiv b_0 \neq 0.$$

The third ODE in (7.201) implies that

$$a(t) = a_0 \mathrm{e}^{\mu t}, \quad \text{with } \mu = 2(m-1)(N-1)b_0 = -\tfrac{4(N-1)}{N}\, b_0.$$

The fifth equation reads $\psi' = 0$, so $\psi =$ constant, and we set

$$\psi(t) \equiv 0 \tag{7.202}$$

by translation. The second ODE gives

$$\varphi(t) = \varphi_0 \mathrm{e}^{\nu t}, \quad \text{with } \nu = 4[(m-1)N + 1]b_0 - \mu = -\tfrac{4}{N}\, b_0.$$

From the first equation in (7.201) we find

$$A(t) = -\tfrac{N}{4(N-1)}\, \mathrm{e}^{2\nu t}(\mu + 2\nu + 8b_0)|\varphi_0|^2. \tag{7.203}$$

The fourth ODE for A now has the form

$$A' = -4b_0 A - \tfrac{N-2}{N}\, a_0 \mathrm{e}^{\mu t}, \tag{7.204}$$

and we want (7.203) to satisfy it. There are two possibilities.

Subcase I: $N = 2$, i.e., $m = 0$. Since $-4b_0 = 2\nu$, (7.203) solves the ODE (7.204), and the exact solution is obtained. This determines another invariant property of the remarkable operator (7.188) in $I\!R^2$ besides those in Section 6.3.

Subcase II: $N = 3$, i.e., $m = \frac{1}{3}$. If $N \neq 2$, we need

$$2\nu = \mu, \quad \text{or} \quad 2\big(-\tfrac{4}{N}\, b_0\big) = -\tfrac{4(N-1)}{N}\, b_0,$$

which yields the only dimension

$$N = 3.$$

Substituting (7.203) into (7.204) yields the consistency condition

$$a_0 = 4b_0^2 |\varphi_0|^2.$$

The corresponding explicit solution of the PME (7.182) in $I\!R^3 \times I\!R_+$ with $m = \frac{1}{3}$ is as follows:

$$u(x,t) = \sqrt{4b_0|\varphi_0|^2 e^{-\frac{8}{3}b_0 t} \sum \left(x_i - \varphi_{i0} e^{-\frac{4}{3}b_0 t}\right)^2 + b_0|x|^2 - \tfrac{1}{2} b_0|\varphi_0|^2 e^{-\frac{8}{3}b_0 t}}.$$

Algebraic case: $\beta + m - 1 \neq 0$. The last equation in (7.201) yields

$$b(t) = c_0 t^{-1}, \quad \text{where } c_0 = -\tfrac{1}{2(\beta+m-1)}.$$

Similarly, from the fifth ODE, $\psi_i' = 0$, so (7.202) holds. The third equation yields

$$a(t) = a_0 t^{\rho_1}, \quad \text{where } \rho_1 = -\tfrac{(m-1)(3N-1)+4}{(m-1)N+2}.$$

We now use the second ODE of (7.201) to get that

$$\varphi_i' + \rho_1 t^{-1} \varphi_i = 4[(m-1)N+1]c_0 t^{-1} \varphi_i.$$

Integrating this linear first-order ODE gives

$$\varphi_i(t) = \varphi_{i0} t^{\rho_2}, \quad \text{where } \rho_2 = \tfrac{(m-1)(N-1)+2}{(m-1)N+2}.$$

Next, let us determine A from the first equation,

$$A(t) = \rho_3|\varphi_0|^2 t^{2\rho_2 - 1}, \quad \text{where } \rho_3 = \tfrac{1}{2[(m-1)N+2]}. \tag{7.205}$$

Finally, consider the fourth ODE that takes the form

$$A' = 2(m-1)Nc_0 t^{-1} A + [(m-1)N - m + 2]a_0 t^{\rho_1}, \tag{7.206}$$

which is assumed to possess solution (7.205), i.e., on substitution,

$$\begin{aligned}&(2\rho_2 - 1)\rho_3|\varphi_0|^2 t^{2\rho_2-2}\\ &= 2(m-1)Nc_0\rho_3|\varphi_0|^2 t^{2\rho_2-2} + [(m-1)N - m + 2]a_0 t^{\rho_1}.\end{aligned} \tag{7.207}$$

Subcase I: $(m-1)N - m + 2 = 0$, i.e.,

$$m = \tfrac{N-2}{N-1}. \tag{7.208}$$

Then (7.205) always solves (7.206).

Subcase II: $(m-1)N - m + 2 \neq 0$. One needs

$$\rho_1 = 2\rho_2 - 2 \quad \Longrightarrow \quad m = \tfrac{3N-7}{3(N-1)},$$

as well as the following relation between constants that is obtained from (7.207):

$$a_0 = \left[\tfrac{3(N-1)}{2(N-3)}\right]^2 |\varphi_0|^2 \quad (N \geq 4).$$

In both subcases, the corresponding explicit solutions of the PME (7.182) for $m = \frac{N-2}{N-1}$ and $m = \frac{3N-7}{3(N-1)}$ are given by

$$u(x,t) = \sqrt{a_0 t^{\rho_1} \sum (x_i - \varphi_{i0} t^{\rho_2})^2 + \tfrac{c_0}{t}\,|x|^2 + \rho_3|\varphi_0|^2 t^{2\rho_2-1}},$$

with the corresponding hypotheses on the parameters.

On solutions in $I\!R^K$. Using a lower-dimensional quadratic form

$$f(x,t) = a(t) \textstyle\sum_{i=1}^{K} (x_i - \varphi_i(t))^2, \quad \text{with some } K < N, \tag{7.209}$$

in the separation problem (7.190) will lead to $g(x,t)$ containing extra linear terms

$$g(x,t) = b(t)\sum_{i=1}^{K}(x_i - \psi_i(t))^2 + \sum_{i=K+1}^{N} C_i(t)x_i + A(t). \tag{7.210}$$

The principal calculations for deriving the corresponding ODE system (7.201), the results, and formulae remain the same, with N replaced by K. Moreover, the ODEs for extra coefficients C_i immediately imply that $C_i = 0$ for all $i \geq K+1$. Therefore, explicit solutions with functions (7.209) and (7.210) truly belong to $I\!R^K$ and this representation cannot supply us with new solutions.

Remarks and comments on the literature

The current notion of invariant sets on linear subspaces W_n (partial invariance) was introduced in [239], where various applications to PDEs with quadratic and cubic operators can be found (we have reflected some of the applications in this chapter).

§ 7.1. Solutions (7.4) were obtained in [397], where the solution structure (7.25) was also proposed for a nonlinear second-order parabolic PDE.

§ 7.2. Quasilinear KS-type models were considered in [570], where some usual compactons (7.37) with $p = 0$, were constructed.

§ 7.3. We use a slightly different version of the GSV introduced in [238], some other results are taken from [239]. By the transformation

$$v = c\,\frac{w_x}{w},$$

equation (7.86) reduces to a PDE with cubic nonlinearities, to which Baker–Hirota transformations apply, [100, 329]. Solutions (7.87) of equation (7.86) were also obtained by another approach based on evolutionary invariant sets, [317].

§ 7.5. We mainly follow [225]. Notice that the orbits of ODEs (7.143) satisfy the comparison principle, so that, for all the PDEs under consideration, the flows on sets M_0 are ordered with respect to initial data. This is a natural property for the second-order parabolic equations (in view of the Maximum Principle), but is rather surprising for higher-order PDEs, though possibly it just reflects a simple geometric structure of these sets. The solution structure (7.139) corresponds to the *functional separation of variables*; see Remarks to Section 8.4 for history and references.

Further applications of evolutionary invariant sets to TFEs were given in [478], where the extension of (7.132) in the form of the Bernoulli ODE

$$u_x = \tfrac{1}{x}\,u + u^n$$

was used (note that almost all explicit solutions therein admit polynomial representation on partially invariant subspaces); see also [479] for further extensions concerning sets (7.147). There exist attempts to treat general sets

$$u_x = G(x,t,u)$$

which often lead to complicated calculus; see [440] and Section 8.3 for parabolic equations.

Applications to the KdV-type equations

$$u_t = u_{xxx} + \psi(u)u_x \tag{7.211}$$

and to more general quasilinear ones, can be found in [223]. For (7.211), the ODE for H is

$$H(HH')'' - 6HH'' - 3(H')^2 + 9H' - 6 = 0, \quad \text{where } \psi(u) = \exp\Big\{\int_1^u \frac{\mathrm{d}z}{H(z)}\Big\}. \tag{7.212}$$

It is curious that (7.212) admits two linear solutions, (i) $H_1(u) = u$, i.e., the KdV equation, $u_t = u_{xxx} + uu_x$, and (ii) $H_2(u) = 2u$, corresponding to

$$\boxed{u_t = u_{xxx} + \sqrt{u}\, u_x,}$$

possessing the explicit blow-up solution

$$u(x,t) = \tfrac{x^2}{(-t)^2}.$$

Therefore, the KdV equation plays the role of a "limit" one (for $u \gg 1$) in the family of PDEs (7.211) admitting M_0; see details in [223]. Let us point out another KdV-type equation

$$\boxed{u_t = u_{xxx} + \ln u\, u_x,}$$

that admits explicit solutions

$$u(x,t) = \mathrm{e}^{C_1(t)+C_2(t)x}$$

on a simple linear subspace with

$$\begin{cases} C_1' = C_1C_2 + C_2^3, \\ C_2' = C_2^2. \end{cases}$$

§ 7.6. Exact solutions (7.197) in $I\!R^2$ and $I\!R^3$ were constructed in [57] by the direct similarity reduction (in $I\!R^2$)

$$u(x,t) = arf(\theta), \quad \text{where} \qquad (7.213)$$

$$\begin{cases} r^2 = x_1^2 + (x_2 - Vt)^2, \\ \theta = \tan^{-1} \frac{x_1}{x_2 - Vt}. \end{cases}$$

The resulting ODE for f admits the explicit solution

$$f(\theta) = \tfrac{1}{2}(1 - \cos\theta).$$

Similar solutions were constructed [58] for the fourth-order TFE

$$u_t = -\nabla \cdot (u^n \nabla \Delta u) \quad \text{in} \quad I\!R^2 \times I\!R_+, \qquad (7.214)$$

but, in this case, the ODE for f in (7.213) cannot be solved explicitly for $n = 3$. This confirms the general conclusion that the solution structure (7.184) does not apply to higher-order PDEs. Substituting this into (7.214) with $n = 1$ (or $n = 2$) leads to the analysis on linear subspaces of dimension larger than two that makes the PDE system more overdetermined and the consistency much more suspicious.

The form of solutions (7.184), (7.186) was proposed by the authors of [501] (see earlier references therein), where systems (7.187) and that for $p = \frac{1}{2}$ were derived and partially analyzed (our more complete analysis is different). Exact solutions with the quadratic $g(x,t)$ for $N = 2$, $m = 0$ can be found in [502] (other solutions in $I\!R^N$ constructed therein for which functions f and g are quadratic on 2D subspaces only, represent the same case). Some solutions in the case of (7.208) were detected in [501] by a sophisticated computational approach; as we have shown, other solutions there with quadratic forms in $I\!R^K$ for all dimensions $K < N$ are of the same nature and are not new.

Open problems

• In particular, some open problems are given in Sections 7.2, 7.4.4, 7.4.5, and 7.5.1.

CHAPTER 8

Sign-Invariants for Second-Order Parabolic Equations and Exact Solutions

We underline a new important aspect of invariant subspaces for nonlinear operators. Our goal is to show that, for second-order parabolic PDEs, proper families of exact solutions may define the so-called sign-invariants (SIs) of parabolic flows, which are nonlinear operators preserving their signs, "$\geq$" and "$\leq$," on evolution orbits. Vice versa, the known SIs may specify the corresponding exact solutions. We will mainly concentrate on the related backward problem: by constructing sign-invariants describe subspaces or sets of the corresponding solutions. Many SIs are generated by differential constraints, which thus preserve their signs. This is an important (and not always well understood) feature of differential constraints for parabolic equations obeying the Maximum Principle.

Our basic model is a *quasilinear heat equation* of the general form

$$u_t = F[u] \equiv \nabla \cdot (k(u)\nabla u) + f(u) \quad \text{in } I\!R^N \times I\!R_+, \tag{8.1}$$

where $k(u) \geq 0$ and $f(u)$ are given smooth functions. We will consider the Cauchy problem for (8.1) with given sufficiently smooth initial data $u(x,0) = u_0(x)$ in $I\!R^N$, and assume that there exists a unique sufficiently smooth solution $u = u(x,t)$, at least locally in time.

We say that the first-order operator $\mathcal{H}[u] = H(x, u, \nabla u, u_t)$ is a *sign-invariant* (SI) of the equation (8.1) if, for any solution $u(x,t)$, the following holds:

$$(1)\ \mathcal{H}[u(x,0)] \geq 0 \text{ in } I\!R^N \implies \mathcal{H}[u(x,t)] \geq 0 \text{ in } I\!R^N \text{ for } t > 0,$$

$$(2)\ \mathcal{H}[u(x,0)] \leq 0 \text{ in } I\!R^N \implies \mathcal{H}[u(x,t)] \leq 0 \text{ in } I\!R^N \text{ for } t > 0,$$

i.e., *both* signs of $\mathcal{H}[u]$ are preserved in evolution. Such an *evolution invariance* of signs is controlled by the Maximum Principle (MP). Since, by definition, any sign-invariant $\mathcal{H}[u]$ is also the *zero-invariant*, i.e.,

$$\mathcal{H}[u(x,0)] = 0 \text{ in } I\!R^N \implies \mathcal{H}[u(x,t)] = 0 \text{ in } I\!R^N \text{ for } t > 0, \tag{8.2}$$

this makes it possible to construct exact solutions of equations (8.1) if we know how to integrate the first-order PDE in (8.2). We will derive finite and, in some cases, infinite-dimensional sets of equations (8.1) possessing solutions that are expressed in terms of dynamical systems or algebraic relations. It turns out that these solutions often belong to some linear invariant subspaces or sets.

Using these connections with SIs, a number of higher-order PDEs admitting similar subspaces and solutions will be found. In view of the lack of the MP, we then lose the SI properties, but the rest of the analysis, including exact solutions, remains valid.

8.1 Quasilinear models, definitions, and first examples

Thus, in order to avoid posing suitable boundary conditions, which are not essential in the present context, we consider the Cauchy problem with smooth initial data

$$u(x,0) = u_0(x) \quad \text{in } I\!R^N, \tag{8.3}$$

and assume that there exists a unique local-in-time solution $u(x,t)$. According to classical parabolic theory, this requires some extra hypotheses on the coefficients of (8.1) and on the class of initial functions u_0.

8.1.1 *On weak and proper solutions*

We illustrate our main results by taking quasilinear heat equations (8.1), which are widely used in applications in diffusion, combustion, and filtration theory. Since we are going to use the MP, let us be more specific about the nonlinear coefficients of the PDEs under consideration. The functions $k(u)$ and $f(u)$ are assumed to be sufficiently smooth. For solvability of parabolic PDEs, we impose the *parabolicity condition*

$$k(u) \geq 0.$$

For $k(u) \equiv 1$, there occurs the *semilinear heat equation*

$$u_t = F_1[u] \equiv \Delta u + f(u) \quad \text{in } I\!R^N \times I\!R_+, \tag{8.4}$$

where Δ is the Laplace operator in $I\!R^N$. For smooth functions $f(u)$, the Cauchy problem for such equations admits a unique classical solution. Even for uniformly parabolic equations, for non-Lipschitz or singular absorption terms, such as

$$f(u) = -\,u^p, \quad \text{with the exponent } p \in (-1, 1) \tag{8.5}$$

(we have dealt with models like that before, especially for $f(u) = -1$ for $p = 0$), the nonnegative solutions may exhibit finite interfaces and need smooth approximations as weak or maximal/minimal proper solutions. Furthermore, we do not hesitate to consider degenerate PDEs (8.1) with $k(0) = 0$, e.g., the PME, where $k(u) = u^\sigma$ with a fixed exponent $\sigma > 0$. For the PME, weak solutions $u(x,t)$ of the Cauchy problem can be compactly supported and are not smooth at the interfaces, where $u = 0$, and the PDE is degenerate. See basics of PME theory in [245, Ch. 2]. In this case, to justify manipulations with derivatives, we impose a conventional assumption that we actually deal with regular approximations of weak solutions. We assume that the weak solution $u(x,t)$ is determined as the limit of a sequence $\{u_n\}$ of smooth solutions,

$$u(x,t) = \lim u_n(x,t) \quad \text{as } n \to \infty.$$

Here, each solution $u_n(x,t)$ solves a regularized uniformly parabolic PDE (8.1) with the heat conduction coefficient $k(u)$ replaced by its strictly positive approximation $k_n(u)$ satisfying $k_n(u) \geq \frac{1}{n} > 0$ for all u, and $k_n \to k$ as $n \to \infty$ uniformly on bounded u-intervals. For instance,

$$k_n(u) = \sqrt{k^2(u) + \tfrac{1}{n^2}}.$$

For non-smooth absorption terms (8.5), the uniformly Lipschitz (analytic) approximation

$$f_n(u) = \left(\tfrac{1}{n^2} + u^2\right)^{\frac{p-1}{2}} u$$

is also used. If necessary, initial data are replaced by bounded smoother truncations $u_{0n}(x) \to u_0(x)$ as $n \to \infty$ uniformly on compact subsets.

The PDE for $u_n(x,t)$ is uniformly parabolic with smooth coefficients, so $u_n(x,t)$ is regular enough for our manipulations. The approximation (regularization) techniques lie in the heart of modern theory of quasilinear singular parabolic PDEs of arbitrary order; see references in [226, Sect. 6.2]

8.1.2 Maximum Principle and first examples of sign-invariants

The MP is the cornerstone of classical theory of second order parabolic PDEs, as explained in many well-known books and monographs, [148, 164, 205, 472, 530]. The MP and various order-preserving comparison techniques and results are associated with the fact that the Laplacian Δu has definite signs "$\geq$" or "$\leq$" at an extremum in x for C_x^2 smooth solutions $u(x,t)$. For instance, as a typical simple application of the MP for equations (8.1) and (8.4) with smooth coefficients and $f(0) = 0$, we have the comparison with the trivial solutions $u = 0$, i.e.,

$$u_0(x) \geq 0 \;\; (\leq 0) \quad \text{in } I\!R^N \implies u(x,t) \geq 0 \;\; (\leq 0) \quad \text{in } I\!R^N \text{ for } t > 0. \tag{8.6}$$

This is the evolution invariance of the *sign* of the solutions $u(x,t)$. On the other hand, a slightly modified comparison implies that the *monotonicity with time* property holds

$$u_t(x,0) \geq 0 \;\; (\leq 0) \text{ in } I\!R^N \implies u_t(x,t) \geq 0 \;\; (\leq 0) \text{ in } I\!R^N \text{ for } t > 0. \tag{8.7}$$

This is the invariance with time of the sign of the derivative u_t (in fact, this represents a first simple SI). Bearing in mind necessary hypotheses on the coefficients and initial data, here the MP in the following form is used. Let a smooth function $J(x,t)$ satisfy a linear parabolic PDE

$$J_t = \mathcal{M}[J] \equiv A\Delta J + \mathbf{B} \cdot \nabla J + CJ \quad \text{in } I\!R^N \times I\!R_+, \tag{8.8}$$

where $A \geq 0$, $\mathbf{B} = (B_1, ..., B_N)$, and C are given bounded coefficients, depending on x, t, and possibly u (the dot "$\cdot$" denotes the scalar product in $I\!R^N$). Then,

$$J(x,0) \geq 0 \;\; (\leq 0) \text{ in } I\!R^N \implies J(x,t) \geq 0 \;\; (\leq 0) \text{ in } I\!R^N \text{ for } t > 0.$$

As above, a rigorous proof uses suitable hypotheses on the coefficients of (8.8) and also on the behavior of $J(x,t)$ as $|x| \to \infty$.

Therefore, property (8.6) follows immediately from the MP, since equations (8.1) and (8.4) have been already written down in the form of (8.8) for the function $J = u$. To show (8.7), one needs to differentiate the PDE with respect to t to obtain (8.8) for $J = u_t$, assuming, as usual, that the regularity of all the functions and coefficients is enough for such manipulations. For instance, differentiating (8.1) yields, for $J = u_t$,

$$J_t = k(u)\Delta J + 2\nabla k(u) \cdot \nabla J + [\Delta k(u) + f'(u)]J, \tag{8.9}$$

which is precisely (8.8) with $A = k(u)$, $\mathbf{B} = 2\nabla k(u)$, and $C = \Delta k(u) + f'(u)$.

In addition, since the first x_i-derivative $J = u_{x_i}$ for any $i = 1, 2, ..., N$ satisfies the same equation (8.9), the MP also implies the following *space* monotonicity property (cf. (8.7)):

$$(u_0)_{x_i} \geq 0 \ \ (\leq 0) \text{ in } I\!R^N \implies u_{x_i}(x,t) \geq 0 \ \ (\leq 0) \text{ in } I\!R^N \text{ for } t > 0.$$

This means the evolution invariance of the sign of u_{x_i}. We next consider examples of more complicated and practical SIs.

8.1.3 First-order sign-invariants

We turn to the general problem of finding first-order SIs for a general fully nonlinear parabolic PDE of the form

$$\mathcal{P}[u] \equiv u_t - L(x, u, \nabla u, \Delta u) = 0 \quad \text{in } I\!R^N \times I\!R_+, \tag{8.10}$$

where $L(x, u, p, q)$ is a given C^∞-function satisfying the parabolicity condition

$$L'_q(x, u, p, q) \geq 0 \quad \text{in } I\!R^N \times I\!R \times I\!R^N \times I\!R. \tag{8.11}$$

As above, consider the Cauchy problem for (8.10) with initial data (8.3). For convenience, let us introduce the set of proper initial functions and solutions of (8.10),

$$\omega_{\mathcal{P}} = \{u_0(x) \in C^2 : \ \exists \ \text{ a unique solution } u(x,t) \in C^\infty\},$$

$$\Omega_{\mathcal{P}} = \{u \in C^2 : \ u(x,t) \text{ solves (8.10) with } u_0 \in \omega_{\mathcal{P}}\}.$$

Consider a general nonlinear first-order Hamilton–Jacobi operator of the form

$$\mathcal{H}[u] \equiv H(x, u, \nabla u, u_t), \tag{8.12}$$

where $H(x, u, p, q)$ is a C^∞-function. We will study the sign of $\mathcal{H}[u]$ on the evolution orbits from $\Omega_{\mathcal{P}}$. Therefore, consider the *reduced* operator

$$\mathcal{H}_{\mathcal{P}}[u] \equiv H(x, u, \nabla u, L(x, u, \nabla u, \Delta u)) \quad \text{in } \Omega_{\mathcal{P}},$$

where u_t is replaced by $L(x, u, \nabla u, \Delta u)$ via (8.10). This gives the following sets:

$$S^+_{\mathcal{H},\mathcal{P}} = \left\{v(x) \in C^2 : \ \mathcal{H}_{\mathcal{P}}[v] \geq 0 \text{ in } I\!R^N\right\} \cap \omega_{\mathcal{P}}, \tag{8.13}$$

$$S^-_{\mathcal{H},\mathcal{P}} = \left\{v(x) \in C^2 : \ \mathcal{H}_{\mathcal{P}}[v] \leq 0 \text{ in } I\!R^N\right\} \cap \omega_{\mathcal{P}}. \tag{8.14}$$

Definition 8.1 The first-order operator (8.12) is said to be a *sign-invariant* of the equation (8.10) if both signs of $\mathcal{H}$ are preserved with time,

$$\forall\, u_0 \in S^+_{\mathcal{H},\mathcal{P}} \ \ (S^-_{\mathcal{H},\mathcal{P}}) \implies u(\cdot,t) \in S^+_{\mathcal{H},\mathcal{P}} \ \ (S^-_{\mathcal{H},\mathcal{P}}) \ \text{ for } \ t > 0. \tag{8.15}$$

If the sign of the operator (8.12) on evolution orbits $\{u(x,t),\ t > 0\}$, i.e., the inequality

$$\mathcal{H}_{\mathcal{P}}[u(x,t)] \geq 0 \ \ (\text{or } \leq 0) \ \text{ in } I\!R^N \text{ for } t > 0,$$

is known, integrating it yields estimates for solutions of the nonlinear parabolic PDE (8.10). Very often such estimates are an important part of general parabolic theory, where the structural properties of possible operators $\mathcal{H}[u]$ play a key role. In subsequent sections, we present special approaches to finding nontrivial pairs of the operators $\{\mathcal{P}, \mathcal{H}\}$ such that (8.15) holds.

8.1.4 *Sign-invariants, zero-invariants, exact solution, and differential constraints*

Using (8.13) and (8.14), it is natural to introduce the set

$$S^0_{\mathcal{H},\mathcal{P}} = S^+_{\mathcal{H},\mathcal{P}} \cap S^-_{\mathcal{H},\mathcal{P}} = \left\{v(x) \in C^\infty : \ \mathcal{H}_{\mathcal{P}}[v] = 0 \text{ in } I\!R^N\right\} \cap \omega_{\mathcal{P}}.$$

It follows from (8.15) that any SI (8.12) also becomes a *zero-invariant* of (8.10), in the sense that the set $S^0_{\mathcal{H},\mathcal{P}}$ is *evolutionary* invariant under the flow generated by (8.10), i.e.,

$$\forall\, u_0 \in S^0_{\mathcal{H},\mathcal{P}} \quad \Longrightarrow \quad u(\cdot, t) \in S^0_{\mathcal{H},\mathcal{P}} \quad \text{for } t > 0.$$

This implies that the parabolic equation (8.10) restricted to the invariant set $S^0_{\mathcal{H},\mathcal{P}}$ is equivalent to the Hamilton–Jacobi equation

$$H(x, u, \nabla u, u_t) = 0, \quad \text{with } u(\cdot, t) \in S^0_{\mathcal{H},\mathcal{P}} \quad \text{for } t > 0. \tag{8.16}$$

It is easier to solve the first-order equation (8.16) than the second-order parabolic PDE (8.10), and, in some cases, this can be done explicitly. In several cases, such exact solutions can be treated from the point of view of linear finite-dimensional subspaces that are (partially) invariant under certain nonlinear operators.

In terms of zero-invariants, (8.16) represents a *differential constraint* associated with the nonlinear PDE (8.10). In general, the problem of determining differential constraints (in our case, the function $H(\cdot)$) is reduced to a complicated nonlinear PDE for H (a compatibility condition), depending on the operator L. This problem cannot be solved even in simpler particular cases. We find several examples, showing how to find a first-order differential constraint. In this analysis, we use some known approaches via the MP coming from qualitative theory of nonlinear parabolic PDEs. In particular, we essentially use results from the theory of *blow-up* solutions of quasilinear heat equations, which was the origin of a number of new ideas and techniques.

8.2 Sign-invariants of the form $u_t - \psi(u)$

Let us apply such SIs to general fully nonlinear parabolic PDEs

$$\mathcal{P}[u] \equiv u_t - L(u, |\nabla u|, \Delta u) = 0 \quad \text{in } I\!R^N \times I\!R_+, \tag{8.17}$$

where $L(u, p, q)$ is smooth and satisfies the parabolicity condition like (8.11). Therefore, we may assume that there exists a smooth function $\ell_0(u, p, s)$ such that

$$L(u, p, \ell_0(u, p, s)) \equiv s \quad \text{for } (u, p, s) \in I\!R \times I\!R_+ \times I\!R. \tag{8.18}$$

We will look for SIs of (8.17) of the form

$$\mathcal{H}[u] \equiv u_t - \psi(u), \tag{8.19}$$

where $\psi(u)$ is a smooth unknown function. Set

$$L^{-1}(u, p) = \ell_0(u, p, \psi(u)), \tag{8.20}$$

and denote $L_u(u, p, q) = \frac{\partial L}{\partial u}$ and $L_p(u, p, q) = \frac{\partial L}{\partial p}$, ..., where the argument q is replaced by $\ell_0(u, p, \psi(u))$.

Theorem 8.1 *Operator* (8.19) *is a sign-invariant of the equation* (8.17) *if* ψ *satisfies the following identity for* $(u, p) \in I\!R \times I\!R_+$*:*

$$\mathcal{F}(u, p) = (L_u - \psi')\psi + L_p\psi' p + L_q(\psi'' p^2 + \psi' L^{-1}) \equiv 0. \tag{8.21}$$

Proof. Setting $J = u_t - \psi(u)$ yields, on differentiation,

$$J_t = u_{tt} - \psi' u_t. \tag{8.22}$$

Calculating u_{tt} from (8.17) implies

$$u_{tt} = L_u u_t + L_p(\nabla u \cdot \nabla u_t)\tfrac{1}{|\nabla u|} + L_q \Delta u_t. \tag{8.23}$$

Since $u_t = J + \psi(u)$, J solves a linear parabolic PDE of the form

$$J_t = \mathcal{M}[J] + \mathcal{F}, \tag{8.24}$$

where $\mathcal{M}$ is an elliptic operator as in (8.8) with $A = L_q$ and coefficients B and C from (8.23). By the MP, it follows from (8.24) that (8.19) is an SI if (8.21) holds. □

Assuming a one-sided partial differential inequality for $\mathcal{F}$, say,

$$\mathcal{F}(u, p) = (L_u - \psi')\psi + L_p\psi' p + L_q(\psi'' p^2 + \psi' L^{-1}) \geq 0, \tag{8.25}$$

the MP guarantees that precisely the sign "$\geq$" is only preserved for the operator (8.19), i.e.,

$$u_t - \psi(u) \geq 0 \quad \text{for } t > 0 \text{ in } I\!R^N, \tag{8.26}$$

provided that the same inequality holds for the initial data at $t = 0$. In classical parabolic theory, such one-sided inequalities are typically associated with the *barrier techniques*. Of course, if only a one-sided estimate is necessary, this essentially widens the set of possible solutions $\{\psi\}$ of the partial differential inequality (PDI) (8.25) and gives other estimates by integrating the PDI (8.26). The one-sided approach is not associated with exact solutions, and will not be considered in this text. One-sided estimates have a range of important applications in blow-up theory of reaction-diffusion PDEs; different applications are described in [509, Ch. 7] and [245, Ch. 10].

Let us return to the main identity (8.21) that is a nonlinear PDE for ψ, which is difficult to analyze for general operators L in (8.17). Our study is now restricted to the class of quasilinear heat equations.

8.2.1 Quasilinear heat equations and higher-order extensions

As the first application of Theorem 8.1, consider quasilinear equations (8.1) for which (8.21) reduces to a system of two ODEs.

Corollary 8.2 *Operator* (8.19) *is a sign-invariant of equation* (8.1) *iff* $\psi(u)$ *satisfies the ODE system*

$$\left[\tfrac{(k\psi)'}{k}\right]' = 0, \quad k'\psi^2 - f^2\left(\tfrac{k\psi}{f}\right)' = 0. \tag{8.27}$$

Proof. By (8.1), $L(u, p, q) = kq + k'p^2 + f$, so that

$$L^{-1} = \tfrac{1}{k}(\psi - f - k'p^2), \quad L_p = 2k'p, \quad L_q = k,$$
$$\text{and} \quad L_u = \left[k'' - \tfrac{(k')^2}{k}\right]p^2 + f' + \tfrac{k'}{k}(\psi - f). \tag{8.28}$$

Plugging (8.28) into (8.21) yields

$$\mathcal{F}(u, p) = k\left[\tfrac{(k\psi)'}{k}\right]' p^2 + \tfrac{1}{k}\left[k'\psi^2 - f^2\left(\tfrac{k\psi}{f}\right)'\right].$$

Since the variables u and p are independent, identity (8.21) is valid, provided that (8.27) holds. □

The system of ODEs (8.27) for two unknowns $\{\psi, f\}$ can be easily integrated. This yields the following family of equations.

Example 8.3 **(Invariant subspaces: blow-up)** Let $\varphi(u)$ be an arbitrary smooth function such that the inverse φ^{-1} exists. Denote $k(u) = \varphi'(u) \geq 0$. Solving (8.27) gives the quasilinear heat equation

$$u_t = \Delta\varphi(u) + \tfrac{a\varphi(u)+b}{\varphi'(u)} + [a\varphi(u) + b]c, \tag{8.29}$$

where $a^2 + b^2 \neq 0$. By (8.19), the PDE (8.29) admits the following SI:

$$\mathcal{H}[u] \equiv u_t - \tfrac{a\varphi(u)+b}{\varphi'(u)} \equiv H(u, u_t). \tag{8.30}$$

The ordinary differential operator (8.30) is also a zero-invariant of the PDE (8.29). This means that if, for a solution $u(x, t)$,

$$H(u, u_t) \equiv \Delta\varphi(u) + [a\varphi(u) + b]c = 0 \quad \text{in } I\!R^N \text{ for } t = 0, \tag{8.31}$$

then

$$H(u, u_t) = 0 \quad \text{in } I\!R^N \text{ for } t > 0. \tag{8.32}$$

Equation (8.32), (8.30) is integrated as a standard ODE $(\varphi(u))'_t = a\varphi(u) + b$. This yields the following exact solutions $u(x, t)$ of (8.29).

(i) If $a \neq 0$, then

$$\varphi(u(x, t)) = \tfrac{1}{a}\,[\rho(x)e^{at} - b], \tag{8.33}$$

where $\rho(x)$ is an arbitrary solution of the linear elliptic PDE

$$\Delta\rho + ac\rho = 0 \quad \text{in } I\!R^N. \tag{8.34}$$

(ii) If $a = 0$ $(b \neq 0)$, then

$$\varphi(u(x, t)) = \rho(x) + bt, \quad \text{where } \Delta\rho + bc = 0 \quad \text{in } I\!R^N.$$

Besides such exact solutions, the SI (8.31) makes it possible to estimate more solutions by using the inequality

$$u_t - \tfrac{a\varphi(u)+b}{\varphi'(u)} \geq 0 \quad (\text{or } \leq 0) \quad \text{for } t \geq 0.$$

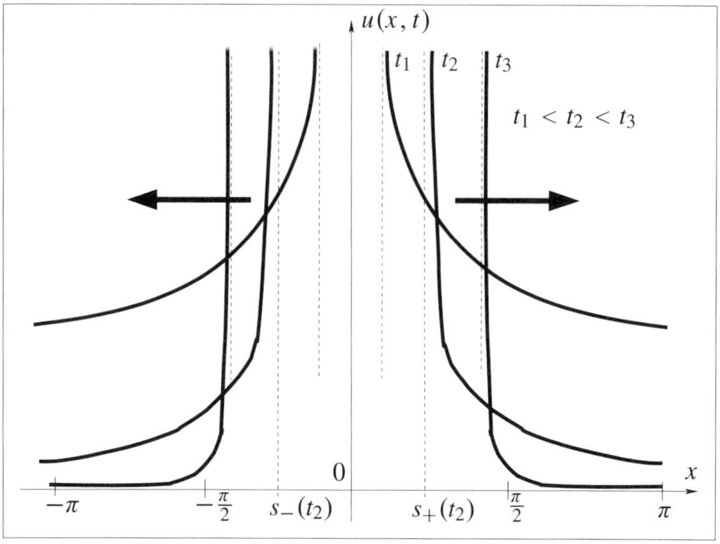

Figure 8.1 Stabilizing blow-up interfaces of the solution (8.36) of the PDE (8.35).

Blow-up interfaces. In the fast-diffusion case $\varphi(u) = -u^{-m}$, $m > 0$, with $a = b = c = 1$ and $N = 1$, we obtain from (8.29) the 1D equation

$$u_t = -(u^{-m})_{xx} + \frac{1}{m}(u - u^{m+1}) + 1 - u^{-m}. \tag{8.35}$$

According to (8.33), taking $\rho(x) = \cos x$ yields the explicit 2π-periodic solution

$$u(x,t) = (1 - e^t \cos x)^{-\frac{1}{m}}, \tag{8.36}$$

with blow-up interfaces propagating as follows:

$$s_{\pm}(t) = \pm \cos^{-1}(e^{-t}) \to \pm(\tfrac{\pi}{2})^{\mp} \quad \text{as } t \to \infty.$$

Such unusual blow-up behavior is illustrated in Figure 8.1. The FBP with blow-up interfaces and their dynamic equations are rigorously justified by the Sturmian intersection comparison approach, [226, Ch. 7].

Invariant subspaces. It is easy to interpret these exact solutions in terms of invariant subspaces. Consider the case $a \neq 0$. Using in (8.29) the *Kirchhoff-type transformation* $v = a\varphi(u) + b$ yields the following quasilinear equation:

$$\boxed{v_t = F[v] \equiv \Phi(v)(\Delta v + acv) + av,} \tag{8.37}$$

with the coefficient $\Phi(v) = \varphi'(\varphi^{-1}(\frac{v-b}{a}))$. Let us introduce the subspace

$$W = \{\rho(x) : \ \rho \text{ solves } (8.34)\}. \tag{8.38}$$

The existence of explicit solutions (8.33) is then a straightforward consequence of the fact that W is invariant under F, i.e., $F[W] \subseteq W$, so this falls into the scope of Chapters 1 and 2. Notice that $\dim W = \infty$ if $N > 1$, and $\dim W = 2$ if $N = 1$.

Substituting in (8.37) yields

$$v(x,t) = C(t)\rho(x) \in W \implies C' = aC, \tag{8.39}$$

which gives the solutions (8.33). For $\varphi(u) = u^{\sigma+1}$ in (8.29), there occurs the PME or the fast diffusion equation with reaction-absorption,

$$\boxed{u_t = \Delta u^{\sigma+1} + \tfrac{a}{\sigma+1}\, u + \tfrac{b}{\sigma+1}\, u^{-\sigma} + acu^{\sigma+1} + bc} \quad (\sigma \neq 0, -1).$$

For $\varphi(u) = \tan u$, we find the parabolic equation

$$\boxed{u_t = (\tan u)_{xx} + (a \tan u + b)(c + \cos^2 u).} \tag{8.40}$$

Example 8.4 The quasilinear heat equation (8.40) with $a = c = 1$ and $b = 0$ admits the SI

$$\mathcal{H}[u] = u_t - \tfrac{1}{2}\sin 2u.$$

It is also a zero-invariant, from which we obtain the explicit solution

$$u(x,t) = \tan^{-1}(\mathrm{e}^t \cos x).$$

The existence of such a solution becomes trivial if we set $v = \tan^{-1} u$, which yields solution $v(x,t) = \mathrm{e}^t \cos x$ of the PDE (cf. (8.37))

$$\boxed{v_t = (1 + v^2)(v_{xx} + v) + v.}$$

The analogy with the invariant subspace (8.38) reveals many extensions of such exact solutions to hyperbolic and higher-order PDEs, though, of course, any connection with SIs is then lost.

Example 8.5 (**Hyperbolic equations**) The hyperbolic PDE

$$\boxed{v_{tt} = F[v] \equiv \Phi(v)(\Delta v + acv) + av}$$

admits the same solutions (8.39) with the ODE $C'' = aC$.

Example 8.6 (**Higher-order PDEs**) The fourth-order parabolic equation

$$\boxed{v_t = F[v] \equiv \Phi(v)(-\Delta^2 v - ac\Delta v) + av,}$$

with an arbitrary function $\Phi(v)$, has solutions (8.39). One can extend the list of PDEs with operators preserving such linear subspaces.

Example 8.7 In a similar fashion, the fourth-order quasilinear hyperbolic PDE

$$\boxed{v_{tttt} = (1 + v^2)(v_{xxxx} - v) + v}$$

admits explicit solutions

$$v(x,t) = \mathrm{e}^{t+x} + \mathrm{e}^{-(t+x)} + \cos(x - t).$$

For convenience, we combine some generalizations in the following statement.

Proposition 8.8 *Let $Q(D_x)$ ($D_x = \nabla_x$ in $I\!R^N$) be a linear differential operator with constant coefficients, and $\varphi(u)$ be a smooth increasing function. The equation*

$$\boxed{u_t = Q(D_x)[\varphi(u)] + \tfrac{a\varphi(u)+b}{\varphi'(u)} + d} \tag{8.41}$$

with constants $a \neq 0$, b, and d such that

$$ad = Q(D_x)[b], \tag{8.42}$$

admits explicit solution (8.33), where $\rho(x)$ solves

$$Q(D_x)[\rho] = 0 \quad \text{in } I\!R^N.$$

Proof. If $a \neq 0$, setting $a\varphi(u) + b = v$ yields

$$v_t = \varphi'(u)\big\{Q(D_x)[v] + ad - Q(D_x)[b]\big\} + av.$$

In view of (8.42), the solution is $v(x,t) = \rho(x)e^{at}$. □

If $a = 0$ in (8.41), setting $v = \varphi(u)$ gives

$$v_t = \varphi'(u)\{Q(D_x)[v] + d\} + b.$$

Hence, under the hypothesis $Q(D_x)1 = 0$ (see (8.42) with $a = 0$), it admits the solution $v(x,t) = \rho(x) + bt$, where ρ solves $\Delta\rho + d = 0$ in $I\!R^N$.

Example 8.9 Taking in (8.41) the operator

$$Q(D_x)[w] = \Delta w + \alpha(\nabla w \cdot n) + \beta w$$

with $ad = \beta b$, Proposition 8.8 yields exact solutions of the following quasilinear heat equation with a nonlinear convective term:

$$\boxed{u_t = \Delta\varphi(u) + \alpha(\nabla\varphi(u) \cdot n) + \tfrac{a\varphi(u)+b}{\varphi'(u)} + \beta\varphi(u) + d.}$$

Extensions to $2m$th-order quasilinear parabolic PDEs are straightforward, but then operator (8.30) is not an SI for any $m \geq 2$, and remains a zero-invariant.

Example 8.10 As a next extension of the same idea, we introduce the following equation with two linear operators $Q_i(D_x)$, and arbitrary functions $\Phi_i(u)$:

$$\boxed{u_t = \Phi_1(u)Q_1(D_x)[\varphi(u)] + \Phi_2(u)Q_2(D_x)[\varphi(u)] + \tfrac{a\varphi(u)+b}{\varphi'(u)},}$$

where $Q_1(D_x)[1] = Q_2(D_x)[1] = 0$. There exist exact solutions (8.33), with $a \neq 0$, and (8.27), with $a = 0$, if $\rho(x)$ solves a system in $I\!R^N$,

$$\begin{cases} Q_1(D_x)[\rho] = 0, \\ Q_2(D_x)[\rho] = 0. \end{cases}$$

8.3 Stationary sign-invariants of the form $H(r, u, u_r)$

Consider the SIs which do not contain the time-derivative u_t. Important examples of such SIs come from blow-up theory; see Remarks. Consider the general 1D parabolic equation in radial geometry

$$\mathcal{P}[u] \equiv u_t - L(r, u, u_r, u_{rr}) = 0, \tag{8.43}$$

where $r > 0$ denotes $|x|$, and $L(r, u, p, q)$ satisfies the parabolicity condition (8.11). We take the SI in the general form

$$\mathcal{H}[u] = H(r, u, u_r), \tag{8.44}$$

where smooth $H(r, u, p)$ satisfies $H_p > 0$. By $H^*(r, u, s)$ we denote the inverse function such that

$$H(r, u, H^*(r, u, s)) \equiv s \quad \text{for } (r, u, s) \in I\!R_+ \times I\!R \times I\!R.$$

We will set $h(r, u) = H^*(r, u, 0)$ and use other notations from the previous section. In particular, in the functions L, H, and their derivatives L_r, L_u,..., H_r, H_u,..., variables p and q are replaced by $h(r, u)$ and $h_r + h_u h$ respectively.

Theorem 8.11 *Operator* (8.44) *is a sign-invariant of the equation* (8.43) *if* H *satisfies the following identity for all* $(r, u) \in I\!R_+ \times I\!R$:

$$\begin{aligned} \mathcal{F}(r, u) &= -h_u L + L_r + L_u h + L_p(h_r + h_u h) \\ &+ L_q\big[h_{rr} + 2h_{ru}h + h_{uu}h^2 + h_u h_r + (h_u)^2 h\big] \equiv 0. \end{aligned} \tag{8.45}$$

Proof. Let $J = H(r, u, u_r)$, so $u_r = H^*(r, u, J)$ and J satisfies

$$J_t = H_u u_t + H_p u_{tr}\,. \tag{8.46}$$

By (8.43), $u_t = L$ and hence,

$$u_{tr} = L_r + L_u H^* + L_p u_{rr} + L_q u_{rrr}\,. \tag{8.47}$$

Let us derive for J a linear parabolic PDE of the form (8.24). Since the coefficient $\mathcal{F}$ in this equation is eventually derived by the standard linearization of the equation about $J = 0$, this allows us to calculate the derivatives in (8.46) and (8.47) from the equality $u_r = h(r, u)$. In this case,

$$u_{rr} = h_r + h_u h \quad \text{and} \quad u_{rrr} = h_{rr} + 2h_{ru}h + h_{uu}h^2 + h_u h_r + (h_u)^2 h.$$

Substituting into (8.46) yields (8.24) with $\mathcal{F} = 0$ given by (8.45). By the MP, this completes the proof. □

8.3.1 Quasilinear heat equations

Consider the equation (8.1) with $k = \varphi'$, which, for radial solutions $u = u(r, t)$ using the radial Laplacian in $I\!R^N$, is written as

$$\boxed{u_t = (k(u)u_r)_r + \tfrac{N-1}{r}\, k(u)u_r + f(u).} \tag{8.48}$$

We take the SI (8.44) in the semilinear form

$$\mathcal{H}[u] = u_r - g(r)\Phi(u), \tag{8.49}$$

where $g(r)$ and $\Phi(u)$ are unknown functions. Identity (8.45) then becomes simpler,

$$\begin{aligned} \mathcal{F}(r,u) &= g^3(r)[\Phi^2(k\Phi)''] \\ &+ g(r)g'(r)[3k'\Phi^2 + 2k\Phi\Phi'] + g(r)(f'\Phi - f\Phi') \\ &+ g''(r)k\Phi + \tfrac{N-1}{r}\left[g^2(r)k'\Phi^2 + \left(g'(r) - \tfrac{g(r)}{r}\right)k\Phi\right] \equiv 0. \end{aligned} \tag{8.50}$$

In general, the right-hand side belongs to a 6D subspace. For $N = 1$, the last term vanishes. We studied a similar identity and obtained special invariant sets and exact solutions in Section 7.5 by using the ODE (7.158) for g (in order to match the results, g should be replaced by $-g$ here). In the present more difficult case of dimensions $N > 1$, we restrict our attention to the three most interesting cases:

(I) $g(r) = r$, **(II)** $g(r) = \frac{1}{r}$, and **(III)** $g(r) = r^{1-N}$ in dimension $N \geq 3$.

(I) $g(r) = r$.

Proposition 8.12 *Operator*

$$\mathcal{H}[u] = u_r - r\Phi(u)$$

is a sign-invariant of (8.48) *if* $\bar{\Phi}(u) = k(u)\Phi(u) \not\equiv 0$ *satisfies the system*

$$\begin{cases} I_1(u) \equiv \bar{\Phi}'' = 0, \\ I_2(u) \equiv \frac{k'}{k}\left(N + \frac{f}{\Phi}\right) + \left(2\ln|\bar{\Phi}| + \frac{f}{\Phi}\right)' = 0. \end{cases} \tag{8.51}$$

Proof. Identity (8.50) with $g(r) = r$ reads $\mathcal{F}(r,u) = r^3\frac{\bar{\Phi}^2}{k^2}I_1 + r\frac{1}{k}I_2$, and system (8.51) follows. □

The first ODE in (8.51) implies that $\bar{\Phi}(u) = au + b$ is linear, and solving the second equation for $f(u)$ yields PDEs possessing blow-up exact solutions.

Example 8.13 **(Focusing blow-up interfaces)** Consider the PDE

$$\boxed{u_t = \Delta\varphi(u) - (au+b)\left[N + \tfrac{2a}{\varphi'(u)}G(u)\right],} \tag{8.52}$$

where $\varphi(u)$ is arbitrary, $\varphi'(u) \geq 0$, and $a \neq 0$. Set

$$G(u) = \int \tfrac{\varphi'(u)\,\mathrm{d}u}{au+b},$$

assuming that the inverse function G^{-1} exists. In the class of radially symmetric solutions $u = u(r,t)$, with $r = |x|$, this equation admits the following SI.

$$\mathcal{H}[u] \equiv u_r - r\tfrac{au+b}{\varphi'(u)}. \tag{8.53}$$

The semilinear operator (8.53) is also a zero-invariant of (8.52), so $\mathcal{H}[u(r,t)] = 0$ for $t > 0$, provided that $\mathcal{H}[u_0] = 0$. This implies that $(G(u))_r - r = 0$ holds, and integrating yields $G(u) = C_1(t) + \frac{1}{2}r^2$. Plugging into (8.52) gives the explicit solution

$$G(u(x,t)) = \tfrac{1}{2}|x|^2 - \mathrm{e}^{-2at}. \tag{8.54}$$

Setting $a = 1$ and $b = 0$, and assuming that φ satisfies *Osgood's criterion*

$$\int^{\infty} \tfrac{\varphi'(u)}{u}\,\mathrm{d}u < \infty,$$

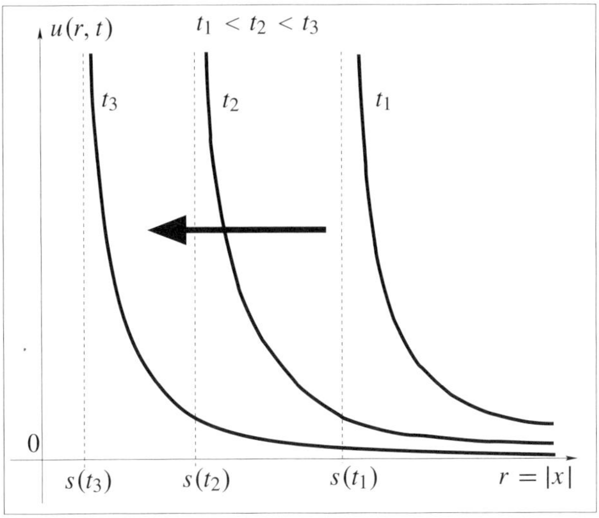

Figure 8.2 Blow-up solution (8.55) with focusing singular interface.

we obtain the blow-up solution

$$u(r,t) = G^{-1}\big(\tfrac{1}{2}\,|x|^2 - \mathrm{e}^{-2t}\big). \tag{8.55}$$

This has the blow-up interface (surface) focusing at the origin in infinite time,

$$|x| = s(t) \equiv \sqrt{2}\,\mathrm{e}^{-t} \to 0 \quad \text{as } t \to \infty,$$

as shown in Figure 8.2.

Partially invariant 2D subspace. Let us present an invariant interpretation of such solutions. Setting $G(u) = v$ in (8.52) yields

$$\boxed{v_t = F[v] \equiv \Phi(v)(\Delta v - N) + a(|\nabla v|^2 - 2v),} \tag{8.56}$$

where $\Phi(v) = \varphi'(G^{-1}(v))$. Consider (8.56) on the 2D subspace $W_2 = \mathcal{L}\{1, |x|^2\}$,

$$v(x,t) = C_1(t) + C_2(t)\,|x|^2.$$

Obviously, W_2 *is not* invariant under the operator F, since

$$F[v] = \Phi(v)N(2C_2 - 1) + a\big[2C_2|x|^2(2C_2 - 1) - 2C_1\big], \tag{8.57}$$

so $F[W_2] \not\subseteq W_2$. But it follows from (8.57) that F admits the invariant set (i.e., W_2 is partially invariant)

$$M = \big\{C_1 + C_2|x|^2 : \ C_2 = \tfrac{1}{2}\big\}.$$

Indeed, $F[M] \subseteq W_2$, and more precisely, $F[M] \subseteq W_1 = \mathcal{L}\{1\}$. Therefore, as in several examples in Chapter 7, the PDE (8.56) on M is reduced to an overdetermined DS, which has a solution. Then, setting $C_2 = \frac{1}{2}$ in (8.57), i.e., looking for the solution

$$v(x,t) = C_1(t) + \tfrac{1}{2}\,|x|^2,$$

yields $F[v] = -2aC_1$. Hence, (8.56) on M becomes an elementary linear ODE,

$$C_1' = -2aC_1,$$

which yields the solution (8.54). Setting in (8.52) $\varphi(u) = u^2$, $a = b = 1$, and $G(u) = 2[u - \ln(1+u)]$ yields the PDE

$$\boxed{u_t = \Delta u^2 + \tfrac{2(1+u)}{u} \ln(1+u) - (N+2)(1+u).}$$

If $\varphi(u) = \ln(1+u^2)$, $a = 1$, and $b = 0$, the PDE is

$$\boxed{u_t = \Delta \ln(1+u^2) - 2(1+u^2)\tan^{-1} u - Nu.}$$

Example 8.14 (**Fourth-order parabolic equation**) We easily extend this simple partial invariance analysis to higher-order PDEs, e.g., to the quasilinear equation

$$\boxed{v_t = \Phi(v)[-\Delta^2 v + 4N(N+2)] + \alpha\big[(\Delta v)^2 - 8(N+2)^2 v\big]}$$

possessing the solution on a set from $\mathcal{L}\{1, |x|^4\}$,

$$v(x,t) = \tfrac{1}{2}\,|x|^4 + \mathrm{e}^{-8(N+2)^2 \alpha t}.$$

Indeed, since $\Delta |x|^4 = 4(N+2)|x|^2$ and $\Delta^2 |x|^4 = 8N(N+2)$, the first term in the equation vanishes, while, in the second term, we observe cancellation of both the $|x|^4$-projections.

Example 8.15 (**Quasilinear KdV-type PDE**) The following third-order nonlinear dispersion equation:

$$\boxed{v_t = \Phi(v)(v_{xxx} - 6) + \alpha(v_x v_{xx} - 18v)}$$

admits the solution on $M \subset \mathcal{L}\{1, x^3\}$,

$$v(x,t) = x^3 + \mathrm{e}^{-18\alpha t}.$$

It is easy to reconstruct other higher-order models, including two and more nonlinear terms annihilating on suitable sets $M \subset W$.

(II) $g(r) = \frac{1}{r}$.

Proposition 8.16 *Operator*

$$\mathcal{H}[u] = u_r - \tfrac{1}{r}\,\Phi(u) \tag{8.58}$$

is a sign-invariant of the equation (8.48) *iff* $\Phi(u) \not\equiv 0$ *satisfies the system*

$$I_1(u) \equiv f'\Phi - f\Phi' = 0, \tag{8.59}$$

$$I_2(u) \equiv \Phi(k\Phi)'' + (N-4)k'\Phi - 2k\Phi' - 2(N-2)k = 0. \tag{8.60}$$

Proof. Setting $g(r) = \frac{1}{r}$ in (8.50) yields $\mathcal{F}(r,u) = \frac{1}{r}\, I_1(u) + \frac{\Phi(u)}{r^3}\, I_2(u)$. □

The first ODE in the system (8.59), (8.60) implies that

$$f(u) = \lambda \Phi(u) \tag{8.61}$$

for some constant λ. The next second-order ODE (8.60) for $\Phi(u)$ cannot be solved explicitly in general. We will show how to construct explicit solutions of (8.48) by using the fact that, under hypotheses (8.61) and (8.60), the operator (8.58) is a zero-invariant of the above quasilinear heat equation.

Example 8.17 Consider (8.48) with $f(u)$ given by (8.61),

$$\boxed{u_t = k(u)\big(u_{rr} + \tfrac{N-1}{r}\, u_r\big) + k'(u)(u_r)^2 + \lambda\Phi(u).} \tag{8.62}$$

Using the zero-invariance of the operator (8.58) means that, for $t > 0$,

$$u_r - \tfrac{1}{r}\,\Phi(u) = 0 \quad \Longrightarrow \quad v \equiv \int^u \tfrac{dz}{\Phi(z)} = \ln r + C_1(t). \tag{8.63}$$

It follows from (8.62) that $v(r,t)$ solves

$$v_t = k(u)\big(v_{rr} + \tfrac{N-1}{r}\, v_r\big) + (k\Phi)'_u(v_r)^2 + \lambda.$$

Substituting $v(r,t)$ from (8.63) yields

$$C_1' = \tfrac{1}{r^2}\, G(v) + \lambda, \quad \text{where} \quad G(v) = (k\Phi)'_u(u) + (N-2)k(u). \tag{8.64}$$

Equation (8.60) implies that there exists a constant $d \in I\!R$ such that

$$G(v) = d\, e^{2v}. \tag{8.65}$$

Indeed, it follows from (8.63) that $u'_v = \Phi(u)$, and (8.64) yields

$$G'_v = \Phi(k\Phi)''_{uu} + (N-2)k'\Phi.$$

Therefore, by (8.60), $G'_v = 2G$, so (8.65) holds. Finally, by (8.64), $C_1' = d\, e^{2C_1} + \lambda$, so, on integration,

$$C_1(t) = \tfrac{1}{2}\, \ln\big(\tfrac{\lambda e^{2\lambda t}}{1 - d e^{2\lambda t}}\big). \tag{8.66}$$

By (8.63), this determines the explicit solution of (8.62).

Consider some particular cases:

(i) In the case of the semilinear heat equation with $k(u) \equiv 1$ in (8.62),

$$\boxed{u_t = \Delta u + \lambda\Phi(u),}$$

the SI has the form $\mathcal{H}[u] = u_r - \frac{1}{r}\,\Phi(u)$, where $\Phi(u)$ solves the following nonlinear ODE (cf. (8.60)):

$$\Phi\Phi'' - 2\Phi' - 2(N-2) = 0.$$

The corresponding explicit solution is again given by (8.63) and (8.66).

(ii) For $N = 2$, (8.60) can be rewritten as $\Phi(k\Phi)'' - 2(k\Phi)' = 0$. Setting $k\Phi = \Psi$, and hence, $\Phi = \frac{2\Psi'}{\Psi''}$, from (8.62) with $\lambda = 1$, we obtain the quasilinear equation

$$u_t = \tfrac{1}{r}\big[r\big(\tfrac{\Psi\Psi''}{2\Psi'}\big)(u)\, u_r\big]_r + 2\big(\tfrac{\Psi'}{\Psi''}\big)(u)$$

admitting the sign and zero-invariant $\mathcal{H}[u] = u_r - \frac{2}{r}\frac{\Psi'(u)}{\Psi''(u)}$. The corresponding explicit solution is $\Psi'(u(r,t)) = C_1(t)r^2$, where $C_1(t)$ satisfies

$$C_1' = 2C_1(C_1 + 1).$$

(iii) For the function

$$\Phi(u) = (2-N)u \quad (N \neq 2), \tag{8.67}$$

which simplifies the ODE (8.60), $k(u)$ has to satisfy $(2-N)(ku)'' = (4-N)k'$. Integrating yields the heat conduction coefficients $k(u) = a + b\,u^{\frac{2}{2-N}}$, i.e., the PDE

$$\boxed{u_t = \nabla \cdot \big[\big(a + bu^{\frac{2}{N-2}}\big)\nabla u\big] + (2-N)u.}$$

(iv) Trying in (8.60) the constant function $\Phi(u) \equiv \gamma \neq 0$ yields a linear ODE for $k(u)$, $\gamma^2 k'' + (N-4)\gamma k' - 2(N-2)k = 0$. Hence, $k(u) = a\mathrm{e}^{\frac{2-N}{\gamma}u} + b\mathrm{e}^{\frac{2}{\gamma}u}$, which yields the PDE with $\gamma = 1$,

$$\boxed{u_t = \nabla \cdot \big[\big(a\mathrm{e}^{(2-N)u} + b\mathrm{e}^{2u}\big)\nabla u\big] + 1.}$$

(v) Finally, trying in (8.60) the linear function (cf. (8.67)) $\Phi(u) = \gamma u$, $\gamma \neq 0$, yields the second-order Euler ODE

$$u^2 k'' + \big(2 + \tfrac{N-4}{\gamma}\big)uk' - \tfrac{2(\gamma+N-2)}{\gamma^2}\,k = 0. \tag{8.68}$$

Here, setting $k(u) = u^\rho$ determines the characteristic equation

$$\rho^2 + \big(1 + \tfrac{N-4}{\gamma}\big)\rho - \tfrac{2(\gamma+N-2)}{\gamma^2} = 0.$$

Therefore, $\rho_1 = \frac{2}{\gamma}$ and $\rho_2 = -1 - \frac{N-2}{\gamma}$. If $\gamma \neq -N$, $\rho_1 \neq \rho_2$, and the general solution of (8.68) has the form $k(u) = au^{\rho_1} + bu^{\rho_2}$, i.e., the PDE with $\gamma = 1$,

$$\boxed{u_t = \nabla \cdot [(au^2 + bu^{1-N})\nabla u] + u.}$$

If $\gamma = -N$, the general solution is $k(u) = u^{-\frac{2}{N}}(a \ln u + b)$, giving the following parabolic equation:

$$\boxed{u_t = \nabla \cdot \big(u^{-\frac{2}{N}} \ln u\, \nabla u\big) - Nu.}$$

It follows from (8.63) with $\Phi(u) = \gamma u$ that, in both cases, the explicit solution is $u(r,t) = r^\gamma \mathrm{e}^{\gamma C_1(t)}$, where $C_1(t)$ solves

$$C_1' = \begin{cases} a(N+\gamma)\mathrm{e}^{2C_1} + \lambda & \text{for } \gamma \neq -N, \\ -aN\mathrm{e}^{2C_1} + \lambda & \text{for } \gamma = -N. \end{cases}$$

(III) $g(r) = r^{1-N}$**.** In this case, (8.50) implies that

$$\begin{aligned} \mathcal{F}(r,u) &= r^{3(1-N)}\Phi^2(k\Phi)'' \\ &- 2(N-1)r^{1-2N}\,\Phi(k\Phi)' + r^{1-N}(f'\Phi - f\Phi'). \end{aligned}$$

By Theorem 8.11, $\mathcal{H}[u] = u_r - r^{1-N}\Phi(u)$ ($N \geq 3$) is an SI of (8.48) if

$$\begin{cases} (k\Phi)' = 0, \\ f'\Phi - f\Phi' = 0. \end{cases}$$

This yields quasilinear heat equations, which have been considered in Section 8.2 in a more general setting; cf. (8.29).

In the 1D case $N = 1$, there exists another trivial choice $g(r) \equiv 1$, where the SI (8.49) has the form

$$\mathcal{H}[u] = u_r - \Phi(u).$$

Then (8.50) implies that f solves the ODE

$$\Phi^2(k\Phi)'' + f'\Phi - f\Phi' = 0.$$

Integrating once yields $f(u) = [a - (k\Phi)']\Phi$. Using the operator $\mathcal{H}[u]$ gives the standard traveling wave solutions $u = \theta(x - \lambda t)$.

8.4 Sign-invariants of the form $u_t - m(u)(u_x)^2 - M(u)$

In this section, we find more general SIs and explicit solutions of quasilinear heat equations with various nonlinearities, including the following two models:

$$\boxed{u_t = u_{xx} + d\,u\big(2\sqrt{\ln u} + \tfrac{1}{\sqrt{\ln u}}\big) + u(2\ln u + 1),}$$

$$\boxed{u_t = (\sqrt{u}\,)_{xx} + \big(\tfrac{1}{\tan^{-1}\sqrt{u}} + 1\big)\,(1 + 4\sqrt{u}\tan^{-1}\sqrt{u}\,)(1+u).} \tag{8.69}$$

As in Section 8.2, consider first the general nonlinear parabolic PDE

$$\mathcal{P}[u] \equiv u_t - L(u, u_x, u_{xx}) = 0 \quad \text{in } I\!R^N \times I\!R_+, \tag{8.70}$$

where the function $L(u, p, q)$ satisfies (8.11). Set (cf. (8.19))

$$\mathcal{H}[u] = u_t - \psi(u, u_x), \tag{8.71}$$

where $\psi(u, p)$ is a smooth function. Assume that $\ell_0(u, p, s)$ is well defined by (8.18). Let $L^{-1}(u, p)$ denote the function (8.20). In the notation of Section 8.2, the following result is derived.

Theorem 8.18 *Operator* (8.71) *is a sign-invariant of the equation* (8.70) *if* $\psi(u, p)$ *satisfies the identity* (*cf.* (8.21))

$$\begin{aligned} \mathcal{F}(u, p) = (L_u - \psi_u)\psi + (L_p\psi_u - L_u\psi_p)p \\ + L_q\big[\psi_{uu}p^2 + 2\psi_{up}L^{-1}p + \psi_{pp}(L^{-1})^2 + \psi_u L^{-1}\big] \equiv 0. \end{aligned} \tag{8.72}$$

Proof. This is the same as in Section 8.2. By calculating J_t from a slightly different formula than (8.22),

$$J_t = u_{tt} - \psi_u u_t - \psi_p u_{xt},$$

u_{tt} from (8.70), and also u_{tx} and u_{txx} by differentiating $J = u_t - \psi$, we derive a PDE such as (8.24). We also use the identities, following from (8.18),

$$L_u + L_q(\ell_0)_u \equiv 0, \;\; L_p + L_q(\ell_0)_p \equiv 0, \;\; \text{and} \;\; L_q(\ell_0)_s \equiv 1.$$

This completes the proof. □

Let us solve the nonlinear PDE (8.72) for $\psi(u, p)$ in some particular cases.

8.4.1 Quasilinear heat equations

Consider the 1D quasilinear equation (8.1). Without loss of generality, we may take it in the form of

$$\mathcal{P}[u] \equiv u_t - [\varphi(u)u_{xx} + f(u)] = 0, \tag{8.73}$$

with a smooth function $\varphi(u) \geq 0$. The equation (8.1) is reduced to (8.73) by the transformation $\hat{u} = \int^u k(s)\,\mathrm{d}s$. Set

$$\mathcal{H}[u] \equiv u_t - \big[m(u)(u_x)^2 + M(u)\big], \tag{8.74}$$

where $m(u)$ and $M(u)$ are smooth unknown functions.

Corollary 8.19 *Operator* (8.74) *is a sign-invariant of equation* (8.73) *if the functions* $m(u)$ *and* $M(u)$ *satisfy the following system of ODEs:*

$$I_1(u) \equiv m''\varphi + 4mm' - m^2\tfrac{\varphi'}{\varphi} + \tfrac{2m^3}{\varphi} = 0, \tag{8.75}$$

$$\begin{aligned} I_2(u) \equiv \varphi M'' + 4m'M - mf' \\ - 5m'f + \tfrac{4m^2}{\varphi}(M - f) + m\tfrac{\varphi'}{\varphi}f = 0, \end{aligned} \tag{8.76}$$

$$I_3(u) \equiv f'M - fM' + \tfrac{2m}{\varphi}(M - f)^2 + M\tfrac{\varphi'}{\varphi}(M - f) = 0. \tag{8.77}$$

Proof. ODEs (8.75)–(8.77) follow from the identity

$$\mathcal{F}(u, p) \equiv I_1 p^4 + I_2 p^2 + I_3,$$

which is easily derived from (8.72). □

We now reduce system (8.75)–(8.77) to a single equation. Let us introduce the new function $v(x, t)$ by setting

$$u = E(v), \tag{8.78}$$

where E is a monotone solution of the ODE

$$E'' = \tfrac{m(E)}{\varphi(E)}(E')^2. \tag{8.79}$$

In terms of v, (8.73) has the form

$$\tilde{\mathcal{P}}[v] \equiv v_t - \big[\varphi(E)v_{xx} + \varphi(E)\tfrac{E''}{E'}(v_x)^2 + \tfrac{f(E)}{E'}\big], \tag{8.80}$$

and the SI (8.74) becomes

$$\tilde{\mathcal{H}}[v] \equiv v_t \quad \big[E'm(E)(v_x)^2 + \tfrac{M(E)}{E'}\big]. \tag{8.81}$$

It follows from (8.79) that the coefficients of $(v_x)^2$ in (8.80) and (8.81) coincide, and (8.75) implies that these are linear functions of v. By differentiating,

$$(E'm(E))''_{vv} \equiv \tfrac{(E')^3}{\varphi(E)} I_1(E) = 0.$$

Thus setting

$$\varphi(E)\tfrac{E''}{E'} = E'm(E) = av + b, \tag{8.82}$$

where a and b are arbitrary constants, and

$$\varphi(E(v)) = \tilde{\varphi}(v), \quad \tfrac{f(E(v))}{E'(v)} = \tilde{f}(v), \quad \tfrac{M(E(v))}{E'(v)} = \tilde{M}(v), \tag{8.83}$$

yields the following parabolic PDE:

$$\tilde{\mathcal{P}}[v] \equiv v_t - \big[\tilde{\varphi}(v)v_{xx} + (av+b)(v_x)^2 + \tilde{f}(v)\big] = 0, \quad \text{with} \tag{8.84}$$

$$\tilde{\mathcal{H}}[v] = v_t - \big[(av+b)(v_x)^2 + \tilde{M}(v)\big]. \tag{8.85}$$

Denoting for convenience

$$G(v) = \tfrac{\tilde{f}(v)-\tilde{M}(v)}{\tilde{\varphi}(v)}, \tag{8.86}$$

equations (8.76) and (8.77) are translated into

$$\tilde{M}'' - (av+b)G' - 5aG = 0, \tag{8.87}$$

$$G'\tilde{M} - G\tilde{M}' + 2(av+b)G^2 = 0, \tag{8.88}$$

respectively. The last ODE (8.88) can be integrated to give

$$\tilde{M}(v) = 2P(v)G(v), \tag{8.89}$$

where $P(v)$ is a quadratic polynomial

$$P(v) = \tfrac{1}{2}\,av^2 + bv + c. \tag{8.90}$$

Substituting (8.90) into (8.87) yields the following linear *hypergeometric* ODE for the function G in (8.86):

$$(av^2 + 2bv + 2c)G'' + 3(av+b)G' - 3aG = 0, \tag{8.91}$$

which plays a key role in the further construction of sign and zero-invariants for our parabolic problem. We call (8.91) the *generating equation*, since actually, by formulae (8.86) and (8.83), it generates the coefficients in the parabolic PDE (8.80) and operator (8.81). In particular, (8.84) and (8.85) imply the following result.

Proposition 8.20 *Under the given notation, the sign-invariant of* (8.84) *is*

$$\tilde{\mathcal{H}}_*[v] = v_{xx} + G(v), \tag{8.92}$$

where G is an arbitrary solution of the ODE (8.91).

Unfortunately (8.91) cannot be solved explicitly in general. Below we use its particular solutions.

8.4.2 *Exact solutions*

Since, by Proposition 8.20, operator (8.92) is a zero-invariant of (8.84) with G satisfying (8.91), it is seen that, for an arbitrary solution $v(x,t)$ of (8.84),

$$v_{xx} + G(v) = 0 \quad \text{for } x \in I\!R,\ t > 0, \tag{8.93}$$

if the same is valid for $t = 0$. Of course, operator (8.85) is also a zero-invariant, i.e., in view of (8.86) and (8.89),

$$v_t = P'(v)(v_x)^2 + 2P(v)G(v) \quad \text{for } t > 0. \tag{8.94}$$

Then (8.93) and (8.94) yield exact solutions of the PDE (8.84). Let a given solution $G(v)$ of (8.91) be well-defined in a neighborhood of $v = 1$. Set

$$Y(v) = \textstyle\int_1^v G(\eta)\,\mathrm{d}\eta\,.$$

It follows from (8.91) that Y solves the linear ODE

$$\mathcal{B}[Y] \equiv 2PY''' + 3P'Y'' - 3P''Y' = 0. \tag{8.95}$$

Multiplying (8.93) by v_x, and integrating over $(0, x)$ yields

$$(v_x)^2 = 2[B(t) - Y(v)], \tag{8.96}$$

where $B(t)$ is a smooth function to be determined later.

Theorem 8.21 *Let G be a solution of (8.91) such that $Y(v)$ exists. Then the quasilinear equation (8.84) admits the explicit solution $v(x,t)$ given by*

$$\int_1^{v(x,t)} \frac{\mathrm{d}z}{\sqrt{B(t)-Y(z)}} = \sqrt{2}\,x + C(t), \tag{8.97}$$

where the functions $B(t)$ and $C(t)$ satisfy the dynamical system

$$\begin{cases} B' = 4aB^2 + \alpha_1 B + \alpha_2, \\ C' = \alpha_3\sqrt{B} + \alpha_4 \frac{1}{\sqrt{B}}, \end{cases} \tag{8.98}$$

$$\alpha_1 = 4P(1)G'(1) + 2P'(1)G(1), \quad \alpha_2 = 2P(1)G^2(1),$$
$$\alpha_3 = 2P'(1), \quad \alpha_4 = 2P(1)G(1).$$

Proof. Integrating (8.96) again yields (8.97). Differentiating (8.97) with respect to t and x implies

$$\begin{aligned} v_t &= \tfrac{1}{2}\sqrt{S(t,v)}\,B'(t)\int_1^v \frac{\mathrm{d}z}{[S(t,z)]^{3/2}} + C'(t)\sqrt{S(t,v)}, \\ v_x &= \sqrt{2\,[S(t,v)}, \quad \text{where} \quad S(t,v) = B(t) - Y(v). \end{aligned} \tag{8.99}$$

Plugging (8.96) and (8.99) into (8.94), we obtain the identity with respect to v and t, which are treated here as the independent variables,

$$\tfrac{1}{2}B'(t)\int_1^v \frac{\mathrm{d}z}{[S(t,z)]^{3/2}} + C'(t) \equiv 2P'(v)\sqrt{S(t,v)} + \frac{2P(v)G(v)}{\sqrt{S(t,v)}}. \tag{8.100}$$

Setting $v = 1$ yields the second ODE in (8.98). Then, differentiating (8.100) with respect to v, and multiplying by $S^{3/2}$ gives

$$\frac{B'(t)-2PG^2}{2S} = 2P''S + 2PG' + P'G. \tag{8.101}$$

It follows from the first ODE (8.98) that (8.101) is valid identically for $v = 1$. Differentiating (8.101) again and using equation (8.91) yields

$$\left(\frac{B'(t)-2PG^2}{2S}\right)'_v = 2P''G.$$

Finally, the identity holds,

$$2aB^2 - \tfrac{1}{2}B' + J_1[v]B + J_2[v] \equiv 0, \quad \text{where} \tag{8.102}$$

$$J_1[v] = 2PY'' + P'Y' - 4P''Y, \quad \text{and}$$
$$J_2[v] = -2PYY'' + P(Y')^2 - PYY' + 2P''Y^2.$$

It is easy to see that, by (8.95), $J_1' = \mathcal{B}[Y] = 0$ and $J_2' = -Y\mathcal{B}[Y] = 0$. Hence, J_1 and J_2 are constants, $J_1[v] \equiv J_1(1)$ and $J_2[v] \equiv J_2(1)$. Then (8.102) is exactly equation (8.98) for $B(t)$. This completes the proof. □

We next describe solutions, corresponding to some particular G satisfying (8.91).

Example 8.22 (**Quasilinear parabolic PDEs**) For any fixed $a, b \in I\!R$, the hypergeometric equation (8.91) has the linear solution

$$G(v) = P'(v) = av + b.$$

Then (8.93) has the form

$$v_{xx} + av + b = 0 \quad \text{for} \quad x \in I\!R, \ t > 0. \tag{8.103}$$

Let $a > 0$. Integrating (8.103) yields

$$v(x,t) = -\tfrac{b}{a} + C_1(t)\cos(\lambda x) + C_2(t)\sin(\lambda x), \tag{8.104}$$

where $\lambda = \sqrt{a}$. Substituting (8.104) into the Hamilton–Jacobi PDE (8.94),

$$v_t = (av + b)\big[(v_x)^2 + (av^2 + 2bv + 2c)\big],$$

we derive the following cubic DS for the coefficients $\{C_1, C_2\}$:

$$\begin{cases} C_1' = [a^2(C_1^2 + C_2^2) + 2ac]C_1, \\ C_2' = [a^2(C_1^2 + C_2^2) + 2ac]C_2, \end{cases} \tag{8.105}$$

possessing blow-up solutions. This determines exact solutions (8.104) of the following quasilinear heat equation (cf. (8.84)):

$$\boxed{\begin{aligned} v_t = \mathcal{D}[v] \equiv \tilde{\varphi}(v)v_{xx} + (av + b)(v_x)^2 \\ + (av + b)\big[\tilde{\varphi}(v) + av^2 + 2bv + 2c\big], \end{aligned}} \tag{8.106}$$

where $\tilde{\varphi}(v) \equiv \varphi(E(v)) \geq 0$ is arbitrary. For solutions $u(x,t) = E(v(x,t))$, where E is given by (8.82), there occurs the PDE

$$\begin{aligned} u_t = \varphi(u)u_{xx} + [aE^{-1}(u) + b] \\ \times \big[\varphi(u) + a(E^{-1}(u))^2 + 2bE^{-1}(u) + 2c\big]\, E'(E^{-1}(u)), \end{aligned}$$

which can easily be rewritten in the divergent form (8.1).

Example 8.23 (**Semilinear heat equations**) Consider

$$u_t = u_{xx} + \tfrac{\alpha R(u) + \beta}{R'(u)}\,, \tag{8.107}$$

where $R(u)$ satisfies the following nonlinear ODE:

$$R' = \exp\big\{-\tfrac{aR^2}{2} - bR - c\big\}. \tag{8.108}$$

The SI is

$$\mathcal{H}[u] = u_t - (aR + b)R'(u_x)^2 - \tfrac{\alpha R(u) + \beta}{R'(u)}\,, \tag{8.109}$$

which is also a zero-invariant of (8.107). To calculate the corresponding solutions, we set $u = E(v)$ in (8.107), where $E(v) = R^{-1}(v)$, to derive

$$\boxed{v_t = F[v] \equiv v_{xx} + (av + b)(v_x)^2 + \alpha v + \beta.} \tag{8.110}$$

Obviously, operator F admits the 2D subspace of linear functions $W_2 = \mathcal{L}\{1, x\}$. Hence, the solutions of (8.110) are

$$v(x,t) = C_1(t) + C_2(t)x \in W_2,$$

$$\begin{cases} C_1' = aC_1C_2^2 + bC_2^2 + \alpha C_1 + \beta, \\ C_2' = aC_2^3 + \alpha C_2, \end{cases}$$

where the DS can be integrated. Adding the term $-v_{xxxx}$ to the right-hand side of (8.110) gives a fourth-order parabolic PDE, which admits the same explicit solutions but, of course, makes the SI (8.109) nonexistent.

Example 8.24 Setting $\tilde{\varphi}(u) \equiv 1, b = 0$ in (8.106) yields

$$u_t = u_{xx} + \frac{a^2R^3(u)+a(2c+1)R(u)}{R'(u)},$$

where $R(u)$ is given by (8.108) with $a \neq 0$ and $b = 0$. It admits the SI

$$\mathcal{H}[u] = u_t - aRR'(u_x)^2 - \frac{a^2R^3+2acR}{R'}.$$

In order to find explicit solutions, setting $u = R^{-1}(v)$ gives

$$\boxed{v_t = F_1[v] \equiv v_{xx} + av(v_x)^2 + a^2v^3 + a(2c+1)v.} \tag{8.111}$$

Assume that $a > 0$, and set $\lambda = \sqrt{a}$. Operator F_1 of this PDE admits the 1D subspace $W_1 = \mathcal{L}\{\cos(\lambda x)\}$. Thus, setting $v(x,t) = C(t)\cos(\lambda x) \in W_1$ yields

$$C' = a^2C^3 + 2acC.$$

If $a < 0$, the subspace is $W_1 = \mathcal{L}\{\cosh(\lambda x)\}$, with $\lambda = \sqrt{|a|}$.

As we have shown in Chapter 1, operator F_1 in (8.111) admits a more general 2D subspace

$$W_2 = \mathcal{L}\{\cos(\lambda x),\ \sin(\lambda x)\}, \tag{8.112}$$

that does not give new exact solutions of (8.111) because of its translational invariance with respect to x. For higher-order evolution PDEs, e.g., for the hyperbolic one

$$v_{tt} = F_1[v], \tag{8.113}$$

using subspace (8.112) yields new solutions via

$$v(x,t) = C_1(t)\cos(\lambda x) + C_2(t)\sin(\lambda x) \in W_2,$$

$$\begin{cases} C_1'' = [a^2(C_1^2 + C_2^2) + 2ac]C_1, \\ C_2'' = [a^2(C_1^2 + C_2^2) + 2ac]C_2. \end{cases}$$

Example 8.25 (**Fourth-order KS-type equation**) There are easy extensions to higher-order PDEs (the SIs are lost). For instance, the fourth-order semilinear KS-type equation with cubic nonlinearities

$$\boxed{v_t = -v_{xxxx} + \alpha\big[v(v_x)^2 + v^3\big] + \beta v}$$

admits the explicit solution

$$v(x,t) = C(t)\cos x, \quad \text{with } C' = \alpha C^3 + (\beta - 1)C.$$

Example 8.26 (**KdV-type equation**) The third-order PDE with the same cubic nonlinearities

$$\boxed{v_t = v_{xxx} + \alpha\big[v(v_x)^2 + v^3\big] + \beta v}$$

has the periodic soliton-type solution

$$v(x,t) = C(t)\cos(x-t), \quad \text{where} \;\; C' = \alpha C^3 + \beta C.$$

Example 8.27 (**Quasilinear heat equations**) Let us return to SIs for the quasilinear parabolic PDEs. In this example, we consider the generating equation (8.91) in the case where $a = c = 0$ and $b = 1$. Then,

$$2vG'' + 3G' = 0 \quad \Longrightarrow \quad G(v) = \tfrac{\alpha}{\sqrt{v}} + \beta, \tag{8.114}$$

where $\alpha \neq 0$ and β are free constants. It follows from (8.86), (8.89), and (8.114) that the corresponding PDE (8.84) is

$$\boxed{v_t = \tilde{\varphi}(v)v_{xx} + (v_x)^2 + \left(\tfrac{\alpha}{\sqrt{v}} + \beta\right)[\tilde{\varphi}(v) + 2v].} \tag{8.115}$$

Exact solutions of (8.115) have been given in Theorem 8.21. The corresponding original equation (8.73) for $u = E(v)$ is now

$$u_t = \varphi(u)u_{xx} + \left[\tfrac{\alpha}{\sqrt{E^{-1}(u)}} + \beta\right]\left[\varphi(u) + 2E^{-1}(u)\right] E'(E^{-1}(u)). \tag{8.116}$$

In the case where $\tilde{\varphi} = 1$, setting $v = \ln u$ yields the semilinear heat equation

$$u_t = u_{xx} + u\left(\tfrac{\alpha}{\sqrt{\ln u}} + \beta\right)(1 + 2\ln u), \tag{8.117}$$

possessing the above solutions $u = e^v$. For $\alpha = 1$ and $\beta = 0$, we find

$$u_t = F_2[u] \equiv u_{xx} + u\left(\tfrac{1}{\sqrt{\ln u}} + 2\sqrt{\ln u}\right), \tag{8.118}$$

which admits the following SI:

$$\mathcal{H}[u] = u_t - \tfrac{1}{u}(u_x)^2 - 2u\sqrt{\ln u}. \tag{8.119}$$

Since (8.119) is also a zero-invariant of (8.118), we obtain solutions $u(x,t) = e^{v(x,t)}$, where $v(x,t)$ solves the algebraic equation

$$t\sqrt{t - \sqrt{v(x,t)}} - \tfrac{1}{3}\sqrt{[t - \sqrt{v(x,t)}\,]^3} = \tfrac{1}{2}x\,. \tag{8.120}$$

For $\alpha = 0$, (8.117) gives the well-known equation reduced to the PDE

$$u_t = u_{xx} + u\ln u,$$

possessing a 4D symmetry Lie algebra [10, p. 135]; see also Example 9.9 for more details. For $\alpha \neq 0$, (8.117) does not have any extra group symmetries.

Example 8.28 (**On higher-order extensions**) It is easy to derive higher-order models possessing precisely the same exact solutions, though some PDEs can be rather artificial. The simple rules of construction are as follows. For instance, using the zero-invariant (8.119) in the PDE (8.118) yields that, for exact solutions, $u_{xx} = \frac{(u_x)^2}{u} - \frac{u}{\sqrt{\ln u}}$, and, on differentiation,

$$u_{xxx} = F_3[u] \equiv \tfrac{2u_x u_{xx}}{u} - \tfrac{(u_x)^3}{u^2} - \tfrac{u_x}{\sqrt{\ln u}} + \tfrac{u_x}{\ln u}.$$

Obviously, the third-order PDE

$$u_t = \psi(u, ...)(u_{xxx} - F_3[u]) + F_2[u],$$

where $\psi(u, ...)$ is an arbitrary function, admits the same solutions $u = e^v$ given by (8.120). One can reconstruct other, less artificial versions of such PDEs. Similarly, calculating u_{xxxx} denoted again by $F_3[u]$ yields a fourth-order PDE, etc.

Example 8.29 (**Quasilinear parabolic PDEs**) We return to quasilinear heat equations and determine other exact solutions and SIs. Recall that function E given by (8.78) (it transforms (8.116) into (8.73), and eventually into (8.1) with $N = 1$) is determined from (8.79) and (8.82), so

$$E'' = \frac{E'}{\varphi(E)} \cdot \tag{8.121}$$

Consider first the coefficient

$$\varphi(u) = \frac{1}{2u}. \tag{8.122}$$

Then (8.121) yields

$$E'' = 2EE', \tag{8.123}$$

and there exist two possibilities.

(i) Take $E(v) = \tan v$. Substituting E into (8.116) yields

$$u_t = \frac{1}{2u} u_{xx} + \big(\frac{\alpha}{\sqrt{\tan^{-1} u}} + \beta\big)\big(\frac{1}{2u} + 2\tan^{-1} u\big)(1 + u^2).$$

Here, setting $u^2 = U$, we arrive at

$$U_t = (\sqrt{U}\,)_{xx} + \big(\frac{\alpha}{\tan^{-1}\sqrt{U}} + \beta\big)\big(1 + 4\sqrt{U}\tan^{-1}\sqrt{U}\,\big)(1 + U), \tag{8.124}$$

which has solutions

$$U(x,t) = \tan^2 v(x,t),$$

where $v(x,t)$ is given in Theorem 8.21. The equation (8.69) is (8.124), with $\alpha = \beta = 1$.

(ii) Take another solution of the ODE (8.123), $E(v) = -\frac{e^{2v}+1}{e^{2v}-1}$. Setting $u^2 = U$ in (8.116) yields

$$U_t = (\sqrt{U}\,)_{xx} + \big[\alpha\big(\tfrac{1}{2}\ln\frac{\sqrt{U}-1}{\sqrt{U}+1}\big)^{-\frac{1}{2}} + \beta\big]\big(1 + 2\sqrt{U}\ln\frac{\sqrt{U}-1}{\sqrt{U}+1}\big)(U - 1),$$

which admits solutions $U(x,t) = E^2(v(x,t))$.

(iii) If we take, for instance,

$$\varphi(u) = \cos^2 u, \tag{8.125}$$

then (8.121) admits the solution $E(v) = \sin^{-1} e^v$. Then (8.116) has the form

$$u_t = \cos^2 u\, u_{xx} + \big(\frac{\alpha}{\sqrt{\ln\sin u}} + \beta\big)\big(\cos^2 u + 2\ln\sin u\big)\tan u.$$

Setting $\tan u = U$ yields the quasilinear heat equation

$$U_t = \big(\tan^{-1} U\big)_{xx} + \big[\alpha\big(\ln\frac{U}{\sqrt{1+U^2}}\big)^{-\frac{1}{2}} + \beta\big]\big[1 + 2(1+U^2)\ln\frac{U}{\sqrt{1+U^2}}\big]U,$$

which has the solution $U(x,t) = \frac{e^{v(x,t)}}{\sqrt{1-e^{2v(x,t)}}}$, with $v(x,t)$ given in Theorem 8.21.

It is easy to find other coefficients $\varphi(u)$ such that (8.121) admits a simple function $E(v)$ determining solutions of the corresponding quasilinear equation (8.116).

(iv) For instance, for $\varphi(u) = \frac{\sqrt{1+u^2}}{u}$, $E(v) = \frac{e^{2v}-1}{2e^v}$, and $U = \sqrt{1+u^2}$, we find

$$U_t = \left(\sqrt{U^2-1}\right)_{xx} + \left[\alpha\left(\ln\left(U+\sqrt{U^2-1}\right)\right)^{-\frac{1}{2}} + \beta\right] \times \left[U + 2\sqrt{U^2-1}\,\ln\left(U+\sqrt{U^2-1}\right)\right],$$

possessing the solution $U(x,t) = \frac{e^{2v(x,t)}+1}{2e^{v(x,t)}}$.

(v) In the case where $\varphi(u) = \frac{1}{\cos u}$, $E(v) = 2\sin^{-1}\frac{e^v}{\sqrt{1+e^{2v}}}$, and $U = \sin u$, we have

$$U_t = \left(\sin^{-1}U\right)_{xx} + \left[\alpha\left(\ln\frac{U}{1+\sqrt{1-U^2}}\right)^{-\frac{1}{2}} + \beta\right]\left(1 + 2\sqrt{1-U^2}\,\ln\frac{U}{1+\sqrt{1-U^2}}\right)U$$

with the solution $U(x,t) = \frac{2e^{v(x,t)}}{e^{2v(x,t)}+1}$.

Example 8.30 Consider now the last case $a = 1$ and $b = c = 0$, so (8.91) is

$$v^2G'' + 3vG' - 3G = 0,$$

with the general solution $G(v) = \alpha v + \beta v^{-3}$. From (8.86) and (8.89), we have, by Theorem 8.21, that the parabolic equation (8.84),

$$\boxed{v_t = \tilde{\varphi}(v)v_{xx} + v(v_x)^2 + (\alpha v + \beta\tfrac{1}{v^3})[\tilde{\varphi}(v) + v^2],}$$

possesses such exact solutions. Setting $v = R(u)$, where R solves (8.108) (R is the inverse function to E given by (8.79)), yields the quasilinear heat equation

$$u_t = \varphi(u)u_{xx} + \frac{(\alpha R^4+\beta)(\varphi(u)+R^2)}{R^3R'}.$$

In the case where $\varphi(u) \equiv 1$, $\alpha = 0$, and $\beta = 1$, this reduces to

$$u_t = u_{xx} + \frac{1+R^2(u)}{R^3(u)R'(u)},$$

where $R(u)$ is given by (8.108) with $a = 1$, $b = c = 0$, and the SI is $\mathcal{H}[u] = u_t - RR'(u_x)^2 - \frac{1}{RR'}$. In particular, there exists the solution

$$R(u(x,t)) = \sqrt{\frac{4t^2-x^2}{2t}}.$$

If $\varphi(u) \equiv 1$ and $\alpha = \beta = 1$, we arrive at the PDE

$$u_t = u_{xx} + \frac{(R^4(u)+1)(1+R^2(u))}{R^3(u)R'(u)},$$

where R is given above, with the SI

$$\mathcal{H}[u] = u_t - RR'(u_x)^2 - \frac{R^4+1}{RR'}$$

and the solution

$$R(u(x,t)) = \sqrt{\frac{\sin 4t + \sin 2x}{\cos 4t}}.$$

In the particular quasilinear case of (8.122), setting $u^2 = U$ yields

$$U_t = \left(\sqrt{U}\right)_{xx} + \tfrac{(\alpha R^4+\beta)(1+2\sqrt{U}\,R^2)}{R^3 R'}$$

(here $R = R(\sqrt{U})$), which admits solutions $U(x,t) = E^2(v(x,t))$, where v is given in Theorem 8.21. In the case of (8.125), setting $\tan u = U$ gives

$$U_t = (\tan^{-1} U)_{xx} + \tfrac{(\alpha R^4+\beta)[1+(1+U^2)R^2]}{R^3 R'},$$

with $R = R(\tan^{-1} U)$, and the solution is $U(x,t) = \tan E(v(x,t))$.

8.5 General first-order Hamilton–Jacobi sign-invariants

In this section, we perform a more general and detailed analysis of first-order SIs for the quasilinear heat equation

$$\boxed{u_t = \psi(u) u_{xx} + q(u) \quad \text{in } I\!R \times I\!R_+,} \tag{8.126}$$

where $\psi(u) \ge 0$ and $q(u)$ are given smooth functions. For convenience, let us perform a smooth transformation $u = \Phi(v)$ to obtain the PDE

$$v_t = \varphi(v) v_{xx} + m(v)(v_x)^2 + f(v), \tag{8.127}$$

with $\varphi(v) \ge 0$. We will deal with a more general (than in Section 8.4) quadratic Hamilton–Jacobi operator of the form

$$\mathcal{H}[v] = v_t - H[v] \equiv v_t - \big[m(v)(v_x)^2 + s(v) v_x + M(v)\big], \tag{8.128}$$

with three free coefficients m, s, and M. Without loss of generality, in (8.128), we set the coefficient of $(v_x)^2$ to be equal to $m(v)$, which is the same function, as in (8.127). In terms of the original solutions $u(x,t)$, the SI has a similar form

$$\tilde{\mathcal{H}}[u] = u_t - \big[h_2(u)(u_x)^2 + h_1(u) u_x + h_0(u)\big].$$

Our goal is to show that more general SIs (8.127) provide us with extra exact solutions and yield other possibilities to derive estimates of solutions.

It is convenient to perform basic calculations for the new coefficients

$$g(v) = -\tfrac{s(v)}{\varphi(v)} \quad \text{and} \quad G(v) = \tfrac{f(v)-M(v)}{\varphi(v)}. \tag{8.129}$$

As in the previous section, $P(v)$ is the quadratic polynomial (8.90), $P(v) = \frac{1}{2} a v^2 + bv + c$. We next state the first main result on the existence of the SI.

Theorem 8.31 *Operator* (8.128) *is a sign-invariant of the PDE* (8.127) *if the coefficients $g(v)$ and $G(v)$ satisfy the following system of ODEs:*

$$2Pg'' + 3P'g' - 8P''g = 0, \tag{8.130}$$

$$2PG'' + 3P'G' - 3P''G = \tfrac{2}{3}(Pg^2)', \tag{8.131}$$

and other coefficients are given by

$$m = P', \quad s' = \tfrac{2}{3}(2P'g + Pg'), \quad \text{and} \quad M = 2PG. \tag{8.132}$$

The linear ODEs (8.130) and (8.131) are of the hypergeometric type. The second equation is non-homogeneous and its right-hand side depends on the solutions of the first equation. Theses are the *generating equations,* which determine the class of parabolic PDEs under consideration. As the next step, let us find the representation of the corresponding exact solutions, since by Theorem 8.31, the operator (8.128) is also a zero-invariant of (8.127), i.e.,

$$\mathcal{H}[v]=0 \quad \text{for all } t>0, \text{ if } \quad \mathcal{H}[v]=0 \quad \text{for } t=0.$$

For fixed constants B and C, let $v = V(x;B,C)$ be the solution of the Cauchy problem for the ODE

$$v_{xx}+g(v)v_x+G(v)=0 \quad \text{for } x>0, \text{ with } v(0)=B,\ v'(0)=C. \tag{8.133}$$

We assume that this ODE is locally well-posed for arbitrary $B, C \in I\!R$. The equation (8.133) is associated with the PDE (8.127) and with the corresponding Hamilton–Jacobi equation

$$\mathcal{H}[v]=0, \tag{8.134}$$

in the sense that, altogether, these three equations are linearly dependent. Clearly, the stationary, time-independent equation (8.133) reduces to a first-order one and sometimes can be integrated in quadratures. We will present such examples below. The simpler case $g=0$, for which (8.133) is always integrated in quadratures, was studied in the previous section, where a different general structure of exact solutions is derived. Finite-dimensional dynamics of exact solutions is as follows:

Theorem 8.32 *Under hypotheses of Theorem* 8.31, *the set of solutions* $\{v(x,t)=V(x;B(t),C(t))\}$ *via zero-invariance of operator* (8.128) *is governed by the following second-order dynamical system for coefficients* $B(t)$ *and* $C(t)$*:*

$$\begin{cases} B' = P'(B)C^2 + s(B)C + 2P(B)G(B), \\ C' = P''(B)C^3 + [s'(B) - 2P'(B)g(B)]C^2 \\ \qquad + [2P(B)G'(B) - s(B)g(B)]C - s(B)G(B). \end{cases} \tag{8.135}$$

The stationary equation (8.133) defines an evolutionary invariant set M of the parabolic flow induced by (8.127), in the sense that

$$v(\cdot,0)\in M \implies v(\cdot,t)\in M \quad \text{for } t>0.$$

The DS (8.135) is the parabolic PDE (8.127) restricted to the set M.

8.5.1 Proofs of main theorems

Proposition 8.33 *Operator* (8.128) *is a sign-invariant of the equation* (8.127) *iff the coefficients satisfy the system*

$$m'' - 0, \tag{8.136}$$

$$I_1(v) \equiv s'' - 4m'g - g'm = 0, \tag{8.137}$$

$$I_2(v) \equiv M'' - 2gs' - 5m'G - mG' + 2mg^2 = 0, \tag{8.138}$$

$$I_3(v) \equiv -3s'G + g'M + 4mgG = 0, \tag{8.139}$$

$$I_4(v) \equiv -\left(\tfrac{M}{G}\right)' + 2m = 0. \tag{8.140}$$

Proof. Set, as usual, $J = \mathcal{H}[v]$. Differentiating this equality with respect to t,

$$J_t = v_{tt} - (H[v])_t,$$

substituting the second derivative v_{tt} from (8.127) differentiated in t, and evaluating other lower-order ones, v_t, v_{tx} and v_{txx}, we arrive at the parabolic PDE

$$J_t = \mathcal{M}[J] + \mathcal{F}, \tag{8.141}$$

where $\mathcal{M}$ is a linear second-order operator that is elliptic in the parabolicity domain of (8.127). By (8.136)–(8.140), the lower-order term is trivial,

$$\mathcal{F} = \varphi\big[m''(v_x)^4 + I_1(v_x)^3 + I_2(v_x)^2 + I_3 v_x + I_4 G^2\big] \equiv 0. \tag{8.142}$$

Hence, J solves the homogeneous linear parabolic PDE

$$J_t = \mathcal{M}[J],$$

and the result follows from the MP. If $\mathcal{F} \not\equiv 0$ in (8.141), then obviously at least one sign of (8.128) cannot be preserved with time on the corresponding set of solutions (the proof is by direct constructing suitable initial data). □

Remark 8.34 Consider the identity (8.142) as a fourth-order algebraic equation for v_x, meaning that the set $\{1, v_x, (v_x)^2, (v_x)^3, (v_x)^4\}$ is linearly dependent. Therefore, there exists the representation $v_x = R(v)$ and, on integration, this kind of zero-invariance gives

$$\int \tfrac{dv}{R(v)} = x + C(t).$$

Plugging into (8.127) gives the elementary ODE $C' =$ constant, which leads to the standard *traveling wave* solution $v(x,t) = \theta(x - \lambda t)$. Therefore, in order to obtain a nontrivial result in the proof, we must assume that all the five terms in (8.142) are linearly independent and this gives the five ODEs (8.136)–(8.140).

Proof of Theorem 8.31. Equation (8.136) defines linear functions $m(v) = av + b \equiv P'(v)$; cf. the first equation (8.132). It follows from (8.140) that

$$\tfrac{M}{G} = 2\int m(v)\, dv \equiv 2P(v); \tag{8.143}$$

cf. the third equation (8.132). Substituting M from (8.143) into (8.139) yields

$$3s' - 4P'g - 2Pg' = 0, \tag{8.144}$$

and the second equation (8.132) follows. The system of ODEs (8.137), (8.144) gives the first generating equation (8.130). Finally, the second one, (8.131), is the result of substituting M from (8.143) and s' from (8.144) into (8.138). □

Proof of Theorem 8.32. Setting $x = 0$ in (8.134) and in $v_{xt} = (H[v])_x$, and using (8.133), (8.132) yields the DS (8.135). □

8.5.2 $P(v)$ is linear polynomial: exact solutions

General structure of SIs. Here, we study the SIs in the *linear* case $a = 0$, where the quadratic polynomial is $P(v) = bv + c$ with $b \neq 0$. Set $c = 0$, corresponding to translation. Moreover, it follows from the generating ODEs (8.130) and (8.131) that the constant b does not play any role in the analysis, so we set $b = 1$. Therefore, consider (8.130) and (8.131) with the linear function

$$P(v) = v. \tag{8.145}$$

The first equation (8.130) is then easily integrated:

$$g(v) = \mu \frac{1}{\sqrt{v}} + \nu \quad (v > 0), \tag{8.146}$$

with arbitrary constants μ and ν.

Let us begin with a simpler case $\nu \neq 0$ and consider the *critical* case $\nu = 0$ later on. Substituting (8.146) into the right-hand side of (8.131) yields

$$G(v) = \alpha \frac{1}{\sqrt{v}} + \beta + K_1\sqrt{v} + K_2 v, \tag{8.147}$$

where $K_1 = \frac{2}{3}\mu\nu$ and $K_2 = \frac{2}{9}\nu^2$. Next, using the general solutions (8.146) and (8.147) of the generating ODEs, we will study the structure of the evolutionary invariant set given by the stationary equation (8.133) and the DS (8.135).

Linearization of the stationary equation. This consists of a few steps.

(i) Set

$$v = \frac{1}{w^2} \tag{8.148}$$

in the ODE

$$\begin{aligned} v_{xx} + \left(\mu \frac{1}{\sqrt{v}} + \nu\right) v_x + \alpha \frac{1}{\sqrt{v}} + \beta + K_1\sqrt{v} + K_2 v = 0 \implies \\ w w_{xx} - 3(w_x)^2 + (\mu w + \nu) w w_x \\ - \tfrac{K_2}{2} w^2 - \tfrac{K_1}{2} w^3 - \tfrac{\beta}{2} w^4 - \tfrac{\alpha}{2} w^5 = 0. \end{aligned} \tag{8.149}$$

(ii) Set

$$w = z_x, \tag{8.150}$$

so that (8.149) reads

$$\begin{aligned} z_x z_{xxx} - 3(z_{xx})^2 + (\mu z_x + \nu) z_x z_{xx} \\ - \tfrac{K_2}{2}(z_x)^2 - \tfrac{K_1}{2}(z_x)^3 - \tfrac{\beta}{2}(z_x)^4 - \tfrac{\alpha}{2}(z_x)^5 = 0. \end{aligned} \tag{8.151}$$

(iii) We now introduce the inverse function with respect to the space variable

$$x = \Psi(y, t) \equiv z^{-1}(y, t), \tag{8.152}$$

so that $z(\Psi(y, t), t) \equiv y$, and

$$z_x = \frac{1}{\Psi_y}, \quad z_{xx} = -\frac{\Psi_{yy}}{(\Psi_y)^3}, \quad z_{xxx} = -\frac{\Psi_{yyy}}{(\Psi_y)^4} + 3\,\frac{(\Psi_{yy})^2}{(\Psi_y)^5}.$$

Plugging these derivatives into (8.151) yields

$$\Psi_{yyy} + (\mu + \nu\Psi_y)\Psi_{yy} + \tfrac{K_2}{2}(\Psi_y)^3 + \tfrac{K_1}{2}(\Psi_y)^2 + \tfrac{\beta}{2}\Psi_y + \tfrac{\alpha}{2} = 0.$$

(iv) Let

$$\Psi_y = V. \tag{8.153}$$

Then V solves

$$V_{yy} + (\mu + \nu V)V_y + \tfrac{K_2}{2} V^3 + \tfrac{K_1}{2} V^2 + \tfrac{\beta}{2} V + \tfrac{\alpha}{2} = 0. \tag{8.154}$$

(v) Finally, let us apply the transformation (used in hyperelliptic function theory since the nineteenth century)

$$V = \kappa \, \frac{\rho_y}{\rho} \tag{8.155}$$

to obtain an equation with cubic nonlinearities,

$$\rho^2\big(\rho_{yyy} + \mu\rho_{yy} + \tfrac{\beta}{2}\rho_y + \tfrac{\alpha}{2\kappa}\rho\big) + (\nu\kappa - 3)\big[\rho\rho_y\rho_{yy} + \tfrac{1}{9}(\nu\kappa - 6)(\rho_y)^3 + \tfrac{\mu}{3}\rho(\rho_y)^2\big] = 0.$$

Therefore, choosing in (8.155)

$$\kappa = \tfrac{3}{\nu} \tag{8.156}$$

gives the linear ODE

$$\rho_{yyy} + \mu\rho_{yy} + \tfrac{\beta}{2}\rho_y + \tfrac{\alpha\nu}{6}\rho = 0. \tag{8.157}$$

The characteristic equation is derived by substituting $\rho = \mathrm{e}^{\lambda y}$,

$$\lambda^3 + \mu\lambda^2 + \tfrac{1}{2}\beta\lambda + \tfrac{1}{6}\alpha\nu = 0. \tag{8.158}$$

At last, the general solution takes the form

$$\rho(y) = H_1\rho_1(y) + H_2\rho_2(y) + H_3\rho_3(y),$$

where ρ_1, ρ_2, and ρ_3 are linearly independent solutions of the ODE (8.157). In view of the representation (8.155) of the solution V, we may set $H_1 = 1$. The trivial choice $H_1 = 0$ leads to a significant simplification of the solution. We write the general solution of (8.157) in the form of

$$\rho(y) = \rho_1(y) + H_2\rho_2(y) + H_3\rho_3(y). \tag{8.159}$$

Linearization of (8.127). Consider (8.128), where the coefficients are constructed from (8.132) with the functions (8.145)–(8.147). This yields

$$m(v) = 1, \quad s(v) = 2\mu\sqrt{v} + \tfrac{4}{3}\nu v + K_3, \quad M(v) = 2vG(v),$$

where K_3 is a free constant. The rest of the coefficients of the quasilinear equation (8.127) are obtained from (8.129),

$$\varphi(v) = -\frac{s(v)}{g(v)} = -\frac{2\mu\sqrt{v}+\frac{4}{3}\nu v+K_3}{\frac{\mu}{\sqrt{v}}+\nu} \quad \text{and} \quad f(v) = [\varphi(v) + 2v]G(v). \tag{8.160}$$

Substituting $G(v)$ from (8.133) into (8.127) yields the PDE

$$v_t = -2vv_{xx} + v_x^2 + \big(K_3 - \tfrac{2}{3}\nu v\big)v_x, \tag{8.161}$$

which is backward parabolic (recall that $v > 0$ by (8.146)). We now apply transformations (i)–(v) to (8.161). The first one, (8.148), yields

$$w_t = -2\big(\tfrac{1}{w^2}\, w_x\big)_x + \big(K_3 w + \tfrac{2\nu}{3w}\big)_x. \tag{8.162}$$

The properties of the famous quasilinear parabolic PDE,

$$w_t = \big(\tfrac{1}{w^2}\, w_x\big)_x,$$

have been known for a long time. In 1951, Storm [537] reduced it to the linear heat equation by the transformation, (8.150) and (8.152). Moreover, the same can be done for (8.162), [194] (the linear convection term $K_3 w_x$ is easily eliminated). See [128] on more general approaches to linearizations of the PDEs. Let us present the corresponding simple calculation for (8.162). Integrating it over $(0, x)$ yields

$$\frac{\partial}{\partial t}\left[\int_0^x w(s,t)\mathrm{d}s + \int_0^t \left(-\frac{2w_x}{w^2} + K_3 w + \frac{2\nu}{3w}\right)(0,\tau)\,\mathrm{d}\tau\right]$$
$$= -\frac{2w_x}{w^2} + K_3 w + \frac{2}{3}\frac{\nu}{w}.$$

Let z denote the function in the square brackets on the left-hand side, so that (8.150) holds. In this case, z solves

$$z_t = -\frac{2z_{xx}}{(z_x)^2} + K_3 z_x + \frac{2}{3}\frac{\nu}{z_x}. \tag{8.163}$$

Next, (8.152) with $\Psi = \Psi(y,t)$ $\left(\text{hence, } z_t = -\frac{\Psi_t}{\Psi_y}\right)$ reduces (8.163) to

$$\Psi_t = -2\Psi_{yy} - K_3 - \frac{2}{3}\nu(\Psi_y)^2.$$

It follows from (8.153) that $V(y,t) = \Psi_y(y,t)$ solves *Burgers' equation*

$$V_t = -2V_{yy} - \frac{4}{3}\nu V V_y. \tag{8.164}$$

One can easily check that transformation (8.155) with coefficient (8.156) reduces (8.164) to the heat equation

$$\rho_t = -2\rho_{yy} + N_0(t)\rho, \tag{8.165}$$

with a function $N_0(t)$ to be determined via the consistency condition with (8.157).

Substituting the general solution (8.159) into (8.165) gives a linear DS for the coefficient $H_2(t)$ and $H_3(t)$.

Example 8.35 Let $\alpha = -\frac{6}{\nu}$, $\beta = 2$ and $\mu = -1$ in (8.147). The characteristic equation (8.158) has the form $(\lambda - 1)(\lambda^2 + 1) = 0$, which yields the general solution

$$\rho(y,t) = \mathrm{e}^y + H_2(t)\cos y + H_3(t)\sin y. \tag{8.166}$$

Plugging (8.166) into (8.165) implies $N_0(t) \equiv 2$ and gives the linear DS

$$\begin{cases} H_2' = 4H_2, \\ H_3' = 4H_3. \end{cases}$$

Therefore, from (8.155), we obtain the explicit solution of both (8.154) and (8.164)

$$V(y,t) = \frac{3}{\nu}\frac{\mathrm{e}^y + \mathrm{e}^{4t}(-A\sin y + B\cos y)}{\mathrm{e}^y + \mathrm{e}^{4t}(A\cos y + B\sin y)},$$

where A and B are arbitrary constants, to be transformed by (iv)–(i) into the solution of the corresponding PDE (8.127).

Example 8.36 Set now $\alpha = 0$, $\beta = 8$, and $\mu = -5$ in (8.147). In this case, from (8.159),

$$\rho(y,t) = 1 + H_2\mathrm{e}^y + H_3\mathrm{e}^{4y},$$

and substituting into (8.165) yields $N_0 = 0$ and the solution

$$V(y,t) = \frac{3}{\nu}\frac{A\mathrm{e}^{y-2t} + 4B\mathrm{e}^{4y-32t}}{1 + A\mathrm{e}^{y-2t} + B\mathrm{e}^{4y-32t}}.$$

8.5.3 The linear critical case

Let $\nu = 0$. Then the final transformation (8.155) with the coefficient (8.156) does not make sense, and, as a result, the exact solutions are not of the exponential-trigonometric soliton-type structure. Recall that (8.145)–(8.147) and (8.160) imply SIs and exact solutions of the following quasilinear heat equation:

$$\boxed{v_t = -\big(2v + \tfrac{K_3}{\mu}\sqrt{v}\,\big)v_{xx} + (v_x)^2 - \tfrac{K_3}{\mu}\sqrt{v}\,\big(\tfrac{\alpha}{\sqrt{v}} + \beta\big).} \tag{8.167}$$

For $\nu = 0$, integrating the stationary ODE yields

$$v_{xx} + \tfrac{\mu}{\sqrt{v}}\,v_x + \tfrac{\alpha}{\sqrt{v}} + \beta = 0 \implies v_x = (\alpha + \beta\sqrt{v})Y(v), \tag{8.168}$$

with $Y(v)$ to be determined. We then obtain the first-order ODE

$$\sqrt{v}(\alpha + \beta\sqrt{v})\tfrac{\mathrm{d}Y}{\mathrm{d}v} = -\tfrac{1}{2Y}(\beta Y^2 + 2\mu Y + 2). \tag{8.169}$$

There exist two cases:

(i) $\beta \ne 0$. Let

$$q^2 = \tfrac{2\beta-\mu^2}{\beta^2} > 0, \quad \text{and set} \quad p = \tfrac{\mu}{\beta}.$$

Integrating the ODE (8.169) yields

$$\begin{aligned} R(Y) &\equiv \tfrac{1}{2\beta}\ln[(Y+p)^2 + q^2] - \tfrac{\mu}{q\beta^2}\tan^{-1}\big(\tfrac{Y+p}{q}\big) \\ &= -\tfrac{1}{\beta}\ln(\alpha + \beta\sqrt{v}) + E \equiv \eta(v), \end{aligned}$$

where $E = E(t)$ is the constant of integration. It then follows from (8.168) that

$$v_x = (\alpha + \beta\sqrt{v})R^{-1}(\eta(v)), \tag{8.170}$$

and hence, solutions $v = v(x,t)$ of (8.167) are given by the quadrature

$$\int_0^{v(x,t)} \frac{\mathrm{d}z}{(\alpha+\beta\sqrt{z})R^{-1}(\eta(z))} = x + H(t), \tag{8.171}$$

with $H = H(t)$ being yet a free function. Differentiating (8.171) in t yields

$$v_t = (\alpha + \beta\sqrt{v})R^{-1}(\eta)\big[H'(t) - E'(t)\int_0^v \big(\tfrac{1}{R^{-1}(\eta(z))}\big)'_\eta \, \tfrac{\mathrm{d}z}{(\alpha+\beta\sqrt{z})}\big]. \tag{8.172}$$

Plugging the derivatives (8.170) and (8.172) into the corresponding Hamilton–Jacobi PDE

$$v_t = (v_x)^2 + (2\mu\sqrt{v} + K_3)v_x + 2\alpha\sqrt{v} + 2\beta v,$$

yields the identity

$$\begin{aligned} H'(t) - E'(t)\textstyle\int_0^v(\cdot) &= (\alpha + \beta\sqrt{v})R^{-1}(\eta(v)) \\ &\quad + 2\mu\sqrt{v} + K_3 + \tfrac{2\sqrt{v}}{R^{-1}(\eta(v))}, \end{aligned} \tag{8.173}$$

where the integral on the left-hand side is the same as in (8.172). This identity should hold for all $t \in I\!R$ and $v > 0$. Setting $v = 0$ in (8.173) yields the first ODE

$$H'(t) = \alpha R^{-1}\big(E(t) - \tfrac{1}{\beta}\ln\alpha\big) + K_3. \tag{8.174}$$

Without loss of generality, we are assuming that $\alpha > 0$. Otherwise, if $\alpha = 0$, we integrate in (8.171) from the limit 1. Differentiating (8.173) with respect to v, one can check that this gives the identity if $E'(t) = 1$, which is the second ODE.

(ii) $\beta = 0$. This case is easier. Integrating (8.169) yields

$$v_x = \alpha R^{-1}(\eta(v)), \quad \eta(v) = -\tfrac{1}{\alpha}\sqrt{v} + E(t), \quad \text{where}$$

$$R(Y) = \tfrac{1}{2\mu}\left[Y - \tfrac{1}{\mu}\ln(\mu Y + 1)\right] \quad \text{and} \quad Y(v) = \tfrac{1}{\alpha}v_x.$$

Therefore, instead of (8.171), we obtain solutions of (8.167) in the form of

$$\int_0^{v(x,t)} \frac{dz}{R^{-1}(\eta(z))} = \alpha x + H(t). \tag{8.175}$$

In a similar fashion, the following DS is derived (cf. (8.174)):

$$H'(t) = \alpha^2 R^{-1}(E(t)) + \alpha K_3, \quad \text{where } E'(t) = 1. \tag{8.176}$$

Theorem 8.37 *Under the given hypotheses, the exact solutions* (8.171) *or* (8.175) *of the PDE* (8.167) *are governed by the ODE* (8.174) *or* (8.176) *with* $E(t) = t$.

8.5.4 The degenerate case

1. We now consider the simplest case of the fully degenerate quadratic polynomial (8.90), where $a = b = 0$, and

$$P(v) = c \neq 0.$$

Then $P'' = P' = 0$ and the first generating ODE (8.130) yields $g'' = 0$, which gives us $g = dv + h$, $d \neq 0$. Setting $h = 0$ by translation gives $g(v) = dv$. From (8.131), we get $G''(v) = \frac{2}{3}d^2 v$, and hence,

$$G(v) = \tfrac{1}{9}d^2 v^3 + K_1 v + K_2, \tag{8.177}$$

with arbitrary constants K_1 and K_2. It then follows from (8.132) that

$$m(v) = 0, \quad s(v) = \tfrac{2}{3}cdv + K, \quad \text{and} \quad M(v) = 2cG(v), \tag{8.178}$$

where K is arbitrary. Finally, from (8.129), we derive the following class of *quasilinear* heat equations:

$$\boxed{v_t = -\tfrac{1}{dv}\left(\tfrac{2}{3}cdv + K\right)v_{xx} + \tfrac{1}{dv}\left(\tfrac{4}{3}cdv - K\right)G(v),} \tag{8.179}$$

which admit the SI

$$\mathcal{H}[v] = v_t - \left(\tfrac{2}{3}cdv + K\right)v_x - 2cG(v).$$

2. The construction of exact solutions of (8.179) is also easier than in the previous case. The corresponding stationary ODE

$$v_{xx} + dvv_x + \tfrac{1}{9}d^2 v^3 + K_1 v + K_2 = 0 \tag{8.180}$$

is linearized by the transformation

$$v = \tfrac{3}{d}\tfrac{\rho_x}{\rho}. \tag{8.181}$$

This yields the linear-third order ODE

$$\rho_{xxx} + K_1\rho_x + \tfrac{1}{3}\,dK_2\rho = 0, \tag{8.182}$$

which results in the linear representation (8.159) of the general solution.

3. Substituting $G(v)$ (as given in (8.177)) from (8.180) into (8.179) yields the classical Burgers equation

$$v_t = -2cv_{xx} - \left(\tfrac{4}{3}\,cdv - K\right)v_x,$$

which, by (8.181), reduces to the linear heat equation with lower-order terms

$$\rho_t = -2c\rho_{xx} + K\rho_x + N_0(t)\rho. \tag{8.183}$$

The rest of the analysis of the consistent system (8.182), (8.183) is similar.

Example 8.38 Setting $K_1 = -1$ and $K_2 = 0$ in (8.182) gives the solution

$$\rho = 1 + H_2\mathrm{e}^x + H_3\mathrm{e}^{-x}$$

of the linearized stationary ODE. Substituting into (8.183) yields $N_0 = 0$ and the explicit solution of (8.179),

$$v(x,t) = \frac{3}{d}\,\frac{A\mathrm{e}^{x+(K-2c)t} - B\mathrm{e}^{-x-(K+2c)t}}{1+A\mathrm{e}^{x+(K-2c)t} + B\mathrm{e}^{-x-(K+2c)t}}\,.$$

Example 8.39 Now let $K_1 = 0$ and $dK_2 = -3$. Then,

$$\rho = \mathrm{e}^x + \mathrm{e}^{-\frac{x}{2}}\left(H_2 \sin\tfrac{\sqrt{3}x}{2} + H_3\cos\tfrac{\sqrt{3}x}{2}\right).$$

Substituting into the heat equation (8.183) implies that $N_0 = 2c - K$, and that the coefficients $H_2(t)$ and $H_3(t)$ solve the linear DS

$$\begin{cases} H_2' = 3\left(c - \frac{K}{2}\right)H_2 - \sqrt{3}\left(c + \frac{K}{2}\right)H_3, \\ H_3' = \sqrt{3}\left(c + \frac{1}{2}\right)H_2 + 3\left(c - \frac{1}{2}\right)H_3, \end{cases}$$

which is easily integrated. In particular, for $K = -2c$, we have the solution of (8.179)

$$v = \frac{3}{2d}\,\frac{2\mathrm{e}^x + \mathrm{e}^{6ct-\frac{x}{2}}\left[(A\sqrt{3}-B)\cos\frac{\sqrt{3}x}{2} - (B\sqrt{3}+A)\sin\frac{\sqrt{3}x}{2}\right]}{\mathrm{e}^x + \mathrm{e}^{6ct-\frac{x}{2}}\left(A\sin\frac{\sqrt{3}x}{2} + B\cos\frac{\sqrt{3}x}{2}\right)}\,.$$

8.5.5 *$P(v)$ is a quadratic polynomial*

Finally, we study the most interesting case $a \neq 0$, where (8.90) is a quadratic polynomial.

1. Single-term polynomial. Consider first the case of the single-term quadratic polynomial

$$P(v) = \tfrac{1}{2}\,av^2, \quad \text{with } a \neq 0 \ \ (\text{i.e., } b = c = 0). \tag{8.184}$$

Then the generating equations (8.130) and (8.131) are solved explicitly,

$$g(v) = \mu v^2 + \tfrac{\nu}{v^4} \ \text{ and } \ G(v) = \tfrac{\alpha}{v^3} + \beta v + \tfrac{1}{16}\left(\mu^2 v^5 - \tfrac{\nu^2}{v^7}\right), \tag{8.185}$$

where $G_0(v) = \frac{\alpha}{v^3} + \beta v$ is the general solution of (8.131).

Let us begin with the structure of exact solutions in the case where $\alpha = \beta = 0$ in (8.185). The stationary ODE (8.133),

$$v_{xx} + \tfrac{1}{v^4}(\mu v^6 + \nu)v_x + \tfrac{1}{16v^7}(\mu v^6 - \nu)(\mu v^6 + \nu) = 0,$$

can then be integrated in quadratures. Introducing the new function Y by

$$v_x = \tfrac{1}{16}\, \tfrac{\mu v^6 - \nu}{v^3}\, Y(v), \tag{8.186}$$

gives the first-order ODE

$$vY\, \tfrac{\mathrm{d}Y}{\mathrm{d}v} = -\tfrac{\mu v^6 + \nu}{\mu v^6 - \nu}\,(Y+4)(3Y+4). \tag{8.187}$$

Integrating yields

$$Y = R^{-1}(\eta(v)), \quad \text{where} \;\; \eta(v) = \tfrac{v^2}{(\mu v^6 - \nu)^{2/3}}\, E, \tag{8.188}$$

E is a free constant and R^{-1} is the inverse function to

$$R(Y) = \tfrac{Y+4}{(3Y+4)^{1/3}}. \tag{8.189}$$

Integrating (8.186) with Y from (8.188) gives the following solutions $v(x,t)$:

$$16 \int_1^{v(x,t)} \tfrac{z^3}{\mu z^6 - \nu}\, \tfrac{\mathrm{d}z}{R^{-1}(\eta(z))} = x + H. \tag{8.190}$$

Here, $E = E(t)$ and $H = H(t)$ are some smooth functions to be determined by substituting (8.190) into the corresponding Hamilton–Jacobi PDE with the coefficients calculated by (8.132),

$$\mathcal{H}[v] \equiv v_t - \big[av(v_x)^2 + \big(\tfrac{1}{2}\, a\mu v^4 + K\big)v_x + \tfrac{a}{16}\,\big(\mu^2 v^7 - \tfrac{\nu^2}{v^5}\big)\big] = 0. \tag{8.191}$$

Using (8.129), (8.184), and (8.185) yields the quasilinear heat equation

$$\boxed{\begin{aligned} v_t = {} & -v^4\, \tfrac{\frac{1}{2}a\mu v^4 + K}{\mu v^6 + \nu}\, v_{xx} + av(v_x)^2 \\ & + \tfrac{1}{16v^5}(\mu v^6 - \nu)\big(\tfrac{1}{2}\, a\mu v^6 + a\nu - Kv^2\big). \end{aligned}} \tag{8.192}$$

Theorem 8.40 *The parabolic PDE (8.192) admits the sign-invariant (8.191) and solutions (8.190), where the functions $E(t)$ in (8.188) and $H(t)$ satisfy the DS*

$$\begin{cases} H' = \tfrac{1}{16}\, a(\mu - \nu)g_E(t) + \tfrac{1}{2}a\mu + K + \tfrac{a(\mu+\nu)}{g_E(t)}, \\ E' = \tfrac{1}{2}\, a\mu\nu E + \tfrac{1}{128}aE^4, \end{cases} \tag{8.193}$$

where $g_E(t) = R^{-1}(E(t)(\mu - \nu)^{-2/3})$.

Proof. We follow the same lines, as in the linear critical case. Calculating derivative v_t from (8.190),

$$v_t - \tfrac{\mu v^6 - \nu}{v^3}\, R^{-1}(\eta(v))\big[\tfrac{1}{16}\, H' - E' \textstyle\int_1^v \tfrac{z^5}{(\mu z^6 - \nu)^{5/3}}\big(\tfrac{1}{R^{-1}(\eta(z))}\big)'_\eta\, \mathrm{d}z\big],$$

and substituting into (8.191), together with v_x from (8.186), we derive the identity

$$\tfrac{1}{16}H' - E' \textstyle\int_1^v (\cdot) = \tfrac{1}{16}\Big[\tfrac{a}{16}\, \tfrac{\mu v^6 - \nu}{v^2} R^{-1} + \tfrac{1}{2}a\mu v^4 + K + a\, \tfrac{\mu v^6 + \nu}{v^2}\, \tfrac{1}{R^{-1}}\Big]. \tag{8.194}$$

Here, setting $v = 1$ yields the first ODE (8.193). Differentiating (8.194) with respect to v and using (8.188) implies the following ODE:

$$vY\,\frac{dY}{dv} = -\frac{2(2\mu v^6+\nu)Y^2\big[Y+\frac{4(2\mu v^6-\nu)}{2\mu v^6+\nu}\big](Y+4)}{(\mu v^6-\nu)Y^2-\big[\frac{256\gamma v^6}{\mu v^6+\nu}+16(\mu v^6+\nu)\big]}, \quad \text{with} \quad \gamma = -\frac{E'}{2aE}. \tag{8.195}$$

Let us compare ODEs (8.195) and (8.187), which must admit a common solution given by (8.188) and (8.189). The right-hand sides of both equations (8.195) and (8.187) coincide if $Y = Y(v)$ solves the cubic algebraic equation

$$Y^3 + 12Y^2 + \big[16 + \tfrac{64v^6(4\gamma+\mu\nu)}{(\mu v^6-\nu)^2}\big](3Y+4) = 0. \tag{8.196}$$

On the other hand, expressions (8.188) and (8.189) can be written in the form of the following cubic equation:

$$Y^3 + 12Y^2 + \big[16 - \tfrac{E^3v^6}{(\mu v^6-\nu)^2}\big](3Y+4) = 0. \tag{8.197}$$

The last coefficients of the term $(3Y+4)$ of (8.196) and (8.197) coincide if $256\gamma = -64\mu\nu - E^3$, from which comes the second ODE in (8.193). □

2. A solution for general polynomial. Consider another solution of the generating ODEs. We again assume that $a \neq 0$, and, by translation, take the quadratic polynomial in the most general form

$$P(v) = \tfrac{1}{2}(av^2 + 2c), \tag{8.198}$$

where $c \neq 0$. Then the system (8.130), (8.131) has the general solution

$$g(v) = \tfrac{1}{4}(2av^2 + c) \quad \text{and}$$
$$G(v) = \tfrac{1}{64}(a^2v^5 + 2acv^3) + \alpha G_0(v) + \beta v,$$

where α and β are arbitrary constants, and $G_0(v)$ denotes the second solution of the homogeneous equation (8.131) such that $G_0(v)$ and v are linearly independent.

Let us restrict ourselves to the *quadrature* case

$$\alpha = 0 \quad \text{and} \quad \beta = \tfrac{3}{256}\,c^2.$$

Then, setting in the corresponding stationary equation (cf. (8.186))

$$v_x = \tfrac{1}{192}(2av^3 + 3cv)Y(v) \tag{8.199}$$

yields the ODE

$$Y\,\frac{dY}{dv} = -\frac{6av^2+3c}{2av^3+3cv}\,(Y+4)(Y+12), \quad \text{so} \tag{8.200}$$

$$Y = R^{-1}(\eta(v)),\ \ \eta(v) = \tfrac{E}{144}\,(2av^3+3cv)^2,\ \ R(Y) = \tfrac{Y+4}{(Y+12)^3}. \tag{8.201}$$

Integrating (8.199) yields the following solutions:

$$192\int_1^{v(x,t)} \frac{1}{2az^3+3cz}\,\frac{dz}{R^{-1}(\eta(z))} = x + H. \tag{8.202}$$

The Hamilton–Jacobi PDE is now

$$\begin{aligned}\mathcal{H}[v] \equiv v_t - \big[&av(v_x)^2 + \tfrac{1}{4}\,(a^2v^4 + 2acv^2 + K)v_x \\ &+ \tfrac{1}{256}(av^2+2c)(4a^2v^5 + 8acv^3 + 3c^2v)\big] = 0.\end{aligned} \tag{8.203}$$

The quasilinear heat equation takes the following (rather awkward) form:

$$\boxed{\begin{aligned} v_t = &-\tfrac{a^2v^4+2acv^2+K}{2av^2+c}\, v_{xx} + av(v_x)^2 \\ &+ \tfrac{1}{256}\,\tfrac{a^2v^4+3acv^2+2c^2-K}{2av^2+c}\,(4a^2v^5 + 8acv^3 + 3c^2v), \end{aligned}} \tag{8.204}$$

where a, c, and K are arbitrary constants.

Theorem 8.41 *PDE* (8.204) *admits the sign-invariant* (8.203) *and solutions* (8.202), *where the coefficients* $E(t)$ *in* (8.201) *and* $H(t)$ *satisfy the DS*

$$\begin{cases} H' = \frac{a}{192}(2a+3c)g_E(t) + \frac{1}{4}(a^2+2ac+K) + \frac{3(a+2c)(2a+c)}{4g_E(t)}, \\ E' = -\frac{a}{128} - \frac{3}{64}c^3E, \end{cases} \tag{8.205}$$

with $g_E(t) = R^{-1}(\frac{1}{144}E(t)(2a+3c)^2)$.

Proof. Calculating v_t from (8.202), and substituting both derivatives v_x and v_t into (8.203) (where setting $v = 1$ in the identity gives the first ODE (8.205)), after differentiating in v, we obtain

$$Y\,\frac{\mathrm{d}Y}{\mathrm{d}v} = -\frac{2av(4av^2+3c)Y^2\left(Y+12\frac{4av^2+5c}{4av^2+3c}\right)(Y+12)}{av^2(2av^2+3c)Y^2+\frac{12288\gamma}{2av^2+c}-144(av^2+2c)(2av^2+c)},$$

where $\gamma = -\frac{E'}{2E}$. It admits solution (8.201) satisfying (8.200) if $\frac{1}{E} - \frac{256\gamma}{a} + \frac{6c^3}{a} = 0$, from which comes the second ODE in (8.205). □

Due to Theorem 8.32, the DS (8.135) is always given explicitly in all the cases (as well as in other cases of sets described by the stationary equation (8.133)). In the case of (8.198), the corresponding solutions $v(x,t)$, in general, do not have explicit quadrature representation, excluding the cases considered.

8.5.6 Interpretation via invariant subspaces and sets

It is not often easy to interpret some of the obtained solutions by using the current ideology of linear subspaces that are (partially) invariant under nonlinear operators. As usual, such interpretations help to reconstruct extensions to higher-order PDEs, avoiding technical manipulations.

Example 8.42 (**Semilinear equation**) Consider the simplest quasilinear case. Fix $a = 1$ in (8.184) and set $\mu = \nu = 0$ in (8.185) (i.e., $g(v) \equiv 0$) and $\alpha = 1$, $\beta = 0$ in (8.185) (i.e., $G(v) = \frac{1}{v^3}$). Next, from (8.132), take $m(v) = v$, $s(v) = 0$, and $M(v) = \frac{1}{v}$, and finally find from (8.129) that $\varphi(v) = 1$ and $f(v) = \frac{1+v^2}{v^3}$. We then obtain the following semilinear heat equation:

$$v_t = v_{xx} + v(v_x)^2 + \frac{1+v^2}{v^3} \tag{8.206}$$

that possesses the explicit solution of the typical self-similar form

$$v_*(x,t) = \sqrt{2t - \tfrac{1}{2t}x^2}\,. \tag{8.207}$$

But (8.207) is not self-similar, in the sense that it cannot be constructed by symmetry group analysis. In view of (8.207), we introduce $V = v^2$, so that

$$\boxed{V_t = F_1[V] + F_2[V] \equiv \big[V_{xx} - \tfrac{1}{2V}(V_x)^2 + \tfrac{2}{V}\big] + \big[\tfrac{1}{2}(V_x)^2 + 2\big].} \tag{8.208}$$

The corresponding explicit solution is

$$V_*(x,t) \equiv (v_*)^2 = 2t - \tfrac{1}{2t}x^2 \in W_2 = \mathcal{L}\{1, x^2\}, \tag{8.209}$$

for all $t > 0$, where the 2D subspace W_2 is invariant under the second Hamilton–Jacobi operator F_2 in (8.208), but not under the operator F_1. Indeed, we have, for $V = C_1 + C_2x^2 \in W_2$,

$$F_2[V] = 2 + 2C_2^2x^2 \in W_2 \quad \text{and} \quad F_1[V] = \tfrac{2(C_1C_2+1)}{C_1+C_2x^2}.$$

We next introduce the following set on W_2:

$$M = \big\{V = C_1 + C_2x^2 : \;\; F_1[V] = 0, \;\; \text{i.e.,} \;\; C_1C_2 + 1 = 0\big\},$$

which is invariant on W_2 under the full operator $F_1 + F_2$ in (8.208). Substituting $V(x,t) = C_1(t) + C_2(t)x^2$, implies that (8.208) on M is the overdetermined DS, which gives (8.209),

$$\begin{cases} C_1' = 2, \\ C_2' = 2C_2^2, \\ C_1C_2 + 1 = 0. \end{cases}$$

Example 8.43 (**Higher-order PDEs**) As the first obvious extension, note that the explicit solution (8.209) satisfies any of the PDEs of arbitrary order

$$\boxed{V_t = \psi(V, V_x, V_{xx}, \ldots)F_1[V] + F_2[V],} \tag{8.210}$$

with an arbitrary function $\psi(\cdot)$. For a more subtle example, consider the following fourth-order parabolic PDE:

$$\boxed{V_t = \big[-V_{xxxx} + \tfrac{1}{6V}(V_{xx})^2 - \tfrac{24}{V}\big] + \big[\tfrac{1}{6}(V_{xx})^2 + 24\big].} \tag{8.211}$$

Then, introducing a similar set, $M \subset W_2$, and looking for solutions $V_*(x,t) = C_1(t) + C_2(t)x^4$ on $W_2 = \mathcal{L}\{1, x^4\}$ yields the consistent overdetermined DS,

$$\begin{cases} C_1' = 24, \\ C_2' = 24C_2^2, \\ C_1C_2 + 1 = 0, \end{cases}$$

i.e., $C_1(t) = 24t$ and $C_2(t) = -\frac{1}{24t}$, from which comes the solution of (8.211),

$$V_*(x,t) = 24t - \tfrac{1}{24t}x^4.$$

Example 8.44 (**Semilinear heat equation**) Setting again $a = 1$ and $\mu = \nu = 0$, we now take $\alpha = \beta = 1$, i.e., according to (8.185), $G(v) = v + \frac{1}{v^3}$. Then v and $V = v^2$ solve the equations

$$v_t = v_{xx} + v(v_x)^2 + (1 + v^2)\big(v + \tfrac{1}{v^3}\big) \implies$$

$$\boxed{V_t = \left[V_{xx} - \tfrac{1}{2V}(V_x)^2 + \tfrac{2}{V} + 2V\right] + \left[\tfrac{1}{2}(V_x)^2 + 2V^2 + 2\right],} \tag{8.212}$$

with the explicit solution

$$V_*(x,t) = \tan 4t + \tfrac{1}{\cos 4t} \sin 2x \in \tilde{W}_2 = \mathcal{L}\{1, \sin 2x\}, \tag{8.213}$$

where $\tilde{W}_2$ is invariant under the second operator F_2 in (8.212). The set $M \subset \tilde{W}_2$ on which the first operator vanishes, $F_1[V] = 0$ on M, takes the form

$$M = \left\{V = C_1 + C_2 \sin 2x : \;\; C_1^2 - C_2^2 + 1 = 0\right\}.$$

Plugging $V = C_1(t) + C_2(t) \sin 2x \in \tilde{W}_2$ into (8.212) gives the overdetermined DS

$$\begin{cases} C_1' = 2(C_1^2 + C_2^2) + 2, \\ C_2' = 4C_1C_2, \\ C_1^2 - C_2^2 + 1 = 0, \end{cases} \tag{8.214}$$

which generates the nontrivial solution (8.213).

Example 8.45 **(Fourth-order parabolic equation)** Clearly, (8.213) solves the general equation (8.210) with operators $F_{1,2}$ from (8.212). Exactly the same solution (8.213) with the same DS (8.214) occurs for the fourth-order PDE

$$\boxed{V_t = -\tfrac{1}{4} V_{xxxx} - \tfrac{1}{2V}(V_x)^2 + \tfrac{2}{V} + 2V + \tfrac{1}{2}(V_x)^2 + 2V^2 + 2,}$$

since $V_{xx} = -\frac{1}{4} V_{xxxx}$ on $\tilde{W}_2$.

Remarks and comments on the literature

The ideas and methodology of differential constraints are classical in PDE theory and initially appeared for first-order equations. Lagrange applied differential constraints to determine total integrals of nonlinear equations with two independent variables,

$$F(x, y, u, u_x, u_y) = 0.$$

Darboux extended this technique to the second-order PDEs [139], which previously had been studied by Monge and Ampère by means of first integrals. A full history of the early years of differential substitution and constraint approaches can be found in Goursat [260] and Forsyth [196]. See also a survey in Kaptsov [326] devoted to some generalizations of the classical Darboux method. Differential constraints are known to lead to difficult systems of PDEs. A general theory of such overdetermined systems was developed by Riquier, Cartan, Ritt, Spencer, and others; see Pommaret [468] for a detailed survey.

Some modern versions of differential constraint approaches are related to the classical *semi-inverse method* that is well known in Continuum Mechanics and more applied areas; see a survey paper by Neményi [433]. The link between the semi-inverse method and symmetries of PDEs was first noted by Birkhoff [61] in application to hydrodynamic problems. In particular, Birkhoff introduced the concept of equations, written in *separate form*, i.e., in the form of a linear combination

$$\textstyle\sum_{i=1}^{m} \Lambda_i(\chi)\Gamma_i(\mu),$$

where the coefficients $\Lambda_i(\chi)$ depend on the set of variables $\chi = \{\chi_1, ..., \chi_s\}$ and functions of these variables, whereas $\Gamma_i(\mu)$ depend on different variables $\mu = \{\mu_1, ..., \mu_r\}$ and functions of these variables. Reducing PDEs to such separate forms makes the method successful. See further comments in the survey in [474]. Later on, in the 1960s, a systematic approach to differential constraints in gas dynamics was proposed by Yanenko [582], which is reflected in the book [525]; see also other applications and references in [13].

In this chapter, we use the basic notions of SIs introduced in [219], where a number of presented results can be found.

§ 8.1. Let us mention again that, in many fundamental problems for quasilinear parabolic PDEs, deriving suitable one-sided estimates (or SIs) plays a key role for the existence, uniqueness, regularity, and different asymptotic results. This is explained in many books; see [33, 205, 245, 509, 530, 533, 550].

§ 8.2. We follow [219] and use SIs proposed in [215]; see more references in [509, Ch. 5]. Corollary 8.2 can be found in [215] and [509, p. 303]. Kirchhoff's transformations in heat conduction theory have been used since 1894, [348].

We mention again that some ideas of the SIs had the origin in blow-up singularity analysis of combustion reaction-diffusion PDEs. The first concept of SI analysis is associated with the notion of the *ψ-criticality* of solutions of parabolic equations playing an important role in blow-up theory; see details and references in [509, Ch. 5]. In general settings, we deal with PDIs of the type

$$\mathcal{H}[u] \equiv u_t - \psi(x, t, u, \nabla u) \geq 0 \quad \text{in } I\!R^N \text{ for } t > 0, \tag{8.215}$$

with *a priori* unknown functions $\psi(x, t, u, \nabla u)$ to be determined from the invariance condition. According to (8.215), the zero-criticality of the solution $u(x, t)$, i.e., the ψ-criticality with $\psi(\cdot) \equiv 0$, implies that $u_t \geq 0$ holds for $t > 0$ if $u_t(x, 0) \geq 0$ initially. In combustion problems, the last inequality is known to characterize initial temperature of the *critical ignition*. Therefore, (8.215) is a natural extension of the critical property, such as the ψ-criticality with respect to a given function ψ. General ψ-critical conditions (8.215) of solutions of parabolic PDEs were introduced in [215]; see extended results and references in [509, Ch. 5]. Estimates of the type (8.215) with different functions ψ have been used in several problems for quasilinear heat equations (8.1), such as heat localization and blow-up behavior, [509, Ch. 5, 7]. A similar idea to derive estimates of the type (8.215) with $\psi = \psi(u)$ for some class of quasilinear heat equations was used in [533], where the applications to blow-up problems are also given. For the semilinear heat equation (8.4), the ψ-criticality (8.215) with $\psi = \delta f(u)$, $\delta =$ constant, has been employed in [207].

§ 8.3. The main results are taken from [219]. Function $g(r) = r$ in the SI (8.49) was introduced in [207], two other cases, $g(r) = \frac{1}{r}$ and $g(r) = r^{1-N}$, were established in [219].

This concept of SI analysis uses the idea by Friedman and McLeod [207]. In the study of single point blow-up for semilinear heat equations (8.4) with nonlinearities $f(u) = e^u$ (the *Frank–Kamenetskii equation* or *solid fuel model*, [594]) and $f(u) = u^p$, they proposed deriving the so-called *gradient estimate* by using a sign-type analysis of the following first-order operator for radial solutions $u = u(r, t)$, $r = |x|$:

$$\mathcal{H}_*[u] \equiv u_r + r\,\Phi(u) \quad \text{for } t > 0. \tag{8.216}$$

A suitable choice of the function Φ provides us with an optimal estimate for blow-up solutions and depends on f. For quasilinear PDEs (8.1), the SI operator takes the form $\mathcal{H}_*[u] = k(u)u_r + r\,\Phi(u)$. A crucial problem is then to choose an *optimal* function $\Phi(u)$ satisfying a nonlinear *ordinary differential inequality*, depending on the coefficients $k(u)$ and $f(u)$ of the parabolic equation (8.1). Various generalizations of this approach to quasilinear heat equations

(8.1) was performed in [233, 243]; see [245, Sect. 10.4], where the most general computations are presented.

This approach has been extended [233] to PDEs with the p-Laplace operator

$$u_t = \nabla \cdot (|\nabla u|^\sigma \, \nabla u) + f(u), \tag{8.217}$$

where the corresponding SIs are $\mathcal{H}_*[u] \equiv |u_r|^\sigma \, u_r + r \, \Phi(u)$.

There exists a direct relation between the results on SIs and a geometric Sturmian theory of nonlinear 1D parabolic equations, [226, Ch. 7]. Geometric theory of such PDEs uses the fact that proper, complete (i.e., sufficiently dense in a natural geometric sense) sets B of particular exact solutions can determine a property of the *B-convexity/concavity* that is defined with respect to the given functional set B. This property is then preserved with time and reduces to the time-invariance of the sign of a certain nonlinear operator (actually, it is a SI) on evolution orbits generated by the parabolic equation.

For second-order parabolic equations, some results on generalized conditional symmetries can be translated to sign invariants; see e.g., [476] and [481], where exact solutions are constructed for equations $u_t = \nabla \cdot (B(u)\nabla u) + A(x, u)$.

§ 8.4. The main results are obtained in [219]. Examples 8.22, 8.23 and 8.24 represent rather elementary equations possessing solutions on invariant subspaces or on sets of lower dimensions 1, 2, or 3. Various extensions of similar differential constraints to parabolic PDEs with non-constant coefficients, extra convection, and gradient diffusivity terms in Sections 8.1–8.4 can be found in [474], together with a quality survey on the relation between semi-inverse methods and symmetries of PDEs.

The 1D invariant subspaces correspond to the *functional separation of variables*, where the additively separable solutions

$$u(x, t) = \phi(A(x) + B(t))$$

are studied. Necessary and sufficient conditions for such a general additive separation of a PDE were obtained in [310], being an extension of the classical *Stäckel form* (1893) [534]. For this separation technique, the results by Steuerwald (1936) [535] have been found effective for finding standing waves. See applications to the sine-Gordon (Enneper) equation, $u_{tx} = \sin u$, in [521], where solutions of the form $u(x, t) = F(f_1(t)f_2(x))$ were described. Such a structure of solutions was earlier derived by Seeger (1953) by integrating Darboux's equation (1894) [142] for Enneper surfaces (pseudo-spherical surfaces with at least one set of planar lines of curvature), and was shown to belong to the manifold of solutions obtained by Steuerwald [535]; see [521].

This functional separation of variables was applied in [316] (see also [13, Ch. 5]) to the 2D equation of vortex structures of an inviscid fluid,

$$\Delta u \equiv u_{xx} + u_{yy} = F(u), \tag{8.218}$$

where u is stream function and $F(u)$ is given (equation (8.218) was derived by H. Lamb [368]; see [29]). According to this method of *generalized separation of variables* (GSV), it was shown [316] that (8.218) admits nontrivial solutions

$$u(x, y) = h(f(x) + g(y)), \tag{8.219}$$

where h, f, and g are some unknown functions to be determined (cf. another application in [423]) iff the right-hand side F takes one of the following five forms:

$$\begin{gathered} A\mathrm{e}^u + B\mathrm{e}^{-2u}, \;\; Au \ln u + Bu, \;\; A \sin u + B(\sin u \ln|\tan \tfrac{u}{4}| + 2\sin \tfrac{u}{2}), \\ A \sinh u + B(\sinh u \ln|\tanh \tfrac{u}{4}| + 2\sinh \tfrac{u}{2}), \\ A \sinh u + B(\sinh u \tan^{-1} \mathrm{e}^{u/4} + 2\cosh \tfrac{u}{2}), \end{gathered}$$

where A and B are constants. All these *Kaptsov's solutions* of the equation (8.218) are given by explicit formulae. Two equations from this list, the *Tzitzéica equation*,*

$$\Delta u = e^u - e^{-2u}, \tag{8.220}$$

and the *Bonnet equation* [71], $\Delta u = \sin u$, are well known and exhibit special symmetry properties and solutions; see [10, Sect. 12.3], [321] for N-soliton solutions of (8.220) and also a survey in [138]. Two Lax pairs of (8.220) are known, and the first one was found by Tzitzéica himself in 1908, and the second pair by Mikhailov seventy years later, [138, p. 2094]. Similar separation results for (8.218) were also obtained in [423], where more general aspects of this functional separation are presented and applied to hyperbolic and elliptic PDEs (in Riemannian and pseudo-Riemannian spaces) with the right-hand side $F = F(x, y, u)$.

Setting in (8.219) $f = \ln \hat{f}$ and $g = \ln \hat{g}$ yields

$$u(x, y) = h(\ln(\hat{f}(x)\hat{g}(y))) \equiv \hat{h}(\hat{f}(x)\hat{g}(y)),$$

exhibiting a typical structure of 1D subspaces with unknown basic functions, so the solutions of this type are naturally associated with invariant subspaces or sets. On the other hand, differentiating in x yields the typical SI structure $u_x = f'(x)H(u)$ studied in Section 8.3. As a next step, for solutions on 2D subspaces $u(x, y) = h(f_1(x) + f_2(x)g(y))$, the corresponding differential constraint takes the form $u_x = g(x)h(y)H(u)$ (cf. [474]), or, which is the same, $[\frac{u_x}{g(x)H(u)}]_x = 0$. One can modify such a constraint to include solutions on three or more-dimensional subspaces, though computations then become much more technical and possibly unbearable.

Similar ideas apply to the *Grad–Shafranov equation* that describes axisymmetric steady flows of an inviscid fluid [29]

$$u_{xx} + u_{rr} + \tfrac{1}{r}\, u_r = r^2 G(u) + F(u),$$

where F and G are arbitrary functions; see [317, 322, 421]. This approach gives a similar classification for the semilinear *Klein–Gordon equation*,

$$u_{tt} - u_{xx} = F(u),$$

(see also [319] where the differential constraints applied) and can be extended to the semilinear and quasilinear heat equations $u_t - \varphi(u)u_{xx} = F(u)$; see a general classification in [482]. In [173], this method was applied to generalized p-Laplacian equations with source, $u_t = (\varphi(u)(u_x)^n)_x + F(u)$, occurring in the theory of turbulent diffusion and non-Newtonian, dilatable, or pseudo-plastic fluids. In [172], the GSV was used for general quasilinear wave equations

$$u_{tt} = (\varphi(u)u_x)_x + F(u).$$

The algorithm of the GSV-method admits a generalization, where the derivative u_x is included in the separation formula (cf. (8.219)) $f(u, u_x) = a(t) + b(x)$; see [597] (and earlier results in [118]) applied to second-order parabolic PDEs $u_t = A(u, u_x)u_{xx} + B(u, u_x)$, and [596] for the KdV-type equations $u_t = u_{xxx} + A(u, u_x)u_{xx} + B(u, u_x)$. See also [479, 483, 480] for a generalization of separation techniques applied to PDEs $u_t = (D(u)u_x)_x + B(x)Q(u)$ and

* This equation was found by Romanian geometer Georges Tzitzéica in 1907, [564]. It is used in the field of geometry and is concerned with surfaces, on which the total curvature at each point is proportional to the fourth power of the distance from a fixed point to the tangent plane. He established its invariance under a Bäcklund transformation, and also constructed, in 1910, a linear representation of solutions incorporating a spectral parameter, which was rediscovered in the soliton theory seventy years later.

other equations. Paper [598] contains a complete classification of the nonlinear wave equations in $I\!R^2 \times I\!R$,

$$u_{tt} = A(u)u_{xx} + B(u)u_{yy} + C(u, u_x, u_y),$$

admitting solutions $f(u) = X(x) + Y(y) + T(t)$.

In particular, the GSV approach easily gives the solution in Example 8.24. On the other hand, the GSV cannot detect elementary solutions in Examples 8.22 and 8.23 dealing with subspaces of dimensions three and two respectively. A generalization of the GSV method of [316] to two and higher-dimensional subspaces seems to be an interesting and challenging problem. The main difficulty is to check whether in 2D, the method would lead to such a simple, closed algorithm of separation as it does in 1D.

Some of the solutions on invariant subspaces and sets can be obtained by constructing suitable differential constraints of higher order. See [325], where, in addition, a new family of exact solutions was constructed for the fast diffusion equation

$$u_t = (u^{-\frac{1}{2}} u_x)_x \, ,$$

by using a third-order nonlinear differential constraint. The solutions are

$$u(x,t) = -\frac{2X'(x)T'(t)}{[X(x)+T(t)]^2}, \quad \text{where}$$
$$(X')^3 = (c_0 + c_1 X + c_2 X^2 + c_3 X^3)^2,$$
$$(T')^3 = A(c_0 - c_1 T + c_2 T^2 - c_3 T^3)^2$$

(X are T are expressed in terms of the Weierstrass function $\wp$) and c_i and A are constants. Another special parabolic equation,

$$v_t = \frac{v_{xx}}{v_x + (v_x)^2} \, ,$$

admits the exact solution $v = v(x,t)$ given implicitly by

$$-\tfrac{1}{8}[\ln|\sinh(3v + 2x)| - \ln|\cos(v + 2x)|] = t,$$

which was constructed in [66] by using nonclassical potential symmetries. Notice another implicitly given solution of the equation of *superslow diffusion*,

$$u_t = (\mathrm{e}^{-1/u} u_x)_x, \quad \text{where } u = -\frac{1}{\ln v}, \quad v = \tfrac{1}{2t}(c^2 - w^2)_+,$$
$$\text{and } |x| = [2 + \ln(2t)]w + (c - w)\ln(c - w) - (c + w)\ln(c + w)$$

($\mathrm{e}^{-1/u}$ is associated with the famous *Arrhenius law* in thermodynamics; here $c > 0$ is a constant). This solution is key in asymptotic theory of superslow diffusion, [245, Ch. 3].

Let us briefly focus on another, rather unusual applications of SIs. Consider the Aronson–Bénilan *semiconvexity* estimate [17] for weak nonnegative solutions of the PME

$$u_t = \Delta u^m \quad \text{in } I\!R^N \times I\!R_+.$$

This plays a key role in general PME theory (see details and references in [245, Ch. 2]), and has the form

$$\Delta u^{m-1} \ge -\frac{C}{t}, \quad \text{where } C = \frac{(m-1)N}{m[N(m-1)+2]}.$$

It can be treated as the ψ-criticality (8.215). Since $\Delta u^{m-1} \equiv \frac{m-1}{mu}(u_t - mu^{m-2}|\nabla u|^2)$, the above estimate is exactly (8.215) with the function ψ, depending on the time-variable t, $\psi(t, u, |\nabla u|) = mu^{m-2}|\nabla u|^2 - \frac{m}{m-1}\frac{c_0}{t}\, u$.

§ 8.5. We mainly follow [236]. In [480], similar Hamilton–Jacobi SIs and solutions were obtained for the quasilinear heat equation with lower-order reaction-convection x-dependent terms, $u_t = (D(u)u_x)_x + Q(x,u)u_x + P(x,u)$.

Equations (8.126) have a wide field of applications in combustion theory, plasma physics, biophysics, and mechanics of porous media. Such quasilinear evolution PDEs are popular in group-theoretical, classical, and nonclassical methods of finding exact solutions, which are responsible for different nonlinear phenomena. There are many directions of mathematical theory of the quasilinear heat equations concerning existence, uniqueness, regularity, asymptotics, and other aspects of the qualitative properties of the solutions. See surveys in [245, 509]. Construction of appropriate one-sided SIs (*barrier techniques*) plays a fundamental role in qualitative theory of nonlinear parabolic PDEs. For instance, classical interior regularity bounds of Bernstein or similar type in the theory of parabolic equations are derived via upper bounds on the term $\xi(x)[(\rho(v))_x]^2$ ($\xi \in C_0^\infty$ is a cut-off function, ρ depends on the coefficients of the PDE), satisfying a parabolic differential inequality. See Oleinik–Kruzhkov [442].

A general description of explicit solutions from Section 8.5.2 was performed in [130] and here we present a couple of examples, showing some links with the SIs. In the particular case, $K = 0$ and $c = -\frac{3}{2}$, where (8.179) is the semilinear heat equation

$$v_t = v_{xx} - 2G(v), \tag{8.221}$$

the evolutionary invariant set governed by the corresponding stationary ODE was introduced in [317]. These explicit solutions have a strong exponential-type nature and have been constructed in [100, 329] by the Penlevé expansion (computationally related to the Baker–Hirota bilinear method). A complete analysis of the types of such explicit solutions of (8.221) with the cubic function (8.177) can be found in [130]. A general classification of exact solutions from [130] applies to the quasilinear equation (8.179). Here, we have considered a few illustrative examples only. In addition, let us mention [595], where one- (a traveling wave) and two-soliton solutions of the third-order non-integrable PDEs with various nonlinear terms (see (4.244) in Chapter 4) were obtained by a variant of the classical dressing method.

SIs and the corresponding solutions of parabolic PDEs with the p-Laplacian operator

$$u_t = \nabla \cdot (|\nabla u|^\sigma \nabla u) + f(u) \quad \text{in } I\!R^N \times I\!R_+ \tag{8.222}$$

are described in [237]. Such types of diffusion-like operators appear, for instance, in the theory of non-Newtonian liquids and in some turbulence problems; see Barenblatt [25] and references therein. Such models are common in combustion problems related to solid fuels. SIs were also detected in [237] for parabolic PDEs with a general gradient-dependent nonlinearity,

$$u_t = h(|\nabla u|)\Delta u + f(u). \tag{8.223}$$

Such nonlinearities are used in mean curvature flow equations related to differential geometry, and to problems of motion of closed hypersurfaces in $I\!R^N$ by its mean curvature, [79, 107]. These PDEs are important in phase transition phenomena, [271]. The 1D equation (8.222) corresponds to $h(p) = (\sigma + 1)p^\sigma$ in (8.223). A typical example is

$$u_t = \frac{u_{xx}}{1+(u_x)^2} - \frac{1}{u} \quad \text{in } I\!R \times I\!R_+,$$

which describes, after a surface parameterization, the evolution of cylindrically symmetric hypersurfaces moving by mean curvature in $I\!R^3$, [178, 532]. Notice also *generalized Burgers' equation* (see [362] and extensions in [115])

$$u_t + uu_x = \pm\nu \, \frac{1-(u_x)^2}{[1+(u_x)^2]^2} \, u_{xx} \, ,$$

which contains a gradient nonlinearity and describes strongly nonlinear processes governing high-amplitude and gradient phenomena. For small gradients $|u_x| \ll 1$, this leads to *Burgers' equation* $u_t + uu_x = \nu u_{xx}$ introduced by I.M. Burgers in 1948.

CHAPTER 9

Invariant Subspaces for Discrete Operators, Moving Mesh Methods, and Lattices

In this chapter, we apply basic techniques of linear invariant subspaces and sets to nonlinear discrete operators and equations. Symmetry analysis of finite-difference equations is a classical subject of group theory; see Handbook of Group Analysis [10, Ch. 17] for key results and references. Concerning general theory of difference equations and literature, we refer to Lakshmikantham–Trigiante [367].

Our main goal concerns finite-dimensional reductions of nonlinear difference equations by using the methodology and the experience that were gained when dealing with invariant subspaces for differential operators. As in the continuous case, we formulate the Main Theorem, which establishes a general representation of the pth-order nonlinear difference operators admitting a given finite-dimensional invariant subspace. This solves the backward problem of invariant subspaces. Some aspects of the forward problem are also considered. We describe subspaces for quadratic first and some higher-order operators.

We next apply these results to moving mesh methods (MMMs) for nonlinear parabolic and hyperbolic PDEs. It is shown that there exist special approximations of differential operators, which preserve finite-dimensional evolution on 2D subspaces. This guarantees that the MMMs under consideration may preserve not only some of the group-invariant solutions (this is the subject of classical Lie symmetry analysis), but also exact solutions on invariant subspaces. For second-order parabolic PDEs, such exact solutions can be used for comparison, for establishing various asymptotic results, and in stabilization analysis.

In the final section, we present examples of exact solutions of lattice dynamical systems as discrete counterparts of parabolic, compacton, and thin film PDEs. In particular, we discuss periodic breathers on anharmonic lattices and detect oscillatory, changing sign behavior for higher-order parabolic models.

9.1 Backward problem of invariant subspaces for discrete operators

Let $\mathbf{Z} = \{0, \pm 1, \pm 2, ...\}$ be the set of integers, and let V be a linear space of real-valued functions defined on $\mathbf{Z}$. Given a function $u : \mathbf{Z} \to I\!R$, the notation $u_i = u(i)$, $i \in \mathbf{Z}$, is used. Fix a natural p and consider a nonlinear difference operator $F : V \to V$ given by

$$F[u]_i \equiv \phi(u_i, u_{i+1}, ..., u_{i+p}, i), \quad i \in \mathbf{Z}, \tag{9.1}$$

where $\phi : I\!R^{p+1} \times \mathbf{Z} \to I\!R$ is a real-valued function. Therefore, (9.1) is a pth-order difference operator. Given a function $g \in V$, we study the difference equation

$$F[u] = g \quad \text{on} \quad \mathbf{Z}, \tag{9.2}$$

which is an infinite system of nonlinear algebraic equations with solutions $u \in V$. Using techniques of linear invariant subspaces, we will describe equations (9.2), which can be reduced to finite systems of algebraic equations.

Usual finite-difference approximations of nonlinear ODEs and PDEs lead to systems such as (9.2). The order p of the difference operator then depends on the order of the approximation of the differential operators. For instance, $p = 2$ corresponds to a symmetric three-point approximation of the second derivative,

$$u_{xx} \sim \tfrac{1}{h^2}\big(u_i - 2u_{i+1} + u_{i+2}\big),$$

where $h > 0$ is a constant step of the discretization (u_i denotes the value of u at the ith point of the grid). We have $p = 2$ also for a three-point symmetric approximation of the first derivative u_x of the form $\frac{1}{2h}\big(u_{i+2} - u_i\big)$.

The preliminaries of our invariant approach to discrete operators is the same as in Section 2.1. In what follows, we are looking for solutions of (9.2) belonging to a linear n-dimensional subspace $W_n \subset V$ given by the span

$$W_n = \mathcal{L}\{f_1, f_2, ..., f_n\} \tag{9.3}$$

of linearly independent functions $f_j \in V$, $j = 1, ..., n$. Suppose that this set of functions represents a *fundamental set of solutions* [367, Ch. 2] of a linear difference equation of the nth-order with the linear operator $L : V \to V$,

$$L[u]_i \equiv \textstyle\sum_{k=0}^{n} a_k(i)u_{i+k} = 0, \quad i \in \mathbf{Z}, \tag{9.4}$$

where $a_k \in V$ for $k = 0, 1, ..., n$ with $a_n(i) \neq 0$ and $a_0(i) \neq 0$ on $\mathbf{Z}$.

We say that a function $I(u_i, u_{i+1}, ..., u_{i+n-1}, i) : I\!R^N \times \mathbf{Z} \to I\!R$ is a *first integral* of the linear equation (9.4) if, for any solution u of (9.4),

$$I(u_i, u_{i+1}, ..., u_{i+n-1}, i) \equiv \text{constant} \quad \text{on } \mathbf{Z}.$$

The notation $I[u]_i$ and $I[u]$ are also used. Linear first integrals $I[u]$ of the nth-order are defined by linear functions,

$$I[u]_i \equiv I(u_i, u_{i+1}, ..., u_{i+n-1}, i) = \textstyle\sum_{k=0}^{n-1} l_k(i)u_{i+k}, \quad i \in \mathbf{Z}, \tag{9.5}$$

where $l_k \in V$ for $k = 0, 1, ..., n$ with $l_{n-1}(i) \neq 0$ on $\mathbf{Z}$. Suppose that the nth-order equation (9.4) admits n linearly independent first integrals $I_1[u], ..., I_n[u]$.

As usual, subspace (9.3) is said to be *invariant* under the operator F if

$$F[W_n] \subseteq W_n.$$

We next state the Main Theorem on the *backward problem* (**Problem II** in our classification in Section 2.1) of invariant subspaces for nonlinear operators: given a linear subspace W_n, describe the set of all the nonlinear difference operators admitting W_n.

Theorem 9.1 (**"Main Theorem"**) *Operator F in* (9.1) *with $p \leq n - 1$ admits the subspace* (9.3) *iff F has the form*

$$F[u] = \textstyle\sum_{j=1}^{n} A_j(I_1[u], ..., I_n[u])f_j, \tag{9.6}$$

where $A_j : I\!R^n \to I\!R$ for $j = 1, ..., n$ are arbitrary functions.

We assume that $p \leq n-1$, since, by (9.4) with $a_n(i) \neq 0$, u_{i+n} is expressed in terms of u_{i+n-1}, ..., u_i. Clearly, for any operator (9.1) restricted to W_n, all the higher-order variables u_{i+k} with $k \geq n$ are excluded by means of the subspace equation (9.4). The set of operators admitting a given subspace is a linear space. In a way similar to that applied in the continuous differential case in Section 2.1, these operators can be considered as the generators of the higher-order symmetries of linear discrete equations. Theorem 9.1 then gives a complete description of such symmetries.

The first application of the theorem concerns problems (9.2) with F preserving a subspace W_n and $g \in W_n$. In this case, every such equation on the subspace W_n, i.e., on the set of functions

$$u = \sum_{i=1}^{n} C_i f_i, \tag{9.7}$$

reduces to a finite-dimensional system of algebraic equations for coefficients $\{C_i\}$. Plugging (9.7) into (9.2) and (9.6) and using the linear properties of the functionals $\{I_i[u]\}$ in (9.5), we obtain the following

Corollary 9.2 *Let*

$$g = \sum_{k=1}^{n} \alpha_k f_k \in W_n,$$

where α_k for $k = 1, ..., n$ are constants. Given an operator F of the form (9.6), *equation* (9.2), *restricted to the subspace W_n of functions* (9.7), *is equivalent to the system of n algebraic equations for coefficients $\{C_1, ..., C_n\}$:*

$$A_j(\Psi_1, ..., \Psi_n) = \alpha_j, \quad j = 1, ..., n, \quad \textit{where} \tag{9.8}$$

$$\Psi_i(C_1, ..., C_n) = \sum_{m=1}^{n} C_m I_i[f_m], \quad i = 1, ..., n. \tag{9.9}$$

As a second application, consider an infinite-dimensional DS

$$\tfrac{\mathrm{d}u}{\mathrm{d}t} = F[u], \tag{9.10}$$

with solutions $u(x,t) : \mathbf{Z} \times I\!R \to I\!R$. If F admits the subspace (9.3) and is represented in the form (9.6), equation (9.10) possesses solutions (9.7) with C_j, depending on t and satisfying the DS

$$\tfrac{\mathrm{d}}{\mathrm{d}t} C_j = A_j(\Psi_1, ..., \Psi_n), \quad j = 1, ..., n.$$

A similar reduction exists in the completely discrete case where the continuous derivative $\frac{\mathrm{d}}{\mathrm{d}t}$ is replaced by the corresponding discrete approximation. A similar finite-dimensional reduction exists for higher-order equations

$$P(\tfrac{\mathrm{d}}{\mathrm{d}t})u = F[u], \tag{9.11}$$

where P is an arbitrary linear polynomial $\big($e.g., $P(\frac{\mathrm{d}}{\mathrm{d}t}) = \frac{\mathrm{d}^2}{\mathrm{d}t^2}\big)$. Equation (9.11) can be considered as a discrete model for the nonlinear PDE

$$P(\tfrac{\partial}{\partial t})u = F_0[u],$$

where F is a discrete approximation of a nonlinear differential operator F_0.

9.1.1 Proof of Theorem 9.1

The proof is straightforward and has a natural extension to other types of nonlinear integro-differential-difference operators.

Proof. The sufficiency is obvious. Let us prove the necessity. The invariance of the subspace (9.3) means that, for any $u \in W_n$,

$$F[u] = \sum_{j=1}^{n} \tilde{A}_j(u) f_j,$$

where the coefficients $\tilde{A}_j$ for $j = 1, ..., n$ are constant for any given u. Note that every $u \in W_n$, being a solution of (9.4), can be uniquely represented by the values of the linearly independent first integrals (9.5) $\{I_k[u],\ k = 1, ..., n\}$. Hence, one can write $\tilde{A}_j(u) \equiv A_j(I_1, ..., I_n)$, which completes the proof. □

9.1.2 First integrals and operators for $n = 2$

We begin with $n = 2$, i.e., consider second-order linear difference equations (9.4),

$$L[u]_i \equiv a_i u_{i+2} + b_i u_{i+1} + c_i u_i = 0. \tag{9.12}$$

Introducing the standard formal scalar product in V,

$$(u, v) = \sum_{(k \in \mathbf{Z})} u_k v_k,$$

the adjoint operator $M[v]$ satisfying

$$(L[u], v) = (u, M[v]) \quad \text{for any} \;\; u,\, v \in V$$

gives the adjoint (conjugate) equation

$$M[v]_i \equiv c_i v_i + b_{i-1} v_{i-1} + a_{i-2} v_{i-2} = 0. \tag{9.13}$$

Proposition 9.3 *Any function of the form*

$$I[u]_i = a_{i-1} v_{i-1} u_{i+1} - c_i v_i u_i, \tag{9.14}$$

where v is a solution of the adjoint equation (9.13), *is the first integral of* (9.12).

Proof. For functions (9.14), the condition

$$I[u]_i = I[u]_{i+1} \quad \text{in} \;\; \{u \in V :\; L[u] = 0\}$$

reduces to $u_{i+1} M[v]_{i+1} = 0$, which completes the proof. □

Example 9.4 Consider the linear equation

$$L[u]_i \equiv u_{i+2} - \tfrac{1}{4} u_i = 0 \tag{9.15}$$

($a_i = 1$, $b_i = 0$, and $c_i = -\frac{1}{4}$). The characteristic equation is $\lambda^2 - \frac{1}{4} = 0$, from which we get the following subspace W_2 of solutions of (9.15):

$$W_2 = \mathcal{L}\{2^{-i}, (-2)^{-i}\}. \tag{9.16}$$

The adjoint equation takes the form $M[v]_i \equiv -\frac{1}{4} v_i + v_{i-2} = 0$ and possesses two linearly independent solutions $v_{1,i} = 2^i$ and $v_{2,i} = (-2)^i$. By Proposition 9.3, there exist two linearly independent first integrals of (9.15),

$$I_1[u]_i = 2^i\left(u_{i+1} + \tfrac{1}{2}u_i\right) \quad \text{and} \quad I_2[u]_i = (-2)^i\left(u_{i+1} - \tfrac{1}{2}u_i\right).$$

It follows from Theorem 9.1 that the family of all the first-order nonlinear operators admitting the 2D subspace (9.16) is given by

$$F[u]_i = A_1(I_1[u]_i, I_2[u]_i)2^{-i} + A_2(I_1[u]_i, I_2[u]_i)(-2)^{-i},$$

where $A_1, A_2 : I\!R^2 \to I\!R$ are arbitrary functions. Due to Corollary 9.2, for any $g \in W_2$, equation $F[u] = g$ is equivalent to a system of two algebraic equations.

For applications, it is important to derive all the *autonomous* operators admitting W_2, i.e., those that do not depend on the independent variable $i \in \mathbf{Z}$. Denoting $2^i(u_{i+1} + \frac{1}{2}u_i) = a$ and $(-2)^i(u_{i+1} - \frac{1}{2}u_i) = b$, we infer that such operators, which are invariant under translation $i \to i+1$, satisfy

$$\begin{aligned}&[A_1(2a, -2b) - 2A_1(a, b)]2^{-i-1}\\ &+[A_2(2a, -2b) + 2A_2(a, b)](-2)^{-i-1} = 0\end{aligned}$$

for all $i \in \mathbf{Z}$ and $a, b \in I\!R$. In view of the linear independence of the functions,

$$A_1(2a, -2b) = 2A_1(a, b) \quad \text{and} \quad A_2(2a, -2b) = -2A_2(a, b) \tag{9.17}$$

for all $a, b \in I\!R$. Thus the autonomous operators have the form

$$F[u]_i = A_1\left(u_{i+1} + \tfrac{1}{2}u_i, u_{i+1} - \tfrac{1}{2}u_i\right) + A_2\left(u_{i+1} + \tfrac{1}{2}u_i, u_{i+1} - \tfrac{1}{2}u_i\right),$$

where A_1 and A_2 are arbitrary functions satisfying (9.17). For instance, the operators

$$F_1[u]_i = \frac{(u_{i+1}+\frac{1}{2}u_i)^3}{(u_{i+1}-\frac{1}{2}u_i)^2} \quad \text{and} \quad F_2[u]_i = \frac{(u_{i+1}+\frac{1}{2}u_i)^2}{(u_{i+1}-\frac{1}{2}u_i)}$$

admit W_2. A more general operator preserving W_2 is given by

$$F[u]_i = C_1F_1[u]_i + C_2F_2[u]_i + \tilde{L}[u]_i,$$

where C_1 and C_2 are constants, and $\tilde{L}$ is any linear operator admitting W_2.

9.2 On the forward problem of invariant subspaces

As in the differential case in Chapter 2, the general *forward* **Problem I** of invariant subspaces: given a nonlinear finite-order difference operator, determine all the linear subspaces that it admits, remains OPEN.

In this section, as a key example, we illustrate the forward problem for the simplest quadratic first-order operator

$$F[u]_i = (u_{i+1} - u_i)^2. \tag{9.18}$$

The forward problem for the corresponding *differential* Hamilton–Jacobi operator

$$F_0[u] = (u_x)^2$$

((9.18) is its difference approximation $[\frac{1}{h}(u_{i+1} - u_i)]^2$ with $h = 1$) has been solved;

see Section 2.2. To be precise, up to translations in x, the operator (9.18) admits: (i) the 1D subspaces $W_1 = \mathcal{L}\{x^2\}$ and $W_1 = \mathcal{L}\{1\}$, (ii) 2D subspaces $W_2 = \mathcal{L}\{1, x\}$ and $W_2 = \mathcal{L}\{1, x^2\}$, given by the ODEs

$$u_{xx} = 0 \quad \text{and} \quad u_{xx} = \tfrac{1}{x}\, u_x,$$

and (iii) a single 3D subspace $W_3 = \mathcal{L}\{1, x, x^2\}$ ($u_{xxx} = 0$).

For convenience, step by step, we give a classification of the invariant subspaces W_n for the discrete operator (9.18) with $n \le 2$, $n = 3$, and $n \ge 4$ respectively.

Case I: $n \le 2$.

Theorem 9.5 *Operator* (9.18) *admits* W_n *with* $n \le 2$ *given by the linear equation with constant coefficients*

$$L[u]_i \equiv u_{i+1} - Bu_i - Cu_{i-1} = 0 \quad (B, C \in I\!R), \tag{9.19}$$

in the following four cases only:

(i) $B = C = 0$ (*equation* $u_{i+1} = 0$), $W_0 = \{0\}$;
(ii) $B = 1,\;\; C = 0$ ($u_{i+1} - u_i = 0$), $W_1 = \mathcal{L}\{1\}$;
(iii) $B = 0,\;\; C = 1$ ($u_{i+1} - u_{i-1} = 0$), $W_2 = \mathcal{L}\{1, (-1)^i\}$;
(iv) $B = 2,\;\; C = -1$ ($u_{i+1} - 2u_i + u_{i-1} = 0$), $W_2 = \mathcal{L}\{1, i\}$.

The subspace in (iii) containing 2-periodic functions is not available in the classification of the differential operators.

Proof. Consider the invariance criterion $L[F[u]] = 0$ if $L[u] = 0$, or

$$(u_{i+2} - u_{i+1})^2 - B(u_{i+1} - u_i)^2 - C(u_i - u_{i-1})^2 \equiv 0$$

on solutions of (9.19). Using (9.19) to eliminate u_{i+1} and u_{i+2}, and taking into account that u_i^2, $u_i u_{i-1}$, and u_{i-1}^2 are linearly independent, we arrive at the system

$$\begin{cases} (B^2 + C - B)^2 = B(B-1)^2 + C, \\ C(B^2 + C - B)(B-1) = C[B(B-1) - 1], \\ C^2(B-1)^2 = C(BC+1). \end{cases} \tag{9.20}$$

Consider two cases:

1. Case $C = 0$. The first equation yields $B(B-1)^3 = 0$, so $B = 0$ or 1.

2. Case $C \neq 0$. Then, substituting C from the second equation (9.20), $C = -B(B-2) - \frac{1}{B-1}$ ($B \neq 1$), yields the following two equations for B:

$$\begin{cases} \left(B - \frac{1}{B-1}\right)^2 - B(B-1)^2 = -B(B-2) - \frac{1}{B-1}, \\ -[(B-1)^2 - B]\left[B(B-2) + \frac{1}{B-1}\right] = 1, \end{cases}$$

that are reduced to $B^2(B-2)(B^2 - 4B + 5) = 0$ and $B^2(B-2)^2(B-1) = 0$, so either $B = 2$, $C = -1$, or $B = 0$, $C = 1$. □

Case II: $n = 3$. A similar result holds.

Theorem 9.6 *Operator* (9.18) *admits a 3D subspace* W_3 *given by*

$$L[u]_i \equiv u_{i+3} - Bu_{i+2} - Cu_{i+1} - Du_i = 0 \quad \textit{iff}$$

(i) $B = 3,\ C = -3,\ D = 1, W_3 = \mathcal{L}\{1, i, i^2\}$;
(ii) $B = C = 0,\ D = 1, W_3 = \mathcal{L}\{1, \cos\frac{2\pi i}{3}, \sin\frac{2\pi i}{3}\}$; *and*
(iii) $B = C = 1,\ D = -1, W_3 = \mathcal{L}\{1, i, (-1)^i\}$.

Case III: $n \geq 4$. Finally, consider linear subspaces given by equations

$$L[u]_i \equiv u_{i+n} - \sum_{k=0}^{n-1} a_k u_{i+k} = 0, \quad n \geq 4 \quad (a_k \in I\!R). \tag{9.21}$$

Theorem 9.7 *The only subspaces W_n with $n \geq 4$ invariant under* (9.18) *are*
(i) $u_{i+n} = u_{i+n-1} + u_{i+1} - u_i$, $W_n = \mathcal{L}\{1, i, e^{2\pi i I/(n-1)}, ..., e^{2\pi i I(n-2)/(n-1)}\}$; *and*
(ii) $u_{i+n} = u_i$, $W_n = \mathcal{L}\{1, e^{2\pi i I/n}, ..., e^{2\pi i I(n-1)/n}\}$, *where* $I^2 = -1$.

A real-valued representation of these subspaces is straightforward.

Proof. Consider the invariance criterion: for any solutions of (9.21),

$$(u_{i+n+1} - u_{i+n})^2 = \sum_{k=0}^{n-1} a_k (u_{i+k+1} - u_{i+k})^2. \tag{9.22}$$

We exclude the variables

$$\begin{aligned} u_{i+n+1} &= \sum_{k=0}^{n-1} a_k u_{i+k+1} = a_{n-1} u_{i+n} + \sum_{k=1}^{n-1} a_{k-1} u_{i+k} \\ &= \sum_{k=1}^{n-1} (a_{n-1} a_k + a_{k-1}) u_{i+k} + a_{n-1} a_0 u_i, \end{aligned}$$

and u_{i+n} given by (9.21), so that

$$\begin{aligned} u_{i+n+1} - u_{i+n} &= \sum_{k=1}^{n-1} [(a_{n-1} - 1) a_k + a_{k-1}] u_{i+k} + (a_{n-1} - 1) a_0 u_i, \\ u_{i+n} - u_{i+n-1} &= (a_{n-1} - 1) u_{i+n-1} + \sum_{k=1}^{n-2} a_k u_{i+k} + a_0 u_i. \end{aligned}$$

For convenience, setting

$$\beta = a_{n-1} - 1, \tag{9.23}$$

and substituting the representations given above into (9.22) yields

$$\begin{aligned} &\Big[\sum_{k=1}^{n-1} (\beta a_k + a_{k-1}) u_{i+k}\Big]^2 + 2\beta a_0 u_i \sum_{k=1}^{n-1} (\beta a_k + a_{k-1}) u_{i+k} \\ &\quad + \beta^2 a_0^2 u_i^2 = a_{n-1}\Big[\big(\beta u_{i+n-1} + \sum_{k=1}^{n-2} a_k u_{i+k}\big)^2 \\ &\quad + 2a_0 u_i \big(\beta u_{i+n-1} + \sum_{k=1}^{n-2} a_k u_{i+k}\big) + a_0^2 u_i^2\Big] \\ &+ a_{n-2}(u_{i+n-1} - u_{i+n-2})^2 + a_{n-3}(u_{i+n-2} - u_{i+n-3})^2 \\ &\quad + ... + a_1(u_{i+2} - u_{i+1})^2 + a_0(u_{i+1} - u_i)^2. \end{aligned} \tag{9.24}$$

Equating the coefficients of the factors u_i^m in both sides for $m = 0$, 1, and 2, respectively, we obtain the following. The equation corresponding to u_i^0 has the form

$$\begin{aligned} &\Big[\sum_{k=1}^{n-1} (\beta a_k + a_{k-1}) u_{i+k}\Big]^2 = a_{n-1}\big(\beta u_{i+n-1} + \sum_{k=1}^{n-2} a_k u_{i+k}\big)^2 \\ &+ a_{n-2}(u_{i+n-1} - u_{i+n-2})^2 + ... + a_1(u_{i+2} - u_{i+1})^2 + a_0 u_{i+1}^2. \end{aligned} \tag{9.25}$$

The equation for u_i is

$$\begin{aligned} &a_0 \beta \sum_{k=1}^{n-1} (\beta a_k + a_{k-1}) u_{i+k} \\ &= a_0\big[a_{n-1}\big(\beta u_{i+n-1} + \sum_{k=1}^{n-2} a_k u_{i+k}\big) - u_{i+1}\big]. \end{aligned} \tag{9.26}$$

Since $a_0 \neq 0$, analyzing the coefficients in (9.26) we obtain:
(i) for u_{i+n-1},

$$\beta[(\beta - 1) a_{n-1} + a_{n-2}] = 0; \tag{9.27}$$

(ii) for u_{i+k} with $2 \le k \le n-2$,

$$\beta(\beta a_k + a_{k-1}) = a_{n-1} a_k; \tag{9.28}$$

and (iii) for u_{i+1},

$$\beta(\beta a_1 + a_0) = a_{n-1} a_1 - 1. \tag{9.29}$$

Finally, considering the coefficients of u_i^2 in (9.24) yields

$$\beta^2 a_0 = a_{n-1} a_0 + 1. \tag{9.30}$$

I. If $\beta = 0$, i.e., $a_{n-1} = 1$, the system (9.27)–(9.30) takes the form

$$a_k = 0 \quad \text{for } 2 \le k \le n-2, \quad a_1 = 1, \ \ a_0 = -1$$

(where (9.25) becomes the identity), and we arrive at the case (i) of the theorem. The subspace is determined from the characteristic equation $\lambda^n = \lambda^{n-1} + \lambda - 1$, so that $\lambda = 1$ or $\lambda^{n-1} = 1$, i.e., denoting $\mathrm{I}^2 = -1$, we have $\lambda_k = \mathrm{e}^{2\pi k \mathrm{I}/(n-1)}$ for $k = 1, 2, ..., n-2$, and $\lambda_{n-1} = \lambda_n = 1$.

II. Let $\beta \ne 0$, i.e., $a_{n-1} \ne 1$. Then the system (9.27)–(9.30) reads

$$(\beta - 1)a_{n-1} + a_{n-2} = 0, \ \ (\beta^2 - a_{n-1})a_k + \beta a_{k-1} = 0, \ \ k = 2, ..., n-2,$$
$$(\beta^2 - a_{n-1})a_1 + \beta a_0 = -1, \ \ (\beta^2 - a_{n-1})a_0 = 1.$$

Denoting

$$\alpha = \beta^2 - a_{n-1} = \beta^2 - \beta - 1, \tag{9.31}$$

we obtain the solution

$$a_0 = \tfrac{1}{\alpha} \ \ (\alpha \ne 0), \quad a_1 = -\tfrac{\alpha+\beta}{\alpha^2}, \quad a_k = \left(-\tfrac{\beta}{\alpha}\right)^{k-1} a_1, \tag{9.32}$$

where $k = 2, ..., n-2$ and $a_{n-2} = 1 - \beta^2$, since $a_{n-1} = 1 + \beta$ by (9.23).

Consider (9.25). The coefficient of u_{i+1}^2 is $(\beta a_1 + a_2)^2 = a_{n-1} a_1^2 + a_1 + a_0$, which yields, by (9.23) and (9.32), $[\beta(\alpha+\beta) - \alpha]^2 = (1+\beta)(\alpha+\beta)^2 - \beta\alpha^2$. Substituting α from (9.31) yields $\beta^2(\beta+1)(\beta-2)(\beta^2-\beta+1) = 0$. If $\beta = -1$, then $\alpha = 1$ by (9.31) that gives the subspace in (ii).

If $\beta = 2$, then $\alpha = 1$ and $a_{n-1} = 1 + \beta = 3$, $a_0 = 1$, $a_1 = -3$ and $a_k = -3(-2)^{k-1}$ for $k = 2, ..., n-2$. Then the last equation in (9.32) implies that $(-2)^{n-3} = 1$ holds, which results in $n = 3$, contradicting the assumption. This completes the proof. □

It is obvious that if $u_{i+n} = u_i$ as in (ii), then

$$F[u]_{i+n} = F[u]_i \tag{9.33}$$

for any operator F that does not depend explicitly on i. The linear subspace W_n given in (ii) is invariant under such an arbitrary operator.

On the maximal dimension of invariant subspaces. In the differential case, the order p of a *nonlinear* ordinary differential operator F is known to satisfy the following estimate (Theorem 2.8)

$$n \le 2p + 1, \tag{9.34}$$

where n is the maximal dimension of an invariant subspace. If $n > 2p+1$, F is a linear operator. It follows from Theorem 9.7 that this is not the case for the discrete operators, where the maximal dimension n can be arbitrarily large, even for the first-order quadratic operator (9.18). Actually, this again emphasizes the obvious fact that the set of difference operators is wider than that of the differential ones. Each pth-order differential operator

$$F_0[u] = F(u, u_x, u_{xx}, ..., D_x^p u)$$

can be obtained from a one-parameter family of the difference operators

$$F_h[u] = F(u, u_x, u_{xx}, ..., D_x^p u),$$

where the arguments are discrete approximations of the derivatives, for instance,

$$u_x = \tfrac{1}{2h}\big(u_{i+1} - u_{i-1}\big), \quad u_{xx} = \tfrac{1}{h^2}\big(u_{i-1} - 2u_i + u_{i+1}\big), \quad \dots . \tag{9.35}$$

This is understood in the sense of a formal limit, $F_0 = \lim F_h$ as $h \to 0$. Without fear of confusion, we will use the same notation for derivatives in continuous and discrete cases.

Example 9.8 **(5D subspace)** The discrete second-order quadratic operator

$$F_h[u] = uu_{xx} - \tfrac{3}{4}(u_x)^2,$$

with the derivative approximations (9.35), admits the same 5D subspace as the differential operator $F_0[u]$ (Example 1.14)

$$W_5 = \mathcal{L}\{1, x, x^2, x^3, x^4\}.$$

Both continuous and discrete parabolic equations, $u_t = F_0[u]$ and $u_t = F_h[u]$ on W_5, reduce to the fifth-order DS for the coefficients $\{C_k(t)\}$ of the solution

$$u(x,t) = \textstyle\sum_{k=0}^4 C_k(t)x^k.$$

Similarly, the hyperbolic second-order equations, $u_{tt} = F_0[u]$ and $u_{tt} = F_h[u]$ on W_5, reduce to tenth-order DSs.

9.3 Invariant subspaces for finite-difference operators

We perform a more detailed analysis of linear subspaces that are invariant under operators, corresponding to finite-difference approximations of differential operators, with polynomial nonlinearities. As a result, a certain structural stability of invariant subspaces and sets of nonlinear differential operators of reaction-diffusion type, with respect to their spatial discretization, is established, and lower-dimensional reductions of the finite-difference solutions on the invariant subspaces are constructed.

We study invariant properties and exact solutions of finite-difference schemes for nonlinear evolution PDEs

$$u_t = F[u] \quad \text{in } I\!R^N \times I\!R_+, \tag{9.36}$$

where F is an ordinary differential (elliptic) operator with polynomial-like nonlinearities. We illustrate the results using the quasilinear parabolic equations from nonlinear diffusion and combustion theory with operators

$$F[u] \equiv \varphi(u)u_{xx} + \psi(u)(u_x)^2 + f(u), \tag{9.37}$$

where φ, ψ, and f are given sufficiently smooth functions, and $\varphi \geq 0$ (the parabolicity condition).

One of the basic tools of solving the PDE (9.36) numerically is the method of finite differences. Consider first the case of discretization in both the spatial and time-variables. Using a given uniform time-grid, $T_\tau = \{n\tau : n = 0, 1, ...\}$ with the fixed, sufficiently small step $\tau > 0$, and an infinite uniform space-grid, $X_h = \{x = kh : k = 0, \pm 1, ...\}$ with the fixed small step $h > 0$, we replace (9.36), (9.37) by the following implicit finite-difference equation

$$u_t = F_h[u] \equiv \varphi(u)u_{xx} + \psi(u)(u_x)^2 + f(u) \tag{9.38}$$

for $(x, t) \in Q_{h\tau} = X_h \times T_\tau$. We again keep the same notation, u_t, for discrete and continuous derivatives. As usual, $u = u(x, t)$ belongs to the space of grid functions on $Q_{h\tau}$, and the derivatives denote

$$\begin{aligned} u_t &= \tfrac{1}{\tau}\big[u(x, t+\tau) - u(x, t)\big], \\ u_x &= \tfrac{1}{2h}\big[u(x+h, t) - u(x-h, t)\big], \\ u_{xx} &= \tfrac{1}{h^2}\big[u(x+h, t) + u(x-h, t) - 2u(x, t)\big], \end{aligned} \tag{9.39}$$

so that $F_h[u] = F[u]$ with the derivatives replaced by those given in (9.39). We consider the Cauchy problem on X_h with prescribed initial data, so we do not need boundary conditions. The PDE is then replaced by a discrete, infinite-dimensional DS. In the case of the spatial discretization only (the *method of lines*), there occurs the continuous time-derivative on the left-hand side of (9.38) that is a *continuous* DS.

In the construction of the finite-difference approximation of the problem, the elliptic differential operator $F_0 : C^2(I\!R) \to C(I\!R)$ is replaced by the corresponding finite-difference operator $F_h : V \to V$, where $V = \{u : X_h \to I\!R\}$ denotes the space of grid functions on X_h. The type of discretization depends on the properties of the differential model, which it is necessary to preserve. Namely, these may be conservation laws (conservation of the mass, or of the first moment, corresponding to the diffusion operator, or other higher-order moments), preservation of a fixed symmetry or scaling invariance of the equation, etc. In what follows, a standard and, sometimes, non-divergent type of discretization is chosen in order to preserve the *invariant subspace* property of the discretized nonlinear operator and the PDE.

We will establish that the finite-difference approximation may preserve the property of the nonlinear operator F to admit a subspace. We show that, in some cases, if F admits a finite-dimensional linear subspace W, such that $F[W] \subseteq W$, the approximating operator F_h can also have a subspace $W_h \subset V$ (i.e., $F_h[W_h] \subseteq W_h$) of the same dimension and with similar basis functions. In the discretized problem, there is also a reduction of the dimension, which gives solutions on the lower-dimensional subspaces. The same effect is proved to be true for the partially discretized equation (9.38) with the continuous derivative u_t, so that (9.38) is a system of nonlinear ODEs. In the discrete case, the space V is infinite-dimensional (as well as in the differential case), so by the exact solutions we mean those that can be expressed in terms of a finite-dimensional discrete or continuous DS.

In some cases, discrete operators may admit invariant sets on linear subspaces for which the exact solutions are obtained from overdetermined DSs. We show that reductions to lower-order systems via invariant subspaces and sets exist under the standard approximations of the differential operator F on the uniform constant grid, provided that F admits an invariant subspace. Since for parabolic problems, the standard comparison theorem upon initial data is true for the implicit schemes such as (9.38), [509, Ch. 7] (or for the corresponding DSs approximation), particular solutions on subspaces or on sets can be used to estimate more general solutions. This makes it possible to describe the asymptotic behavior and attractors for equations (9.38).

9.3.1 First examples

We begin with a special example, exhibiting some common features of the group-invariant and invariant subspace solutions.

Example 9.9 (**Semilinear heat equation**) Consider the parabolic PDE

$$v_t = v_{xx} + v \ln v,$$

which admits a four-parameter Lie group of symmetries with operators [10, p. 135]

$$X_1 = \tfrac{\partial}{\partial t}, \quad X_2 = \tfrac{\partial}{\partial x}, \quad X_3 = 2\mathrm{e}^t \tfrac{\partial}{\partial x} - \mathrm{e}^t x v \tfrac{\partial}{\partial v}, \quad X_4 = \mathrm{e}^t v \tfrac{\partial}{\partial v},$$

and, therefore, has a wide class of invariant solutions. There is a simple invariant subspace interpretation of those solutions generated by the symmetry $\varepsilon_0 X_2 + X_3$ with a constant ε_0, [10, p. 140]. The pressure-like function $u = \ln v$ satisfies

$$\boxed{u_t = F_0[u] \equiv u_{xx} + (u_x)^2 + u.} \tag{9.40}$$

The quadratic operator F admits a simple subspace $W_2 = \mathcal{L}\{1, x^2\}$, so that, for

$$u(x,t) = C_1(t) + C_2(t)x^2, \tag{9.41}$$

the following holds: $F_0[u] = C_1 + 2C_2 + \big(C_2 + 4C_2^2\big)x^2 \in W_2$. Plugging (9.41) into (9.40) yields a nonlinear DS

$$\begin{cases} C_1' = C_1 + 2C_2, \\ C_2' = C_2 + 4C_2^2. \end{cases} \tag{9.42}$$

This is easily solved explicitly and gives exact solutions; see precise expressions in [10, p. 140]. The standard finite-difference scheme actually given by (9.40) with discrete derivatives also admits solutions (9.41), with a similar action of the operator and the same DS.

Example 9.10 The corresponding N-dimensional PDE

$$v_t = \Delta v + v \ln v$$

by the same transformation $v = e^u$ is reduced to

$$\boxed{u_t = F_0[u] \equiv \Delta u + |\nabla u|^2 + u \quad \text{in } I\!R^N \times I\!R_+.} \tag{9.43}$$

This quadratic PDE possesses the solutions

$$u(x,t) = C_0(t) + \textstyle\sum_{k=1}^N C_k(t)x_k^2 \tag{9.44}$$

on the subspace $W_{N+1} = \mathcal{L}\{1, x_1^2, ..., x_N^2\}$ that is easily seen to be also admitted by the operator of the corresponding scheme on a uniform grid X_h,

$$u_t = F_h[u] \equiv \Delta_h u + |\nabla_h u|^2 + u, \tag{9.45}$$

with the standard discrete approximations

$$\Delta_h u = \textstyle\sum_{k=1}^N u_{x_k x_k} \quad \text{and} \quad |\nabla_h u|^2 = \sum_{k=1}^N (u_{x_k})^2. \tag{9.46}$$

Substituting (9.44) into (9.45) yields the system

$$\begin{cases} C_0' = C_0 + 2\sum_{(i)} C_i, \\ C_k' = C_k + 4C_k^2, \ k = 1, ..., N. \end{cases}$$

The DS for solutions (9.44) of the parabolic PDE (9.43) is the same with the difference derivatives $\frac{\mathrm{d}}{\mathrm{d}t}$.

Remark 9.11 (Hyperbolic PDEs) Invariant reductions to lower-order systems exist for discretized *hyperbolic* equations

$$\boxed{u_{tt} = F_h[u]}$$

with the quadratic operators from (9.40) or (9.45), and for other operators to be considered. Observe that, in the case of higher-order time-derivatives, the set of solutions on the 3D subspace $W_3 = \mathcal{L}\{1, x, x^2\}$ contains new nontrivial solutions (for the parabolic case (9.40) such a generalization reveals nothing new).

Example 9.12 (Semilinear heat equation with blow-up) Consider the semilinear PDE from Example 1.11,

$$v_t = v_{xx} + v \ln^2 v.$$

Setting $v = e^u$ yields the PDE possessing blow-up solutions,

$$\boxed{u_t = F[u] \equiv u_{xx} + (u_x)^2 + u^2,} \tag{9.47}$$

with the quadratic operator F preserving the subspace $W_2 = \mathcal{L}\{1, \cos x\}$.

Let us state an analogy of the invariant property for equation (9.47), where all the derivatives are discrete.

Proposition 9.13 *The discrete operator F in (9.47) admits*

$$W_2^h = \mathcal{L}\{1, \cos(\lambda x)\}, \quad \textit{where } \lambda = \tfrac{1}{h}\sin^{-1} h \quad (0 < h \le 1). \tag{9.48}$$

Proof. Using simple formulae

$$(\cos(\lambda x))_x = -l \sin \lambda x \quad \text{and} \quad (\cos(\lambda x))_{xx} = -m \cos \lambda x, \tag{9.49}$$

where $l = \frac{1}{h}\sin(\lambda h)$ and $m = \frac{4}{h^2}\sin^2(\frac{\lambda h}{2})$, we find that, for any $u = C_1 + C_2\cos(\lambda x) \in W_2^h$,

$$F[u] = (mC_2 + 2C_1C_2)\cos(\lambda x) + \left[\cos^2(\lambda x) + l^2\sin^2(\lambda x)\right]C_2^2 + C_1^2.$$

Then $F[u] \in W_2^h$ implies $l^2 = 1$, so, without loss of generality, we set $l = 1$, which yields $\lambda = \frac{1}{h}\sin^{-1} h$, while $m = 2/(1+\sqrt{1-h^2})$. □

Hence, there exist the following solutions of (9.47) in the discrete case:

$$u(x,t) = C_1(t) + C_2(t)\cos(\lambda x), \tag{9.50}$$
$$\begin{cases} C_1' = C_1^2 + C_2^2, \\ C_2' = mC_2 + 2C_1C_2. \end{cases}$$

If $h \to 0$, then $\lambda,\ m \to 1$ and, more precisely, $\lambda(h) = 1 + \frac{1}{6}h^2 + O(h^4)$, so that subspace (9.48) coincides with $W_2 = \mathcal{L}\{1, \cos x\}$ at $h = 0$.

Remark 9.14 **(KS-type equations)** Similar solutions on W_2^h exist for the discretized fourth-order Kuramoto–Sivashinsky type equation

$$\boxed{u_t = -u_{xxxx} + F[u].}$$

Example 9.15 **(The PME)** Consider the PME with lower-order terms

$$v_t = (v^\sigma v_x)_x + v^{\sigma+1} + av^{1-\sigma} + bv,$$

where $\sigma \neq 0, -1$ and a and b are given constants. The pressure function $u = v^\sigma$ solves

$$\boxed{u_t = F[u] \equiv uu_{xx} + \tfrac{1}{\sigma}(u_x)^2 + \sigma u^2 + \sigma a + \sigma bu.} \tag{9.51}$$

The quadratic operator F is known to admit the 2D subspaces (see Section 1.4)

$$W_2 = \begin{cases} \mathcal{L}\{1, \cos(\lambda x)\}, & \text{if } \sigma > -1, \\ \mathcal{L}\{1, \cosh(\lambda x)\}, & \text{if } \sigma < -1, \end{cases}$$

where $\lambda = \sigma/\sqrt{|\sigma+1|}$. Consider, for instance, the case of $\sigma > -1$, where there exist solutions (9.50), with the DS

$$\begin{cases} C_1' = \sigma C_1^2 + \frac{\sigma}{\sigma+1}C_2^2 + \sigma bC_1 + \sigma a, \\ C_2' = \frac{\sigma(\sigma+2)}{\sigma+1}C_1C_2 + \sigma bC_2. \end{cases} \tag{9.52}$$

Let (9.51) now be a standard finite-difference scheme. Similarly to the previous example, we establish the following:

Proposition 9.16 *The discrete operator F in* (9.51) *has the subspace*

$$W_2^h = \mathcal{L}\{1,\ \cos(\lambda x)\}, \quad \textit{if} \quad l^2 + m\sigma - \sigma^2 = 0, \tag{9.53}$$

where m and l are as given in (9.49).

For $\sigma > -1$, on the subspace (9.53), i.e., for functions (9.50) restricted to X_h, the PDE (9.51) is equivalent to the following DS:

$$\begin{cases} C_1' = \sigma C_1^2 + \frac{l^2}{\sigma}C_2^2 + \sigma bC_1 + \sigma a, \\ C_2' = (2\sigma - m)C_1C_2 + \sigma bC_2. \end{cases} \tag{9.54}$$

Setting $k = \cos(\lambda h)$, let us write down the algebraic equation in (9.53) in the form

$$k^2 + 2\sigma k + \sigma^2 h^2 - 2\sigma - 1 = 0.$$

It is easy to clarify for which σ and h this equation possesses roots $k \in [-1, 1]$. Avoiding a detailed analysis, we indicate, for example, that, for $\sigma \in \left(-\frac{1}{1+h}, 0\right) \cup \left(0, \frac{4}{h^2}\right)$, there exists the root $k = -\sigma + \sqrt{(\sigma+1)^2 - \sigma^2 h^2}$, and, for the corresponding $\lambda = \frac{1}{h}\cos^{-1} k$, one obtains that

$$\lambda(h) = \tfrac{\sigma}{\sqrt{\sigma+1}} + \tfrac{\sigma^3(\sigma+4)}{24(\sigma+1)^{5/2}}\, h^2 + O(h^4) \quad \text{for small } h > 0. \tag{9.55}$$

This means convergence as $h \to 0$ to the solution (9.50), with the DS (9.52), of the PDE (9.51). There are other branches of $\lambda(h)$ that do not have a differential analogy. Subspaces of the hyperbolic $\cosh(\lambda x)$ function are studied in a similar fashion.

Example 9.17 (**Quasilinear heat equations**) Let $\varphi(v)$ be an arbitrary smooth monotone function with φ^{-1} inverse. Consider the PDE from Example 8.3,

$$v_t = (\varphi(v))_{xx} + \tfrac{a\varphi(v)+b}{\varphi'(v)} + [a\varphi(v) + b]c.$$

Setting $u = a\varphi(v) + b$ yields

$$\boxed{u_t = F[u] \equiv \Phi(u)(u_{xx} + acu) + au,} \tag{9.56}$$

where $\Phi(u) = \varphi'(\varphi^{-1}(\frac{1}{a}(u-b)))$. Operator F was shown to admit the 2D subspace

$$W_2 = \mathcal{L}\{\rho_1(x), \rho_2(x)\}, \tag{9.57}$$

where ρ_1 and ρ_2 are arbitrary linearly independent solutions of $\rho'' + ac\rho = 0$. Therefore, (9.56) possesses exact solutions

$$u(x,t) = C_1(t)\rho_1(x) + C_2(t)\rho_2(x), \quad \text{where } C_1' = aC_1, \;\; C_2' = aC_2. \tag{9.58}$$

It is easy to see that the same construction applies to the discrete equation (9.56), and there exist exact solutions (9.58) with ρ satisfying the same linear difference equation. Finally, we obtain a discrete linear DS.

9.3.2 N-dimensional quasilinear operators

Let us present further examples of subspaces for discretized nonlinear operators in $I\!R^N$. We borrow some samples of such differential operators from Section 6.1 and Proposition 6.1 therein.

Example 9.18 (**Non-symmetric blow-up and extinction**) Consider positive solutions of the PME with an extra lower-order term

$$v_t = \nabla \cdot (v^\sigma \nabla v) + \delta v^{1-\sigma} \quad \text{in } I\!R^N \times I\!R_+, \tag{9.59}$$

where $\sigma \neq 0$ is a fixed constant. Here, $\delta = 1$ corresponds to the source term, generating for $\sigma < 0$ non-symmetric blow-up solutions. For $\delta = -1$, (9.59) becomes a quasilinear heat equation with absorption that, for $\sigma > 0$, describes non-symmetric finite-time extinction phenomena; see [235] for details. The pressure function $u = v^\sigma$ solves

$$\boxed{u_t = F_0[u] \equiv u \Delta u + \tfrac{1}{\sigma} \, |\nabla u|^2 + \sigma \delta.}$$

Operator F preserves the $(N{+}1)$-dimensional subspace

$$W_{N+1} = \mathcal{L}\{1, x_1^2, ..., x_N^2\}. \tag{9.60}$$

The same subspace is admitted by the discrete approximation of F_0 in the PDE

$$u_t = F_h[u] \equiv u \Delta_h u + \tfrac{1}{\sigma} \, |\nabla_h u|^2 + \sigma \delta,$$

with difference operators from (9.46). This implies that, for solutions (9.44),

$$\begin{cases} C_0' = 2C_0 M + \sigma \delta, \quad M = \sum_{i=1}^N C_i, \\ C_k' = 2C_k M + \frac{4}{\sigma} \, C_k^2, \;\; k = 1, 2, ..., N. \end{cases} \tag{9.61}$$

Due to the subspace (9.60), the DS (9.61) coincides with the DS in the differential case.

Example 9.19 (**Blow-up for exponential PDEs**) Consider the quasilinear heat equation with exponential nonlinearities,

$$v_t = \Delta \mathrm{e}^v + \mathrm{e}^v + a \mathrm{e}^{-v} + b.$$

Setting $u = \mathrm{e}^v$ yields the quadratic PDE

$$\boxed{u_t = F[u] \equiv u \Delta u + u^2 + a + bu.} \tag{9.62}$$

F admits the subspace (see Proposition 6.9)

$$W_2 = \mathcal{L}\{1, \; f(x)\}, \tag{9.63}$$

where f is a solution of the linear elliptic equation

$$\Delta f + f = 0 \quad \text{in} \;\; I\!\!R^N. \tag{9.64}$$

Therefore, substituting $u(x, t) = C_1(t) + C_2(t) f(x)$ leads to the DS

$$\begin{cases} C_1' = C_1^2 + bC_1 + a, \\ C_2' = C_1 C_2 + bC_2, \end{cases} \tag{9.65}$$

that can be solved explicitly. The corresponding discretized equation

$$u_t = F_h[u] \equiv u \Delta_h u + u^2 + a + bu$$

possesses the same solutions on the subspace (9.63), provided that f solves the linear discrete equation $\Delta_h f + f = 0$. The discrete system for expansion coefficients coincides with (9.65).

Example 9.20 (**On quasilinear hyperbolic PDEs**) Consider the evolution PDE

$$\boxed{u_{tt} = F[u] \equiv u \Delta u + u^2 + a + bu.} \tag{9.66}$$

Fixing an arbitrary finite number n of linearly independent solutions $f_1, ..., f_n$ of (9.64) ($n = 2$ if $N = 1$) yields the subspace

$$W_{n+1} = \mathcal{L}\{1, f_1, ..., f_n\}. \tag{9.67}$$

Plugging $u(x,t) = C_0(t) + C_1(t) f_1 + ... + C_n(t) f_n$ gives the DS

$$\begin{cases} C_0'' = C_0^2 + bC_0 + a, \\ C_k'' = C_0 C_k + bC_k, \end{cases}$$

for $k = 1, ..., n$. Similar solutions exist for the corresponding discrete equation.

For the parabolic problem (9.62), last ODEs of the DS are

$$C_k' = C_k(b + C_0) \quad \Longrightarrow \quad \frac{C_k(t)}{C_l(t)} = \text{constant}$$

for all $k, l = 1, ..., n$. Hence, the solutions on the extended subspace (9.67) coincide with those on the 2D subspace (9.63), where f is a linear combination of $f_1, ..., f_n$. This is not the case for the hyperbolic PDE (9.66).

Example 9.21 (**Fast diffusion**) Consider the PDE

$$v_t = \nabla \cdot \left(v^{-\frac{4}{N+2}} \nabla v\right) + b v^{\frac{N+6}{N+2}} + cv$$

in $I\!R^N \times I\!R_+$ with $N \geq 2$. The pressure variable $u = v^{-\frac{4}{N+2}}$ satisfies

$$\boxed{u_t = F_0[u] - \tfrac{4}{N+2} b - \tfrac{4}{N+2} cu, \quad \text{where} \quad F_0[u] = u \Delta u - \tfrac{N+2}{4} |\nabla u|^2.}$$

In Example 6.25, we have shown that F admits $W_3 = \mathcal{L}\{1, |x|^2, |x|^4\}$, and have constructed the corresponding solutions. Consider now the discretized radially symmetric operator

$$F_h[u] = u\left(u_{rr} + \tfrac{N-1}{r} u_r\right) - \tfrac{N+2}{4}(u_r)^2, \quad \text{where } r = |x|, \tag{9.68}$$

that is defined in the space of functions on the uniform grid $X_h^+ = X_h \cap \{r > 0\}$, with the standard approximation of the derivatives u_r and u_{rr}. The subspace of functions restricted to X_h^+ is then invariant under the operator (9.68). Hence,

$$\begin{aligned} u = C_1 + C_2 r^2 + C_3 r^4 \quad &\Longrightarrow \\ F_h[u] = 2NC_1C_2 + 4(N-1)&C_1C_3 h + 2C_1C_3h^2 \\ + \{4(N+2)C_1C_3 + (N&-2)C_2^2 - 12C_2C_3 h \\ + [2C_2C_3 - 4(N&+2)C_3^2]h^2\} r^2 \\ + [2NC_2C_3 - 4(N+5)&C_3^2 h + 2C_3^2 h^2] r^4 \in W_3. \end{aligned} \tag{9.69}$$

Therefore, the discrete equation admits solutions (9.69), which, for $h, \tau = 0$, coincide with the corresponding differential ones.

9.3.3 Five-dimensional invariant subspaces for $N = 1$

1. Polynomial subspaces. We begin with the 1D quasilinear heat equation from Example 1.14,

$$v_t = \left(v^{-\frac{4}{3}} v_x\right)_x + a v^{\frac{7}{3}} + bv.$$

The pressure function $u = v^{-\frac{4}{3}}$ satisfies the quadratic PDE

$$\boxed{u_t = F[u] \equiv u u_{xx} - \tfrac{3}{4}(u_x)^2 - \tfrac{4}{3} a - \tfrac{4}{3} bu.} \tag{9.70}$$

F is known to preserve the 5D subspace, $W_5 = \mathcal{L}\{1, x, x^2, x^3, x^4\}$, so that (9.70) possesses solutions

$$u(x,t) = C_1(t) + C_2(t)x + C_3(t)x^2 + C_4(t)x^3 + C_5(t)x^4, \tag{9.71}$$

$$\begin{cases} C_1' = 2C_1C_3 - \frac{3}{4}C_2^2 - \frac{4}{3}a - \frac{4}{3}bC_1 \equiv \Psi_1, \\ C_2' = 6C_1C_4 - C_2C_3 - \frac{4}{3}bC_2 \equiv \Psi_2, \\ C_3' = 12C_1C_5 + \frac{3}{2}C_2C_4 - C_3^2 - \frac{4}{3}bC_3 \equiv \Psi_3, \\ C_4' = 6C_2C_5 - C_3C_4 - \frac{4}{3}bC_4 \equiv \Psi_4, \\ C_5' = 2C_3C_5 - \frac{3}{4}C_4^2 - \frac{4}{3}bC_5 \equiv \Psi_5. \end{cases} \tag{9.72}$$

For the corresponding discretized equation (9.70), the same subspace is obtained.

Proposition 9.22 *The discrete operator F in* (9.70) *admits W_5.*

Proof. Substituting (9.71) restricted to X_h into (9.70) and using formulae

$$(x)_x = 1, \;\; (x^2)_x = 2x, \;\; (x^3)_x = 3x^2 + h^2, \;\; (x^4)_x = 4x^3 + 4xh^2, \tag{9.73}$$

we find that (9.70) has exact solutions (9.71), where the coefficients solve the DS

$$\begin{cases} C_1' = \Psi_1 + (2C_1C_5 - \frac{3}{2}C_2C_4)h^2 - \frac{3}{4}C_4^2h^4, \\ C_2' = \Psi_2 - (4C_2C_5 + 3C_3C_4)h^2 - 6C_4C_5h^4, \\ C_3' = \Psi_3 - (10C_3C_5 + \frac{9}{2}C_4^2)h^2 - 12C_5^2h^4, \\ C_4' = \Psi_4 - 22C_4C_5h^2, \\ C_5' = \Psi_5 - 22C_5^2h^2. \end{cases} \tag{9.74}$$

Passing to the limit as $h, \tau \to 0$ in (9.74) yields the DS (9.72). □

Example 9.23 (**Extended polynomial subspaces**) All the polynomial subspaces remain to exist for discrete higher-order operators. For instance, Proposition 3.2 established that the fourth-order quadratic thin film equation (TFE)

$$\boxed{u_t = F[u] \equiv -uu_{xxxx} + \beta u_x u_{xxx}\,,}$$

where the basic subspace is known to be $W_5 = \mathcal{L}\{1, x, x^2, x^3, x^4\}$, and admits solutions on the extended subspaces

$$W_6 = \mathcal{L}\{1, x, x^2, x^3, x^4, x^5\}, \quad \text{if } \beta = \tfrac{2}{5},$$
$$W_7 = \mathcal{L}\{1, x, x^2, x^3, x^4, x^5, x^6\}, \quad \text{if } \beta = \tfrac{1}{2}.$$

The same subspaces can be used for the discretized operator, $F[u]$, with any standard approximation of the derivatives, and lead to similar discrete DSs.

2. Trigonometric subspaces. Next consider the quasilinear heat equation with absorption from Example 1.15,

$$v_t = \left(v^{-\frac{4}{3}}v_x\right)_x - v^{-\frac{1}{3}}.$$

The pressure transformation $u = v^{-\frac{4}{3}}$ reduces it to

$$\boxed{u_t = F_0[u] \equiv uu_{xx} - \tfrac{3}{4}(u_x)^2 + \tfrac{4}{3}u^2,} \tag{9.75}$$

where the quadratic operator F admits

$$W_5 = \mathcal{L}\{1, \cos(\lambda x), \sin(\lambda x), \cos(\tfrac{\lambda x}{2}), \sin(\tfrac{\lambda x}{2})\}, \quad \text{with } \lambda = \tfrac{4}{\sqrt{3}}. \tag{9.76}$$

Hence, (9.75) has solutions

$$u = C_1 + C_2 \cos(\lambda x) + C_3 \sin(\lambda x) + C_4 \cos(\tfrac{\lambda x}{2}) + C_5 \sin(\tfrac{\lambda x}{2}). \tag{9.77}$$

The DS is given in Example 1.15.

Unlike in the case of polynomial subspaces, for trigonometric ones, the invariance of the nonlinear operator is no longer true for the discretization taken in the same form as in (9.75). For preserving the subspace in the discrete case, we consider a slightly different approximation of the operator,

$$u_t = F_h[u] \equiv u u_{xx} + \alpha (u_x)^2 + \beta u^2, \tag{9.78}$$

where the parameters α and β depend on h, and conclude as follows:

Proposition 9.24 *Operator F_h in (9.78) admits subspace (9.76) if the constants α, β and λ satisfy the system*

$$\begin{cases} m\big[1 + \alpha \cos^2(\tfrac{\lambda h}{2})\big] = \beta, \\ m\big\{1 + 4\cos^2(\tfrac{\lambda h}{4})[1 + \alpha \cos(\tfrac{\lambda h}{2})]\big\} = \beta \end{cases} \tag{9.79}$$

(here $m = \frac{4}{h^2} \sin^2(\frac{\lambda h}{2})$ as in (9.49)).

In the case of the quasilinear heat operator (cf. (9.75) with $\sigma = -\frac{4}{3}$),

$$F_h[u] = u u_{xx} + \tfrac{1}{\sigma}(u_x)^2 - \sigma u^2,$$

i.e., we still fix $\alpha = \frac{1}{\sigma}$ and $\beta = -\sigma$ with some $\sigma = \sigma(h)$, (9.79) gives the following asymptotic values of the parameters σ and λ for which the 5D subspace (9.76) exists: as $h \to 0$,

$$\sigma = -\tfrac{4}{3} + \tfrac{68}{27} h^2 + O(h^4) \quad \text{and} \quad \lambda^2 = \tfrac{16}{3} + \tfrac{32}{27} h^2 + O(h^4).$$

For $h = 0$, we obtain the differential case $\sigma = -\frac{4}{3}$, $\lambda = \frac{4}{\sqrt{3}}$, and subspace (9.76).

Finally, let us present the discrete DS for the coefficients of the solution (9.77) of the discrete equation (9.78). Under the hypotheses of Proposition 9.24, this is

$$\begin{cases} C_1' = \beta C_1^2 + \gamma_1(C_2^2 + C_3^2) + \gamma_2(C_4^2 + C_5^2), \\ C_2' = \gamma_3 C_1 C_2 + \gamma_4(C_4^2 - C_5^2), \\ C_3' = \gamma_3 C_1 C_3 + 2\gamma_4 C_4 C_5, \\ C_4' = \gamma_5 C_1 C_4 + \gamma_6(C_2 C_4 + C_3 C_5), \\ C_5' = \gamma_5 C_1 C_5 - \gamma_6(C_2 C_5 - C_3 C_4), \end{cases}$$

where the coefficient $\{\gamma_i\}$ are

$$\begin{gathered} \gamma_1 = \beta - 4\rho_1, \ \gamma_{2,4} = \tfrac{1}{2}(\beta \pm \alpha\rho_1 - 4\rho_2), \ \gamma_3 = 2(\beta - 2\rho_1), \\ \gamma_5 = 2(\beta h^2 - 2\rho_2), \ \gamma_6 = \tfrac{2}{h^2}\alpha \sin(\tfrac{\lambda h}{2}) \sin(\lambda h), \end{gathered} \tag{9.80}$$

with $\rho_1 = \frac{1}{h^2} \sin^2(\frac{\lambda h}{2}) = \frac{m}{4}$ and $\rho_2 = \frac{1}{h^2} \sin^2(\frac{\lambda h}{4})$. For $h = 0$, coefficients (9.80) determine those in the differential system.

Example 9.25 (**The TFE**) It follows from Proposition 3.16 that the TFE

$$\boxed{u_t = F_0[u] \equiv -uu_{xxxx} + \tfrac{15}{22}\, u_x u_{xxx} + \tfrac{56}{11}\, u^2}$$

admits solutions on $W_5 = \mathcal{L}\{1, \cos x, \sin x, \cos 2x, \sin 2x\}$. In order to keep this subspace invariant in the discretized case, the operator is chosen to contain two parameters, depending on h,

$$u_t = F_h[u] \equiv -uu_{xxxx} + \beta u_x u_{xxx} + \delta u^2,$$

where $\beta \to \frac{15}{22}$ and $\delta \to \frac{56}{11}$ as $h \to 0$. The corresponding discrete DS also converges to the continuous one.

9.3.4 Exact solutions on partially invariant subspaces

Example 9.26 (**Quasilinear heat equations**) We take solutions from Example 8.13, and consider the PDE

$$v_t = (\varphi(v))_{xx} - (av + b)\big[1 + \tfrac{2a}{\varphi'(v)} G(v)\big], \quad \text{where} \quad G(v) = \int^v \tfrac{\varphi'(z)\,\mathrm{d}z}{az+b}.$$

Then the function $u = G(v)$ solves

$$\boxed{u_t = F[u] \equiv \Phi(u)(u_{xx} - 1) + a[(u_x)^2 - 2u],} \tag{9.81}$$

with $\Phi(u) = \varphi'(G^{-1}(u))$. Consider (9.81) on the subspace $W_2 = \mathcal{L}\{1, x^2\}$ which is not invariant under F. Setting $u = C_1 + C_2 x^2 \in W_2$ yields

$$F[u] = \Phi(u)(2C_2 - 1) + a\big[2C_2(2C_2 - 1)x^2 - 2C_1\big],$$

so that $F[u] \notin W_2$ for all $u \in W_2$. We have shown that, for F, W_2 is partially invariant and there exists the invariant set (an affine subspace in W_2),

$$M = \big\{u = C_1 + C_2 x^2 : \ C_2 = \tfrac{1}{2}\big\}, \quad \text{such that} \quad F[M] \subseteq W_2.$$

On M, (9.81) reduces to an overdetermined DS which is consistent. Substituting into (9.81) yields the solution

$$u(x,t) = C_1(t) + \tfrac{1}{2} x^2 \in M, \quad \text{where } C_1' = -2aC_1. \tag{9.82}$$

The same analysis applies to the discrete equation (9.81), since the discrete operator F admits the set M. Therefore, there exists the solution (9.82) on M, with the coefficient $C_1(t)$ satisfying the discrete linear equation $C_1' = -2aC_1$ in T_τ.

The results on partially invariant polynomial subspaces are easy to extend to higher-order PDEs similar to those in Example 8.7 and Proposition 8.8. In the next PDE associated with Example 8.44, the conclusion for the discrete equation is not so straightforward, since we deal with trigonometric subspaces for which the consistency of overdetermined DSs is not easy to check.

Example 9.27 (**Semilinear parabolic model**) Consider the PDE

$$\boxed{u_t = F_1[u] + F_2[u] \equiv \big[u_{xx} - \tfrac{1}{2}\tfrac{(u_x)^2}{u} + \tfrac{2}{u} + 2u\big] + \big[\tfrac{1}{2}(u_x)^2 + 2u^2 + 2\big].} \tag{9.83}$$

For convenience, we briefly repeat the argument from Example 8.44. Subspace $W_2 = \mathcal{L}\{1, \sin 2x\}$ is invariant under F_2 and, for any

$$u(x,t) = C_1(t) + C_2(t)\sin 2x, \tag{9.84}$$

$F_2[u] = 2\big(C_1^2 + C_2^2\big) + 2 + 4C_1C_2 \sin 2x \in W_2$, but W_2 is not invariant under F_1, where

$$F_1[u] = \tfrac{2(C_1^2 - C_2^2 + 1)}{C_1 + C_2 \sin 2x} \quad \text{for } u \in W_2. \tag{9.85}$$

There exists the set M on W_2 (a hyperbola in the variables $\{C_1, C_2\}$),

$$M = \big\{u = C_1 + C_2 \sin 2x : \; C_1^2 - C_2^2 + 1 = 0\big\} \tag{9.86}$$

on which $F_1[u] = 0$. Plugging (9.84) into (9.83) yields the overdetermined DS

$$\begin{cases} C_1' = 2\big(C_1^2 + C_2^2\big) + 2, \\ C_2' = 4C_1C_2, \\ C_1^2 - C_2^2 + 1 = 0, \end{cases} \tag{9.87}$$

which yields the solution of (9.83), $u(x,t) = \tan 4t + \frac{1}{\cos 4t} \sin 2x \in M$.

Consider the corresponding discretized equation

$$u_t = F_h[u] \equiv F_{1h}[u] + F_{2h}[u] \tag{9.88}$$

on the subspace $W_2^h = \mathcal{L}\{1, \sin(\lambda x)\}$. Operator F_{2h} is as shown in (9.83), and

$$F_{1h}[u] = \alpha u_{xx} - \tfrac{1}{2u}(u_x)^2 + \tfrac{2}{u} + 2u,$$

where $\alpha(h) \to 1$ (and $\lambda(h) \to 2$) as $h \to 0$.

Proposition 9.28 (i) *W_2^h is invariant under F_{2h} iff*

$$\tfrac{1}{h}\sin(\lambda h) = 2. \tag{9.89}$$

(ii) *If, in addition, $\alpha = \cos^2(\frac{\lambda h}{2})$, then $F_h[M] \subseteq W_2^h$.*

Since $F_{1h}[u] = 0$ on M and, as in the differential case,

$$F_{2h}[u] = 2\big(C_1^2 + C_2^2\big) + 2 + 4C_1C_2 \sin(\lambda x),$$

(9.88) on M is equivalent to the overdetermined discrete system (9.87), where the continuous derivatives $(\cdot)'$ are replaced by discrete ones. It can be shown that, unlike in the differential case, for both the implicit and explicit finite-difference schemes, the discrete dynamical system does not have a solution satisfying $C_1^2 - C_2^2 + 1 = 0$ on T_τ, so our result is negative. On the other hand, if the time is not discretized, system (9.87) gives the exact solutions of $u_t = F_h[u]$. Similarly, we can analyze invariant sets and consistency of the DSs for discretized higher-order PDEs studied in Examples 8.43–8.45.

9.4 Invariant properties of moving mesh operators and applications

The moving mesh methods (MMMs) are known to be an efficient approach to solving the nonlinear evolution PDEs where the solutions exhibit essentially non-stationary and singular behavior, such as blow-up, extinction, or quenching. We will describe some basics of MMM theory and refer to [293, 292, 99] for general principles and applications.

We apply the results on invariant subspaces for discrete operators to some MMMs for typical parabolic or hyperbolic PDEs. We show that, for a class of nonlinear 1D evolution equations, the MMMs may preserve linear invariant subspaces of nonlinear differential operators under a special approximation of the *spatial gradient operator*. The corresponding discrete evolution on such subspaces may coincide with that for fixed meshes, and even with the continuous ones, regardless of the fast deformation of the moving mesh.

9.4.1 Introduction: MMMs and invariant subspaces

We deal with a general nonlinear evolution PDE

$$u_t = F[u], \quad x \in S \subseteq I\!R, \;\; t > 0, \tag{9.90}$$

where $F[u] = F\big(u, u_x, ..., D_x^k u\big)$, $F \in C^\infty$ is an kth-order ordinary differential operator, and S is an interval. Assume that, being endowed with suitable boundary conditions on the lateral boundary of $Q = S \times I\!R_+$ (if $S \neq I\!R$), the equation (9.90) defines a smooth flow for bounded initial data u_0. For $S = I\!R$, we mean the Cauchy problem with a given initial function $u_0(x)$.

The MMMs approximate (9.90) on a moving mesh (MM) $\Omega = \{x = X_i(t), \; i \in I, \; t > 0\} \subset Q$, $I = \{1, 2, ..., K\}$, with *a priori* unknown curves $x = X_i(t)$, depending on the solution $u(x,t)$ under consideration. Using identity

$$\tfrac{\mathrm{d}}{\mathrm{d}t}\, u(X_i, t) = u_t(X_i, t) + u_x(X_i, t) X_i',$$

the differential equation for $U = \{U_i(t) = u(X_i(t), t)\}$ reads

$$U_i' - u_x(X_i, t) X_i' = F[u(X_i, t)], \quad t > 0; \quad i \in I.$$

Discretizing this equation yields a dynamical system,

$$U_i' - [D_h U]_i X_i' = F_h[U] \quad \text{for } t > 0, \tag{9.91}$$

where F_h is a suitable approximation of the nonlinear operator F, and D_h approximates the gradient operator $D_x = \frac{\mathrm{d}}{\mathrm{d}x}$. The evolution equations for the moving mesh Ω are added to (9.91), and these, all together, give a finite-dimensional DS for the functions $\{U_i(t), \; i \in I\}$. Such approaches make it possible to eventually concentrate the MM close to crucial singularities of the solutions. Giving such advantages in computing the singular behavior, the extra linear discrete gradient operator D_h that appears in (9.91) may strongly change the discrete PDE.

Even for semilinear second-order parabolic PDEs, equation (9.91) no longer obeys the Maximum Principle (MP). On the contrary, if $X_i'(t) \equiv 0$ for all i, i.e., the mesh is fixed (stationary), the MP holds, provided that the approximation F_h preserves such a *positivity* property. Therefore, moving meshes may destroy the positivity or monotonicity features of solutions that are expected to be inherited by the discrete equation from the original parabolic PDE. Mathematical analysis of MMMs is harder than for the corresponding explicit or implicit difference schemes with stationary meshes. This includes the questions of convergence of MMMs and their asymptotic behavior. Comparison of solutions of MMMs for parabolic PDEs becomes a complicated procedure, because the solutions are defined on the distinct MMs, depending on solutions that are unknown *a priori*.

We will study the principal question of *invariant linear subspaces* generated by the MM operators (MMOs). It was shown in the previous section how to choose a linear subspace W that is invariant under a discrete nonlinear operator F_h. In contrast, for typical DSs (9.91) occurring for the MMMs, extra work should be done. We will need to prove the necessary and sufficient condition for the gradient approximation D_h to preserve the invariance of a given subspace W. This guarantees that the invariant MMO admits the same subspace and, moreover, the discrete dynamics on W coincides with the differential one, regardless of the structure of the MM.

For parabolic PDEs, we establish the necessary and sufficient conditions for the positivity of the solutions (the weak MP), and also prove comparison of solutions of MMMs. As a general conclusion, for typical semilinear parabolic PDEs, these important properties remain valid if *the MM does not move too fast.*

We will also consider invariant MMOs for higher-order evolution PDEs,

$$D_t^m u = F[u] \equiv F\big(u, u_t, ..., D_t^{m-1} u, u_x, ..., D_x^k u\big), \tag{9.92}$$

where $m \geq 2$ and $m \geq k$. In particular, we consider the second-order hyperbolic PDEs

$$u_{tt} = F(u, u_x, u_{xx}).$$

9.4.2 *Invariant subspaces for differential and discrete operators*

For convenience, we list below a number of already known differential and corresponding discrete quadratic operators preserving invariant subspaces.

1. Invariant subspaces of differential operators. Let us state the main invariance hypothesis on the nonlinear operator F.

Hypothesis (*Inv*). The operator F admits a *2D subspace* $W_2 = \mathcal{L}\{1, f(x)\}$, where f is a given smooth function, i.e.,

$$F[W_2] \subseteq W_2. \tag{9.93}$$

Under hypothesis (*Inv*), the PDE (9.90) on W_2 is equivalent to a nonlinear second-order DS. In view of (9.93), for any $u = C_1 + C_2 f \in W_2$,

$$F[C_1 + C_2 f] = \Psi_1(C_1, C_2) + \Psi_2(C_1, C_2) f \in W_2. \tag{9.94}$$

Substituting (9.94) into (9.90) yields the DS

$$\begin{cases} C_1' = \Psi_1(C_1, C_2), \\ C_2' = \Psi_2(C_1, C_2). \end{cases} \tag{9.95}$$

The following basic examples of operators and invariant subspaces are taken from Sections 1.3–1.5 and 3.1.

(i) Quadratic operators with 2D subspaces:

$$F_1[u] = u_{xx} + (u_x)^2 + u, \;\; W_2 = \mathcal{L}\{1, x^2\}, \tag{9.96}$$

$$F_2[u] = u_{xx} + (u_x)^2 + u^2, \;\; W_2 = \mathcal{L}\{1, \cos x\}, \tag{9.97}$$

$$F_3[u] = uu_{xx} + \tfrac{1}{\sigma}(u_x)^2 + \alpha u + \beta, \;\; W_2 = \mathcal{L}\{1, x^2\}, \tag{9.98}$$

$$F_4[u] = \;uu_{xx} + \tfrac{1}{\sigma}(u_x)^2 + \sigma u^2, \;\; W_2 = \mathcal{L}\{1, \cos(\lambda x)\}, \;\lambda = \tfrac{\sigma}{\sqrt{\sigma+1}}, \tag{9.99}$$

$$F_5[u] = -u_{xxxx} - (u_x)^2 + u^2, \;\; W_2 = \mathcal{L}\{1, \cosh x\}, \tag{9.100}$$

$$F_6[u] = -uu_{xxxx} + \alpha u_x u_{xxx} + \beta (u_{xx})^2 + \gamma u + \delta, \;\; W_2 = \mathcal{L}\{1, x^4\}. \tag{9.101}$$

The operators $F_1 - F_4$ correspond to the semilinear and quasilinear second-order parabolic PDEs of reaction-diffusion type, where F_3 and F_4 contain the differential operator $uu_{xx} + \frac{1}{\sigma}(u_x)^2$ of the porous medium ($\sigma > 0$) or fast diffusion ($\sigma < 0$) type. The operator F_5 reproduces a fourth-order parabolic equation, which is a Kuramoto–Sivashinsky-type PDE with the extra lower-order source term u^2 (see Section 3.8), while F_6 occurs in TFE theory (Section 3.1). PDEs

$$\boxed{u_{tt} = F_3[u] + \gamma\, u_{xxxx} + \delta\, u_{xx}} \quad \text{and} \quad \boxed{u_{tt} = F_4[u] + \gamma\, u_{xxxx} + \delta\, u_{xx}}$$

are Boussinesq-type equations from water wave theory (Section 5.3). We next discuss their exact solutions in greater detail.

Example 9.29 (**PME with source**) The quasilinear heat equation with a source,

$$\boxed{u_t = F_4[u] \equiv uu_{xx} + \tfrac{1}{\sigma}(u_x)^2 + \sigma u^2, \quad \sigma > 0,} \tag{9.102}$$

plays a key role in the study of localization of blow-up in combustion, [509, Ch. 4]. The quadratic operator F_4 admits subspace W_2 given in (9.99), so that this PDE has the solutions

$$u(x,t) = C_1(t) + C_2(t)\cos(\lambda x) \in W_2, \tag{9.103}$$

$$\begin{cases} C_1' = \sigma C_1^2 + \frac{\sigma}{\sigma+1}\, C_2^2, \\ C_2' = \frac{\sigma(\sigma+2)}{(\sigma+1)}\, C_1 C_2. \end{cases} \tag{9.104}$$

Such solutions describe localization and blow-up phenomena. In particular, the Sturmian intersection comparison with such solutions establishes [509, p. 249] that the interface $s(t)$ of any weak solution $u(x,t) \ge 0$ of (9.102) with the connected compact support satisfies the localization estimate

$$|s(t) - s(0)| \le \tfrac{1}{2}\, L_S = \pi\, \tfrac{\sqrt{\sigma+1}}{\sigma} \quad \text{for any } t \in [0, T)$$

(L_S is the *fundamental length*), where T is the blow-up time of the solution. If $C_1(t) \equiv C_2(t)$ in (9.104), we obtain the ODE $C_1' = \alpha\, C_2^2$ with $\alpha = \frac{\sigma(\sigma+2)}{\sigma+1}$. Hence, $C_1(t) = [\alpha(T-t)]^{-1}$, and this gives the separable solution

$$u(x,t) = \tfrac{1}{\alpha(T-t)}\, 2\cos^2(\tfrac{\lambda x}{2}) \ge 0. \tag{9.105}$$

Therefore, subspace W_2 contains the 1D manifold (with the parameter $T \in I\!R$) of similarity solutions (9.105). It is important that these solutions are asymptotically stable; see the stability analysis of blow-up dynamics in Example 3.17.

Example 9.30 **(Quasilinear hyperbolic equation)** The PDE

$$\boxed{u_{tt} = F_4[u] \equiv (uu_x)_x + u^2,}$$

which is hyperbolic in the positivity domain $\{u > 0\}$, admits solutions

$$u(x,t) = C_1(t) + C_2(t)\cos\big(\tfrac{x}{\sqrt{2}}\big) \in W_2 = \mathcal{L}\big\{1, \cos\big(\tfrac{x}{\sqrt{2}}\big)\big\},$$
$$\begin{cases} C_1'' = C_1 + \frac{1}{2}\,C_2^2, \\ C_2'' = \frac{3}{2}\,C_1C_2. \end{cases}$$

For $C_1 = C_2$, we find the ODE $C_1'' = \frac{3}{2}\,C_1^2$, which is integrated in quadratures with the solutions given by elliptic functions. Taking $C_1(t) = 4(T-t)^{-2}$ gives the separable blow-up solution

$$u(x,t) = \tfrac{4}{(T-t)^2}\,\big[1 + \cos\big(\tfrac{x}{\sqrt{2}}\big)\big].$$

Its asymptotic stability on W_2 can be checked, as done in Example 5.1.

(ii) Five and more-dimensional subspaces exist for special quadratic operators:

$$F_7[u] = uu_{xx} - \tfrac{3}{4}(u_x)^2, \quad \text{with } W_5 = \mathcal{L}\{1, x, x^2, x^3, x^4\}$$

(notice that (9.101) also admits W_5), and

$$F_8[u] = uu_{xx} - \tfrac{3}{4}(u_x)^2 + \tfrac{4}{3}\,u^2,$$

with W_5 given in (9.76). Note that F_7 also admits 2D subspaces $\mathcal{L}\{1, x^2\}$ and $\mathcal{L}\{1, x^4\}$. Operator

$$F_9[u] = uu_{xx} - \tfrac{2}{3}(u_x)^2$$

admits $W_2 = \mathcal{L}\{1, x^3\}$, which is extended to the 4D $W_4 = \mathcal{L}\{1, x, x^2, x^3\}$; see Example 1.13. Fourth-order thin film operators from Section 3.1 are the origin of many invariant subspaces. For instance, in the previous section we have used

$$F_{10}[u] = -uu_{xxxx} + \tfrac{2}{5}\,u_xu_{xxx}\,,$$

admitting $W_6 = \mathcal{L}\{1, x, x^2, x^3, x^4, x^5\}$ and the 2D restrictions $\mathcal{L}\{1, x^5\}$ and $\mathcal{L}\{1, x^4\}$. Operator

$$F_{11}[u] = -uu_{xxxx} + \tfrac{1}{2}\,u_xu_{xxx}$$

admits $W_7 = \mathcal{L}\{1, x, x^2, x^3, x^4, x^5, x^6\}$, as well as $\mathcal{L}\{1, x^6\}$ and $\mathcal{L}\{1, x^4\}$.

2. Discrete operators. In Section 9.3, we have shown that invariant subspaces can be preserved and are stable with respect to standard discretizations of operators on fixed meshes. In other words, the corresponding discrete operators, F_{kh}, with sufficiently small steps of uniform discretization $h > 0$ can admit subspaces with the basis functions being $O(h)$-perturbations of those for differential operators. For the 2D subspace, $W_2 = \mathcal{L}\{1, f(x)\}$, this implies that

$$F_h[C_1 + C_2 f] \equiv \Psi_{1h}(C_1, C_2) + \Psi_{2h}(C_1, C_2) f \in W_2, \tag{9.106}$$

so that the discrete equation on a fixed mesh,

$$U_t = F_h[U],$$

has the solutions $U(x,t) = C_1(t)+C_2(t)f(x)$ on W_2 with the coefficients satisfying a DS, which is similar to (9.95),

$$\begin{cases} C_1' = \Psi_{1h}(C_1, C_2), \\ C_2' = \Psi_{2h}(C_1, C_2). \end{cases} \tag{9.107}$$

By approximation, Ψ_{1h} and Ψ_{2h} converge as $h \to 0$ to the coefficients of the DS (9.95).

Temporarily assuming that the mesh is fixed and uniform, with a given $h > 0$, we introduce the natural approximations of the spatial derivatives

$$u_x = \tfrac{1}{2h}\big(U_{i+1} - U_{i-1}\big), \quad u_{xx} = \tfrac{1}{h^2}\big(U_{i+1} + U_{i-1} - 2U_i\big), \dots . \tag{9.108}$$

Substituting those into operators $F_1 - F_{11}$ yields the corresponding discrete operators $F_{1h} - F_{11h}$. As was shown in the previous section, the polynomial invariant subspaces then remain the same for discrete operators. Concerning trigonometric and exponential subspaces, operator F_{2h} in (9.97) admits

$$W_{2h} = \mathcal{L}\{1, \cos(\lambda x)\}, \quad \text{with} \quad \tfrac{1}{h^2}\sin^2(\lambda h) = 1, \tag{9.109}$$

and F_{4h} in (9.99) admits the same subspace as in (9.109), where λ satisfies

$$\tfrac{4}{h^2}\sin^2\big(\tfrac{\lambda h}{2}\big)\big[1 + \tfrac{1}{\sigma}\cos^2\big(\tfrac{\lambda h}{2}\big)\big] = \sigma. \tag{9.110}$$

The exponential subspace

$$\mathcal{L}\{1, \cosh(\lambda x)\}, \quad \text{with} \quad \lambda = \lambda(h), \tag{9.111}$$

exists for the fourth-order operator F_{5h} in (9.100), etc.

Example 9.31 Consider the discrete equation (9.102). On W_{2h} given by (9.109), it reduces to the DS (9.54), with $a = b = 0$, for the coefficients $\{C_1, C_2\}$ of the expansion (9.103). For $h \ll 1$, the parameter $\lambda(h)$ of the basis function satisfies (9.55), so that $\lambda(h) \to \frac{\sigma}{\sqrt{\sigma+1}}$ as $h \to 0^+$, and, in the limit, the discrete DS (9.54) coincides with the continuous system (9.104).

For standard approximations of derivatives (9.108), subspaces (9.109), (9.111) can be invariant on uniform fixed meshes only. In order to preserve this invariance on non-uniform MMs, different approximations have to be used.

Invariance of moving mesh operators. We are going to show that invariant subspaces can be preserved by MM operators. Consider a general MMM for (9.90). Assume that, given initial data u_0, we approximate the evolution orbit $\{u(t),\ t > 0\}$ by the approximate solution $U(t) = \{U_i(t),\ i \in I\}$, $I = \{1, 2, ..., K\}$, defined on the finite moving mesh $\Omega(U) = \{X_i(t),\ i \in I;\ t > 0\}$. Here, U satisfies the *moving mesh equation* (MME)

$$\big[\mathbf{B_h}[U]\big]_i = U_i' - [D_h U]_i X_i' - F_h[U] = 0, \tag{9.112}$$

where $\mathbf{B_h}$ is the *moving mesh operator* (MMO) defined at the internal points $i \in I_m = \{2, ..., K-1\} \subset I$. Assume that suitable initial data are prescribed at $t = 0$, and boundary conditions are given on the lateral boundary $\Gamma(U)$ of $\Omega(U)$, $i \in I_m$ (if $S \neq I\!R$), so that the solution U is well-defined locally in time on $\Omega(U)$. For a fixed

$\tau \geq 0$, we denote $\Omega_\tau(U) = \Omega(U) \cap \{t = \tau\}$. By $\partial\Omega(U)$ we denote the *parabolic boundary* of $\Omega(U)$, i.e., that part of the boundary where initial and boundary data are prescribed.

We will study the invariant properties of the MMO $\mathbf{B_h}$. We do not take into account the mechanism of generation of the moving mesh $\Omega(U)$, which is crucial for construction of MMMs, [293, 292, 99]; see also applications to blow-up problems in [88, 87]. We construct MMEs with an *arbitrary* MMs $\Omega(U)$, which possess solutions on the subspaces of the discrete nonlinear operators F_h. A special choice of MMs can improve invariant properties of the MMMs. There are examples, for which the requirement of optimal approximations of invariant subspaces (forming a stable attractor of the given flow) plays an important role in the correct description of evolution properties of PDEs. Moreover, this may be the decisive principle of construction of the corresponding MMOs.

We restrict our attention to the problem of invariant subspaces of the MMO (9.112). We will deal with 2D subspaces introduced above.

Theorem 9.32 *Assume that, given moving mesh $\Omega(U)$, the discrete operator F_h on $\Omega(U)$ admits the 2D subspace $W_{2h} = \mathcal{L}\{1, f(x)\}$, with a smooth function f. Let the discrete gradient operator satisfy*

$$[D_h f(X)]_i = f'(X_i) \quad \text{on} \ \ \Omega(U). \tag{9.113}$$

Then $\mathbf{B_h} : W_{2h} \to W_{2h}$ *and the MME* (9.112) *on* $\Omega(U)$ *is equivalent to a nonlinear 2D dynamical system that coincides with* (9.107) *for the corresponding fixed mesh.*

For instance, we can use the following discrete gradient operator satisfying (9.113):

$$[D_h U]_i = \tfrac{U_{i+1}-U_{i-1}}{f(X_{i+1})-f(X_{i-1})}\, f'(X_i). \tag{9.114}$$

On the mesh satisfying $|X_{i+1} - X_{i-1}| = O(h)$, it approximates $\frac{\mathrm{d}u}{\mathrm{d}x}$, with at least order $O(h)$.

Proof of Theorem 9.32. We are looking for a solution of (9.112) in the form of

$$U_i = U(X_i(t), t) = C_1(t) + C_2(t) f(X_i(t)) \in W_{2h}. \tag{9.115}$$

Plugging (9.115) into (9.112) yields

$$\begin{aligned} C_1' + C_2' f(X_i) + C_2\big\{f'(X_i) - [D_h f(X)]_i\big\}X_i' \\ = F_h[C_1 + C_2 f(X) \in W_{2h}. \end{aligned} \tag{9.116}$$

In view of (9.113), the term depending on X_i' on the left-hand side of (9.116) vanishes, and, due to (9.106), the DS (9.107) is obtained. □

We consider evolution PDEs with the above nonlinear operators, $F_1 - F_{11}$, and show how to construct suitable discretizations F_h of the operators to get solutions of the MMEs on the invariant subspaces.

9.4.3 *Invariant subspace* $L\{1, x^m\}$

We have presented a number of such subspaces with $m = 2,\ 3$, or 4.

Semilinear operator F_1**.** Consider the PDE (9.40) with the operator (9.96). The corresponding MME (9.112) has the form

$$U_i' - [D_h U]_i X_i' = F_{1h}[U] \equiv U_{XX} + (U_X)^2 + U_i \quad \text{on } \Omega(U), \tag{9.117}$$

with the approximation F_{1h} to be determined later. In view of (9.114) with the function $f(x) = x^2$, the gradient operator is

$$[D_h U]_i = \frac{U_{i+1} - U_{i-1}}{X_{i+1}^2 - X_{i-1}^2}\, 2X_i, \tag{9.118}$$

so $[D_h X^2]_i = 2X_i$ by (9.113). Consider the discrete operator F_{1h} admitting W_2 if the discrete derivatives are given by

$$U_X = D_h U \quad \text{and} \quad U_{XX} = d_h D_h U, \tag{9.119}$$

where d_h denotes the standard symmetric gradient approximation,

$$d_h U = \frac{U_{i+1} - U_{i-1}}{X_{i+1} - X_{i-1}}, \quad \text{so that} \quad d_h D_h X^2 = 2. \tag{9.120}$$

Finally, given the mesh function $U = C_1(t) + C_2(t) X^2 \in W_2$, the following holds:

$$F_{1h}[U] = C_1 + 2C_1 + C_2(1 + 4C_2) X^2 \in W_2, \tag{9.121}$$

so that equation (9.117) on W_2 is the DS (9.42). Hence, the discrete evolution on W_2 coincides with the differential one induced by the PDE (9.40) on W_2.

On conditionally invariant MMOs. Observe that, in general, the natural approximation $U_{XX} = D_h^2 U$ does not satisfy the second condition in (9.120). One can calculate that

$$\left[D_h^2 X^2\right]_i = \frac{4X_i}{X_{i+1} + X_{i-1}},$$

and, therefore, taking the condition $D_h^2 X^2 = 2$ as one of the generating conditions of the MMs, we arrive at $X_i = \frac{1}{2}(X_{i+1} + X_{i-1})$, i.e., the MM has to be uniform. Hence, on any uniform MM $\Omega(U)$, the MME (9.117) with $U_{XX} = D_h^2 U$ admits solutions on W_2. This MME is *conditionally* invariant, i.e., not for every MM. On the other hand, the invariance of the standard approximation $U_{XX} = d_h^2 U$ implies that

$$\left[d_h d_h X^2\right]_i \equiv \frac{X_{i+2} - X_{i-2}}{X_{i+1} - X_{i-1}} = 2,$$

which is true for uniform MMs $\Omega(U)$. Finally, the invariant MMO in (9.117) leaves the extended subspace $W_3 = \mathcal{L}\{1, x, x^2\}$ invariant on any uniform mesh, since the discrete gradient operator correctly differentiates the linear function x on such meshes, $[D_h X]_i = \frac{2X_i}{X_{i+1} - X_{i-1}} = 1$.

Porous medium operators F_3**,** F_7**, and** F_9**.** Consider the PDE

$$\boxed{u_t = u u_{xx} + \tfrac{1}{\sigma}(u_x)^2 + \alpha u + \beta \quad \text{on } W_2 = \mathcal{L}\{1, x^2\}.} \tag{9.122}$$

Using (9.118) and (9.119), we introduce the MME on $\Omega(V)$

$$U_i' - [D_h U]_i X_i' = F_{3h}[U] \equiv U_i d_h D_h U + \tfrac{1}{\sigma}(D_h U)^2 + \alpha U_i + \beta,$$

which admits solutions $U = C_1(t) + C_2(t) X^2$. The expansion coefficients $\{C_1, C_2\}$

solve the same DS as in the differential case,

$$\begin{cases} C_1' = 2C_1C_2 + \alpha C_1 + \beta, \\ C_2' = \frac{2(\sigma+2)}{\sigma}\, C_2^2 + \alpha C_2. \end{cases}$$

The invariance result is true for the extension W_3 on any uniform mesh. The subspace W_2 remains invariant if we add any linear differential operator $L[u] = \gamma\, u_{xx} + \delta u_x : W_2 \to W_2$ to the right-hand side of (9.122).

The fourth-order operator F_6**.** For the quadratic operator (9.101), the invariant approximation of the gradient operator is

$$[D_h U]_i = \frac{U_{i+1} - U_{i-1}}{X_{i+1}^4 - X_{i-1}^4}\, 3X_i^3.$$

Then the discrete DS is the same as for the PDE $u_t = F_6[u]$ on W_2. Exact solutions of both the continuous and discrete DS are $u = C_1 + C_2 x^4$ and $U = C_1 + C_2 X^4$, where

$$\begin{cases} C_1' = -24C_1C_2 + \gamma\, C_1 + \delta, \\ C_2' = (-24 + 96\alpha + 144\beta)C_2^2 + \gamma\, C_2. \end{cases}$$

Any operator $F_6 + L$, for any linear higher-order operator L with constant coefficients, admits the same subspace. This also applies to other higher-order operators on polynomial subspaces.

9.4.4 Invariant subspace $L\{1, \cos x\}$

Unlike the polynomials, the trigonometric (and exponential) functions need special discrete approximations.

Semilinear operator F_2**.** The MME, corresponding to the evolution PDE

$$\boxed{u_t = u_{xx} + (u_x)^2 + u^2 \quad \text{on } W_2 = \mathcal{L}\{1, \cos x\},} \tag{9.123}$$

has the form

$$U_i' - [D_h U]_i X_i' = F_{2h}[U] \equiv U_{XX} + (U_X)^2 + U_i^2 \quad \text{on } \Omega(U), \tag{9.124}$$

where, by (9.114),

$$[D_h U]_i = \frac{U_{i+1} - U_{i-1}}{\cos(X_{i+1}) - \cos(X_{i-1})}\, (-\sin(X_i)). \tag{9.125}$$

Then $[D_h \cos X]_i = -\sin(X_i)$ by (9.113), and we choose $U_X = D_h U$. In order to keep invariant properties of the second derivative, we take the approximation

$$U_{XX} = \overline{D}_h D_h U, \quad \text{with } [\overline{D}_h U]_i = \frac{U_{i+1} - U_{i-1}}{\sin(X_{i+1}) - \sin(X_{i-1})}\, \cos(X_i). \tag{9.126}$$

Therefore, $[\overline{D}_h D_h \cos X]_i = -\cos(X_i)$. Finally, the invariant MMO (9.124) preserves the continuous evolution on W_2. For the solutions

$$U = C_1(t) + C_2(t) \cos X \in W_2, \tag{9.127}$$

the DS is the same as for PDE (9.123),

$$\begin{cases} C_1' = C_1^2 + C_2^2, \\ C_2' = -C_2 + 2C_1C_2. \end{cases}$$

The DS is not explicitly solved, but is easily studied on the phase-plane. The equation (9.123) describes regional blow-up, where the blow-up sets of bell-shaped solutions have measure 2π; see [245, Ch. 9]. On any uniform MM,

$$[D_h \sin X]_i = \cos X_i \quad \text{and} \quad [D_h \cos X]_i = -\sin X_i,$$

so that the operator F_{2h} with $U_X = D_h U$ and $U_{XX} = D_h^2 U$ admits the extended subspace $W_3 = \mathcal{L}\{1, \cos x, \sin x\}$.

Quasilinear operators F$_4$ and F$_5$. As the next example, consider the equation (9.90), (9.99) in the case where F_4 admits $W_2 = \mathcal{L}\{1, \cos x\}$, so that

$$\lambda = \tfrac{\sigma}{\sqrt{\sigma+1}} = 1, \quad \text{i.e.,} \ \sigma = \sigma_\pm = \tfrac{1}{2}(1 \pm \sqrt{5})$$

(note that $-\sigma_-$ is the *Golden Mean*). The MME on $\Omega(U)$ is

$$U_i' - [D_h U]_i X_i' = F_{4h}[U] \equiv U_i \overline{D}_h D_h U + \tfrac{1}{\sigma}(D_h U)^2 + \sigma U_i^2,$$

where the derivatives are given in (9.125) and (9.126). This MME exactly reproduces the continuous solutions (9.127) and the corresponding DS is

$$\begin{cases} C_1' = \sigma C_1^2 + \frac{1}{\sigma} C_2^2, \\ C_2' = (2\sigma - 1) C_1 C_2. \end{cases}$$

On uniform MMs, the MMO approximates the corresponding extended subspace $W_3 = \mathcal{L}\{1, \cos x, \sin x\}$. A similar analysis applies to the Kuramoto–Sivashinsky equation, $u_t = F_5[u]$, and to many other PDEs in Section 3.1 on trigonometric subspaces.

9.4.5 *Higher-order evolution equations*

We describe invariant properties of the equation (9.92). Setting

$$D_t^j u = v_j \quad \text{for} \ j = 1, ..., m-1,$$

reduces it to the system of PDEs

$$\begin{cases} u_t = v_1, \\ (v_j)_t = v_{j+1}, \ \ j = 1, ..., m-2, \\ (v_{m-1})_t = F\big(u, v_1, ..., v_{m-1}, u_x, ..., D_x^k u\big). \end{cases}$$

Introducing the variable $W = (u, v_1, ..., v_{m-1})^T \in I\!R^m$ yields the DS

$$W_t = \tilde{F}[W], \quad t > 0.$$

The corresponding MME for W takes the standard form in $I\!R^m$ similar to (9.112),

$$W_i' - [D_h W]_i X_i' = \tilde{F}_h[W].$$

By checking the invariant property of the gradient operator, we see that Theorem 9.32 is true, and the gradient approximation (9.114) preserves invariance of W_2.

Example 9.33 (**Boussinesq-type PDE**) Consider the PDE with operator F_3,

$$u_{tt} = u u_{xx} + \tfrac{1}{\sigma}(u_x)^2 + \alpha u + \beta.$$

It is equivalent to the system

$$\begin{cases} u_t = v, \\ v_t = F_3[u] + \alpha u + \beta, \end{cases}$$

and the corresponding MME is

$$\begin{cases} U_i' - [D_h U]_i X_i' = V_i, \\ V_i' - [D_h V]_i X_i' = F_{3h}[U] + \alpha U_i + \beta. \end{cases}$$

We define the gradient operator by (9.118), and then, by Theorem 9.32, the MMO admits the subspace $W_2 = \mathcal{L}\{1, x^2\}$.

9.4.6 *Positivity and comparison*

In order to use solutions on invariant subspaces, we first need to establish the Maximum Principle aspects of the MMMs for the second-order parabolic PDEs. Let us begin with the positivity property.

Weak Maximum Principle. We return to the MMO (9.112) and consider second-order equations of parabolic type, assuming a natural three-point approximation of the nonlinear operator F, so

$$[F_h[U]]_i = F_h(U_{i-1}, U_i, U_{i+1}, i, X).$$

Theorem 9.34 *Let $U(t) \ge 0$ on $\partial\Omega(U)$. Then the inequality*

$$\big\{ [D_h U]_i X_i' + F_h[U] \big\}\big|_{U_i=0} \ge 0 \tag{9.128}$$

for all $i \in I_m = \{2, ..., K-1\}$ and $U_j \ge 0$ for $i \ne j$, $t > 0$, is necessary and sufficient for the positivity of $U(t)$ with arbitrary nonnegative data on $\partial\Omega(U)$.

Proof. It follows from (9.112) that, at any instant, when $U_i = 0$ at an internal point of $\Omega(U)$, we have to have $U_i' \ge 0$, which gives the desired result. □

This applies to other homogeneous boundary conditions, including Neumann's,

$$U_X = 0 \quad \text{on the lateral boundary of } \Omega(U).$$

The criterion in the theorem can easily be checked for invariant or standard MMOs.

Example 9.35 **(Semilinear parabolic PDEs)** Consider

$$u_t = u_{xx} + g(x, t, u, u_x),$$

where g satisfies $g(x, t, 0, u_x) \equiv 0$. We take the standard MME on the MM $\Omega(U)$,

$$U_i' - [D_h U]_i X_i' = U_{XX} + g(X_i, t, U_i, U_X).$$

The difference operators are

$$U_X \equiv [D_h U]_i = \tfrac{U_{i+1} - U_{i-1}}{h_i + h_{i-1}} \quad \text{and} \quad U_{XX} = \tfrac{2}{h_i + h_{i-1}} \big(\tfrac{U_{i+1} - U_i}{h_i} - \tfrac{U_i - U_{i-1}}{h_{i-1}} \big),$$

where $h_i = X_{i+1} - X_i > 0$. Then the condition (9.128) reads

$$\tfrac{U_{i+1} - U_{i-1}}{h_i + h_{i-1}} X_i' + \tfrac{2}{h_i + h_{i-1}} \big(\tfrac{U_{i+1}}{h_i} + \tfrac{U_{i-1}}{h_{i-1}} \big) \ge 0, \quad \text{or}$$

$$U_{i+1}\big(X_i' + \tfrac{2}{h_i}\big) + U_{i-1}\big(\tfrac{2}{h_{i-1}} - X_i'\big) \ge 0$$

for any $U_{i\pm1} \geq 0$. This gives the *criterion* of the (internal) positivity of the solution

$$-\tfrac{2}{h_i} \leq X_i' \leq \tfrac{2}{h_{i-1}} \quad \text{for } i \in I_m, \; t > 0.$$

Both inequalities are valid if

$$|X_i'(t)| \leq 2\min\left\{\tfrac{1}{h_i},\, \tfrac{1}{h_{i-1}}\right\} \quad \text{for } i \in I_m, \; t > 0, \tag{9.129}$$

i.e., if the MM does not move very fast. Thus any MM moving sufficiently slow preserves positivity of the solution just as the stationary mesh does. This is a natural continuity property of discrete parabolic operators that describe a transitional behavior from fixed meshes to the MMs.

Comparison principle. This is a consequence of the positivity result.

Theorem 9.36 *Let U and W be two solutions of the MME with $\Omega(U) = \Omega(W)$. Let $U \geq W$ on $\partial\Omega$. The inequality*

$$\left\{ [D_h(U - W)]_i X_i' + F_h[U] - F_h[W] \right\}\Big|_{U_i=W_i} \geq 0 \tag{9.130}$$

for all $i \in I_m$ and $U_j \geq W_j$ for $i \neq j$, $t > 0$, is necessary and sufficient for the comparison $U \geq W$ in Ω for arbitrary ordered data on $\partial\Omega$.

Proof is straightforward, and the result is equivalent to the positivity of the difference $U - W \geq 0$ in Ω satisfying a linearized MME. Since the proof is essentially of the interior nature, comparison is true for the homogeneous Neumann conditions or monotone nonlinear Neumann conditions preserving ordering of solutions on the lateral boundary.

Example 9.37 Consider the semilinear heat equation

$$u_t = u_{xx} + g(x, t, u).$$

Taking the standard MME as given in Example 9.35, one obtains that (9.129) guarantees comparison of solutions of this MME on any such coinciding MMs.

On asymptotic properties of MMEs. The standard comparison of different solutions U and W assumes that the corresponding MMs, $\Omega(U)$ and $\Omega(W)$, coincide. Since the MMs depend on the solutions, this makes the comparison a delicate matter. Using invariant MMOs makes such a comparison not only possible, but also is an effective way to study asymptotic properties of discrete equations. By invariant MMOs, we actually compare general solutions with exact solutions taken on the same MMs. The main conclusion directly follows from the invariance Theorem 9.32 and the comparison Theorem 9.36.

Theorem 9.38 *Let the hypotheses of Theorem 9.32 hold. Fix a solution U on $\Omega(U)$ and an exact invariant solution $W(t) \in W_{2h}$ for $t \geq 0$ such that*

$$U \geq W \;\; (U \leq W) \quad on \;\; \partial\Omega(U).$$

Assume that the hypothesis (9.130) is true for solutions that are defined on the same MM $\Omega = \Omega(U)$. Then,

$$U \geq W \;\; (U \leq W) \quad in \;\; \Omega(U).$$

Example 9.39 (Comparison and asymptotic behavior) Consider the semilinear equation (9.40). Let U be a discrete solution of the MME on an MM $\Omega(U)$,

$$U_i' - [D_h U]_i X_i' = U_{XX} + (U_X)^2 + U_i,$$

with the gradient approximation D_h given in (9.118), and invariant approximations (9.119) of the derivatives U_X and U_{XX}. The quadratic operator F_{1h} on the right-hand side admits the same subspace as in (9.96), with the basis function $f(X) = X^2$. Let $W(t) \in W_2$ be an exact solution. By Theorem 9.32, this solution satisfies the same MME on the MM $\Omega(U)$. The comparison condition then reduces to

$$\begin{aligned}(U_{i+1} - W_{i+1})\big[\Delta_i X_i' + \tfrac{1}{H_i h_i} + \tfrac{1}{2H_i}(U_X + W_X)\big]\\ + (U_{i-1} - W_{i-1})\big[-\Delta_i X_i' + \tfrac{1}{H_i h_{i-1}} - \tfrac{1}{2H_i}(U_X + W_X)\big] \geq 0\end{aligned}$$

for all $U_{i\pm1} \geq W_{i\pm1}$, where

$$\Delta_i = \tfrac{f'(X_i)}{f(X_{i+1}) - f(X_{i-1})},$$

$h_i = X_{i+1} - X_i$, and $H_i = \frac{1}{2}(h_i + h_{i-1})$. This gives the following condition of comparison:

$$|H_i \Delta_i X_i'(t)| \leq \min\big\{\tfrac{1}{h_i}, \tfrac{1}{h_{i-1}}\big\} - \tfrac{1}{2}\sup_{\Omega_t}(|U_X + W_X|),$$

i.e., if, for sufficiently regular solutions with bounded derivatives U_X and W_X, the MM does not move faster than $O\big(\frac{1}{h}\big)$ (similar to the positivity (9.129)).

The above comparison is true for any solution on W_2. Choose the solution

$$W(x,t) = C_1(t) + C_2(t)x^2 \in W_2,$$

where the coefficients $\{C_1, C_2\}$ satisfy the DS (9.42). Assume that $U \geq W$ on the lateral boundary of $\Omega(U)$, so that

$$U(X,0) \geq W(X,0) \equiv C_1(0) + C_2(0)X^2 \quad \text{in } \Omega_0(U).$$

Then, by comparison,

$$U(X,t) \geq W(X,t) \equiv C_1(t) + C_2(t)X^2 \quad \text{in } \Omega(U),$$

provided that the MM satisfies the above comparison condition. A similar estimate from below is also proved. Finally, solving the DS (9.42), we obtain the exact asymptotic behavior as $t \to \infty$ of a wide class of solutions of this invariant MME. In this case, as $t \to \infty$, the discrete solutions converge to the principle separable solution, as in the easy continuous case of the parabolic PDE, [509, p. 93].

More examples on the comparison with solutions on invariant subspaces and asymptotic behavior can be constructed for parabolic PDEs and invariant approximations of other quadratic operators given above.

9.5 Applications to anharmonic lattices

In this section, we present examples of exact solutions of nonlinear lattices. The dynamics of such discrete systems have been extensively studied since the 1950s, starting with the celebrated work of Fermi–Pasta–Ulam [183]. In fact, the soliton concept appeared for the first time in the context of nonlinear lattices, before becoming widely used in many areas of physics and mechanics; see a history review in Remoissenet [487]. Another classical example is the *Toda lattice*

$$u_n'' = \mathrm{e}^{u_{n-1}-u_n} - \mathrm{e}^{u_n-u_{n+1}}.$$

It was introduced by M. Toda in 1967 and had an earlier geometric origin detected by Darboux (1887) via iterations of, in modern terminology, Laplace–Darboux transformations, [141]. The Toda lattice was used to describe anharmonic oscillations in a 1D crystal lattice. This is known to be an integrable Hamiltonian system admitting a Lax pair, N-soliton solutions given by Hirota's expression, *etc.*; see Toda [556]. The related discrete sine-Gordon (FK) model was discussed in Example 5.5.

Lattice dynamical systems occur in a variety of applications, where the spatial structure has a discrete character. These apply in chemical reaction theory, cellular neural networks, image processing, and pattern recognition, in different areas of material science, in electrical engineering in laser physics, etc. Discrete breathers, i.e., time-periodic patterns of lattice dynamics with spatially localized oscillations, are supported by many nonlinear lattices, and occur in several physical models, such as dynamical properties of crystals, Josephson junction arrays, DNA denaturation, vibration dynamics of ionic crystals, etc. See [304] for further references and a survey on mathematical results in this area of lattice theory.

Example 9.40 (Parabolic quadratic equation) Consider a quadratic operator, which was studied in Example 5.6 in the differential case,

$$F_2[u] = 2|u_x|u_{xx} + u_n^2. \tag{9.131}$$

In computations, we use the following symmetric expressions for discrete derivatives (later on, we often set $h = 1$):

$$\begin{aligned} u_x = \tfrac{1}{2h}\big(u_{n+1} - u_{n-1}\big), \quad u_{xx} = (u_x)_x \equiv \tfrac{1}{h^2}\big(u_{n-1} - 2u_n + u_{n+1}\big), \\ u_{xxx} = (u_{xx})_x \equiv \tfrac{1}{2h^2}\big(-u_{n-2} + 2u_{n-1} - 2u_{n+1} + u_{n+2}\big), \\ u_{xxxx} = (u_{xxx})_x, \quad u_{xxxxx} = (u_{xxxx})_x, \quad \text{etc.} \end{aligned} \tag{9.132}$$

The results can easily be recalculated for other derivative approximations. Some approaches are extended to uniform lattices in $I\!R^N$, e.g., for the operator

$$F_2[u] = 2|\nabla u|\Delta u + u_n^2,$$

with natural definitions of the discrete gradient and the Laplacian operators.

The discrete operator F_2 in (9.131) is associated with the 2D subspace

$$W_2 = \mathcal{L}\{1, f(x)\}, \quad \text{with } f_n = f(n), \tag{9.133}$$

where f solves the equation

$$F_2[f] - f_n = 0. \tag{9.134}$$

Consider the corresponding discrete parabolic equation

$$u_n' = F_2[u]. \tag{9.135}$$

Plugging the expression

$$u_n(t) = C_1(t) + C_2(t) f_n \in W_2, \quad \text{with } C_2 \ge 0 \tag{9.136}$$

(in fact, the cone, $K_+ = \{C_{1,2} \ge 0\} \subset W_2$, is invariant under F_2) into (9.135) yields the same DS as in the differential case,

$$\begin{cases} C_1' = C_1^2, \\ C_2' = C_2^2 + 2C_1C_2. \end{cases}$$

Setting $C_1(t) \equiv 0$ yields $C_2' = C_2^2$ and the separate variable solution

$$u_n(t) = \tfrac{1}{T-t}\, f_n,$$

where $T > 0$ is the blow-up time. Such separation of variables in finite-difference equations is well known in blow-up theory; see [509, p. 491] and references therein. Choosing a nontrivial $C_1(t)$ from the first ODE, $C_1(t) = -\frac{1}{t}$, we obtain more interesting blow-up solutions on such a uniform grid,

$$u_n(t) = -\tfrac{1}{t} + \tfrac{T}{(T-t)t}\, f_n.$$

On infinite propagation in the lattice. It is easy to see that implicit schemes such as (9.135), with operator (9.131), cannot describe processes with finite propagation. For the stationary ("elliptic") equation (9.134), this means that $f_n = f(n)$ is not compactly supported, unlike in the differential case in Example 5.6. The asymptotic decay of the monotone fast decreasing solution $\{f_n > 0\}$ for large n is easy to obtain from the following asymptotic representation of (9.134):

$$f_n = F_2[f] = f_{n-1}^2(1 + O(f_n)) \quad \text{for } n \gg 1,$$

where we have used sharp estimates $f_x = \frac{1}{2}(f_{n+1} - f_{n-1}) \approx -\frac{1}{2} f_{n-1}$ and $f_{xx} = f_{n-1} - 2f_n + f_{n+1} \approx f_{n-1}$. Hence, for some constants $c_* \in (0,1)$ and $C > 0$,

$$f_n \sim C(c_*)^{2^n} \quad \text{as } n \to \infty.$$

We thus observe a fast double-exponential decay of the function $f(n)$.

Concerning the discrete parabolic equation with a gradient dependent diffusion (we omit the quadratic term u_n^2 that is negligible on such decay asymptotics)

$$u_n' = |u_{n+1} - u_{n-1}|(u_{n-1} + u_{n+1} - 2u_n), \tag{9.137}$$

assuming that the initial data $\{u_{0n}\}$ are compactly supported, we find that, for small $t > 0$ as $n \to \infty$, the sharp asymptotic form of the equation is given by

$$u_n' = u_{n-1}^2(1 + O(u_n)). \tag{9.138}$$

Let $n_0 > 0$ be the interface point of u_{0n}, so that $u_{0n_0} = 0$ and $c_0 = u_{0n_0-1} > 0$. In this case, (9.137) at $n = n_0$ yields that

$$u_{n_0}(t) = c_0 t(1 + o(1)) \quad \text{as } t \to 0. \tag{9.139}$$

Further iterations of the equation (9.138) leads to the following asymptotic behavior as $k = n - n_0 \to \infty$:

$$u_{n_0+k}(t) = \left[6 \prod_{i=0}^{k} (2^{i+2} - 1)^{-2^{k-i}}\right] c_0^{2^k} t^{2^k+1} + \dots. \tag{9.140}$$

Therefore, the formula

$$u_{n_0+k}(t) \sim \mathrm{e}^{2^k \ln t} \to 0$$

gives the sharp asymptotic rate of convergence of the solution as $t \to 0$ to compactly supported initial data u_{0n}. This has an easy relation to asymptotics (1.61) of the delay ODE (1.60).

Example 9.41 (**Oscillatory behavior near "interfaces"**) Consider the higher-order *parabolic lattice* (9.135) with the operator

$$F_2[u] = -(|u_{xx}|u_{xx})_{xx} + u^2,$$

which uses the central differences as in (9.132). This model is associated with the differential analogy in Example 5.7 (for $n = 1$), where we have detected a specific oscillatory behavior of weak solutions near interfaces. Apparently, there occur the same subspace and a DS on W_2, where f solves (9.134).

Studying the decay behavior for $n \gg 1$, we have, instead of (9.138), the following asymptotic equation for small solutions:

$$u_n' = -|u_{n-2}|u_{n-2}. \tag{9.141}$$

Posing, similar to (9.139), the positive conditions at two "boundary" points

$$u_{n_0}(t) = c_0 t, \quad u_{n_0+1}(t) = d_0 t \quad \text{as } t \to 0 \quad (0 < d_0 < c_0),$$

we obtain from (9.141) the following asymptotic behavior as $t \to 0$:

$$\begin{gathered} u_{n_0+2}(t) = -\tfrac{c_0^2}{3}\, t^3 < 0, \;\; u_{n_0+3}(t) = -\tfrac{d_0^2}{3}\, t^3 < 0, \\ u_{n_0+4}(t) = \left(\tfrac{c_0^2}{3}\right)^2 \tfrac{1}{7}\, t^7 > 0, \;\; u_{n_0+5}(t) = \left(\tfrac{d_0^2}{3}\right)^2 \tfrac{1}{7}\, t^7 > 0, \\ u_{n_0+6}(t) = -\left(\tfrac{c_0^2}{3}\right)^{2\cdot 2} \left(\tfrac{1}{7}\right)^2 \tfrac{1}{15}\, t^{15} < 0, \;\; u_{n_0+7}(t) = -\left(\tfrac{d_0^2}{3}\right)^{2\cdot 2} \left(\tfrac{1}{7}\right)^2 \tfrac{1}{15}\, t^{15} < 0, \end{gathered}$$

etc. This shows a clear oscillatory structure of solutions, and, writing the general formula as in (9.140) (with $2^k \mapsto 2^{k/2}$), we will observe a rescaled 2-periodic behavior as $n \to \infty$. This is a discrete analogy of solutions of changing sign in the differential counterpart in Example 5.7.

Example 9.42 (**Quadratic hyperbolic equation**) Consider the *hyperbolic lattice* with the second-order operator (9.131),

$$u_n'' = F_2[u] - u_n.$$

Using the subspace (9.133), we obtain solutions (9.136) with the DS

$$\begin{cases} C_1'' = C_1^2 - C_1, \\ C_2'' = C_2^2 + 2C_1C_2 - C_2 \end{cases}$$

($C_2 \geq 0$). The analysis of infinite propagation with compactly supported initial data $\{u_{0n},\, u_{1n}\}$ is performed in a similar fashion, where the governing equation is

$$u_n'' = u_{n-1}^2(1 + O(u_n)) \quad \text{for } n \gg 1,$$

that admits asymptotics as in (9.140).

Example 9.43 (**Cubic hyperbolic equation**) The next lattice is associated with the PDE (5.52) possessing exact solutions on the invariant subspace

$$W_3 = \mathcal{L}\{1, \cos(\lambda x), \sin(\lambda x)\}, \quad \text{with } \lambda > 0.$$

We now take the corresponding lattice in the form of

$$u_n'' = F_3[u] \equiv u_n^2(u_{n-1} + u_{n+1} - 2u_n) - \alpha\, u_n\big(\tfrac{u_{n+1}-u_{n-1}}{2}\big)^2 + \beta u_n^3. \tag{9.142}$$

Proposition 9.44 *The discrete operator F_3 in* (9.142) *preserves W_3 if*

$$\alpha = \tfrac{1}{1+\cos\lambda} \ \text{ and } \ \beta = 1 - \cos\lambda \quad (\lambda \neq \pi(2k+1)).$$

Therefore, the dynamics on W_2 is as follows:

$$u_n(t) = C_1(t) + C_2(t)\cos(\lambda n) + C_3(t)\sin(\lambda n),$$
$$C_i'' = (1 - \cos\lambda)C_i(C_1^2 - C_2^2 - C_3^2), \quad i = 1, 2, 3.$$

For convenience, we present the results of calculations for the general discrete cubic operator in Proposition 1.17. Here $\lambda = 1$.

Proposition 9.45 *The discrete operator* (1.57) *admits $W_3 = \mathcal{L}\{1, \cos x, \sin x\}$ if*

$$b_1 = \tfrac{1}{2}\big(b_5 Q_1 + b_7 R_1^2 Q_1 - b_{10} Q_1^3\big), \ b_2 = \tfrac{1}{2}\, b_6 Q_1,$$
$$b_3 = \tfrac{1}{2R_1^2}\big(-b_5 Q_1 + 3b_7 R_1^2 Q_1 + 2b_8 Q_1^2 - 3b_{10} Q_1^3\big),$$
$$b_4 = \tfrac{1}{R_1^2}\big(b_7 Q_1^2 - \tfrac{1}{2}\, b_6 Q_1\big),$$

where $R_1 = \frac{1}{h}\sinh h$ (= $\sin 1$ for $h = 1$) and $Q_1 = \frac{4}{h^2}\sin^2(\frac{h}{2})$ (= $4\sin^2(\frac{1}{2})$).

Example 9.46 (**5th-order PDE from compacton theory**) Consider the nonlinear dispersion lattice from compacton theory in Section 4.3 (equation (4.97), $\alpha = 1$)

$$u_t = F_5[u] + \delta u + \varepsilon \equiv (u^2)_{xxxxx} + \beta(u^2)_{xxx} + \gamma\,(u^2)_x + \delta u + \varepsilon. \tag{9.143}$$

We will first study it on the same subspace, W_3, as in Proposition 9.45. For convenience of passing to the limit, we introduce the mesh parameter $h > 0$ in the definition of discrete operators in (9.132). Then the following results hold:

(i) W_3 is invariant if

$$Q_2^2 - Q_2\beta + \gamma = 0, \quad \text{where now } Q_\lambda = \tfrac{4}{h^2}\sin^2(\tfrac{\lambda h}{2}). \tag{9.144}$$

Therefore, $Q_2 \to 4$ as $h \to 0$, so, in the limit, the invariance condition (9.144) coincides with the differential one (4.12) with $\alpha = 1$.

(ii) The DS for the lattice (9.143) on W_3 is (4.98) with the coefficients $\mu = 2R_1(Q_2 - Q_1)(\beta - Q_2 - Q_1)$ (we again use the step h), where $R_\lambda = \frac{1}{h}\sin(\lambda h)$. Hence, $\mu \to 6(\beta - 5)$ as $h \to 0$ as in (4.98). The exact solutions are the same as in (4.99) and (4.100), so these are not presented here.

Concerning 5D subspaces, let us mention that, as in Example 4.13, the discrete equation (9.143) possesses solutions

$$u_n(t) = C_1(t) + C_2(t)\cos n + C_3(t)\sin n + C_4(t)\cos 2n + C_5(t)\sin 2n,$$

provided that $\gamma = Q_3 Q_4$ and $\beta = Q_3 + Q_4$ (here $Q_3 \to 9$ and $Q_4 \to 16$ as $h \to 0$), and the DS for $\delta = \varepsilon = 0$ is

$$\begin{cases} C_1' = 0, \\ C_2' = \mu(C_2C_5 - C_3C_4 + 2C_1C_3), \\ C_3' = -\mu(C_2C_4 + C_3C_5 + 2C_1C_2), \\ C_4' = 2\nu(2C_1C_5 + C_2C_3), \\ C_5' = \nu\big(C_3^2 - C_2^2 - 4C_1C_4\big). \end{cases}$$

The coefficients μ and ν satisfy $\mu = (Q_1 - Q_3)(Q_1 - Q_4) \to 120$ and $\nu = (Q_2 - Q_3)(Q_2 - Q_4) \to 60$ as $h \to 0$, so in the limit $h = 0$, we obtain the DS (4.104) for the continuous PDE. The dynamical evolution properties of such solutions on W_5 are quite similar in both continuous and discrete cases, so one can observe the discrete TW (4.105).

Example 9.47 (4th-order PDE from thin film theory) In this last example, we consider the lattice counterpart of the thin film PDE in Example 3.17,

$$u_t = -uu_{xxxx} - u_x u_{xxx} + \beta u^2.$$

Then $\mathcal{L}\{1, \cos x, \sin x\}$ is invariant if $\beta = Q_1(Q_1 + R_1^2) \to 2$ as $h \to 0$, and for solutions

$$u_n(t) = C_1(t) + C_2(t)\cos n + C_3(t)\sin n,$$

we obtain the DS (cf. (3.117) for $h = 0$)

$$\begin{cases} C_1' = Q_1\big[(Q_1 + R_1^2)C_1^2 + R_1^2 C_2^2 + R_1^2 C_3^2\big], \\ C_2' = \mu C_1 C_2, \quad C_3' = \mu C_1 C_3, \end{cases} \tag{9.145}$$

where $\mu = Q_1(Q_1^2 + 2R_1^2) \to 3$. For $C_1 = C_2$ and $C_3 = 0$, we have the ODE $C_1' = \mu C_1^2$ and the following localized blow-up pattern on this lattice:

$$u_n(t) = \tfrac{2}{\mu(T-t)} \cos^2\big(\tfrac{n}{2}\big).$$

Remark 9.48 1. (Other approximations) Using another discrete derivative u_x, instead of (9.132), slightly changes the coefficients of the resulting DSs according to the easy rule

$$u_x = \tfrac{1}{h}\big(u_{n+1} - u_n\big) \quad \Longrightarrow \quad (\sin \lambda x)_x = R_\lambda \cos x - \tfrac{h}{2}\, Q_\lambda \sin x.$$

2. (Exponential functions) For subspaces composed of exponential or hyperbolic functions, the calculations yield

$$(\mathrm{e}^{\lambda x})_x = Q_\lambda \mathrm{e}^{\lambda x}, \quad \text{where } Q_\lambda = \tfrac{1}{2h}\big(\mathrm{e}^{\lambda h} - \mathrm{e}^{-\lambda h}\big).$$

3. (Polynomial subspaces) These were especially important for a number of PDE examples in the previous chapters. All the computations are easily translated to the discrete case using the rule

$$(x^n)_x = \tfrac{1}{2h}\big[(x+h)^n - (x-h)^n\big] = \tfrac{1}{2}\textstyle\sum_{k=1}^n C_n^k\big[1 + (-1)^k\big]x^{n-k}h^{k-1}.$$

Of course, this affects the coefficients of the DSs.

Remarks and comments on the literature

There are many classical texts on difference equations; see references in [367, p. 34, p. 72]. Concerning applications of Lie group theory to difference equations, see lists of references in [151, 152, 490].

§ 9.1, 9.2. The main results are taken from [242].

§ 9.3. Some of the results were published in [221]. There exists a group-invariant finite-difference scheme for (9.40), which admits explicit solutions, and then the corresponding space grid depends on the time variable (and on the solution itself); see [23]. Some solutions in this section can be justified by generalized conditional symmetries, [119].

It has been known for a long time that, for a number of fully integrable PDEs, there exist techniques of their discretization without destroying integrability, where the discrete equations admit N-soliton solutions that are obtained by similar methods over linear spaces of exponential functions. The idea of such discretization techniques was proposed by Hirota in 1973; see [286], where the extra references can be found. Algebro-geometric and elliptic solutions of the fully discretized KP and 2D Toda equations were constructed in [357] by using Baker functions. One can expect that, for non-integrable PDEs possessing one and two-soliton solutions similar to those in Section 8.5 and in [100, 329, 595], there exist discretizations preserving such exact solutions constructed by similar manipulations with exponential functions.

§ 9.4. The main results are available in [222]. The basics of MMM theory are given in [88, 293, 292, 99, 98], some applications to blow-up solutions of higher-order parabolic PDEs can be found in [87].

§ 9.5. Discrete "almost compact" breathers of lattices associated with cubic PDEs were studied in [500]. More recent references on exact TW, and other solutions of lattice differential equations can be found in [291]. References to main results and applications of lattice dynamical systems in chemical reaction theory, image processing, pattern recognition, material sciences, and biology are given in [124]. The most well-studied exact solutions of lattice parabolic PDEs are TWs; see a more recent survey [408] and also [404], where non-local models were treated. Concerning integrable cases, *complexiton-type* exact solutions of the Toda lattice were constructed in [406]. Notice a relation of *embedded solitons* in dynamical lattices [258] to typical invariant subspaces and sets.

For higher-order linear and nonlinear PDEs, using various approximating discrete dynamical systems via *Saint-Venant's principle* (originally formulated by A.-J.-C. Barré de Saint-Venant in linear elasticity theory in 1855 [508]) is an effective tool for existence, uniqueness, and asymptotic analysis. This discrete approach leads to concepts of energy solutions, and can be applied in general singular nonlinear cases. We refer to the fundamental Oleinik–Radkevich paper [443]; see also a survey on more recent results and applications in [240].

Open problems

- The general forward **Problem I** of invariant subspaces (and several other related aspects): given a nonlinear finite-order difference operator, determine all the linear subspaces which it admits; see Section 9.2.

References

[1] M.J. Ablowitz, S. Chakravarty, and R.G. Halburg, Integrable systems and reductions of the self-dual Yang-Mills equations, *J. Math. Phys.*, **44** (2003), 3147–3173.

[2] M.J. Ablowitz and P.A. Clarkson, *Solitons, Nonlinear Evolution Equations and Inverse Scattering*, Cambridge Univ. Press, Cambridge, 1991.

[3] M.J. Ablowitz and P.A. Clarkson, Solitons and symmetries, *J. Engrg. Math.*, **36** (1999), 1–9.

[4] M.J. Ablowitz and H. Segur, *Solitons and the Inverse Scattering Transform*, SIAM, Philadelphia, 1981.

[5] G.B. Airy, Tides and waves, *Encyclopedia Metropolitana*, Vol. **5**, London, 1845, pp. 241–396.

[6] M.S. Alber, R. Camassa, Yu.N. Fedorov, D.D. Holm, and J.E. Marsden, On billiard solutions of nonlinear PDEs, *Phys. Lett.*, **264A** (1999), 171–178.

[7] E. Alfinito, V. Grassi, R.A. Leo, G. Profilo, and G. Soliani, Equations of the reaction–diffusion type with a loop algebra structure, *Inverse Problems,* **14** (1998), 1387–1401.

[8] S. Alinhac, *Blow-up for Nonlinear Hyperbolic Equations,* Birkhäuser Boston, Inc., Boston, MA, 1995.

[9] H. Amann, *Linear and Quasilinear Parabolic Problems*, Vol. I. *Abstract Linear Theory*, Birkhäuser Boston, Inc., Boston, MA, 1995.

[10] W.F. Ames, R.L. Anderson, V.A. Dorodnitsyn, E.V. Ferapontov, R.K. Gazizov, N.H. Ibragimov, and S.R. Svirshchevskii, *CRC Handbook of Lie Group Analysis of Differential Equations*, Vol. **1**: *Symmetries, Exact Solutions and Conservation Laws*, CRC Press, Boca Raton, FL, 1994.

[11] I.M. Anderson, N. Kamran, and P.J. Olver, Internal, external and generalizaed symmetries, *Adv. Math.,* **100** (1993), 53–100.

[12] R.L. Anderson and N.H. Ibragimov, *Lie -Bäcklund Transformations in Applications*, SIAM Studies in Appl. Math., 1, SIAM, Philadelphia, Pa., 1979.

[13] V.K. Andreev, O.V. Kaptsov, V.V. Pukhnachov, and A.A. Rodionov, *Applications of Group-Theoretical Methods in Hydrodynamics,* Kluwer Acad. Publ., Dordrecht, 1998.

[14] L. Ansini and L. Giacomelli, Doubly nonlinear thin-film equations in one dimension, *Arch. Rational Mech. Anal.,* **173** (2004), 89–131.

[15] P. Appel, Mémoire sur déblais et les remblais des systèmes continus ou discontinus, Mémoires présentées par divers savants à l'Académie des Sciences de l'Institut de France, I. N. **29**, Paris, 1887, pp. 1–208.

[16] V.A. Arkad'ev, A.K. Pogrebkov, and M.K. Polivanov, Singular solutions of the KdV equation and the method of the inverse problem, *Zap. Nauchn. Sem. LOMI,* **133** (1984), 17–37 (in Russian).

[17] D.G. Aronson and P. Bénilan, Régularité des solutions de l'équation des milieux poreux dan $\mathbf{R}^N$, *C. R. Acad. Sci. Paris Sér. A-B,* **288** (1979), A103–A105.

[18] D.J. Arrigo, Group properties of $u_{xx} - u_y^m u_{yy} = f(u)$, *Internat. J. Non-Linear Mech.,* **26** (1991), 619–629.

[19] C. Athorne, J.C. Eilbeck, and V.Z. Enolskii, Identities for the classical genus two $\wp$ function, *J. Geom. Phys.*, **48** (2003), 354–368.

[20] C. Athorne, J.C. Eilbeck, and V.Z. Enolskii, A SL(2) covariant theory of genus 2 hyperelliptic functions, *Math. Proc. Cambidge Philos. Soc.*, **136** (2004), 269–286.

[21] Yu.Yu. Bagderina and A.P. Chupakhin, Invariant and partially invariant solutions of the Green–Naghdi equations, *J. Appl. Mech. Techn. Phys.*, **46** (2005), 791–799.

[22] H.F. Baker, On a system of differential equations leading to periodic functions, *Acta Math.*, **27** (1903), 135–156.

[23] M.I. Bakirova and V.A. Dorodnitsyn, An invariant difference model for the equation $u_t = u_{xx} + \delta u \ln u$, *Differ. Equat.*, **30** (1994), 1565–1570.

[24] J. Bao, J. Chen, B. Guan, and M. Ji, Liouville property and regularity of a Hessian quotient equation, *Amer. J. Math.*, **125** (2003), 301–316.

[25] G.I. Barenblatt, *Scaling, Self-Similarity, and Intermediate Asymptotics*, With a foreword by Ya.B. Zeldovich, Cambridge Univ. Press, Cambridge, 1996.

[26] G.I. Barenblatt, Self-similar turbulence propagation from an instantaneous plane source, In: *Nonl. Dynam. Turbulence,* G.I. Barenblatt, G. Iooss and D.D. Joseph, Eds, Pitman, Boston, MA, 1983, pp. 48–60.

[27] G.I. Barenblatt, N.L. Galerkina, and M.V. Luneva, Evolution of a turbulent burst, *Inzhenerno-Fizicheskii Zh.*, **53** (1987), 733–740 (in Russian).

[28] G.I. Barenblatt and Ya.B. Zel'dovich, On dipole-type solutions in problems of nonstationary filtration of gas under polytropic regime, *Prikl. Mat. Meh.*, **21** (1957), 718–720 (in Russian).

[29] G.K. Batchelor, *An Introduction to Fluid Dynamics*, Second Edition, Cambridge Univ. Press, Cambridge, 1999.

[30] R. Beals, D.H. Sattinger, and J. Szmigielski, Acoustic scattering and the extended Korteweg-de Vries hierarchy, *Adv. Math.*, **140** (1998), 190–206.

[31] R. Beals, D.H. Sattinger, and J. Szmigielski, Multipeakons and the classical moment problem, *Adv. Math.*, **154** (2000), 229–257.

[32] R. Beals, D.H. Sattinger, and J. Szmigielski, Continued fractions and integrable systems, *J. Comput. Appl. Math.*, **153** (2003), 47–60.

[33] J. Bebernes and D. Eberly, *Mathematical Problems from Combustion Theory*, Appl. Math. Sci., **83**, Springer-Verlag, New York, 1989.

[34] J. Becker and G. Grün, The thin-film equation: recent advances and some new perspectives, *J. Phys.: Condens. Matter*, **17** (2005), S291–S307.

[35] E.D. Belokolos, A.I. Bobenko, V.Z. Enolskii, A.R. Its, and V.B. Matveev, *Algebro-Geometric Approach to Nonlinear Integrable Equations*, Springer, Berlin, 1994.

[36] E.D. Belokolos and V.Z. Enolskii, Reduction of abelian functions and algebraically integrable systems. I, II, *J. Math. Sci. (New York)*, **106** (2001), 3395–3486; **108** (2002), 295–374.

[37] Ph. Bénilan and M. Crandall, The continuous dependence on φ of solutions of $u_t - \Delta\varphi(u) = 0$, *Indiana Univ. Math. J.*, **30** (1981), 161–177.

[38] T.B. Benjamin, J.L. Bona, and J.J. Mahoney, Model equations for long waves in nonlinear dispersive systems, *Philos. Trans. Roy. Soc. London Ser. A*, **272** (1972), 47–78.

[39] D.J. Benney, Long waves on liquid films, *J. Math. and Phys.*, **45** (1966), 150–155.

[40] D.J. Benney and J.C. Luke, On the interactions of permanent waves of finite amplitude, *J. Math. Phys.*, **43** (1964), 309–313.

[41] M. Berger, *Nonlinearity and Functional Analysis*, Acad. Press, New York, 1977.

[42] R. Berker, Intégration des équations du mouvement d'un fluide visqueux incompressible, *Handbuch der Physik*, **8** (1963), 1–384; Springer, Berlin.

[43] A.S. Berman, Laminar flow in channels with porous walls, *J. Appl. Phys.,* **24** (1953), 1232–1235.

[44] F. Bernis and A. Friedman, Higher order nonlinear degenerate parabolic equations, *J. Differ. Equat.,* **83** (1990), 179–206.

[45] F. Bernis, J. Hulshof, and J.R. King, Dipoles and similarity solutions of the thin film equation in the half-line, *Nonlinearity,* **13** (2000), 413–439.

[46] F. Bernis, J. Hulshof, and F. Quirós, The "linear" limit of thin film flows as an obstacle-type free boundary problem, *SIAM J. Appl. Math.,* **61** (2000), 1062–1079.

[47] F. Bernis and J.B. McLeod, Similarity solutions of a higher order nonlinear diffusion equation, *Nonl. Anal.,* **17** (1991), 1039–1068.

[48] F. Bernis, L.A. Peletier, and S.M. Williams, Source type solutions of a fourth order nonlinear degenerate parabolic equation, *Nonl. Anal.,* **18** (1992), 217–234.

[49] F. Bernis, J. Hulshof, and J.L. Vázquez, A very singular solution for the dual porous medium equation and the asymptotic behaviour of general solutions, *J. reine angew. Math.,* **435** (1993), 1–31.

[50] A.J. Bernoff and A.L. Bertozzi, Singularities in a modified Kuramoto-Sivashinsky equation describing interface motion for phase transition, *Phys. D,* **85** (1995), 375–404.

[51] S. Bernstein, Sur la généralisation du probléme de Dirichlet. Deuxième partie, *Math. Ann.,* **69** (1910), 82–136.

[52] A.L. Bertozzi, Symmetric singularity formation in lubrication-type equations for interface motion, *SIAM J. Appl. Math.,* **56** (1996), 681–714.

[53] A.L. Bertozzi, G. Grün, and T.P. Witelski, Dewetting films: bifurcations and concentrations, *Nonlinearity,* **14** (2001), 1569–1592.

[54] A.L. Bertozzi and M. Pugh, The lubrication approximation for thin viscous films: regularity and long-time behaviour of weak solutions, *Commun. Pure Appl. Math.,* **49** (1996), 85–123.

[55] A.L. Bertozzi and M.C. Pugh, Long-wave instabilities and saturation in thin film equations, *Commun. Pure Appl. Math.,* **51** (1998), 625–661.

[56] M. Bertsch, R. Dal Passo, and R. Kersner, The evolution of turbulent bursts: $b - \varepsilon$ model, *European J. Appl. Math.,* **5** (1994), 537–557.

[57] S. Betelú, A two-dimensional corner solution for a nonlinear diffusion equation, *Appl. Math. Lett.,* **13** (2000), 119–123.

[58] S.I. Betelú and J.A. Diez, A two-dimensional similarity solution for capillary driven flows, *Phys. D,* **126** (1999), 136–140.

[59] S.I. Betelú and J. King, Explicit solutions of a two-dimensional fourth-order nonlinear diffusion equation, *Math. Comp. Modelling,* **37** (2003), 395–403.

[60] R. Beutler, Positon solutions of the sine-Gordon equation, *J. Math. Phys.,* **34** (1993), 3098–3109.

[61] G. Birkhoff, *Hydrodynamics. A Study in Logic, Fact, and Similitude*, Princeton Univ. Press, Princeton, N.J., 1950.

[62] M.S. Birman and M.Z. Solomjak, *Spectral Theory of Self-Adjoint Operators in Hilbert Space*, D. Reidel Publ. Comp., Dordrecht/Tokyo, 1987.

[63] H. Blasius, Grenzchichten in Flüssigkeiten mit kleiner Reibung, *Z. Math. Phys.,* **56** (1908), 1–37.

[64] P.A. Bleher, J.L. Lebowitz, and E.R. Speer, Existence and positivity of solutions of a fourth-order nonlinear PDE describing interface fluctuations, *Commun. Pure Appl. Math.,* **47** (1994), 923–942.

[65] G.W. Bluman and S. Kumei, *Symmetries and Differential Equations*, Springer-Verlag,

New York, 1989.

[66] G.W. Bluman and Z. Yan, Nonclassical potential solutions of partial differential equations, *European J. Appl. Math.*, **16** (2005), 239–261.

[67] S. Bochner, Über Sturm–Liouvillesche Polynomsysteme, *Math. Zeit.*, **29** (1929), 730–736.

[68] L.V. Bogdanov and V.E. Zakharov, The Boussinesq equation revisited, *Phys. D*, **165** (2002), 137–162.

[69] J.L. Bona, Y. Lie, and M.M. Tom, The Cauchy problem and stability of solitary-wave solutions for RLW–KP-type equations, *J. Differ. Equat.*, **185** (2002), 437–482.

[70] J.L. Bona and F.B. Weissler, Similarity solutions of the generalized Korteweg–de Vries equation, *Math. Proc Camb. Phil. Soc.*, **127** (1999), 323–351.

[71] O. Bonnet, Mémoire sur la théorie des surfaces applicables sur une surface donnée, *J. l'École Impériale Polytech.*, **XXV** (1867), 1.

[72] L.A. Bordag and V.B. Matveev, Selfsimilar solutions of the Korteweg-de Vries equation and potentials with a trivial S-matrix, *Theoret. Math. Phys.*, **34** (1978), 272–275.

[73] M. Born and L. Infeld, Foundation of a new field theory, *Proc. Roy. Soc. London A*, **144** (1934), 425–451.

[74] J. Boussinesq, Théorie générale des mouvements qui sont propagés dans un canal rectangulaire horizontal, *C. R. Acad. Sci. Paris*, **73** (1871), 256–260.

[75] J. Boussinesq, Théories des ondes et des remous qui se propagent le long d'un canal rectangulaire horizontal, en communiquant au liquide contenu dans ce canal des vitesses sensiblement pareilles de la surface au fond, *J. Math. Pures Appl.*, **17**, No. 2 (1872), 55–108.

[76] J. Boussinesq, Essai sur la théorie des eaux courantes, *Mémoires présentées par divers savants à l'Académie des Sciences de l'Institut de France*, I. N. **23**, Paris, 1877, pp. 1–680.

[77] J. Boussinesq, Recherches théorique sur l'écoulement des nappes d'eau infiltrées dans le sol et sur le débit des sources, *J. Math. Pures Appl.*, **10**, No. 1 (1904), 5–78.

[78] M. Bowen, J. Hulshof, and J.R. King, Anomaluous exponents and dipole solutions for the thin film equation, *SIAM J. Appl. Math.*, **62** (2001), 149–179.

[79] K.A. Brakke, *The Motion of a Surface by its Mean Curvature*, Princeton Univ. Press, Princeton, N.J., 1978.

[80] Y. Brenner, Hydrodynamic structure of the augmented Born-Infeld equations, *Arch. Rational Mech. Anal.*, **172** (2004), 65–91.

[81] A. Bressan, *Hyperbolic Systems of Conservation Laws. The One Dimensional Cauchy Problem*, Oxford Univ. Press, Oxford, 2000.

[82] P. Broadbridge and P. Tritscher, An integrable fourth-order nonlinear evolution equation applied to thermal grooving of metal surfaces, *IMA J. Appl. Math.*, **53** (1994), 249–265.

[83] L.W. Bruch, M.W. Cole, and E. Zaremba, *Physical Adsorption: Forces and Phenomena*, Oxford Univ. Press, Oxford, 1997.

[84] M. Bruschi and F. Calogero, On the integrability of certain matric evolution equations, *Phys. Lett. A*, **273** (2000), 167–172.

[85] M.S. Bruzón, P.A. Clarkson, M.L. Gandarias, and E. Medina, The symmetry reductions of a turbulence model, *J. Phys. A*, **34** (2001), 3751–3760.

[86] V.M. Buchstaber, V.Z. Enolskii, and D.V. Leikin, Kleinian functions, hyperelliptic Jacobians and applications, In: *Review in Mathematics and Mathematical Physics* (London), S.P. Novikov and I.M. Krichever, Eds, India: Gordon & Breach, 1997, pp. 1–125.

[87] C.J. Budd, V.A. Galaktionov, and J.F. Williams, Self-similar blow-up in higher-order

semilinear parabolic equations, *SIAM J. Appl. Math.*, **64** (2004), 1775–1809.

[88] C.J. Budd, W. Huang, and R.D. Russell, Moving mesh methods for problems with blow-up, *SIAM J. Sci. Comput.*, **17** (1996), 305–327.

[89] G.I. Burde, Potential symmetriex of the nonlinear wave equation $u_{tt} = (uu_x)_x$ and related exact and approximate solutions, *J. Phys. A,* **34** (2001), 5355–5371.

[90] L.A. Caffarelli and A. Friedman, The blow-up boundary for nonlinear wave equations, *Trans. Amer. Math. Soc.,* **297** (1986), 223–241.

[91] J.W. Cahn, C.M. Elliott, and A. Novick-Cohen, The Cahn-Hilliard equation with a concentration dependent mobility: motion by minus the Laplacian of the mean curvature, *European J. Appl. Math.,* **7** (1996), 287–301.

[92] J.W. Cahn and J.E. Hilliard, Free energy of a nonuniform system. I. Interfacial free energy, *J. Chem. Phys.,* **28** (1958), 258–267.

[93] F. Calogero and A. Degasperis, Nonlinear evolution equations solvable by the inverse spectral transform. I, *Nuovo Cimento B,* **32** (1976), 201–242.

[94] R. Camassa and D.D. Holm, An integrable shallow water equation with peaked solitons, *Phys. Rev. Lett.,* **71** (1993), 1661–1664.

[95] R. Camassa, J.M. Hyman, and B.P. Luce, Nonlinear waves and solitons in physical systems, *Phys. D,* **123** (1998), 1–20.

[96] R. Camassa and A.I. Zenchuk, On the initial value problem for a completely integrable shallow water wave equation, *Phys. Lett. A,* **281** (2001), 26–33.

[97] H.-D. Cao and X.-P. Zhu, A complete proof of the Poincaré and geometrization conjectures – Application of the Hamilton-Perelman theory of the Ricci flow, Asian J. Math., **10** (2006), 165-492.

[98] W. Cao, R. Carretero-González, W. Huang, and R.D. Russel, Variational mesh adaptation methods for axisymmetric problems, *SIAM J. Numer. Anal.,* **41** (2003), 235–257.

[99] W. Cao, W. Huang, and R.D. Russel, A moving mesh method based on the geometric conservation law, *SIAM J. Sci. Comput.,* **24** (2002), 118–142.

[100] F. Cariello and M. Tabor, Painlevé expansions for nonintegrable evolution equations, *Phys. D,* **39** (1989), 77–94.

[101] A.S. Carstea, Extension of the bilinear formalism to supersymmetric KdV-type equations, *Nonlinearity*, **13** (2000), 1645–1656.

[102] S. Chakravarty, M.J. Ablowitz, and L.A. Takhtajan, Self-dual Yang-Mills equation and new special functions in integrable systems, In: *Nonlinear Evolution Equations and Dynamical Systems*, M. Boiti, L. Martina, and F. Pempinelli, Eds, World Sci. Publ., River Edge, NJ, 1992.

[103] S.A. Chaplygin, *Gas jets*, Ucenye Zapiski Imperat. Moskov. Univ., Otd. Fiz.-Mat., **21** (1904), 1–121; GITTL, Moscow/Leningrad, 1949; Tech. Memos. Nat. Adv. Comm. Aeronaut., no. 1063, 112 pp., 1944.

[104] J. Chazy, Sur les équations différentielles dont l'intégrale générale est uniforme et admet des singularities essentielles mobiles, *C. R. Acad. Sci. Paris,* **149** (1909), 563–565.

[105] Q. Chen, Maximum principles, uniqueness and existence for harmonic maps with potential and Landau-Lifshitz equations, *Calc. Var. PDEs,* **48** (1999), 91–107.

[106] Y. Chen, Existence and singularities for the Dirichlet boundary value problems of Landau–Lifshitz equations, *Nonl. Anal.*, **48** (2002), 411–426.

[107] Y.G. Chen, Y. Giga, and S. Goto, Uniqueness and existence of viscosity solutions of generalized mean curvature flow equations, *J. Differ. Geom.,* **33** (1991), 749–786.

[108] S.Y. Cheng and S.-T. Yau, Complete affine hyperspheres. I. The completeness of affine metrics, *Commun. Pure Appl. Math.,* **39** (1986), 839–866.

[109] I. Cherednik, *Basic Methods of Soliton Theory*, World Sci. Publ. Co., Inc., River Edge, NJ, 1996.

[110] R.M. Cherniha, New non-Lie ansätze and exact solutions of nonlinear reaction-diffusuion-convection equations, *J. Phys. A,* **31** (1998), 8179–8198.

[111] R.M. Cherniha, New exact solutions of a nonlinear equation in mathematical biology and their properties, *Ukranian Math. J.,* **53** (2001), 1712–1727.

[112] R.M. Cherniha, Nonlinear Galilei-invariant PDEs with infinite-dimensional Lie symmetry, *J. Math. Anal. Appl.,* **253** (2001), 126–141.

[113] R. Cherniha and J.R. King, Nonlinear reaction-diffusion systems with variable diffusivities: Lie symmetries, ansätze and exact solutions, *J. Math. Anal. Appl.,* **308** (2005), 11–35.

[114] R. Cherniha and M. Serov, Nonlinear systems of the Burgers-type equations: Lie and Q-conditional symmetries, ansätze and solutions, *J. Math. Anal., Appl.,* **282** (2003), 305–328.

[115] A. Chertock, A. Kurganov, and P. Rosenau, Formation of discontinuities in flux-saturated degenerate parabolic equations, *Nonlinearity,* **16** (2003), 1875–1898.

[116] K.-S. Chou, A blow-up criterion for the curve shortening flow by surface diffusion, *Hokkaido Math. J.,* **32** (2003), 1–19.

[117] K.-S. Chou, D. Geng, and S.-S. Yan, Critical dimensions of a Hessian equation involving critical exponent and a related asymptotic result, *J. Differ. Equat.,* **129** (1996), 79–110.

[118] K.-S. Chou and C. Qu, Symmetry groups and separation of variables of a class of nonlinear diffusion–convection equations, *J. Phys. A,* **32** (1999), 6271–6286.

[119] K.-S. Chou and C. Qu, Generalized conditional symmetries of nonlinear differential-difference equations, *Phys. Lett. A,* **280** (2001), 303–308.

[120] K.-S. Chou and C. Qu, Integrable equations arising from motions of plane curves, *Phys. D,* **162** (2002), 9–33.

[121] K.-S. Chou and C. Qu, Optimal systems and group classification of (1+2)-dimensional heat equation, *Acta Appl. Math.,* **83** (2004), 257–287.

[122] K.-S. Chou and X.-J. Wang, A logarithmic Gauss curvature flow and the Minkowski problem, *Ann. Inst. H. Poincaré, Anal. Non Linéaire,* **17** (2000), 733–751.

[123] K.-S. Chou and X.-J. Wang, A variational theory of the Hessian equation, *Commun. Pure Appl. Math.,* **54** (2001), 1029–1064.

[124] S.-N. Chow, J. Mallet-Paret, and W. Shen, Travelling waves in lattice dynamical systems, *J. Differ. Equat.,* **149** (1998), 248–291.

[125] K.W. Chow, W.C. Lai, C.K. Shek, and K. Tso, Positon-like solutions of nonlinear evolution equations in $(2+1)$ dimensions, *Chaos, Solitons Fractals,* **9** (1998), 1901–1912.

[126] P.L. Christiansen, J.C. Eilbeck, V.Z. Enolskii, and N.A. Kostov, Quasi-periodic and periodic solutions for coupled nonlinear Schrödinger equations of Manakov type, *Proc. Roy. Soc. London A,* **456** (2000), 2263–2281.

[127] P.A. Clarkson, Nonclassical symmetry reductions of the Boussinesq equation, *Chaos, Solitons Fractals,* **5** (1995), 2261–2301.

[128] P.A. Clarkson, A.S. Fokas, and M. Ablowitz, Hodograph transformastions of linearizable partial differential equations, *SIAM J. Appl. Math.,* **49** (1989), 1188–1209.

[129] P.A. Clarkson and S. Hood, Nonclassical symmetry reductions and exact solutions of the Zabolotskaya–Khokhlov equation, *European J. Appl. Math.,* **3** (1992), 381–414.

[130] P.A. Clarkson and E.L. Mansfield, Symmetry reductions and exact solutions of a class of nonlinear heat equations, *Phys. D,* **70** (1994), 250–288.

[131] G.M. Coclite and K.H. Karlsen, On the well-posedness of the Degasperis–Procesi equation, J. Funct. Anal., **233** (2006), 60–91.

[132] E.A. Coddington and N. Levinson, *Theory of Ordinary Differential Equations*, McGraw-Hill Book Company, Inc., New York/London, 1955.

[133] J.D. Cole and L.P. Cook, *Transonic Aerodynamics,* North-Holland, Amsterdam, 1986.

[134] A. Constantin and J. Escher, Well-posedness, global existence, and blow-up phenomena for a periodic quasi-linear hyperbolic equation, *Commun. Pure Appl. Math.,* **51** (1998), 475–504.

[135] A. Constantin and H.P. McKean, A shallow water equation on the circle, *Commun. Pure Appl. Math.,* **52** (1999), 949–982.

[136] A. Constantin and L. Molinet, Orbital stability of solitary waves for a shallow water equation, *Phys. D,* **157** (2001), 75–89.

[137] R. Conte, Exact solutions of nonlinear partial differential equations by singularity analysis, In: *Direct and Inverse Methods in Nonlinear Evolution Equations*, *Lect. Notes in Phys.,* Vol. **632**, Springer, Berlin, 2003, pp. 1–83.

[138] R. Conte, M. Musette, and A.M. Grundland, Bäcklund transformation of partial differential equations from the Painlevé-Gambier classification. II. Tzitzéica equation, *J. Math. Phys.*, **40** (1999), 2092–2106.

[139] G. Darboux, Sur la théorie des équations aux dérivées partielles, *C. R. Acad. Sci. Paris,* **70** (1870), 746–749.

[140] G. Darboux, Sur la théorie des coordonnées curvilignes et les systémes orthogonaux, *Ann. Ec. Normale Supér.,* **7** (1878), 101–150.

[141] G. Darboux, Lecons sur la théorie générale des surfaces, Gauthier-Villars, Paris, 1887.

[142] G. Darboux, Lecons sur la théorie générale des surfaces et les applications géometriques du calcul infinitésimal. Troisiéme partie, Gauthier-Villars, Paris, 1894.

[143] P. Daskalopoulos and R. Hamilton, The free boundary in the Gauss curvature flow with flat sides, *J. reine angew. Math.,* **510** (1999), 187–227.

[144] P. Daskalopoulos and K.-A. Lee, Worn stones with flat sides all time regularity of the interface, *Invent. Math.,* **156** (2004), 445–493.

[145] A. Degasperis, D.D. Holm, and A.N.W. Hone, A new integrable equation with peakon solutions, *Theoret. Math. Phys.,* **133** (2002), 1463–1474.

[146] P. Degond, F. Méhats, and C. Ringhofer, Quantum energy-transport and drift-diffusion models, *J. Stat. Phys.,* **118** (2005), 625–667.

[147] B. Dey, Compacton solutions for a class of two parameter generalized odd-order Korteweg–de Vries equations, *Phys. Rev. E*, **57** (1998), 4733–4738.

[148] E. DiBenedetto, *Degenerate Parabolic Equations,* Universitext, Springer-Verlag, New York, 1993.

[149] G.C. Dong, *Nonlinear Partial Differential Equations of Second Order,* Transl. Math. Monogr., **95**, Amer. Math. Soc., Providence, RI, 1991.

[150] V.A. Dorodnitsyn, On invariant solutions of the equation of non-linear heat conduction with a source, USSR *Comput. Math. Math. Phys.,* **22** (1982), 115–122.

[151] V.A. Dorodnitsyn, Symmetry of finite-difference equations, In: *CRC Handbook of Lie Group Analysis of Differential Equations*, Vol. **1**. *Symmetries, Exact Solutions and Conservation Laws*, CRC Press, Boca Raton, FL, 1994, pp. 365–403.

[152] V. Dorodnitsyn, R. Kozlov, and P. Winternitz, Continuous symmetries of Lagrangians and exact solutions of discrete equations, *J. Math. Phys.,* **45** (2004), 336–359.

[153] J. Drach, Sur les groupes complexes de rationalité et sur l'intégration par quadratures, *C. R. Acad. Sci. Paris,* **167** (1918), 743–746.

[154] J. Drach, Sur l'intégration par quadratures de l'équation différentielle $\frac{d^2y}{dx^2} = [\varphi(x) + h]y$, *C. R. Acad. Sci. Paris*, **168** (1919), 337–340.

[155] B.R. Duffy and H.K. Moffatt, A similarity solution for viscous source flow on a vertical plane, *European J. Appl. Math.*, **8** (1997), 37–47.

[156] Yu.A. Dubinskii, *Analytic Pseudo-Differential Operators and their Applications*, *Math. and Appl.* (Soviet Ser.), 68, Kluver Acad. Publ. Group, Dordrecht, 1991.

[157] B.A. Dubrovin, *Geometry of 2D Topological Field Theories*, Lect. Notes in Math., Vol. **1620**, Springer, Berlin, 1996, pp. 120–348.

[158] M. Dunaiski, A class of Einstein–Weyl spaces associated to an integrable system of hydrodynamic type, J. Geom. Phys., **51** (2004), 126–137.

[159] Y. Ebihara, H. Fukuda, and M. Kurokiba, On degenerate parabolic equations with quadratic convection, *Fukuoka Univ. Sci. Rep.*, **26** (1996), 79–87.

[160] Y. Ebihara and J. Kameda, On quasilinear bidegenerate parabolic equations, *Math. Appl. Comput.*, **14** (1995), 3-1–315.

[161] Y. Ebihara and T. Kitada,On the behavior of explicit solutions of quasilinear hyperbolic-elliptic equations with convective term, *Adv. Math. Sci. Appl.*, **7** (1997), 225–243.

[162] Y. Ebihara, T. Kitada, and M. Kurokiba, Explicit solutions of quasilinear hyperbolic-elliptic equations with quadratic nonlinear terms, *Fukuoka Univ. Sci. Rep.*, **24** (1994), 39–48.

[163] Yu.V. Egorov, V.A. Galaktionov, V.A. Kondratiev, and S.I. Pohozaev, Global solutions of higher-order semilinear parabolic equations in the supercritical range, *Adv. Differ. Equat.*, **9** (2004), 1009–1038.

[164] S.D. Eidelman, *Parabolic Systems,* North-Holland Publ. Co., Amsterdam, 1969.

[165] J.C. Eilbeck and V.Z. Enolskii, Bilinear operators and the power series for the Wierstrass σ function, *J. Phys. A*, **33** (2000), 791–794.

[166] J.C. Eilbeck, V.Z. Enolskii, and H. Holden, The hyperelliptic ζ-function and the integrable massive Thirring model, *Roy. Soc. Lond. Proc. Ser. A Math. Phys. Eng. Sci.*, **459** (2003), 1581–1610.

[167] C.M. Elliott and H. Garcke, On the Cahn–Hilliard equation with degenerate mobility, *SIAM J. Math. Anal.*, **27** (1996), 404–423.

[168] R. Emden, *Gaskugeln, Anwendungen der mechanischen Warmentheorie auf Kosmologie und meteorologische Probleme*, Chap. XII, Teubner, Leipzig, 1907.

[169] A. Enneper, Über asymptotische Linien, *Nachr. Königl. Gesellsch. d. Wissenschaften Göttingen*, 1870, pp. 493–511.

[170] V.P. Ermakov, Differential equations of second order. Conditions of integrability in final form, *Izvestiya Kievskogo Univ. III*, **9** (1880), 1–25 (in Russian).

[171] F.J. Ernst, New formulation of the axially symmetric gravitational field problem, *Phys. Rev.*, **167** (1968), 1175–1178.

[172] P.G. Estévez and C.-Z. Qu, Separation of variables in nonlinear wave equations with variable wave speed, *Theoret. Math. Phys.*, **133** (2002), 1490–1497.

[173] P.G. Estévez, C. Qu, and S. Zhang, Separation of variables of a generalized porous medium equation with nonlinear source, *J. Math. Anal. Appl.*, **275** (2002), 44–59.

[174] J.D. Evans, V.A. Galaktionov, and J.R. King, Source-type solutions of the fourth-order unstable thin film equation, *European J. Appl. Math.*, to appear.

[175] J.D. Evans, V.A. Galaktionov, and J.R. King, Unstable sixth-order thin film equation. I. Blow-up similarity solutions; II. Global similarity patterns, *Nonlinearity*, submitted.

[176] J.D. Evans, J.R. King, and A.B. Tayler, Finite length mask effects in the isolation oxidation of silicon, *IMA J. Appl. Math.*, **58** (1997), 121–146.

[177] J.D. Evans, M. Vynnycky, and S.P. Ferro, Oxidation-induced stresses in the isolation oxidation of silicon, *J. Engrg. Math.*, **38** (2000), 191–218.

[178] L.C. Evans and J. Spruck, Motion of level sets by mean curvature. I, *J. Differ. Geom.*, **33** (1991), 635–681.

[179] A.V. Faminskii, On the mixed problem for quasilinear equations of the third order, *J. Math. Sci.*, **110** (2002), 2476–2507.

[180] E. Fan, An algebraic method for finding a series of exact solutions to integrable and nonintegrable nonlinear evolution equations, *J. Phys. A*, **36** (2003), 7009–7026.

[181] E.V. Ferapontov, Decomposition of higher-order equations of Monge–Ampère type, *Lett. Math. Phys.*, **62** (2002), 193–198.

[182] E.V. Ferapontov, Hypersurfaces with flat centroaffine metric and equations of associativity, *Geom. Dedicata*, **103** (2004), 33–49.

[183] E. Fermi, J. Pasta, and S. Ulam, Studies in nonlinear problems, Los Alamos Sci. Lab. Report LA-1940, 1955; Reprinted in *Lec. Appl. Math.*, **15**, 143–156.

[184] M.C. Ferreira, R.A. Kraenkel, and A.I. Zenchuk, Soliton–cuspon interaction for the Camassa–Holm equation, *J. Phys. A*, **32** (1999), 8665–8670.

[185] R. Ferreira and F. Bernis, Source-type solutions to thin-film equations in higher dimensions, *European J. Appl. Math.*, **8** (1997), 507–524.

[186] D. Feyel and A.S. Üstünel, Monge-Kantorovitch measure transportation and Monge-Ampère equation on Wiener space, *Probab. Theory Related Fields*, **128** (2004), 347–385.

[187] P.C. Fife, *Mathematical Aspects of Reacting and Diffusing Systems*, Lect. Notes in Biomath., **28**, Springer-Verlag, Berlin/New York, 1979.

[188] F. Finkel and N. Kamran, The Lie algebraic structure of differential operators admitting invariant spaces of polynomials, *Adv. Appl. Math.*, **20** (1998), 300–322.

[189] J.C. Flitton and J.R. King, Moving-boundary and fixed-domain problems for a sixth-order thin-film equation, *European J. Appl. Math.*, **15** (2004), 713-754.

[190] V.A. Florin, Earth compaction and seepage with variable porosity taking into account the inflience of bound water, *Izvestiya Akad. Nauk SSSR, Otdel. Tekhn. Nauk*, No. **11** (1951), 1625–1649.

[191] A.S. Fokas and Q.M. Liu, Nonlinear interaction of traveling waves of nonintegrable equations, *Phys. Rev. Lett.*, **72** (1994), 3293–3296.

[192] A.S. Fokas and Q.M. Liu, Generalized conditional symmetries and exact solutions of non-integrable equations, *Theoret. Math. Phys.*, **99** (1994), 571–582.

[193] A.S. Fokas and Q.M. Liu, Asymptotic integrability of water waves, *Phys. Rev. Lett.*, **77** (1996), 2347–2351.

[194] A.S. Fokas and Y.C. Yortsos, On the exactly solvable equations $S_t = [(\beta S + \gamma)^{-2} S_x]_x + \alpha(\beta S + \gamma)^{-2} S_x$ occurring in two-phase flow in porous media, *SIAM J. Appl. Math.*, **42** (1982), 318–332.

[195] B. Fornberg and G.B. Whitham, A numerical and theoretical study of certain nonlinear wave phenomena, *Phil. Trans. Roy. Soc. London Ser. A*, **289** (1978), 373–404.

[196] A.R. Forsyth, *Theory of Differential Equations*. 5,6. *Partial Differential Equations*, Dover Publications, Inc., New York, 1959.

[197] M.V. Foursov, On integrable coupled KdV-type systems, *Inverse Problems*, **16** (2000), 259–274.

[198] M.V. Foursov, Towards the complete classification of homogeneous two-component integrable equations, *J. Math. Phys.*, **44** (2003), 3088–3096.

[199] M.V. Foursov and E.M. Vorob'ev, Solutions of the nonlinear wave equation $u_{tt} = (uu_x)_x$ invariant under conditional symmetries, *J. Phys. A*, **29** (1996), 6363–6373.

[200] R.H. Fowler, The form near infinity of real continuous solutions of a certain differential equation of the second order, *Quart. J. Math.*, **45** (1914), 289–305.

[201] M.L. Frankel, On the nonlinear evolution of a solid-liquid interface, *Phys. Lett. A*, **128** (1988), 57–60.

[202] M.L. Frankel and G.I. Sivashinsky, On the nonlinear thermal diffusive theory of curved flames, *J. Phys.*, **48** (1987), 25–28.

[203] M.L. Frankel and G.I. Sivashinsky, On the equation of a curved flame front, *Phys. D*, **30** (1988), 28–42.

[204] N.C. Freeman and J.J.C. Nimmo, Soliton solutions of the Korteweg-de Vries and Kadomtsev-Petviashvili equations: the Wronskian technique. A method of obtaining the N-soliton solution of the Boussinesq equation in terms of a Wronskian, *Phys. Lett. A*, **95** (1983), 1–3, 4–6.

[205] A. Friedman, *Partial Differential Equations of Parabolic Type*, Prentice Hall Inc., Englewood Cliffs, NJ, 1964.

[206] A. Friedman, *Variational Principles and Free-Boundary Problems*, A Wiley-Interscience Publ., New York, 1982.

[207] A. Friedman and B. McLeod, Blow-up of positive solutions of semilinear heat equations, *Indiana Univ. Math. J.*, **34** (1985), 425–447.

[208] K. Friedrichs, Über die ausgeseichnete Randbedingung in der Spektraltheorie der halbbeschränkten gewöhnlichen Differentialoperatoren zweiter Ordnung, *Math. Ann.*, **121** (1935/36), 1–23.

[209] B. Fuchssteiner, The Lie algebra structure of nonlinear evolution equations and infinite-dimensional abelian symmetry groups, *Progr. Theoret. Phys.*, **65** (1981), 861–876.

[210] B. Fuchssteiner, The Lie algebra structure of degenerate Hamiltonian and bi-Hamiltonian systems, *Progr. Theoret. Phys.*, **68** (1982), 1082–1104.

[211] B. Fuchssteiner, Some tricks from the symmetry-toolbox for nonlinear equations: generalizations of the Camassa–Holm equation, *Phys. D*, **95** (1996), 229–243.

[212] B. Fuchssteiner and A.S. Fokas, Symplectic structures, their Bäcklund transformations and hereditary symmetries, *Phys. D*, **4** (1981), 47–66.

[213] W. Fushchich and R.M. Cherniha, The Galilean relativistic principle and nonlinear partial differential equations, *J. Phys. A*, **18** (1985), 3491–3503.

[214] M. Gage and R.S. Hamilton, The heat equation shrinking convex plane curves, *J. Differ. Geom.*, **23** (1986), 69–96.

[215] V.A. Galaktionov, Two methods of comparison of solutions of parabolic equations, *Soviet Physics Dokl.*, **25** (1980), 250–251.

[216] V.A. Galaktionov, Conditions for global non-existence and localization of solutions of the Cauchy problem for a class of nonlinear parabolic equations, *USSR Comput. Math. Math. Phys.*, **23** (1983), 36–44.

[217] V.A. Galaktionov, On new exact blow-up solutions for nonlinear heat conduction equations with source and applications, *Differ. Integr. Equat.*, **3** (1990), 863–874.

[218] V.A. Galaktionov, Best possible upper bound for blow-up solutions of the quasilinear heat conduction equation with source, *SIAM J. Math. Anal.*, **22** (1991), 1293–1302.

[219] V.A. Galaktionov, Quasilinear heat equations with first-order sign-invariants and new explicit solutions, *Nonl. Anal.*, **23** (1994), 1595–1621.

[220] V.A. Galaktionov, Invariant subspaces and new explicit solutions to evolution equations with quadratic nonlinearities, *Proc. Roy. Soc. Edinburgh, Sect. A*, **125** (1995), 225–246.

[221] V.A. Galaktionov, On invariant subspaces for nonlinear finite-difference operators,

Proc. Royal Soc. Edinburgh, Sect. A, **128** (1998), 1293–1308.

[222] V.A. Galaktionov, Invariant and positivity properties of moving mesh operators for nonlinear evolution equations, Preprint 99/17, School Math. Sci., Univ. of Bath, 1999.

[223] V.A. Galaktionov, Ordered invariant sets for nonlinear evolution equations of KdV-type, *Comput. Maths Math. Phys.*, **39** (1999), 1499–1505.

[224] V.A. Galaktionov, Invariant solutions of two models of evolution of turbulent bursts, *European J. Appl. Math.*, **10** (1999), 237–249.

[225] V.A. Galaktionov, Groups of scalings and invariant sets for higher-order nonlinear evolution equations, *Differ. Integr. Equat.*, **14** (2001), 913–924.

[226] V.A. Galaktionov, *Geometric Sturmian Theory of Nonlinear Parabolic Equations and Applications*, Chapman & Hall/CRC, Boca Raton, FL, 2004.

[227] V.A. Galaktionov, On interfaces and oscillatory solutions of higher-order semilinear parabolic equations with nonlipschitz nonlinearities, *Stud. Appl. Math.*, **117** (206), 353–389.

[228] V.A. Galaktionov, V.A. Dorodnitsyn, G.G. Elenin, S.P. Kurdyumov, and A.A. Samarskii, A quasilinear equation of heat conduction with a source: peaking, localization, symmetry, exact solutions, asymptotic behavior, structures, *J. Soviet Math.*, **41** (1988), No. 5, pp. 1222–1292.

[229] V.A. Galaktionov, S.P. Kurdyumov, S.A. Posashkov, and A.A. Samarskii, A non-linear elliptic problem with a complex spectrum of solutions, *USSR Comput. Math. Math. Phys.*, **26** (1986), 48–54.

[230] V.A. Galaktionov and L.A. Peletier, Asymptotic behaviour near finite-time extinction for the fast diffusion equation, *Arch. Ration. Mech. Anal.*, **139** (1997), 83–98.

[231] V.A. Galaktionov and S.I. Pohozaev, On similarity solutions and blow-up spectra for a semilinear wave equation, *Quart. Appl. Math.*, **61** (2003), 583–600.

[232] V.A. Galaktionov and S.A. Posashkov, New exact solutions of parabolic equations with quadratic nonlinearities, *USSR Comput. Math. Math. Phys.*, **29** (1989), 112–119.

[233] V.A. Galaktionov and S.A. Posashkov, Single point blow-up for N-dimensional quasilinear equation with gradient diffusion and source, *Indiana Univ. Math. J.*, **40** (1991), 1041–1060.

[234] V.A. Galaktionov and S.A. Posashkov, Exact solutions and invariant spaces for nonlinear equations of gradient diffusion, *Comput. Maths Math. Phys.*, **34** (1994), 313–321.

[235] V.A. Galaktionov and S.A. Posashkov, Examples of nonsymmetric extinction and blow-up for quasilinear heat equations, *Differ. Integr. Equat.*, **8** (1995), 87–103.

[236] V.A. Galaktionov and S.A. Posashkov, New explicit solutions of quasilinear heat equations with general first-order sign-invariants, *Phys. D*, **99** (1996), 217–236.

[237] V.A. Galaktionov and S.A. Posashkov, Maximal sign-invariants of quasilinear parabolic equations with gradient diffusivity, *J. Math. Phys.*, **39** (1998), 4948–4964.

[238] V.A. Galaktionov, S.A. Posashkov, and S.R. Svirshchevskii, Generalized separation of variables for differential equations with polynomial nonlinearities, *Differ. Equat.*, **31** (1995), 233–240.

[239] V.A. Galaktionov, S.A. Posashkov, and S.R. Svirschevskii, On invariant sets and explicit solutions of nonlinear evolution equations with quadratic nonlinearities, *Differ. Integr. Equat.*, **8** (1995), 1997–2024.

[240] V.A. Galaktionov and A.E. Shishkov, Saint Venant's principle in blow-up for higher-order quasilinear parabolic equations, *Proc. Royal Soc. Edinburgh, Sect. A*, **133** (2003), 1075–1119.

[241] V.A. Galaktionov, S.I. Shmarev, and J.L. Vazquez, Behaviour of interfaces in a diffusion-absorption equation with critical exponents, *Interfaces Free Bound.*, **2** (2000),

425–448.

[242] V.A. Galaktionov and S.R. Svirshchevskii, Nonlinear difference operators admitting invariant subspaces, Preprint 99/07, School Math. Sci., University of Bath, 1999.

[243] V.A. Galaktionov and J.L. Vazquez, Blow-up for quasilinear heat equations described by means of nonlinear Hamilton-Jacobi equations, *J. Differ. Equat.*, **127** (1996), 1–40.

[244] V.A. Galaktionov and J.L. Vazquez, Blow-up of a class of solutions with free boundaries for the Navier–Stokes equations, *Adv. Differ. Equat.*, **4** (1999), 297–321.

[245] V.A. Galaktionov and J.L. Vázquez, *A Stability Technique for Evolution Partial Differential Equations. A Dynamical Systems Approach, Progr. in Nonl. Differ. Equat. and their Appl.*, **56**, Birkhäuser Boston, Inc., MA, 2004.

[246] M.L. Gandarias and M.S. Bruzón, Nonclassical symmetries for a family of Cahn-Hilliard equations, *Phys. Lett. A,* **34** (1999), 331–337.

[247] M.L. Gandarias, M.S. Bruzón, and J. Ramirez, Classical symmetry reductions of the Schwarz-Korteweg-de Vries equation in the dimension 2+1, *Theoret. Math. Phys.,* **134** (2003), 62–71.

[248] X. Geng and C. Cao, Explicit solutions of the 2+1-dimensional breaking soliton equation, *Chaos, Solitons Fractals,* **22** (2004), 683–691.

[249] P.G. de Gennes, Wetting: statics and dynamics, *Rev. Mod. Phys.*, **57** (1985), 827–863.

[250] F. Gesztesy and R. Weikard, Elliptic algebro-geometric solutions of the KdV and AKNS hierarchies – an analytic approach, *Bull. Amer. Math. Soc. (N.S.),* **35** (1998), 271–317.

[251] S. Ghosh and D. Sarma, Bilinearization of $N = 1$ supersymmetric modified KdV equations, *Nonlinearity*, **16** (2003), 411–418.

[252] J. Gibbons and S.P. Tsarev, Conformal maps and reductions of the Benney equations, *Phys. Lett. A,* **258** (1999), 263–271.

[253] D. Gilbarg and N.S. Trudinger, *Elliptic Partial Differential Equations of Second Order,* Springer-Verlag, Berlin, 2001.

[254] L. Gilles, S.C. Hagness, and L. Vázquez, Comparison between staggered and unstaggered finite-difference time-domain grid for few-cycle temporal optical soliton propagation, *J. Comput. Phys.,* **161** (2000), 379–400.

[255] R.T. Glassey, J.K. Hunter, and Y. Zheng, Singularities in a variational wave equation, *J. Differ. Equat.,* **129** (1996), 49–78.

[256] J. Goard and P. Broadbridge, Solutions to nonlinear partial differential equations from symmetry-enhancing and symmetry-preserving constraints, *J. Math. Anal. Appl.,* **238** (1999), 369–384.

[257] D. Gomez-Ullate, N. Kamran, and R. Milson, Structure theorems for linear and nonlinear differential operators admitting invariant polynomial subspaces, Preprint, 2006 (arXiv:nlin.SI/0604070 v1).

[258] S. González-Pérez-Sandi, J. Fujioka, and B.A. Malomed, Embedded solitons in dynamical lattices, Phys. D, **197** (2004), 86–100.

[259] H. Görtler, Zum Übergang von Unterschall zu Überschallgescwindigkeiten in Düsen, *Z. angew. Math. Mech.,* **19** (1939), 325.

[260] E. Goursat, *Leçons sur l'intégration des équations aux dérivées partielles du second ordre a deux variables indépendantes*, T. II, Paris, 1898 (Librairie sci. A. Hermann).

[261] L.V. Govor, J. Parisi, G.H. Bauer, and G. Reiter, Instability and droplet formation in evaporating thin films of a binary solution, *Phys. Rev. E*, **71**, 051603 (2005).

[262] A.E. Green and P. Naghdi, A derivation of equations for wave propagation in water of variable depth, *J. Fluid Mech.,* **78** (1976), 237–246.

[263] H.P. Greenspan, On the motion of a small viscous droplet that wets a surface, *J. Fluid*

Mech., **84** (1978), 125–143.

[264] G. Grün, Degenerate parabolic equations of fourth order and a plasticity model with non-local hardening, *Z. Anal. Anwendungen,* **14** (1995), 541–573.

[265] G. Grün, Droplet spreading under weak slippage – existence for the Cauchy problem, *Commun. Partial Differ. Equat.,* **29** (2004), 1697–1744.

[266] A.M. Grundland, A.J. Hariton, and V. Hussin, Group-invariant solutions of relativistic and nonrelativistic models in field theory, *J. Math. Phys.,* **44** (2003), 2874–2890.

[267] K.G. Guderley, *Theorie Schallnaher Strömungen*, Berlin/Heidelberg, 1957; *The Theory of Transonic Flow*, Pergamon Press, Oxford/Frankfurt, 1962.

[268] F. Güngör, V.I. Lahno, and R.Z. Zhdanov, Symmetry classification of KdV-type nonlinear evolution equations, *J. Math. Phys.,* **45** (2004), 2280–2313.

[269] B. Guo, Y. Han, and G. Yang, Blow up problem for Landau-Lifshitz equations in two dimensions, *Commun. Nonlinear Sci. Numer. Simul.,* **5** (2000), 43–44.

[270] B. Guo and G. Yang, Some exact nontrivial global solutions with values in unit sphere for two-dimensional Landau-Lifshitz equations, *J. Math. Phys.,* **42** (2001), 5223–5227.

[271] M.E. Gurtin, Toward a nonequilibrium thermodynamics of two-phase materials, *Arch. Ration. Mech. Anal.,* **100** (1988), 275–312.

[272] C.E. Gutiérrez, *The Monge-Ampère Equation,* Birkhäuser Inc., Boston, MA, 2001.

[273] C.E. Gutiérrez and Q. Huang, A generalization of a theorem by Calabi to the parabolic Monge-Ampère equation, *Indiana Univ. Math. J.,* **47** (1998), 1459–1480.

[274] C.E. Gutiérrez and Q. Huang, $W^{2,p}$ estimates for the parabolic Monge-Ampère equation, *Arch. Ration. Mech. Anal.,* **159** (2001), 137–177.

[275] S. Gutierrez and L. Vega, Self-similar solutions of the localized induction approximation: singularity formation, *Nonlinearity,* **17** (2004), 2091–2136.

[276] J. Hale, *Theory of Functional Differential Equations,* Springer-Verlag, New York/ Heidelberg, 1977.

[277] G. Halphen, Sur un systéme d'équations différentielles, *C. R. Acad. Sci. Paris,* **92** (1881), 1101–1103.

[278] R.S. Hamilton, Three-manifolds with positive Ricci curvature, *J. Differ. Geom.,* **17** (1982), 255–306.

[279] P.S. Hammond, Nonlinear adjustement of a thin annular film of viscous fluid surrounding a thread of another within a circular cylindrical pipe, *J. Fluid Mech.,* **137** (1983), 363–384.

[280] G.H. Hardy, *Note on a theorem of Hilbert, Math. Z.*, **6** (1920), 314–317.

[281] J. Harris, *Algebraic Geometry,* Springer-Verlag, New York, 1992.

[282] A.C. Hern, *REDUCE. User's Manual,* RAND Publ. CP78, Santa Monica, CA, 1991.

[283] A.A. Himonas and G. Misiolek, Well-posedness of the Cauchy problem for a shallow water equation on the circle, *J. Differ. Equat.,* **161** (2000), 479–495.

[284] R. Hirota, Exact solutions of the Korteweg-de Vries equation for multiple collisions of solitons, *Phys. Rev. Lett.,* **27** (1971), 1192–1194.

[285] R. Hirota, Exact N-soliton solutions of the wave equation of long waves in shallow-water and in nonlinear lattices, *J. Math. Phys.,* **14** (1973), 810–814.

[286] R. Hirota, Discretization of coupled modified KdV equations, *Chaos, Solitons Fractals,* **11** (2000), 77–84.

[287] R. Hirota, X. B. Hu, and X. Y. Tang, A vector potential KdV equation and vector Ito equation: soliton solutions, bilinear Bäcklund transformations and Lax pairs, *J. Math. Anal. Appl.,* **288** (2003), 326–348.

[288] R. Hirota and J. Satsuma, Soliton solutions of a coupled Korteweg-de Vries equation, *Phys. Lett. A,* **85** (1981), 407–408.

[289] T. Hocherman and P. Rosenau, On KS-type equations describing the evolution and rupture of a liquid interface, *Phys. D*, **67** (1993), 113–125.

[290] D.D. Holm, J.E. Marsden, and T.S. Ratiu, The Euler–Poincaré equations and semidirect products with applications to continuum theories, *Adv. Math.*, **137** (1998), 1–81.

[291] J. Huang, G. Lu, and S. Ruan, Travelling wave solutions in delayed lattice differential equations with partial monotonicity, *Nonl. Anal.*, **60** (2005), 1331–1350.

[292] W. Huang, Y. Ren and R.D. Russell, Moving mesh methods based on moving mesh partial differential equations, *J. Comput. Physics*, **113** (1994), 279–290.

[293] W. Huang and R.D. Russell, A moving collocation method for solving time dependent partial differential equations, *Appl. Numer. Math.*, **20** (1996), 101–116.

[294] J.K. Hunter and R. Saxton, Dynamics of director fields, *SIAM J. Appl. Math.*, **51** (1991), 1498–1521.

[295] J.K. Hunter and Y.X. Zheng, On a completely integrable nonlinear hyperbolic variational equation, *Phys. D*, **79** (1994), 361–386.

[296] J.M. Hyman and P. Rosenau, Pulsating multiplet solutions of quintic wave equations, *Phys. D*, **123** (1998), 502–512.

[297] N.H. Ibragimov, *Transformations Groups Applied to Mathematical Physics*, D. Reidel Publ. Co., Dordrecht, 1985.

[298] N.H. Ibragimov and S.R. Svirshchevskii, Lie-Bäcklund symmetries of submaximal order of ordinary differential equations, *Nonlinear Dynam.*, **28** (2002), 155–166.

[299] N.H. Ibragimov, M. Torrisi, and A. Valenti, Differential invariants of nonlinear equations $v_{tt} = f(x, v_x)v_{xx} + g(x, v_x)$, *Commun. Nonlinear Sci. Numer. Simul.*, **9** (2004), 69–80.

[300] Y.H. Ichikawa, K. Konno, and M. Wadati, Nonlinear transverse oscillation of elastic beams under tension, *J. Phys. Soc. Japan*, **50** (1981), 1799–1802.

[301] S. Iona and F. Calogero, Integrable systems of quartic oscillators in ordinary (three-dimensional) space, *J. Phys. A*, **35** (2002), 3091–3098.

[302] M. Ito, Symmetries and conservation laws of a coupled nonlinear wave equation, *Phys. Lett. A*, **91** (1982), 335–338.

[303] N.M. Ivochkina and O.A. Ladyzhenskaya, Parabolic equations generated by symmetric functions of the eigenvalues of the Hessian or by the principal curvature of a surface I. Parabolic Monge-Ampère equations, *St. Petersb. Math. J.*, **6** (1995), 575–594.

[304] G. James and P. Noble, Breathers on diatomic Fermi–Pasta–Ulam lattices, Phys. D, **19** (2004), 124–171.

[305] F. John, Blow-up of solutions of nonlinear wave equations in three space dimensions, *Manuscripta Math.*, **28** (1979), 235–268.

[306] R.S. Johnson, Camassa-Holm, Korteweg-de Vries and related models for water waves, *J. Fluid Mech.*, **455** (2002), 63–82.

[307] A. Jüngel and D. Matthes, An algorithmic construction of entropies in higher-order nonlinear PDEs, *Nonlinearity*, **19** (2006), 633–659.

[308] B.B. Kadomtsev and V.I. Petviashvili, On the stability of solitary waves in weakly dispersing media, *Sov. Physics Dokl.*, **15** (1970), 539–541.

[309] A.S. Kalashnikov, Some problems of the qualitative theory of second-order nonlinear degenerate parabolic equations, *Russian Math. Surveys*, **42** (1987), 169–222.

[310] E.G. Kalnins and W. Miller Jr., Differential-Stäckel matrices, *J. Math. Phys.*, **26** (1985), 1560–1565.

[311] A.M. Kamchatnov, On the Baker-Akhiezer function in the AKNS scheme, *J. Phys. A*, **34** (2001), L441–L446.

[312] N. Kamran, R. Milson, and P.J. Olver, Invariant modules and the reduction of nonlinear

partial differential equations to dynamical systems, *Adv. Math.*, **156** (2000), 286–319.

[313] L.V. Kantorovich, On the transfer of masses, *Dokl. Acad. Nauk SSSR*, **37** (1942), 227–229.

[314] L.V. Kantorovich, On a problem of Monge, *Uspekhi Matem. Nauk*, **3** (1948), 225–226.

[315] P.L. Kapitza and S.P. Kapitza, Wave flow of thin layers of a viscous fluid, *Zh. Eksp. Teor. Fiz.*, **19** (1949), 105 (in Russian).

[316] O.V. Kaptsov, New solutions of two-dimensional steady Euler equations, *J. Appl. Math. Mech.*, **54** (1990), 337–342.

[317] O.V. Kaptsov, Invariant sets of evolution equations, *Nonl. Anal.*, **19** (1992), 753–761.

[318] O.V. Kaptsov, Construction of exact solutions of systems of diffusion equations, *Mat. Model.*, **7** (1995), 107–115.

[319] O.V. Kaptsov, B-determining equations: applications to nonlinear partial differential equations, *European J. Appl. Math.*, **6** (1995), 265–286.

[320] O.V. Kaptsov, Construction of exact solutions of the Boussinesq equation, *J. Appl. Mech. Tech. Phys.*, **39** (1998), 389–392.

[321] O.V. Kaptsov, Some classes of two-dimensional stationary vortex structures in an ideal fluid, *J. Appl. Mech. Tech. Phys.*, **39** (1998), 524–527.

[322] O.V. Kaptsov, Linear determining equations for differential constraints, *Sbornik: Math.*, **189** (1998), 1839–1854.

[323] O.V. Kaptsov, Involute distributions, invariant manifolds, and defining equations, *Siberian Math. J.*, **43** (2002), 428–438.

[324] O.V. Kaptsov and A.V. Schmidt, Linear determining equations for differential constraints, *Glasgow Math. J., Sect. A*, **47** (2005), 109–120.

[325] O.V. Kaptsov and I.V. Verevkin, Differential constraints and exact solutions of nonlinear diffusion equations, *J. Phys. A*, **36** (2003), 1401–1414.

[326] O.V. Kaptsov and A.V. Zabluda, Characteristic invariants and Darboux's method, *J. Phys. A*, **38** (2005), 3133–3144.

[327] Th. von Kármán, Ueber laminare und turbulente Reibung, *ZAMM*, **1** (1921), 233–252.

[328] T. Kato, Blow-up of solutions of some nonlinear hyperbolic equations, *Commun. Pure Appl. Math.*, **33** (1980), 501–505.

[329] T. Kawahara and M. Tanaka, Interactions of travelling fronts: an exact solution of a nonlinear diffusion equation, *Phys. Lett. A*, **97** (1983), 311–314.

[330] S. Kawamoto, An exact transformation from the Harry Dym equation to the modified KdV equation, *J. Phys. Soc. Japan*, **54** (1985), 2055–2056.

[331] J.B. Keller, On solutions of $\Delta u = f(u)$, *Commun. Pure Appl. Math.*, **10** (1957), 503–510.

[332] J.B. Keller, On solutions of nonlinear wave equations, *Commun. Pure Appl. Math.*, **10** (1957), 523–530.

[333] R. Kersner, On the behaviour of temperature fronts in media with non-linear heat conductivity under absorption, *Moscow Univ. Math. Bull.*, **33** (1978), 35–41.

[334] R. Kersner, Some properties of generalized solutions of quasilinear degenerate parabolic equations, *Acta Math. Acad. Sci. Hungar.*, **32** (1978), 301–330.

[335] B. Khesin and G. Misiolek, Euler equations on homogeneous spaces and Virasoro orbits, *Adv. Math.*, **176** (2003), 116–144.

[336] I.T. Kiguradze and T. Kusano, Periodic solutions of nonautonomous ordinary differential equations of higher order, *Differ. Equat.*, **35** (1999), 71–77.

[337] K.Y. Kim, R.O. Reid, and R.E. Whitaker, On an open radiational boundary condition for weakly dispersive tsunami waves, *J. Comput. Phys.*, **76** (1988), 327–348.

[338] J.R. King, *Mathematical Aspects of Semiconductor Process Modelling*, PhD Thesis,

University of Oxford, Oxford, 1986.

[339] J.R. King, The isolation oxidation of silicon: the reaction-controlled case, *SIAM J. Appl. Math.,* **49** (1989), 1064–1080.

[340] J.R. King, Mathematical analysis of a model for substitutional diffusion, *Proc. Roy. Soc. London Ser. A,* **430** (1990), 377–404.

[341] J.R. King, Some non-self-similar solutions to a nonlinear diffusion equation, *J. Phys. A,* **25** (1992), 4861–4868.

[342] J.R. King, Exact multidimensional solutions to some nonlinear diffusion equations, *Quart. J. Mech. Appl. Math.,* **46** (1993), 419–436.

[343] J.R. King, Exact polynomial solutions to some nonlinear diffusion equations, *Phys. D,* **64** (1993), 35–65.

[344] J.R. King, Self-similar behaviour for the fast diffusion equation of fast nonlinear diffusion, *Phil. Trans. Roy. Soc. London Ser. A,* **343** (1993), 337–375.

[345] J.R. King, Two generalisations of the thin film equation, *Math. Comput. Modelling,* **34** (2001), 737–756.

[346] J.R. King and J.M. Oliver, Thin-film modelling of poroviscous free surface flows, *European J. Appl. Math.,* **15** (2005), 1–35.

[347] J.R. King and S.M. Cox, Asymptotic analysis of the steady-state and time-dependent Berman problem, *J. Engrg. Math.,* **39** (2001), 87–130.

[348] G. Kirchhoff, *Vorlesungen über die Theorie der Warme*, Barth, Leipzig, 1894.

[349] Y. Kivshar, Compactons in discrete lattices, *Nonlinear Coherent. Struct. Phys. Biol.,* **329** (1994), 255–258.

[350] P.Ya. Kochina, The Zhukovskii function and some problems in filtration theory, *J. Appl. Math. Mech.,* **61** (1997), 153–155.

[351] V. Kolmanovskii and A. Myshkis, *Applied Theory of Functional-Differential Equations,* Kluwer Acad. Publ. Group, Dordrecht, 1992.

[352] A.N. Kolmogorov and S.V. Fomin, *Elements of the Theory of Functions and Functional Analysis*, Fourth edition, revised, Nauka, Moscow, 1976.

[353] A.N. Kolmogorov, I.G. Petrovskii, and N.S. Piskunov, Study of the diffusion equation with growth of the quantity of matter and its application to a biological problem, *Byull. Moskov. Gos. Univ., Sect. A,* **1** (1937), 1–26. English. transl.: [457], pp. 105–130.

[354] D.J. Korteweg and G. de Vries, On the change of form of long waves advancing in a rectangular channel, and on a new type of long stationary waves, *Phil. Mag.,* **39**, No. 5 (1895), 422–442.

[355] M.A. Krasnosel'skii and P.P. Zabreiko, *Geometrical Methods of Nonlinear Analysis*, Springer-Verlag, Berlin/Tokyo, 1984.

[356] I.M. Krichever and S.P. Novikov, Holomorphic vector bundles over Riemann surfaces and the Kadomtsev-Petviashvili equation. I, *Funct. Anal. Appl.,* **12** (1978), 276–286.

[357] I. Krichever, P. Wiegmann, and A. Zabrodin, Elliptic solutions to difference non-linear equations and related many-body problems, *Commun. Math. Phys.,* **193** (1998), 373–396.

[358] M. Kruskal, Nonlinear wave equations, In: *Dynamical Systems, Theory and Applications,* Rencontres, Battelle Res. Inst., Seattle, Wash., 1974) , J. Moser Ed., *Lect. Notes in Phys.,* Vol. **38**, Springer, Berlin, 1975, pp. 310–354.

[359] N.V. Krylov, Sequences of convex functions, and estimates of the maximum of the solution of a parabolic equation, *Siberian Math. J.,* **17** (1976), 226–236.

[360] C.N. Kumar and P.K. Panigrahi, Compacton-like solutions for modified KdV and other nonlinear equations, 23 April 1999, REVTeX electr. manuscr.

[361] B.A. Kupershmidt, A super Korteweg-de Vries equation: an integrable system, *Phys.*

Lett. A, **102** (1984), 213–215.
[362] A. Kurganov, D. Levy, and P. Rosenau, On Burgers-type equations with nonmonotonic dissipative fluxes, *Commun. Pure Appl. Math.*, **51** (1998), 443–473.
[363] V.P. Kuznetsov, Equations of nonlinear acoustics, *Soviet Phys. Acoustics,* **16** (1970), 467–470.
[364] K.-H. Kwek, H. Gao, W. Zhang, and C. Qu, An initial boundary value problem of Camassa-Holm equation, *J. Math. Phys.*, **41** (2000), 8279–8285.
[365] O.A. Ladyzhenskaya, V.A. Solonnikov, and N.N. Ural'tseva, *Linear and Quasilinear Equations of Parabolic Type,* Amer. Math. Soc., Providence, R.I., 1967.
[366] V. Lahno and R. Zhdanov, Group classification of nonlinear wave equation, *J. Math. Phys.,* **46** (2005), 053301-1–37.
[367] V. Lakshmikantham and D. Trigiante, *Theory of Difference Equations: Numerical Methods and Applications*, Second Edition, Marcel Dekker, Inc., New York, 2002.
[368] H. Lamb, *Hydrodynamics,* Sixth Edition (First Edition, 1879), Cambridge Univ. Press, Cambridge, 1932.
[369] L.D. Landau and E.M. Lifshitz, On the theory of the dispersion of magnetic permeability in ferromagnetic bodies, *Soviet J. Phys.,* **8** (1935), 153 (Reproduces in: *Collected Papers of L.D. Landau*, Pergamon Press, New York, 1965, pp. 101–114.)
[370] L.D. Landau and E.M. Lifshitz, *Fluid Mechanics*, Pergamon Press, London, 1959.
[371] L.D. Landau and E.M. Lifshitz, *Electrodynamics of Continuous Media*, Pergamon Press, Oxford/Paris, 1960.
[372] L.D. Landau and E.M. Lifshitz, *Theory of Elasticity*, Third Edition, Pergamon Press, Oxford, 1986.
[373] S. Lang, *Algebra*, Addison-Wesley Publ. Comp., Reading/Tokyo, 1984.
[374] M.Ya. Lankerovich, Ordinary differential equations admitting a group of maximal dimension, *Dinamika Sploshn. Sredy*, **37** (1978), 133–138 (in Russian).
[375] N. Larkin, On the problem of transonic gas dynamics, *Mat. Cont.,* **15** (1998), 169–186.
[376] R.S. Laugesen and M.C. Pugh, Energy levels of steady states for thin-film-type equations, *J. Differ. Equat.,* **182** (2002), 377–415.
[377] M.H. Lee, Pseudodifferential operators of several variables and Baker functions, *Lett. Math. Phys.,* **60** (2002), 1–8.
[378] L.S. Leibenzon, *Motion of Natural Fluids and Gases in a Porous Medium*, GITTL, Moscow–Leningrad, 1947.
[379] J. Lenells, Travelling wave solutions of the Camassa–Holm equation, *J. Differ. Equat.,* **217** (2005), 393–430.
[380] J. Lenells, Conservation laws of the Camassa–Holm equation, *J. Phys. A,* **38** (2005), 869–880.
[381] J. Levandosky, Smoothing properties of nonlinear dispersive equations in two spatial dimensions, *J. Differ. Equat.,* **175** (2001), 275–352.
[382] H.A. Levine, The role of critical exponents in blow-up problems, *SIAM Rev.,* **32** (1990), 262–288.
[383] B. Lewis and G. Elbe, On the theory of flame propagation, *J. Chem. Phys.,* **2** (1934), 537–546.
[384] A.-M. Li and F. Jia, A Bernstein property of affine maximal hypersurfaces, *Ann. Global Anal. Geom.,* **23** (2003), 359–372.
[385] A.M. Li, U. Simon, and G.S. Zhao, *Global Affine Differential Geometry of Hypersurfaces,* Walter De Gruyter, Berlin, 1993.
[386] Yi.A. Li, Linear stability of solitary waves of the Green-Naghdi equations, *Commun.*

Pure Appl. Math., **54** (2001), 501–536.

[387] Yi.A. Li and P.J. Olver, Well-posedness and blow-up solutions for an integrable non linearly dispersive model wave equations, *J. Differ. Equat.*, **162** (2000), 27–63.

[388] Y.A. Li and P.J. Olver, Convergence of solitary-wave solutions in a perturbed bi-Hamiltonian dynamical system. I, II, *Discrete Contin. Dynam. Systems*, **3** (1997), 419–432; **4** (1998), 159–191.

[389] S. Lie, Über die Integration durch bestimmte Integrale von einer Klasse lineare partiellen Differentialgleichungen, *Arch. Math.*, **6**, Heft 3 (1881), 328–368.

[390] S. Lie, Classification und Integration von gewöhnlichen Differentialgleichungen zwischen x, y, die eine Gruppe von Transformationen gestatten, *Arch. Math. Naturv., Christiania*, **9** (1883), 371–393.

[391] S. Lie, Algemeine Untersuchungen über Differentialgleichungen, die eine continuirliche, endliche Gruppe gestatten, *Math. Ann.*, **25** (1885), 71–151.

[392] S. Lie, Zur allgemeinen Theorie der partiellen Differentialgleichungen beliebieger Ordnung, *Leipz. Berichte*, **47** (1895), 53–128.

[393] S. Lie, *Geometrie der Berührungstransformationen/ Dargestellt von Sophus Lie und Georg Scheffers*, B.G. Teubner, Leipzig, 1896.

[394] E.M. Lifshitz and L.P. Pitaevskii, *Physical Kinetics*, Pergamon Press, New York, 1981.

[395] C.C. Lin, E. Reissner, and H.S. Tsien, On two-dimensional non-steady motion of a slender body in a compressible fluid, *J. Math. Phys.*, **27** (1948), 220–231.

[396] J.-L. Lions, *Quelques méthodes de résolution des problèmes aux limites non linéaires*, Dunod, Gauthier-Villars, Paris, 1969.

[397] Q.M. Liu, Exact solutions to nonlinear equations with quadratic nonlinearities, *J. Phys. A*, **34** (2001), 5083–5088.

[398] X. Liu, S. Jiang, and Y. Han, New explicit solutions to the n-dimensional Landau–Lifshitz equations, *Phys. Lett. A.*, **281** (2001), 324–326.

[399] Z. Liu and Y. Mao, Existence theorems for periodic solutions of higher order nonlinear differential equations, *J. Math. Anal. Appl.*, **216** (1997), 481–490.

[400] C. Loewner and L. Nirenberg, Partial differential equations invariant under conformal or projective transformations, In: *Contributions to Analysis*, Acad. Press, New York, 1974, pp. 245–272.

[401] Ya.B. Lopatinskii, On a method of reducing boundary problems for a system of differential equations of elliptic type to regular integral equations, *Ukrain. Math. Zh.*, **5** (1953), 123–151.

[402] D.K. Ludlow, P.A. Clarkson, and A.P. Bassom, New similarity solutions of the unsteady incompressible boundary-layer equations, *Quart. J. Mech. Appl. Math.*, **53** (2000), 175–206.

[403] A. Lunardi, *Analytic Semigroups and Optimal Regularity in Parabolic Problems*, Progr. in Nonl. Differ. Equat. Appl., Vol. **16**, Birkhäuser Verlag, Basel, 1995.

[404] S. Ma and X. Zou, Propagation and its failure in a lattice delayed differetial equation with global interaction, *J. Differ. Equat.*, **212** (2005), 129–190.

[405] W.X. Ma, Complexiton solutions to the Korteweg-de Vries equation, *Phys. Lett. A.*, **301** (2002), 35–44.

[406] W.-X. Ma and K.-I. Maruno, Complexiton solutions of the Toda lattice equation, *Phys. A.*, **343** (2004), 219–237.

[407] W.-X. Ma and Y. You, Solving the Korteweg-de Vries equation by its bilinear form: Wronskian solutions, *Trans. Amer. Math. Soc.*, **357** (2004), 1753–1778.

[408] J. Mallet-Paret, Travelling waves in spatially discrete dynamical systems of diffusive type, *Dynamical Systems*, *Lect. Notes in Math.*, Vol. **1822**, Springer, Berlin, 2003,

pp. 231–298.

[409] S.V. Manakov, On the theory of two dimensional stationary self-focusing of electromagnetic waves, *Sov. Phys. JETP*, **38** (1974), 248–253.

[410] Yu.I. Manin and A.O. Radul, A supersymmetric extension of the Kadomtsev-Petviashvili hierarchy, *Commun. Math. Phys.*, **98** (1985), 65–77.

[411] P. Manneville, The Kuramoto-Sivashinsky equation: a progress report, In: *Propagation in Systems Far from Equilibrium*, J. Weisfreid et all, Eds, Springer, Berlin, 1988, pp. 265–280.

[412] A.S. Markus, *Introduction to the Spectral Theory of Polynomial Operator Pencils*, Transl. Math. Mon., Vol. **71**, Amer. Math. Soc., Providence, RI, 1988.

[413] L. Martina, M.B. Sheftel, and P. Winternitz, Group foliation and non-invariant solutions of the heavenly equation, *J. Phys. A*, **34** (2001), 9243–9263.

[414] L.K. Martinson, Propagation of heat wave in a nonlinear medium with absorption, *Zh. Prikl. Mekh. i Tekhn. Fiz.*, No. **4** (1979), 36–39 (in Russian); English translation: *J. Appl. Mech. Techn. Phys.*, **20** (1979).

[415] S. Matsutani, Hyperelliptic solutions of KdV and KP equations: re-evaluation of Baker's study on hyperelliptic sigma functions, *J. Phys. A*, **34** (2001), 4721–4732.

[416] S. Matsutani, Close loop solitons and sigma functions: classical and quantized elasticas with genera one and two, *J. Geom. Phys.*, **39** (2001), 50–61.

[417] S. Matsutani, Hyperelliptic solutions of modified Korteweg-de Vries equation of genus g: essentials of the Miura transformation, *J. Phys. A*, **35** (2002), 4321–4333.

[418] V.B. Matveev, Generalized Wronskian formula for solutions of the KdV equation: first applications, *Phys. Lett. A*, **166** (1992), 205–208.

[419] V.B. Matveev, Positons: slowly decreasing analogues of solitons, *Theoret. Math. Phys.*, **131** (2002), 483–497.

[420] D. McKenzie, The generation and compaction of partially molten rock, *J. Petrol*, **25** (1984), 713–765.

[421] E.Yu. Meshcheryakova, New steady and self-similar solutions of the Euler equations, *J. Appl. Mech. Tech. Phys.*, **44** (2003), 455–460.

[422] Th. Meyer, Dissertation, Göttingen, 1908.

[423] W. Miller, Jr. and L.A. Rubel, Functional separation of variables for Laplace equations in two dimensions, *J. Phys. A*, **26** (1993), 1901–1913.

[424] R. von Mises, Bemerkungen zur Hydrodynamik, ZAMM, **7** (1927), 425–531.

[425] E. Mitidieri and S.I. Pohozaev, *Apriori Estimates and Blow-up of Solutions to Nonlinear Partial Differential Equations and Inequalities*, Proc. Steklov Inst. Math., Vol. **234**, Intern. Acad. Publ. Comp. Nauka/Interperiodica, Moscow, 2001.

[426] R.M. Miura, Korteweg-de Vries equation and generalizations. I. A remarkable explicit nonlinear transformation, *J. Math. Phys.*, **9** (1968), 1202–1204.

[427] G. Monge, Mémoire sur la théorie des déblais et des remblais, *Histoire de l'Académie Royale des Sciences*, Paris, 1781, pp. 666–704.

[428] W.W. Mullins, Theory of thermal grooving, *J. Appl. Phys.*, **28** (1957), 333–339.

[429] M. Muskat, *The Flow of Homogeneous Fluids Through Porous Media*, McGraw-Hill, New York, 1937.

[430] T.G. Myers, Thin films with high surface tension, *SIAM Rev.*, **40** (1998), 441–462.

[431] T.G. Myers, J.P.F. Charpin, and S.J. Chapman, The flow and solidification of a thin fluid film on an arbitrary three-dimensional surface, *Phys. Fluids*, **14** (2002), 2788–2803.

[432] M.A. Naimark, *Linear Differential Operators*, Part II. *Linear Differential Operators in Hilbert Space*, Frederick Ungar Publ. Co., New York, 1968.

[433] P.F. Neményi, Recent developments in inverse and semi-inverse methods in the Mechanics of Continua, *Adv. Appl. Mech.*, **2** (1951), 123–151.

[434] V.V. Nemytskii and V.V. Stepanov, *Qualitative Theory of Differential Equations*, New York, Dover, 1989.

[435] G.V. Nesterenko, The use of pseudosymmetries to construct exact solutions of differential equations, *Differ. Equat.*, **26** (1990), 855–859.

[436] A.C. Newell, *Solitons in Mathematics and Physics*, CBMS-NSF Regional Conf. Ser. Appl. Math., 48, SIAM, Phil., PA, 1985.

[437] W.I. Newman, Some exact solutions to a nonlinear diffusion problem in population genetics and combustion, *J. Theoret. Biol.*, **85** (1980), 325–334.

[438] A. Novick-Cohen and L.A. Segel, Nonlinear aspects of the Cahn-Hilliard equation, *Phys. D,* **10** (1984), 277–298.

[439] M.C. Nucci, Iterating the nonclassical symmetries methods, *Phys. D*, **78** (1994), 124–134.

[440] M.C. Nucci, Nonclassical symmetries as special solutions of heir equations, *J. Math. Anal. Appl.,*, **279** (2003), 168–179.

[441] O.A. Oleinik, Discontinuous solutions of non-linear differential equations, *Uspehi Mat. Nauk.*, **12** (1957), 3–73; *Amer. Math. Soc. Transl.* (2), **26** (1963), 95–172.

[442] O.A. Oleinik and S.N. Kruzhkov, Quasi-linear second-order parabolic equations with several independent variables, *Russian Math. Surveys,* **16** (1961), 105–146.

[443] O.A. Oleinik and E.V. Radkevich, The method of introducing a parameter for the investigation of evolution equations, *Russian Math. Surveys,* **33** (1978), 7–84.

[444] O.A. Oleinik and V.N. Samokhin, *Mathematical Models in Boundary Layer Theory*, Chapman & Hall/CRC, Boca Raton, London, 1999.

[445] P.J. Olver, Symmetry and explicit solutions of partial differential equations, *Appl. Numer. Math.*, **10** (1992), 307–324.

[446] P.J. Olver, *Applications of Lie Groups to Differential Equations*, Second Edition, Graduate Texts in Mathematics, Vol. **107**, Springer-Verlag, New York, 1993.

[447] P.J. Olver, Direct reduction and differential constraints, *Proc. Roy. Soc. London Ser. A,* **444** (1994), 509–523.

[448] P.J. Olver and P. Rosenau, The construction of special solutions to partial differential equations, *Phys. Lett. A,* **114** (1986), 107–112.

[449] P.J. Olver and P. Rosenau, Tri-Hamiltonian duality between solitons and solitary-wave solutions having compact support, *Phys. Rev. E* (3), **53** (1996), 1900–1906.

[450] A. Oron, Nonlinear dynamics of three-dimensional long-wave Marangoni instability in thin liquid films, *Phys. Fluids,* **12** (2000), 1633–1645.

[451] A. Oron, S.H. Davies, and S.G. Bankoff, Long-scale evolution of thin liquids films, *Rev. Modern Phys.*, **69** (1997), 931–980.

[452] A. Oron and O. Gottlied, Nonlinear dynamics of temporally excited falling liquid films, *Phys. Fluids,* **14** (2002), 2622–2636.

[453] A. Oron and P. Rosenau, Some symmetries of the nonlinear heat and wave equations, *Phys. Lett. A,* **118** (1986), 172–176.

[454] F. Otto, Lubrication approximation with prescribed non-zero contact angle: an existence result, *Commun. Partial Differ. Equat.*, **23** (1998), 2077–2164.

[455] L.V. Ovsiannikov, *Investigation of Gas Flows with Straight Transition Line*, PhD Thesis, Leningrad, 1948.

[456] L.V. Ovsiannikov, *Group Analysis of Differential Equations*, Acad. Press, Inc., New York/London, 1982.

[457] *Dynamics of Curved Fronts,* P. Pelcé, Ed., Acad. Press, Inc., New York, 1988.

[458] R.L. Pego and J.R. Quintero, Two-dimensional solitary waves for a Benney–Luke equation, *Phys. D*, **132** (1999), 476–496.

[459] L.A. Peletier and W.C. Troy, *Spatial Patterns. Higher Order Models in Physics and Mechanics*, Birkhäuser, Boston/Berlin, 2001.

[460] L. Perko, *Differential Equations and Dynamical Systems*, Springer, New York, 1991.

[461] D.H. Peregrine, Calculations on the development of an undular bore, *J. Fluid Mech.*, **25** (1966), 321–330.

[462] M. Pinl and H.W. Trapp, Stationäre Krümmungsdichten auf Hyperflächen des euklidischen R_{n+1}, *Math. Ann.*, **176** (1968), 257–272.

[463] J.F. Plebański, Some solutions of complex Einstein equations, *J. Math. Phys.*, **16** (1975), 2395–2402.

[464] S.I. Pohozaev, On the eigenfunctions of the equation $\Delta u + \lambda f(u) = 0$, *Soviet Math. Dokl.*, **6** (1965), 1408–1411.

[465] S.I. Pohozaev, On the eigenfunctions of quasilinear elliptic problems, *Math. USSR Sbornik*, **11** (1970), 171–188.

[466] S.I. Pohozaev, On the problem by L.V. Ovsiannikov, *Prikl. Mehan. i Tekhn. Fiz.*, **2** (1989), 5–10 (in Russian).

[467] S.I. Pohozaev, Personal communication.

[468] J.E. Pommaret, *Systems of Partial Differential Equations and Lie Pseudogroups*, Gordon & Breach Sci. Publ., New York, 1978.

[469] A.V. Porubov and M.G. Velarde, Strain kinks in an elastic rod embedded in a viscoelastic medium, *Wave Motion*, **35** (2002), 189–204.

[470] L. Prandtl, Über Flüssigkeitsbewegung bei sehr kleiner Reibung, In: *Verhandlungen des dritten Internationalen Mathematiker Kongresses*, Heidelberg, 1904, Teubner, Leipzig, 1905, pp. 484–491.

[471] L.J. Pratt and M.E. Stern, Dynamics of potential vorticity fronts and eddy detachment, *J. Phys. Oceanogr.*, **16** (1986), 1101–1120.

[472] M.H. Protter and H.F. Weinberger, *Maximum Principles in Differential Equations*, Springer-Verlag, New York, 1984.

[473] I. Proudman and K. Johnson, Boundary-layer growth near a rear stagnation point, *J. Fluid Mech.*, **12** (1962), 161–168.

[474] E. Pucci and G. Saccomandi, Evolution equations, invariant surface conditions and functional separation of variables, *Phys. D*, **139** (2000), 28–47.

[475] V.V. Pukhnachov, On a problem of viscosity strip deformation with a free boundary, *C. R. Acad. Sci. Paris, Sér. I Math.*, **328** (1999), 357–362.

[476] C. Qu, Group classification and generalized conditional symmetry reduction of the nonlinear diffusion–convection equation with a nonlinear source, *Stud. Appl. Math.*, **99** (1997), 107–136.

[477] C. Qu, Classification and reduction of some systems of quasilinear partial differential equations, *Nonl. Anal.*, **42** (2000), 301–327.

[478] C. Qu, Symmetries and solutions to the thin film equations, J. Math. Anal. Appl., **317** (2006), 381–397.

[479] C. Qu and P.G. Estévez, Extended rotation and scaling groups for nonlinear evolution equations, *Nonl. Anal.*, **52** (2003), 1655–1673.

[480] C. Qu and P.G. Estébvez, On nonlinear diffusion equations with x dependent convection and absorption, *Nonl. Anal.*, **57** (2004), 549–577.

[481] C. Qu, L. Ji, and J. Dou, Exact solutions and generalized conditional symmetries to $(n + 1)$-dimensional nonlinear diffusion equations with source term, *J. Phys. A*, **343** (2005), 139–147.

[482] C. Qu, S. Zhang, and R. Liu, Separation of variables and exact solutions to quasilinear diffusion equations with nonlinear source, *Phys. D,* **144** (2000), 97–123.

[483] C.-Z. Qu and S.-L. Zhang, Group foliation method and functional separation of variables to nonlinear diffusion equations, *Chin. Phys. Lett.,* **22** (2005), 1563–1566.

[484] J. Ramírez, M.S. Bruzón, and M.L. Gandarias, The Schwarzian Korteweg-de Vries equation in $(2+1)$ dimensions, *J. Phys. A,* **36** (2003), 1467–1484.

[485] C. Rasinariu, U. Sukhatme, and A. Khare, Negaton and positon solutions of the KdV and mKdV hierarchy, *J. Phys. A*, **29** (1996), 1803–1823.

[486] R.C. Reilly, On the Hessian of a function and the curvatures of its graph, *Michigan Math. J.,* **20** (1973), 373–383.

[487] M. Remoissenet, *Waves Called Solitons*, 3rd ed., Springer-Verlag, Heidelverg, 1999.

[488] D. Riabouchinsky, Quelques considérations sur les mouvements plans rotationnels d'un liquide, *C. R. Hebd. Acad. Sci.,* **179** (1924), 1133–1136.

[489] L.A. Richard's, Capillary conduction of liquids through porous mediums, *Physics,* **1** (1931), 345–357.

[490] M.A. Rodríguez and P. Winternitz, Lie symmetries and exact solutions of first-order difference schemes, *J. Phys. A,* **37** (2004), 6129–6142.

[491] P. Rosenau, Nonlinear dispertion and compact structures, *Phys. Rev. Lett.,* **73** (1994), 1737–1741.

[492] P. Rosenau, On solitons, compactons, and Lagrange maps, *Phys. Lett. A,* **211** (1996), 265–275.

[493] P. Rosenau, On a class of nonlinear dispersive-dissipative interactions, *Phys. D,* **123** (1998), 525–546.

[494] P. Rosenau, Compact and noncompact dispersive patterns, *Phys. Lett. A,* **275** (2000), 193–203.

[495] P. Rosenau, Personal communication, 2005.

[496] P. Rosenau and J.M. Hyman, Compactons: solitons with finite wavelength, *Phys. Rev. Lett.,* **70** (1993), 564–567.

[497] P. Rosenau, J.M. Hyman, and M. Staley, On multi-dimensional compactons, *Phys. Rev. Lett.,* to appear.

[498] P. Rosenau and S. Kamin, Thermal waves in an absorbing and convecting medium, *Phys. D,* **8** (1983), 273–283.

[499] P. Rosenau and D. Levy, Compactons in a class of nonlinearly quintic equations, *Phys. Lett. A,* **252** (1999), 297–306.

[500] P. Rosenau and S. Schochet, Almost compact breathers in anharmonic lattices near the continuum limit, *Phys. Rev. Lett.,* **94**, 045503 (2005).

[501] G.A. Rudykh and E.I. Semenov, Existence and construction of anisotropic solutions to the multidimensional equation of nonlinear diffusion. I, II, *Siberian Math. J.,* **41** (2000), 940–959; **42** (2001), 157–175.

[502] G.A. Rudykh and E.I. Semenov, Exact non-self-similar solutions of the equation $u_t = \Delta \ln u$, *Math. Notes*, **70** (2001), 714–719.

[503] J.S. Russell, On waves, In: *Report of the* 14th *Meeting*, British Assoc. Adv. Sci., London, John Murrey, 1845, pp. 311–390.

[504] O.S. Ryzhov and S.A. Khristianovitch, On nonlinear reflection of weak shock waves, *J. Appl. Math. Mech.,* **22** (1958), 826–843.

[505] O.S. Ryzhov and G.M. Shefter, On unsteady gas flows in Laval nozzles, *Soviet Physics Dokl.,* **4** (1959), 939–942.

[506] G. Saccomandi, Elastic rods, Weierstrass' theory and special travelling waves solutions with compact support, *Int. J. Non-Lin. Mech.,* **39** (2004), 331–339.

[507] T. Sadakane, Soliton solutions of XXZ lattice Landau-Lifshitz equation, *J. Math. Phys.*, **42** (2001), 5457–5471.

[508] A.-J.-C. Barré de Saint-Venant, *De la Torsion des Prismes*, Imprimére Impériale, Paris, 1855.

[509] A.A. Samarskii, V.A. Galaktionov, S.P. Kurdyumov, and A.P. Mikhailov, *Blow-up in Quasilinear Parabolic Equations,* Walter de Gruyter & Co., Berlin, 1995.

[510] A.A. Samarskii, N.V. Zmitrenko, S.P. Kurdyumov, and A.P. Mikhailov, Thermal structures and fundamental length in a medium with non-linear heat conduction and volumetric heat sources, *Soviet Phys. Dokl.*, **21** (1976), 141–143.

[511] J. Sander and K. Hutter, On the development of the theory of the solitary wave. A historical essay, *Acta Mech.,* **86** (1991), 111–152.

[512] J.A. Sanders and J.P. Wang, On the integrability of second order evolution equations with two components, *J. Differ. Equat.*, **203** (2004), 1–27.

[513] A. Sansom, J.R. King, and D.S. Riley, Degenerate-diffusion models for the spreading of thin non-isothermic gravity currents, *J. Engrg. Math.,* **48** (2004), 43–68.

[514] D.C. Sarocka and A.J. Bernoff, An intrinsic equation of interfacial motion for the solidification of a pure hypercooled melt, *Phys. D,* **85** (1995), 348–374.

[515] J. Satsuma, A Wronskian representation of N-soliton solutions of nonlinear evolution equations, *J. Phys. Soc. Japan,* **46** (1979), 359–360.

[516] O.C. Schnürer and K. Smoczyk, Neumann and second boundary value problems for Hessian and Gauss curvature flows, *Ann. Inst. H. Poincaré, Anal. Non Linéaire,* **20** (2003), 1043–1073.

[517] H. Schamel, A modified Korteweg-de Vries equation for ion acoustic waves due to resonant electrons, *J. Plasma Phys.,* **9** (1973), 377–387.

[518] R.W. Schrage, *A Theoretical Study of Interphase Mass Transfer*, Columbia Univ. Press, New York, 1953.

[519] D.R. Scott and D.J. Stivenson, Magma solitons, *Geophys. Res. Lett.,* **11** (1984), 1161–1164.

[520] A. Seeger, H. Donth, and A. Kochendörfer, Theorie der Versetzungen in eindimensionalen Atomreihen. III. Versetzungen, Eigenbewegungen und ihre Wechselwirkung, *Z. Phys.*, **134** (1953), 173–193.

[521] A. Seeger and Z. Wesolowski, Standing-wave solutions of the Enneper (sine-Gordon) equation, *Internat. J. Engr. Sci.,* **19** (1981), 1535–1549.

[522] H. Segur and M.D. Kruskal, Nonexistence of small-amplitude breather solutions in ϕ^4 theory, *Phys. Rev. Lett.,* **58** (1987), 747–750.

[523] A. Sergeyev, Constructing conditionally integrable evolution systems in (1+1) dimensions: a generalization of invariant modules approach, *J. Phys. A,* **35** (2002), 7653–7660.

[524] A.E. Shishkov, Dead cores and instantaneous compactification of the supports of energy solutions of quasilinear parabolic equations of arbitrary order, *Sbornik: Math.*, **190** (1999), 1843–1869.

[525] A.F. Sidorov, V.P. Shapeev, and N.N. Yanenko, *Method of Differential Constraints and its Applications in Gas Dynamics*, Nauka, Sibirsk. Otdel., Novosibirsk, 1984.

[526] G.I. Sivashinsky, Nonlinear analysis of hydrodynamic instability in laminar flames, *Acta Astronaut.,* **4** (1977), 1177–1206.

[527] R. Slipic and D.J. Benney, Lump interaction and collapse in the modified Zakharov–Kuznetsov equation, *Studies Appl. Math.,* **105** (2000), 385–403.

[528] F.T. Smith and R.J. Bodonyi, Nonlinear critical layers and their development in streaming-flow stability, *J. Fluid Mech.,* **118** (1982), 165–185.

[529] P.C. Smith, A similarity solution for slow viscous flow down an inclined plane, *J. Fluid Mech.*, **58** (1973), 275–288.
[530] J. Smoller, *Shock Waves and Reaction-Diffusion Equations,* Springer-Verlag, New York/Berlin, 1983.
[531] N.F. Smyth and J.M. Hill, High-order nonlinear diffusion, *IMA J. Appl. Math.*, **40** (1988), 73–86.
[532] H.M. Soner and P.E. Souganidis, Singularities and uniqueness of cylindrically symmetric surfaces moving by mean curvature, *Commun. Partial Differ. Equat.*, **18** (1993), 859–894.
[533] R.P. Sperb, *Maximum Principles and their Applications*, Acad. Press, New York/ London, 1981.
[534] P. Stäckel, Über die Bewegung eines Punktes in eines n-fachen Mannigfaltigkeit, *Mat. Ann.*, **42** (1893), 537.
[535] R. Steuerwald, Über Ennepersche Flächen und Bäcklundsche Transformation, *Abhandlungen der Bayerischen Akademie der Wissenschaften,* Mathem.-naturwissenschaftl. Abtlg., Neue Folge, Heft 40, München, 1936.
[536] G. Stokes, On the theory of oscillatory waves, *Trans. Cambridge Phil. Soc.,* **8** (1847), 441–455.
[537] M.L. Storm, Heat conduction in simple metals, *J. Appl. Phys.*, **22** (1951), 940–951.
[538] C. Sturm, Mémoire sur une classe d'équations à différences partielles, *J. Math. Pures Appl.,* **1** (1836), 373–444.
[539] M. Svendsen and H.C. Fogedby, Phase shift analysis of the Landau-Lifshitz equation, *J. Phys. A,* **26** (1993), 1717–1730.
[540] S.R. Svirshchevskii, *Group Properties of Heat Conduction Models*, PhD Thesis, Keldysh Inst. Appl. Math., Moscow, 1984.
[541] S.R. Svirshchevskii, High-order symmetries of linear differential equations and linear spaces invariant under nonlinear operators, Preprint No. 14, Inst. Math. Modell., Moscow, 1993.
[542] S.R. Svirshchevskii, Group classification and invariant solutions of nonlinear polyharmonic equations, *Differ. Equat.,* **29** (1993), 1538–1547.
[543] S.R. Svirshchevskii, Lie-Bäcklund symmetries of linear ODE and invariant linear spaces, In: *Modern Group Analysis*, Inter-Univ. Proceedings., Moscow Phys.-Techn. Inst., Moscow, 1993, pp. 75–82.
[544] S.R. Svirshchevskii, Lie-Bäcklund symmetries of linear ODEs and generalized separation of variables in nonlinear equations, *Phys. Lett. A,* **199** (1995), 344–348.
[545] S.R. Svirshchevskii, First- and second-order nonlinear differential operators that have invariant linear spaces of maximal dimension, *Theoret. Math. Phys.,* **105** (1995), 1346–1353.
[546] S.R. Svirshchevskii, Invariant linear spaces and exact solutions of nonlinear evolution equations, In: *Symm. in Nonl. Math. Phys.*, Vol. **2** (Kiev, 1995); *J. Nonl. Math. Phys.,* **3** (1996), 164–169.
[547] S.R. Svirshchevskii, Ordinary differential operators possessing invariant subspaces of polynomial type, *Commun. Nonlinear Sci. Numer. Simul.,* **9** (2004), 105–115.
[548] S. Täklind, Sur les classes quasianalytiques des solutions des équa-tions aux dérivees partielles du type parabolique, *Nova Acta Regalis Societatis Scientiarum Uppsaliensis* (4), **10**, No. 3 (1936), 3–55.
[549] J. Tanthanuch and S.V. Meleshko, On definition of an admitted Lie group for functional differential equations, *Commun. Nonlinear Sci. Numer. Simul.,* **9** (2004), 117–125.
[550] M.E. Taylor, *Partial Differential Equations III. Nonlinear Equations,* Springer-Verlag,

New York, 1996.
[551] W.E. Thirring, A soluble relativistic field theory, *Ann. Physics,* **3** (1958), 91–112.
[552] A.N. Tikhonov, Uniqueness theorem for the equation of heat conduction, *Mat. Sb.,* **42** (1935), 199–215.
[553] S.S. Titov, On solutions of nonlinear partial differential equations of the form of a simple variable, *Chisl. Met. Mech. Sploshnoi Sredy,* **8** (1977), 586–599 (in Russian).
[554] S.S. Titov, On transonic gas flow around thin bodies, in: *Analytical Methods in Continues Medium Mechanics*, ed. A.F. Sidorov, Sverdlovsk, 1979, p. 65 (in Russian).
[555] S.S. Titov, A method of finite-dimensional rings to solve nonlinear equations of mathematical physics, in: *Aerodynamics of Plane and Axis-Symmetric Flows of Liquids*, Saratov University, Saratov, 1988, 104–109 (in Russian).
[556] M. Toda, *Theory of Nonlinear Lattices*, Springer, Berlin, 1989.
[557] K. Toda and S.-J. Yu, The investigation into the Schwarz-Korteweg-de Vries equation and the Schwarz derivative in (2+1) dimensions, *J. Math. Phys.,* **41** (2000), 4747–4751.
[558] S. Tomotika and T. Tamada, Studies on two-dimensional transonic flows of compressible fluid.–Part I, *Quart. Appl. Math.,* **7** (1949), 381–397.
[559] R.A. Trasca, M.W. Cole, and R.D. Diehl, Systematic model behavior of adsorption on flat surfaces, *Phys. Rev. E*, **68**, 041605 (2003).
[560] N.S. Trudinger and X. Wang, Bernstein-Jörgens theorem for a fourth order partial differential equation, *J. Partial Differ. Equat.,* **15** (2002), 78–88.
[561] K. Tso, Remarks on critical exponents for Hessian operators, *Ann. Inst. H. Poincaré, Anal. Non Linéaire,* **7** (1990), 113–122.
[562] T. Tsuchida and T. Wolf, Classification of polynomial integrable systems of mixed scalar and vector evolution equations. I, J. Phys. A, **38** (2005), 7691–7733.
[563] A. Turbiner, Quasi-exactly-solvable problems and sl(2) algebra, *Commun. Math. Phys.,* **118** (1988), 467–474.
[564] G. Tzitzéica, Sur une nouvelle classe de surfaces, *C. R. Acad. Sci., Paris,* **144** (1907), 1257–1259.
[565] M.A. Vainberg and V.A. Trenogin, *Theory of Branching of Solutions of Non-Linear Equations*, Noordhoff Int. Publ., Leiden, 1974.
[566] E.M. Vorob'ev, Reduction and quotient equations for differential equations with symmetries, *Acta Appl. Math.,* **23** (1991), 1–24.
[567] E.M. Vorob'ev, N.V. Ignatovich, and E.O. Semenova, Invariant and partially invariant solutions of boundary value problems, *Soviet Physics Dokl.,* **34** (1989), 505–507.
[568] M. Wadati, Y.H. Ichikawa, and T. Shimizu, Cusp soluton of a new integrable nonlinear evolution equation, *Progr. Theoret. Phys,* **64** (1980), 1959–1967.
[569] J.R. Ward, Asymptotic conditions for periodic solutions of ordinary differential equations, *Proc. Amer. Math. Soc.,* **81** (1981), 415–420.
[570] A.M. Wazwaz, A study on compacton-like solutions for the modified KdV and fifth order KdV-like equations, *Appl. Math. Comput.,* **147** (2004), 439–447.
[571] A.M. Wazwaz, An analytic study of compacton solutions for variants of Kuramoto–Sivashinsky equation, *Appl. Math. Comput.,* **148** (2004), 571–585.
[572] G. Webb, M.P. Sorensen, M. Brio, A.R. Zakharian, and J.V. Moloney, Variational principles, Lie point symmetries, and similarity solutions of the vector Maxwell equations in non-linear optics, *Phys. D,* **191** (2004), 49–80.
[573] K. Weierstrass, Beitrag zur Theorie der Abel'schen Integrale, In: *Jahreber. Königl. Katolischen Gymnasium zu Braunsberg in dem Schuljahre* 1848/49, 1849, pp. 3–23.
[574] G.B. Whitham, Non-linear dispersive waves, *Proc. Roy. Soc. London Ser. A,* **283** (1965), 238–261.

[575] G.B. Whitham, *Linear and Nonlinear Waves,* Wiley Interscience, New York, 1974.

[576] T.P. Witelski, A.J. Bernoff, and A.L. Bertozzi, Blow-up and dissipation in a critical-case unstable thin film equation, *European J. Appl. Math.,* **15** (2004), 223–256.

[577] L.-F. Wu, The Ricci flow on complete $I\!R^2$, *Commun. Anal. Geom.,* **1** (1993), 439–472.

[578] Z. Wu, J. Zhao, J. Yin, and H. Li, *Nonlinear Diffusion Equations,* World Scientific Publ. Co., Inc., River Edge, NJ, 2001.

[579] Z. Xin and P. Zhang, On the weak solutions to a shallow water equation, *Commun. Pure Appl. Math.,* **53** (2000), 1411–1433.

[580] Z. Yan, Constructing exact solutions for two-dimensional nonlinear dispersion Boussinesq equation. II, *Chaos, Solitons Fractals,* **18** (2003), 869–880.

[581] Z. Yan and G. Bluman, New compacton solutions and solitary patterns solutions of nonlinearly dispersive Boussinesq equations, *Comp. Physics Commun.,* **149** (2002), 1–18.

[582] N.N. Yanenko, The compatability theory and methods of integration of systems of nonlinear partial differential equations, In: *Proceedings of All-Union Math. Congress,* Vol. **2**, Nauka, Leningrad, 1964, pp. 613–621 (in Russian).

[583] R.-X. Yao and Z.-B. Li, Conservation laws and new exact solutions for the generalized seventh order KdV equation, *Chaos, Solitons Fractals,* **20** (2004), 259–266.

[584] Y. Yao, New type of exact solutions of nonlinear evolution equations via the new Sine–Poisson equation expansion method, *Chaos, Solitons Fractals,* **26** (2005), 1081–1086.

[585] M. Yoshino, Global solvability of Monge-Ampère type equations, *Commun. Part. Differ. Equat.,* **25** (2000), 1925–1950.

[586] S.-J. Yu, K. Toda, and T. Fukuyama, N-soliton solutions to a (2+1)-dimensional integrable equation, *J. Phys. A,* **31** (1998), 10181–10186.

[587] E.A. Zabolotskaya and R.V. Khohlov, Quasi-plane waves in the nonlinear acoustic confined beams, *Soviet Phys. Acoustics,* **15** (1969), 35–40.

[588] E.A. Zabolotskaya and R.V. Khohlov, Convergent and divergent sound beams in nonlinear media, *Soviet Phys. Acoustics,* **16** (1970), 39–43.

[589] N.J. Zabusky, Exact solution for the vibrations of a nonlinear continuous model string, *J. Math. Phys.,* **3** (1962), 1028–1039.

[590] N.J. Zabusky and M.D. Kruskal, Interaction of solitons in a collisionless plasma and the recurrence of initial states, *Phys. Rev. Lett.,* **15** (1965), 240–243.

[591] V.E. Zakharov and E.A. Kuznetsov, On three-dimensional solitons, *Sov. Phys. JETP,* **39** (1974), 285–286.

[592] R.Z. Zhdanov, Conditional Lie-Bäcklund symmetry and reduction of evolution equations, *J. Phys. A,* **28** (1995), 3841–3850.

[593] R.Z. Zhdanov, Higher conditional symmetry and reduction of initial value problems, *Nonlinear Dynam.,* **28** (2002), 17–27.

[594] Ya.B. Zel'dovich, G.I. Barenblatt, V.B. Librovich, and G.M. Makhviladze, *The Mathematical Theory of Combustion and Explosions*, Consultants Bureau [Plenum], New York, 1985.

[595] A.I. Zenchuk, On the construction of particular solutions to (1+1)-dimensional partial differential equations, *J. Phys. A,* **35** (2002), 1791–1803.

[596] S.-L. Zhang and S.-Y. Lou, Derivative-dependent functional separable solutions for the KdV-type equations, *Phys. A,* **335** (2004), 430–444.

[597] S.-L. Zhang, S.-Y. Lou, and C.-Z. Qu, New variable separation approach: application to nonlinear diffusion equations, *J. Phys. A,* **36** (2003), 1223–1242.

[598] S.-L. Zhang, S.-Y. Lou, and C.-Z. Qu, Functional variable separation for extended (1+2)-dimensional nonlinear wave equations, *Chin. Phys. Lett.,* **22** (2005), 2731–2734.

List of Frequently Used Abbreviations

1D – one-dimensional
AKNS – Ablowitz–Kaup–Newell–Segur
BI – Born–Infeld
CP – Cauchy problem
DS – dynamical system
FW – Fornberg–Whitham
FBP – free-boundary problem
FK – Frenkel–Kontorova
FFCH – Fuchssteiner–Fokas–Camassa–Holm
GSV – generalized separation of variables
GT – Gibbons–Tsarev
GN – Green–Naghdi
HE – heat equation
KP – Kadomtsev–Petviashvili
KFG – Kármán–Fal'kovich–Guderley
KS – Kuramoto–Sivashinsky
KdV – Korteweg–de Vries
KPPF – Kolmogorov–Petrovskii–Piskunov–Fisher
LL – Landau–Lifshitz
LRT – Lin–Reissner–Tsien
MP – Maximum Principle
M-A – Monge-Ampère
MMM – moving mesh method
ODE – ordinary differential equation
PDE – partial differential equation
PDI – partial differential inequality
PBBM – Peregrine–Benjamin–Bona–Mahoney
PME – porous medium equation
RH – Rosenau–Hyman
sdYM – self-dual Yang–Mills
SI – sign-invariant
TFE – thin film equation
TW – traveling wave
ZKB – Zel'dovich–Kompaneetz–Barenblatt

Index